Physics

A Practical Approach

O.N. Bishop *B. Sc., (Bristol.) B. Sc., (Oxon.)*

Ⓜ MACMILLAN
EDUCATION

First Published 1981

Published by
Macmillan Education Limited
London and Basingstoke
Companies and representatives throughout the world

ISBN 0 333 22593 7

Printed in Hong Kong

Acknowledgements

The author and publishers wish to thank the following
sources for photographs used within this book:

Architect's Department of the Greater London Council
British Airways
British Hovercraft Corporation Limited
Camera Press/NASA
Dr J. Hannay, Imperial College of Science, London
International Computers Limited
Keystone
Kratos
Materials Reclamation
By Courtesy of the Post Office
TRRL photo
C. James Webb
Zoological Society of London
Arthur Mechen and John Newman/STL – cover

The publishers have made every effort to trace the
copyright holders of all illustrations but where they
have failed to do so they will be pleased to make the
necessary arrangements at the first opportunity.

Contents

About this book

Physics – A Practical Approach is written specifically for students taking General Certificate of Education (Ordinary Level) and School Certificate examinations. It covers all the major syllabuses including those of Cambridge, London, AEB, the East African Examinations Council, the West African Examinations Council, and the Malaysian Examinations Syndicate. In the UK it covers most CSE syllabuses as well.

Physics – A Practical Approach blends the best of both the enquiry and traditional methods of learning. The facts and methods are up-to-date and are presented in a clear style, with particular attention paid to language level. All the information necessary for the examination is included and is balanced by a large amount of enquiry material. Itemised experiments and investigations with easy-to-follow instructions, are given throughout the book and questions relating to them reinforce understanding of the facts, principles and processes involved. Aided by the comprehensive Answers and Discussion section at the end of the book the student can readily use the exercises to assess his or her progress. In this way the book is suitable for students working largely on their own.

The book is well illustrated with large labelled diagrams and photographs. Many summary tables are included for quick and easy reference and revision. SI units are used throughout and an introductory chapter provides the rules and explanation necessary for a rigorous and systematic approach to measurement.

As far as a practical approach is concerned the section 'Physics in Action' is of particular interest. This is a broad and stimulating look at the applications of physics to modern industry and technology in areas such as optics, power generation, electronics, transport and communications. In addition there are suggestions for project work to build electronic relays, switches, automatic systems, radio receivers – devices which the students can construct in their own time and use in the home.

The keynote of the book is relevance – throughout, physics is presented through analogies and situations with which the average student will be familiar and will encounter in everyday life.

O.N. Bishop
Wiseton

Introduction

The practical activities of this course are listed under 4 headings:

1 Instructions for laboratory work. These give you information and detailed guidance. Work with a partner or in a small group. Complete as many of these items as time and facilities allow, beginning each item when you come to it in the text. The instructions cover several different kinds of laboratory activity. Some tell you how to use scientific apparatus and how to make measurements of physical quantities. Some demonstrate the important ideas of physics. Some are the starting point from which we get ideas that are to be discussed later in the text. Some are *experiments* – we do these to find out if our ideas about physics are true. Scientific discoveries are made by people who think of new ideas and then do experiments to test them. A few of the demonstrations require special skills or expensive apparatus. These will normally be done for you by your teacher. Detailed instructions for these are not given, but an outline of the method and its results are described in the main text.

2 Questions about the laboratory work or about topics that are under discussion in the text. Work all these questions as you come to them. They give you practice in using new ideas. Some introduce other aspects of the topic, so answering the questions is an essential part of the course.

3 Projects. At the end of some chapters there is a list of interesting projects. These give you outlines of things you might do as a class activity, as displays for Open Days or Science Fairs, as a hobby activity at home or in your Science Club. Choose the activities that interest you most and preferably, add some original ideas of your own.

4 Revision questions at the end of the book. These are to give you practice for examination work after you have completed each chapter.

Finding your way through the book:

The book is divided into 7 main parts, A to G. In each part we look at physics from a different angle. It is best to study the parts in alphabetical order. In Parts A, B, D, E, and F it is best to study the chapters in numerical order. In part C you may study the chapters in any order you like. In Part G you may study the chapters in any order except that it is best to study chapter 24 before chapter 25; chapter 26 before chapter 29; chapter 31 before chapter 32.

The instructions for laboratory work include many questions. Answers to these and to the groups of questions in the text are given in the Answers and Discussion section at the end of the book. When working through a set of instructions or a group of questions, it is best to attempt the questions one at a time, checking the answer after each question. The answer to one question often includes some discussion which helps you to answer the next question.

PART A MEASURING

1 Measurement of length, mass, time

1.00 Units for measuring

Most scientists make measurements, and they use the International System of Units. This is often called SI, for short. We use SI in this book. A few of its units are commonly used in everyday life. SI has seven *base units* for the seven physical quantities. We shall use three base units in this chapter, and two others later in the book.

1.01 Measuring length

The SI unit of length is the **metre.** From this we get other useful lengths:

Length	Name	Symbol
1000 metres	kilometre	km
1 metre	**metre**	**m**
1/10 metre	decimetre	dm
1/100 metre	centimetre	cm
1/1 000 metre	millimetre	mm

There are units of length smaller than the millimetre. We shall study these later.

Instructions: measuring lengths

1 Practise using instruments for measuring lengths (see figs. 1.1 to 1.5). Measure all kinds of objects that are in the laboratory or classroom: books, pens, coins, wire, furniture, yourself, the room, and many other things.

2 *Before* you make each measurement, you should *estimate* what you think the length will be. Then use an instrument to find out if your estimate is correct.

3 When a person walks or runs, his steps are not all exactly the same length. Find a way of measuring the *average* length of the steps. When you have thought about this and have found a way, turn to the Answers and Discussion section, beginning on page 333.

4 A sheet of paper or metal foil is so thin that we cannot accurately measure its thickness, even with a micrometer screw gauge. Find an accurate way of measuring the thickness of paper or foil.

1.02 Measuring mass

The SI unit for a quantity of matter (or mass) is the kilogram. This gives us several other units of mass:

Mass	Name	Symbol
1 kilogram	**kilogram**	**kg**
1/1 000 kilogram	gram	g
1/1 000 000 kilogram	milligram	mg

Always in everyday life, and almost always in the laboratory, we find the mass of an object by using a balance or scales. We usually call this *'weighing'* the object, but the *weight* of the object is the downward **force** with which the Earth pulls on the object. You can feel this downward force if you hold the object in your hands. The greater the mass of the object, the greater the force. Thus balances are used to measure or compare masses and the units used are g or kg.

When we need to know the *weight* of an object, we measure it in **newtons.** 1 newton is the force which will give a mass of 1 kg an acceleration of 1 m/s². 1 kg mass, free-falling, has an acceleration of $g = 9.81$ m/s², so it has a force of g newtons acting on it.

The weight W of 1 kg mass, then, is given by

$$W = 1 \text{ kg} \times 9.81 \text{ m/s}^2 = 9.81 \text{ newtons.}$$

and g can be written as 9.81 newtons/kilogram (N/kg) or, for convenience, 10 N/kg. Therefore, to find the weight of an object, multiply its mass (in kg) by 10 N/kg. For example, a 100 g brass mass used with a beam balance has a mass of 100 g (or 0.1 kg); however, its weight is 0.1 kg × 10 N/kg = 1.0 N. Thus a lump of brass of mass 100 g can

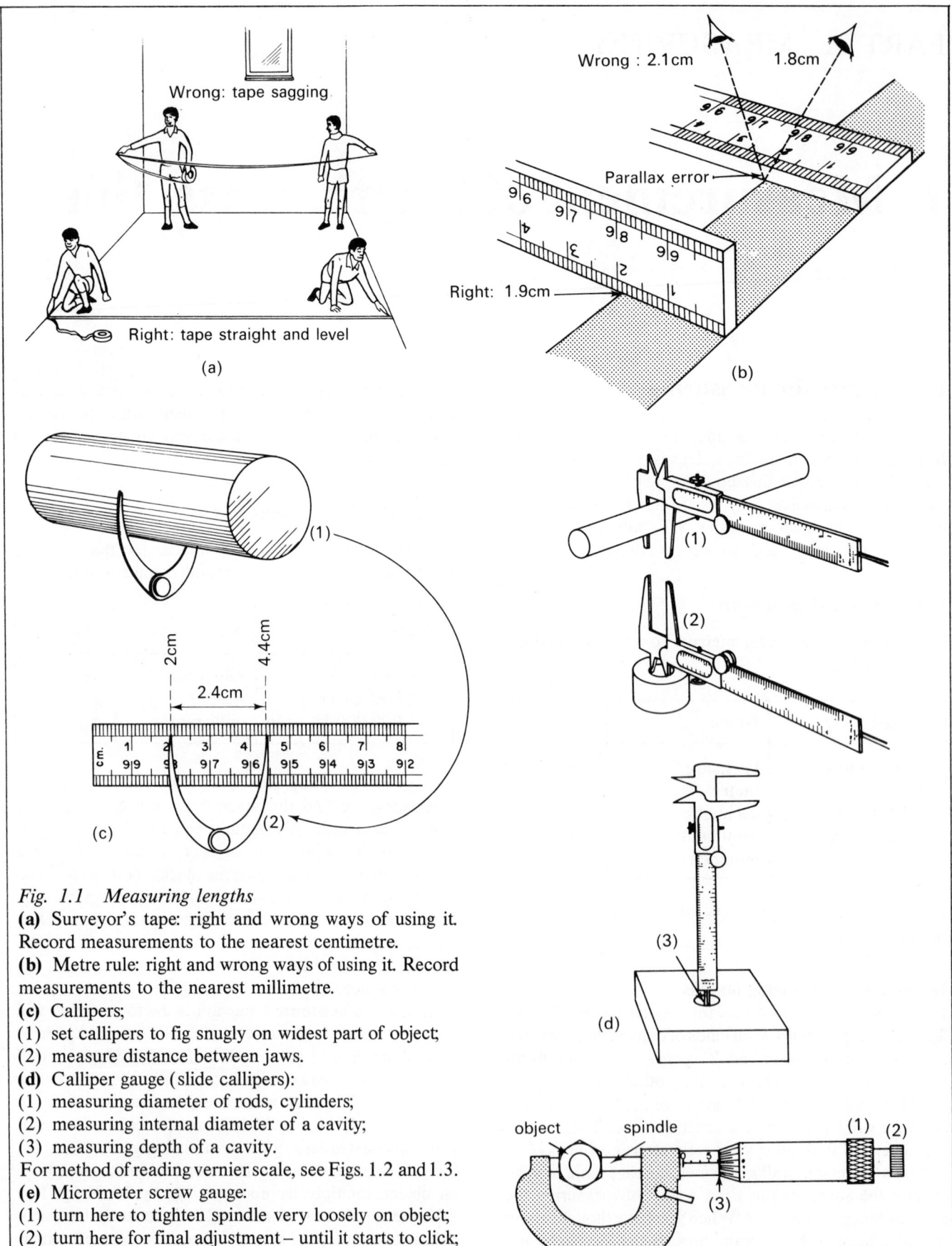

Fig. 1.1 Measuring lengths

(a) Surveyor's tape: right and wrong ways of using it. Record measurements to the nearest centimetre.

(b) Metre rule: right and wrong ways of using it. Record measurements to the nearest millimetre.

(c) Callipers;

(1) set callipers to fig snugly on widest part of object;

(2) measure distance between jaws.

(d) Calliper gauge (slide callipers):

(1) measuring diameter of rods, cylinders;

(2) measuring internal diameter of a cavity;

(3) measuring depth of a cavity.

For method of reading vernier scale, see Figs. 1.2 and 1.3.

(e) Micrometer screw gauge:

(1) turn here to tighten spindle very loosely on object;

(2) turn here for final adjustment – until it starts to click;

(3) read measurement here – see Figs. 1.4 and 1.5.

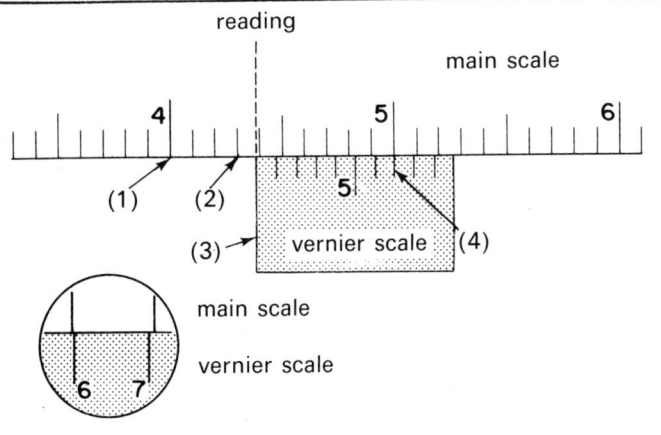

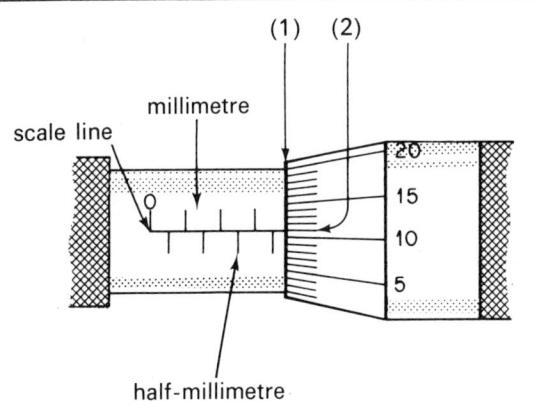

Fig. 1.2 Reading a vernier

(a) (1) count centimetres along main scale; write . 4
(2) count millimetres; write second figure 43
(3) *as a check* estimate tenths of millimetre
(estimate; 'between 6 and 8 tenths')
(4) look along vernier scale until you find a mark that is
exactly in line with one of the marks on the main scale.
The number of this mark, *counting along the vernier
scale*, is the third figure of the result . . . . . 43.7 mm
(b) If at step **4 no** marks are exactly opposite, take the
nearest. Use a lens if necessary. In the figure the nearest
mark is '6', so third figure of result is '6'.

Fig. 1.4 Reading a micrometer screw gauge:
Divisions on the rotating scale represent 0.01 mm each.
There are 50 divisions on the scale, so one turn moves the
spindle 50 × 0.01 mm = 0.5 mm. To read the gauge:
(1) count millimetres and half-millimetres
from zero of scale to this edge . . . . . . 3.50 mm
(2) read division comes level with
the scale line . . . . 0.11 mm
(3) add these two . . . . 3.61 mm
If divisions are not exactly opposite the scale line, read the
nearest. If they are equally distant, read the higher.

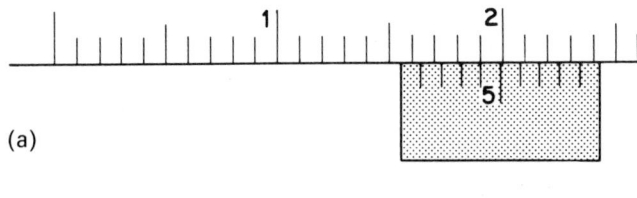

(a)

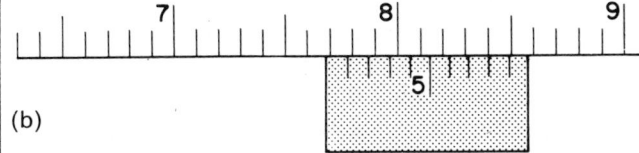

(b)

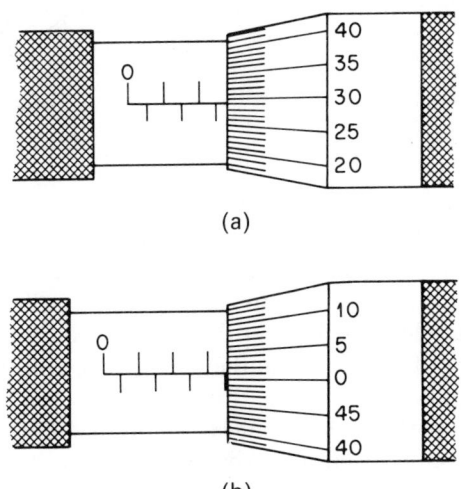

(a)

(b)

*Fig. 1.5 Test yourself – can you read a micrometer
screw gauge? (See Answers and Discussion section)*

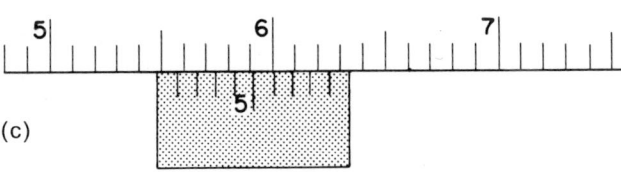

(c)

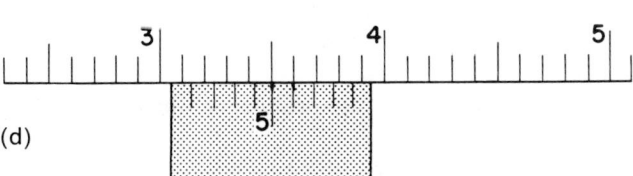

(d)

*Fig. 1.3 Test yourself – can you read a vernier? (see
Answers and Discussion section)*

be used in the laboratory to exert a force of 1 N on some
other object because of its weight. Clearly, mass and
weight have different units and are not the same or even
similar quantities.

Instructions: finding the mass
1 Practise using balances for finding the mass of objects,
(see fig. 1.6). Find the mass of all kinds of things that are in
the laboratory or classroom.

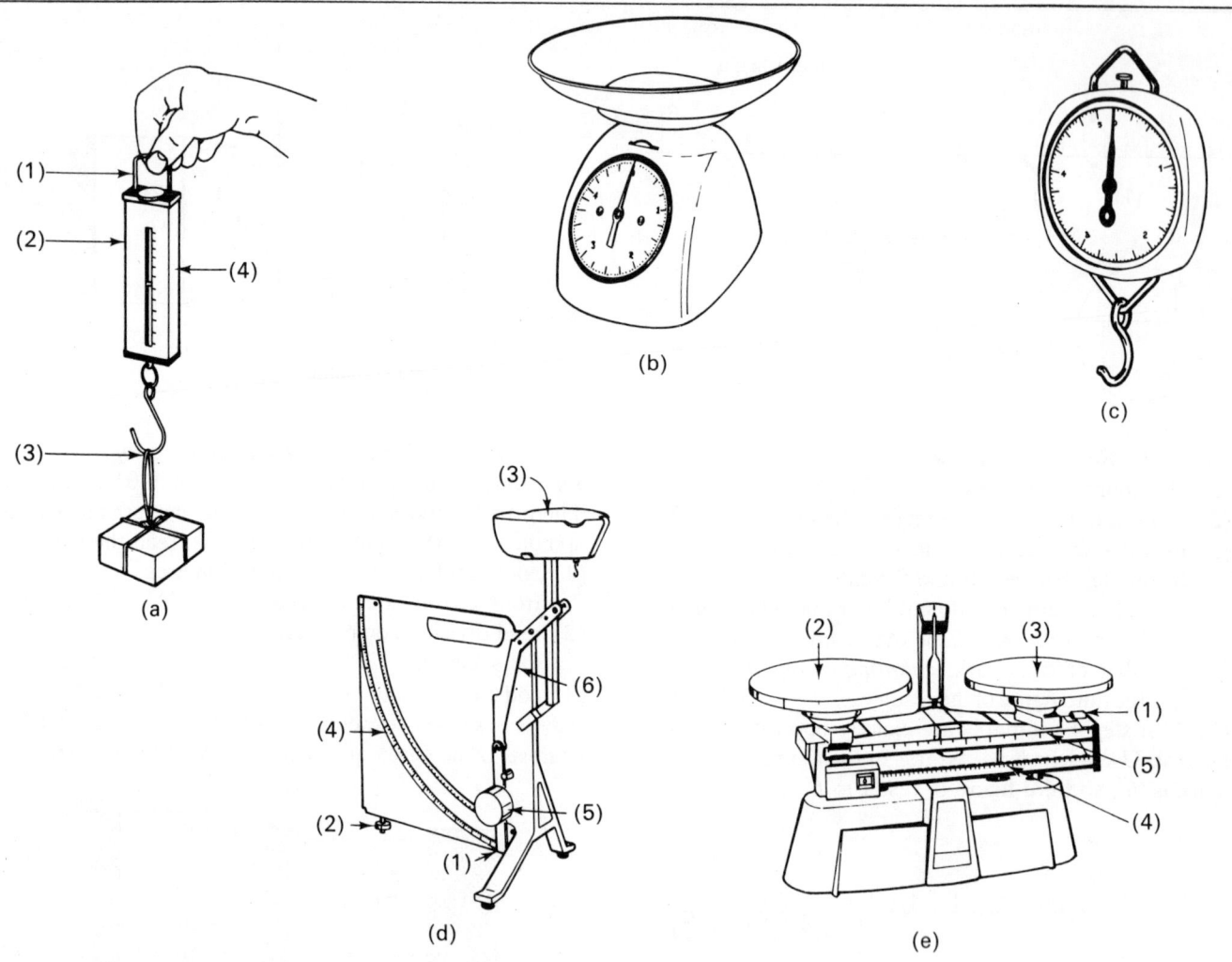

Fig. 1.6 Measuring mass

(a) Spring balance:
(1) let the balance hang down freely from the ring;
(2) check that it reads zero when nothing is on the hook; if not, adjust the knob at the top of the balance;
(3) hang the object from the hook, using thin thread if necessary;
(4) record the reading: the scale mark nearest the pointer.

(b) (c) Other kinds of spring balance.

(d) Lever balance:
(1) check that the balance reads zero when nothing is on the pan;
(2) if not, adjust this screw leg;
(3) put the object on the pan;
(4) record the reading: the scale mark nearest the pointer;
(5) most lever balances work on two ranges; if the large weight is *here*, as illustrated, read on the 'high' scale (usually up to 1 000 g);
(6) if the weight is here, read on the 'low' scale (usually up to 250 g).

(e) Beam balance:
(1) check that the beam swings freely and evenly about its central position when there is nothing on the pans; if

not, adjust this nut (or the one at the other end, if there is one);
(2) put the object on this pan;
(3) if you are using separate weights, put them on this pan until the beam swings evenly about its central position. Add up the weights and record the total.
(4) if you are using sliding weights, start with both sliders at the left-hand end (NOTE: check that they were both at that end when you balanced the beam at Step 1). Now slide this weight along, step by step to the right, until the right-hand side of the balance goes down; then slide the weight back *one* step to the left: the left-hand side goes down;
(5) slide this weight (the smaller one), step by step to the right, until the right-hand side of the balance goes down. Then slide it slightly back to the left, until the balance swings level;
(6) add the readings of the positions of the two weights, and record the total.
(7) Some balances have three sliding weights: step the largest first, then the middle weight, and finally the smallest. These balances often do not have a pan for weights on the right-hand side.

2 First make an estimate. Then, check your estimate against the actual reading shown on the balance.

3 Loose powders, chemicals and liquids must not be put directly on a balance pan. How can you find the mass of substances like these?

4 Some objects, such as thumb-tacks, are too light to 'weigh' on the kinds of balance you are using for this activity. How can you find the mass of light objects, using one of the kinds of balance from fig. 1.6?

1.03 Measuring time

The SI unit of time is the **second**. From this we get other units:

Time	Name	Symbol
24 hours (= 86 400 seconds)	day	d
60 minutes (= 3 600 seconds)	hour	h
60 seconds	minute	min
1 second	**second**	s
1/1 000 second	millisecond	ms

For times much longer than 60 seconds we use minutes, hours and days. They are convenient units and we are familiar with them. We use the 3-letter symbol 'min' for minute, so as not to confuse it with the symbol for metre.

Instructions: timing

1 Practise using a stop-watch or stop-clock. Your teacher will show you how to operate it.

2 Estimate how long it takes for you to run 50m, or how long it takes for you to read 100 words. Then measure the time, using the stop-watch. Repeat for several other activities. It is a good habit to estimate lengths, weights and times *before* measuring them – *why* is it a good habit?

1.10 Some derived SI units

The base units of SI can be combined to give other useful units. These units are *derived* from the base units. We shall study a few of these in the remainder of this chapter, and meet many more in later chapters.

1.11 Units of area and volume

The SI unit of area is the **square metre** (symbol, **m²**). The SI unit of volume is the **cubic metre** (symbol, **m³**). For small areas and volumes we can use square millimetre (mm²), square centimetre (cm²), cubic millimetre (mm³) and cubic centimetre (cm³), converting to SI units in the

final answer. There is also a unit of volume called the litre (symbol, *l*), which is equal to 1 000cm³. For large areas we can use square kilometre (km²), and a special unit often used for areas of land, the hectare (equal to 10 000m²).

Instructions: measuring volumes

1 Measure volumes of quantities of liquid by using a graduated beaker or a measuring cylinder (fig. 1.7). Keep your eye level with the *lowest* part of the liquid surface. In fig. 1.7, the volume should be recorded as 175cm³.

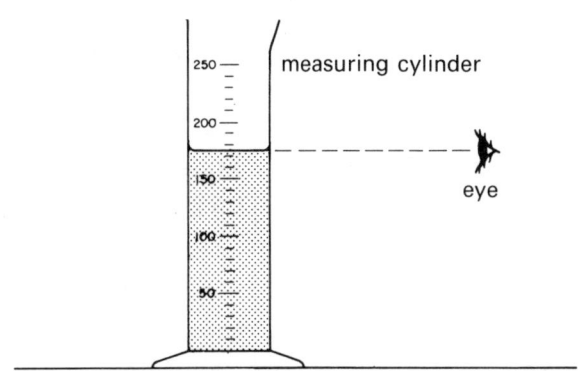

Fig. 1.7 Measuring the volume of liquid in a measuring cylinder.

2 For measuring volumes of irregular solids (lumps of rock, pieces of metal of complicated shape), use the *displacement* method. You can use a special can for this (fig. 1.8). Fill the can so that the water is level with the bottom of the spout. Then place the object in the can, collecting the water which overflows; measure its volume; this equals the volume of the object.

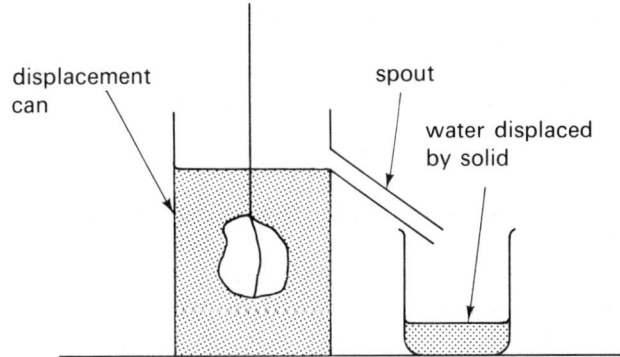

Fig. 1.8 Measuring the volume of an irregularly shaped object.

3 If you do not have a displacement can, use a measuring cylinder instead. Fill it about half-full with water; read and record the volume of water, V_1. Put the object in the water so that it is completely covered; read and record the total volume of water and object, V_2. Calculate the volume of the object, V_2-V_1.

4 If the object is soluble in water, use a liquid in which it is insoluble, such as oil or ethanol.

1.12 Density

We can measure the length, breadth and height of an iron block; we can also measure its mass. These *four* measurements can be combined in an interesting way.

Instructions: measuring a property of iron
1 You need about 10 pieces of iron: several rectangular blocks of different sizes and shapes, discs and rods, and a few irregularly shaped pieces such as nuts and bolts.
2 Draw a table.

Piece No	Volume V (in cm³)	Mass m (in g)	m/V
1			
2			
3			
etc.			

3 Measure the volume of each piece. You will find it convenient to measure the 'lengths' of the regular shapes in centimetres and calculate the volumes in cubic centimetres (cm³).
4 Find the mass of each piece. Again you will find it more convenient to find the mass in grams (g).
5 For each piece, divide its mass (m) by its volume (V) and write the answer in the last column of the table. Of course, some of these divisions do not work out exactly. You may get a long row of figures after the decimal point. You do not need these figures. Work out all answers to just *one* figure after the decimal point.
6 What do you notice about the figures in the last column of the table? Compare your results with those of others in your class – what do you notice?

Instructions: measuring the same property for other materials
1 Do we obtain the same kind of result if we use other materials? To answer this, repeat the instructions above, using other materials, such as aluminium, brass, wood, glass, stone, paraffin wax. Use several different blocks of each kind of material.
2 When you have measured volumes and masses, calculate m/V for each material.
3 What do you notice about the values of m/V? Compare your results with those of others in your class.

The value of m/V is a property of the material. It tells us something about the material. m/V is high for materials such as lead and iron. It is low for materials such as wood and wax. In everyday life we sometimes *say* that iron and lead are 'heavy'. We also *say* that wood is 'light'. We *mean* that a fairly small piece of iron or lead has a large mass compared to that of, say, a piece of wood of similar size. However we should never use the words 'heavy' and 'light' to describe materials in that way. After all, a large plank of wood may well be heavier than a small lump of lead.

The correct way to contrast these materials is to say that iron or lead is *denser* than wood or wax. The value of m/V is a way of expressing their *density*. Density is a property of a material. Iron has a particular density; it does not matter if the iron is in the form of a block, a disc, a nut or a bolt, it still has the same density. The value of m/V for iron is the same, no matter what shape the piece of iron has. For high-density materials, m/V is large; for low-density materials m/V is small.

So far we have thought of m/V simply as a number. Now we need to find some units for it. Suppose we have a block of cast iron (fig. 1.9) which we have 'weighed' to find its mass and measured to find its volume. To calculate its density we divide its mass, m (in g) by its volume, V (in cm³). The answer to this division is 7.0. To show how we got this number, we can write something after it: for cast iron, density = 7.0 grams divided by cubic centimetres. This is rather a long way of writing it, so let us use a shorter way: for cast iron, density = 7.0g/cm^3. From this statement, anyone can see that we divided a mass by a volume. The statement is *read* like this. 'For cast iron, density equals seven grams per cubic centimetre'. The word 'per' means 'for each'. However, the SI unit of density is kilograms per cubic metre (kg/m^3) and hence we need to know how to convert g/cm^3 obtained in practical experiments into SI units of kg/m^3.

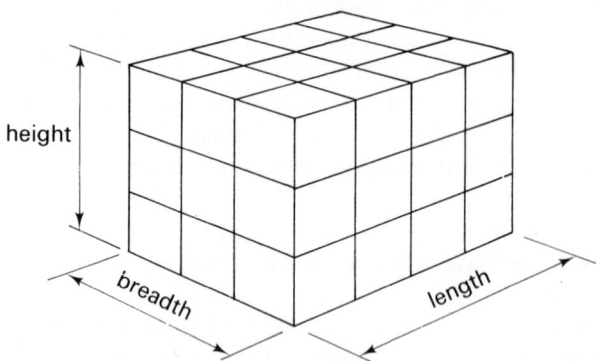

Fig. 1.9 A block of cast iron. Its faces are marked in centimetre squares; $V = 4 \times 3 \times 3 = 36 \text{ cm}^3$; $W = 252$ g.

If 1 kg = 1 000 g and 1 m³ = 1 000 000 cm³

then $\quad 1\dfrac{\text{kg}}{\text{m}^3} = \dfrac{1\,000}{1\,000\,000}\ \dfrac{\text{g}}{\text{cm}^3}$

$\quad 1\ \text{kg/m}^3 = \dfrac{1}{1\,000}\ \text{g/cm}^3$

$$1\ 000\ \text{kg/m}^3 = 1\ \text{g/cm}^3$$

thus $7\text{g/cm}^3 = 7 \times 1\ 000\ \text{kg/m}^3 = 7\ 000\text{kg/m}^3$

If your answer for density is a number in g/cm³ you can convert the units to kg/m³ provided you multiply the number by 1 000.

You can look at the idea of density in another way. The block of iron in fig. 1.9 can be thought of as 36 separate cubes, each with a volume of 1cm³. If we break the block into 36 separate cubes, each cube has a mass of 7.0g. The iron has a mass of 7.0g per cube, or cubic centimetre.

Questions
1 What is the mass of 10 cubes of cast iron, if each cube has a volume of 1cm³?
2 What is the mass of 258 such cubes?
3 What is the mass of a rectangular block of cast iron, measuring 15cm x 3cm x 2cm?
4 The density of bamboo is 0.4g/cm³. (Note: read this density as '0.4 grams per cubic centimetre'). What is the mass of a solid cylinder of bamboo, radius 2cm and 150cm long? (Take $\pi = \frac{22}{7}$)
5 The density of nylon is 1.14g/cm³. What is the mass of a ball of nylon, radius 2cm?

The calculations show how useful it is to know the density of a material. If you know the size of an object you can work out its mass without actually measuring that mass.

Density is used so often in physics that we have given it a special symbol. Instead of writing 'm/V' we usually write the single Greek letter, ρ, rho, to mean the same thing.

We write: for cast iron, $\rho = 7\ 000\ \text{kg/m}^3$

We read: 'For cast iron, density equals 7 000 kilograms per cubic metre'.

We have now expressed density in base units.

Instructions: measuring the density of a liquid
1 Place a graduated beaker or measuring cylinder on the pan of a balance. Find the mass of it empty. Record the empty mass m_1, in grams.
2 Pour water (or other liquid) into the beaker until you have some convenient quantity. Record the volume of the liquid, V, in cubic centimetres.
3 Find the mass of the beaker or cylinder with the liquid in it. Record the mass of beaker plus liquid, m_2, in grams.
4 Calculate the mass of the liquid. $m = m_2 - m_1$
5 Calculate the density of the liquid. $\rho = m/V$
6 The units in which density is calculated are g/cm³. As explained above, density can be expressed in base units, using the same figures. Write your result in kg/m³.
7 Repeat the experiment using different liquids such as: water, glycerine, various oils, methylated spirit and turpentine. After finishing the experiment, look in the Answer and Discussion section, to see what results you might expect to get.

In the activity above you probably used a beam balance to find the mass of the liquid and a measuring cylinder to find its volume. A good beam balance can measure to the nearest 0.1g. If you know the mass of the empty cylinder to the nearest 0.1g, and the mass of the cylinder with liquid in it to the nearest 0.1g, you can calculate the mass of the liquid to the nearest 0.2g. In other words, there can be an error of up to 0.1g in *both* measurements. When we subtract one from the other there can be an error of up to $0.1 + 0.1 = 0.2$g in the answer. If you had 100cm³ of liquid and this weighed 100g, the error in measuring its *mass* is 0.2g, or *0.2%*.

A measuring cylinder of 100cm³ is graduated to the nearest 1cm³. Your measurement of the volume of the liquid is accurate to the nearest 1cm³: a possible error of *1%*. You know the mass much more accurately than you know the volume. When you calculate m/V, the answer you get is no more accurate than the *least* accurate of the two measurements used in calculating it. So your calculated density has a possible error of up to 1%.

If we want to get a more accurate measurement of density, it is a waste of time to use a more accurate balance. The mass is already known with an accuracy five times that of the volume measurement. We must use a more accurate method to measure the volume. The set of instructions explains this method.

Instructions: a more accurate method for measuring the density of a liquid
1 Use a burette, or pipette, to deliver a measured volume of the liquid into an empty beaker of known mass.
2 Repeat the procedure given above to find the mass of the 'known' volume of the liquid.
3 Calculate the density of the liquid. $\rho = m/V$

The *relative density* of a substance is the density of the substance relative to the density of water.

$$\text{relative density of a liquid} = \frac{\text{density of the liquid}}{\text{density of water}}$$

$$\text{r.d.} = \frac{\text{mass of liquid}}{\text{volume of liquid}} \div \frac{\text{mass of water}}{\text{volume of water}}$$

If the *same* volume of liquid and water is used, then, relative density of a liquid = mass of liquid ÷ mass of water

$$\text{r.d. of a liquid} = \frac{\text{mass of liquid}}{\text{mass of an } equal \text{ volume of water}}$$

Note also since weight is proportional to mass the relative density can also be expressed by

$$\text{r.d. of liquid} = \frac{\text{weight of liquid}}{\text{weight of an } equal \text{ volume of water.}}$$

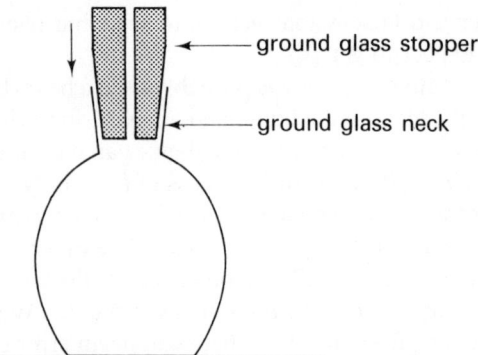

Fig. 1.10 A density bottle.

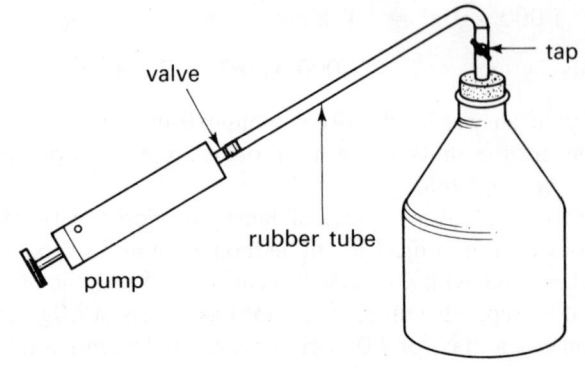

Fig. 1.11 Finding the density of air.

Instructions: an accurate method for measuring the relative density of a liquid

1 Use a density bottle (fig. 1.10). When this is correctly filled, it should always contain exactly the same volume of liquid. Make sure that the bottle is clean and dry. Find and record its mass, m_1, including the stopper.

2 Remove the stopper and fill the bottle up to the neck with distilled water. Slowly put the stopper in. Water should rise up the narrow tube and overflow at the top. Use a piece of paper tissue to wipe across the top of the stopper to remove any small drop of water standing up above the level of the top of the stopper. Also wipe the bottle carefully outside, to remove any drops of water. Find the mass of the bottle, with water in it. Record its mass, m_2.

3 Remove the stopper, pour out the water. Use a paper tissue to wipe water from the bottle and the stopper. Rinse the bottle with some other liquid for which you wish to find the density; use methylated spirit, glycerine or turpentine. Pour this out, and rinse again with another small quantity of the liquid. This is to make sure all traces of water are gone.

4 Repeat step 2, using the other liquid. Record the mass of this liquid, m_3.

5 The mass of water in the bottle is $m_2 - m_1$.

6 The mass of other liquid is $m_3 - m_1$. The volume of other liquid is the same volume as the water, because the bottle contains exactly the same volume each time. Calculate the relative density of the other liquid.

$$\text{relative density} = \frac{m_3 - m_1}{m_2 - m_1}$$

Instructions: measuring the density of air

1 You need a large plastic container and a pump (fig. 1.11). The container has a tap. The pump has a non-return valve, like the valve on a football inflator.

2 Find and record the mass, m_1, of the empty container. Is the container truly empty?

3 Pump air into the container for as long as you can. Find the mass of the container, which now has *extra* air inside

it. If the mass has not increased very much, try pumping in some more air. Find the mass again. Record the final mass, m_2.

4 To find out how much air is in the container, we let it out slowly and measure its volume (fig. 1.12). Use a beaker or jar of which you know the volume. Count how many

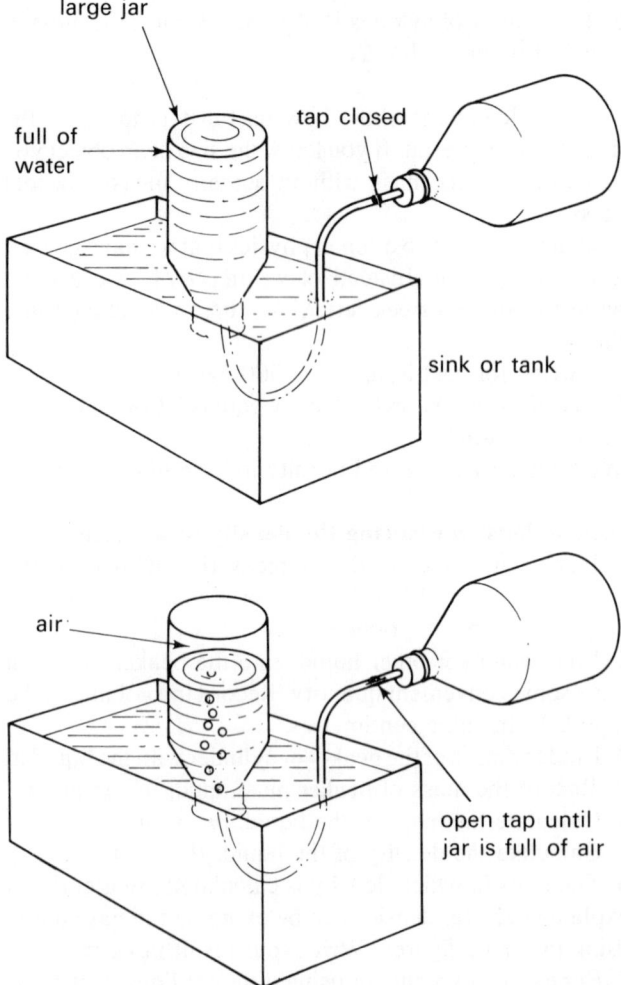

Fig. 1.12 Measuring the volume of air in the container.

8

beakersful of air you collect. At the end, when no more air comes from the container, you have measured all the *extra* air. The container now just has inside it as much air as it had to begin with. Calculate the total volume of extra air, V.

5 Calculate the mass of the extra air, $m = m_2 - m_1$.

6 Calculate the density of air, $\rho = m/V$. Compare your result with the figures given in the Answers and Discussion section.

1.13 Units of pressure

Instructions: discovering the effect of pressure

1 You need the apparatus shown in fig. 1.13. The card should be stiff, about 2mm thick; use the same kind of card for all the squares. For plastic foam use the soft flexible kind, often used for packing breakable equipment and for putting in cushions.

2 Draw a table.

Area, $A/10\,000$ in m^2	Weight, W in N	W/A in N/m^2
1		
4		
9		
16		
25		

(*Note* 1/10 000 m² is 1 cm², hence the numbers in the Area column are effectively in square centimetres but the multiplying factor of 1/10 000 enables the column to have SI units of m²).

3 Put a card on the foam. Pile 'weights' on the card until the card sinks into the foam. The top surface of the card must be exactly level with the top surface of the foam. It is best to put your eye level with the bench, and look along the top of the foam, to judge if foam and card are level. Use balance 'weights', or small pieces of metal, glass or stone. Arrange these carefully on the card so that the card sinks evenly into the foam. It must not tilt to one side.

4 Add up all the 'weights' on the card. If you are using pieces of metal or other materials, put them all together on a balance and 'weigh' them. As explained earlier, a 100g (0.1kg) mass has a weight W given by $W = 0.1\,\text{kg} \times 10\,\text{N/kg} = 1\,\text{N}$. Hence if you know the mass of the

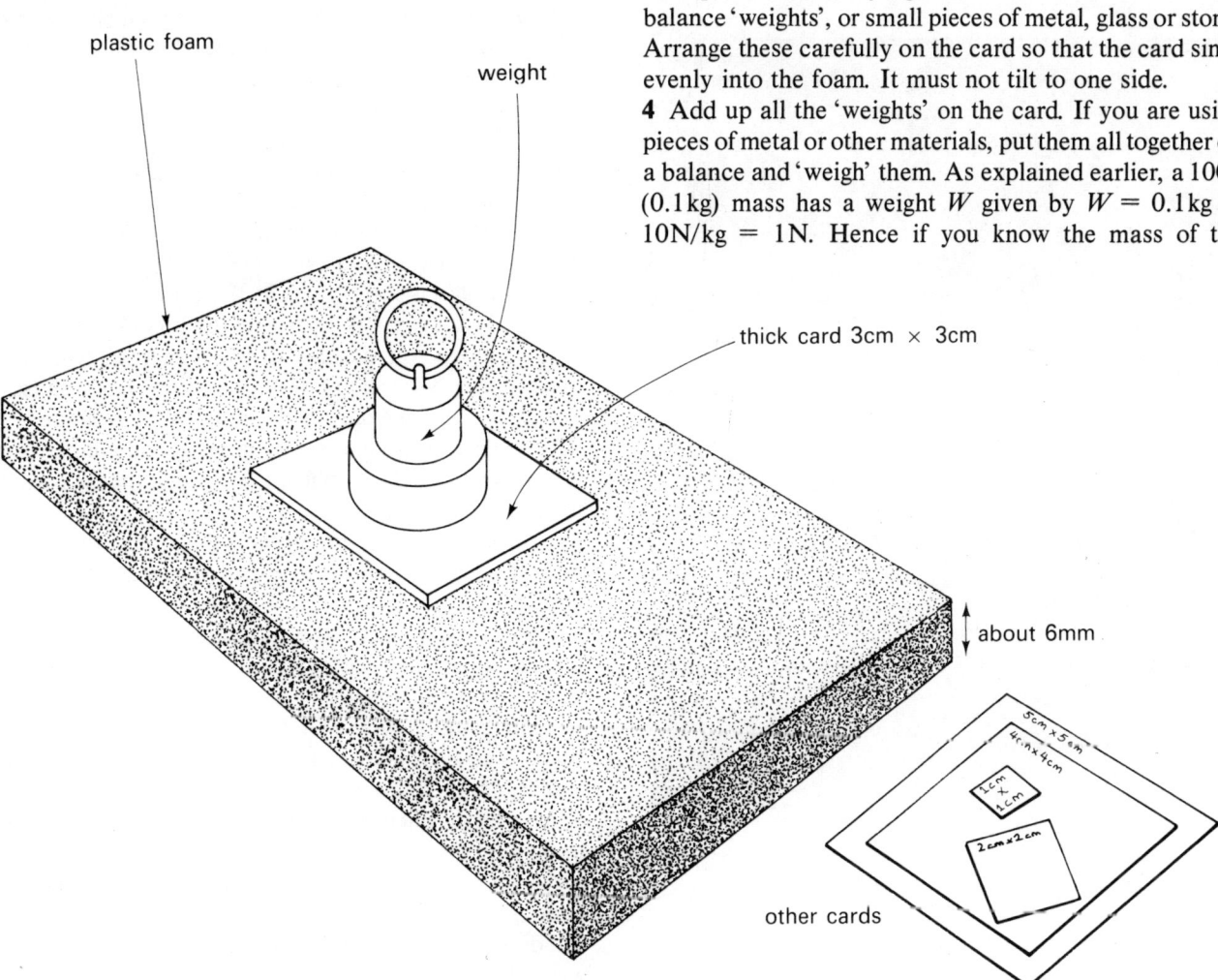

plastic foam

weight

thick card 3cm × 3cm

about 6mm

5cm × 5cm
4cm × 4cm
1cm × 1cm
2cm × 2cm

other cards

Fig. 1.13 Apparatus for investigating pressure.

'weights', pieces of metal or other materials you can calculate the weight W in newtons. Record the weight, W. You are using the 'objects' to exert force on the card and foam, so the units of measurement are newtons.

5 Repeat steps 3 and 4, using the other squares of card.
6 For each card, calculate W/A. Write the result in the last column of the table. What do you notice about the values of W/A? Do other people in your class find the same?

We have derived another unit. This is a unit for pressure. To measure the pressure on the foam we divided a weight (in N) by an area (in m²). The pressure is expressed in newtons per square metre (N/m²); however, the correct SI name for this unit is the **pascal** (Pa).

The amount that the foam is compressed depends on the pressure acting on it. If we put a heavy weight on a small card, pressure is high. The card sinks deeply into the foam. If we put a heavy weight on a large card, its weight is spread over a large area of foam. Pressure is low and the foam can support it.

Fig. 1.14 Viking I lander; note the large pads on each leg.

When the spacecraft *Viking* was sent to land on Mars, scientists were not certain what the surface of Mars would be like. They thought it might be a soft soil and the *Viking* would sink deeply into the soil after landing. To prevent this, the legs of the lander had large pads on them (fig. 1.14) to spread the weight of the lander over a large area of soil. This reduced the pressure on the soil, making the lander less likely to sink into soft soil.

Questions

1 A fish-tank measures 40 cm × 20 cm × 25 cm inside. What is the mass of water it holds when it is full to the brim?

2 The same fish-tank is filled with sea-water. The mass of the sea-water is 20.6 kg. Calculate the density of the sea-water.

3 A swimming-pool is 50 m long, 20 m broad, and 2 m deep. When it is full to the brim with water, what is the mass of the water?

4 Two beakers of equal mass are put on the pans of a beam balance, which swings level. 190 cm³ of ethanol ($\rho = 800$ kg/m³) are put in one beaker. How much castor oil ($\rho = 950$ kg/m³) must be put into the other beaker to make the balance swing level again?

5 Calculate the mass of air in your laboratory or class-room.

6 An average person breathes about 400 cm³ of air in and out at each breath. What is the mass of one breath?

7 One of the world's largest airships was a *Zeppelin*, with a gas capacity of almost 200 000 m³. The density of hydrogen gas is 0.09 kg/m³ at the pressure used in the airship. What mass of hydrogen did it contain?

8 A scuba-diver has his cylinder charged with compressed air. After charging, the cylinder has a mass of 2.34 kg more than it did when 'empty'. What volume of air will the diver be able to get from this cylinder if he swims just below the water surface? (Density of the air when he breathes it = 1.2 kg/m³).

9 What is your weight? What is the area of the soles of your two feet? What pressure do you exert on the ground when standing?

10 A woman weighs 630 N. The area of the soles of both feet is 300 cm². What pressure does she exert on the ground when standing?

11 The same woman (Q.10) wears shoes with narrow heels. The heel area is 1 cm² on each shoe. When she stands, almost her whole weight is supported on her heels. What pressure does she exert on the floor when standing in these shoes?

12 A rectangular block of gold measures 10 cm × 4 cm × 1 cm. The density of gold is 19 500 kg/m³. The block is standing on a flat table. What is the pressure on the table when the block is standing on (a) one of its largest faces, (b) one of its long edge faces, and (c) one of its end faces?

Fig. 1.15   Reindeer. Their large feet spread their weight evenly so they are able to walk on the ice and snow.

Project

Some kinds of animal that live in sandy places or spend a lot of time on soft mud have feet which have a large area. This reduces the pressure on the ground as they walk. They do not sink into the soft sand or mud. They can travel about easily. Find out the names of animals like this; collect photographs or make drawings showing their feet. Try to find out the weights of the animals and the area of their feet. Calculate the pressures. Make a display of your work and put it on the wall of the classroom or laboratory.

2 Using measurements

2.00 Levers

The simplest kind of lever is a straight bar which is usually made of a strong material such as metal or wood. A lever has: *a pivot* which is some kind of axle or bearing around which it can turn, *a load* which we are usually trying to move or exert a force on, by using the lever and *an effort* which is the force we apply to the lever to try to move the load. The arrow in fig. 2.1 points in the direction of the force (W) produced by the weight of the load. In the type of lever shown in fig. 2.1 we are trying to move the load in the direction opposite to this force.

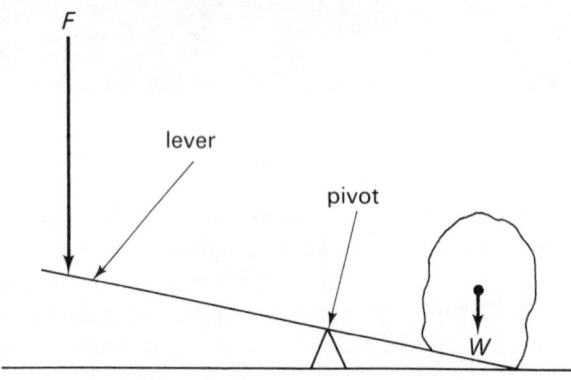

Fig. 2.1   The forces acting on a typical lever.

Instructions: balancing load and effort
1, 2, 3 Figure 2.2 shows you how to balance load and effort, and what measurements to make. Repeat the balancing several times, using different weights as load and effort, and placing them at different distances from the pivot.
4 What can you say about xL and yE when the forces are balanced?

Instructions: a lever with a pivot at one end
1 Fig. 2.3 shows you how to balance load and effort. Use loads of different weights and at different distances, and measure the effort required to balance the load. With this lever the effort is an upward force, measured by the spring balance.
2 What can you say about xL and yE when the forces are balanced?
3 Is there an error in balancing this kind of lever? What is the cause of it?
4 Which force is the greater, E or L?

Instructions: a third kind of lever
1 Fig. 2.4 shows you how to balance load and effort. Use different weights as load, and put load and efforts at different distances from the fulcrum (x always greater than y).
2 What can you say about xL and yE when the forces are balanced?
3 What error occurs in balancing?
4 Which force is greater, E or L?

2.10 Pendulums

Instructions: factors affecting the period of swing (T) of a pendulum
1 Figure 2.5 shows you how to make a pendulum. To begin with, make l exactly 50cm.
2 Measure the period T, the time for one complete swing from centre, to left, to centre, to right and back to centre. It is best to time, say 20 swings and divide the total time, t, by 20 to find the period of 1 swing, T.
3 Repeat for different values of l, between 10cm and 100cm. Make up a rule (in words) that gives the relation between T and l.
4 With $l = 50$cm, find out if T is affected by the size of the amplitude, a. If so, make up a rule to give the relation between T and a.
5 With $l = 50$cm, find out if T is affected by the mass of the bob, m. If so, make up a rule to give the relation between T and m.

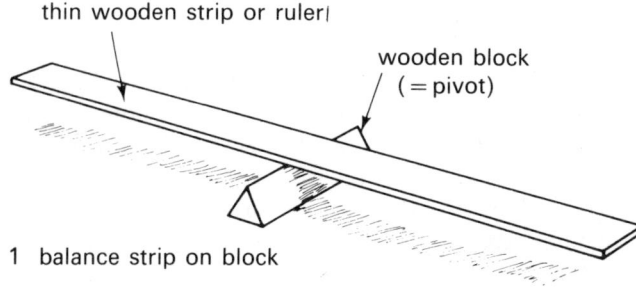

1 balance strip on block

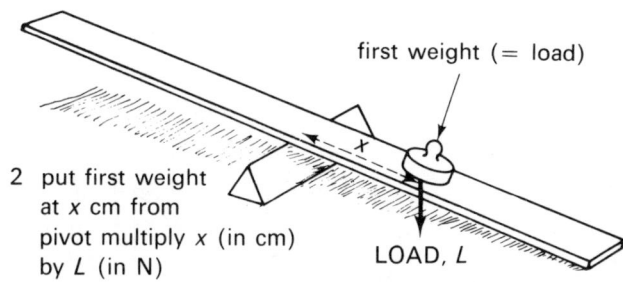

2 put first weight at x cm from pivot multiply x (in cm) by L (in N)

first weight (= load)

LOAD, L

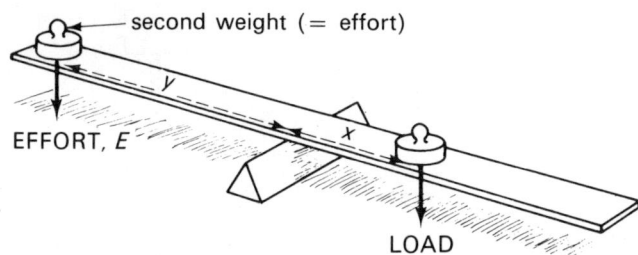

second weight (= effort)

EFFORT, E

LOAD

3 balance load by effort multiply y (in cm) by E (in N)

Fig. 2.2 Apparatus for balancing load and effort.

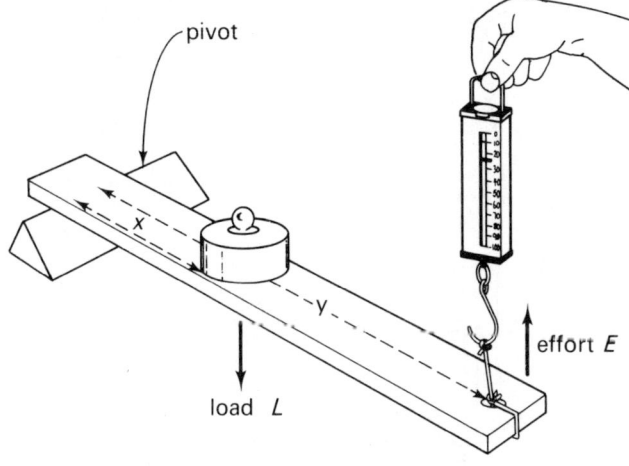

pivot

effort E

load L

1 balance load by effort
2 multiply x by load L
3 multiply y by effort E

Fig. 2.3 Apparatus for investigating the action of a lever with a pivot at one end.

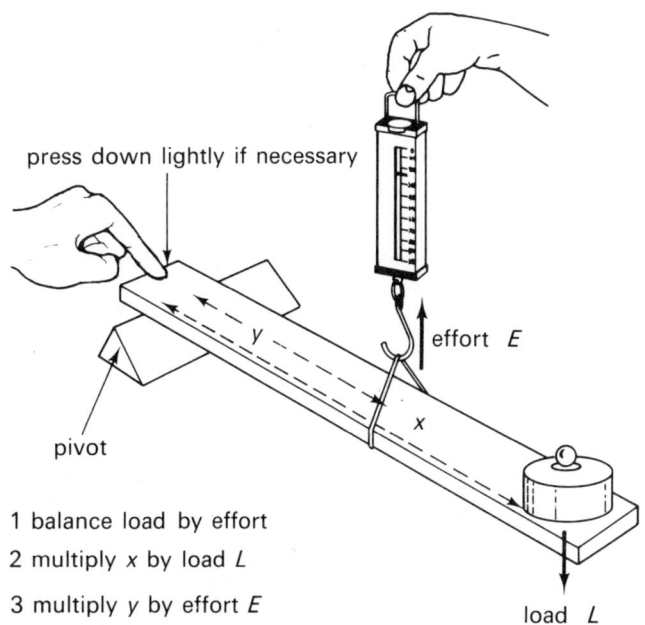

press down lightly if necessary

pivot

effort E

x

load L

1 balance load by effort
2 multiply x by load L
3 multiply y by effort E

Fig. 2.4 A third kind of lever.

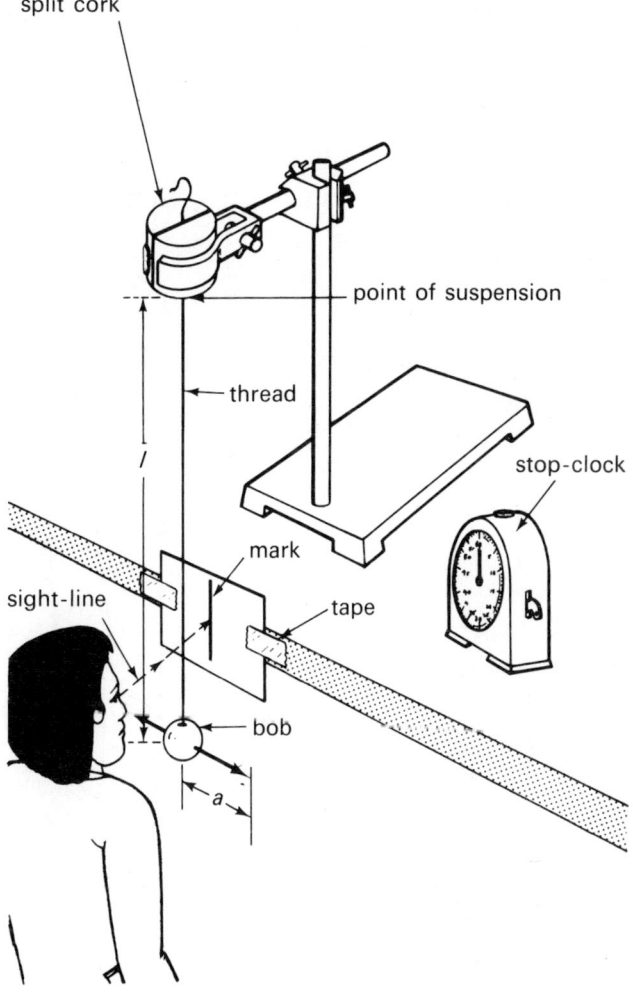

split cork

point of suspension

thread

l

mark

sight-line

tape

stop-clock

bob

a

Fig. 2.5 Apparatus for investigating pendulums.

2.20 Pulleys

Instructions: using systems of pulleys

1 Figure 2.6 shows some ways of using pulleys. Make up each of these systems, one at a time. Use different loads and find how much effort *just* begins to lift the load.

2 Find the value of L divided by E (L/E) for each pulley system. Try to discover a rule that tells you what L/E will be.

3 Design a pulley system for which $L/E = 5$.

2.30 Air pressure

Solid objects press on surfaces (see 1.13). Liquids and gases press on surfaces too. The Earth's atmosphere is around you and above you. It is several kilometres high. The force produced by the weight of this air is pressing on you, and on all the objects in the room. A good way to demonstrate this is to take a metal can and put a little water in it. With the cap of the can off, the can is heated until the water boils. The can becomes filled with steam,

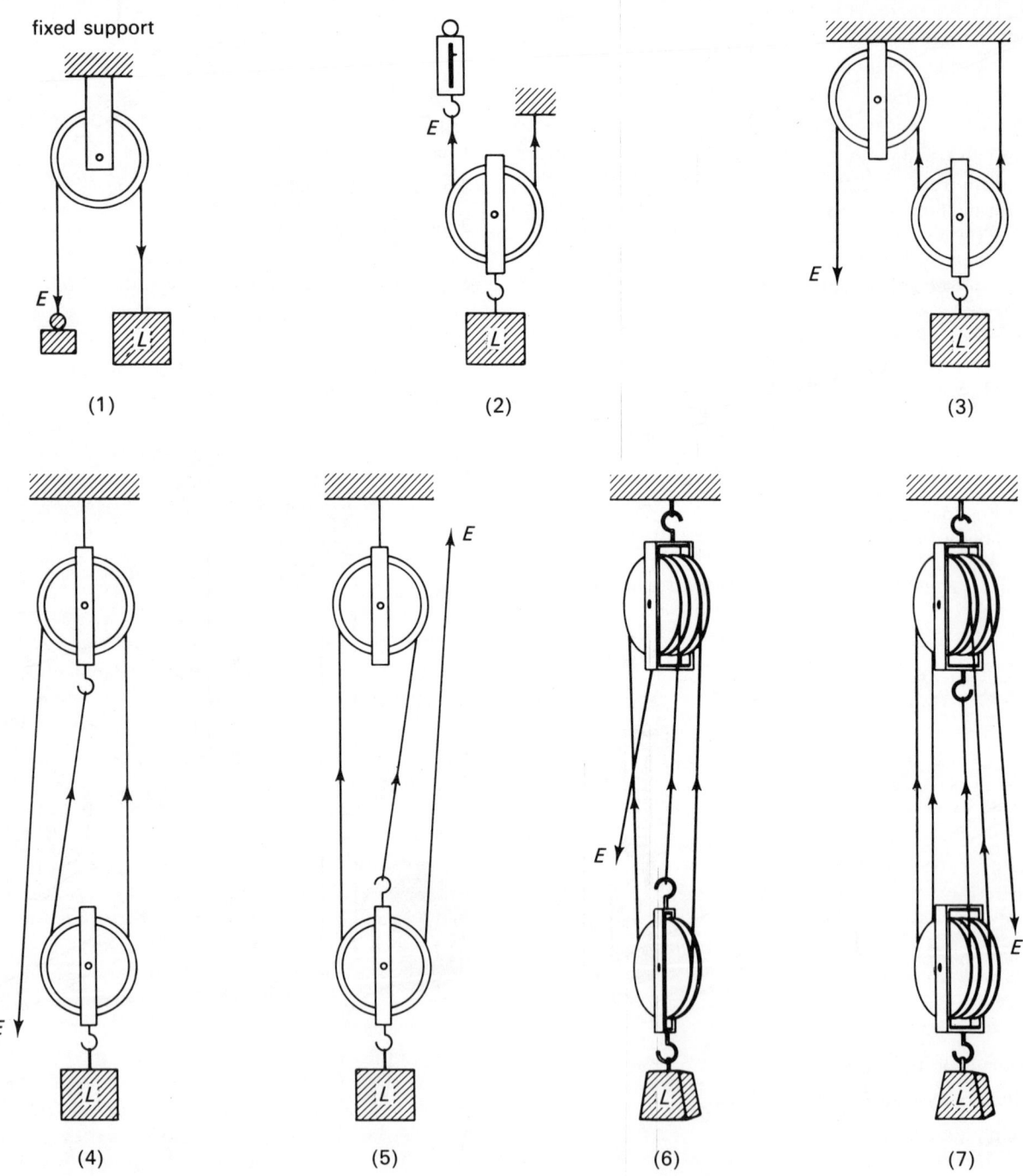

Fig. 2.6 Pulley systems.

and all the air is driven out of it. Then the cap is screwed tightly on. The can is taken away from the heat. Cold water is poured over it so that the steam inside is cooled and turns back into water. Water takes up *much* less space than steam. The can has almost nothing inside it. Now the pressure of the atmosphere is pressing *inwards* on the outside of the walls of the can, but there is very little gas inside to press *outwards* on the walls of the can. As soon as the steam turns into water, the can is squashed almost flat. This shows the great pressure of the atmosphere.

Another demonstration was performed in 1654 by Otto von Guericke, the Mayor of Magdeburg, in Germany. He used two hollow hemispheres (half spheres) made of bronze. They were sealed together with a ring of greased leather. One hemisphere had a tap (fig. 2.7). The two hemispheres were held together while the air between them was sucked away. He used a vacuum pump which he had invented. Then the tap was closed. The air pressure on each hemisphere pushed them tightly together. Next, a team of 8 horses was harnessed to each hemisphere. They were driven in opposite directions, to try to pull the hemispheres apart. They failed to do so. The pressure of the atmosphere was stronger than 16 horses!

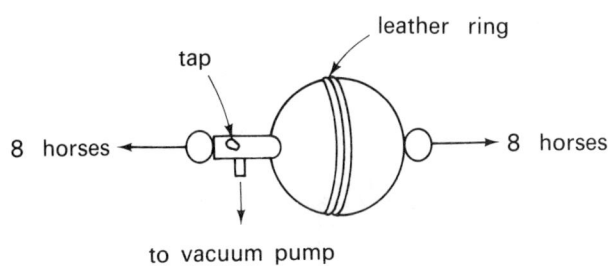

Magdeburg hemispheres

Fig. 2.7 Magdeburg hemispheres.

Your school laboratory may have a small model of the Magdeburg hemispheres. If so, you can repeat von Guericke's demonstration on a small scale, using students instead of horses.

Instructions: finding the direction of air pressure
1 Fill a jar with water to the brim.
2 Place a flat card across the mouth of the jar, leaving no bubbles of air inside.
3 Put your finger on the centre of the card to hold it in place. Carefully invert the jar (fig. 2.8).
4 Take your finger away. What happens to the card and water? Can you explain this?

2.31 Measuring atmospheric pressure

Atmospheric pressure is large. At sea level it is usually a little over 100 000 Pa. If a can has a surface area of about

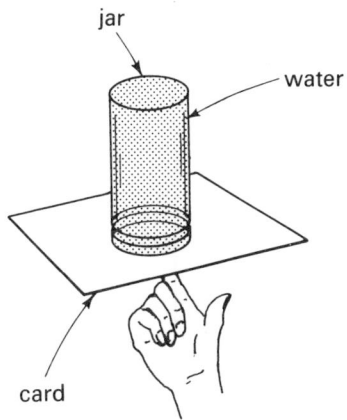

Fig. 2.8   The direction of air pressure.

0.2 m², the total force on the can is equivalent to more than 20 000 N. No wonder that it is crushed almost flat!

Questions
1 If you have seen the demonstration with the can (p. 14), what was the surface area of the can? What was the total force acting on it?
2 Why did the can not collapse under atmospheric pressure *before* the demonstration was tried?
3 What is the total atmospheric pressure acting on your body? (Assume that the total area of your skin is 2 m².) Why are you not pushed down on to the floor? Why does your body not become crushed under this great pressure?

At altitudes higher than sea level, the amount of air above us is less, so atmospheric pressure is less. For example, at 1 500 m above sea level it is about 85 000 Pa. At 10 000 m above sea level it is only about 26 000 Pa – roughly one quarter of the pressure at sea level. Atmospheric pressure varies slightly from day to day and even from hour to hour. These variations are connected with changes in weather or seasonal climate. To predict the weather, for example, the arrival of a hurricane, we need to measure atmospheric pressure and its changes. One instrument used for measuring atmospheric pressure is the *mercury barometer* (fig. 2.9). We use a piece of thick-walled glass tubing about 850 mm long, sealed at one end. We fill it with mercury. Then we invert the tube over a small dish, which contains more mercury. The mercury does not run out of the tube completely. Some runs out into the dish, but most of the mercury remains in the tube, as in the drawing. If we put a ruler beside the tube we can measure the length of the column of mercury in the tube. The height, *h*, of the column is nearly always between 700 and 800 mm, and most often it is between 750 mm and 770 mm. Within these limits the height varies slightly from time to time. In some parts of the world it rises and falls a few millimetres daily. In other parts the rise and fall is irregular and is connected with changes in local weather.

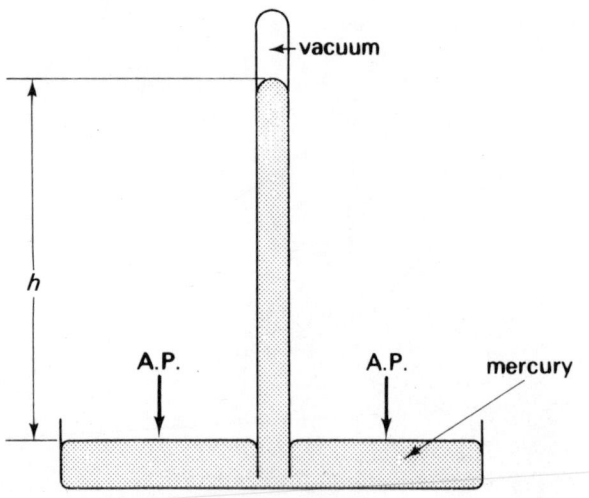

Fig. 2.9 A simple mercury barometer.

When the tube is inverted, some mercury runs out. A space is left above the mercury. Nothing can enter this space. It is a vacuum. The pressure in it is zero. This means that the pressure on the top of the mercury column is zero. Further down the column the pressure increases, due to the weight of mercury above. At the bottom of the column (level with the surface of the mercury on the dish) the pressure is that due to the weight of a column of mercury, of height h.

The pressure at the surface of the mercury in the dish is atmospheric pressure (A.P.). The pressure here and the pressure at the bottom of the column are *equal*, for the surface of the mercury in the dish, and the bottom of the column are at the *same level*. At the bottom of the column the pressure due to the column exactly balances the pressure due to the atmosphere. As the atmospheric pressure changes, the height of the column changes too, to keep the balance.

To read a barometer we measure the height, h. We can plot graphs of daily or hourly readings. A convenient unit for measuring and recording atmospheric pressure is 'millimetre of mercury'. This is simply the height of the column of the barometer as measured in millimetres. On this scale of measurement, the standard atmospheric pressure is 760 millimetres of mercury, or 760mm Hg for short.

For accurate measurement we use a special type of mercury barometer, known as a *Fortin barometer* (fig. 2.10). It hangs from a bearing, so that it is vertical. Its glass tube is enclosed in a brass tube, except at the upper end, where there is an opening. Through this opening the top of the mercury column can be seen. Behind it is a mirror. One edge of the opening has a scale, graduated in millimetres. The distances on this scale give the height of the column, measured from a zero point. The zero point is the tip of the ivory peg in the container at the bottom of the barometer. For accurate readings of height, there is a

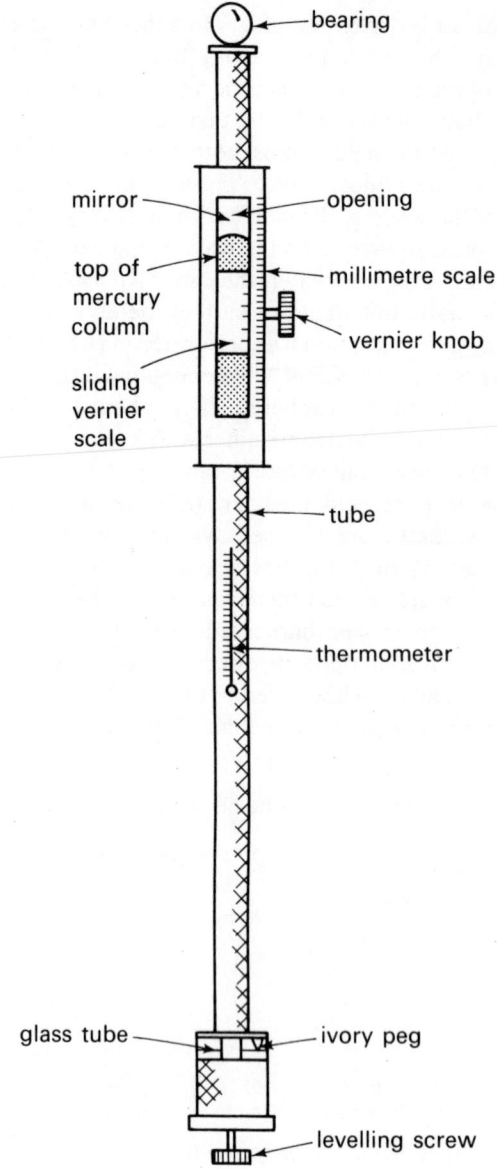

Fig. 2.10 A Fortin barometer.

vernier scale (fig. 1.2) which can be moved up and down by an adjusting knob.

One problem with a barometer is that as the mercury level in the tube rises, the level in the dish falls. In a Fortin barometer, the dish or container has a bottom made of leather, which can be raised or lowered by turning a levelling screw.

To measure atmospheric pressure, the levelling screw is adjusted so that the mercury in the container just touches the tip of the ivory peg. Then we know that the lower end of the mercury column is at the zero of the scale. The vernier scale is adjusted until the top of the scale, the top of the column and the top of the reflection of the vernier scale in the mirror, are all in line. The use of the mirror avoids error due to parallax (fig. 1.1b). The height of the column can then be read to the nearest tenth of a millimetre.

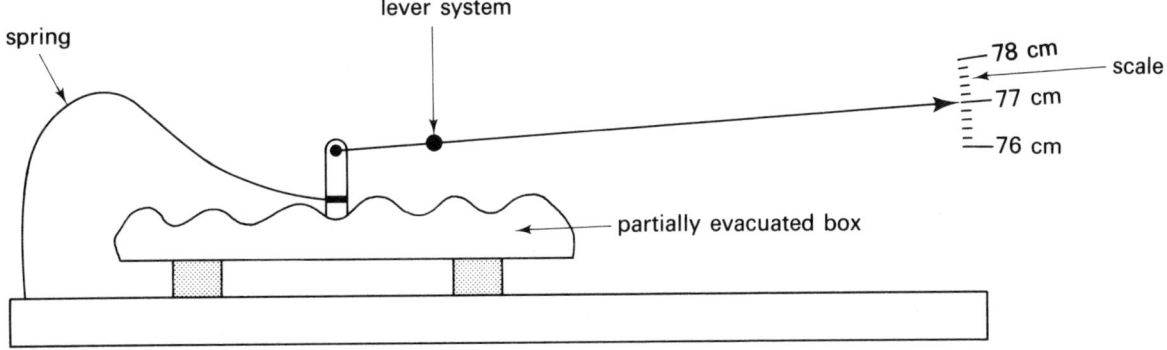

Fig. 2.11   An aneroid barometer.

The thermometer attached to the barometer is for measuring the temperature of the barometer. As temperature increases, the brass tube expands – in other words, the measuring scale expands. A correction can be made to the reading, to allow for this. Also the density of mercury decreases as temperature increases, and a correction can be made for this effect too. These corrections are needed only for specially accurate work.

The other common type of barometer works in an entirely different way. The *aneroid barometer* (fig. 2.11) has a flat circular box made of thin flexible metal. Most of the air is sucked out of this and it is sealed. If atmospheric pressure increases, the box is squashed flatter. If air pressure decreases the box bulges outward. The changes in shape of the box are too small for you to notice. One side of the box is connected to a system of levers which magnifies any movements made by the side of the box. The levers make a pointer rotate over a scale. The scale can be marked to show atmospheric pressure in centimetres and millimetres of mercury, or it can be marked in Pa (N/m²) or some other units of pressure. In the diagram the lever system has been drawn very simply, but in most aneroid barometers it is more complicated than this. The advantages of the aneroid barometer compared with the Fortin barometer are its cheapness, its portability and its compactness. There is also the advantage that a stack of several boxes may be connected to a system of levers which moves a pen. This draws a continuous record of pressure on a sheet of graph paper, which is on a rotating drum. By means of this *barograph* variations in pressure may be continuously recorded for a whole week. The disadvantage of the aneroid barometer is that it may get out of adjustment, and its scale will need to be marked again by comparing it with a Fortin barometer in the same room. Also it does not have the accuracy of the Fortin barometer.

Since atmospheric pressure decreases as we rise above sea level, we can use this effect for estimating height above sea level. Mountain climbers carry small aneroid barometers to help them tell how high they have climbed. In an aircraft there is usually a type of aneroid barometer called an *altimeter*. It responds to the changes in atmospheric pressure as the aircraft climbs and dives. The scale is marked to indicate height, in metres. The pointer is set to zero before the aircraft takes off. During the flight the instrument indicates the height of the aircraft above the airfield from which the flight began. When landing at another airfield, allowance must be made for any difference in altitude of the two airfields. Allowance must also be made for any large variation in the altitude of the ground over which the aircraft is flying, particularly mountain ranges. Unfortunately, atmospheric pressure may change rapidly in stormy weather; then the altimeter gives a false reading. If possible, other systems of height measurement are used, such as radar.

2.32   Measuring gas pressure

The *manometer* works on the same principle as the barometer. It is used for measuring the pressure of any gas which is in a container.

Instructions: using a manometer
1 Set up the apparatus in fig. 2.12.
2 Open the clip, so that air can freely enter or leave the flask through tube A. What happens to the water levels in the manometer? Explain this using ideas about pressure.
3 Blow gently into A, and close the clip while still blowing. What happens to the water levels? Explain.
4 Measure the difference between water levels. The water surface is curved; what point of it must be used for the measurements?
5 Open the clip. Suck gently at A and close the clip while still sucking. What happens to the water levels? Explain.
6 Measure the difference between levels. Why should this difference be recorded as a negative height?
7 Connect A to a gas tap in the laboratory. Open the clip. Turn on the gas tap and leave it open. Measure the gas pressure.
8 Turn off the gas tap before removing the manometer from it.
9 If you had used mercury in the manometer instead of

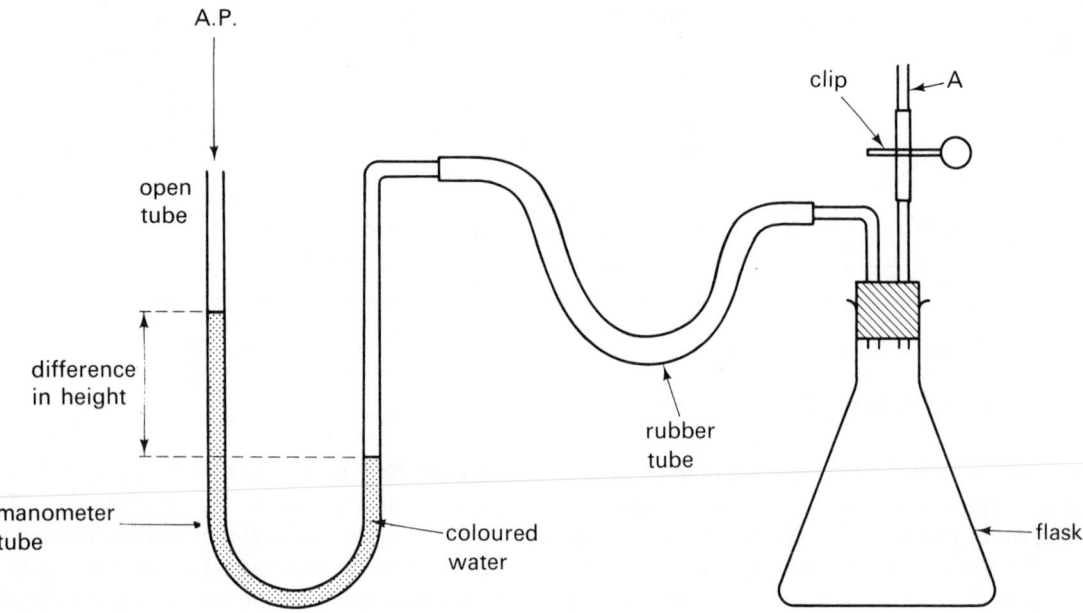

Fig. 2.12 A manometer.

water, what difference would you have noticed? What point on the surface of the mercury must be used for measurements?

A manometer measures a *difference* in pressures. The pressure on one side of the manometer is different from that on the other. Usually one side is open to the atmosphere, so the pressure on this side is atmospheric. We will call this the open side. The other side of the manometer is connected to the gas container; this is the closed side. The pressure on the liquid surface in the closed side is the pressure of the gas in the container. The same pressure is found at the *same level* (13.00) in the open side of the manometer (dashed line in figure). The pressure here is also equal to atmospheric pressure *plus* the pressure due to the weight of the liquid in the tube above, (above the level of the dashed line). If we measure the difference in levels in the two sides we have a way of measuring the difference between atmospheric pressure and the pressure of the gas in the container. This difference may be positive, as in the figure, or negative. Ways of calculating this will be described later.

Gas pressure can also be measured by the *Bourdon gauge*. One of these is used in the apparatus shown in fig. 17.12. The gauge consists of a slightly curved tube, which has one end sealed and the other end connected to the gas container. If gas pressure increases, the extra pressure makes the tube become straighter. The sealed end of the tube is free to move and this movement is magnified by levers. The levers are connected to a pointer which rotates over a scale marked in Pa (N/m²), or some other unit of pressure. Like the manometer, this instrument indicates a difference in pressure between the gas in the container

(pressing on the curved tube from its inside) and atmospheric pressure (pressing on the curved tube from outside). A zero reading on the scale means that the gas in the container is at atmospheric pressure. If you have a Bourdon gauge in your laboratory, investigate how it works; you could use it instead of a manometer and repeat the instructions on page 17.

Projects

1 Look around your home, your school (including the workshops and office) to find as many examples as you can of the use of levers. Make a display to show the different kinds. If possible put some of the levers in the display, and make drawings of those which are too big to display, or which cannot be borrowed. Label each item to show where the pivot is, where the load is and where the effort is made. If possible, measure load and effort for some of these machines (make up your own method, based on the instructions on page 12). Calculate values for *L/E*, and thus see how much each machine helps you to exert a large force with relatively little effort.

2 Make a display of pulley systems. Find out who uses pulleys in your neighbourhood and what they are used for. Make drawings or models (from wood, or using construction sets) to show the many uses for pulley systems.

3 Keep a daily record of atmospheric pressure for several weeks. Plot a graph of your measurements. Does the pressure show any connection with local weather conditions?

4 Examine a syringe (plastic disposable type). Use it to draw up water. From your knowledge of atmospheric pressure, write a short description explaining how it works.

3 Measuring temperature

3.00 Simple thermometers

Human skin contains small organs that are sensitive to temperature. We use these to detect the temperature of our surroundings. For most purposes they work well, but it is possible to trick them. If you have three bowls of water, one as hot as the hand can bear, one icy cold, and the third one at room temperature, you can confuse the heat-sensitive organs of your skin. Put one hand in the hot water and one in the cold, and leave them there for a few minutes. Then put *both* hands in the water that is at room temperature. To the hand that was in hot water, this water feels cold. To the hand that was in cold water, this water feels warm. Both cannot be correct!

This illustrates the importance of using *instruments* to measure temperature. An instrument for measuring temperature is called a *thermometer*. It makes use of some *physical property* that changes with temperature. One such property is that most substances get bigger (or *expand*) when heated.

Instructions: making and using a water thermometer
1 Make a water thermometer like the one in fig. 3.1. To make the water easier to see, add some coloured dye to it.
2 Put the thermometer in different places, some warm, some cool. In each place, leave it for a few minutes until the top of the water column in the tube has reached a steady position. Then mark the card, level with the top of the water. Write the name of the place beside the mark. When you have finished, you will have several marks at different levels.

Questions
1 Which mark indicates the highest temperature? Which indicates the lowest temperature?
2 Which is the hottest place you put the thermometer in? Which is the coolest?

Instructions: making and using an air thermometer
1 Use the same tube, stopper and flask, but have only a short thread of coloured water in the tube (fig. 3.2).

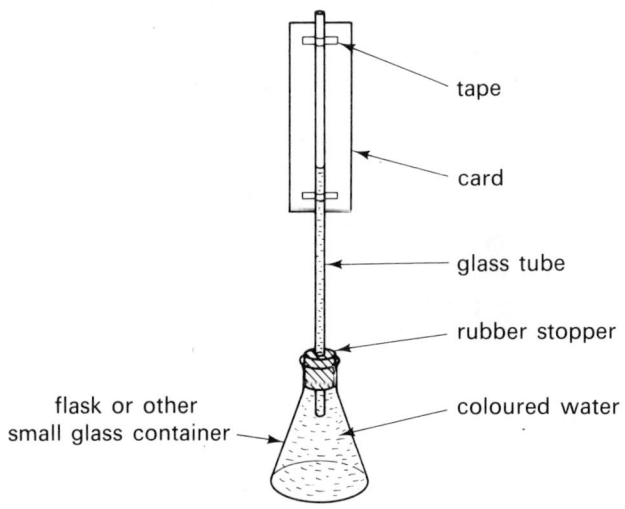

Fig. 3.1 A water thermometer.

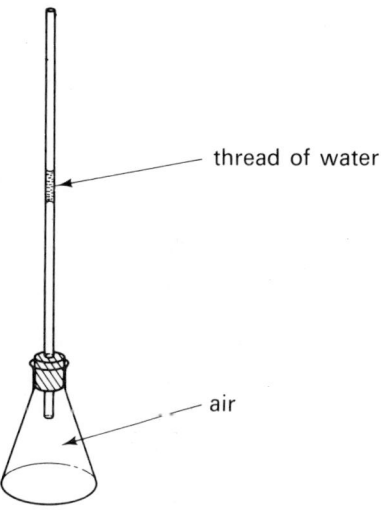

Fig. 3.2 An air thermometer.

2 Hold the flask between your hands, then take your hands away. Repeat this several times.

Questions
1 In what way is the action of this thermometer very different from the water thermometer?
2 Explain why it acts as it does.

Though air thermometers and water thermometers are simple and cheap to make, they have several disadvantages. You can use them to compare temperatures of different places, to say that one place is hotter than another, but you cannot say *how much* hotter, (why not?). You cannot compare temperatures on one thermometer with those on another, (why not?). Both types of thermometer become dried-out very quickly and need to be re-marked frequently.

The mercury-in-glass thermometer is much better. This is made by taking a glass tube with a bulb at one end. The bulb is filled with mercury and the mercury heated so that it expands and fills the tube. Then the tube is sealed. As the mercury cools again it contracts. The end of the thread moves down the tube, leaving almost a vacuum in the space above. The tube is sealed, so there are no problems about evaporation. The tube is of very narrow bore; a small change of temperature and a small change in the volume of the mercury can produce a large change in the position of the end of the mercury column. This makes the thermometer sensitive to small temperature changes.

3.10 The Celsius scale

It is of little use to mark a thermometer with the names of rooms. We need a scale of *numbers* to indicate temperature. The same number should mean the same temperature on any thermometer. We need a temperature scale that is known and used internationally. The *Celsius scale* is the most widely used. It is sometimes called the centigrade scale, but this name is not used in SI. To mark a thermometer with the Celsius scale we find the position of the end of the mercury thread at two fixed temperatures: the *upper* and *lower fixed points* of the scale.

3.11 The fixed points

Instructions: checking the lower fixed point on a thermometer
1 Place a thermometer with its bulb in a funnel containing melting ice (fig. 3.3b). This is the way to find where to put the zero mark, for zero on the Celsius scale is defined as the temperature at which pure ice melts.
2 The zero mark is already on the thermometer you are using. Is it in *exactly* the correct place? When the mercury thread has stopped contracting, check that it is exactly level with the zero mark. We call this temperature 'zero degrees Celsius' and write it '0°C'. The cheaper types of thermometer may not be perfectly accurate; it is useful to

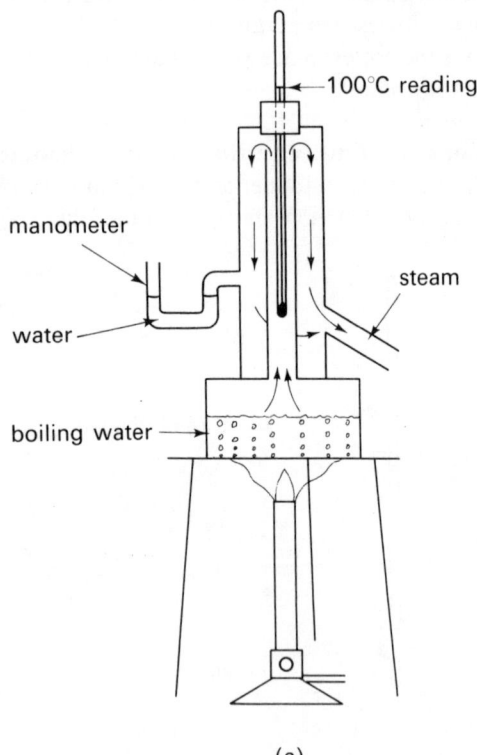

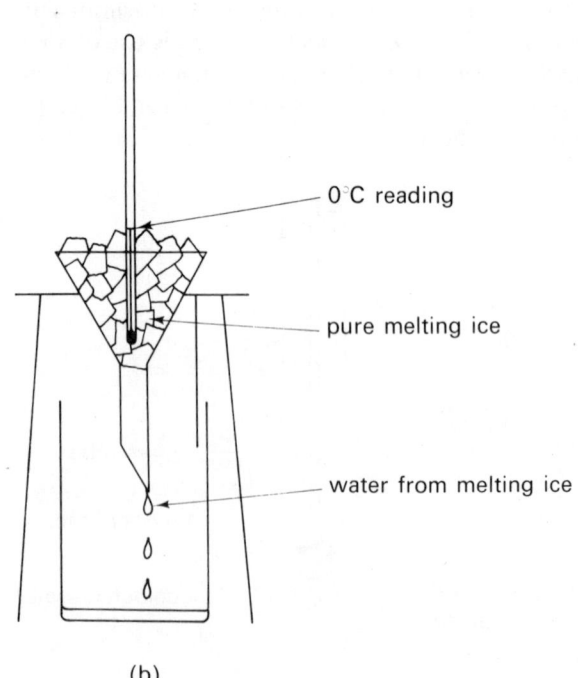

Fig. 3.3 Marking the fixed points on a thermometer: (a) upper (b) lower.

know the error of each thermometer and to keep a note of it.

Instructions: checking the upper fixed point on a thermometer

1 Hang a thermometer with its bulb in steam coming from water boiling in a large beaker. The upper fixed point on the Celsius scale is defined as the temperature of steam from boiling water at standard atmospheric pressure (760 mm Hg).

2 You should find that the mercury column gets longer, until it comes level with '100' mark of the thermometer. Does yours do this?

3 If it does not come to exactly '100', there may be an error in marking the thermometer, but it may be that the pressure of the atmosphere is not standard when you are checking the fixed point. This can affect boiling point, as we shall see later.

When we have marked the lower fixed point (0°C) and upper fixed point (100°C), we simply divide the distance between them into 100 equal parts. These are the *degrees* of the Celsius scale.

The equipment used above for checking the upper fixed point is simple. A more complicated apparatus is used for really accurate marking. This is the *hypsometer* (fig 3.3 a). The double walls help keep the steam at exactly 100 °C. It does not cool and condense, as it is likely to do in an open beaker. Some hypsometers have a manometer (2.32) to show that the pressure inside the hypsometer is atmospheric pressure. A barometer (2.31) is needed at the time of marking the thermometer, to check that atmospheric pressure in the room is exactly 760mm Hg. If it is not, the true position for the 100 °C mark is calculated so as to allow for this.

Questions

1 When the lower and upper fixed points had been marked on a thermometer they were 220mm apart. It was then noted that the end of the mercury column was 49.5mm above the lower fixed point. What was the temperature of the room?

2 On the thermometer described in question 1, where would you place the mark to indicate (a) 60°C, (b) −10°C?

3 A thermometer when checked reads correctly at the lower fixed point, but at the upper fixed point it reads 102°C. When this thermometer reads 56°C, what is the true temperature?

3.20 Some different kinds of thermometer

Most of the thermometers commonly in use are the liquid-in-glass type, similar to the mercury-in-glass type we have already described. Once we have decided on two fixed points and divided the distance between them into 100 equal parts, we have established the temperature difference that is represented by 1 degree on the Celsius scale. It is then possible to continue the scale up and down, to graduate the thermometer for temperatures lower than 0°C and higher than 100°C. Most laboratory thermometers are graduated from −10°C up to 110°C. There is a limit to how far up or down the scale one can go. With mercury we cannot measure temperatures below −39°C, for it freezes at this temperature. The limit to measuring high temperatures is set by the fact that it boils at 357°C. These properties of mercury set a limit to the range of mercury thermometers. In some parts of the world, such as Russia and Canada, temperatures are often lower than −39°C, so a mercury thermometer is useless. Instead we can use ethanol, which has its freezing point at −117°C. However its boiling point is 79°C, which rather restricts its use in the laboratory. It is quite suitable for measuring room temperature and the temperature of glass-houses, as these rarely exceed 50°C. Ethanol has the disadvantage that it wets the glass of the tube easily; the column may often break into drops, which makes readings very inaccurate. Also, at high temperatures, ethanol easily evaporates from the column, and may condense at the upper end of the tube, if that is cooler, causing serious error. Ethanol is colourless, and a dye must be added to it to make it easily visible; there is no such problem with mercury for it is a shiny metal and easily seen. Being a metal, it carries heat more easily than ethanol does. The mercury quickly reaches the temperature of its surroundings and indicates an accurate reading.

The *clinical* thermometer is a special type of mercury thermometer used for measuring body temperature (fig. 3.4). We need to know the temperature of the body accurately, since a change in temperature of only 1 degree may indicate that the person is in ill health. The thermometer must therefore be sensitive to small temperature changes. The normal human temperature is 37°C. It never falls below 35°C or rises above 43°C so temperatures outside this range need not be included on the scale. The thermometer has a short tube. The bore of the tube is extremely fine, so that small changes of temperature cause large changes in the length of the mercury column. When the bulb of the thermometer is left under a person's tongue for a minute, the mercury expands along the tube, indicating the temperature. When the thermometer is removed, the mercury column does not contract into the bulb again. There is a constriction, a narrow kink, in the tube. The column breaks at this point; the mercury below the constriction returns to the bulb, but the mercury above the constriction remains in the tube. The temperature can then be read carefully and recorded, for the mercury column stays at the reading it had when it was in the mouth. After the reading has been taken, the thermometer

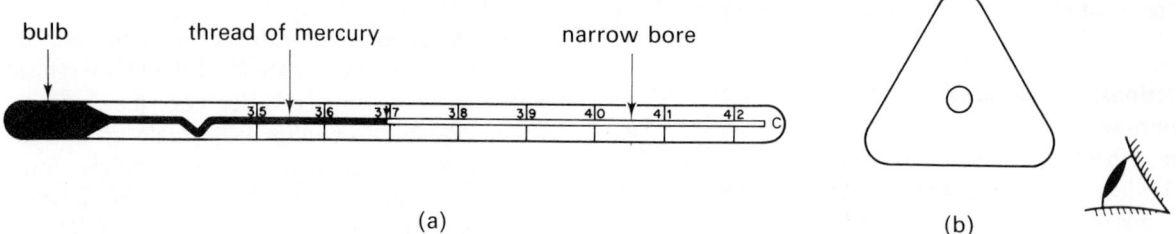

bulb　　　thread of mercury　　　narrow bore

(a)　　　　　　　　　　　　　　　　(b)

Fig. 3.4　A clinical thermometer (a) side view (b) section.

is shaken to force the mercury back past the constriction into the bulb.

The column of mercury is very fine indeed, to give greatest sensitivity. It is therefore rather hard to see. To make it more easily visible, the thermometer tube is specially shaped so as to have the effect of a lens (fig. 3.4b). This makes the column look much thicker.

The thermometers described in this chapter all depend on the *expansion* that occurs when the temperature of a liquid is increased. We can use other physical properties for measuring temperature. Thermometers based on other properties include the bimetallic thermometer, the thermo-couple, the platinum resistance thermometer and the thermistor. These are described in later chapters.

3.30　Melting points

The melting point of ice made from pure water is 0°C. This is actually a *definition* of what we mean by 0°C, for the melting ice is used to *fix* the lower point of the Celsius scale. Ice was chosen for determining the fixed point because pure water is a substance which is easily available. Other substances have different melting points. We have already mentioned ethanol which melts at −117°C. Some substances have melting points lower than this, such as hydrogen which melts from the solid state to the liquid state at −259°C. Some substances have much higher melting points. The melting point of lead is 327°C. This is not an exceptionally high temperature. One of the reasons why lead is suitable for water pipes is that it can easily be melted with a blow-lamp when pipes are to be joined. Solder is an alloy of lead and tin (melting point of tin, 232°C). The melting point of the solder makes it a very useful material for joining certain metals, especially for making good electrical connections. Carbon has a very high melting point (3 500°C) and, close to it, is the metal tungsten (3 380°C). Both these substances have been used for making the filaments of electric lamps. The earliest lamps used carbon filaments, but modern lamps use tungsten. The filaments reach high temperatures and glow brightly without melting.

The melting points quoted above are all for pure substances. With solder, the exact melting point depends on the proportion of lead to tin. By adjusting the propor-tion we can make high-melting point or low-melting point solders to suit special purposes. Similarly, if water has any other substance mixed with it, the melting point is altered.

Instructions: finding the effect of dissolved substances on the melting point of water

1 Make two solutions of sodium chloride (common salt) in water; solution A: 6.5 g sodium chloride in 100 cm³ water, solution B: 13 g sodium chloride in 100 cm³ water.
2 Pour each solution into an ice-making tray and freeze them in the ice compartment of a refrigerator.
3 When the solutions are frozen, break up the salt-water-ice into small pieces, but keep each kind separate.
4 Measure the melting-points of each kind, using the apparatus shown in fig. 3.3b.

Questions
1 What melting points did you find?
2 Write a simple rule to summarise your discoveries.
3 Why is it essential to use ice made from *pure* water when we are marking the lower fixed point on a thermo-meter?

Melting point is also affected by pressure. In fig. 3.5 the heavy weights pulling on the thin wire produce a very high

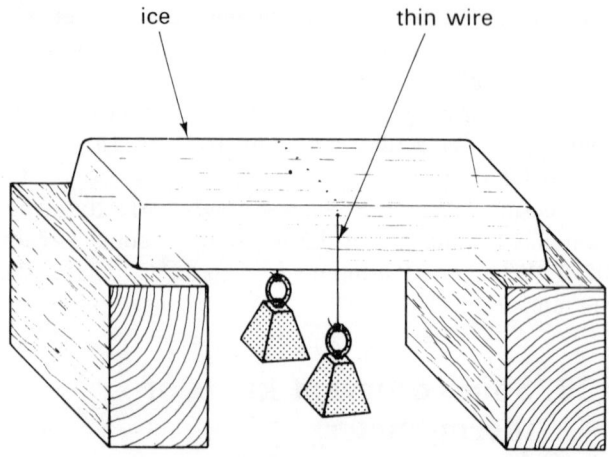

ice　　　　　　　　　　thin wire

Fig. 3.5　Effect of pressure on the melting point of ice (see text).

pressure on the ice. The melting point of the ice just beneath the wire is lowered. The ice cannot remain solid, as its melting point is now lower than the temperature of the block. The ice immediately beneath the wire melts. The wire passes through the water on to the ice surface beneath the melted layer, and the process is repeated. The water above the wire is no longer under pressure so its melting point is restored to normal. It is cold enough to freeze again and does so. Gradually the wire passes through the block, the ice melting beneath it and re-freezing above it. Eventually the wire passes completely through the block, yet the block remains in one piece. Some other substances show the same pressure effect, but there are others for which the application of pressure will raise the melting point.

3.40 Boiling points

The boiling point of water is 100°C. This is a *definition* of the upper fixed point. Boiling points of other substances range from very low (hydrogen, −253°C) to very high (tungsten, 5 700°C). If a liquid has substances dissolved in it, the boiling point is altered.

Instructions: finding the effect of dissolved substances on the boiling point of water
1 Make two solutions of sucrose; solution A: 150g sucrose in 100cm³ water, solution B: 300g sucrose in 100cm³ water.
2 Find the boiling points of these solutions. Boil the solutions in beakers and place the bulb of the thermometer directly *in the boiling solution*. Take readings as soon as boiling begins.
3 Let the solutions boil for a few minutes. Read the boiling points again.

Questions
1 What boiling points did you find?
2 Write a simple rule to summarise your discoveries.
3 Explain the reason for the different boiling points recorded after the solutions had been boiling for several minutes.
4 Suggest a reason why we do not put the bulb of the thermometer in the water when we mark the upper fixed point on a thermometer (fig. 3.3a).

It can be demonstrated that the boiling point is affected by pressure. Some water is boiled in a flask (fig. 3.6) until the flask is full of steam. The stopper is quickly pushed into the neck of the flask and the flask is held upside down under a tap. Cold water cools the upper part of the flask and the steam turns into water. The water takes up much less room than the steam so there is a sudden reduction in pressure inside the flask. The water below is not as hot as 100°C now but it is still hot enough to boil because the reduced pressure has lowered its boiling point. In mountain areas where people live at high altitude the low atmospheric pressure causes problems in cooking. At 1 500m, water boils at approximately 95°C. At higher altitudes the boiling point is even less. Food takes longer to cook and it is not possible to make water hot enough to prepare good tea or coffee. One way over the problem is to use a pressure cooker. This is a cooking vessel which has a lid sealed with a rubber or plastic ring and clamped in place, so that it does not come off when the pressure inside the vessel is higher than the pressure outside. In the lid is a valve which has weights or springs on it. As the food cooks the pressure increases inside the vessel until it is around standard atmospheric pressure. Water inside the cooker boils at 100°C and food is cooked properly. If pressure increases too much, excess pressure causes the valve to open. This type of vessel is very useful at sea level too. The

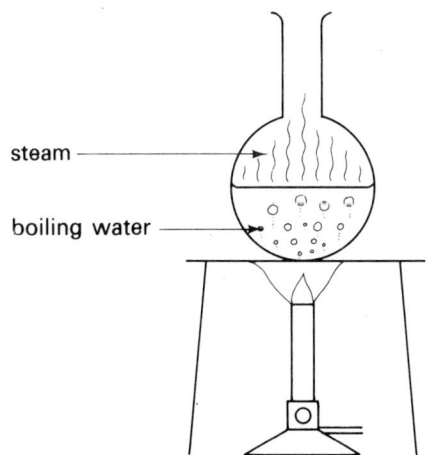

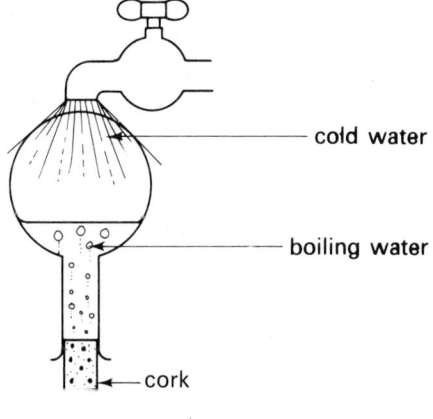

Fig. 3.6   Effect of pressure on the boiling point of water (see text).

usual domestic type has its valve set to regulate the pressure to twice standard atmospheric pressure. At this pressure water boils at 120°C. The high temperature makes cooking faster. This saves fuel and time. It is also said that the food value of some foods cooked in this way is improved.

A larger pressure cooker is used in some laboratories:, it is called an *autoclave*. This works in exactly the same way. It is used for killing germs on glass-ware and surgical instruments and for making sure that there are no living organisms in culture solutions in which bacteria and fungi are to be grown. Many bacteria and fungi (or their spores) can survive in boiling water or steam at 100°C for quite long periods, but most are killed by exposure to 120°C for 20 minutes. An autoclave is an essential piece of equipment for public health laboratories and hospitals.

4 From measurements to laws

4.00 Working scientifically

In science we ask questions. For example, using this lever, how much effort is needed to move that load? To answer a question like that we make measurements. We try to discover rules. We like to discover useful rules:

Effort × (distance between effort and pivot) =
Load × (distance between load and pivot)

You discovered that rule in chapter 2. Now you are able to answer questions about levers, loads and efforts.

You can start to think about rules even *before* you begin making the measurements. You could think of some rules for pendulums:
(a) The longer the pendulum, the longer the period;
(b) The heavier the bob, the longer the period (This is NOT TRUE, but it is a rule you *might* have thought of *before* you did the activity on page 12).

This is going too fast! You cannot call these statements *rules* if you have not tested them to find out IF THEY ARE TRUE. You can call them *hypotheses*.

A hypothesis is an idea that you *believe* is right, until you (or somebody else) prove it is wrong.

Each of the two hypotheses (as we will now call them) seem to be reasonable, we *believe* in them. To be scientific, we must next *test* them. We must do an *experiment*, to find out if our belief holds true. You have already done the experiment and you know that hypothesis 1 is true, but hypothesis 2 is false. We *accept* hypothesis 1; we continue to believe in it, until it can be proved to be false. We *reject* hypothesis 2; we no longer believe in it. If possible, we think of another hypothesis to replace it. This new hypothesis must agree with all the facts we already know about pendulums. You know from your experiment that a good hypothesis is: the mass of the bob makes no difference to the period. This seems true enough to be believed in. Maybe some day, someone will show that even this hypothesis is not true. Perhaps it is not true of a bob of several thousand kilograms mass. Who can

tell, unless they try the experiment? Until somebody proves it is *not* true, we believe in the hypothesis. We *use* it too; people who make pendulum clocks use both these hypotheses, and their clocks keep very good time.

A hypothesis which seems to be true and has been tested can help us to state a rule. This is a short set of working instructions. If you want to know how to calculate load when you know effort and the distances of load and effort from the pivot, the rule is '$xL = yE$'. If you want to know how to alter a pendulum to make it swing faster, the rule is 'make it shorter'.

Some hypotheses have been tested over and over again by many people doing many different kinds of experiments. These hypotheses have always been found true. Gradually scientists have begun to believe that these hypotheses will *never* be proved to be false. Such hypotheses are then called *laws* or *principles*. A law or principle is a statement that summarises our ideas about some important section of science.

Whenever we can, we express a law as an equation. This is usually the clearest way of expressing it. The hypothesis about load and effort in levers has been tested many times; we accept it as a law; we express it as an equation, $xL = yE$.

The equation *on its own* is not enough. We must say exactly what we mean by those symbols. We must say exactly what type of lever it applies to. For example it would not apply if the lever were slightly flexible, so we must say that the lever must be rigid.

The equation, $xL = yE$, is part of a general law about levers called the *law of moments*. We shall come back to this again in chapter 12. In the rest of this chapter we study three new subjects: friction, upthrust, and flotation.

4.10 Friction

Put this book on the table and push it slowly away from you. Can you feel a force which tries to stop the book from sliding? This is friction. When two surfaces (book cover and table top) rub together, friction is the force which opposes the motion.

Questions

1 If the book is resting on the table, and you are not trying to move it, is there friction?

2 If you tilt the table, but the book does not slide, is there friction?

3 If you tilt the table and the book slides, is there friction?

4 Which kind of book slides more easily when pushed (or when the table is tilted), a book with a shiny cover or a book with a rough cover?

5 Which kind of book slides more easily when pushed (or when the table is tilted), a thick, heavy book or a thin, light book. Assume that the shape and size of the pages and cover are the same for both kinds of book.

6 You are standing still, then you take a pace forward. How does friction help you to move?

7 You are walking and suddenly stop. How does friction help you to stop?

8 You have to move a large packing case. It is heavy. It is too large for you to lift completely off the ground. You have to slide it along the ground by pushing it. How does friction make this job harder? How does friction help you do this job?

9 You put the packing case on a trolley with wheels. Why is it easier to push the case now?

10 If you did not have a trolley, in what other ways could you reduce the friction between the trolley and the ground?

11 Why are worn automobile tyres dangerous?

12 You are riding a bicycle. You find that the pedals do not turn easily. What could you do to make it go more easily?

13 You are riding a bicycle. You put on the brakes to stop. In what ways does friction help you?

14 Is friction a nuisance or a help?

In answering the questions you probably found that you know quite a lot about friction already. It is important in everyday life. However to discover laws we need accurate information. We can start from what we already know, and use our knowledge to help us collect more information about friction. We must *measure* friction.

4.11 Measuring friction

In fig. 4.1 a wooden block rests on a surface. We can put weights on the block to increase the force of contact between block and surface. More and more weights are added to the loop of paper at the end of the thread. Find out what is the greatest weight that can be added *before* the block begins to slide. This tells us the greatest amount of friction that can occur between block and surface.

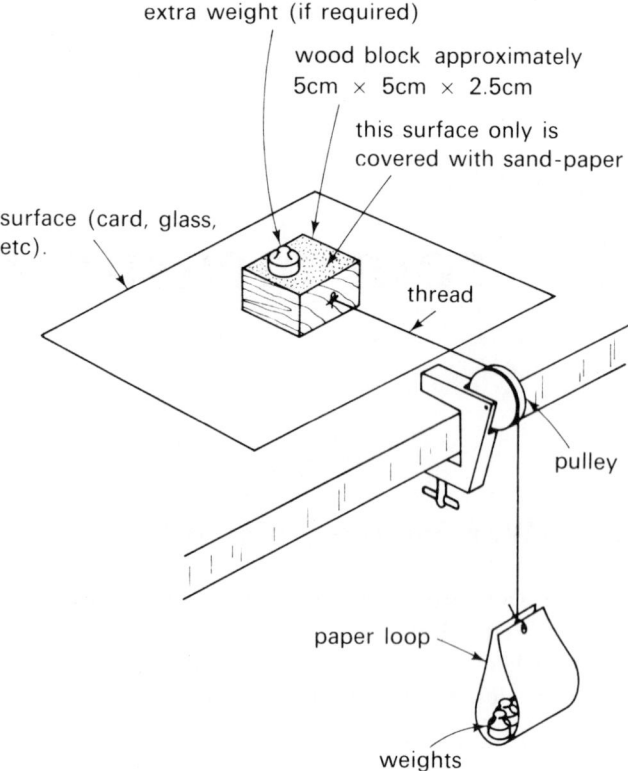

Fig. 4.1   Apparatus for investigating friction.

Instructions: experiments on friction

1 Here are some hypotheses about friction between the block and the surface:

(a) the larger the area of contact, the greater the friction;

(b) the larger the area of contact, the smaller the friction;

(c) area of contact has no effect on friction;

(d) the larger the force of contact, the greater the friction;

(e) the larger the force of contact, the smaller the friction;

(f) force of contact has no effect on friction;

(g) the smoother the surfaces, the greater the friction;

(h) the smoother the surfaces, the smaller the friction;

(i) the type of surface has no effect on friction.

2 Use apparatus like that in fig. 4.1 to test which (if any) of the above hypotheses are true.

When there was no weight on the paper loop there was no friction, for there was no force to oppose. As you increased the weight on the loop, friction increased and was *equal* to that weight. Friction is a self-regulating force. As weight was increased, friction increased, until the block *only just* remained still. This force we measured as the maximum weight that could be added. We call the friction at this stage the *limiting friction*. When limiting friction is reached, any further weights on the loop cause the block to slide.

In all the experiments, you have measured friction when the block was still. We call this *static friction*. If the block is sliding, we call this *sliding friction* or *dynamic friction*.

Instructions: experiments on sliding friction

1 Use the same apparatus as for the previous instructions. Use a rough card surface, place the block with its large wood surface downward. Put an extra 100g weight on the block.

2 Add weights gradually until the block just begins to slide.

3 As soon as the block begins to slide, *quickly* take away the final small weight that was added.

4 Does the block stop sliding? Explain.

4.12 The laws of friction

The experiments on page 26 showed that we can accept hypotheses (c), (d) and (h), provided that by 'friction' we mean 'limiting static friction'. If the applied force by which we are trying to make the block slide is less than limiting static friction, the friction is equal to the applied force. Under these circumstances, hypotheses (c), (f) and (i) are true and may be accepted. Hypotheses (a), (b), (e) and (g) are false under *any* circumstances and are rejected. The accepted hypotheses have been tested many times. We can now think of them as *laws of friction*. We can add to these the hypothesis supported by the instructions above i.e. dynamic friction is less than limiting static friction. Knowing these laws, we can make use of them.

Questions

1 List examples of ways in which we use friction to help us.

2 For some of the examples you have listed, say what is done to make friction as great as possible.

3 List examples of ways in which friction is not helpful.

4 For some of the examples you have listed, say what is done to make friction as small as possible.

In answer to question 4 you will probably have mentioned the use of *oil* on bearings. Oil is a *lubricant*. Its purpose is to reduce the friction between the surfaces as they rub together. In machinery, rubbing produces heat, and causes the surfaces to become worn away.

Many fluids can be used as lubricants. Water can be used but has the disadvantage that it evaporates quickly, especially if it gets hot. Also it may cause metal surfaces to corrode and become rough. This is why oil is generally preferred for lubricating metal machinery. With heavy machinery the pressure on the bearings may be high. This forces the oil away so that the two surfaces come into contact and are not lubricated. To avoid this the oil is forced into the bearings under pressure.

Air is a cheap fluid; it is easily available; it can be used as a lubricant. In machines which run at very high speeds compressed air is fed into the bearings. The surfaces are held apart by air under high pressure. Since air can be compressed by pressure (and liquids like oil and water cannot), air has a cushioning effect. It acts like an air-bed or air-cushion. This helps make the parts turn very smoothly, in spite of small irregularities on the bearing surfaces.

On a much larger scale, compressed air is used to support vehicles such as hovercraft (fig. 4.2). They can

Fig. 4.2 A hovercraft.

move easily over the surface of land or water with very little friction. The fact that the surface is rough (waves or rough ground) makes little difference. The vehicle is supported on a cushion of air produced by the downward-directed air jets around the edge of the vehicle. The force needed to push these vehicles along is slight. Smaller versions of hovercraft are used in factories and construction sites to take the place of trolleys. They are used for carrying machinery and materials. They can be easily pushed by hand. The fact that they can move just as easily over rough surfaces means that they can pass over steps and grooves in the floor or the rough ground on a construction site.

4.20 Upthrust

This is a force which acts on solid objects when they are placed in a fluid; that is, in a liquid or a gas.

4.21 Measuring upthrust

Instructions: investigating the weight of an object in water

1 Prepare the apparatus shown in fig. 4.3 and measure W_1.

2 Suspend the can as in fig. 4.3b. Measure the apparent weight, W_2.

3 Calculate the apparent loss in weight, W_1-W_2. This is equal in value to U, the upthrust of the water on the can (fig. 4.3c).

4 Calculate the volume of the can that is below water level. Since part of the can is below water, some of the water has been pushed to one side to make room for the can. We say that some water has been *displaced*. What volume of water has been displaced? What is the weight of this displaced water?

5 Repeat steps 2 to 4 with the can submerged to the marks at 4 cm, 6 cm, 8 cm and so on.

6 What do you notice about your value for upthrust and your value for the weight of displaced water at each stage of the experiment?

With any liquids, the upthrust is equal to the weight of *liquid* displaced. For liquids denser than water, upthrust is greater than for water. The apparent loss in weight of an object in salt water is greater than its apparent loss in pure water. You could check this statement by designing your own experiment, and then trying it.

4.22 Archimedes' principle

About 2 100 years ago a Greek scientist, Archimedes, did experiments with objects immersed in water and found results similar to those you found in the activity above. As

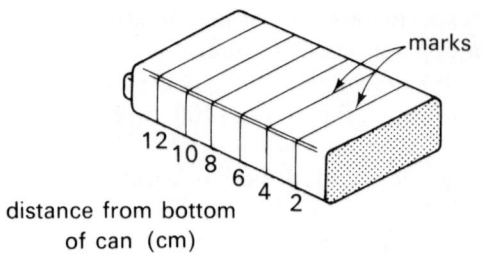

(a)

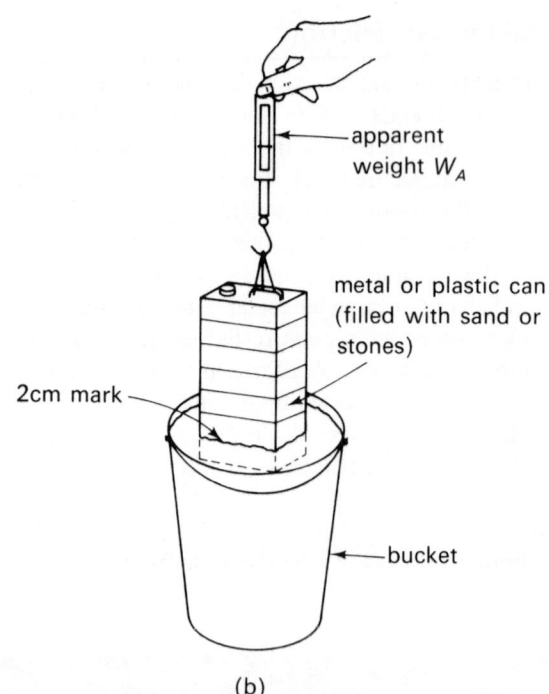

(b)

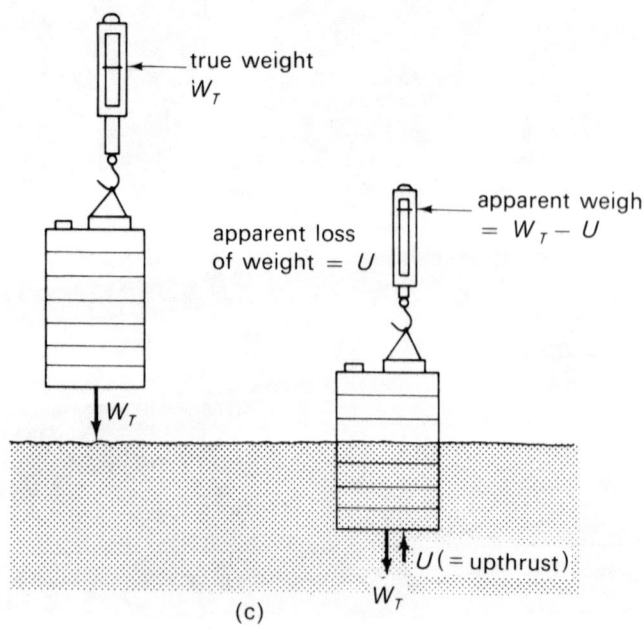

(c)

Fig. 4.3 Measuring upthrust.

28

a result he stated a hypothesis which is nowadays known as Archimedes' principle.

> When an object is immersed (or partly immersed) in a fluid, the upthrust is equal to the weight of the fluid displaced.

In chapter 13 we will see *why* this should happen. For the moment we will *use* this principle in various ways.

Instructions: finding the relative density of an object

1 For the object use a piece of stone or a block of metal. Tie a thread to it and hang it from a spring balance. Record its weight, W_T in newton.
2 Lower the object, still hanging by the thread from the balance, into water. It must not touch the sides or bottom of the water container. When the object is completely under the water surface, record its apparent weight W_A.
3 The upthrust in water is $(W_T - W_A)$.
4 The volume of the water displaced is exactly equal to the volume of the stone. The upthrust is equal to the weight of the water displaced.
5 The relative density of the object is

$$\frac{\text{weight of the object}}{\text{weight of an equal volume of water}}$$

or in this case $\dfrac{\text{weight of the object}}{\text{upthrust in water}} = \dfrac{W_T}{W_T - W_A}.$

Note We do not need to measure the volume of the object. We need to know only its weight in air and its weight in water. This method is very suitable for finding the relative density of an object that has an irregular shape. The only condition is that the object must not be soluble in water.

Instructions: finding the relative density of a liquid

1 Hang a piece of stone or block of metal from a spring balance, as in the previous activity.
2 Record the weight of the object, W_T in newtons.
3 Immerse the object completely in water. Record its apparent weight, W_A (in N).
4 Immerse the object completely in the liquid. Record its apparent weight, W_L (in N). The liquid used can be ethanol or an oil. If you use any other liquid, make sure that it is one which does not dissolve or chemically attack the object.
5 The upthrust in water is $(W_T - W_A)$.
The upthrust in the liquid is $(W_T - W_L)$.
The weight of the water displaced is $(W_T - W_A)$ (Archimedes' principle).
The weight of the liquid displaced is $(W_T - W_L)$ (Archimedes' principle).
The volume of the water and the liquid displaced is the same in each case because it equals the volume of the object. Hence the relative density of the liquid =

$$\frac{\text{weight of the liquid}}{\text{weight of an equal volume of water}} = \frac{W_T - W_L}{W_T - W_A}.$$

Calculate your result, using this equation.
Note We do not need to know or to calculate the volume of the solid. We do not need to know its density.

4.30 Floating

The first investigation uses the same can that you used in the activity on page 28. This time we will make it float and see what happens to its weight.

Instructions: finding the apparent weight of a floating can

1 Empty some of the sand or rocks from the can used on page 28. Adjust the amount of sand or rocks so that the can floats with about three-quarters of its volume below water.
2 Repeat steps 1 to 5 of the activity on page 28. Continue until you reach a level at which the can floats. This will probably not be exactly level with any of the lines already marked on the can. When it floats, draw a line round the can to show the water level. Calculate what volume is below water level. Record the apparent weight when floating.
3 Calculate the upthrust and weight of water displaced for each stage of the experiment, and write the results in the table.

Questions

1 When the can is submerged to the 2 cm mark, is the apparent loss in weight equal to the weight of water displaced? Is this true for all other levels, when the can is *not* floating?
2 When the can is floating, what is the relation between the upthrust and its weight, W_T?

An object in water is acted on by two forces:

its weight – a downward force,
upthrust – an upward force.

If the object is floating, these two forces are equal in size. They are opposite in direction and balance each other. The object neither rises nor sinks. It floats. These ideas can be written as a short sentence which summarises the *law of flotation*.

An object floats in a fluid if the weight of the fluid it displaces is equal to the weight of the object.

This law is really a special application of Archimedes' principle, not an entirely new law. It refers to a *fluid,* so it applies to all liquids and to all gases. We will study the topic of floating in gases in 4.32.

Fig. 4.4 Rafts (see text).

The simplest floating vehicle is a raft. We can make a raft from logs of a low-density wood such as balsa (as used in South America). In fig. 4.4a we have a raft made from five logs with a total volume 0.5 m³. Balsa has $\rho = 200 \text{kg/m}^3$, so the raft weighs only 1 000N. It floats high in the water. If five men of average weight (700N each) ride on the raft, the total weight of men and raft is 4 500N. The raft sinks until it displaces 4 500N of water. This weight of water has volume 0.45 m³, and the raft floats mostly submerged (fig. 4.4b) though the men are safely above water level. If we add just one more man, the total weight becomes 5 200N. Now the raft needs to displace 0.52 m³ if it is to float. Its total volume is 0.5 m³. Even when completely submerged it cannot support six men (fig. 4.4c).

The calculations assume that the wood is dry. In practice, balsa wood gradually absorbs water; its density gradually increases and it will not carry five men far. Bamboo has $\rho = 400 \text{kg/m}^3$, so this too is useful for raft-making. Denser woods such as teak ($\rho = 850 \text{kg/m}^3$) are almost useless. A teak raft of the same size as the balsa raft weighs 4 250N. If there is one man on the raft, the total weight is 4 950N. It displaces 0.495 m³ of water and is almost completely submerged (fig. 4.4d).

Rafts were first made by early men, thousands of years ago. Then they discovered that a hollow log displaces just as much water as a solid one, but weighs much less. It can carry a bigger useful load (man or cargo) without sinking. See if you can work out why.

Questions
1 If a log weighs 1 000N, and the density of the wood is 700 kg/m³, how many boys (average weight 400N) can sit on the log without making it sink?
2 If the same log is hollowed out to make a canoe, and 800N of wood are removed, leaving a canoe weighing 200N, how many boys can sit in the canoe, without making it sink?

In making boats we try to make the hull of the boat as light as possible, but it must be strong enough to stand against the forces of wind, waves and rocks. By using wood, or wood and skin, or wood and canvas we can make a very light hull, strong enough for many purposes. It is cheap, but is liable to damage in rough water and may rot after a while. Steel makes a very strong hull, but needs protection from corrosion (30.10). Recently fibre-glass has been used for small boats. It has low density, is very strong and does not corrode. Hulls can also be made cheaply from reinforced concrete. The main aim is to have a light structure to displace a large volume of water. Then a large useful load can be carried inside the hull.

For safety, large boats and aircraft flying over water

carry inflatable rubber boats. These have very low weight, since most of their volume consists of air. They do not sink, even when full of water. The human body is slightly less dense than water. If we relax in the water we naturally float. In rough water we may not float high enough to breathe easily. Life-jackets, swimming-rings and the many kinds of floats used by people learning to swim all contain a lot of air. They have large volume and low weight, so can give support to the swimmer. Some life-jackets and the floating-boards used by swimmers make use of the low-density plastic material called expanded polystyrene. This is a solid foam, containing a very high proportion of air and has extremely low density.

Instructions: measuring the relative density of expanded polystyrene

1 Use a swimming float made of expanded polystyrene, or a block of expanded polystyrene obtained from the laboratory or a store (it is often used for packing laboratory apparatus, radios, hifi equipment, cameras etc.) You need a weight (called the sinker) that is heavy enough to make the polystyrene block sink completely in water, when tied to the block. Try this out first, to make sure the sinker really does make the block sink.

2 Hang the sinker from a spring balance, so that the sinker is in water. Record its apparent weight, W_1.

3 Tie the polystyrene block to the same thread. With the sinker in water and the polystyrene in air (fig. 4.5), measure their combined weight, W_2.

4 Tie the polystyrene and sinker together. Measure their combined weight when both are submerged in water, W_3.

5 Calculate the weight of the polystyrene in air, $W = W_2 - W_1$.

6 Calculate the apparent loss in weight of the polystyrene when submerged in water.

7 What weight of water is displaced by the polystyrene? The sinker is in the water for all the weighings, so we can forget about the water it displaces.

8 Calculate the relative density of polystyrene from:

$$\frac{W_2 - W_1}{W_2 - W_3}$$

This method for finding the relative density of solids that float in water can be used for other substances. You would repeat this experiment, using cork, paraffin wax or wood. None of these have relative density as low as that of expanded polystyrene.

Note: If you know the relative density of a substance you can easily work out the density by multiplying the number (relative density) by 1 000 and adding the units kg/m³. For example, the relative density of wood is 0.7 and hence the density of wood is $0.7 \times 1\,000\,\text{kg/m}^3 = 700\,\text{kg/m}^3$.

4.31 Hydrometer

Instructions: making a simple hydrometer

1 Use a piece of wood about $1\,\text{cm} \times 1\,\text{cm} \times 15\,\text{cm}$. To make it float vertically, fix a nail or small piece of metal to one end (fig. 4.6a). The rod should float about two-thirds submerged, as in the figure.

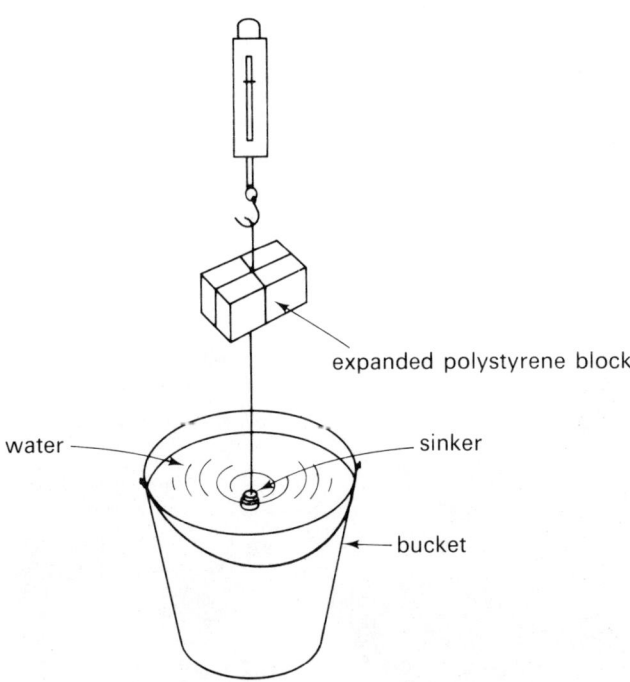

Fig. 4.5   *Finding the relative density of expanded poly-styrene.*

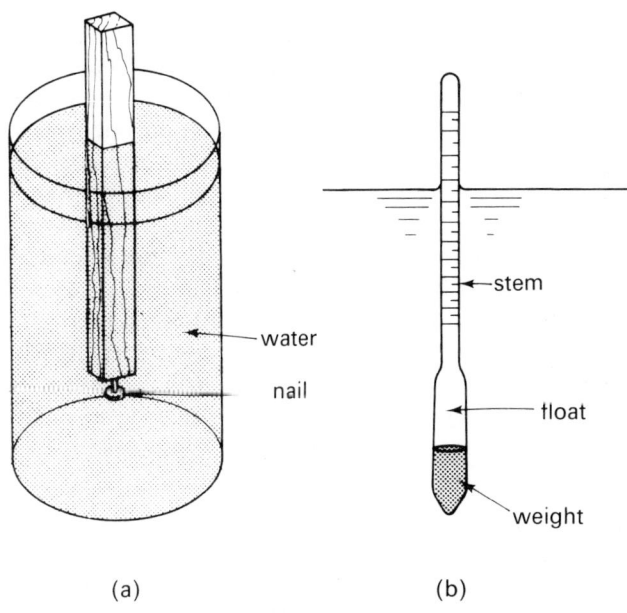

(a)   (b)

Fig. 4.6   *(a) Simple hydrometer (b) Commercially made hydrometer.*

2 Fill a tall jar with water. Float the rod in the water.

3 With a pencil draw a line around the rod, level with the water surface. Write the figure '1 000' beside this mark. What can you say about the volume of the rod (plus nail or metal) below the level of the mark? If you repeat the experiment using oil instead of water, where do you expect to make the new mark?

4 Float the rod in oil (ρ between 800kg/m³ and 900kg/m³), ethanol (ρ = 800kg/m³), 15% sodium chloride solution (ρ = 1 100kg/m³), 45% sucrose solution (ρ = 1 200kg/m³). As you make each mark, write the density of the liquid beside it. Remember to clean the rod thoroughly in between trials.

5 You now have a useful measuring instrument. What can it be used for?

The advantage of the hydrometer is that to measure the density of a liquid you need only float it in the liquid and take one reading; this gives you the density directly. The method used on page 7 requires two measurements of mass and one of volume, and then some calculations. However, remember that the marks on a hydrometer can only be put there by placing the instrument in liquids of known density. The densities of these liquids can only be known by using methods like those on page 7.

An accurate hydrometer works on exactly the same principle as your simple model, but is made in a different way (fig. 4.6b). It is usually made of glass, so that it can be cleaned easily and does not soak up the liquids it is floated in. It is not chemically attacked by liquids such as acids. Its extreme lower end is weighted with lead to make it float really upright. The upper part, or *stem*, is very narrow. Very small differences in density make the hydrometer rise or sink a lot. It is sensitive to small density changes and gives readings to the nearest 0.5kg/m³. A small change of density means a large upward or downward change in floating level, so a hydrometer can measure only a limited range of densities. Hydrometers are often sold in sets, each one covering a small part of the density range, such as 700 to 800kg/m³, 800 to 900kg/m³, 900 to 1 000kg/m³, and so on.

The density of a solution is related to the amount of substance dissolved in it. If we know the density we know how much of a given substance is dissolved in a given volume of the solution. Hydrometers are made with special scales on them to indicate the amount of dissolved substance. They can be marked to indicate the amount of dissolved sucrose, or the percentage of ethanol.

4.32 Floating in air

The laws of flotation apply to all fluids, including air. When we weigh objects we weigh them in air. This is not really a correct thing to do. The object displaces air and there is upthrust acting on it. We measure its apparent weight in air. This is less than its true weight.

32

Fig. 4.7 Hot air balloon.

Questions

1 A block of aluminium has a volume of 10cm³ and weighs 0.27N. The density of air in which it is to be weighed is 1.2kg/m³. What is the upthrust on the block?

2 When the block is weighed, what will be its apparent weight?

3 To find the true weight of the block, under what conditions should we weigh it?

4 For ordinary purposes in the laboratory and in everyday life we can forget about upthrust when weighing objects in air. Why?

5 You have a large balloon, containing 1 000m³ of hydrogen gas, of density 0.09kg/m³. The weight of the fabric of the balloon is 10 000N. What is the total weight of the balloon and the hydrogen in it?

6 We can ignore the volume of the fabric and think of the total volume displaced by the balloon as 1 000m³. If the density of air is 1.2kg/m³, what is the upthrust on the balloon?

7 What will happen to the balloon?

8 What load must be attached to the balloon to keep it at constant height?

9 How could you make the balloon rise?

10 How could you make the balloon come down to Earth again?

11 What is a serious disadvantage of using hydrogen in balloons and airships?

12 What other low-density gases can be used instead? State their advantages and disadvantages compared with hydrogen. (*Hint* see fig. 4.7.)

Projects

1 Find out as much as you can about balloons and airships as methods of transport. Find out the greatest heights reached, the longest distances travelled, the longest times of flight. Find out what uses balloons are put to nowadays. What are their advantages and disadvantages compared with aeroplanes? Collect photographs and drawings and make a display in your laboratory.

2 Make and fly a hot-air balloon. You can make it from segments cut from thin tissue-paper. Sheets from airmail editions of newspapers are very suitable. The balloon should be at least 0.5m in diameter, preferably larger. A Butagas stove makes a good airheater but take great care not to set light to the balloon whilst you are heating the air inside it.

5 Discovering about light

5.00 Rays of light

Light from the Sun and stars travels across millions of kilometres of Space before it reaches Earth. Light comes from fire, from electric lamps and from fire-flies. They are all *sources* of light. Light travels in many directions from a source until it is detected by our eyes, or it activates the film in a camera. What laws does it obey on its journey?

When the Sun shines through a gap in the clouds we can sometimes see the path the light is taking. We see a *beam* of light. We can see a beam when light shines in through the window and the air of the room is dusty or smoky. We see the beams because the light is reflected to our eyes from droplets of water, raindrops, dust or smoke particles in the air. The beam from an automobile headlamp spreads out slightly (fig. 5.1.) Obviously, the light is not all travelling in exactly the same direction. In most beams the light is travelling roughly in the same direction but not in *exactly* the same direction; beams usually get wider or narrower. In fig. 5.1, the direction of travel of light in the beam is indicated by thin lines. They represent the *rays of* light in the beam. In the laboratory we work with *very narrow beams*. We can think of these as rays and in diagrams indicate them by thin lines, with arrow-heads to show the direction in which the light is travelling.

5.01 Shadows

Where there are rays of light we have a beam of light. Where there are no rays of light we have a *shadow*. Shadows are produced by opaque objects, objects that do not let light pass through them. The shadow is formed because of a law which states; light travels in straight lines. This law has been known for many years and until recently there was no reason to disbelieve it. More recently, scientists have found that light does *not always* travel in exactly straight lines. A ray of light passing by a heavy object, such as the Sun, has its path curved by the effect of the Sun's gravity. The effect is very small, and we can forget about it when we are doing experiments with light in the laboratory, but this point is mentioned here to remind us that laws can be broken. Later discoveries may show that a law is not exactly true. Scientists are used to changing their minds, to altering their hypotheses and even their laws, when there is *new* evidence that shows the *old* ideas to be wrong. Sometimes the old ideas are still correct enough to be useful, except in special circumstances. This law about light is a very useful one in everyday life and in laboratory experiments. We shall continue to use it, even though we know that it is not always exactly true.

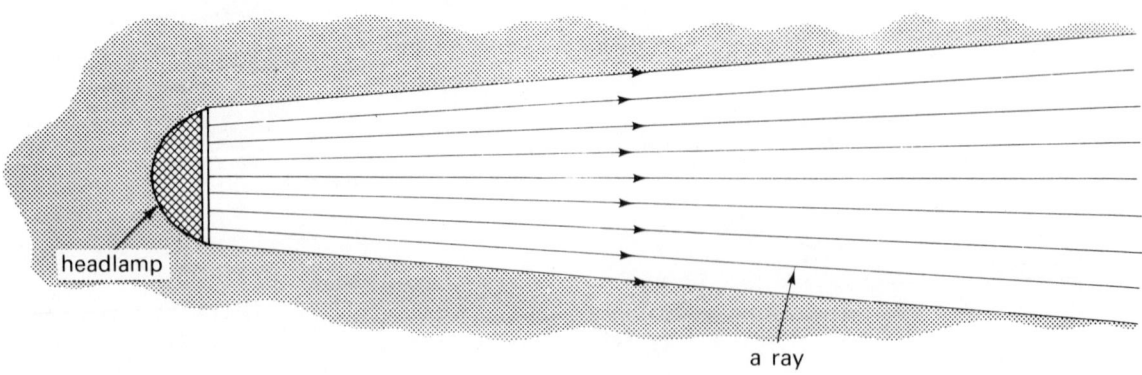

headlamp

a ray

Fig. 5.1 A beam of light.

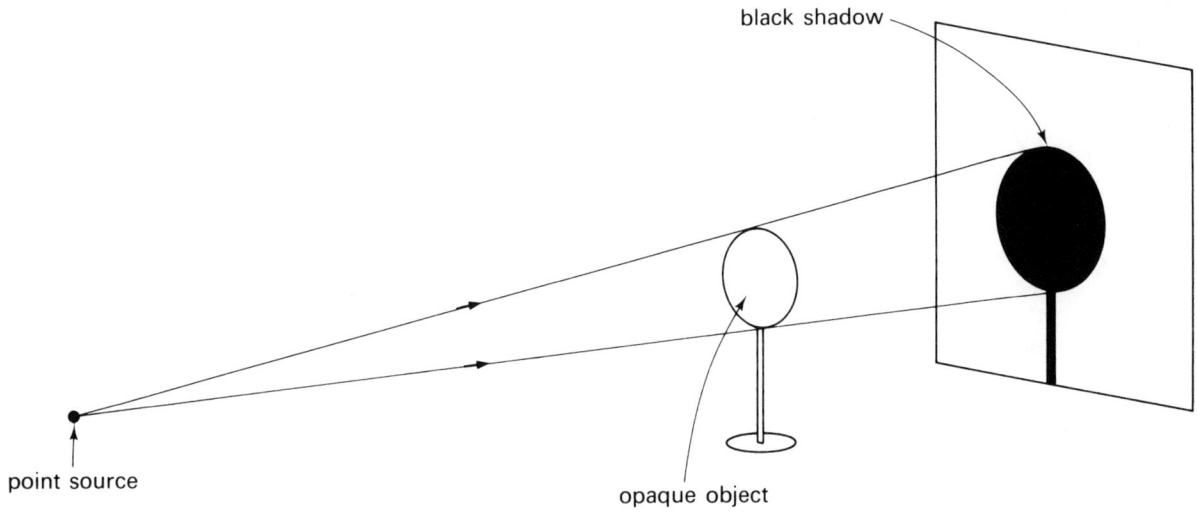

Fig. 5.2   Shadow cast by a point source.

In fig. 5.2 we have a tiny source of light, a *point source*. Between this and the screen is an opaque object, such as a ruler or a coin. In the figure the object is circular and casts a circular shadow. This shadow has a sharp edge. Fig. 5.3 shows why a shadow has a soft edge when the source of light is large. Imagine that you are a fly walking across the screen, from top to bottom. At first, you can see the whole of the lamp from where you stand. The screen there is brightly lit. As you walk downwards, you notice that the object is coming between you and the source. More and more of the lamp is passing out of sight behind the object. You are in the 'grey shadow' region and walking into even darker regions. This region of grey half-shadow is called the *penumbra*. It ranges from almost completely light to almost completely dark. This is why the shadow has a soft outline. In the black shadow (or *umbra*), you cannot see any part of the lamp. It is completely hidden by the object.

No light reaches you. As you walk downwards, you pass into the penumbra again. You gradually see more and more of the lamp reappearing from behind the object as you (the fly) walk along. Near the bottom of the screen you can see the whole lamp, as once more you emerge into full light.

With a large source of light there is a wide penumbra, and shadows are soft. Even with a small light source, like the pinhole, there *is* a penumbra, but it is far too narrow to be noticed.

5.02 Eclipses

In the first sentence of this chapter the Sun and stars were mentioned as sources of light. Nuclear reactions in these bodies produce energy. This travels across Space and reaches us as light, heat and radio waves (21.40, 32.20).

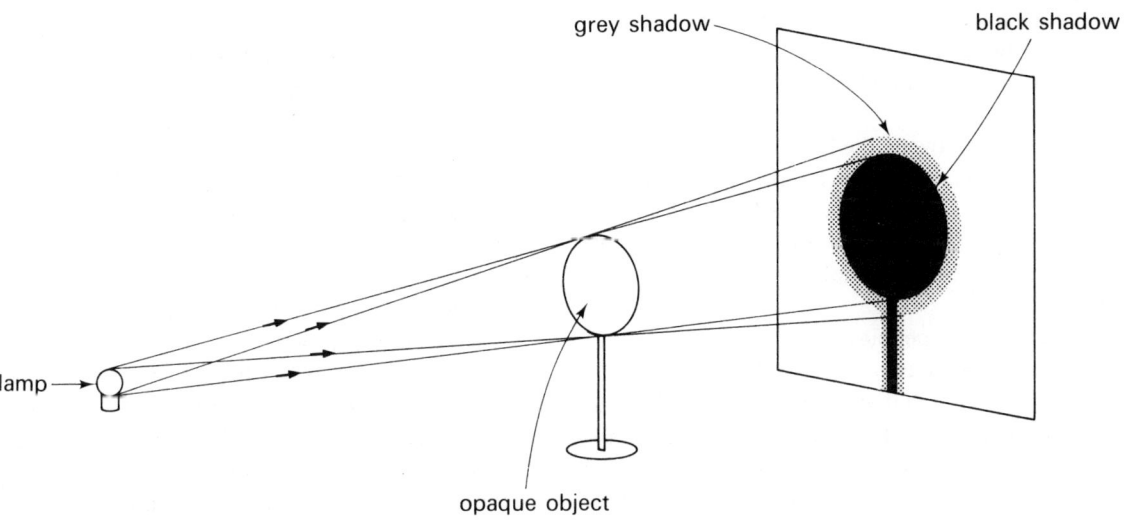

Fig. 5.3   Shadow cast by a large source.

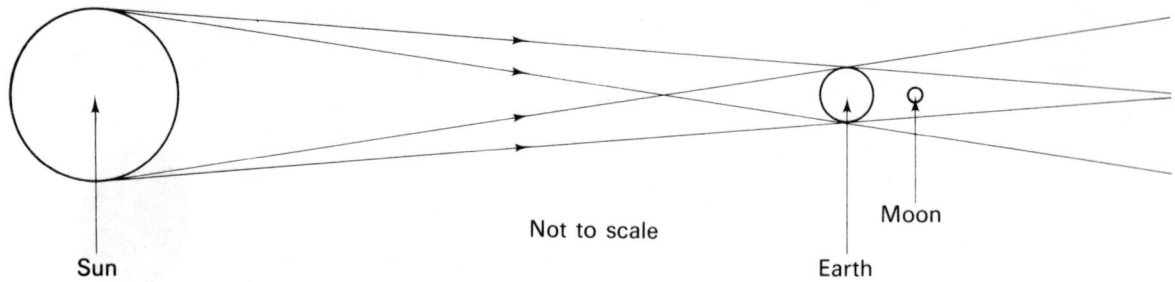

Fig. 5.4 *Eclipse of the Moon.*

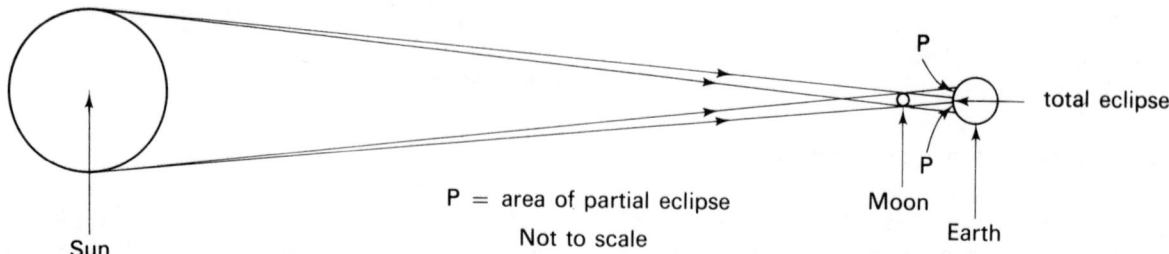

P = area of partial eclipse

Not to scale

Fig. 5.5 *Eclipse of the Sun.*

The Moon is not a source of light. When we look at the Moon we are looking at sunlight reflected from the Moon's surface. The Moon circles around the Earth. When the Sun, the Earth and the Moon are in line, the shadow of the Earth falls on the Moon (fig. 5.4). The Moon receives no light from the Sun and so there is none to reflect. We cannot see the Moon at that time. We say there is an *eclipse of the Moon.*

The Earth is small and far from the Sun (much smaller and much further away than is shown in fig. 5.4). Its umbra is a very narrow cone extending into space. As the Moon circles the Earth it passes close to the umbra every month. The orbit of the Moon is not regular though. Often the Moon passes a little above the umbra, and often it passes a little below it. At these times there is no eclipse. The eclipse takes place only when all three are more or less in line.

When the Moon is between the Earth and the Sun, the shadow of the Moon may fall on the Earth (fig. 5.5.) Then, if we are standing on a part of the Earth that is in the Moon's umbra, we cannot see the Sun. We call this an *eclipse of the Sun.* The Sun is completely out of sight, so it is a *total eclipse.*

At the same time, other people may be in the penumbra. For them the Sun is only partly covered. This is a *partial eclipse.* The Moon is smaller than the Earth, and its umbra is narrow. Often when the Moon is between Earth and Sun, the umbra misses Earth altogether. Then there is no eclipse. At other times the Moon may be further from the Earth, and though the three are in line the umbra does not reach as far as the Earth's surface. Then we see the Sun

covered, except for a bright ring around the edge. This is an *annular eclipse.*

At total eclipse, the brilliant disc of the Sun is covered. We can then see the rim of the Sun's disc (the corona). This is not as bright as the main part of the disc and is normally hard to see. At the eclipse astronomers have a chance to study the activities in the corona and other outer layers of the Sun.

5.03 A pinhole camera

This kind of camera makes use of the fact that light travels in straight lines.

Instructions: making a pinhole camera
1 Use a nail and hammer to make a *small* hole in the centre of the bottom of a metal can.
2 It is best, though not essential, to paint the inside of the can with dull black paint, or to line it with black paper.
3 Cover the other end of the can with thin tracing paper. Fold it round and secure it with a rubber band (fig. 5.6).
4 Wrap a piece of card around the can to form a light-shield and secure this with a rubber band. The pinhole camera is now complete.
5 Point the camera at a brightly-lit scene. Look on the screen to see the picture produced by the camera. The picture may be rather faint, in which case it is best to use the camera indoors, so that little light can fall on your side of the screen. Point the camera towards the window. Look for the outline of the window and the images of brightly-lit objects outdoors.

36

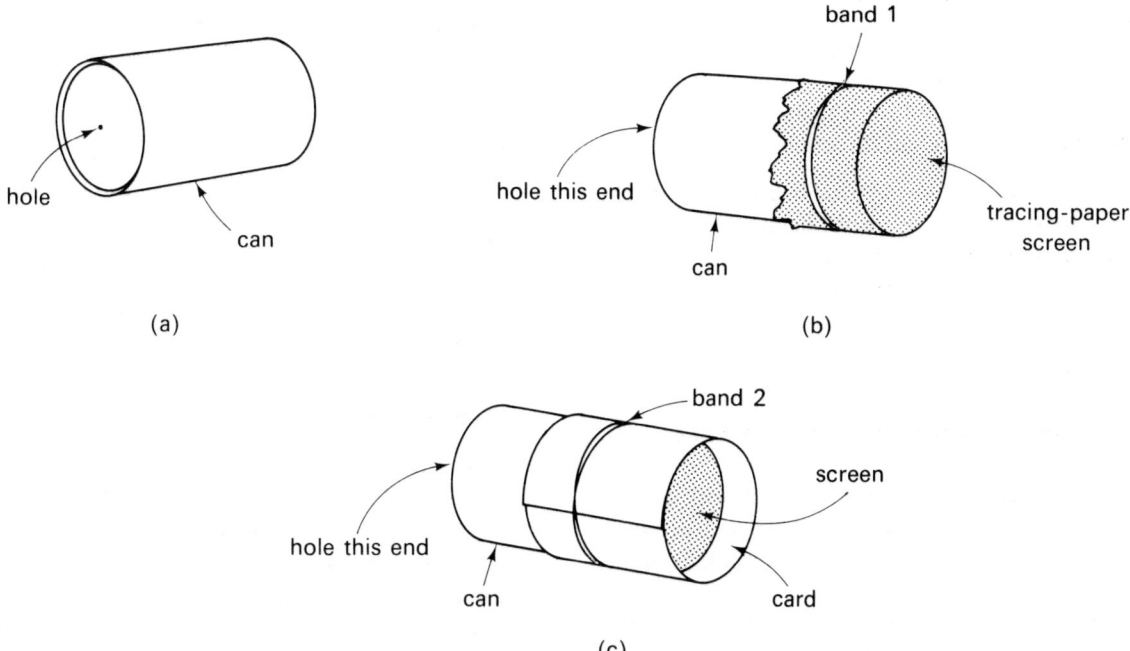

Fig. 5.6 Making a pinhole camera.

Questions

1 What do you notice about the picture on the screen of the camera?

2 Explain how the camera works.

3 To make the picture brighter we could make the pinhole larger. What are the disadvantages of doing this?

We could put a piece of photographic film in place of the tissue paper screen, and use the pinhole camera to take a photograph (fig. 5.7). If you know something about photography, you could try this. The difficulty of this simple and cheap type of camera is that the picture is faint, so the exposure time must be long. If the pinhole is made

bigger, it gives a blurred picture. A photographer's camera has a large hole, but there is a lens to control the light rays passing through it. The way lenses work is described later (chapter 24).

5.10 Reflection

Light travels in straight lines until it comes to the surface of an object. Then it is either:

absorbed – its energy is converted into other forms, usually heat, and it is no longer light.

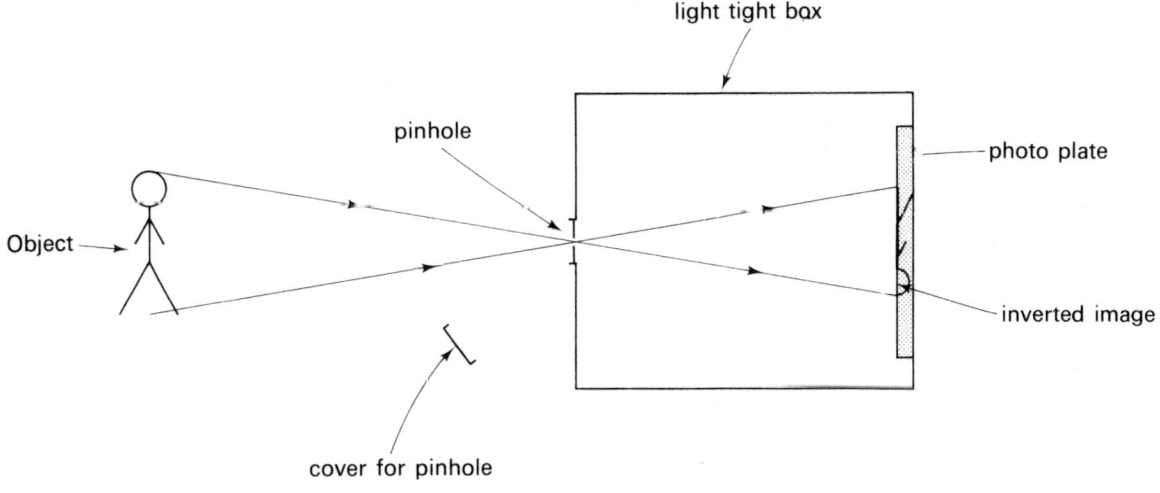

Fig. 5.7 How a pinhole camera works.

transmitted – it passes into the material of the object, as when light passes into water in a bowl. It may change its direction at the surface (5.20).

reflected – it is sent away from the surface again, and now is travelling in a different direction.

In this section we are studying what happens when light is reflected. First we will see what happens when it is reflected by a perfectly flat, perfectly smooth surface, such as the surface of a *plane mirror*.

5.11 The laws of reflection

Instructions: investigating reflection in a plane mirror (1)

1 As a source of light, use a lamp and a card (with a small hole) in front of it. Put a plane mirror on the bench between the source and another card (with a small hole), as in fig. 5.8.

2 Look *down* through the hole at the *reflection* of the source in the mirror. Put a dot of ink on the mirror at the exact point where you can see the source. This dot is exactly on the ray of light travelling from the source to the mirror and then being reflected up and through the second hole to your eye.

3 Carefully thread a piece of cord through the two holes. Hold it against the mirror where the dot is marked. Pull the cord tight; you will need someone to help you.

4 The cord traces the path of a ray from the source to the mirror and then to your eye. The ray from the source to the mirror is called the *incident ray*. After it has been reflected we call it the *reflected ray*.

5 Look down on the apparatus and note the path of the incident and reflected rays (by looking at the cord). What can you say about these paths? If you are not sure what answer to give, repeat the instructions a few times with the source and the other pinhole in different positions. Then try again to answer the question. You have demonstrated one law of reflection:

The incident ray, the normal and the reflected ray all lie in the same plane.

This is shown in fig. 5.9. You can see the normal marked there; it is a line which is perpendicular to (at right angles to) the surface of the mirror at the point where the ray of light is reflected.

If both rays and the normal are in one plane, it is easier to do our experiments in one plane. We can let the rays of light skim across the surface of a sheet of paper so we can see their paths really clearly.

Instructions: investigating reflection in a plane mirror (2)

1 Use a piece of paper about 20 cm × 30 cm. Draw a line across the paper, dividing it into halves (line AB, fig. 5.10).

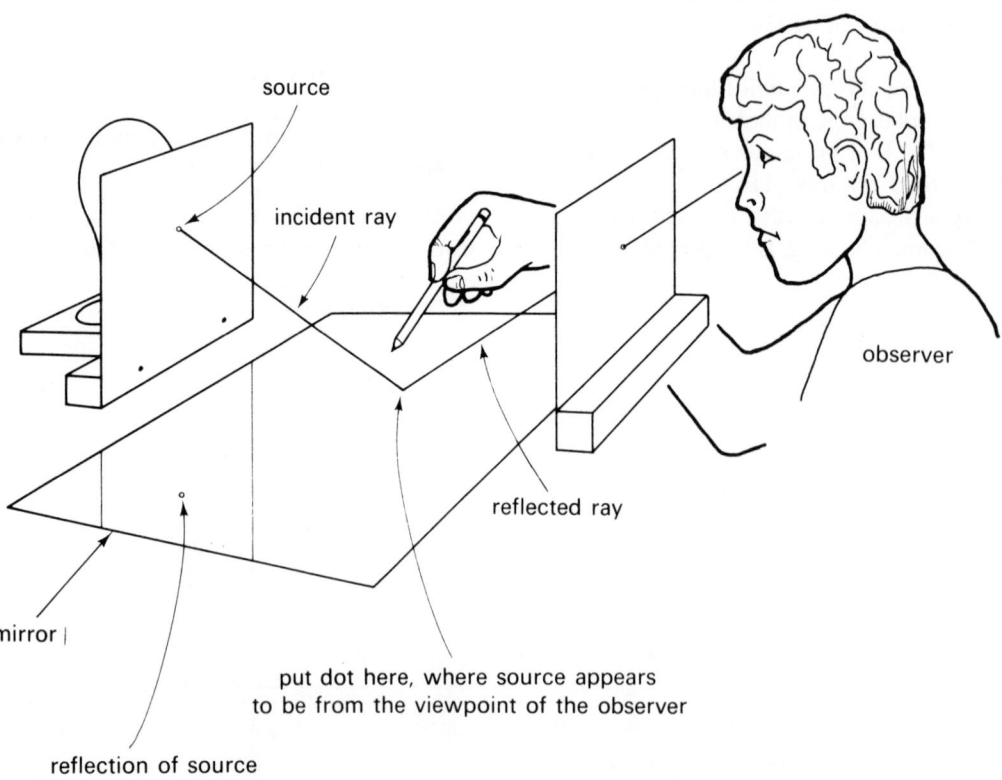

source

incident ray

observer

reflected ray

mirror |

put dot here, where source appears to be from the viewpoint of the observer

reflection of source

Fig. 5.8 Plotting the path of a reflected ray.

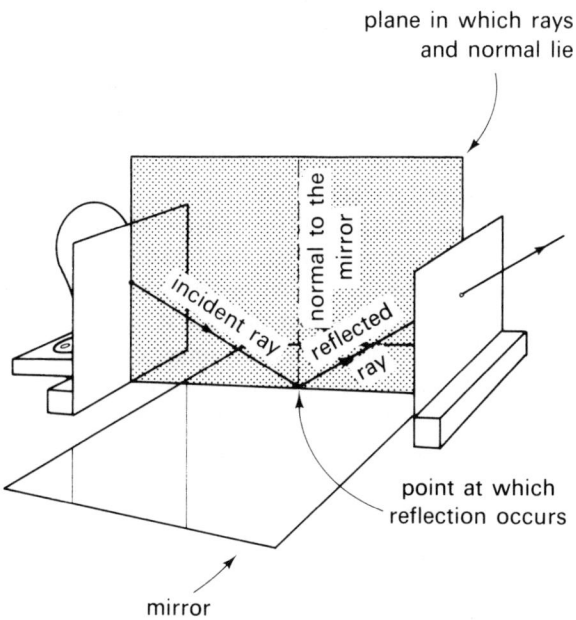

Fig. 5.9 *Interpretation of the result.*

2 Support the mirror in a wooden block. Place it on the line. Reflection occurs at the back of the mirror, not on the front glass surface. Set the mirror with its back surface *exactly* on the line.

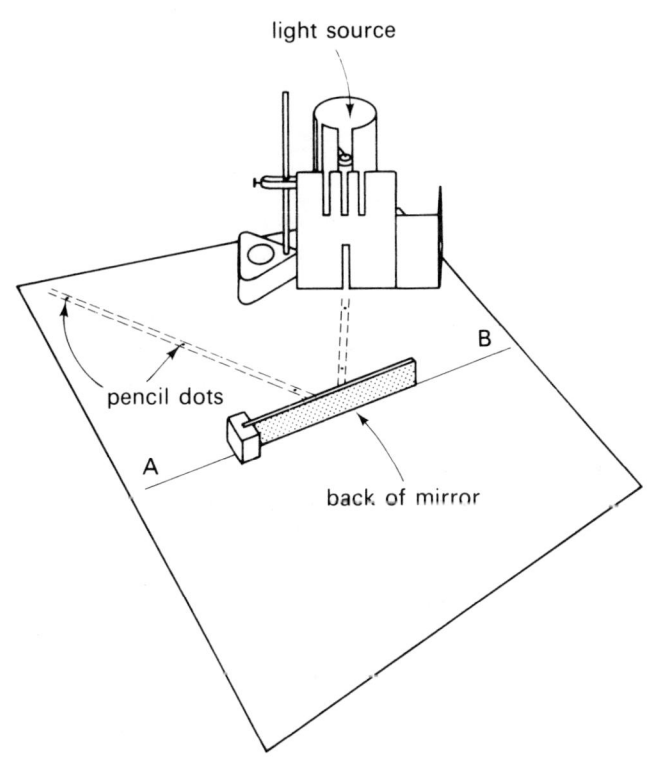

Fig. 5.10 *Reflection in a plane mirror.*

3 The source of light is a lamp with a small filament, or a thin vertical filament. If a screen with a narrow slit is placed in front of it, you get a very thin beam of light. We can think of this as a ray. It skims across the paper and you can follow its path a long way.

4 Aim the ray so that it strikes the mirror.

5 Mark the paper with two or three dots spaced along the incident ray.

6 Mark the paper with two or three dots spaced along the reflected ray.

7 Remove the apparatus from the paper. Join the dots with straight lines (use a ruler). This plots the exact paths of the incident and reflected rays. Where do these rays meet? Label this point with the letter C.

8 Draw the normal at point C; use a set-square or protractor to make the angle between the normal and the mirror (line AB) exactly 90°.

9 Measure the angle between the incident ray and the normal.

10 Measure the angle between the reflected ray and the normal. What do you notice about the two angles you have measured?

11 Repeat steps 4 to 9 three more times. Use the other half of the paper, and the other side of the paper. Let the ray strike the mirror at a different angle each time.

12 Label all the four drawings you have made in the same way as fig. 5.11. Write the sizes of the angles of incidence and reflection on the drawings.

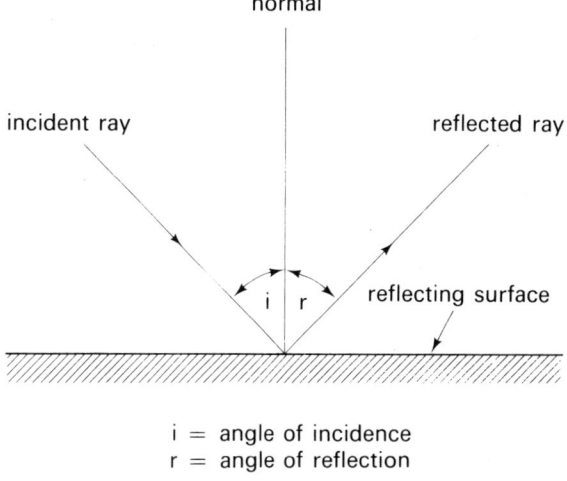

i = angle of incidence
r = angle of reflection

Fig. 5.11 *Reflection at a plane surface.*

13 Make a hypothesis about the relation between the angle of incidence and the angle of reflection.

When you were following the instructions above you will have looked in the mirror and seen the reflection of the slotted screen and lamp. These are *images*, seen in the mirror. The lamp appears to be somewhere behind the mirror. Exactly where does it appear to be?

Instructions: finding the position of the image

1 The apparatus is the same as for the previous experiment except that you need a screen with *three* slits, to give three rays of light.

2 Mark the line AB across the paper.

3 Place the lamp and screen on one side of the line so that the three rays are aimed towards where the mirror will be. Are the rays parallel? From where do they diverge?

4 Put the mirror with its *back* on the line.

5 Mark the three incident rays and the three reflected rays with dots. The three reflected rays diverge (fig. 5.12). From where do they appear to diverge?

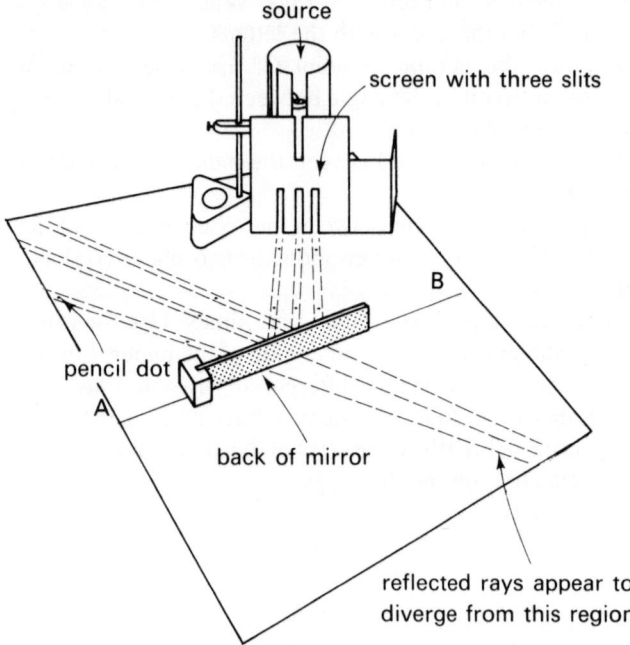

Fig. 5.12 *Finding the position of the image.*

6 To find the exact point from which they diverge, remove all the apparatus from the paper. Join the dots with ruled lines.

7 The incident rays diverge from a point just below where the lamp filament was. Mark this point with an O, (the *object*).

8 The reflected rays appear to diverge from a point behind the mirror. Continue these rays with dashed lines to show where they appear to have come from. Where they meet, mark the point with an I, (the *image*).

9 Join O and I by a straight line. What do you notice about the angle between this line and the line of the mirror, AB? What do you notice about the distance of O and the distance of I from the mirror?

The previous investigation suggests that there is a point behind the mirror from which light rays appear to diverge. This is the image of the filament. If this image is a definite point, we ought to be able to place a second object at this point. Then the image and the second object will appear to be *at the same place.* Let us test this idea by experiment.

Instructions: an experiment to put an object in the same position as an image

1 The image and object appear to be in the same position if they show *no parallax* (fig. 1.1b). Draw line AB, as before.

2 Set up the mirror with its back to the line. Place a pin anywhere in front of the mirror. This is the first object.

3 Place a second pin *behind* the mirror, so that it appears to be in the same position as the image of the first pin (fig. 5.13). As you move your head from side to side, the *top* of the second pin should remain exactly in line with the *bottom* of the reflected image of the first pin, as shown in the figure. If the second pin and the image come out of line as you move your head, there is parallax. Alter the position of the second pin. When you have found the position in which there is no parallax, the second pin and image are always in line, from whichever direction you look. They appear to be in the same place.

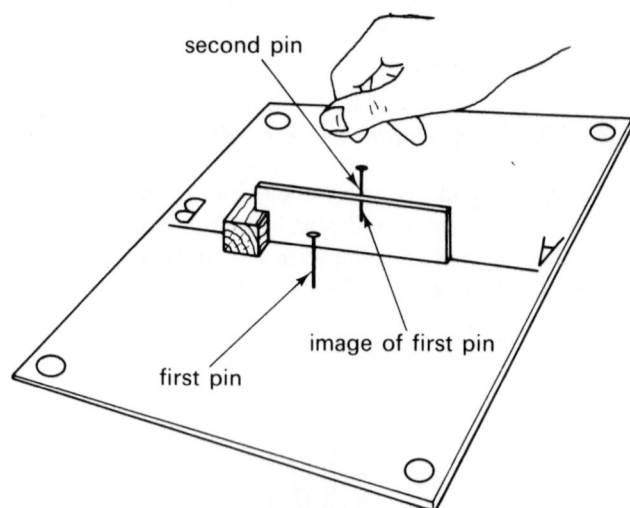

Fig. 5.13 *Finding the position of the image by the method of no parallax.*

4 Mark the position of the second pin. Draw a line from the first pin O to the second pin, which appears to be at the same place as the image I. The line OI cuts the line of the mirror (AB) at H.

5 Check that OI is at right-angles to AB (in other words, it is normal to the mirror).

6 Check that OH = IH. Was the second pin truly in the same position as the image?

Instructions: finding the nature of the reflected image

1 Set up the mirror as in the previous instructions. Draw a large capital letter R on the paper in front of the mirror. Make the letter about 5 cm high.

2 Put a pin in the paper at the top left-hand corner of the letter. Locate the position of the image of this pin, using a second pin and the method of no parallax (page 2). Mark the position of the second pin by a dot drawn on the paper.

3 Repeat, placing the first pin at various positions on the letter, and locating its image each time. In this way you are plotting the position of the image of the letter as a set of dots.
4 When you have finished, join the dots with a pencil line.
5 Compare the drawing of the image with the original letter. What difference can you see? In what way are they alike?
6 Which capital letters could not be used for this experiment?

5.12 Using the laws of reflection

What happens if we have two mirrors at right-angles, with one pin between them. How many reflections can we see, and where are they? We can use our knowledge of the laws of reflection to predict what will happen. Fig. 5.14 shows what happens. Light is reflected in mirror 1. An image is formed at I_1 (fig. 5.14a). It is on the normal to the mirror from O, and equidistant from the mirror. We can find its position by drawing the normal and marking out equal distances. Similarly, we can find the position of the reflection of O in mirror 2, image I_2. Rays of light from the object, reflected in one mirror only are drawn in fig. 5.14a. In fig. 5.14b we see rays that are reflected twice. The eye at A sees a ray that has come from O to mirror 2, then reflected to mirror 1, and then reflected to the eye. Looking into mirror 1, the eye sees the reflection of I_2 in mirror 1; call this image I_{12}. Looking in mirror 2 from position B, the eye sees the reflection of I_1 in mirror 2; call this image I_{21}. It can be seen that images I_{12} and I_{21} occupy exactly the same position. As we move the eye around we see O, I_1, I_2 and I_{12} or I_{21}. We see the object and three images; apparently there are four pins.

Instructions: locating reflections in two mirrors
1 Check the predictions made above by doing this experiment. Find the positions of I_1, I_2, I_{12} and I_{21} by the method of no parallax. Confirm that I_{12} and I_{21} coincide.
2 Predict what will happen if the angle between the mirrors is altered to 60°.
3 Perform an experiment to check your prediction. Do you know of a toy that uses two mirrors set at 60°?

Reflection changes the direction of a ray of light. This can be useful. We can use a mirror for seeing round corners. Mirrors are used in automobiles to allow the driver to view the road behind him. They are used in stores to help the storekeeper view all corners of his store, and watch for thieves. Two mirrors may be used to construct a simple kind of periscope (fig. 5.15). This is used for looking over walls and other obstacles. You can look over the heads of people in front of you when you are at the back of a crowd of spectators. You can easily make a periscope by fixing two mirrors in a long cardboard box or tube.

When a beam of parallel rays falls on a plane mirror, each ray is reflected according to the laws of reflection and therefore all parallel rays are reflected parallel to each other (fig. 5.16a). The beam is reflected and continues in a new direction. If the surface is not highly polished, but is rough, the rays meet it at many different angles of incidence. Each ray obeys the laws of reflection at the point where it strikes the surface. Because the surface is rough, the rays are scattered in all directions (fig. 5.16b). This is irregular reflection. In headlamps, spotlights, and flashlamps, we use highly polished mirrors (sometimes curved, but *not rough*) to give regular reflection and produce a narrow beam of light to illuminate a small area brightly. In the shades on room lights and desk lamps, the reflecting surfaces are usually rough. The light is scattered

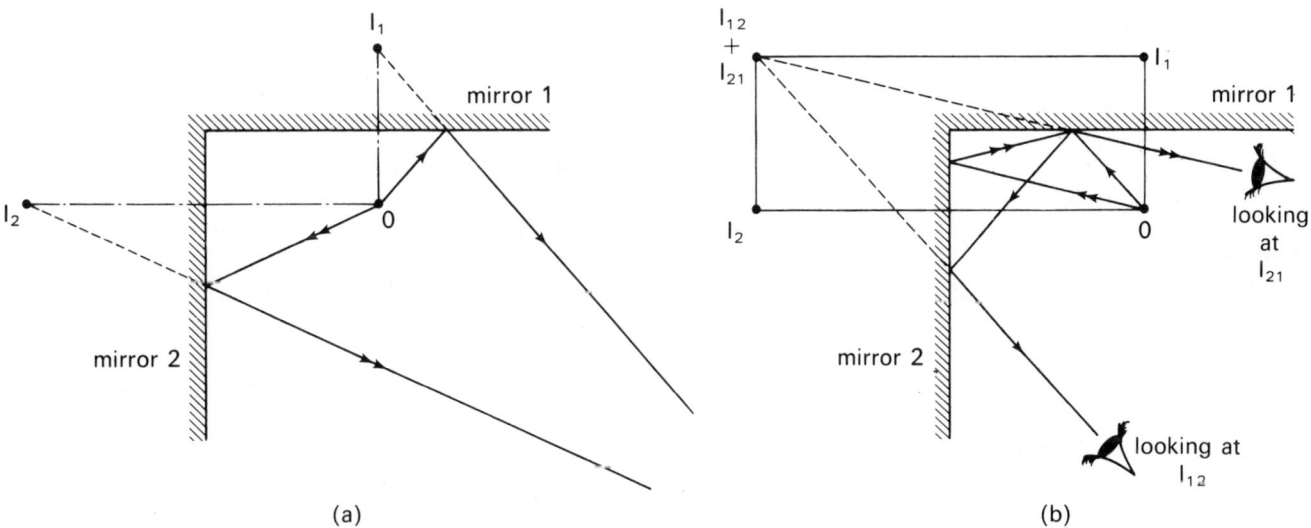

(a)   (b)

Fig. 5.14 Reflection in two perpendicular mirrors: (a) images produced by single reflection; (b) images produced by double reflection.

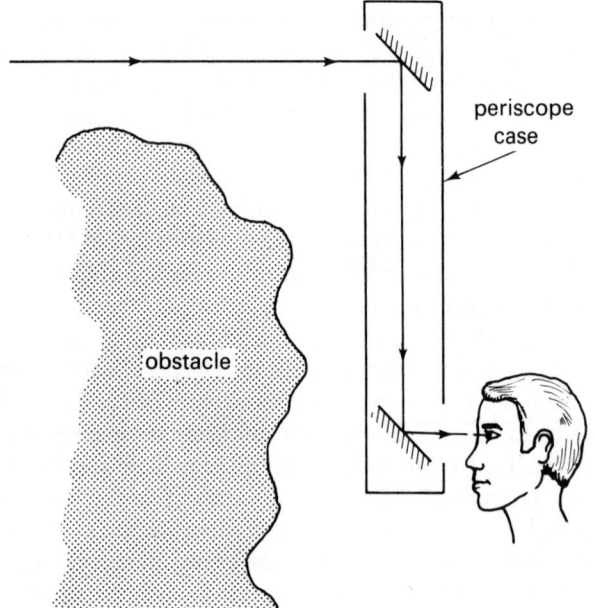

Fig. 5.15   A simple periscope.

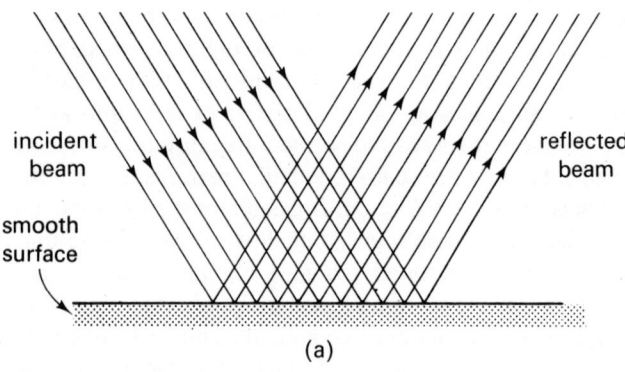

(a)

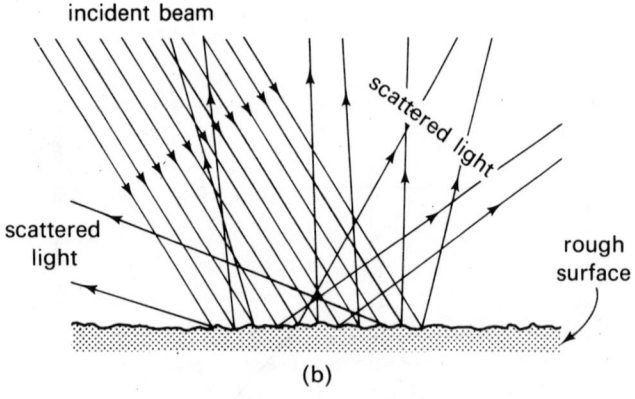

(b)

Fig. 5.16   Reflection at (a) a smooth surface and (b) a rough surface.

over a wide area, giving even illumination that is more comfortable to the eye. There is more about using the laws of reflection in chapter 25.

5.20 Refraction

5.21 The laws of refraction

Refraction occurs when a ray of light strikes the surface of a transparent object, such as a block of glass. The ray *passes through* the surface, and continues its journey *inside* the transparent object. As it passes through the surface, the direction of the ray may be changed. This change of direction is called *refraction*. We can investigate the laws of refraction by studying what happens when a ray is travelling in air (which is a transparent substance) and strikes the surface of a glass block (another transparent substance). Before it reaches the surface, we call it the *incident ray*. This is the same name that we used for a ray reaching a reflecting surface. After the ray has passed through the surface, we call it the *refracted ray* (fig. 5.17). One of the laws of refraction is like one of the laws of reflection: the incident ray, the refracted ray and the normal are all in the same plane. The law that gives the relation between the angle of incidence, i, and the angle of refraction, r, can be discovered by making measurements.

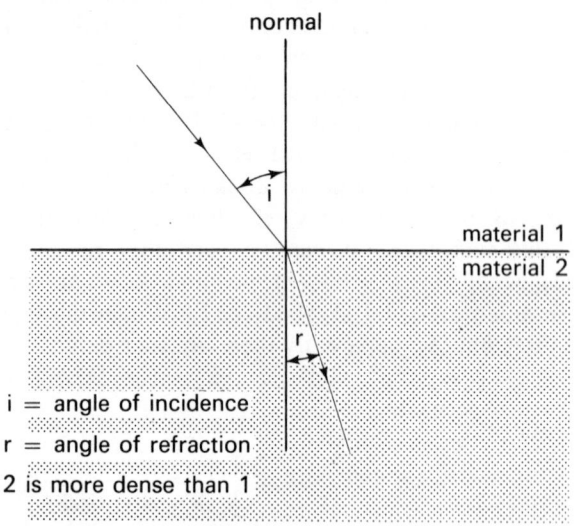

Fig. 5.17   Refraction.

Instructions: investigating refraction at an air-glass surface

1 Put a rectangular block of glass on a piece of paper and draw round it. This is a guide for when you need to remove and replace the block. It is best to use a block that is painted white on one side, and place this side down on the paper.

2 Use a lamp and screen with one slit (fig. 5.10) or a ray-box to produce a single ray. Aim this so that it strikes one long side of the block (fig. 5.18a).

3 Look at the path taken by the ray as it enters the block at one surface. It is refracted, passes through the block, strikes the opposite surface, is refracted *again* and leaves

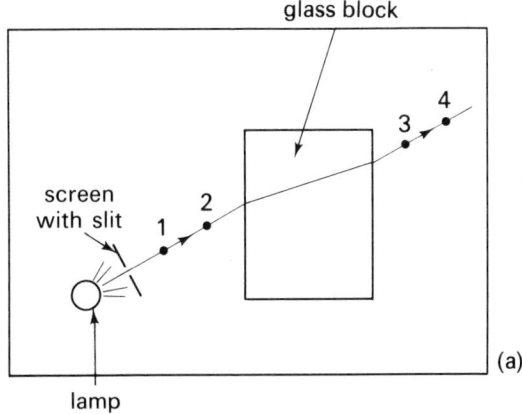
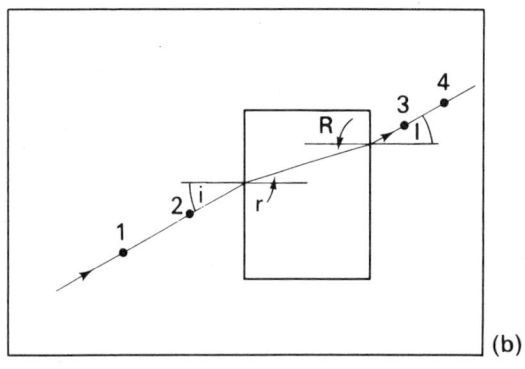

Fig. 5.18 *Refraction at an air-glass surface.*

the block. When it leaves the block we call it the *emergent ray.* If your block is *not* painted white underneath you will not easily see the ray inside the block, but you can easily see the incident and the emergent rays.

4 Alter the angle at which the incident ray strikes the block. What do you notice about the directions of the incident and emergent rays?

5 Mark the incident and emergent rays with four pencil dots (1, 2, 3 and 4 as in the figure).

6 Remove the apparatus from the paper. Join dots 1 and 2 and continue the line to the point where the ray entered the block. Join dots 3 and 4 and continue the line to the point where the ray left the block. Now you can join the incident and emergent rays to find the path of the ray as it travelled through the block (fig. 5.18b).

7 Draw the normal at the point where the incident ray strikes the block. Measure the angle of incidence (i) and the angle of refraction (r).

8 Repeat steps 5 to 7 about 5 times, using rays with different angles of incidence. Make a table of each value of i, and the corresponding value of r. Can you see any relation between i and r? With reflection the relation was very easy to see ($i = r$). With refraction it is not easy to see, and it would probably take you a long time to find it. It was found by a scientist called Snell, and has become known as *Snell's law.* Snell discovered that if we use the *sine* of i and the *sine* of r, there is a relation between these sines. To see how this works, set out your results in a table: the table has five columns, headed i, r, sin i, sin r and sin i/sin r.

9 Look up the sine of each value of i and of each value of r in trigonometrical tables. Write these in your table.

10 Calculate sin i/sin r for each pair of values and write the results in the last column of the table. What do you notice about the figures in this column?

11 Calculate the average value of sin i/sin r.

Instructions: investigating refraction in other materials

1 Repeat the previous instructions, using a block of some other transparent material, such as perspex, or use a thin-walled rectangular tank (made of glass or plastic) such as a small fish-tank. This can be filled with water or another liquid such as glycerine, ethanol, strong sucrose solution or kerosene. If the glass is thin and of good quality we can ignore its effect, and think of the tank simply as a 'block' of water, etc.

2 Calculate the average value of sin i/sin r for each substance.

For reflection, the rule $i = r$ applies for *all* smooth reflecting surfaces: plane glass mirror, polished steel, polished copper, etc. Reflection is not affected by the material from which the reflecting surface is made. Is the same true for refraction?

Your experiments show that the value sin i/sin r depends on the material causing the refraction. It is a property of the material. To be more exact it is a property of two materials: the material in which the incident ray is travelling (in these experiments, air) and the material in which the refracted ray travels (glass, water, perspex etc). For any pair of transparent materials sin i/sin r is a constant. We call it the *refractive index.* Refractive index is usually given the symbol, n, which is much quicker to write than sin i/sin r. Refractive index is obtained by dividing a sine by a sine. It is a simple ratio: it is just a number, with *no units.* Usually we measure refractive index by plotting the path of a ray passing from air into some denser substance, such as glass. We quote the refractive index of glass as $n = 1.5$, or the refractive index of ice as 1.31, meaning that this refers to a ray going from air to glass (or ice). It is also possible to have a ray going from glass to air, or from ice to glass, or from water to ice. If there is any chance of confusion we write small letters before and after the 'n' to indicate which materials are involved and in which direction the ray is going.

$_A n_G$ means the ray is going from air to glass,
$_G n_A$ means the ray is going from glass to air,
$_W n_I$ means the ray is going from water to ice.

43

As shown in fig. 5.18b, a ray can pass in *either direction* along the path. $i = I$ and $r = R$. For a ray going from glass to air, the angle of incidence is R, the angle of refraction is I.

$$_G n_A = \frac{\sin R}{\sin I} \qquad = \frac{\sin r}{\sin i} \qquad = \frac{1}{_A n_G}$$

5.22 The effects of refraction

In the hot season mirages often appear on tarmac roads and sandy areas. This effect is thought to be due to refraction (fig. 5.19). The air near the ground is extremely hot, and therefore less dense than air above it. Rays of light which are directed toward the ground are passing from dense air into air that is less dense. They are refracted away from the normal. The rays are refracted more and more as they get closer to the ground and eventually they are horizontal. Then, if they pass into denser air, they are refracted upwards. When we look at these rays we apparently see a reflection. Our experience tells us that reflections like these occur in pools of water, so we *think* we see a pool of water where none exists.

One of the most important applications of refraction is the use of lenses. This topic will be dealt with in chapter 24. Another use is the splitting of white light into its component colours, (Instructions, page 46).

When a straight stick is placed half in water, it appears to be bent (fig. 5.20). This can be explained by refraction. As rays of light leave the water surface they are refracted, and the effect is to make the parts of the stick below the water appear to be less deep in the water than they really are. So the stick appears to be bent upwards towards the surface. For the same reason, a pool of water or swimming pool appears to be less deep than it really is. If you have used a face-mask or goggles when swimming underwater, you will have noticed that fish and other objects appear to be nearer to you than they really are. We make use of this effect in a method for measuring refractive index.

Instructions: measuring the refractive index of a substance by its real and apparent depth
1 Draw a line on a sheet of paper and stand a glass block on this line. If you look down through the block, you can see the line. It *appears* not to be as far from the top surface of the block as it really is. (Compare with fig. 5.20.)

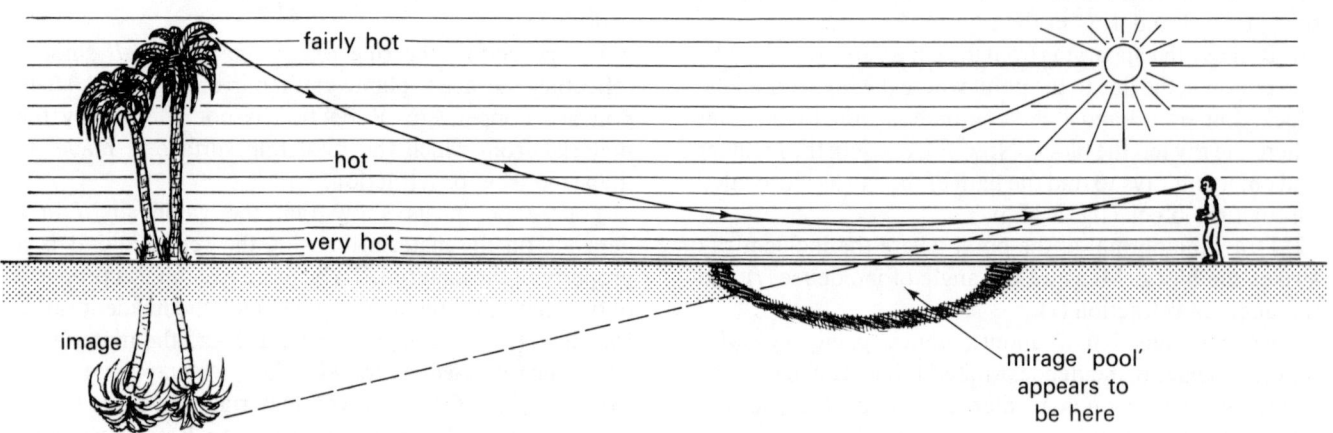

Fig. 5.19   A mirage.

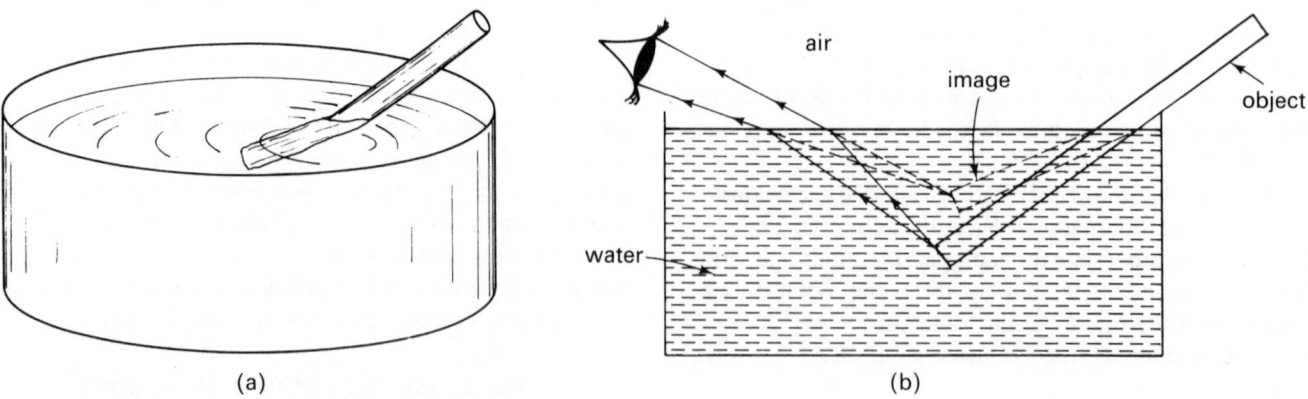

Fig. 5.20   The illusion of the bent stick; (a) the illusion; (b) the explanation.

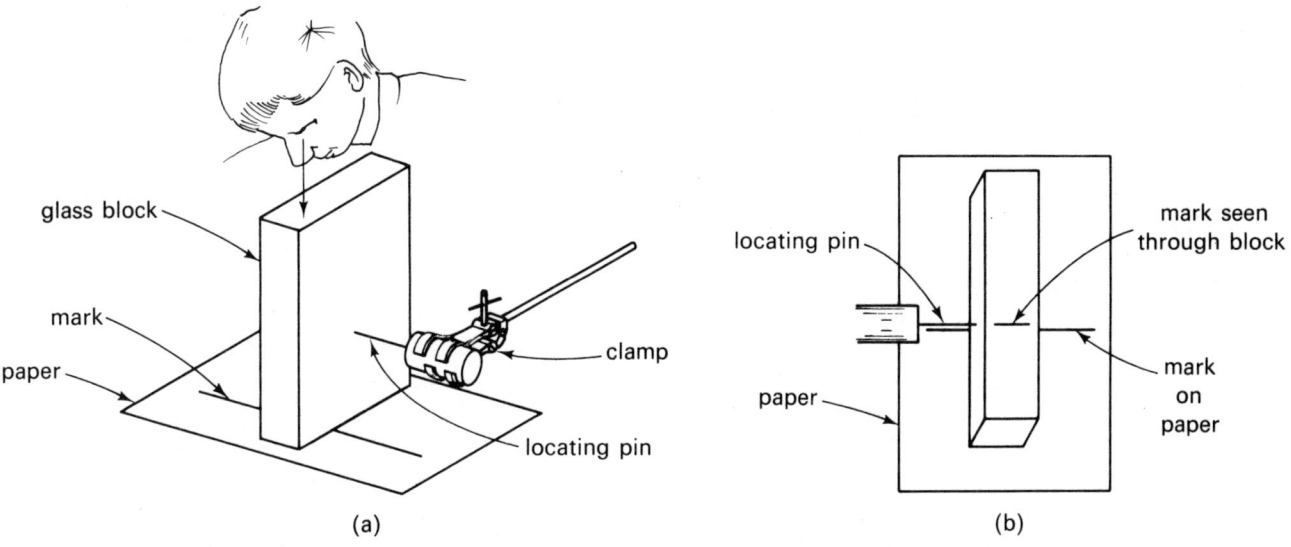

Fig. 5.21 *Refractive index by the method of real and apparent depth; (a) the apparatus; (b) view of mark and locating pin, seen from above.*

2 Use a locating pin held in a stand (fig. 5.21 a). Move it up or down until it is at the same level as the mark appears to be when seen through the block. Use the method of no parallax (see page 2). As you move your head backwards and forwards, the locating pin and the mark (seen through the block) must stay in line (fig. 5.21 b). If they do not, move the pin up or down until they do.

3 Measure the distance between the pin and the *top* surface of the block. This is the *apparent depth, d_a*.

4 Measure the distance between the top and bottom surfaces of the block. This is the *real depth, d_r*.

5 The method of calculation is explained by fig. 5.22. A ray leaves the mark M and passes vertically up through the block. It leaves the block at F and passes without refraction to the eye of the viewer. Another ray passes up at a very slight angle to the vertical. In the figure the angle is drawn to look bigger than it really is. The ray leaves the block at E, is refracted, and enters the eye. The two rays appear to diverge from M'. The mark is apparently at M'. Refractive index from *glass* to *air* is

$$_G n_A = \frac{\sin i}{\sin r} .$$

In triangle EFM, angle EMF $= i$ (alternate angles)

$$\sin i = \sin \mathrm{EMF} = \mathrm{EF/ME} .$$

The triangle is very long and thin; ME and MF are almost equal, and since

$$\mathrm{MF} = d_r , \text{we can say that } \mathrm{ME} = d_r ,$$

$$\text{and } \sin i = \mathrm{EF}/d_r .$$

In triangle EFM', angle EM'F $= r$ (corresponding angles)

$$\sin r = \sin \mathrm{EM'F} = \mathrm{EF/M'E}$$

The triangle is very long and thin; M'E and M'F are almost equal, and since

$$\mathrm{M'F} = d_a , \text{we can say that } \mathrm{M'E} = d_a ,$$

$$\text{and } \sin r = \mathrm{EF}/d_a .$$

Therefore $_G n_A = \dfrac{\sin i}{\sin r} = \dfrac{\mathrm{EF}}{d_r} \times \dfrac{d_a}{\mathrm{EF}} = \dfrac{d_a}{d_r}$

and $_A n_G = \dfrac{1}{_G n_A} = \dfrac{d_r}{d_a} .$

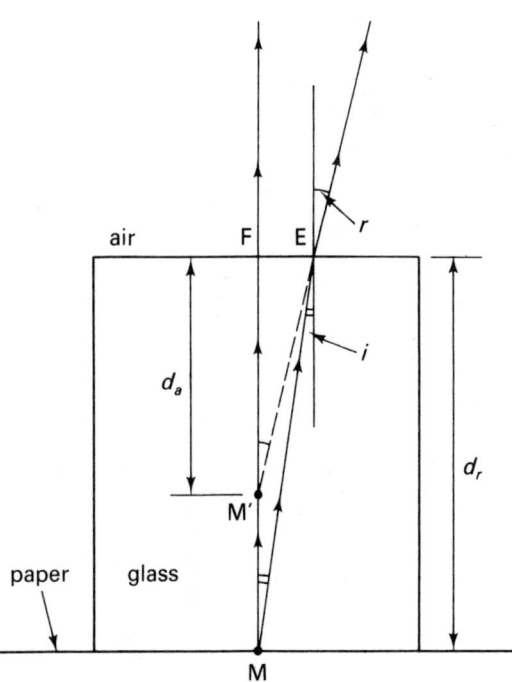

Fig. 5.22 *Explanation of the calculation of refractive index.*

This reasoning gives us a simple rule:

> to calculate refractive index, divide real depth by apparent depth.

Calculate the refractive index, $_An_G$, from the measurements you have made.

6 Repeat the procedure and calculation to find the refractive index of a liquid. Fill a beaker with water, glycerine or kerosine. Stand it on a sheet of paper on which a mark has been drawn. Use a beaker with a thin, flat bottom. The effect of refraction in the glass can be ignored. Distances are measured from the top surface of the liquid.

Instructions: investigating refraction through a triangular glass prism

1 Hold up a sewing needle or shiny pin in sunlight. It should gleam brightly. What colour is the light coming from the needle to your eye?

2 Hold a triangular prism so that it is between the needle

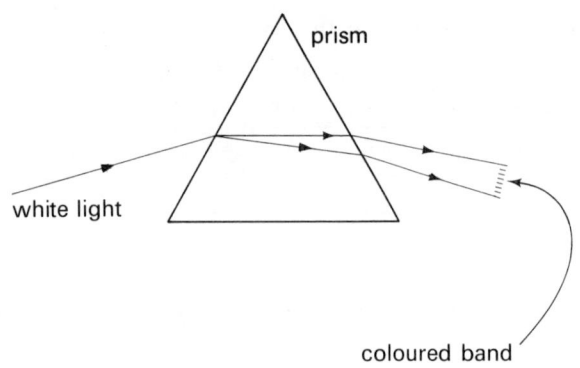

Fig. 5.23   Refraction through a triangular glass prism.

and your eye. Turn the prism around in various positions, and look through it at the needle. What do you see now?

3 Set up a lamp with a one-slit screen in front of it, and place it on paper, as in fig. 5.10.

4 Let the ray of light strike one side of the prism, as in fig. 5.23. Look for the coloured band. Mark the incident ray of white light and the regions of coloured bands, using pencil dots. Can you explain where the colours come from?

5 For which colour is $_An_G$ highest; for which colour is it lowest? On passing through a prism a narrow beam of white light is split or *dispersed*. Even though the beam is narrow, it has a width of a few millimetres, so that the differently coloured beams overlap each other. The colours merge with neighbouring colours. If the slit is very narrow, the colours are better separated, but the light is too faint to see. To demonstrate the colours clearly, use an apparatus similar to that shown in fig. 5.24. The source of white light is bright, with a narrow slit in front of it. The lens focusses the narrow beam so that, when the prism is not there it forms a very narrow image (S) of the slit on a distant screen. This would appear as a narrow band of bright white light. If the prism is put in this focussed beam, each colour is separately focussed on the screen. We get a continuous band of colours. In order, they are red (R), orange, yellow, green, blue, indigo and violet (V). This is called a *spectrum*. In practice the screen is placed a long way from the prism, so that the spectrum may be several centimetres wide and the regions of different colours can easily be seen.

5.23 The spectrum

The colours of the spectrum are pure colours, or *spectral colours*. They are different kinds of light.

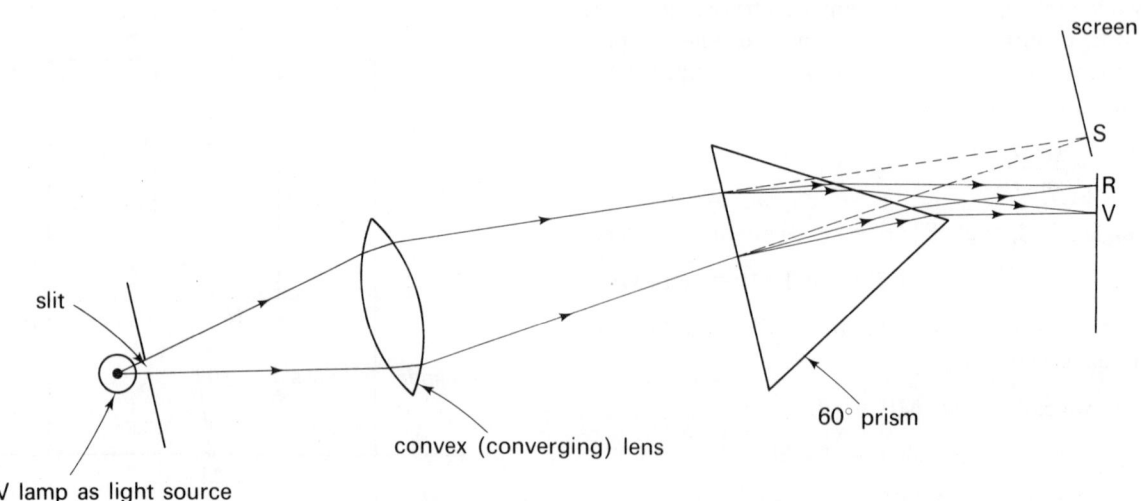

Fig. 5.24   Apparatus for producing a spectrum.

White light is not a pure colour. It is a mixture of all the spectral colours. When we see this mixture, it produces an effect in the brain which we call 'white'. There are many other mixtures of spectral colours, some of which are given special names, such as brown, pink, khaki, cream, olive green, carmine and many more.

When white light shines on a diamond it is refracted and dispersed. Beams of spectral colours leave the diamond in many directions. If a shower of rain falls while the Sun is shining, refraction and dispersion occur in the raindrops. If you stand with the Sun behind you and the shower is in front of you, you see a spectrum in the form of an arc in the sky, a rainbow. You can see the same effect in the spray from a garden hose. Dispersion of white light in the drops of water sends differently coloured light in different directions. From some regions of the shower or spray, your eyes receive red light; from regions next to this they receive orange light, and so on.

5.30 Colour

5.31 Coloured light

We can use the apparatus of fig. 5.24 to discover more about coloured light. A piece of red glass is placed in the beam of white light (between the lens and the prism). Only the red region of the spectrum appears on the screen. The glass lets only red light pass through it; we say it *transmits* red, and *absorbs* light of other colours (page 37). Such pieces of coloured glass or plastic are often used as colour filters. We use them in the lights on the stage of a theatre and in traffic lights, for example. A colour filter may transmit light of only one colour, or it may transmit several colours in various amounts. For example a pale yellow filter transmits a little red light, some orange light, a lot of yellow light and a little green light. The effect of the transmitted mixture is 'pale yellow'.

Instructions: investigating reflection of coloured light
1 Put a colour filter in front of the lens of a slide projector to give a beam of coloured light. Darken the room, hold pieces of paper and coloured pictures in the beam of light. Record what you see.
2 Repeat, with beams of various other colours.
3 What effect do you see when:
(a) Coloured light shines on white paper?
(b) Coloured light shines on paper of the same colour as the light?
(c) Coloured light shines on black paper?
(d) Coloured light shines on paper of colour different from that of the light?

4 Explain the effects you have noticed by saying which colours are reflected by the paper and which colours are absorbed.

Instructions: mixing coloured lights
1 Use three slide projectors, each with a colour filter: red, green or blue, fig. 5.25.

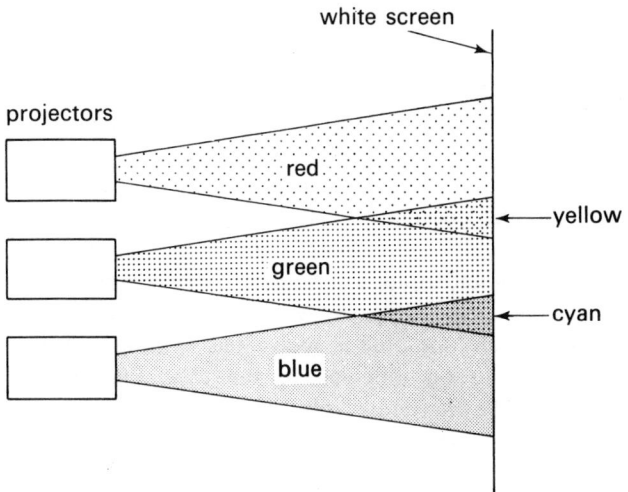

Fig. 5.25   The mixing of coloured lights.

2 In a darkened room, project their beams on a white screen.
3 Overlap two of the beams. What colour can you see where they overlap?
4 Overlap the beams in all possible pairs; then try overlapping all three beams. What colours can you see on the screen?

The three coloured lights, red, green and blue are the three *primary lights*. By mixing two or all three in various amounts we can obtain all the colours that we know. When we mix coloured lights we *add* them together. We begin with a dark screen (no light); if we add red light to green light, the screen appears to be yellow. This is only an effect in the brain; the yellow seen on the screen is *not* yellow light such as we find in the yellow region of a spectrum, but the red-green mixture and pure spectral yellow *appear* alike. This effect is made use of in the theatre, where coloured floodlights are mixed to obtain all the colours needed on the stage. The effect is used in a colour television set. When the set is off, the screen is almost black. The screen is covered with small dots or lines of special paint that glow red or green or blue. By making differently coloured dots glow with different brightness, all possible colours can be made to appear on the screen. If you look closely, you can see the separate dots of primary colours. From the normal viewing distance they merge, our eyes blend them, then a full-colour picture is seen.

5.32 Coloured pigments

The mixing of paints, pigments or inks works by *subtracting* colours from the white mixture. We begin with white paper and shine white light on it. When we put paint or ink on the paper, it absorbs (or subtracts) lights of certain colours from the white mixture and reflects the remainder. If we mix two paints on the same area of paper, both paints subtract light from the white mixture. For example, the effect of a mixture of blue and yellow paints is shown in fig. 5.26. The particles of paint are shown very highly magnified. In the figure, white light strikes a blue particle. Few paints reflect only pure spectral colours, so the light reflected from the blue particle is a mixture of blue and green lights, probably more blue than green. Red light, yellow light and orange light have been absorbed, or subtracted, from white. When this mixture strikes a yellow particle, the blue is subtracted, leaving only the green light to be reflected to the eye. The person looking at the mixture of blue and yellow paints sees green. Artists use mixtures of three basic paints, red, yellow and blue. With these, and perhaps some white paint and black paint for altering brightness, they can mix any other colour they require. However, most artists also use a number of paints of special colours,

as this is a more convenient way in which to work. You could experiment with mixtures of artist's paints to see what colours you can obtain. What would you expect to see if you mixed all three: red, blue and yellow, together?

Because the pigments used in painting are usually not pure spectral colours it is a complicated matter to make a particular colour by mixing the artist's basic pigments. A lot of experience is required.

Printers who produce coloured pictures and photographs in magazines and books use a different set of colours. These are:

magenta – a bluish-red colour that reflects red and blue, but subtracts green,
yellow – reflects red and green, but subtracts blue,
cyan – a bluish-green colour that reflects blue and green, but subtracts red.

Each colour reflects two colours but subtracts (or absorbs) the third.

If you look at a colour photograph in a magazine, and use a lens, you can see that it is made up from tiny dots of ink, of different sizes. These dots are magenta, yellow or cyan. In printing, the paper is run through the press three times. Three separate pictures are printed on top of each other in the three colours. Often a fourth printing is made, using black ink, to improve the contrast between light and shade.

Projects

1 Make a pinhole camera and try taking photographs with it. You will need a dark-room for putting the film in the camera, and for taking it out after the exposure. Instead of film you could use a piece of enlarging paper. This gives you a paper negative. It is not easy to make a positive print from this, but making the negative is interesting.

2 Make a kaleidoscope (5.12).

3 If you are keen on photography, find out how colour photographs are made. Photographic handbooks and the manufacturer's leaflets can tell you how the film is made and processed. The information in 5.30 is helpful. You can find out about the different methods of colour photography: colour transparencies (slides); colour print film and the making of prints from colour negatives; instant colour prints (for example, Polaroid).

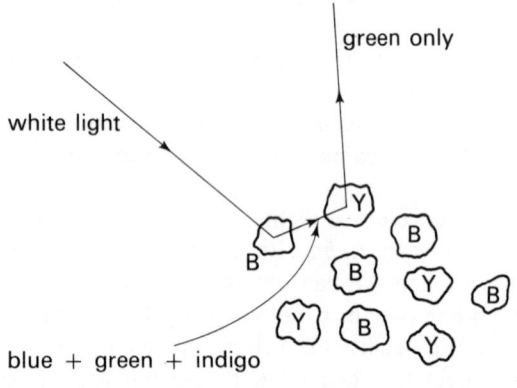

B = blue pigment
Y = yellow pigment

Fig. 5.26 Why a mixture of blue and yellow pigments looks green.

6 Discovering about heat

6.00 What happens to substances when they are heated?

In 3.30 and 3.40 we studied two ways in which a substance may change when it is heated: it may melt or it may boil. Both these changes are called a *change of state*. There are three states of matter: solid, liquid and vapour. Melting is a change from the solid to the liquid state; boiling is a change from the liquid to the vapour state. Gases are substances which are in the vapour state at normal temperatures. Melting and boiling occur at definite temperatures, depending on the substance, the surrounding pressure and the presence of impurities (if any). Liquids can also turn into the vapour state by evaporation. This occurs at temperatures below the boiling point. When a pool of water dries in the sunshine the water does not boil; it evaporates and the pool gradually goes.

If you heat paper, it does not melt; it catches fire. It combines with oxygen from the air and turns to ash, carbon dioxide, water vapour and other vapours. Substances have been formed that were not there before. You cannot make paper by collecting these substances and letting them cool. Heating paper has caused a *chemical change;* new substances are formed; the change cannot be reversed. In contrast, change of state is a *physical change*. No new substances are produced; the process can be reversed.

In chapter 3 we saw a different kind of physical change. We saw that liquids and gases expand when they are heated; they contract when they cool. There is no chemical change during expansion or contraction. Expansion is a physical change which we find very useful. The thermometers described in chapter 3 make use of this kind of change in liquids or gases, but we shall show that solids too expand when heated. Before we do, though, we will study the expansion of liquids in more detail.

6.10 The thermal expansion of liquids

Expansion produced by heating is called thermal expansion to distinguish it from expansion produced by other means.

Instructions: Investigating thermal expansion of liquids
1 Make four identical thermometers (fig. 3.1). Fill one with water. Fill the other three with liquids chosen from this list: ethanol; methylated spirit; acetone; glycerine; xylene.
2 Stand all the thermometers in a large beaker, water-bath or saucepan containing water at room temperature. Leave them there until the level of liquid in each thermometer is steady.
3 Mark the liquid level on the card of each thermometer.
4 Heat the water steadily, and watch the level of the liquid in each thermometer. Stop heating the water when it becomes as hot as your finger can bear, for some of the liquids have boiling points lower than 100°C. Acetone boils at 56.6°C.
5 Stir the water in the bath when it is warm, so that all thermometers are at the same temperature.
6 Mark the level of the liquid in each thermometer.

Questions
1 Assume that the internal diameters of the thermometer tubes are equal. Which liquid expanded the most when heated?
2 List the liquids in order, from the one which expanded most to the one which expanded least.
3 If you had used mercury in a thermometer you would have found that it expanded a little less than the water. From this information, suggest one way in which an ethanol thermometer is better than a mercury thermometer.
4 Why is acetone not much use as a liquid for thermometers?

6.11 The thermal expansion of water

Instructions: investigating the thermal expansion of water (1)

1 Use the water thermometer from the previous investigation.

2 Make a mixture of ice and salt. Crush some ice then mix with it about half its volume of salt (sodium chloride). This gives a freezing mixture with a temperature of about $-18°C$. Instead you could use very cold ice made in a deep-freeze compartment.

3 Place the thermometer in the freezing mixture. Watch the water column as the thermometer is cooled. What is happening to the water?

4 As soon as the water begins to freeze inside the thermometer, take it out of the freezing mixture. Why?

Water is a very common substance, but it has some extremely unusual properties. One of these strange properties is its behaviour when it is cooled. If we begin with water that is warmer than $4°C$, we find that it contracts as it is cooled. This is not unusual. When it has cooled down to $4°C$, it has reached its *minimum volume*. As it is cooled from $4°C$ down to $0°C$ it *expands*. This is very unusual. As it freezes at $0°C$ it expands a lot. Below $0°C$ the ice contracts as it cools, so it is behaving normally again.

The fact that water expands as it is cooled below $4°C$ and freezes has several important results. The force of its expansion is extremely strong. You can show this by putting water in a closed bottle and freezing it in the ice compartment of a refrigerator. In the colder parts of the world, including the mountaintops of the tropics, rain-water runs into cracks in the rocks. As it freezes at night or in winter it expands. This forces the layers of rock apart. The large rocks are cracked into smaller pieces. In this way the expansion of freezing water plays an important part in erosion. In cold countries water may freeze in water-pipes in the home. The water in the radiator of an automobile may freeze, if the vehicle is left standing outdoors. Expansion may burst the water pipes or the radiator, causing expensive damage. Pipes must be protected from cold by enclosing them as much as possible. The water in the automobile radiator should have anti-freeze added to it. This dissolved substance (see page 22) lowers the freezing point of the water.

Instructions: investigating the thermal expansion of water (2)

1 Use the same water thermometer as before. The water in it should be at room temperature.

2 Hold the bulb of the thermometer over a lighted burner. Watch the water level in the tube from the moment you begin heating. Explain what you see.

6.20 The thermal expansion of solids

Solids expand when heated. For a given change of temperature the expansion of solids is much less than that of liquids. It is less easy to see solids expanding, even when heated to high temperatures, but expansion of solids can be important.

6.21 The thermal expansion of metals

Instructions: investigating the thermal expansion of metals

Try one or more of these demonstrations:

1 Use the ball-and-ring apparatus. This consists of a metal ring into which a metal ball fits exactly, when they are both at the same temperature. Begin with both ball and ring at room temperature. Heat the ball strongly for several minutes over a bunsen flame. Now try to pass it through the ring. What has happened to the ball? Support the ring on a tripod, with the ball resting on the ring. Leave the ball to cool. What happens? Explain. In what way does this demonstrate that expansion is a physical change?

2 Use the bar-and-gauge apparatus. This consists of a metal rod on a handle (the bar). The gauge is a piece of metal with a slot which just fits over the length of the bar, and a circular hole that just fits around it. Fit the bar into the slot and hole to make certain that it is a close fit. Heat the bar strongly over a bunsen burner for several minutes. Does it fit into the slot? Does it fit into the hole? Why test by using both slot *and* hole? Leave the bar to cool and test again. In what way does this demonstrate that expansion is a physical change?

3 Use the apparatus illustrated in fig. 6.1. Heat the rod. Record the change in the position of the pointer. Which way does the pointer move? What does this show? Which way does the pointer move when the rod cools? What does this show? In what way does this demonstrate that thermal expansion is a physical change?

Though metal objects do not expand much in proportion to their size, the forces involved are very great. On a sunny day you can hear creaking noises in the roof of a building if it is made from galvanized iron sheets. This is due to the iron expanding as it gets hot. From time to time the sheets slip slightly over each other, making a noise. The reverse happens in the evening, as they cool. Though an iron bar or sheet expands only a fraction of a millimetre, a large metal structure such as a bridge may expand several centimetres in length. This must be allowed for when the bridge is designed. Sometimes one end of the bridge is supported on rollers (fig. 6.2). The other end is fixed. As the bridge expands, the end on rollers can move slightly, enough to avoid strain that would damage it.

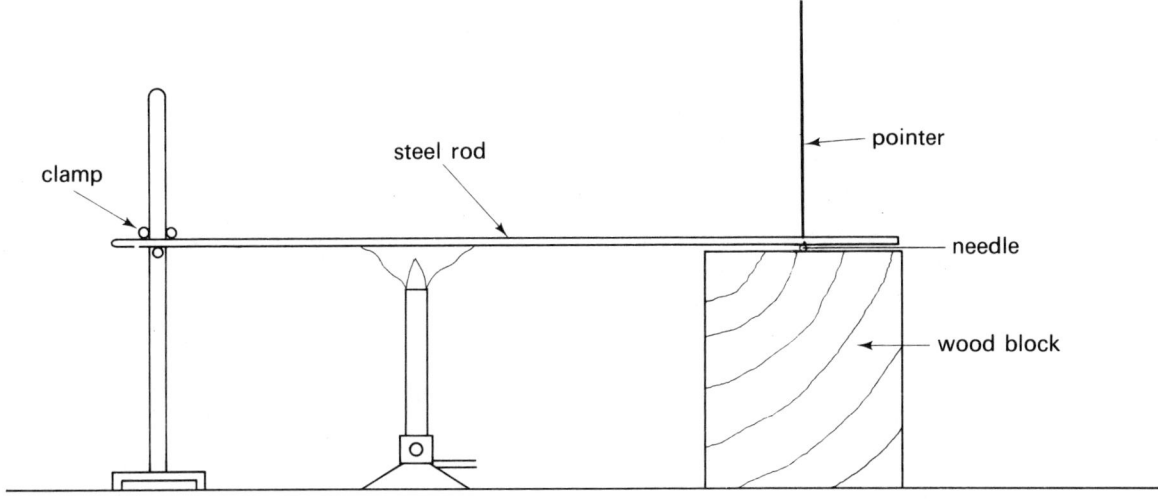

Fig. 6.1   Expansion of a rod when heated.

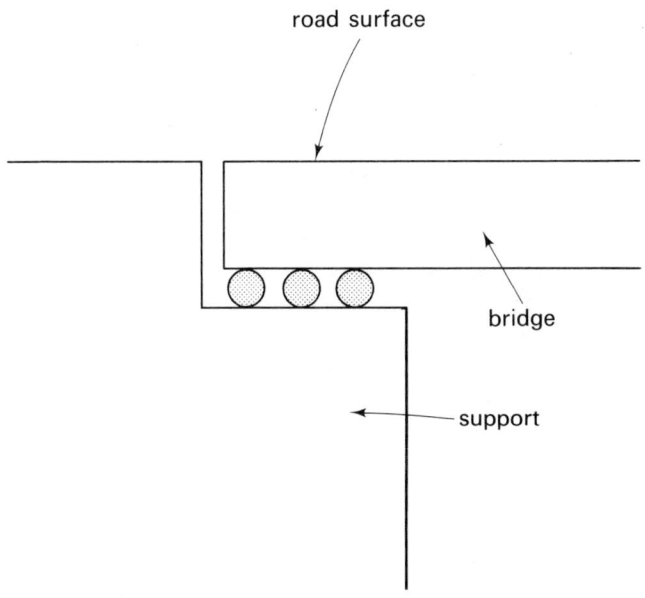

Fig. 6.2   Allowing for expansion in a bridge.

A problem occurs with railway tracks. These are long, yet must be laid exactly end to end. A gap is left between the end of one rail and the next. The ends are held in line with fishplates, which are strips of metal bolted to the ends of the rails by slotted holes. The slots allow the ends of the rails to slide either way as they expand or contract. If the ends were firmly fixed together, with no gap, expansion on a hot day would make the rails bend inwards or outwards. The rails would be distorted and no longer parallel. Trains might be derailed. Another way of allowing for expansion is to taper the rails at each end. The end of one rail overlaps the end of the rail next to it. As the rails expand or contract their ends can slide past one another. This method is used in modern tracks in which rails of normal length are welded together to produce very long lengths of rail, without a join to allow for expansion. The tapering method allows considerable expansion to occur without damage to the rail.

If they are heated the same amount, different liquids expand by different amounts (6.10). This is true of solids too. In general, metals and alloys expand more than other solid materials, but they vary in the amount by which they expand. Aluminium and lead expand about twice as much as iron, when heated the same amount. Invar, an alloy of steel and nickel, expands hardly at all.

Instructions: heating the bimetallic strip
1 The apparatus is shown in fig. 6.3. The two strips of metal are riveted together.
2 Hold the strip in the flame of a bunsen burner for a few minutes. Turn it and move it to make sure that it is evenly heated. What happens to the strip? Explain.
3 Let the strip cool. What happens? Explain.

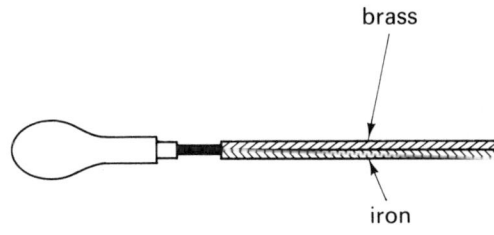

Fig. 6.3   Bimetallic strip.

51

6.22 The thermal expansion of other solids

Ordinary window-glass (soda-glass) expands about a tenth as much as iron. Glass is a brittle substance though; if strong forces act on it, it does not bend (like iron) but cracks. A jug or bottle made from soda-glass is easily cracked by heat. If we pour boiling water into it when it is cold, the inside of the jug becomes hot; the glass expands. The outside of the jug is still cold; it is not expanding. This produces a strong force or strain in the glass, making it crack.

Questions

1 In the laboratory we can use flasks and test-tubes made of soda-glass, if the glass is thin. Explain why thin glass is less likely than thick glass to crack when heated.

2 In the laboratory we can also have glassware made from borosilicate glass, such as Pyrex. This glass expands very little when heated. Why is this glass very resistant to damage by heating? Do you know of any use made of this glass in the home?

6.30 Using expansion in everyday life

If the stopper of a glass bottle becomes stuck, we can use expansion to help us remove it. We stand the bottle in boiling water, making sure that the stopper is not in the water. The bottle expands, but the stopper does not. The stopper comes loose. Another way of heating the neck of the bottle is to wrap a cloth round it, then pull it firmly to and fro many times. The friction heats the neck. This is rather like the ball-and-ring experiment in reverse.

When two sheets of metal are to be permanently joined, we sometimes use rivets. Holes are bored in the two sheets, a very hot rivet is passed through and the rivet is hammered strongly on both sides. This makes a head at each end. The heads hold the sheets together. As the rivet cools, it contracts. This pulls the sheets even more firmly together.

The bimetallic strip has several applications in every-day life. We can use it to make a thermometer. In this the strip is coiled into a spiral. The metal on the outside of the turns is usually brass. The metal on the inside of the turns is usually invar steel. One end of the coil is fixed to the case of the thermometer; the other has a pointer fixed to it. As temperature increases the brass expands more than the invar (which hardly expands at all). The coil becomes more tightly coiled. The pointer moves along the scale to indicate a higher temperature. When the temperature goes down, the coil becomes less tightly coiled and the pointer moves to a lower region of the scale.

Figure 6.4 shows how a bimetallic strip can be used to operate an electric switch. When the strip is touching the contact (C), a current flows through the strip and through the heater. This may be a room heater or a heater in a cooker. When the temperature increases the strip bends downwards. This breaks contact with C. The heater is switched off. Then the temperature begins to fall. The strip becomes straight. It touches the contact. The heater is switched on again. In this way the switch keeps the temperature of the room or cooker more or less constant. The switch can also be used for controlling the temperature of a fish-tank or an electric clothes-iron. Most *thermostatic* switches (as these are called), have a way in which the position of the contact can be adjusted, so that the switch can be set to operate at some chosen temperature.

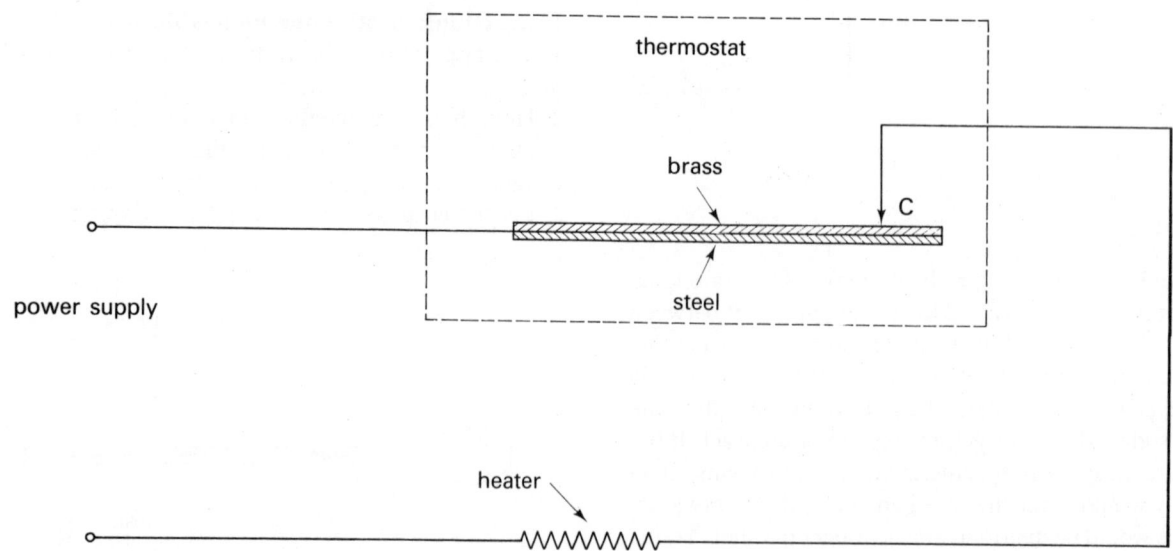

Fig. 6.4 A thermostat switch.

6.40 The transfer of heat

When you go to school in the morning you feel cool. By midday, even indoors you begin to feel hot; at least you do if you live in a warm country. Where does the heat come from? How does it reach you? In other words how is the heat transferred from the source of heat to you?

The main source of heat is the Sun. It is about 150 million kilometres away, yet we can feel the energy coming from it. Some of this energy is in the form of heat, some of it is light. It crosses space as *radiation*. This is one way in which heat can be transferred. The Sun shines on the roof of your school and much of this radiation is absorbed. It passes into the material from which the roof is made. This material becomes hot. The heat passes through this material; we say it is *conducted* through the material of the roof and ceiling. It may then be radiated from the ceiling; it may also be radiated from the walls if the Sun is shining on the outside of the walls. The radiated heat passes across the space between the ceiling or walls and you. The radiation is absorbed in your skin and you feel the heat of it. These are two ways in which heat is transferred: by radiation, in which form it can cross empty space, and by conduction through matter. The third method of transfer of heat is *convection*. If the Sun is shining in through a window at the beginning or end of the day and it strikes the walls and furniture, these become hot. The air in contact with these hot objects also becomes warmed. It circulates around the room, carrying the heat with it.

6.41 Radiation

Instructions: absorbing radiation from the Sun
1 Paint two metal cans or calorimeters on the outside; paint one dull black, the other white (fig. 6.5). Make a lid of expanded polystyrene or cork for each can. Make a small hole in each lid for the thermometers.
2 Stand the cans outdoors side by side on a pad made from a pile of newspaper sheets or on a cork mat.
3 Fill the cans with water at room temperature; the cans must contain equal amounts of water.
4 It is best if the cans are left in bright sunshine for about 3 hours. If there is no sunshine they can be heated by radiation from an electric fire, the type that has a red-hot heating coil and a reflector.
5 Record the temperature of the water in the cans every 15 minutes.

Questions
1 Which can heats more quickly?
2 How can you be sure that the heat is reaching the cans by radiation, not by conduction from the soil or the surface the cans are standing on?
3 Which kind of surface is better at absorbing radiation?

Instructions: investigating radiating surfaces
1 Use the same cans as above. Place them on a table indoors, away from draughts. Stand them on a pad made from newspaper sheets, or a cork mat.
2 Fill each can with equal amounts of very hot water. Put the lids on quickly and begin recording the temperature of the water in the cans every minute.

Questions
1 Which can loses heat more quickly?
2 Much of the heat is being lost by radiation, but some may be lost by conduction to the table. What has been done to make loss of heat by conduction as little as possible?
3 Why are you told to keep the cans away from draughts?
4 What has been done to make loss of heat by convection as little as possible?

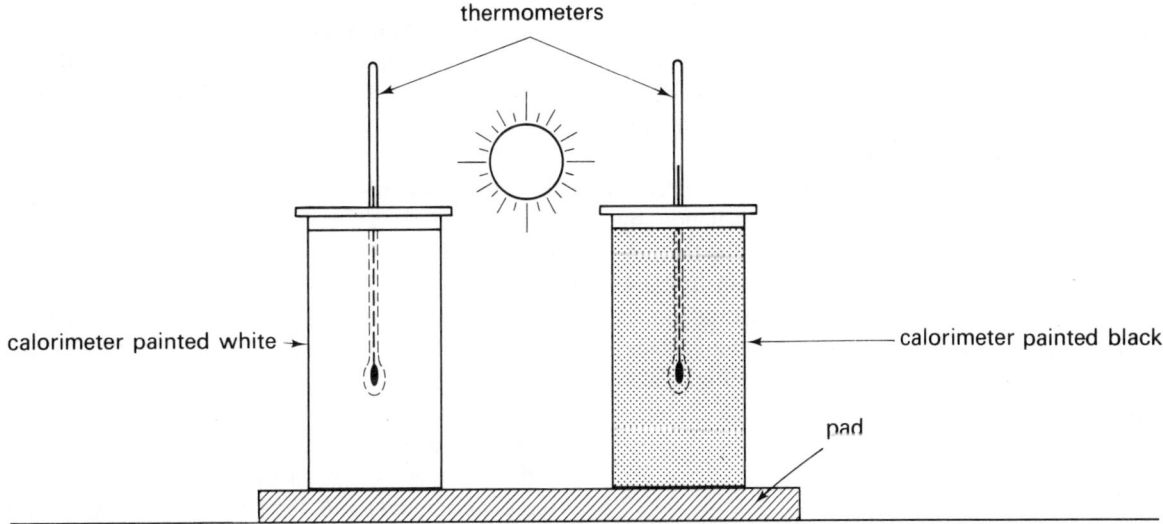

thermometers

calorimeter painted white →   ← calorimeter painted black

pad

Fig. 6.5 Absorbing radiation from the Sun.

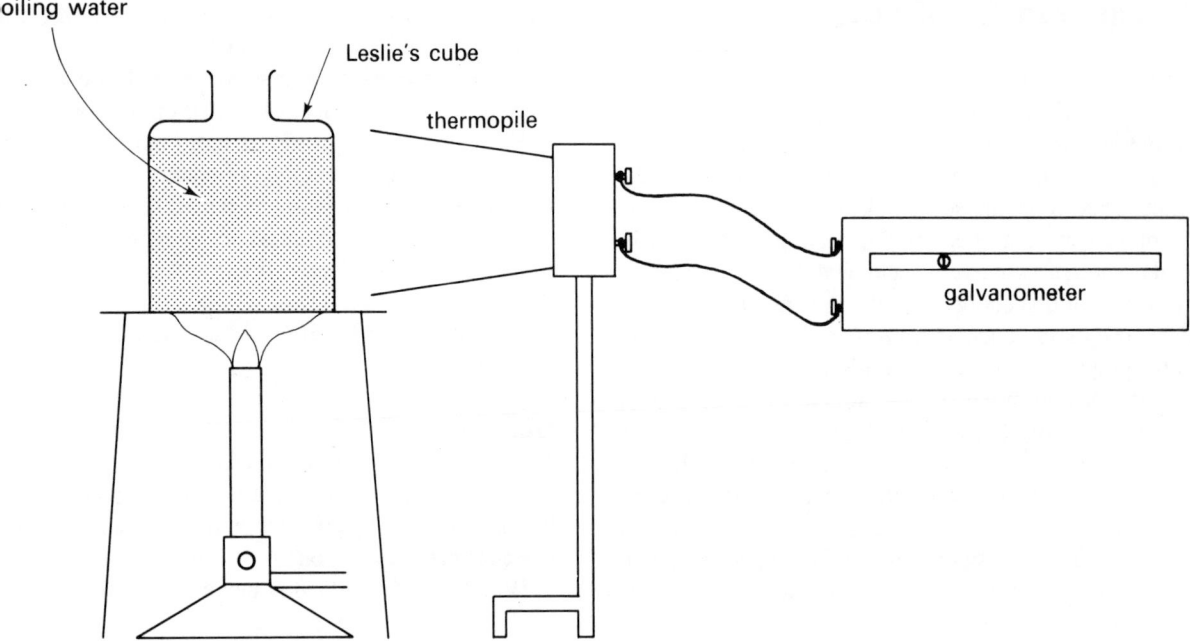

Fig. 6.6 Using a thermopile to detect radiated heat.

5 Which kind of surface is better at radiating heat?

6 Compare the results of this investigation with those of the previous one. What can you say about the heat-radiating properties of surfaces that are good absorbers?

7 What is the advantage of making a kettle, teapot or coffeepot from a shiny metal?

8 What is the advantage of painting the outer walls of a building white?

9 What is the advantage of wearing white clothes in hot climates?

Radiated heat can be detected by an instrument called a thermopile. The way this works is described later in chapter 29. Figure 6.6 shows how to use it to detect heat being radiated from a container made hot by boiling water in it. To demonstrate the effects of different surfaces we can use a special container called *Leslie's cube*. This has four sides and each is different. Usually they are dull black, shiny black, white, and shiny metal (tin). If the thermopile is placed close to and equidistant from each surface in turn we can measure the heat radiation by noting how much electric current comes from the thermopile. The current is detected by a sensitive instrument, called a galvanometer. It is found that the best radiator is the dull black surface. This is followed by shiny black, white and shiny metal, in that order. This effect may also be demonstrated with separate cans.

6.42 Conduction

One way of comparing solids to see which are best at conducting heat is the apparatus shown in fig. 6.7. The rods are of different materials, such as, iron, copper, aluminium, brass, monel metal, bronze, glass and lead. They are of equal diameter and length. When the water is boiling, you can touch the ends of the rods carefully to find which is hottest. A better way of comparing them is to fix a small tack at the end of each rod, using melted candle-wax. This is done before heating begins. When the end of each rod reaches the temperature of the melting-point of wax, the tack drops off. In this way the materials can be listed in order of their ability to conduct heat. Your teacher may be able to demonstrate this for you in the laboratory.

Questions

1 Of the materials listed above, which are the best conductors? Which is the worst?

2 Jars made of thick soda-glass crack easily when hot liquid is poured into them. In what way does the conduction of heat by glass make the glass liable to crack so easily?

3 Why is a kettle or saucepan difficult to use if it has a metal handle? What sort of material is good for making handles of saucepans and kettles?

Metals differ in their ability to conduct heat, but they are all classed as *good conductors*. Many other substances, including glass conduct heat so badly that we call them bad conductors. Or we can call them *good insulators*. These substances conduct heat so badly that very little heat passes along a narrow rod like the rods used in the experiment (fig. 6.7). To compare materials that are good insulators we need to use a disc or mat of material instead of a rod.

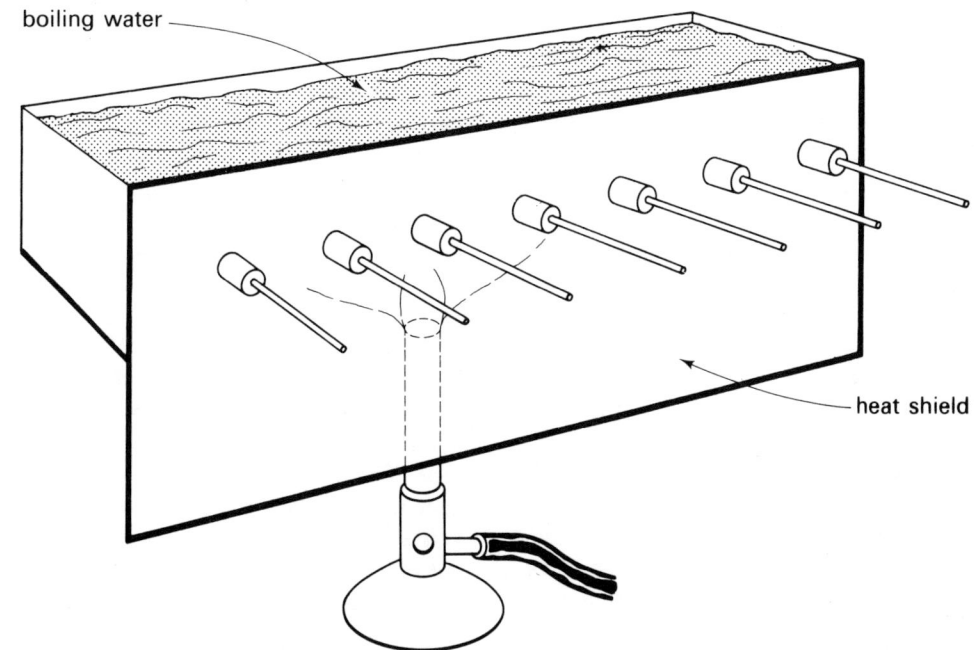

Fig. 6.7 Comparing heat conduction in rods of different materials.

Instructions: testing some good insulators

1 Make three or four mats of insulating materials that you would like to test. Include materials used locally for building houses and for making clothes. Materials which can be tested include glass, cardboard, paper (a pile of newspaper sheets 4 mm thick), kapok or cotton wool, cork, bark, felt, wool fabric, cotton fabric, nylon fabric, terylene fabric, sheets of various plastics including expanded polystyrene, plywood, hardboard, insulating board, rubber, foam rubber, brick, concrete. You could also include a metal mat for comparison with the insulators.

All mats should be about 10 cm square and about 4 mm thick. As far as possible they should be of equal thickness.

2 Put candle wax on the mats (fig. 6.8 b): let four drops of wax run from the candle at each spot; they run together and solidify to make a large spot of wax. If the material is absorbent, let one drop fall first; let this solidify, *then* let three more drops run on top of it.

3 Put a little water in a large saucepan or water-bath. Place a large flat sheet of metal over the pan (fig. 6.8 a). Boil the water, let it simmer gently. The steam heats the sheet which is at approximately 100°C all over.

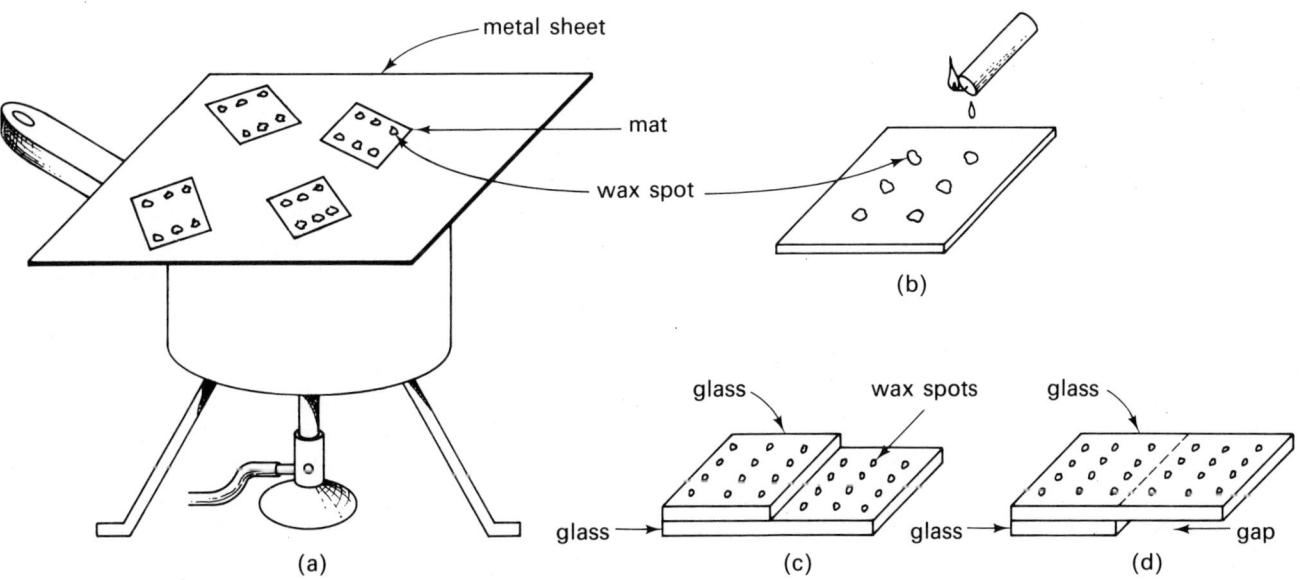

Fig. 6.8 Testing some good insulators.

4 When the metal sheet is hot, place the mats on it. Put them as near to the centre of the sheet as possible, but not touching each other.

5 Watch the wax spots. On which mat do they begin to melt first? By watching the melting of the drops you can place the materials in order, from the poorest insulator to the best.

6 Which were the best insulators among those you tested? In what ways are these materials used in everyday life?

Instructions: finding out whether the thickness of the material affects conduction

1 Make some mats and half-mats, as in fig. 6.8c.

2 Use the hot metal sheet as in the previous experiment. Which gives the better insulation, a single layer or a double layer?

3 Make some mats and half-mats of glass, and arrange them in the two ways shown in fig. 6.8c and d. In both of these the heat passes through one or two thicknesses of glass. In one (d) there is a gap between the hot metal and the single sheet of glass. Put both on the hot metal. Explain any difference you notice about the conduction through the glass. What is in the gap? Is this a good insulator?

Practical tests show that air is a good insulator. It is slightly better than the best of the materials you have tested. The best materials contain a lot of air. Kapok, fabrics and felt all consist of fibres woven or matted together. There is plenty of air between the fibres. This is why these materials are good insulators.

Instructions: testing water as a conductor

1 Clamp a test-tube containing cold water at an angle of about 45° to the vertical.

2 Place a thermometer in the water, with its bulb near the *bottom* of the tube. This too should be in a clamp, so that it does not touch the sides of the tube.

3 Heat the tube with a bunsen flame. Heat only the top of the tube, not the bottom. After the water has been heated for 20 seconds, record the reading on the thermometer.

4 Start again with a cool tube and water. Now clamp the thermometer so that its bulb is at the *top* of the tube, close below the surface of the water.

5 Heat for 20 seconds. This time heat the *bottom* of the tube, not the top. After 20 seconds record the reading on the thermometer.

6 Is water a good conductor? Explain the different results you get from the two methods of heating.

7 If you cannot answer the question in step 6, repeat the whole procedure. This time, put something in the water to show you the way it is moving. You can use sawdust (boil it first in water, so that it sinks instead of floating), or used tea leaves, or used coffee grounds. Watch how the pieces move around while you heat the tube. Then try again to answer the question in step 6.

6.43 Convection

This is the third method of transfer of heat. Water is a very bad *conductor* of heat, but large quantities of heat can be carried in water by convection. Another demonstration of the same effect is shown in fig. 6.9. At the top of the tube the water is boiling, yet ice at the bottom of the tube does not melt for some time. If we reverse the positions of ice and flame, with the ice floating at the top of the tube and the flame heating the bottom, the ice melts in a few seconds. Convection has quickly carried heat from the bottom of the tube to the top. When water is heated from below, it rises, carrying the heat with it. Its place is taken by cool water coming down from above. This too is heated and rises. We get a current of water circulating in the test-tube. This is a *convection current*. It quickly distributes heat through the water. As water is a bad conductor, a kettle of water would take hours to boil if conduction was the only way of transferring heat. With convection, the kettle boils in a few minutes.

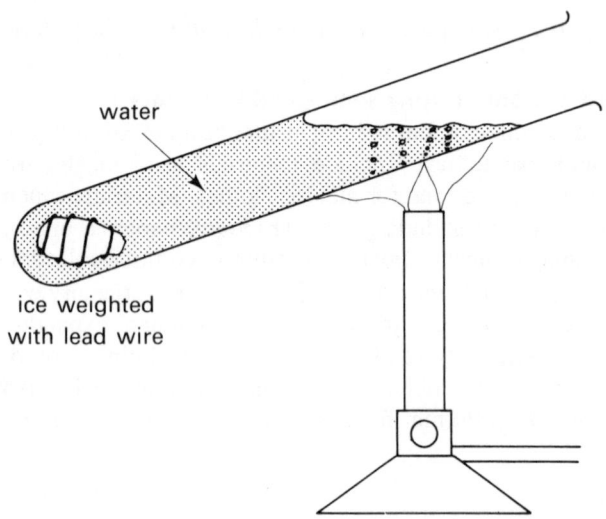

water

ice weighted
with lead wire

Fig. 6.9 A demonstration that water is a poor conductor.

Convection involves movement of warmed water from one region to another. In the apparatus shown in fig. 6.10, we heat the water in the left-hand tube. This rises. It is replaced by cooler water coming from the right-hand tube, along the lower tube. Crystals of potassium manganate (VII) make a red solution when dissolved. We can see the red solution circulating through the tubes in a clockwise direction. This is a clear demonstration of a convection current. This current is made use of in a hot water system at home (fig. 6.11). Water is heated in the boiler and circulates, as shown by the arrows, from the boiler to the storage tank and back to the boiler again. Gradually the water in the tank becomes hotter, as heat is transferred from fire (coal or oil) to the storage tank. The tank may be

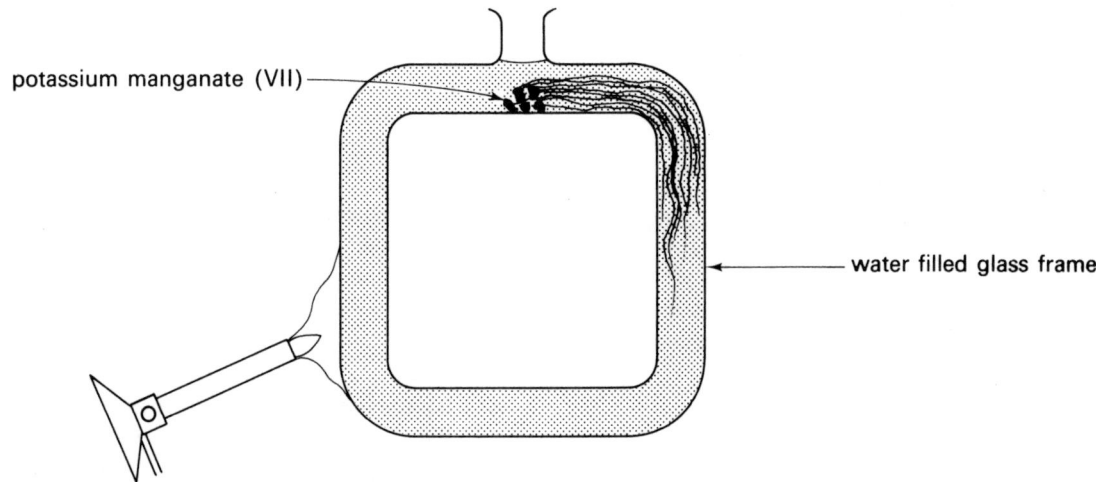

potassium manganate (VII)

water filled glass frame

Fig. 6.10 Convection current.

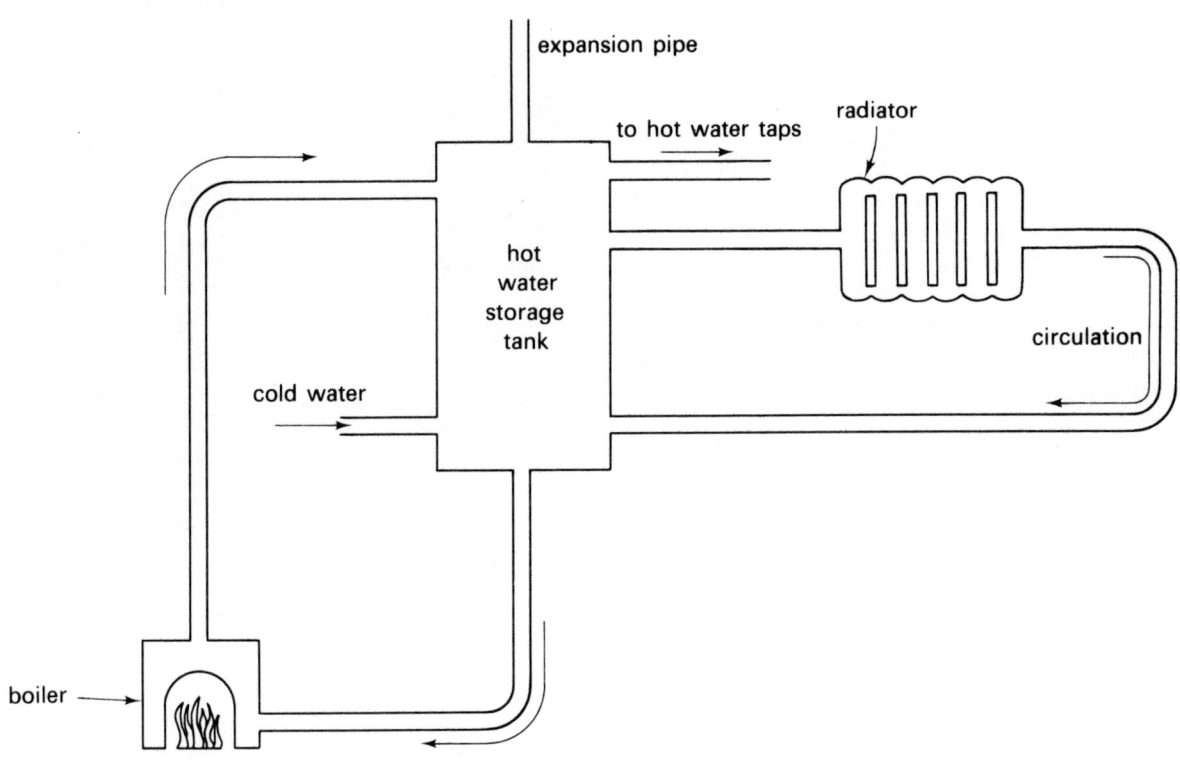

expansion pipe

to hot water taps

radiator

hot water storage tank

circulation

cold water

boiler

Fig. 6.11 Diagram of a hot water system in a house.

covered with insulating material (kapok, felt, or a layer of glass fibres) to prevent loss of heat by convection in the air around the tank. The water is hottest at the top of the tank. Pipes run from the top of the tank to hot water taps in the kitchen or bathroom. Cold water enters where shown to replace the water that is drawn off.

Questions
1 Explain the circulation of water through the radiator and back to the tank.

2 Why is this called a radiator? Do you think that this is a good name?

3 What do you think is the purpose of the expansion pipe in the system?

4 If water is to be heated by electricity, where is the heater usually placed?

5 Convection occurs in all liquids. It can also occur in gases including air. The apparatus which is shown in fig. 6.12 demonstrates convection in air. Explain what is happening.

57

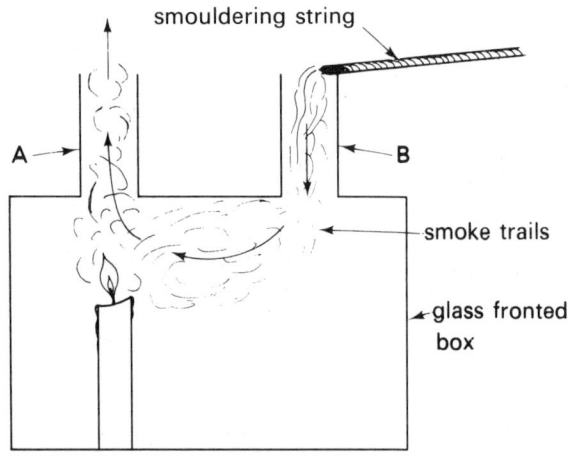

Fig. 6.12 *Demonstration of convection in air.*

6.44 Science and technology

In this chapter you have found out a lot about the ways in which heat is transferred. You know about conduction, convection and radiation. You know about good conductors and good insulators. This knowledge is part of *science*. Using this knowledge in everyday life is *technology*. A good example of the technology of heat transfer is the vacuum flask. The idea of the flask is to make it as difficult as possible for heat to enter or to leave the flask. If hot food is put in the flask, it keeps hot for hours. If cold food is put in the flask, it keeps cold for hours. The vacuum flask makes use of very many of the ideas about heat transfer that you have learned in this chapter. Look at the drawing of a typical flask (fig. 6.13). Try to think how it works, then answer these questions.

Questions

1 List the ways in which the conduction of heat is prevented.

2 List the ways in which the convection of heat is prevented.

3 In what way is the radiation of heat reduced?

4 Instead of a cork we may have another type of stopper. What is this like and in what way is it a good insulator?

Some other examples of the technology of heat transfer (either increasing it, decreasing it, or making use of it) can be found in the projects which follow.

Projects

1 The heat of the Sun can cause convection in the atmosphere. This causes winds. Find out as much as you can about this subject. Your geography teacher may be able to help you find books and charts on this subject.

Fig. 6.14 *The space suit insulates the astronaut against the extreme cold on the Moon.*

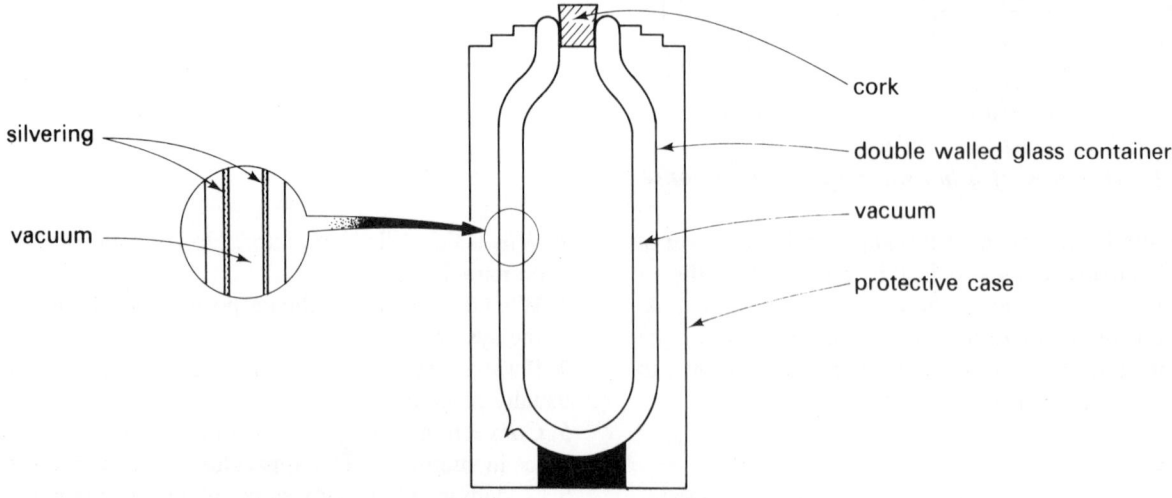

Fig. 6.13 *A sectional view of a vacuum flask.*

Some key-words to help you in your search for information: sea breeze; land breeze; seasonal winds, monsoon; trade winds; windmills, windmill generators.

2 Convection is used in some types of electric fire which have artificial coal or logs. It is also used in some types of table-lamp to make shadows move around the shade in an amusing way. Find out how these work. You could make a small windmill out of paper and balance it on a pin. Then use it to detect rising air currents caused by convection.

3 Sometimes we want to transfer a lot of heat quickly from one place to another. Think of some examples of this, and how they work. Here are some ideas to begin with: metal cooking pots; soldering iron; cooling system of an automobile engine; electric hot-plate; electric room heater; electric clothes iron; kerosene stove or cooker. Note that for some of these examples there are several types, working in different ways. Think of more examples of your own.

4 Sometimes we want to stop the transfer of heat. Think of examples of this, and how they work. Here are some ideas: window shutters; deep verandahs on the sunny side of a house; shiny metal roof; straw or thatch roof; walls made from two layers with air between (cavity walls); ice-box; fur or feathers on mammals and birds; space suits. You will find lots of examples if you study local building methods.

7 Discovering about sound

7.00 Making a noise

There are lots of ways of making a noise. How many can you think of? Try some of these suggestions.
(a) Make a stick with notches in one edge; run a card along the notched edge.
(b) Make a toothed wheel (fig. 7.1); turn the wheel and hold a card against the moving teeth.

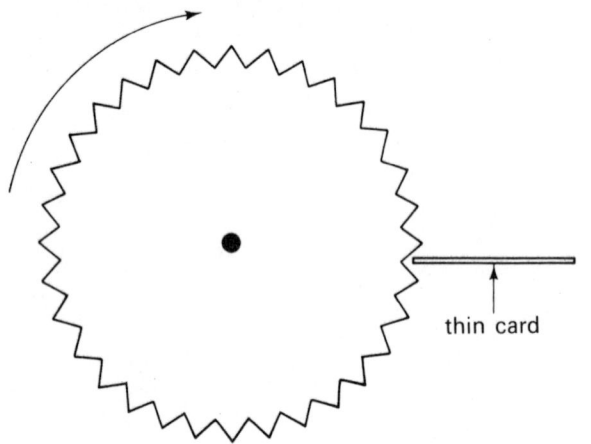

thin card

Fig. 7.1 Making noise: a card held against a rotating toothed wheel.

(c) Turn a bicycle upside down; turn the pedals and hold a card against the tyre on the back wheel.
(d) Hold a hacksaw blade firmly on the edge of the table, with most of the blade projecting over the edge. Push the free end down a little way, then let go.
(e) Hold a steel wire tightly, as in fig. 7.2. Pluck it at C, between the bridges.
(f) Hang a hollow tube, such as a length of bamboo about 50cm long, from a string. Hit it with a hammer.
(g) Operate an electric bell.
(h) Stick corrugated cardboard of several different kinds on a piece of thicker cardboard. Rub your finger-nails gently over the corrugations. This is rather like the 'washboard' used by some jazz musicians years ago.
(i) Connect a piece of wire grid (flyscreen, or wire gauze as used in the laboratory) by wire to one terminal of a loudspeaker; another wire goes from the other loud-speaker terminal to a flashlamp battery. From the other terminal of the battery there is a third wire. Rub the free ends of this wire across the grid.

Think of some more ways of making a noise; then try the experiments below.

Instructions: making noise
1 Try several of the methods listed above, and some of the ways you have thought of too. If the equipment is set out around the laboratory, you can work in pairs or threes and circulate to each set of equipment in turn.
2 With each noise-maker, find out how you can make a loud noise and a soft noise.
3 With each noise-maker, find out how you can make a low-pitch noise (a buzzing sound) and a high-pitch noise (a squeak). With some it is not possible to alter pitch. Can you say why?
4 With some of the noise-makers, try to find out which part the sound comes from. Put your finger-tip *gently* on any part that you think might be giving off noise. What do you feel?
5 When you use noise-makers such as (a), (b), (h) and (i) try working them very slowly: one notch, tooth, grid-wire or corrugation per second. Then try them faster. Then try them as fast as you can.
6 If possible, make the noises close to a microphone which is connected to an oscilloscope. Then you can see the 'shapes' of the sounds you are making.

Questions
1 In each of the noise-makers, something happens which makes the sound. What makes sounds?
2 How does the sound get to your ears?
3 Why are some sounds loud and others soft?

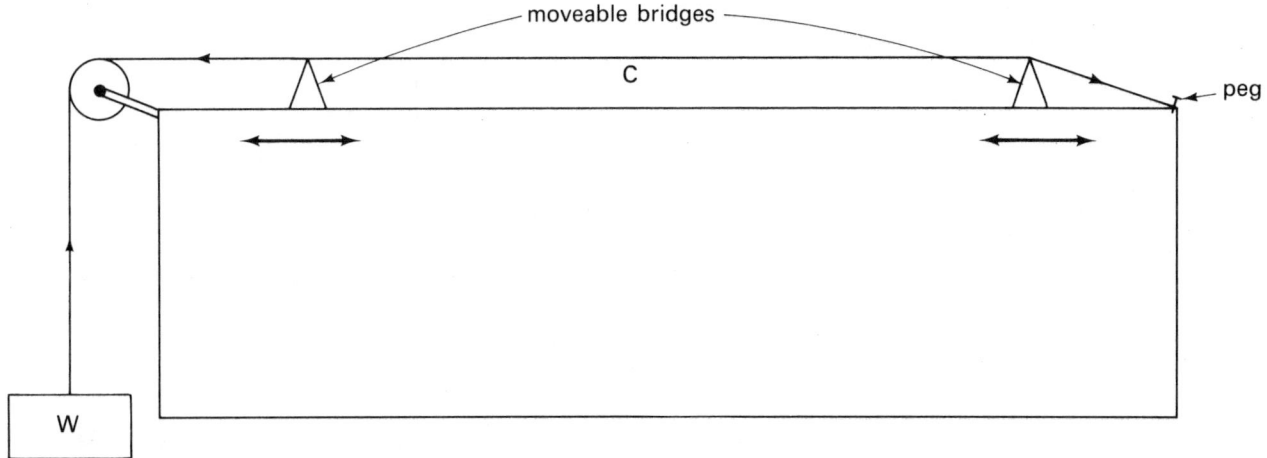

Fig. 7.2 Making noise by plucking a tightly stretched steel wire.

4 Why do some sounds have low pitch and others have high pitch?

Sounds are caused by vibration. A card for example vibrates in the noise-maker; it makes the air next to it vibrate; this air makes the air next to it vibrate; the vibrations spread through the air in all directions; they reach your ear-drum; this vibrates; it makes the mechanism inside your ear vibrate; then the vibrations are detected by the inner ear and messages are sent to the brain; then we hear the vibrations, and call them sound.

The pitch of the sound depends on how fast you operate the noise-maker. If you move the card quickly along the notched stick, or turn the toothed wheel rapidly, you get a high-pitched sound. You are making the card vibrate many more times per second. The number of vibrations per second is called the *frequency* of the sound. If the frequency is as low as one or two vibrations per second we can hear the separate vibrations. Each time the card passes one notch or one tooth we hear one sharp click. If we make about ten vibrations per second, we can still hear the separate clicks. Now they come too fast to be counted, but you can still tell that you are hearing a number of separate clicks, with silence between them. As frequency is increased it becomes more and more difficult to hear the clicks separately. Instead we hear a low-pitch buzzing sound. This happens when frequency, f, is about 30 vibrations per second. As frequency increases above this level we hear a sound of higher and higher pitch.

With some of the noise-makers (for example (d) and (e)) frequency can be altered by altering the length of the wire or strip. Pitch depends on the size of the vibrating object. When the strip or wire is long, it vibrates with low frequency, making a low-pitched sound. When it is short, pitch is high. This is like the behaviour of a pendulum (page 12). A short pendulum has a short period. It vibrates quickly. It has high frequency. A long pendulum has low frequency. Pendulums do not vibrate fast enough to produce sound but the idea is the same. Their natural rate of vibration depends on size (on length, in the case of the pendulum, the vibrating wire, the bamboo tube, the hacksaw blade and so on.) We can sum up the properties of sound:

(a) it is caused by objects vibrating;

(b) the greater the size of the vibrations, the louder the sound;

(c) the greater the frequency of vibration, the higher the pitch of the sound.

There is a third way in which sounds can differ. Different noise-makers make different *kinds* of noise. They do not sound alike, even if they are equally loud and have equal frequency. We say there is a difference of *quality*. We learn to recognise the quality of the sounds we hear. We can learn to recognise the differing qualities of musical instruments, and the various noise-makers and other sources of sound. Quality is affected by the exact way the noise-making part vibrates. A hacksaw blade vibrates in a different way from a piece of card held against a toothed wheel. If you have looked at the shapes of sounds, using an oscilloscope, you may have noticed the many different patterns produced. In some the vibrations are smooth, in some they are very spiky, others produce most complicated patterns of vibration. Notice also the difference between low-pitched sounds and high-pitched sounds, and between loud sounds and soft sounds.

7.10 Making musical sounds

Over the years men have invented many kinds of musical instrument. In all of them something is made to vibrate: a string or wire; a sheet of metal or skin or some other material; a block of wood; a hollow cylinder of wood or metal; a column of air inside a pipe.

Every country has its own musical instruments. Think of as many instruments as you can to put under the

61

headings given above. Perhaps you can think of new headings to add to the list. Some modern musical instruments produce vibrations by electronic circuits; these are converted into sound by a loudspeaker. Can you think of examples?

Instructions: making musical sounds

1 Examine musical instruments obtained locally. Find out what part vibrates when the instrument is played. Examine and use a tuning fork too.

2 Find out how the instruments can be played loudly and softly.

3 Find out how the pitch of the sound can be changed, as tunes are played.

4 Look at the 'shapes' of the sounds, using an oscilloscope.

One way in which musical instruments differ from noise-makers is that their vibrations are more regular and have a steady frequency. Also they are usually played at a number of definite frequencies or notes, and not at frequencies between. There are some exceptions to this, can you name some? An example of a musical note is the one called 'middle C' on the piano. If you have a tuning fork which gives a note of this pitch, you will find that the fork has the number '256' marked on it. In different parts of the world, different sets of notes (musical scales) are used. One that is widely used is the diatonic scale. In one form of this, the frequencies of each note are:

C	D	E	F	G	A	B	C'
256	288	320	341	384	427	480	512

This is the scientific scale. The numbers indicate the number of vibrations per second. The higher the number, the higher the pitch of the note. You could sound the tuning forks that you have in the laboratory to hear what these notes sound like. The top C (C') of the scale has frequency exactly double that of the C at the lower end of the scale. The musical distance between these two notes is called an octave (it includes eight notes). If we continue up the scale we have D' an octave above D, with a frequency of 576 vibrations per second. In this way we can name notes of higher and higher pitch. We can also go down the scale. A note with frequency of 160 vibrations per second is one octave below E. One of frequency 80 vibrations per second is two octaves below E, and so on. For various reasons, the exact frequencies of the scientific scale are not suitable for many kinds of musical instrument, especially for those which have a keyboard. Musicians generally use the even-tempered scale in which the frequencies differ slightly from those given above. The unit of frequency is the number of vibrations per second. For short we write: frequency of middle C, $f = 256/s$.

This unit is used a lot, so it has been given a special name which means 'number of vibrations per second'. We call it a *hertz*. Its symbol is Hz. It is named after Heinrich Hertz, an early pioneer of radio, so the symbol begins with a capital letter. This is a derived SI unit. Like other SI units it can have multiples and submultiples. We say that the frequency of middle C is 256 Hz. The frequency of the high-pitched sound from a jet engine is around 10 kHz (kilohertz). The unit is used for stating the frequency of radio transmitters. Short-wave transmitters operate at very high frequency: you often hear the announcer give the station frequency in megahertz (MHz). 1 MHz is a frequency of 1 million vibrations per second.

7.20 The transmission of sound

Sound is usually transmitted to us through air, but it can be transmitted through liquids and solids too. When you are swimming under water it is easy to hear the sounds of the engines of boats and the noise of the waves. Sounds can pass through solid materials, such as the walls of a room. Wherever there is matter, either solid, liquid or gas, sound can be transmitted. What if there is no matter? Figure 7.3 shows an apparatus designed to answer this question. An electric bell is enclosed in a glass jar. The electric current is turned on and the bell rings. The sound can easily be heard in the room. Then a vacuum pump is connected to the jar. The air is sucked out of the jar. When the air is gone and there is a vacuum in the jar, the sound of the bell cannot be heard. There is no air to transmit the sound from the bell to the jar, from where it can pass through the air of the room, to your ears.

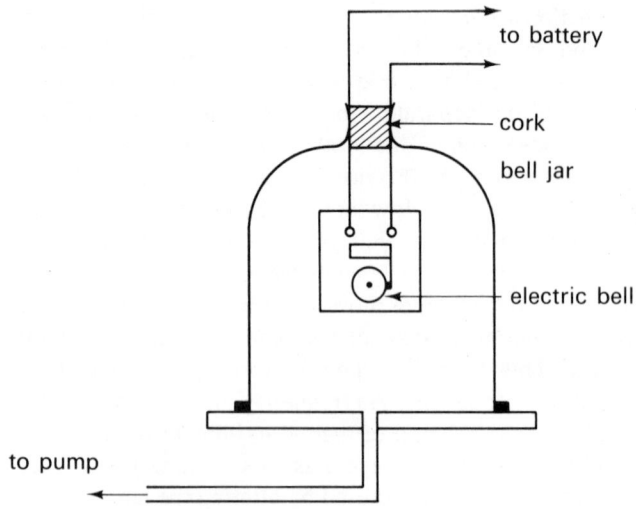

Fig. 7.3 A demonstration of the transmission of sound through a vacuum, see text.

Questions

1 If you cannot hear the bell when there is a vacuum in the jar, how can you be sure that it is still ringing?

2 It is not strictly true to say that the bell cannot be heard. A slight sound of ringing can be heard, even when there is a vacuum in the jar. Explain how this happens.

3 In what way would you improve this apparatus to make it less easy for sound to pass from the bell to the jar?

4 What do you expect to hear after the pump has been disconnected and the air is slowly let back into the jar?

5 Can astronauts hear each other speak while they are on the Moon or in some other place in space where there is no atmosphere?

When light strikes a surface it may be transmitted (pass through into the material of the surface), absorbed, or reflected. Sound behaves in the same way. If we are indoors with windows and doors closed, we can still hear outdoor noises such as traffic. The sound is transmitted through the walls of the room. Some is absorbed, for the traffic does not sound as loud as it does outside, but a lot may be transmitted. How much is absorbed and how much is transmitted depends on the materials from which the walls are made. Soft materials such as fabrics and padding materials such as glass-fibre and foam plastic are good at absorbing sound. They absorb much of the energy of the sound as it passes through. If the walls of a room are covered with fabrics, or if space in the walls is filled with sound-proof padding, the amount of sound entering or leaving a room can be greatly reduced. With light we found that reflection obeyed certain laws. Does sound obey laws when reflected? If so, what are these laws? An experiment may tell us.

Instructions: investigating the reflection of sound

1 Use a hard plaster wall or some other hard surface that is smooth.

2 To make a beam of sound use a cardboard tube. Put a watch at one end, as a source of sound (fig. 7.4).

3 Direct this beam at the wall.

4 To find the reflected beam, use another tube. Place your ear at one end. Move the tube around until the reflected sound is heard as loudly as possible. The tube must be directed at the exact spot on the wall where the beam of sound from the watch is expected to strike the wall.

5 Decide where the beam is striking the wall. Imagine a normal to the wall at this point (fig. 5.12). Measure the angle of incidence of the beam. Measure the angle of reflection.

6 Repeat for several different angles of incidence. What can be said about the values of the angles?

Smooth, hard surfaces reflect sound well. Rooms that have many such surfaces tend to be noisy. Curtains, carpets, soft furniture, people and their clothes, all help to reduce reflection and increase absorption. Such rooms tend to be quiet. A library is a quiet place not just because people are asked not to talk and are expected to move about quietly. The books which line the library walls are excellent absorbers of sound.

Outdoors, sounds are reflected from the walls of buildings and from pavements and roads. If we clap hands we often hear the clap reflected back from some distant surface a fraction of a second later. This is an *echo*. During that short time the sound has travelled from the hands to the surface and back to us again. The greater the distance, the longer the time taken for the echo to return. In a thunderstorm the lightning spark pushes through the air, causing a sharp 'crack'. The sound of this comes directly to us. Immediately afterward we hear echoes of the 'crack' from all the hills and buildings in the district. Such a loud

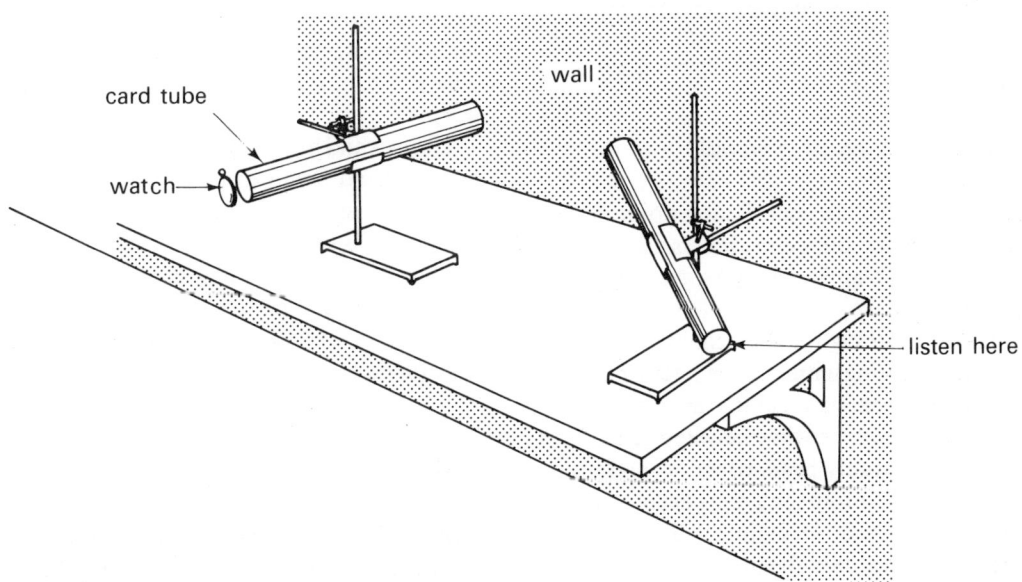

Fig. 7.4 The reflection of sound.

sound can be heard from many kilometres away. Echoes can be heard from hills far away. Several seconds pass while the sound travels to distant hills and is echoed to us. After the first 'crack' we hear a series of echoes, making the irregular, rumbling sound of thunder.

Do you know how to estimate the distance away of a lightning strike? Many people count the number of seconds between the flash and the first crack of thunder. They divide this time by three and the result is the distance, in kilometres. For example, if the time between flash and crack is 6 seconds, the strike is 2km away. The reason behind this calculation is that light and sound travel at very different speeds. The speed of light is so great that for this purpose we can say that the flash reaches us immediately. The sound takes several seconds to arrive. We will say more about the speed of light later (21.40). Next we will measure the speed of sound.

7.30 The speed of sound

Instructions: measuring the speed of sound
1 This is done out of doors. You need a large wall or building to act as a sound reflector (fig. 7.5). You need a stop-watch and a partner to operate it.

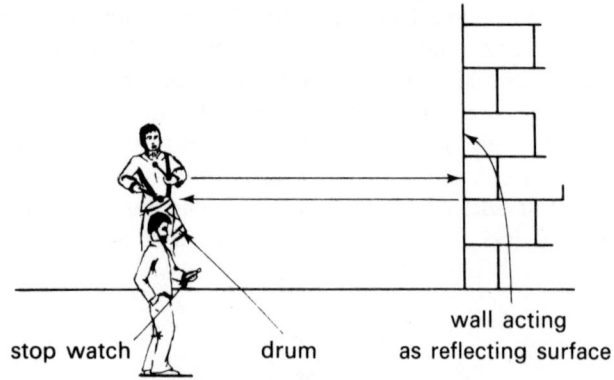

stop watch   drum   wall acting
                                   as reflecting surface

Fig. 7.5   Measuring the speed of sound.

2 Stand about 100m from the wall. Beat a drum or clap your hands and listen for the echo. If this is not loud enough, make a louder clap by clapping two large flat pieces of wood together.
3 Beat a few times, listening for the echoes. Gradually increase your rate of beating until you are beating *in time with* the echoes. Keep beating regularly at this rate.
4 Your partner is to measure the time between beats. This is less than a second. Twenty beats or more (preferably 50) must be measured. To make the timing accurate, your partner begins with a count-down in time with your beats: 'five, four, three, two, one, *zero* (begin timing), one, two, three, . . . , fifty (stop timing)'.
5 Measure your distance from the wall.

6 The time taken for 50 beats is Ts. The time for one beat is $t = T/50$. Calculate t.
7 The distance between you and the wall is dm. The sound travels from you to the wall and back to you. The distance travelled is $2d$m. The time taken is ts. Calculate the speed of sound, c, where $c = \dfrac{2d}{t}$ metres per second.

Questions
1 What result did you obtain for c?
2 If a wind was blowing from you towards the wall, would this make any difference to your result?
3 Convert your result for c into kilometres per hour. What word do we use to describe aeroplanes that fly faster than this?
4 Do you think that loud sounds travel faster than soft sounds? Give reasons for your answer.
5 Do you think that high-pitched sounds travel faster than low-pitched sounds? Give reasons.

Before we say more about the speed of sound we should note that we have derived another SI unit. This is the unit for speed (and for velocity, see later). We divide a distance (in metres) by a time (in seconds). This gives us speed in metres per second. In symbols this is written as m/s. This makes use of the base units of SI. For some purposes it is convenient to express speeds in other units, such as kilometres per hour.

As the answers to questions 4 and 5 show, the speed of sound is not affected by its loudness or pitch. Reasoning of the same kind tells us that it is not affected by quality. We might begin to wonder if anything affects it. Air pressure has no effect, but air temperature does. The speed increases as air temperature increases. At 0°C c is approximately 331m/s. It increases by about 0.6m/s for every degree above 0°C. This is why your result is expected to be higher than 331m/s. The speed of sound in air is also affected by the amount of water vapour in the air. The effect is small, and is too complicated to be stated simply.

If you stand by a long iron fence or framework (a bridge) and a friend some distance away bangs the fence, the sound of the bang reaches you first through the iron. A fraction of a second later the sound reaches you through the air. The reason is that in iron, c is over 5 000m/s. In general sound travels at very high speeds in solids. In liquids it travels less fast, for example in water c is just over four times the speed in air.

7.40 Resonance

Instructions: finding out the nature of resonance
1 Make a pendulum (fig. 2.5) with a heavy bob (at least 100g) and 80cm long. Let the bob hang still.
2 Cut a strip of thin paper about 5cm wide and 15cm long.

Hold this by one end. Hit the bob with the other end. This has only a small effect on the bob.

3 Hit the bob repeatedly with the paper. Always hit from the same direction. Try to start the bob swinging. It seems impossible to make such a heavy bob swing by hitting it with a thin piece of paper. Yet it can be done. What is the secret?

4 Measure the period of the pendulum (chapter 2). How many swings does it make per second? This number of swings is its *frequency*. Record its frequency, f, in hertz.

5 Hit the pendulum with the paper at a regular rate, which is close to f.

6 Hit the pendulum with the paper at a regular rate, which is close to $2f$.

7 Hit the pendulum with the paper at a regular rate, which is close to $\frac{f}{2}$.

8 Hit the pendulum with the paper at an irregular rate.

9 Which rate of hitting (steps 5 to 8) is best for making the pendulum swing?

Instructions: investigating resonance

1 Tie a length of thread horizontally between two supports about 60 cm apart. Make the thread as tight as possible, so that it does not sag. Hang four pendulums from it. Pendulum A is the one you used above with 100 g bob and 80 cm long. The other pendulums have bobs of mass about 10 g each. Their lengths are: B, 75 cm; C, 80 cm; D, 85 cm.

2 Let all pendulums hang still.

3 Pull the bob of pendulum A towards you (about 5 cm), then let it swing. Watch the other three pendulums during the next five minutes.

4 Repeat steps 2 and 3, but set B swinging instead of A.

5 Repeat twice more, but setting C swinging and then setting D swinging. Which pendulums show resonance?

6 Make pendulum B longer: about 78 cm long. Make pendulum D shorter: about 82 cm long. They are not now very different from pendulums A and C.

7 Repeat steps 2 and 3. B, C and D may all begin to swing. In what way does the behaviour of B and D differ from that of C?

We have seen that two systems are in resonance when their frequencies are exactly equal. In the first investigation the two systems were pendulum and paper; in the second, the two systems were pendulums A and C. Because of resonance *energy* can be transferred from one system to the other. Only a very small amount of energy is transferred at each swing. The paper could have only a very small effect on the bob. Only small forces could be transmitted along the thread joining the pendulums. Because there is resonance the small amounts of energy are added together. Over a period of time the *total* amount of energy transferred can add up to a large amount. For example, even a small child can make a heavy adult swing

high on a swing, *if* the pushes are in resonance with the frequency of the swing. If you are jumping up and down on a springy wooden board or diving-board, *and if your rate of jumping is the same as that of the natural frequency of the board*, you can jump higher and higher. You can jump much higher than you could from firm ground. All the energy for these really high jumps has come from you, not from the board. Jump by jump, you have added more and more energy to the two systems (you and the board). You and the board are in resonance. If a party of soldiers is crossing a bridge they may all march in step. If the frequency of their steps is in resonance with one of the natural frequencies of the framework of the bridge, the bridge will resonate. It will vibrate more and more strongly, so strongly that it damages its supports and the bridge collapses. This has actually happened. To avoid this danger soldiers break step (each walks at his own pace) when crossing a bridge.

Sound consists of vibrations. These have a definite frequency. Sound vibrations should show resonance. Try this experiment to confirm that they do.

Instructions: investigating resonance with sound

1 Collect several large jars, of capacity 1 litre (1 000 cm³) or more.

2 Across the mouth of each jar fix a sheet of thin plastic or rubber. Use polythene from a plastic bag, or a piece from a rubber balloon. Hold the sheet in place with a rubber band round the neck of the jar. You can also use a sheet of the very thin plastic film used for food-wrapping ('Glad-wrap', 'Cling-wrap' etc.). This needs no rubber band. Pull it tightly across the mouth of the jar to make it like a drum.

3 Cut some small grains of cork and put five or six of these on each drum.

4 Experiment with one of the jars at a time. Hum loudly, or sing notes with your mouth close to the jar. Take care not to actually blow on the pieces of cork. Try high notes and low notes. Try notes that begin high and slide slowly down to a lower pitch. Watch the cork pieces. Try to find out why they sometimes jump about and why they sometimes lie still. Your ideas on resonance should help you to explain what is happening.

5 Remove the drum-skin; blow across the mouth of the jar. What do you hear?

6 Repeat steps 4 and 5 with some other jars. What is the relation between the size of the jar and the pitch of note that makes the pieces jump?

7 You can reduce the volume of air in the jar by pouring water in it. Then it resonates at higher frequency. Use a tuning fork as a source of sound. With the fork vibrating, touch its stem on the jar. See if the pieces jump. If not, add water or pour some water out and try again. In this way you can *tune* the jar to resonate with the fork. When the jar and fork resonate, and the pieces jump, what can we say about the frequencies of the jar and of the fork?

8 Tune a number of jars to resonate at a number of different frequencies. Stand them in a row close to a radio which is playing a fairly slow piece of music, or stand the jars on a piano which somebody is playing slowly. Watch the pieces jump as different notes are sounded.

Instructions: investigating resonance in strings

1 In this investigation we use a length of wire as in fig. 7.2. To detect when it is vibrating, make a small 'rider'. Cut a piece of paper about 2 mm × 10 mm. Fold this in halves to make a '∧'. Place this on the wire, about half way between the bridges. With resonance the rider jumps about; with strong resonance it may even be thrown off the wire. Now try some investigations of resonance similar to those above using the wire instead of the bottles. Adjust the length of the wire by moving the bridges, so that it resonates with different tuning forks, or with notes played on musical instruments.

2 If you have a piano or some other instrument which has several tuned strings, stand close to it and sing various notes. Keep the 'loud' (or 'sustain') pedal pressed while you do this. After you have sung a note stop and listen. Can you hear a note coming from the instrument? Explain what is happening.

While you were experimenting with the jars, you may sometimes have seen the cork pieces jump when you were *not* making any sound. This was probably due to some sound in the distance, resonating with your jar. This effect can be a nuisance. If a window-pane or a door or some other object in a room resonates with sounds made by passing traffic, it resonates and rattles every time traffic goes by. This is very irritating to people in the room. The resonant frequency of some rooms is close to the frequency of the human voice. When people speak the air in the room resonates. A booming sound is heard when certain voice sounds are made. The effect varies according to the person; a room may boom when a man speaks with a low-pitched voice, but perhaps not if a child with a high-pitched voice is speaking. Continual booming of this type is irritating and it is often hard for others to hear what the speaker is saying. A good lecture theatre is designed to prevent this from happening. The architect plans the room with its walls at the correct angles and with panels of special sound-absorbing materials on some of the walls and often on the ceiling. Then resonance does not occur at or near the frequency of the human voice and the lecturer is clearly heard.

Cheap loudspeakers and loudspeaker cabinets resonate at certain frequencies. The effect is to produce an exceptionally loud note whenever a note of the same frequency appears in the music being reproduced by the speaker. With notes of other frequencies there is no effect. In this way certain notes of a musical piece sound much louder than they should do; there is distortion of the music. The better quality loudspeakers and cabinets are designed to have little resonance, so they reproduce the music without distortion.

Fig. 7.6   The range of notes of this organ is attained by using pipes of many different lengths.

8 Discovering about magnets

8.00 Magnetic poles

Instructions: finding out about the poles of a bar magnet

1 Suspend a bar magnet from a wooden stand, as in fig. 8.1a.

2 The magnet may spin around for a few minutes, as the thread unwinds a little. When it stops spinning, note the direction in which it comes to rest (fig. 8.1b).

3 Turn the magnet at right-angles to that direction (fig. 8.1c), then let it go.

4 Turn the magnet through 180° (fig. 8.1d), then let it go.

5 Turn the magnet through different angles and let it go.

Questions

1 Does it always come to rest pointing in the same direction?

2 Which direction is this?

3 Does it always come to rest with the *same* end pointing in the *same* direction, as in fig. 8.1d?

4 Why must you use a wooden stand in this experiment and not an iron stand?

The experiment shows that the magnet is something more than just a bar of metal. Its magnetic properties make it come to rest in a definite direction. This direction is fixed relative to the Earth. Also we note that there is a *difference*

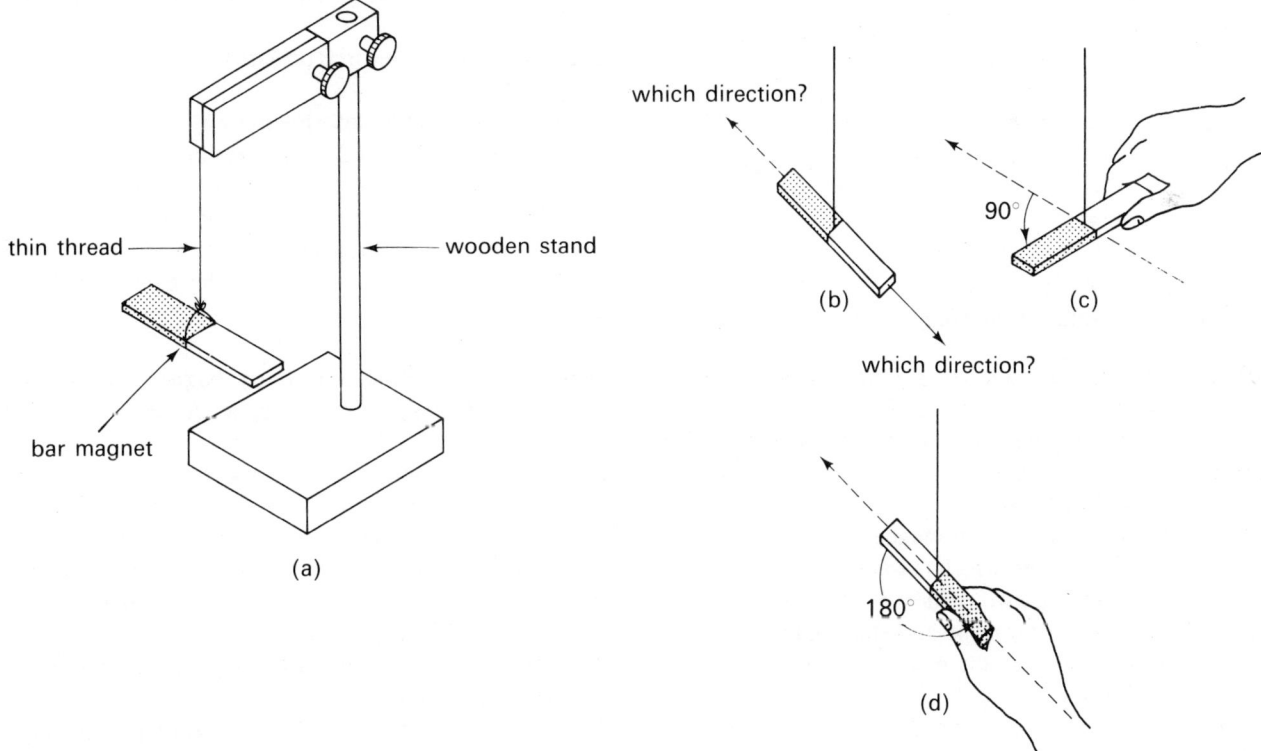

Fig. 8.1   The poles of a bar magnet.

between one end of the magnet and the other: one end always points in one direction, the other end always points in the opposite direction. We say that the magnet has *poles*. For convenience, we call the pole at the north-pointing end the *north pole*, and the other pole the *south pole*. A lot of bar magnets are painted red on the half which has the north pole, and blue on the other half. Others are painted the same colour all over, but the north end is stamped with an 'N' or marked in some other way. However you should always check the magnet by letting it swing as in the experiment above, for magnets that have been used a lot may have had their *polarity* reversed.

Instructions: finding the poles of a bar magnet

1 Fix a bar magnet in a wooden or plastic clamp (fig. 8.2). Do not use a metal clamp.

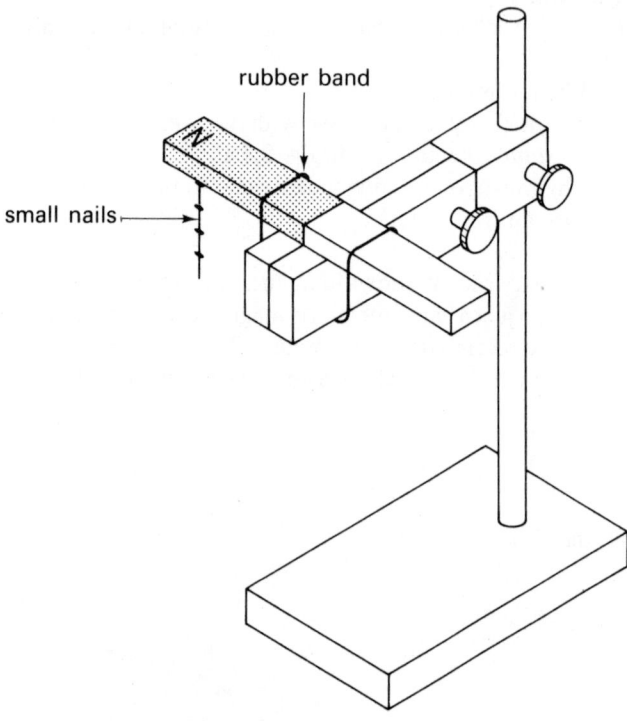

Fig. 8.2 Investigating the poles of a magnet.

2 Place an iron nail (about 1 or 1.5 cm long) at the north end of the magnet. Hang other nails below it, as in the figure. See how many nails you can hang in a single chain from the end of the magnet. This gives you some idea of the strength of the magnet.
3 At this stage compare your magnet with others being used in your class to see which is the strongest. If yours will carry only one or two nails it is too weak for this experiment, get a stronger one and start again.
4 Draw the magnet, with the nails hanging from it.
5 Remove the nails.
6 Place a nail on the magnet exactly 0.5 cm from the north end. See how many nails you can hang on it now.

7 Add a drawing of these nails to your drawing. Your drawing will show several chains of nails hanging on the magnet, but really you hang only one chain at a time. The drawing is a diagram which shows the strength of the magnet at different places along its length.
8 Repeat steps 5 to 7 at distances 1, 1.5, 2, 2.5, and every half centimetre along the magnet as far as the middle. Then work from the south end in a similar way. At the finish, your diagram will show how the strength of the magnet varies along its length.

Questions

1 At what point or points is the magnet strongest?
2 The poles are the places where the magnet is strongest. Mark on your diagram the *exact* positions of the north and south poles of your magnet.

Instructions: investigating the behaviour of magnetic poles

1 Use two bar magnets (we will call them A and B, for short). If you do not know which is the north end of each, hang them on thread as in fig. 8.1. Mark the north end of each, using pencil.
2 Hang magnet A on a thread and let it come to rest, pointing north-south.
3 Hold magnet B in your hand and bring its north pole slowly towards the north pole of magnet A (fig. 8.3a). Observe what happens.
4 Bring the north pole of B up to the south pole of A (fig. 8.3b). What happens?
5 Bring the south pole of B up to the north pole of A. What happens?
6 Bring the south pole of B up to the south pole of A. What happens?
7 Work out a rule to describe the way in which the poles behave. This simple rule has always been found true. It is sometimes known as the *first law of magnetism*. It is a useful rule for, if we know the poles of one magnet, we can use it to test another magnet of which we do not know the poles. But we must be careful, as the next investigation shows.

Instructions: finding out about magnetic attraction

1 Set up a hanging magnet, magnet A, as in the instructions above.
2 Instead of magnet B, use a bar of iron which is *not* magnetised. If you are not sure if it is magnetised or not, hang it on thread to see if it always points north-south. As bars of iron are often slightly magnetised, it is best to check it anyway. Assuming that you find it to be unmagnetised, proceed with the investigation.
3 Bring one end of the bar up to the north pole of magnet A. What happens?
4 Bring one end of the bar up to the south pole of magnet A. What happens?

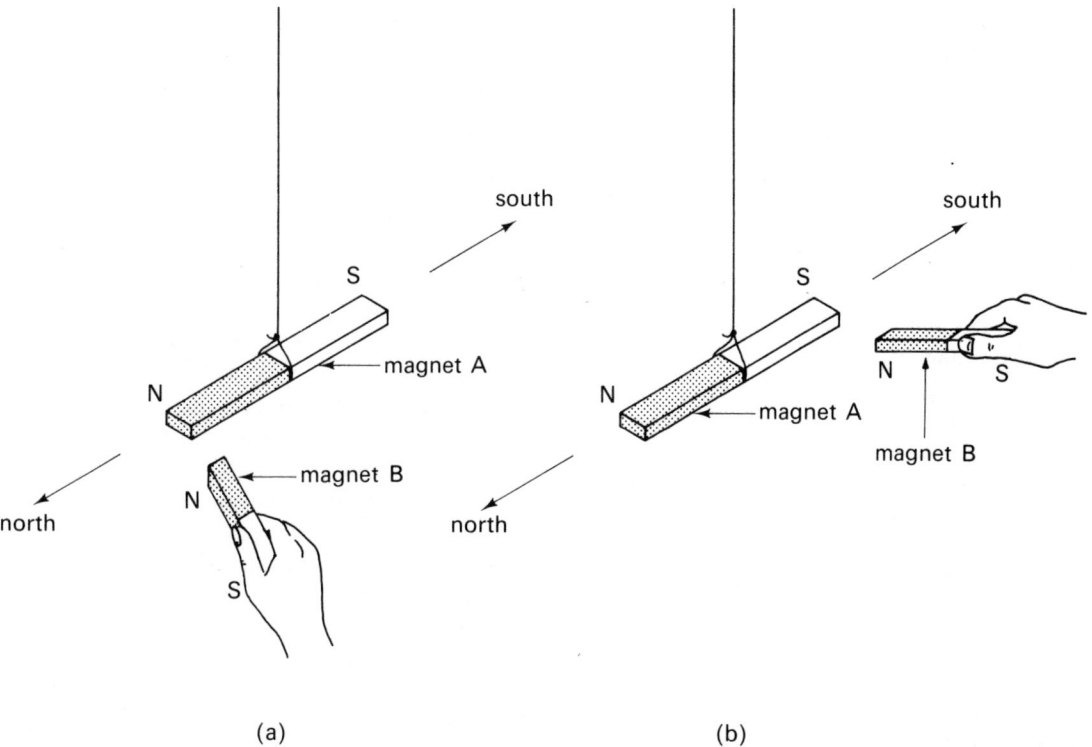

(a)　　　　　　　　　　　　　(b)

Fig. 8.3　The behaviour of magnetic poles.

5 Bring the other end of the bar up to the north and south poles of magnet A in turn. What happens in each case?

Questions
1 What *does not* happen with a bar and magnet that does happen with two magnets?
2 How can you use the results of this investigation to give you a rule for telling if a bar is magnetised or not?

8.10 Making magnets

Instructions: making a magnet by induction
1 Support a bar magnet in a wooden or plastic clamp. In another clamp at the same height support a bar of iron (soft iron such as a large iron nail is very suitable, but use a *new* nail for, if it has been used before for experiments like these it may already be magnetised). At first, the magnet and bar must be at least 30 cm apart (fig. 8.4a).
2 Try hanging small nails on the bar, as in fig. 8.2. Again, use *new* nails, for nails that have been used in experiments before may be magnetised already.
3 Bring the bar magnet close to the iron bar, but do not let them touch (fig. 8.4b).
4 Try hanging nails from the bar.
5 If nails hang, find the position of the poles, as in the previous experiment.
6 Use a hanging bar magnet to find the polarity of the end of the iron bar furthest from the bar magnet.

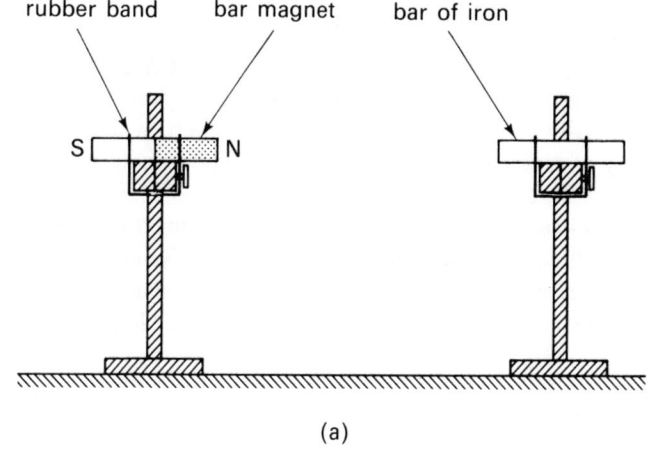

(a)

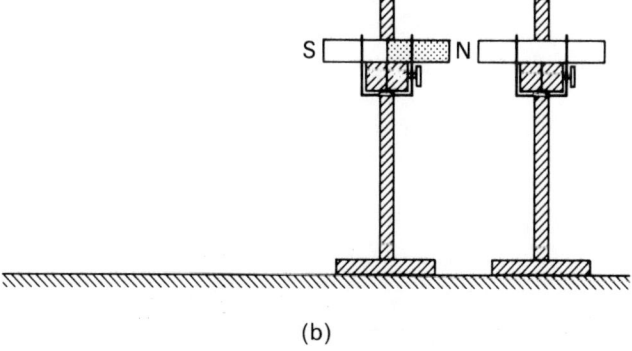

(b)

Fig. 8.4　Making a magnet by induction.

7 Take the bar magnet away. Repeat steps 4 and 5.

8 Now reverse the bar magnet, so that its south pole is next to the iron bar. Repeat steps 3 to 6.

Questions

1 What happens to the iron bar when the bar magnet is put near it?

2 Where are the poles of the iron bar?

3 When the bar and magnet are arranged as in fig. 8.4b, where is the north pole of the bar? Where is the south pole?

4 Work out a rule that tells you where the north and south poles of an induced magnet will be.

5 Is the magnetisation permanent?

6 You probably found that when the magnet was very close to the bar, the two were very strongly attracted together. Possibly the attraction was so strong that the stands were pulled over and you had difficulty in holding the bar and the magnet apart. Explain this, using the rule of induction (question 4, above) and the first law of magnetism.

7 Use the same two rules to explain the result of the experiment on magnetic attraction (p. 68).

Magnetisation by induction is usually a temporary effect, and its main interest is to explain why magnets and pieces of magnetic material are attracted to each other. Note that a *magnetic material* is a substance that is capable of being magnetised, but at any given time it may or may not be magnetised.

If we take a steel bar and induce magnetism in it over and over again, always in the same direction, we can gradually increase the strength of its magnetism. This is a way of making a permanent magnet. The principle is to stroke the bar with one or two magnets (fig. 8.5); we call it magnetisation by touch. As the pole of the magnet passes along the bar it magnetises it, leaving an *unlike* pole at the end where it is removed from the bar. The magnet is then taken some distance from the bar, (the dashed lines of the figure) to bring it into position for the next stroke.

Some of the experiments have suggested that the Earth itself has magnetic properties. We make use of this effect in two other methods of making magnets. One method is to heat a bar of steel until it is red-hot, then leave it to cool while it is resting in a north-south direction. Or place a steel bar in a north-south direction and beat it with a hammer repeatedly. Such methods were widely used once, but are no longer used for the commercial manufacture of magnets. Steel bars that are parts of the frameworks of buildings, bridges or machinery can become magnetised in this way if they lie roughly north-south. Over a long period of time, slight vibrations such as those caused by passing traffic, or the action of the machinery itself, have the same effect as the hammering. The structures become magnetised. This is why it is important for you not to do your experiments on magnetism close to the steel framework of the building or close to iron pipes.

Industrial methods of making magnets use an electric current. The principle of this is illustrated in fig. 8.6. Use a coil of wire consisting of many turns, several hundred if possible. A battery is connected to the coil for a fraction of a second; the magnet will not become stronger even if the battery is connected for longer.

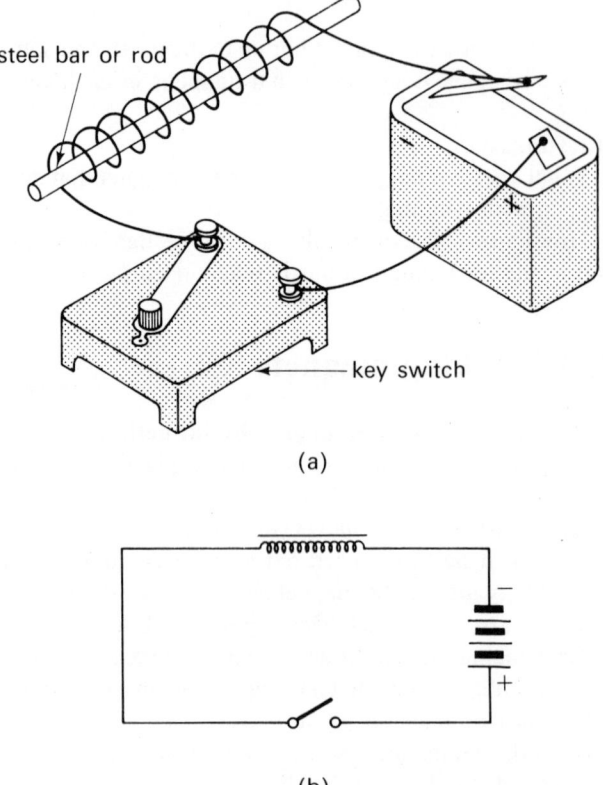

(a)

(b)

Fig. 8.6 Magnetisation by using an electric coil: (a) the apparatus, (b) a diagram of the same circuit, using electrical symbols. Can you see what the symbols mean?

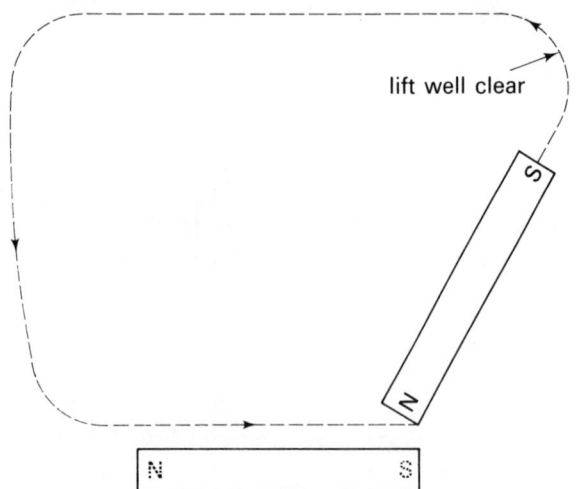

Fig. 8.5 Magnetisation by touch.

Instructions: making magnets

1 Use one or more of the methods described above to magnetise small lengths of steel. Cut lengths of steel rod or use steel bolts or screws, steel sewing-needles, pieces cut from an old clock-spring, or even strips from old razor-blades.

2 When you have made a magnet, test it to find the position of its poles. Do not bring a strong bar magnet too close to your home-made magnet. Your magnet is probably very weak and if a strong bar magnet is brought too close to it, the bar magnet may induce enough magnetism in your magnet to destroy its magnetism or reverse it. Keep the bar magnet as far away as possible and look carefully for *small* effects of repulsion and attraction.

8.20 Magnetic fields

Instructions: investigating the magnetic fields of bar magnets and horseshoe magnets

1 Put a bar magnet on the bench and cover it with a sheet of paper, about 20cm × 30cm.

2 Sprinkle fine iron filings over the paper. It is convenient if you have the filings in a 'pepper-pot' or a small can with several small holes in the lid.

3 Tap the paper gently a few times.

4 Make a sketch of the pattern you see.

5 Repeat with various other arrangements of magnets as shown in fig. 8.7.

6 Compare your results with those shown in the Answer and Discussion section.

The pattern of iron filings shows a large number of lines. These are *lines of magnetic force*. They show the *direction* of the magnetic force at different places around the magnet. What is happening is that the iron filings are each being made into tiny magnets, by induction. These line themselves up with the magnetic field, and attract one another so they tend to lie end-to-end along definite lines. These lines are crowded together near the poles of the magnet indicating that the magnetic field is stronger there (fig. 8.8). Some can be seen to run in a curve from north pole to south pole. Imagine a north pole of a magnet placed close to the north pole of the magnet in fig. 34.4(i). Like poles repel. The north pole is therefore repelled. The force on it acts along one of these lines. The pole is repelled along a line of magnetic force, away from the north pole, until it comes near to the south pole. Then it is attracted, still moving along the line of magnetic force. These lines can be thought of as the path that is followed by a pole which is in the field of the magnet and is free to move in any direction.

Some lines of force seem to run straight out from the poles of the bar magnet; a few centimetres away they are too weak to affect the iron filings, so we lose trace of them. We must imagine these as running out in wide curves and eventually curving round to reach the opposite pole. If other magnets or magnetic materials are nearby they may not do this. In (c) and (d) we see what happens when other magnets are nearby. Lines can run from the north pole of

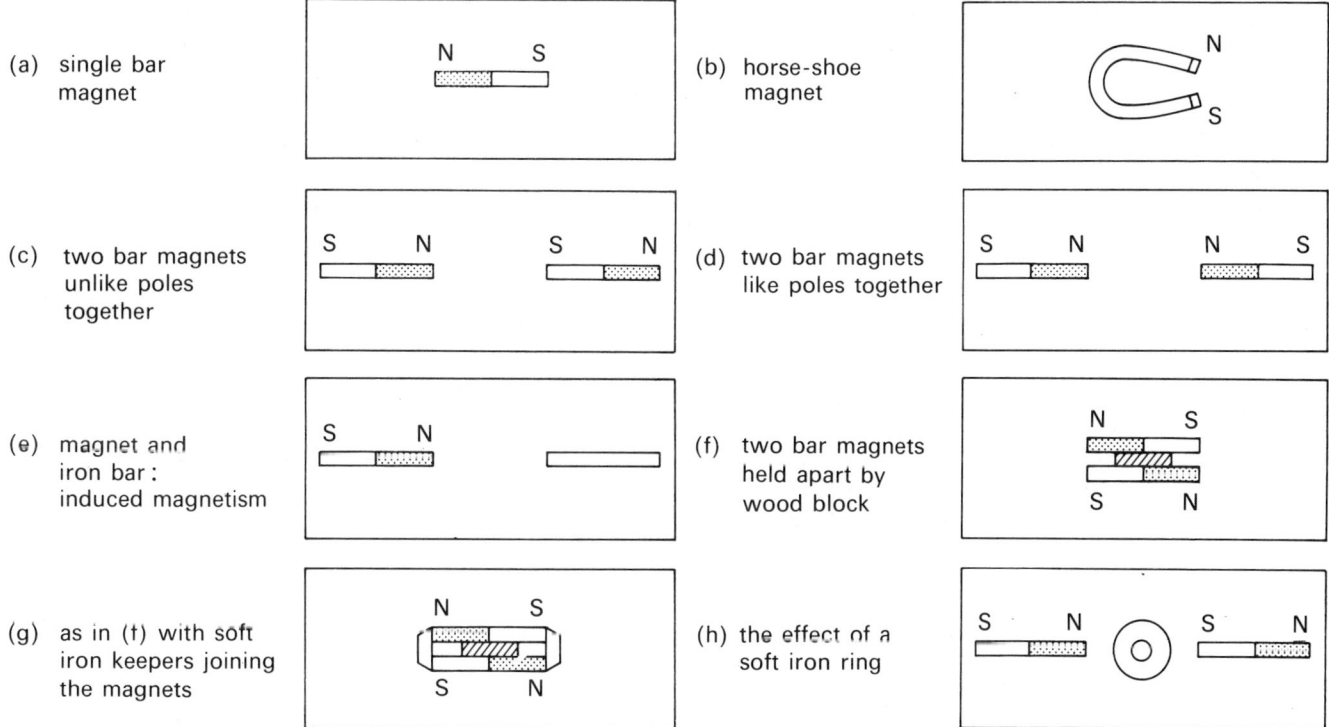

Fig. 8.7 Investigate these magnetic fields (Note: the magnets etc. are underneath the pages)

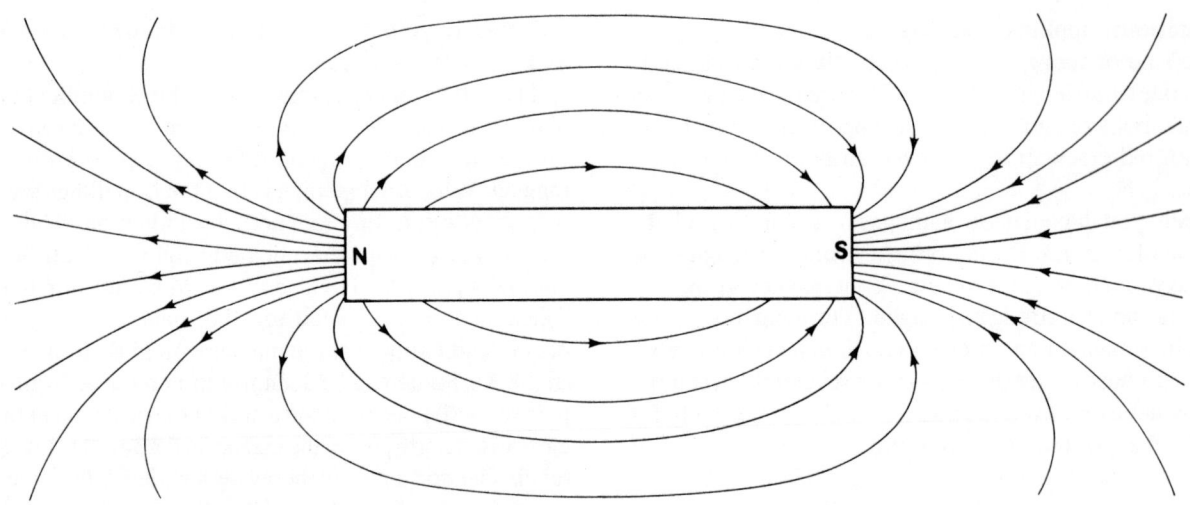

Fig. 8.8   Lines of magnetic force around a bar magnet.

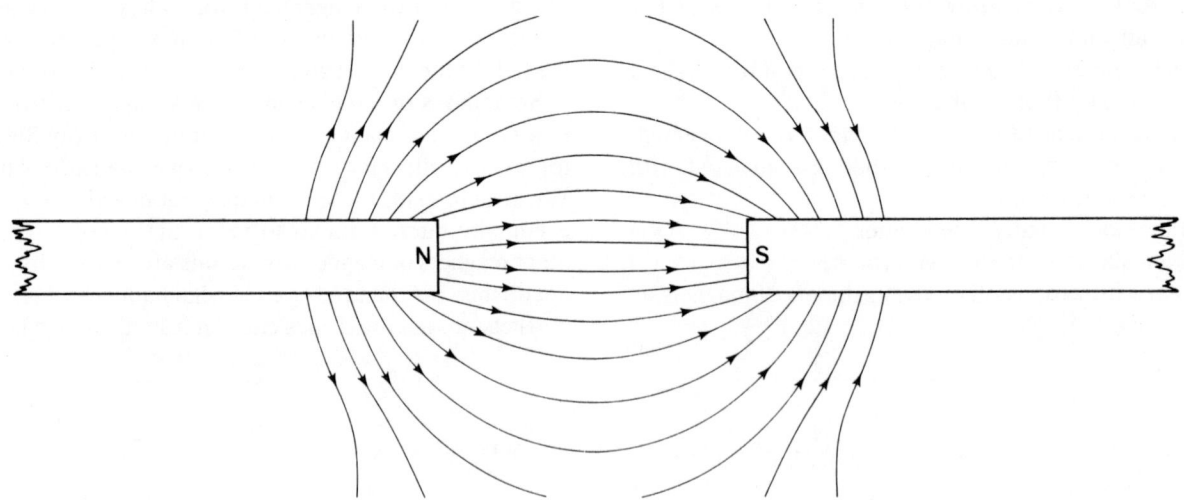

Fig. 8.9   Lines of magnetic force between unlike poles.

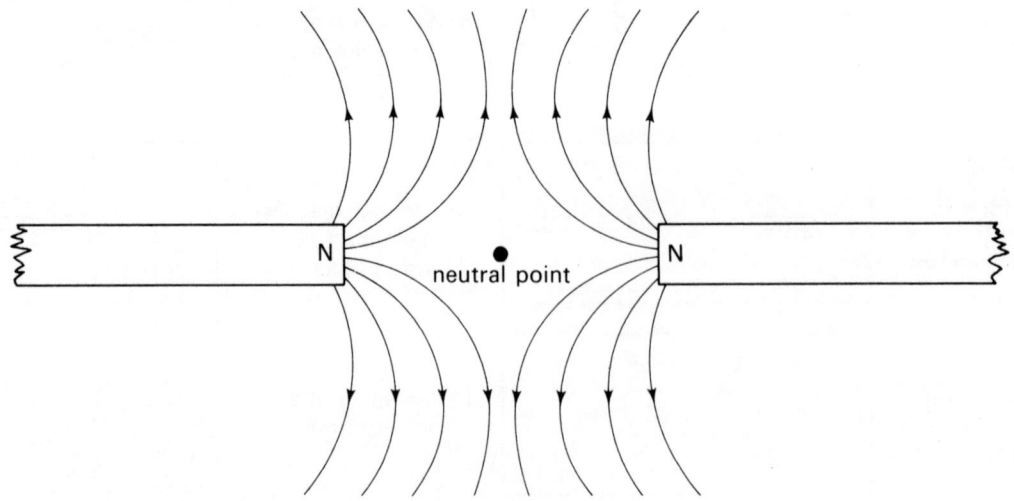

Fig. 8.10   Lines of magnetic force between two like poles.

one magnet to the south pole of a magnet next to it (fig. 8.9). There is attraction between these poles. When like poles are together (as in (d) and fig. 8.10) lines do not cross the space between. The lines from one pole are strongly deflected away from the lines from the other pole, indicating repulsion. A north pole placed at the neutral point is repelled equally by both bar magnets. It does not move. In (e) the lines show that magnetism has been induced in the iron bar. Lines around the bar are fairly far apart, indicating that the induced magnetism is weaker than that of the magnet. Lines of force join the magnet with the nearest end of the bar, showing that the induced pole is unlike the inducing pole and that there is attraction between them.

The two magnets in (f) are strongly attracted and the lines of force at their poles are very close together with hardly any other lines elsewhere. If we place short bars of soft iron across the poles, as in (g) we find very few lines and those found are extremely weak. The iron bars become induced magnets and the lines of force between the magnets pass into and along the bars. Though they are not easy to see, we can think of the lines of force as running round and round inside the closed loop formed by the two magnets and two bars. When bar magnets are stored they should be placed as in (g), using the two short bars (or keepers). If this is done, they retain their strength for longer. For the same reason, a short keeper bar should be placed across the poles of a horse-shoe magnet.

In (h) there is a strong field in the space between the poles of the two magnets. Many lines of force cross this space and they can be seen to pass mainly through the iron ring. The magnetism induced in the ring accounts for this. Note that there are *no lines* crossing the space in the centre of the ring. Because the lines are concentrated in the iron of the ring, the field is strong there, but there is no field in the central space. The ring is acting as a *magnetic shield*, or *magnetic screen*.

8.30 Magnetic materials

A magnetic material is one that, when placed in a magnetic field, becomes strongly magnetised. The most commonly used material is iron. Soft iron is magnetised only as long as it remains in the field. On removal it loses its magnetism. Steel and various alloys of iron with metals such as cobalt and nickel strongly retain their magnetism, and can be used for making very powerful magnets. To understand these properties we need to know more about what happens when a magnetic material is magnetised.

Instructions: how small can a magnet be?
1 Magnetise a piece of thin steel strip about 15 cm long (or a length of clock-spring) by the electrical method of fig. 8.6.

2 Test the magnet to find the position of its poles by placing one end and then the other near to (but not too near to) a hanging bar magnet.
3 Cut the magnet in half. Test the two halves.
4 Cut each half into two. Test all four quarters.
5 Repeat this procedure, cutting the magnet into shorter and shorter lengths.

Questions
1 When the magnet is cut into two, is one half just a south-pole and the other half just a north-pole?
2 How many times can you cut the magnet and still obtain small magnets?

We can cut a large magnet into small pieces and obtain many small magnets. If we cut it into the smallest pieces we can cut with any cutting tool, we still have pieces that are small magnets. The atoms themselves are the magnets and their magnetic effect is caused by their electrons orbiting the nucleus (page 177). Atoms of all elements have orbiting electrons and therefore *all* materials should show magnetic properties. All materials are affected by being placed in a magnetic field but, because of the arrangement of their atoms, the effect is very slight for most of them. It is unlike the magnetic effects we have been discussing in this chapter. One group of materials has special properties; we call them *ferromagnetic* materials. The 'ferro' part of the word refers to 'iron' for these materials have the magnetic properties shown by iron and a few other elements. They respond very strongly to a magnetic field. In these, the molecules are arranged in groups, several million in a group, called a *domain*. All the molecules in a group are set in the same direction, so that a domain is a very tiny magnet. In a bar of iron that is unmagnetised the molecules in a domain are all arranged to produce a magnetic field in one direction but those of different domains give magnetic fields pointing in different directions (fig. 8.11 a). In effect we have millions of small

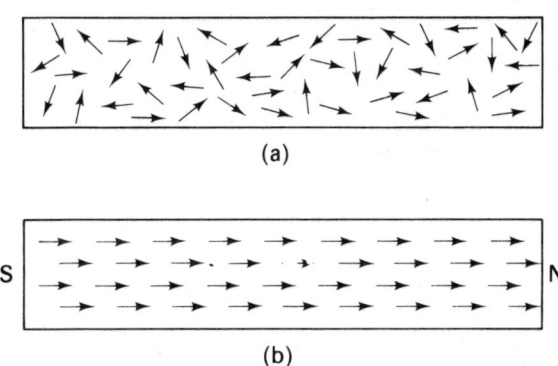

(a)

(b)

Fig. 8.11 *Magnetism in a bar of iron. Each arrow represents a group of iron atoms, all similarly arranged to form a tiny magnetic region in the bar.*

magnets jumbled together and the bar *as a whole* is unmagnetised. If the bar is placed in a magnetic field some of the domains change their direction to lie along the magnetic lines of force of the field. All the molecules in a domain change at once. Not all domains change, but quite a lot do. Then we have a bar that has most domains arranged in the same direction, but a few pointing in other directions. The bar is partly magnetised. As we continue to magnetise the bar, perhaps by putting it in a stronger field, more and more domains line up in the same direction and the bar becomes more strongly magnetised. In the end, when all the domains are lined up, the bar is magnetised as strongly as it can be (fig. 8.11b) and further attempts to magnetise it have no effect.

We can now understand how some of the methods of making magnets work. If a bar is heated to red-heat the molecules of a domain are given extra energy and are more likely to change direction. They are more likely to rearrange themselves to lie along the lines of force of the Earth's magnetic field (fig. 8.13). As the bar cools they remain in this position and the bar is now magnetised. A similar explanation applies to the method of beating the bar with a hammer. The method of induction is caused by the domains lining themselves up in the magnetic field of the inducing magnet. When the magnet is removed some of the fields in the domains take up other positions so the magnetism is wholly or partly lost.

If a magnet can be made by arranging the domains in one main direction, we should be able to demagnetise a material by disarranging the domains. This can be done by heating the bar and allowing it to cool while it is in an *east-west* direction. Or we can hammer it strongly while in this position. In general, any banging of a magnet disturbs the domains and makes the magnet lose strength. Magnets should be handled with care and not dropped.

It might be thought that if a bar is placed east-west and heated or banged it would become a magnet with a north pole along one side and a south pole along the other. This does not happen for the many domains along one side all pointing the same way repel one another and come to lie in many different directions, so the bar becomes demagnetised.

A magnet can also be demagnetised by passing an electric current through a coil. We use an alternating current i.e. one that is changing its direction several times a second. It is like reversing the connections to battery terminals very fast many times every second. The bar becomes magnetised in one direction, then in the other at high frequency. While this is going on we remove the bar from the coil and slowly take it further and further from the coil. As it gets further away the field is weaker. Fewer and fewer domains respond to the changes in the direction of the field. Each domain takes up some position but they are not now all arranged in the same direction, so the bar is demagnetised.

Tape recorders make use of a similar principle. The plastic tape is coated with a compound which contains groups of molecules. The recording head produces a magnetic field which varies in direction and strength according to the musical or other sound vibrations that are being recorded. As the tape passes over the head the field affects the groups of molecules on the tape. They come to lie in a pattern depending on the nature of the recorded sound. When the tape is played, the magnetic fields of these groups of molecules are detected by the playing head and a signal is produced which can be amplified and sent to the loudspeaker. When we no longer need a recording we can demagnetise the tape, so that it can be used again. This can be done in the tape-recorder by a demagnetising head. This produces a rapidly alternating field which disarranges the groups of molecules as the tape passes by the head. It is quicker to demagnetise the whole tape at once by using a special tape demagnetising device. This works rather like the coil method described above. Instead of taking the reel of tape slowly away from the coil, the coil is made to produce a field which is strong at first and gets weaker and weaker as it alternates. The effect is the same: the groups of molecules are disarranged and the tape has been demagnetised. Such effects can happen accidentally if tapes are exposed to strong magnetic fields. Modern loudspeakers have extremely powerful permanent magnets made from special alloys which are capable of being magnetised to very high strength. If tapes are stored close to loudspeakers, the field can partly demagnetise the tape, so spoiling the quality of the recording.

8.40 The Earth's magnetic field

A magnet that is hanging on a thread turns until it comes to rest in one definite direction. This suggests that the Earth must have some of the properties of a magnet. The Earth's magnet has a magnetic field which acts on suspended magnets. The field is used in some of the methods of making magnets. Compared with the field of a bar magnet the Earth's field is very weak. The bar magnet turns only if it is suspended on a thread; the Earth's field is not strong enough to make the magnet rotate when it is lying on the table, for the magnetic force is not enough to overcome friction (page 25). If we sprinkle iron filings on a sheet of paper we see no lines of force, for the Earth's field is not strong enough to affect the filings. We need a more sensitive method. One method is to use a magnetic compass. This is much the same as the bar magnet on a thread but it is more sensitive. Here a light, magnetised needle is balanced on a needle-point so that it can turn very easily. It is in a case which protects it from damage and prevents it from being disturbed by currents of air. A compass needle shows the direction of the Earth's field at the point where the needle is suspended. We can use this

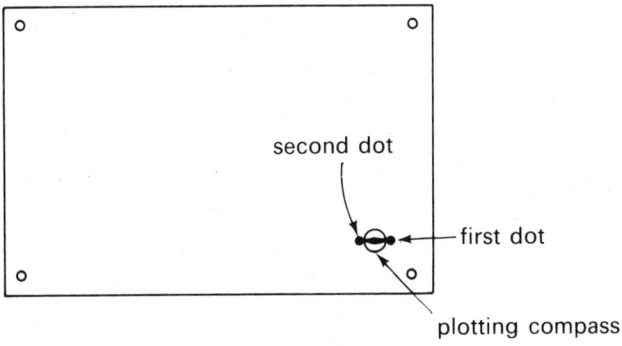

Fig. 8.12 Plotting the Earth's magnetic field.

to help us to find our way about on land or on sea. In the next experiment we use a compass to plot the Earth's field in the laboratory. For this we use a very small compass called a plotting compass.

Instructions: plotting the Earth's field

1 Fix a sheet of paper to the table with sticky tape, or pin it on a drawing-board.

2 Put a pencil dot at one edge of the sheet. Line up the compass so that one end of its needle is over the dot and the other end is over the sheet (fig. 8.12a). You can see the dot through the glass bottom of the compass.

3 Place a second pencil dot close to the case of the compass, in line with the other end of the needle.

4 Shift the compass so that one end of its needle lies over the second dot. Make a third dot in line with the other end of the needle.

5 Continue in this way, making dots across the paper (fig. 8.12b).

6 Join the dots with pencil lines. You have traced one of the lines of force of the Earth's magnetic field.

7 Start again at several more dots spaced along the edge of the paper and draw some more lines of force.

8 Mark the north and south ends of the lines.

9 Draw the direction in which the North Pole lies (true north) on the paper. To find this use an accurately placed wind-vane or a large-scale map which shows the location of your school buildings and has the direction of north marked on it. At night you can locate the Pole Star (Polaris) if you are in the northern hemisphere. The direction of this star is almost due north. Unfortunately there is no conveniently-placed star in the southern hemisphere. Alternatively note the position of the shadow of a vertical stick at noon, *local time*, which can be several minutes different from the time shown on your clock. Your teacher will help you work out local time. At noon, local time, the shadow of the stick points due north or due south, depending on your latitude and the time of year.

10 Measure the angle between the lines of force and the direction of north. This is the *angle of declination* for your district.

We can imagine that the Earth's magnetic field is produced by a short bar magnet set at an angle to the axis on which the Earth spins. Its poles are not directly beneath the North Pole and South Pole of the Earth, but are below places on the Earth's surface that we call the *North Geomagnetic Pole* (fig. 8.13a) and the *South Geomagnetic Pole*. The geomagnetic poles are at present over 1 000 km from the North and South Poles and they have no fixed location. Year by year the imaginary magnet changes position and the locations of the geomagnetic poles change with it; there seems to be no way of predicting the exact amount or direction of movement of these poles.

Since the geomagnetic poles are not at the geographical poles of the Earth, the lines of force do not run due north and south, but at some angle to this direction. This is illustrated in fig. 8.13a, where you can see compasses on different parts of the Earth, all with their needles along the lines of force. The angle between the line of force and the direction of the North Pole is the angle of declination (*d*). In some places (for example A in the figure) the North Geomagnetic Pole lies to the west of the North Pole; in other places (B) it lies to the east. In a few places (C) the geomagnetic and geographical poles are exactly in line, so the angle of declination is zero.

Another complication is that the poles of the imaginary magnet in the Earth lie deep below the surface (fig. 8.13b). The lines of force to not run horizontally over the surface of the Earth but at various angles to the surface. To measure the angle at which the lines of force dip into the surface, (called the *angle of dip*), we use a dip needle (fig. 8.13c). This is simply a compass needle carefully balanced so that it does not tend to turn under its own weight, and supported so that it can rotate in a vertical plane. When set up with its plane of rotation along the direction of a line of force, its needle takes up the angle of dip. If you have such a needle, you could measure the angle of dip in your district. The one shown in the figure is a very simple type; for accurate measurements more complicated mountings are used, though the principle is the same. At the geomagnetic poles the angle of dip is 90°.

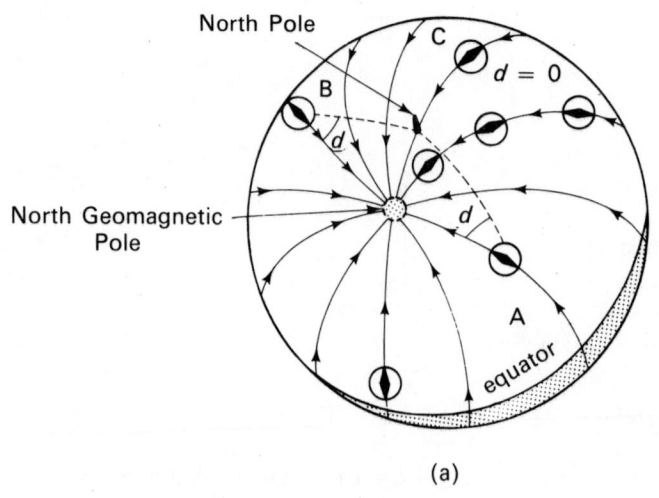

(a)

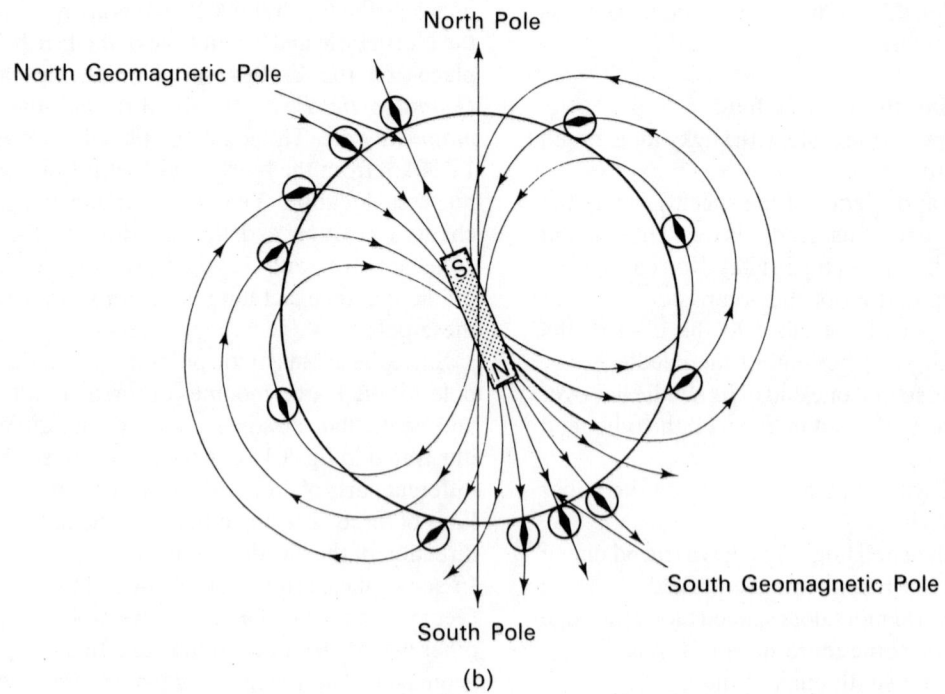

(b)

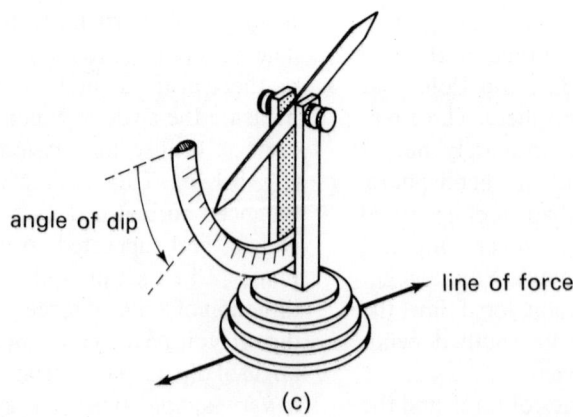

(c)

Fig. 8.13   Earth's magnetic field: (a) as shown by compass needles; (b) as shown by dip needles; (c) a dip needle.

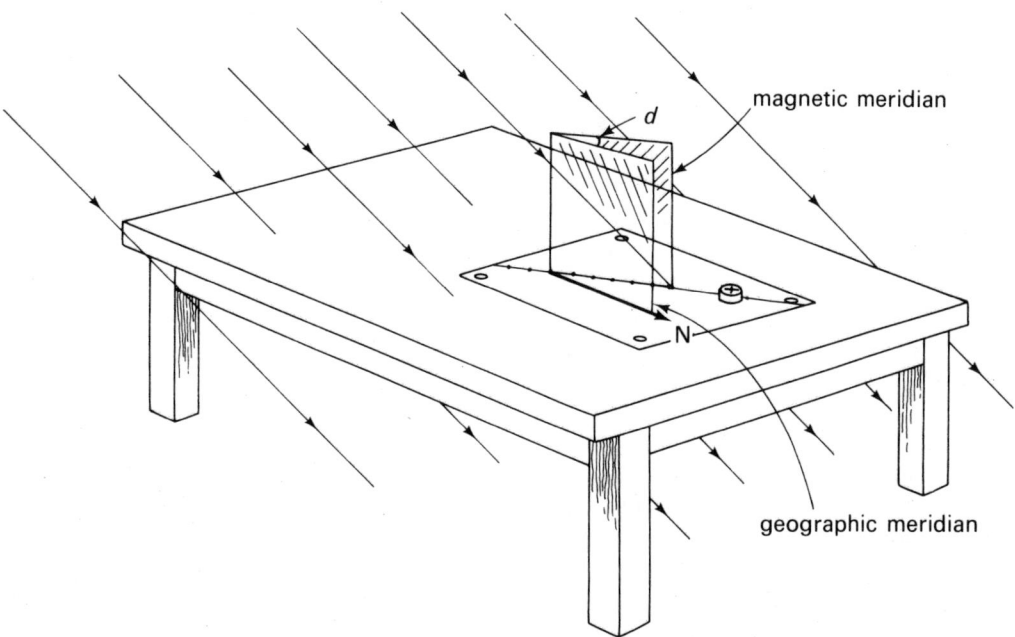

Fig. 8.14 The meridians and the angle of declination (d).

As we go away from the poles the angle is less than 90°, though it remains high over many parts of the Earth because the imaginary magnet is short compared with the Earth's diameter. In equatorial regions the angle of dip is zero or close to zero.

We can visualise the lines of force coming through the laboratory, as in fig. 8.14. The vertical plane in which one of these lines passes down through the laboratory and into the Earth beneath is the *magnetic meridian* at that place. The vertical plane which, if continued, would include the North Pole is the *geographic meridian*. The angle between them is the *angle of declination*. You can see from the figure that when you performed the experiment above you did not actually plot lines of force. Your paper was horizontal but the lines of force are not (unless you are at a place where the angle of dip is zero). The line you plotted is the line where the magnetic meridian cuts into your paper.

The lines of force do not pass in regular curves from one pole of the Earth's imaginary magnet to the other. Their path is irregular in places, owing to concentrations of magnetic materials in the Earth's crust. There is no easy way that we can calculate what the angles of declination and dip should be for any place. They must be measured locally. Maps have been made, showing the angle of declination at different places over the Earth's surface. They also estimate the correction that must be applied to allow for changes in the positions of the geomagnetic poles as time passes. Such maps are useful to navigators and surveyors who use magnetic compasses in their work.

Many rocks have magnetic properties. One of these, called magnetite, is strongly magnetised. Pieces of this rock were suspended by thread and used by sailors long ago in the way that we use compasses nowadays. This rock is sometimes called 'lode-stone'. When rocks form and cool they may become magnetised in rather the same way that a cooling bar of steel becomes magnetised. The direction of their magnetisation is determined by the direction of the Earth's magnetic field at that time. If we know the direction of the Earth's field in the past when the rocks were formed, we can tell in which direction the rocks should be magnetised. Sometimes we find that they are not magnetised in the expected direction. This suggests that the rocks have shifted since they were formed. By careful and very delicate measurements we can find how much the rocks have moved. We use this information to study the past history of the Earth's crust. For example, we can work out the movements of the continents over periods of millions of years.

We have said that the lines of force of the Earth may be affected by large amounts of magnetic materials in the Earth's crust. They are also affected locally by quantities of iron and steel, such as are found in buildings, vehicles and machinery. We might also expect them to be affected by magnets. Before we investigate this practically we must notice one point about fig. 8.13b that was not mentioned before. The imaginary magnet in the Earth is shown with its *south* pole pointing towards the *North* Geomagnetic Pole, and its *north* pole pointing towards the *South* Geomagnetic Pole. This is not a mistake in the diagram! We have defined the north pole of our compass needle and hanging bar magnet as the one which points (roughly) towards the north. It follows from this, and from the first law of magnetism that the northernmost pole of the imaginary magnet must be a south pole in a magnetic sense.

Instructions: investigating the Earth's magnetic field in the region of a bar magnet

1 Place a large (40cm × 50cm) piece of paper on a table, or a drawing-board, so that its shorter edges are parallel to the magnetic meridian. Use a compass to position the paper carefully, then tape it or pin it in position. Make sure that no magnets are nearby to disturb the compass.

2 Place a bar magnet on the paper, with its north pole towards the North Geomagnetic Pole, as in fig. 8.15. We can imagine the Earth's field as a number of parallel lines running across the paper (the dashed lines in the figure). We can also imagine the lines of force of the bar magnet (thin lines). The arrows on the lines indicate the direction from north pole (of a magnet) to south pole. Remember that the north pole of the Earth's imaginary magnet is to the south of you. How do these two magnetic fields combine? We use a plotting compass to find the answer to this question.

3 Begin at one corner of the paper and plot a line of force (a good place to start is shown in the figure). Plot the line all the way across the paper or, if it goes to the magnet, as far as the magnet.

4 Plot other lines, starting from points along the southern or northern edges of the paper.

5 If you see any regions where there appear to be no lines,

place the plotting compass in these regions and try to plot lines of force there.

6 Repeat the whole procedure, with the magnet reversed, so that its north pole is toward the southern edge of the paper. For the repeat, the paper is better arranged with its *long* sides parallel to the magnetic meridian.

Questions

1 In general the magnetic field is strongest where lines of force crowd close together. Where they are drawn far apart it is weakest. What parts of the field are strongest in your first trial?

2 What parts are weakest in your first trial?

3 On your drawing, mark with an 'X' any places where you think the field is very weak or zero. This point is called a *neutral point*. Think of the neutral point as being a point where the field caused by the magnet exactly cancels the field caused by the Earth.

4 Answer questions 1 to 3 again for your second trial and plot the neutral points on your drawing.

5 What would happen if we put a very small compass exactly on a neutral point?

6 What would have happened if you had used a stronger magnet?

7 Do two lines of force ever cross each other?

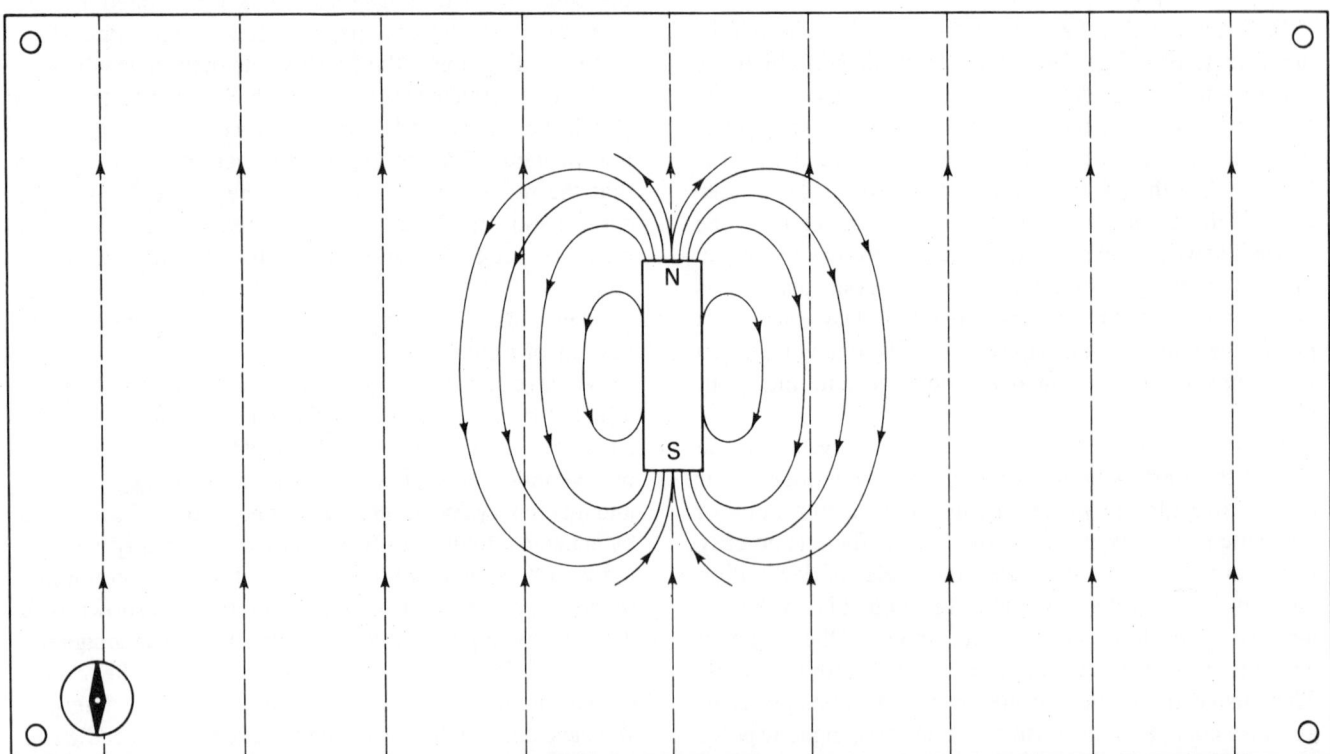

northern edge of paper

southern edge of paper

Fig. 8.15   Investigating the Earth's field in the region of a bar magnet. The Earth's field (dashed lines) and the field of the magnet (thin lines) are drawn as if they were separate; in the investigation, we see how they combine.

9 Discovering about electricity

9.00 Electric cells

There are many ways of obtaining electricity. Several of them are dealt with in other parts of this book. The electric cell is one of the commonest ways: you use it every time you switch on an electric flash-lamp or a portable radio. To make a cell you need three items. Two of these are strips of metal called *electrodes*. One electrode is of one kind of metal, and the other of a different kind of metal. One of the electrodes can be carbon, which is not a metal but has similar properties. The third item is a container in which there is a solution of a substance in water. This solution is called an *electrolyte*. Some examples of electrolytes are given in the experiment which follows.

Instructions: making an electric cell

1 For the electrolyte use one of these: lime-juice or lemon-juice; vinegar; dilute sulphuric acid; a solution of sodium chloride in water; a strong solution of ammonium chloride in water. You need only about $50 cm^3$ of solution, in a small glass or plastic container.

2 For electrodes, use strips or rods of aluminium, carbon, copper, iron, lead or zinc. The strips should be about 1 cm wide and 5 cm long.

3 To make a cell, take any pair of electrodes and place them in the electrolyte so that they do not touch each other (fig. 9.1 a).

4 Wire one electrode (either one, it does not matter which) to the negative terminal of the voltmeter. Connect

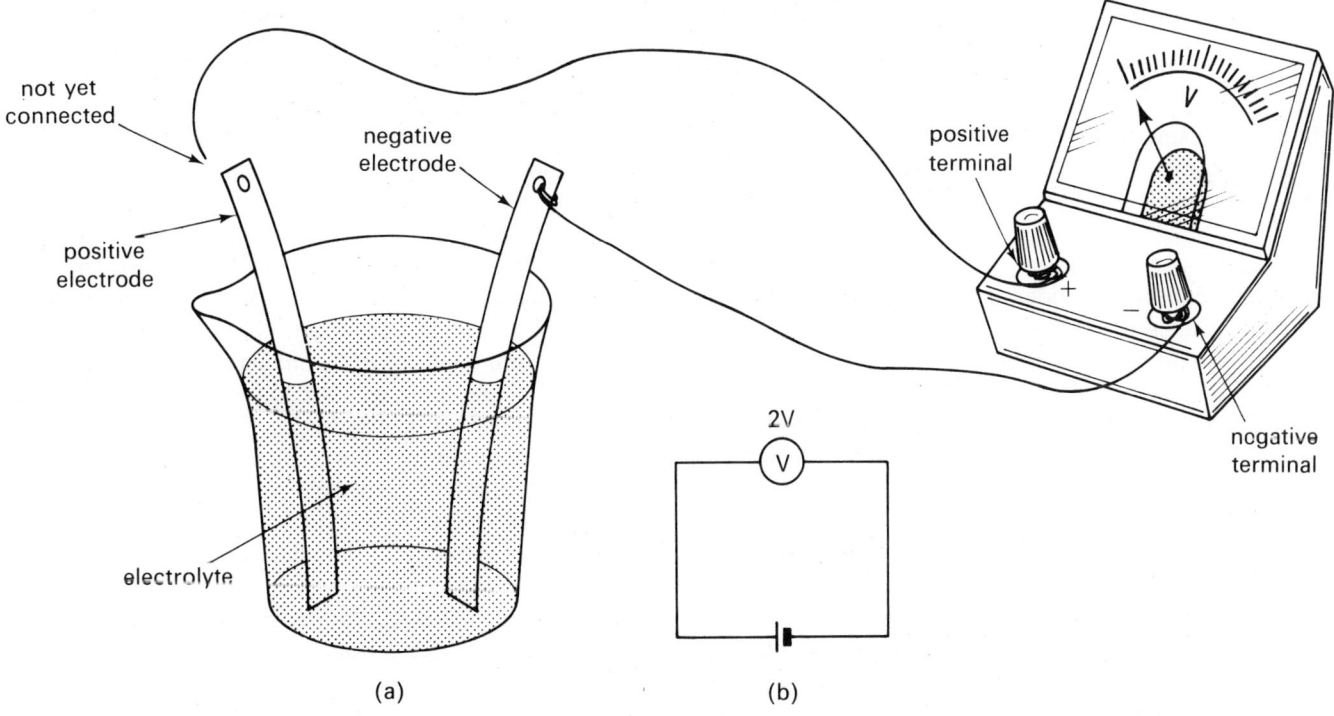

(a)   (b)

Fig. 9.1 A simple cell.

another wire to the positive terminal of the voltmeter, as in the figure.

5 Touch this loose wire against the other electrode. If the needle of the voltmeter moves to the right, your connections are correct; twist the wire firmly around the electrode. This is the positive electrode. If the needle of the voltmeter moves to the left, below zero, the electricity is flowing in the wrong direction. Connect the wires to the opposite electrodes. The needle should then move to the right and you will know that the electrode which is connected to the positive terminal of the voltmeter is the positive electrode.

6 Read the voltage between the electrodes. This is a way of measuring the *force* which drives electricity from the positive electrode, through the wire to the voltmeter, through the voltmeter (making its needle move), through the other wire to the negative electrode and through the electrolyte, back to the positive electrode again. The electricity makes a complete *circuit* (fig. 9.1b) and the force which drives it is called the *electromotive force*. For short, we call it the *e.m.f.*

7 Record your results in a table.

Positive electrode	Negative electrode	e.m.f. (volts)

8 Repeat using other pairs of electrodes. Some metals make a positive electrode when paired with some metals, but make negative electrodes when paired with others.

Questions

1 Which pair of electrodes gave the biggest e.m.f.?

2 Which material was *always* the positive electrode when paired with *any* other material?

3 Are you able to light a torch-bulb with the electricity from this kind of cell? Try it and see. It is best to use a low-voltage bulb (1.25 volt or 2.5 volt) and shade it from bright light.

Measurement of electromotive force has introduced a new SI unit, the *volt*. The symbol is V, which is a capital letter because the unit is named after Alessandro Volta. He was a physicist in the eighteenth century who invented the simple type of cell that you used in the experiment. The volt is a derived SI unit. The way in which it is derived and defined is described later in chapter 15. Like other SI units it can have multiples, such as the kilovolt. The electric generator at a power station may have an e.m.f. of 11 kV, or more. The volt also has sub-multiples, for example the millivolt (mV), a thousandth of a volt. When you speak into a microphone a very small e.m.f. is generated amounting to a few millivolts.

9.01 Leclanché cells

There are many different ways of making cells. Some ways are better than others and there are special types of cell suitable for different purposes. The original type of Leclanché cell (fig. 9.2) was widely used years ago for operating electric bells and similar devices. A disadvantage of the original type was that it was a wet cell, like the cells you made in the previous activity. If a wet cell is shaken or turned over, some or all of the liquid may be spilled, so making a mess, possibly damaging equipment or furniture, and making the cell less effective.

Nowadays we use a dry version of the Leclanché cell. We usually call it simply a 'dry cell'. Its structure is shown in fig. 9.3. The electrolyte is a jelly-like material containing ammonium chloride. The electrodes are made from zinc and carbon. The zinc electrode (negative) is shaped in the form of a cylindrical can which holds the electrolyte and the carbon electrode. The carbon electrode consists of a rod of carbon. Surrounding the rod is a mixture of powdered carbon and manganese (IV) oxide. The latter substance is included to help overcome a disadvantage that most cells possess. This is called *polarization*.

If a cell is connected to an electric bell or other device which takes a large electric current, the cell works well at first, and its e.m.f. is 1.5 V. (Note that this figure is roughly what you obtained for your carbon-zinc cell, and is one of the largest e.m.f.'s obtainable with cells made from cheap materials.) The working of the cell produces hydrogen gas which collects in bubbles around the carbon electrode. You may have noticed this in your experimental cells. This has two serious effects: (a) the bubbles interfere with the passage of electricity through the cell, and (b) the bubbles themselves act as a kind of cell inside the main cell; the force of this hydrogen cell is in the opposite direction from the force of the main cell. In other words, the e.m.f. of the main cell is opposed by a *back e.m.f.* caused by the bubbles of hydrogen. This is what we mean by polarization. As the back e.m.f. gradually increases, it opposes the e.m.f. of the cell more and more and the total e.m.f. available to drive electricity around the circuit becomes less and less. The bell rings less and less strongly.

Manganese (IV) oxide removes the bubbles of hydrogen; it reacts with hydrogen, producing water. So polarization is prevented and the cell produces its full e.m.f.

Sometimes, when a very large amount of electricity is passing around the circuit, the hydrogen is formed more

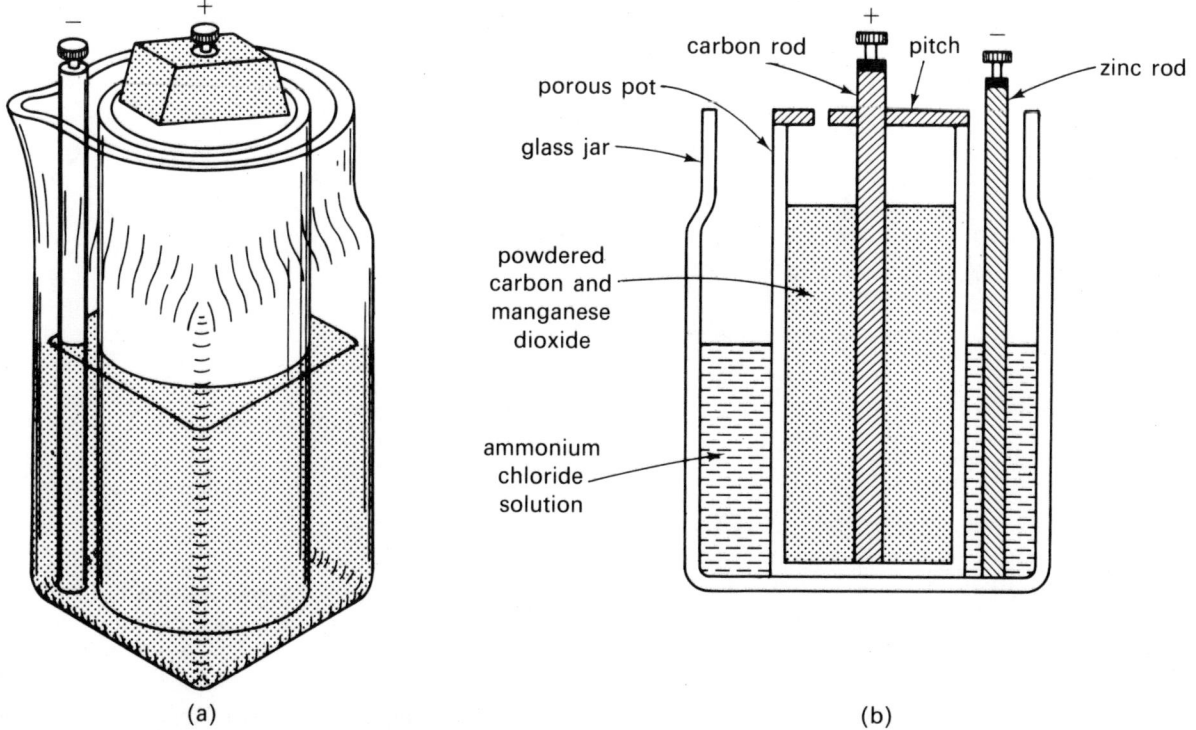

Fig. 9.2 A Leclanché cell (a) ready for use; (b) sectional view.

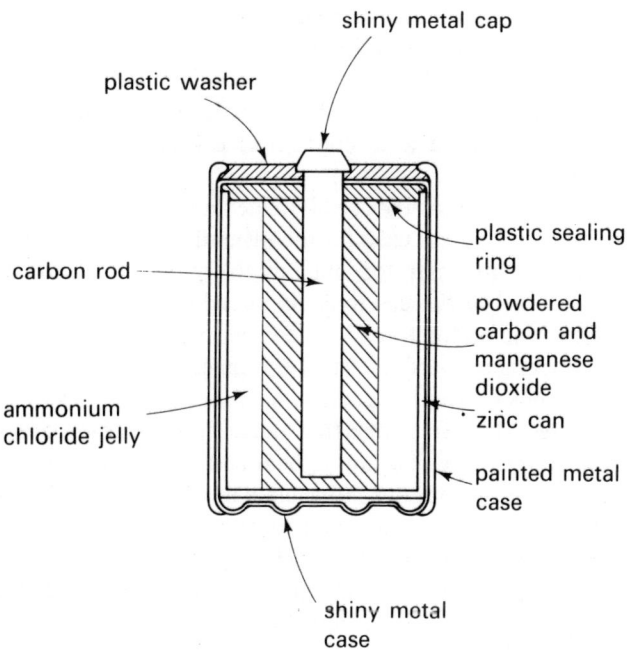

Fig. 9.3 Sectional view of a dry cell.

quickly than the manganese (IV) oxide can remove it. The e.m.f. of the cell falls. However, if the switch is turned off and the cell is left to recover, the hydrogen is gradually removed by the manganese (IV) oxide and the cell regains its full e.m.f. This is why Leclanché cells (wet or dry) are especially useful for electric bells on doors and for alarms. They can deliver a large electric current for a short time and recover during the periods between, when they are not being used.

The fact that the dry cell can be sealed and is unspillable means that it can be used in all kinds of portable equipment, such as electric flashlamps, portable radio sets, pocket calculators, toys and many other devices. Most of the dry cells used today are of the Leclanché type and can be bought as single cells (with an e.m.f. of 1.5 V) or as a number of cells packed together to make a *battery*. The total e.m.f. of the battery may be 3 V, 6 V, 9 V, or more, depending on how many dry cells are connected together inside it. Usually the cell is contained in an outer metal casing, which includes a shiny metal plate at one end and a smaller one at the upper end of the carbon rod, giving good electrical contact with the equipment in which it is used.

9.10 Conductors and insulators

We have met these words before, when we were studying heat (chapter 6). An electrical conductor is a material which allows electricity to pass along it easily. An electrical insulator is a material which does not allow electricity to pass easily.

Questions
1 Recall your investigations with cells. Name some materials which are electrical conductors.
2 Make a list of some materials which are good insulators.

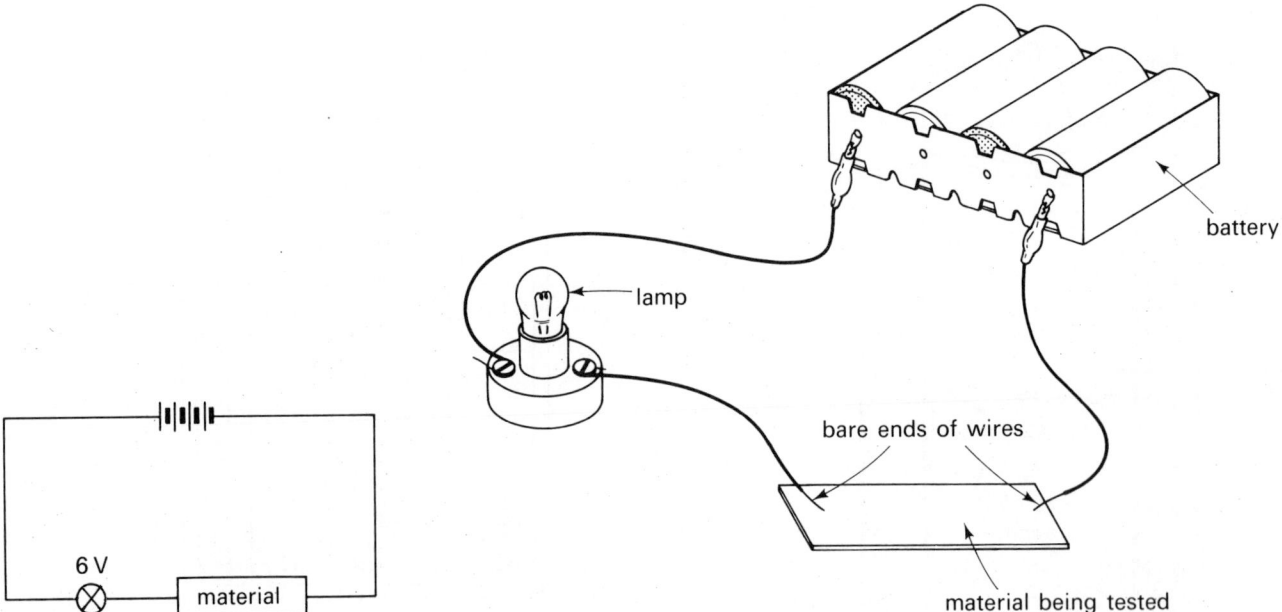

labels on figure: battery, lamp, bare ends of wires, material being tested

6 V ⊗ — material

Fig. 9.4 Testing conductors and insulators.

Instructions: testing conductors and insulators

1 Connect together a testing circuit, as in fig. 9.4. The drawing shows a battery-holder in which there are four dry cells, connected so as to give a total e.m.f. of 6V. You can use any battery you like, giving up to 9V. The lamp is in the circuit to indicate (when it glows) that electricity is passing through the material which is being tested.

2 Collect some materials to test. Testing is simple: just place the bare ends of the wire at either end of the material and see if the lamp lights or not. Materials for testing could include: copper; solder; different kinds of plastics; iron; mica; paraffin wax; zinc; glass; paper; wood; rocks; silver; rubber; carbon rod; brick; aluminium; cotton thread; the 'lead' (graphite) of a pencil; ice; bronze (a coin). Also put some liquids in small beakers and dip the two bare wires in them. Liquids to test are: the electrolytes used in cells; tap water; rain water; ethanol; oils of various kinds. You can also put loose solids (powders) in beakers: sodium chloride; sugar; flour; soils (wet and dry); sulphur.

3 As you test each substance write its name in one of two lists, headed 'Conductors' and 'Insulators'.

Questions

1 Which types of material are good conductors of electricity?

2 Which materials might be *useful* as insulators?

3 What do you notice about materials which are good thermal conductors and those which are good electrical conductors?

The most commonly used electrical conductor is copper. It is used for making wires and most of the conducting parts of electrical equipment. Silver is a better conductor than copper but, though it is used for special purposes, it is too expensive for common use. Nowadays even copper is becoming expensive and it is being replaced by aluminium which is not quite such a good conductor as copper, but is cheaper. Solder, an alloy of lead and tin, is used for making firm joints between copper wires and is a good conductor of electricity. Occasionally the liquid metal, mercury, is used as a conductor, especially in certain types of switch.

Plastics such as PVC, polythene and polystyrene have come to be widely used for the insulating coating around electric wires and for making the parts of electrical equipment which need to be non-conducting. Previously, rubber and cotton thread were used to cover wires. Non-conducting parts of electrical equipment were made from wood, bakelite and a specially hardened rubber known as ebonite. Mica is a rock that occurs in thin transparent sheets. It is very resistant to heat and is used as a former for winding electrical heating elements such as those used in electric clothes irons and toasters. Glass is used for making insulators for carrying power-lines. Boards made of fibre-glass are used to mount the many components that form part of a modern radio set or amplifier. The knobs and switches by which we operate this equipment and the switches, sockets, plugs and lamp-holders that we use at home and school are all made from hard-wearing plastics.

9.20 The magnetic effect of an electric current

When we connect a cell to other pieces of electrical equipment to make a circuit, electricity flows from the cell, through the wires and equipment and back to the cell again. We say that there is an *electric current*, just as we say there is a water current when water flows along a river. We usually think of this electric current as flowing from the positive terminal of the cell and returning to the negative terminal of the cell. In this section we look at the effects that this current has in the region surrounding the wire. The first experiment is based on one performed in the nineteenth century by the Danish physicist, Hans Oersted, the first investigator of electromagnetism.

9.21 The effect of a current in a straight wire

Instructions: Oersted's experiment

1 Suspend a bar magnet from a thread, or use a compass. Allow the magnet or compass needle to come to rest lying along the magnetic meridian (page 77), that is to say, roughly north-south.

2 Arrange a straight piece of wire above the needle, running along the magnetic meridian (fig. 9.5). For supplying the current, use a 2 V rechargeable cell (accumulator) as this can provide a larger current than a dry cell. However, a dry cell can be used if it is more convenient.

3 Press the push-button *for an instant*. Watch the compass needle. Record the direction of its movement in a table. The easiest way to do this is to say which way the north-pointing pole of the needle moves, to east or to west.

4 Reverse the direction of flow of current by exchanging the wires at the terminals of the cell. Press the button. Record the direction of movement of the needle.

5 Next mount the compass on a block, so that the wire is just below the compass. Repeat the experiment with the current running south-to-north and then north-to-south. Complete the table as shown below.

Questions

1 Is the direction of movement of the needle affected by the direction of flow of the current?

2 Is the direction of movement affected by whether the wire is above or below the needle?

3 Make up a rule to help you remember what happens.

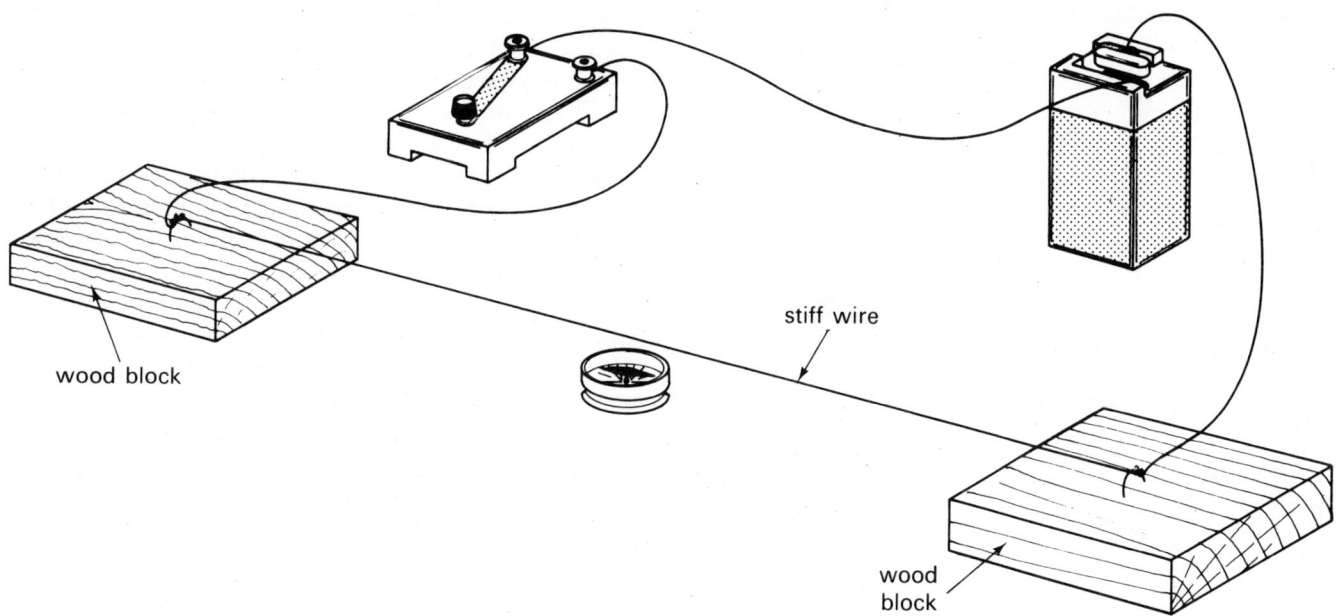

Fig. 9.5 Oersted's experiment.

Position of wire	Direction of flow of current	Direction of movement of north pole of needle
Above the needle	south to north	
	north to south	
Below the needle	south to north	
	north to south	

83

In science we have *laws* which state the most important ideas about the way things behave. We also have *rules* for things that are perhaps less important, and perhaps we do not understand exactly why they happen. The rule is just a useful way of helping us to remember what happens. A useful rule to help you remember the result of this experiment was invented by Ampere, who was another pioneer of the study of electricity. His rule is called *Ampere's swimming rule*. Imagine that you are swimming along the wire in the direction of the current. You are facing the needle i.e. if the wire is above the needle, you are swimming face-down; if the wire is below the needle, you are swimming on your back. When the current passes, the north pole of the needle will be turned towards your left hand. Now we will see why this rule holds true.

The apparatus used to demonstrate the field around a straight wire is shown in fig. 9.6. A straight wire passes vertically through a horizontal wooden platform. To give a bigger effect we sometimes use several pieces of wire running side-by-side and insulated from one another. The wire is connected to cells giving a total e.m.f. of about 6 V. A resistor in the circuit prevents the current from being too large, which would damage the cells.

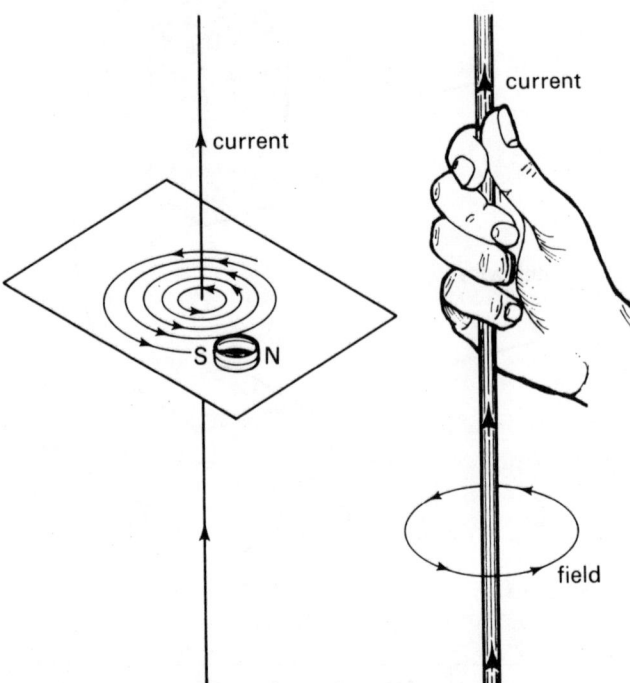

Fig. 9.6 The magnetic field around a straight wire that is carrying an electric current.

Some iron filings are sprinkled on the platform and when the current is switched on the platform is tapped a few times. The iron filings become arranged in circles, all centred on the wire. We can remove the filings and use a plotting compass (as in fig. 9.6) to plot the lines of force close to the wire. We confirm that these are circles. When the current is flowing *up* through the wire, as in the figure, the compass needle always points in an anti-clockwise direction as we move it around the wire. The direction of the magnetic force is anti-clockwise.

Question
What do you think will happen if the current is reversed, and flows *down* through the wire?

A rule has been invented to help you remember the results of this experiment. This is the *right-hand grip rule*. Imagine that you grip the wire with your right hand, as you would grip a piece of rope that you were going to pull. Your thumb lies along the wire. If you are gripping the wire so that your thumb points in the direction that the current flows in, the direction of the field is the same as the direction in which your fingers circle the wire.

9.22 The field around a coil

The ways of demonstrating this are the same in principle as the one already described. The coil, either a long narrow coil or a short wide coil, is made so that it passes through small holes in a wooden platform. A strong current is passed and the field can be detected either by iron filings or by using a plotting compass (fig. 9.7a). The field of the short coil is what might be expected from applying the right-hand grip rule. On the nearer side of the coil in the figure the current is passing upwards. This causes an anti-clockwise field around this bunch of wires. At the farther side, the current is passing down, causing a clockwise field. In between the lines of force are parts of larger circles: the direction of force is as shown.

If we imagine such a coil pulled out to separate the turns, we get the long coil (fig. 9.7b). Lines of force are entering the coil at the end nearer to us and leaving at the far end. The field of this coil is very much like that which we find around the poles of a bar magnet. In both demonstrations the direction of the field is reversed if we reverse the direction of the current. When the current is turned off, the magnetic field disappears completely.

Figure 9.7c is a view of the coil with the platform left out. With the current flowing as shown the magnetic force is away from the left-hand end of the coil. If this were a bar magnet we should say that this is the north pole. The right-hand end is the south pole. By passing a current through a coil we have produced the same effect as making a magnet. This effect is *electromagnetism*. There is another rule to help you remember the details. If you look at the coil from the end which is its *north pole*, you imagine the current flowing around the coil in an *anti-clockwise direction*. The 'N' with arrow-heads on it also has this anti-clockwise look. As we look at the coil from the other end we see the *south pole*, and a *clockwise current*. An 'S' with arrow-heads on it has a clockwise look.

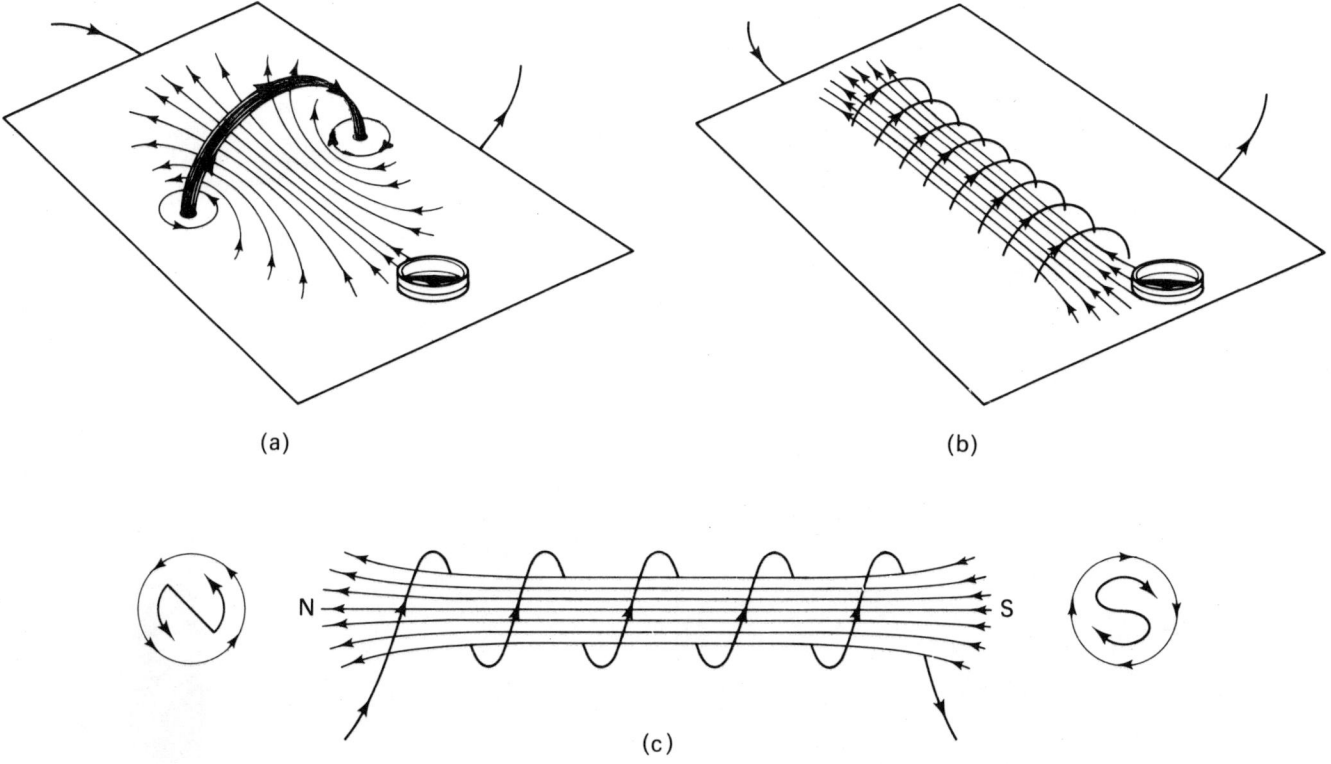

Fig. 9.7 *The magnetic field around a coil: (a) a short, wide coil; (b) a long, narrow coil; (c) a way of remembering the polarity of a coil.*

9.30 Using electromagnetism

Instructions: making and using an electromagnet

1 Cut a piece of stiff paper about 4cm × 8cm. Wrap it around a wooden rod (or test-tube or felt-tip pen) about 1.5cm diameter and fix it with sticky tape (fig. 9.8a).

2 Take a piece of thin plastic-covered wire about 2m long and wind this round the paper. There should be between 60 and 100 turns; the exact number does not matter. There will be more than one layer of turns: keep winding in the same direction. It does not matter if the turns are not evenly spaced (fig. 9.8b).

3 When you have finished winding, twist the ends of the wire around each other to prevent the coil coming unwound and wrap some sticky-tape around it to hold it together (fig. 9.8c).

4 Slide the rod out of the coil which is now ready for use.

5 Allow a compass needle (or a hanging bar magnet) to come to rest in the magnetic meridian. Place the coil near to one end of the magnet. Connect the coil to a 6V dry battery (fig. 9.8d and e). Before you press the button, work out the direction in which the current will flow through the coil. Use fig. 9.7c to decide which end of the coil will be north pole and which end will be south. Press the button, watch how the needle behaves and find out if your prediction is correct.

To get a strong effect we need a strong current. A dry cell delivers a strong current, but not for long. Do not hold the button down for more than one or two seconds or the battery will soon become exhausted. If you are using accumulator cells, you must connect a resistor in the circuit. Your teacher will provide one of the correct value (about 5 ohms) and tell you how to connect it in the circuit. This prevents the current from being too strong and damaging the accumulator.

6 Hold the coil with one end just above a pile of pins or paper-clips (fig. 9.8f). Press the button for one or two seconds while you find out how many pins can be picked up. If you are not successful in picking up pins, do not be surprised, but go on to step 7.

7 Make a soft-iron *core* for the coil. To do this take enough iron nails to make a bundle fat enough to just fit inside the paper tube (fig. 9.8g). The nails should be a little longer than the tube. Wrap some sticky tape round the bundle. Place this core inside the tube.

8 Hold the coil so that one end of the core is just touching a pile of pins or paperclips. Press the button for one or two seconds while you find out how many pins can be picked up. What happens?

9 Press the button to pick up some pins. Raise the coil above the table, then release the button. What happens?

10 Keep the coil and core, ready for the next activity. If you wanted to make a stronger electromagnet, what methods would you try?

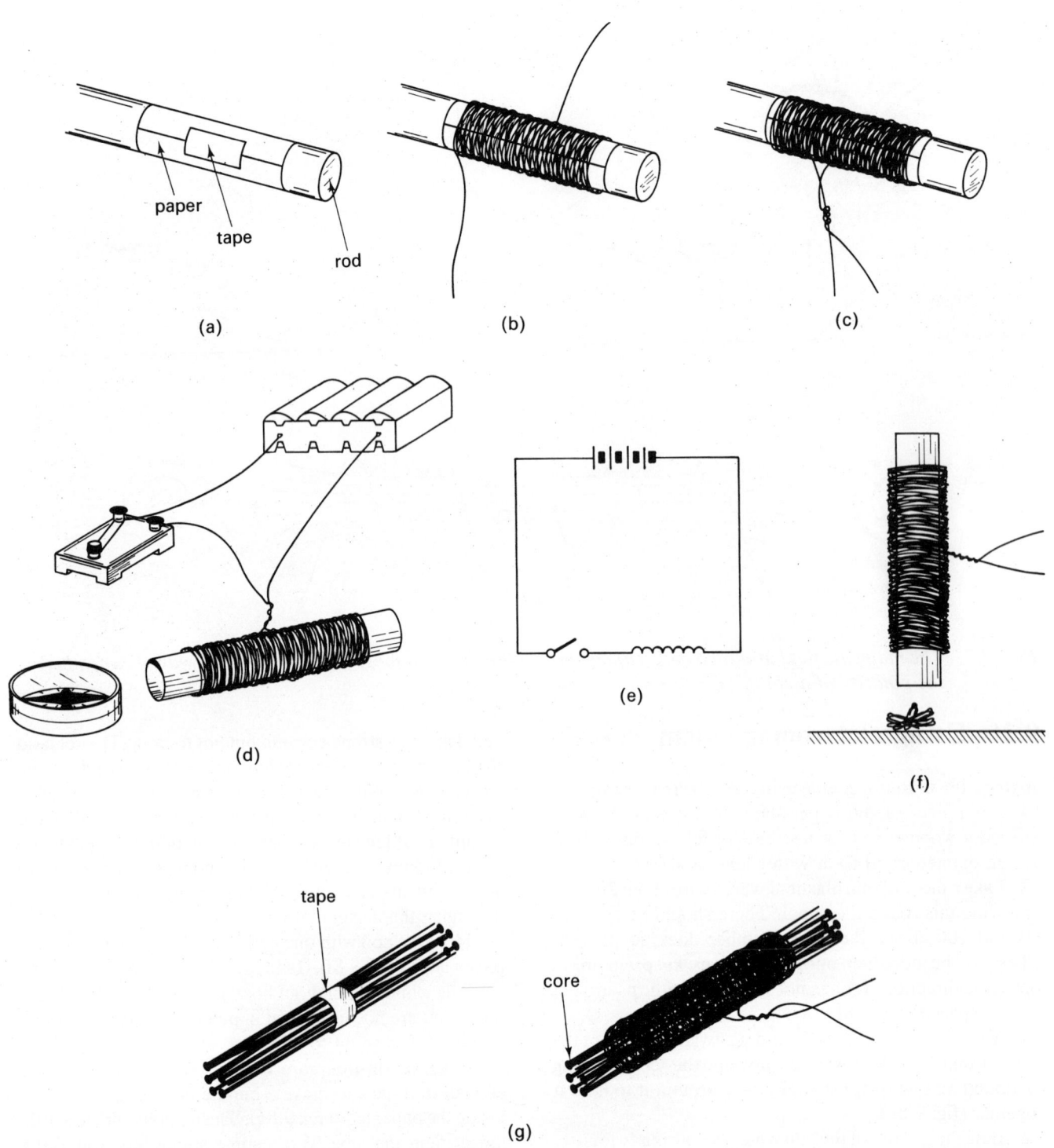

Fig. 9.8 *Making an electromagnet.*

An electromagnet has the one useful property of a bar magnet: it can be used for picking up ferromagnetic objects. In addition it has its own special property: its magnetism can be switched on and off. When it is switched on, it picks up and holds the object; when it is switched off, the object is released. Large electromagnets are used in industry for lifting heavy steel parts, such as girders and sheet metal and for handling scrap iron and steel (fig. 9.9). If you are good at model-building, you could make a model crane and attach your electromagnet to it. Electromagnets are often used in scientific equipment for producing strong magnetic fields. With specially-shaped poles they can be used as 'lenses' to direct beams of charged particles (chapter 23). Small, simple electromagnets have a number of uses in everyday life, as the next activity shows.

Fig. 9.9 Electromagnets have many uses in industry.

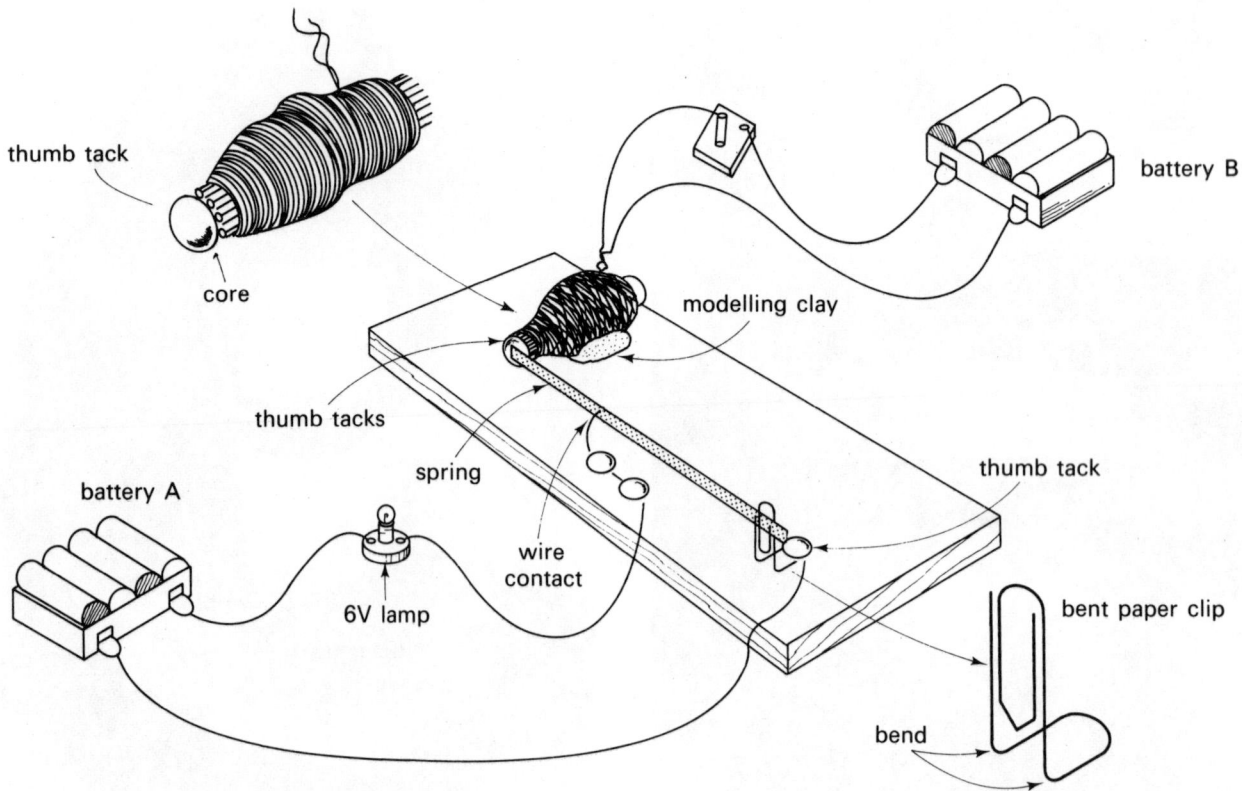

Fig. 9.10　One way of using the electromagnet.

Instructions: using the electromagnet

1 Cut a length of thin clock-spring; rub it on both sides with a file to remove any varnish that might be on it. Clip one end of the spring in a bent paper-clip that is fixed to a piece of wood, with a thumb-tack (fig. 9.10).

2 At the other end of the spring attach a thumb-tack: bore a small hole to accept it or attach it with wax or glue.

3 Make a contact from bent bare wire. This wire is wound round two thumb-tacks pinned to the board, to hold it in place.

4 By turning the paper-clip slightly you can adjust the spring so that it *just* touches the contact. Wire the contact and paper-clip to a lamp and battery, A, as shown in the figure. The spring and contact are making a switch. Since the spring is just touching the contact, the lamp should glow as soon as you connect battery A. How can you turn the lamp off again?

5 The ends of the nails give the core a rough surface at its ends. To make the end smooth, push a large thumb-tack among the nails. Note that though the tack is plated with some non-ferromagnetic metal or alloy, such as chromium or brass, the tack is made of iron. Fix the coil to the board by forcing it into a lump of modelling clay (Plasticine etc.). Adjust its position so that the thumb-tack in the core is exactly opposite the thumb-tack on the spring. The gap between them should be about a millimetre.

6 Connect the push-button switch and battery B to the coil.

7 Press the button. What happens?

8 How could you alter this circuit so that the lamp is off normally, but is switched on when you press the button? Alter your circuit to make this happen.

The device you have made is a very simple kind of *relay*. You can add contacts to it to make it more complicated. With a contact on either side of the spring you can arrange to switch one lamp off and another lamp on, as you press the button. There are many other ways in which devices can be switched by this relay.

A relay (fig. 9.11) is simply a switch that is operated by an electromagnet. The coil needs only a low-voltage supply (usually between 6 and 12 volts) but, if the contacts are suitably made, they can switch on lamps or motors or other devices operating at much higher power. Relays are often used for switching mains-powered lamps, the motors of elevators and the resistors used in controlling the motors of electric locomotives. In a telephone exchange we find very complex relays, which are operated by the electric signals coming from the dial on the telephone set. These relays switch the connections between your telephone and that of the person you wish to speak to, depending on the number you dial. Though relays are still widely used, they are gradually being replaced nowadays by electronic circuits which do the same job more efficiently. Modern telephone exchanges work by electronic switching.

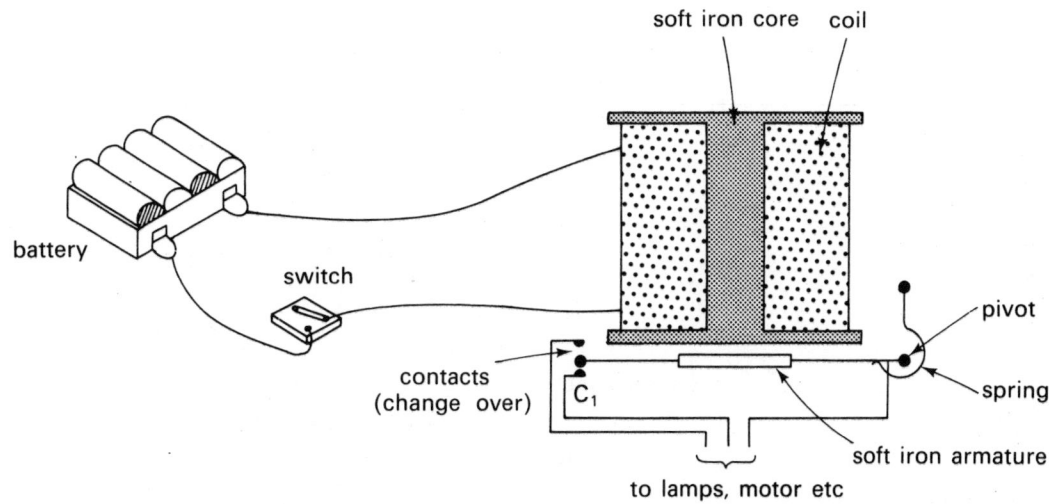

Fig. 9.11 How a relay works: C_1 is a pair of contacts which can be used for switching two separate circuits.

Instructions: making a relay operate itself

1 Make the relay as in fig. 9.10, but connect it so that the relay switch is in the circuit that supplies power to the coil (fig. 9.12).

2 Press the button. Listen for a buzzing sound. You may need to adjust the pressure of the spring against the contact. You may also need to adjust the size of the gap between the two thumb-tacks. The buzzing sound may last only a few seconds. If you look closely you will see sparks between the contact and spring; after a short while their surfaces become oxidised by the sparking and good electrical contact is impossible. When this happens, scrape the contact and spring to clean them.

3 Explain what causes the buzzing sound.

The device you have made is like the electric buzzers that are used in homes and offices. If there was a hammer attached to the spring, and a gong for the hammer to strike, this would make an electric bell. This is another useful signalling device (fig. 9.13). In most types there are two electromagnets wired so that one produces a north pole at its end nearer the spring, and the other produces a south pole. Their other ends are joined by a strip of soft iron, called a yoke. In the diagram the two magnets and yoke are drawn all in one piece to make the drawing simpler. Examine an electric bell to see the details of the way it is made. Another strip of soft iron, the armature, is fixed to

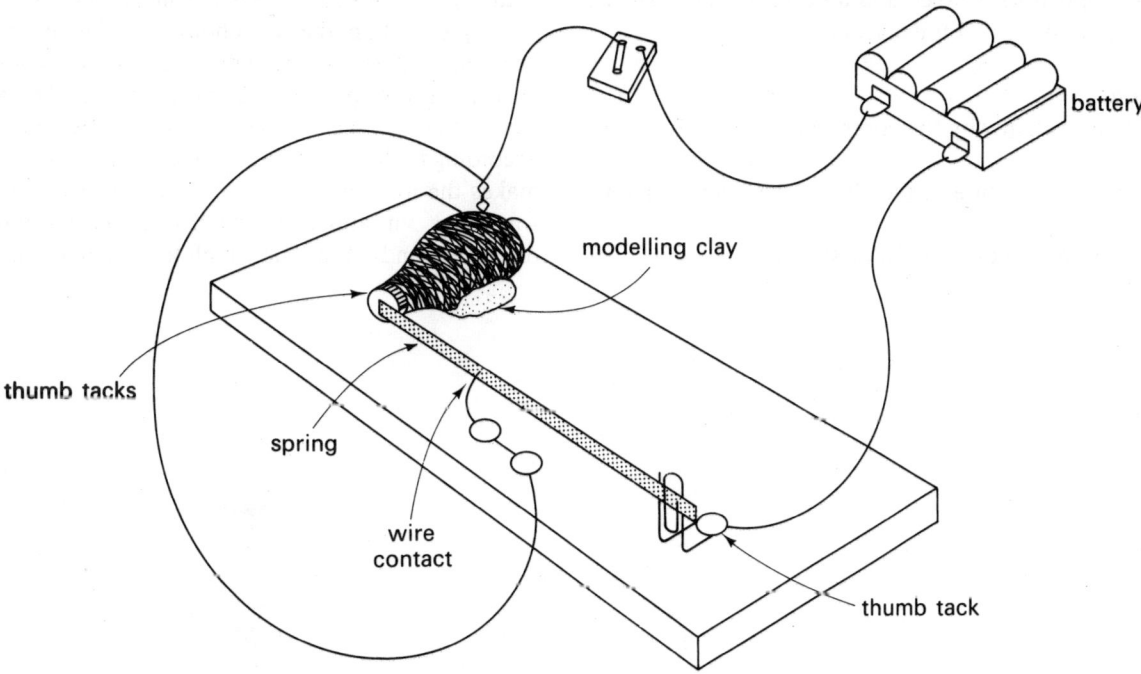

Fig. 9.12 How to connect the relay so that it operates itself.

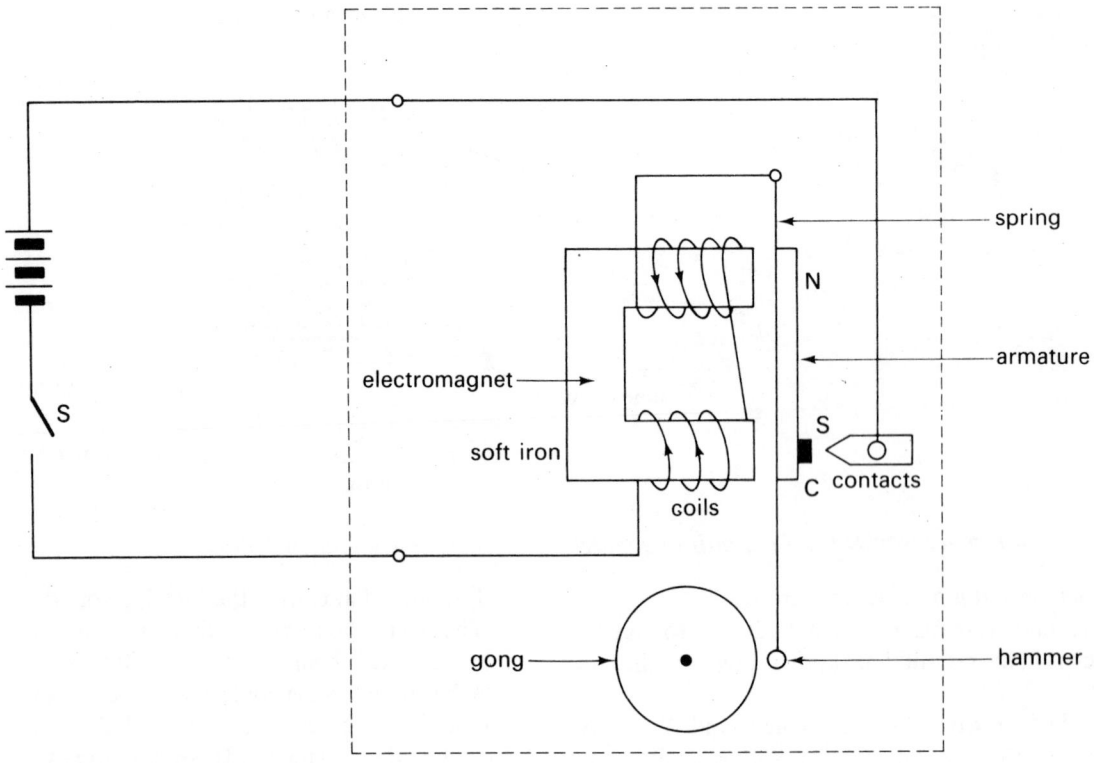

Fig. 9.13   Diagram of an electric bell.

the spring. The coils have many turns and the magnetic field is much stronger than that of your buzzer, so a bell can make a very loud noise. The contact consists of a pillar with a screw that can be adjusted to make contact with the armature at just the right point for best operation. The contact is tipped with silver, and touches against a small silver contact disc, C, on the armature.

Questions

1 Can you explain why the cores of the coils are joined by a yoke?

2 Why is the armature in the electric bell made of soft iron?

3 Why are the contacts made of silver?

If we vary the strength of the current flowing through an electromagnet coil, the strength of the magnetic field varies too. In a telephone earphone there is a small electromagnet. Close to its poles is a very thin sheet of iron. The current flowing along the telephone wires is made to vary in strength when somebody speaks into the mouthpiece of another telephone set. The current varies according to the sound vibrations of the voice. This varying current passes through the coil of the earphone electromagnet. The magnetic field varies in strength according to the vibrations produced by the voice and this makes the iron sheet, or diaphragm, vibrate in the same way. The vibrations of the iron sheet reproduce the original sound vibrations which the listener can hear.

10 Investigating motion and momentum

Momentum is a property of all moving objects. Objects that are not moving have no momentum. Objects that are very light or objects that are moving very slowly (such as a crawling snail or a feathery seed carried in the wind) have very small momentum. Objects that are very heavy or objects that are moving very rapidly (such as a walking rhinoceros or a bullet fired from a gun) have great momentum. You can think of momentum as being related to the force required to stop the object from moving. It is now necessary to make the definition of momentum more precise.

The momentum of an object depends on only two factors:
(a) its velocity
(b) its mass.
We have not yet defined either of these terms. Before we can understand what is meant by momentum, we must first discuss precisely what we mean by velocity and by mass.

10.00 The mathematics of motion

You have probably already heard the word 'velocity' and know that it is something to do with 'speed'. We will begin by discussing speed.

10.01 Speed

Uniform speed is easy to define. If an object moves in a straight line and travels equal distances in equal intervals of time, it has uniform (or constant) speed. We measure its speed by measuring the distance travelled and the time taken. If the distance is measured in metres and the time in seconds, and if we divide the *number of* metres by the *number of* seconds, we obtain another *number*, which is

its speed in metres per second. For example, an object travels at uniform speed for 2 s and the distance travelled is 50 m. Its speed, u, is calculated like this:

$$u = \frac{50}{2} = 25 \, \text{m/s}.$$

The unit of speed, m/s, is a derived SI unit.

We can also measure the speed of an object that is not travelling in a straight line. It can be travelling in a circle or along some irregular zigzag path. As long as we measure the distance travelled *along the path*, we can calculate its speed using the same equation. For example, a bee flies from flower to flower in a very irregular path, and travels a total distance of 20 m *as measured along the path it has flown*, in 9.6 s. Its speed, u, is given by

$$u = \frac{20}{9.6} = 2.08 \, \text{m/s}.$$

A racing car takes 2 min to complete one lap of a racing track. The distance of one lap, measured along the track, is 6 km. Its speed is

$$u = \frac{6}{2} = 3 \, \text{km/min}$$

or $u = 3 \times 60 = 180 \, \text{km/h}$

or $u = \frac{3}{60} \times 1\,000 = 50 \, \text{m/s}.$

Suppose that an object is travelling along a path and we measure its distance from one end of the path every second. The results are shown in a table on the following page.
Do not confuse the letter s, which is the symbol for the SI unit of time, the *second*, with the letter *s* which is the frequently used symbol for the physical quantity of *distance* travelled by a moving object. Care should be taken to distinguish between the two.

Time taken t in seconds	0	1	2	3	4	5	6	7	8	9	10
Distance travelled s in metres	0	15	30	45	60	75	90	105	120	135	150

Simply by looking at this table we can tell that the motion of the object is uniform. The distance travelled in each second is 15 m. A graph to show the relation between these two sets of numbers is called a *distance-time graph* (fig. 10.1, curve *a*). The graph is a straight line, indicating that speed is constant or uniform. We have measured the distances and times on only 11 occasions during the journey of the object. These 11 instants are represented by the 11 points on the graph. The fact that these 11 points lie on a straight line does not in itself, *prove* that speed is uniform. Between one instant and the next speed may have increased and then decreased, making the graph a wavy line which still passes through the 11 points we have plotted. This would be non-uniform speed. There is no evidence to suggest that this has happened, so we *assume* that speed is uniform and we base all our further calculations on this assumption. If we have any reason to think that speed is not uniform (for example, if we know that the moving object is an automobile moving along a rocky,

muddy track) we could check for uniformity by taking measurements more often, say, every tenth of a second. Then we would have a more precise graph of the motion, with 101 points on it. In the discussion which follows we shall assume uniformity of speed.

When speed is uniform we can calculate it by using the distance and time for any part of the journey. During the first second, distance travelled is 15 m and time taken is 1 s, so $u = 15/1 = 15$ m/s. Between the end of the 4th second and the end of the 7th second it travels 45 m, so its speed is $u = 45/3 = 15$ m/s. Look at fig. 10.1. When we divide the length of the line AB by the length of BC, we are also dividing a number representing a distance by a number representing a time. We call this the *gradient* of the curve. The curve is a straight line, so its gradient is the same all the way along its length. This explains why we always obtain the value 15 m/s when we calculate the speed of the object. This leads us to a useful rule: *the gradient of a distance-time graph tells us the speed.*

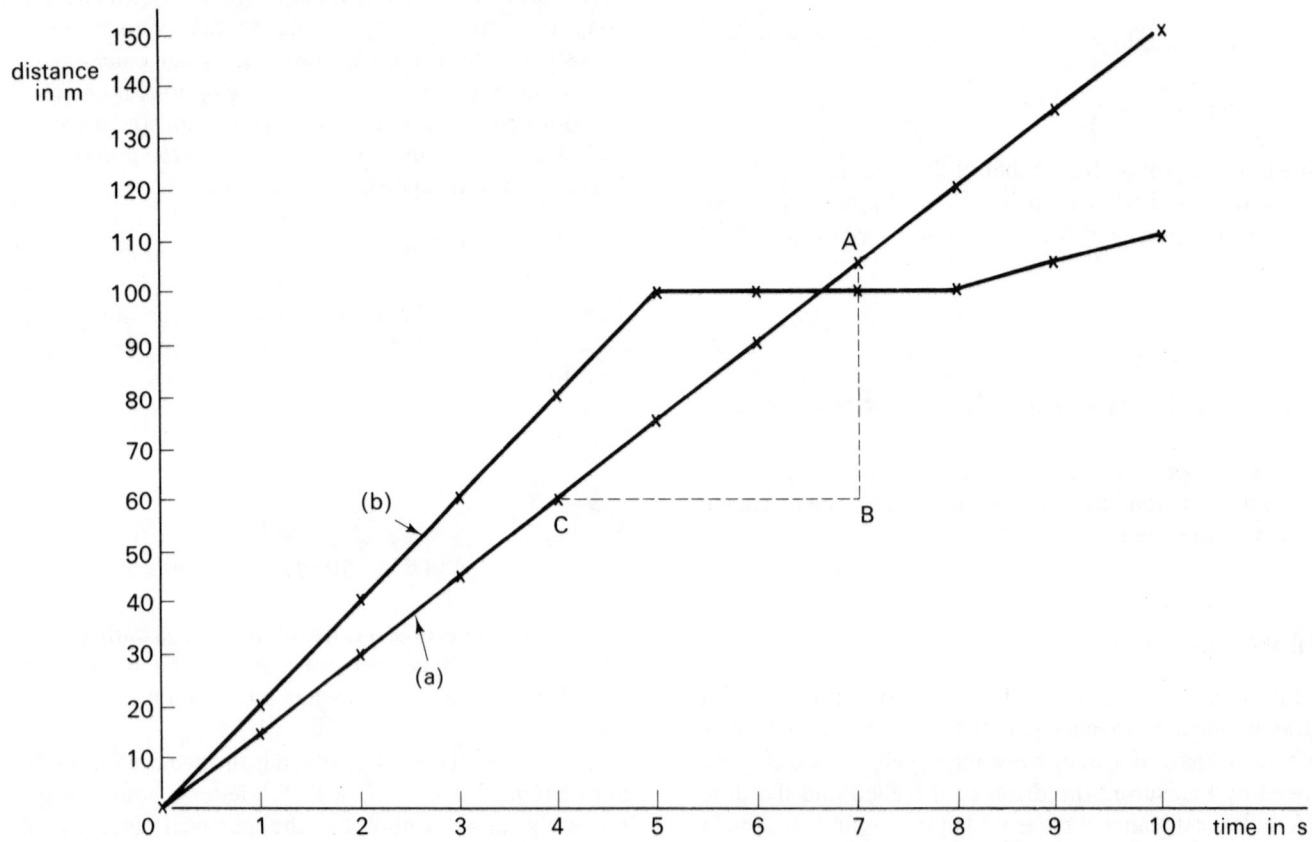

Fig. 10.1 *Distance-time graphs.*

Here is another set of results:

Time taken t in seconds	0	1	2	3	4	5	6	7	8	9	10
Distance travelled s in metres	0	20	40	60	80	100	100	100	100	105	110

This is plotted as curve b in fig. 10.1. Until the end of the 5th second, the object is travelling at uniform speed (straight line), which we can calculate to be 20m/s. This speed is greater than that of the previous object; we can see that during the first 5 seconds the gradient of the graph is steeper. During the next 3 seconds the object remains at the same distance from its starting point. It is not moving, its speed is zero. The gradient of the graph is zero; the line runs parallel with the t-axis. For the final 2 seconds the object moves with a uniform speed of 5m/s. This is less than the speed of the previous object. We can see that the gradient of the graph is less. At each stage of its journey the speed of the object is represented by the gradient of the distance-time graph.

From the numbers in the tables above we can also calculate the *average speed*. The first object moved at uniform speed for its whole journey, so its average speed is the same as its uniform speed, 15m/s. For the second object the average speed for the whole journey is calculated like this:

$$\text{average speed, } u = \frac{\text{total number of metres travelled}}{\text{total number of seconds taken}} = \frac{110}{10} = 11 \text{ m/s.}$$

Though its average speed is 11m/s, the object never at any time travelled at this speed. The uniform speeds for its journey were 20m/s, 0m/s and 5m/s. But an object which started from the same point at the same time and travelled with a uniform speed of 11m/s would have covered an equal distance in 10s.

Question

The motion of an object is recorded in the table shown below.

Plot a distance-time graph of these numbers. Describe the motion of the object, giving its speed at different stages of its journey. What is its average speed for (a) the first 5 seconds of the journey, and (b) for the whole journey?

Speed can be measured in another way. In an automobile we have a speedometer, made to indicate the speed of the vehicle at any instant. At uniform speed the needle of the speedometer stays at the same point on the scale. We can read the speedometer and so measure the speed at intervals of one second or at any other convenient interval during the journey. If the object of the table on page 92 had a speedometer, the reading would have been 15m/s for the whole journey. A graph of the speedometer reading is shown in fig. 10.2, curve a. This is called a *speed-time graph* and since speed is uniform, the curve is a straight line parallel with the t-axis. Between this curve and the t-axis is a rectangular area, shaded in the figure. The object travelled for 10s at a speed of 15m/s, covering a total distance 150m. To calculate the distance we have multiplied the number representing time (10) by the number representing speed (15) and obtained a third number ($10 \times 15 = 150$), which represents distance travelled. On the graph we can multiply a number representing the length of the shaded rectangle by a number representing the breadth of the rectangle, obtaining a third number representing the *area* of the rectangle. This leads us to another useful rule: *the area beneath the curve of a speed-time graph tells us the distance covered.*

Figure 10.2, curve b, shows the speed-time curve for the object of the table above. If this had a speedometer it would indicate 20m/s for the first 5s, then drop instantly to 0m/s for 3s, and finally rise to 5m/s for 2s. We can use the 'area beneath the curve of the graph' rule to find out how far the object travels. In the first 5s the distance travelled is:

$$s_1 = 20 \times 5 = 100\text{m.}$$

For the next 3 seconds the object is stationary. For the final 2 seconds, the distance travelled is:

$$s_2 = 5 \times 2 = 10\text{m.}$$

The total distance is 110m, which agrees with the value in the table.

Time taken t in seconds	0	1	2	3	4	5	6	7	8	9	10
Distance travelled s in metres	0	5	10	15	20	30	40	50	50	52	54

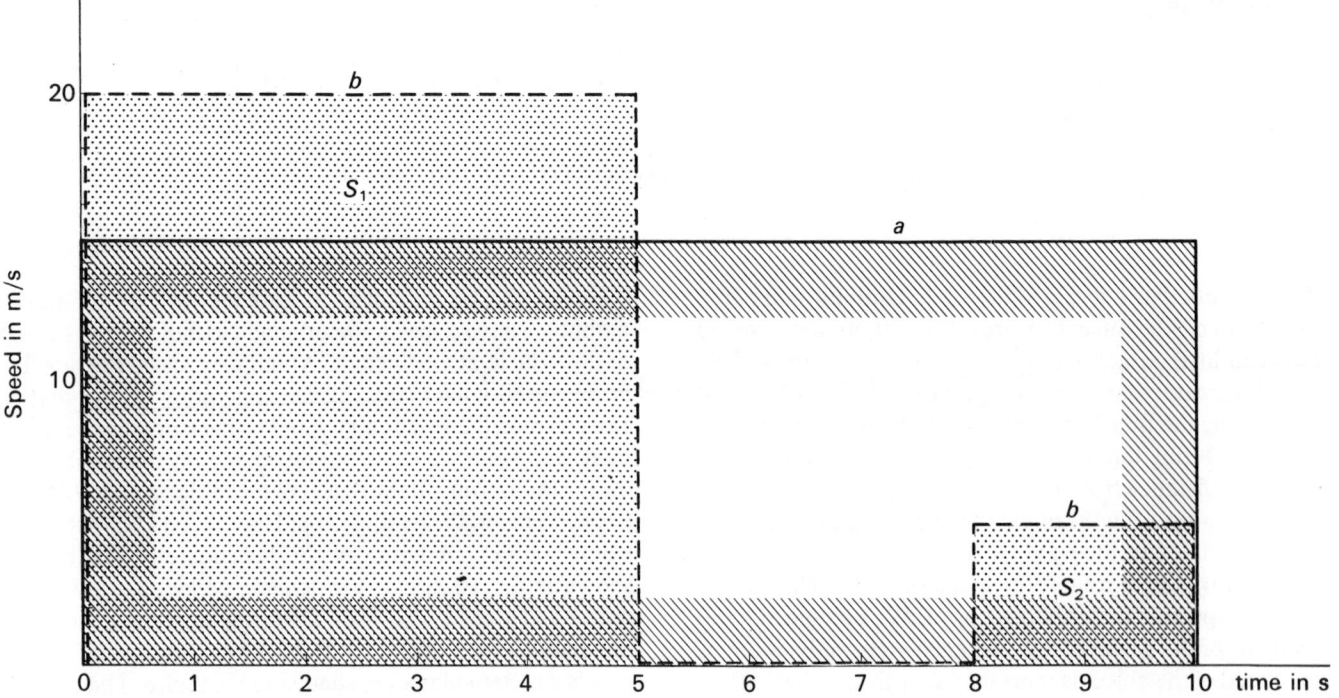

Fig. 10.2 Speed-time graphs.

Question

Plot a speed-time graph for the object of the table at the bottom of page 93. Calculate the areas under the curve and find the total distance travelled.

Unlike the moving objects we have studied so far, real objects do not change speed instantaneously. An automobile, a locomotive or a bullet change speed gradually. Their distance-time graphs do not have sharp corners (like fig. 10.1, curve *b*) and their speed-time graphs are never parallel to the *u*-axis (like parts of the curves of fig. 10.2).

10.02 Scalar and vector quantities

Quantities such as speed and distance can be stated by writing a number and a unit. These are called *scalar quantities*. Other scalar quantities that you have used are time, temperature, volume, pressure and density. They have no particular direction associated with them. In fig. 10.3a we see the path taken by a boy walking on level ground. First he walks 3 m east, then he turns to walk 4 m north. He covers a total distance of 3 + 4 = 7 m in walking from O to A and then to B. However, if we measure how far he is from his starting point, we find that his distance measured in a *straight line* is only 5 m. We call this distance his *displacement.*

To specify his displacement we also have to state the direction of the point he has reached in relation to the point he started out from. The direction is 36°52′ north of east. To specify his displacement we must state *two* quantities –

the distance *and the direction*. Thus his displacement from O is 5 m, 36°52′ north of east. Such a quantity as displacement, that can be stated only by giving both its value and its direction, is known as a *vector quantity*. Some vector quantities that we have studied are friction, magnetic force and weight (which has downward direction). In diagrams we can represent vectors by arrows; often the length of the arrow represents the size (or magnitude) of the vector, and the direction of the arrow corresponds to the direction of the vector.

The value and direction of a vector depend only on the original and final starting points. In fig. 10.3b, the boy wanders from O to A, covering a distance of 4 m, and then wanders from A to B, covering a distance of 6 m. The total distance travelled is 4 + 6 = 10 m, but his displacement is still 5 m 36°52′ north of east.

To further illustrate the difference between distance and displacement we can draw graphs to represent a girl walking at a steady speed of 2 m/s from X to Y and back again to X, see fig. 10.4.

It takes the girl 4 seconds to reach Y and a further 4 seconds to return to X. If we plot a distance-time graph and a displacement-time graph for this journey we get very different results, see fig. 10.5a and fig. 10.5b. The distance travelled by the girl is 16 m as shown in fig. 10.5a and the displacement is zero as shown by fig. 10.5b. Note that the gradient of the distance-time graph gives a speed of 2 m/s but the displacement-time graph gives a velocity of +2 m/s (from X to Y) in the first 4 seconds and a velocity of –2 m/s (from Y to X) in the next 4 seconds.

Fig. 10.3   *Illustrating the differences between distance and displacement (see text).*

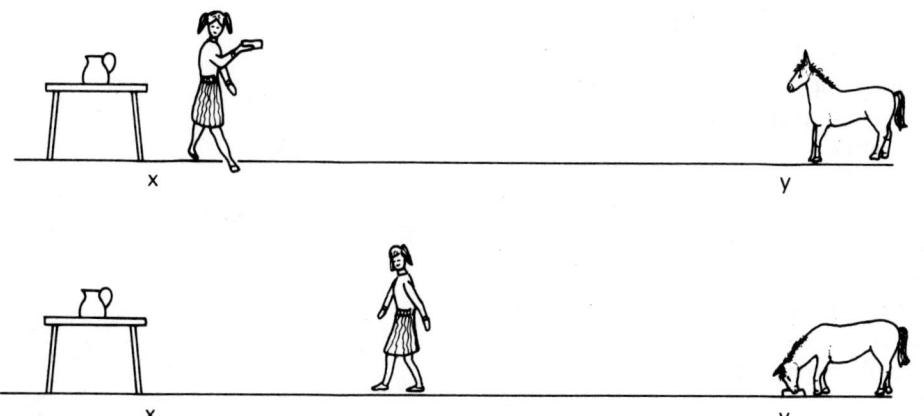

Fig. 10.4   *A journey resulting in zero displacement.*

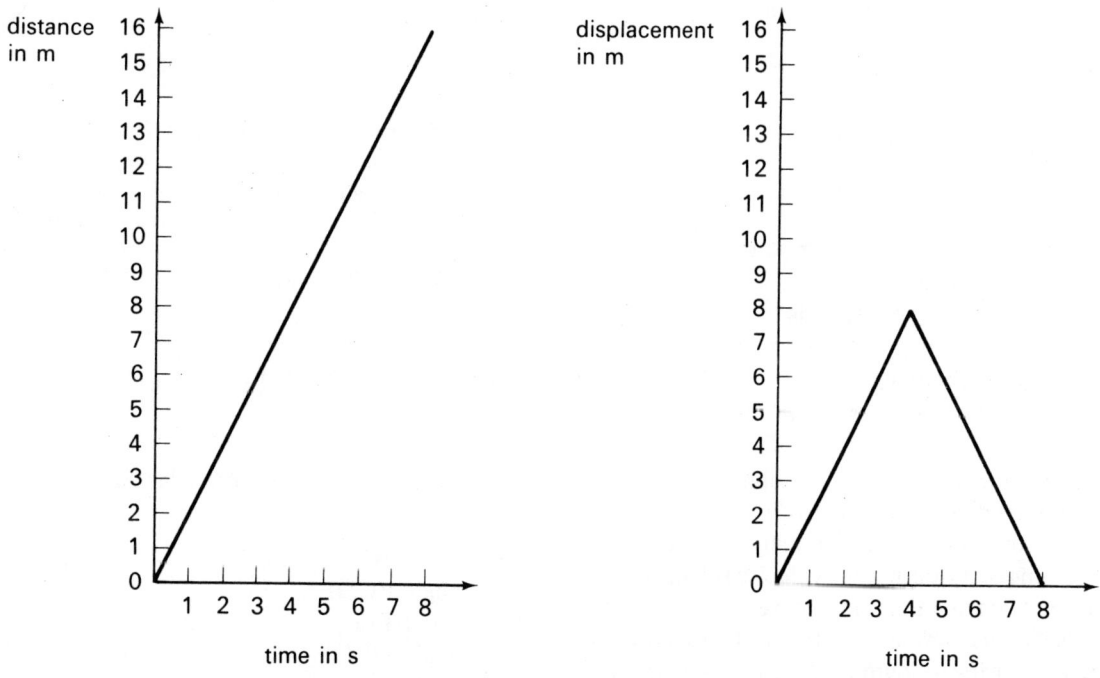

Fig. 10.5   *(a) The distance-time graph (b) the displacement-time graph of the journey of Fig. 10.4.*

10.03 Velocity

Like displacement, velocity is a vector quantity; both magnitude and direction must be stated. Velocity is the rate of change of *displacement* while speed is the rate of change of *distance*. To show the difference between speed and velocity we can study the motion of a ball projected vertically into the air with a velocity of 50m/s. The ball rises for 5 seconds, stops and then falls downwards to reach 50m/s in a further 5 seconds. Figure 10.6a and fig. 10.6b show graphs of speed-time and velocity-time for a 10 second interval. In fig. 10.6a the speed decreases to zero after 5 seconds and then increases to 50m/s again in a further 5 seconds. Note we do not concern ourselves with whether the ball is moving up or down. However, in fig. 10.6b the velocity decreases to zero after 5 seconds and then decreases to −50m/s in a further 5 seconds. The minus sign is to show that the ball is moving downwards because we have chosen upwards to be positive. From the speed-time graph we can find the distance travelled by finding the area under the curve. Area $= \frac{1}{2} \times 5 \times 50 + \frac{1}{2} \times 5 \times 50 = 125 + 125 = 250$. We know that the ball has travelled 125m upwards and then 125m downwards for a total distance of 250m. Also from the gradient of the lines we see that in the first 5 seconds the speed is decreasing at the rate of −10m/s² and during the next 5 seconds the speed is increasing at the rate of +10m/s². Again we ignore the direction of motion when dealing with speed.

From the velocity-time graph we can find the displacement by finding the area under the curve. Area $= \frac{1}{2} \times 5 \times (+50) + \frac{1}{2} \times 5 \times (−50) = \frac{1}{2} \times 5 \times 50 − \frac{1}{2} \times 5 \times 50 = 0$. The ball has returned to its original position. It has no displacement.

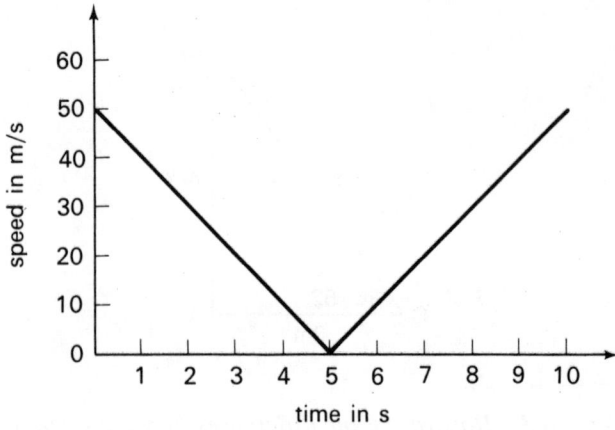

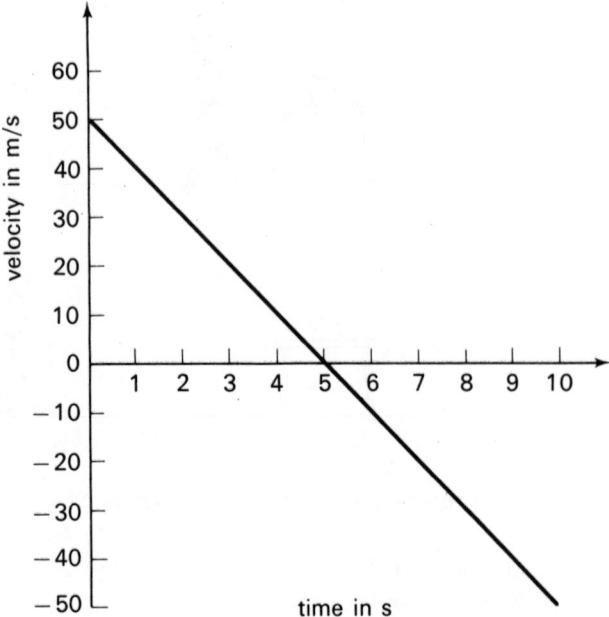

Fig. 10.6 (a) The speed-time graph (b) The velocity-time graph of the motion of a ball thrown vertically into the air.

10.04 Acceleration

Look at this data:

Time taken t in seconds	0 1 2 3 4 5 6 7 8 9 10
Velocity u in metres per second	0 5 10 15 20 20 20 17.5 15 12.5 10

The object, which could be an automobile, starts from rest. Its velocity gradually increases to 20m/s. It remains uniform for 2s, then decreases gradually. Ten seconds after the beginning of its journey the velocity is 10m/s; it is still moving, but from that time onward we have no further data. A velocity-time graph (fig. 10.7) shows how its velocity changes during its journey. We can now calculate the displacement (distance travelled in a straight line in a specified direction). We think of the area below the curve as a collection of shapes : triangles and rectangles. Their areas are:

triangle ABC	$\frac{1}{2} \times 4 \times 20 = 40$
rectangle BCDE	$2 \times 20 = 40$
triangle EFG	$\frac{1}{2} \times 4 \times 10 = 20$
rectangle DFGH	$4 \times 10 = 40$

The total area = 140. Therefore the total displacement (distance travelled in a specified direction) = 140m.

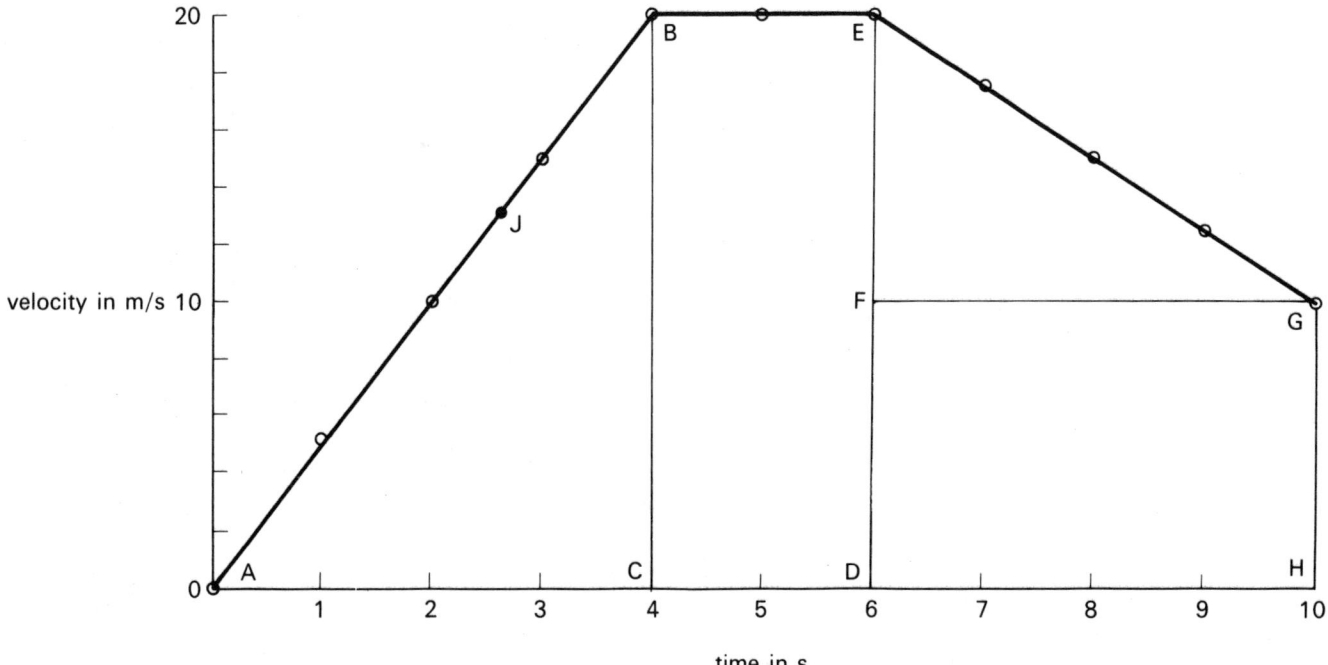

Fig. 10.7 A velocity graph of the numbers from table on page 96.

During the first 4s the velocity of the object increases; it is *accelerating*. During each second its velocity increases by an equal amount, by 5m/s. We call this *uniform acceleration*. We calculate uniform acceleration by dividing the number of metres per second of velocity increase by the number of seconds during which it increases. In this example the velocity increased from 0m/s to 20m/s, an increase of 20m/s. This took 4s. Therefore acceleration $a = 20/4 = 5$ metres per second per second. For clarity and to save writing 'per second' twice, we write the symbol for acceleration as 'm/s²'. We say it as 'metres per second per second' or 'metres per second squared'. In calculating acceleration we have divided a number representing increase in velocity by a number representing time. On the graph we can divide a number representing the length BC by a number representing the length AC. This gives us the gradient of the velocity-time graph. This leads us to the third useful rule: *the gradient of a velocity-time graph tells us acceleration.*

In fig. 10.7, the curve slopes upwards from A to B and is a straight line, showing a uniform gradient and therefore uniform acceleration. From B to E the gradient is zero: acceleration is zero; velocity is uniform. From E to G the gradient is negative; its value is calculated as follows:

$$\text{gradient} = \frac{EF}{FG} = \frac{-10}{4} = -2.5.$$

This represents an acceleration of $-2.5\,\text{m/s}^2$. Velocity decreases at the rate of 2.5m/s every second. This negative acceleration is often called *deceleration* or *retardation*.

Question

Calculate separately the areas beneath the curve of fig. 10.7 for each second of the journey. Make a table to show the displacement during each second of the journey. Find the total displacement.

Figure 10.8 is a displacement-time graph of the motion of fig. 10.7, as calculated in the question above. From A to B the displacement increases and the curve becomes gradually steeper. This shows that velocity is not uniform. The object is accelerating. From B to E the curve is straight, indicating uniform velocity and zero acceleration. From E to G the gradient gradually decreases, showing decreasing velocity and deceleration. By examining a displacement-time graph in this way we can determine the displacement (distance travelled in a given direction) at any instant. For example we can say that after 7.2s the object has a displacement of 103m (point H) that is, the object is 103m from its starting position in some known direction. We can measure the gradient of the graph and calculate the velocity. Along straight parts of the curve this is simple. For example, along the part BE the gradient

is $\dfrac{EJ}{BJ} = \dfrac{40}{2} = 20$. This represents a velocity of 20m/s.

Along parts of the curve that are not straight, velocity is changing continuously. To find the velocity at any instant we draw a tangent to the curve. For example, to find the velocity when $t = 2.6$s, we draw a tangent LM to the curve at point K, for which $t = 2.6$. Completing triangle LMN, we measure its height and base, using the correct scale of

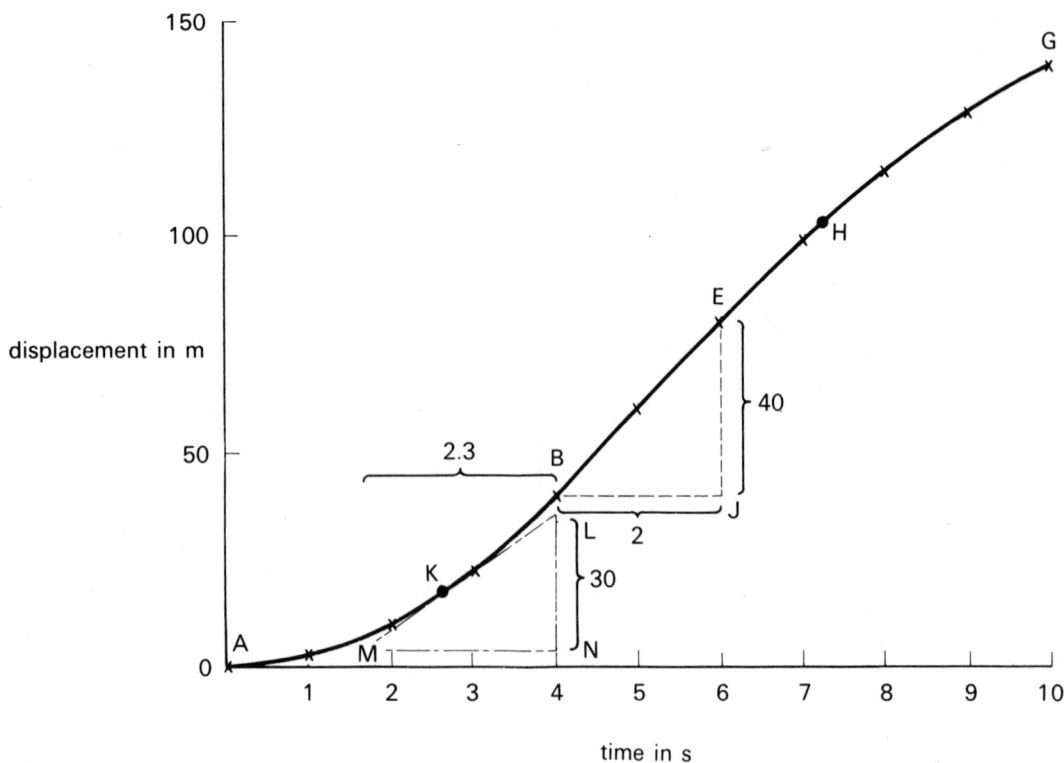

Fig. 10.8   Displacement-time graph of numbers derived from Fig. 10.7.

units of each axis. The gradient of the tangent is 30/2.3 =
13. This tells us that at 2.6 s the velocity of the object is
13 m/s. This result agrees with a value taken from the
velocity-time graph, fig. 10.7, point J. Thus, you can
obtain velocity from either graph by using the appropriate
method. The precision of your results may depend a lot on
your skill at drawing, as when you have to plot a curve and
then draw a tangent to it, but with problems involving
uniform velocity or acceleration it is usually possible to
avoid plotting an accurate graph. Instead the graph may be
sketched as a simple line diagram and gradients and areas
may be calculated directly from the sketch. In most of the
questions which follow, a simple sketch is all that is
needed.

Questions

1 An antelope starts from rest and accelerates uniformly
in a straight line. It reaches a velocity of 20 m/s after it has
travelled 180 m. Find its acceleration and the time it takes
to cover this distance.

2 A train travelling at 60 m/s decelerates to 40 m/s.
During the period of deceleration its displacement is 2 km.
How long does deceleration take and what is the rate of
deceleration?

3 An automobile passes the speed limit sign at the edge of
a town, when it is travelling at 10 m/s. It accelerates at
1 m/s² for 15 s. Then it travels at uniform velocity for 3
minutes. Next the driver sees a level-crossing gate 500 m

ahead. He decelerates uniformly, bringing the automobile
to rest at the gate. What is the highest velocity reached by
the automobile? What is the time spent decelerating?
What is the displacement (distance) between the speed
limit and the level-crossing gate? What is the total time of
the journey? What is the average velocity for the whole
journey?

4 A butterfly flies beside a fence, the posts of which are
1 m apart. The time at which it passes each post is recorded
in the following table:

Post number (displacement) s in metres	0	1	2	3	4	5
Time t in seconds	0	2	4.5	6.25	8.35	10.4

Plot a displacement-time graph. Is the butterfly's velocity
uniform? During which part of the journey is its velocity
greatest? What is its greatest velocity? Is the acceleration
uniform over the whole journey?

5 A rocket is fired vertically upwards. Its displacement
above the Earth is measured every 5 s. At first it is
accelerated uniformly by its motors. After a period the
motors cut out and the rocket decelerates uniformly for the
remainder of its flight. This is recorded as follows:

Time t in seconds	0	5	10	15	20	25	30	35	40
Displacement s in metres	0	125	500	1 125	2 000	3 125	4 300	5 325	6 200

Plot a displacement-time graph of these numbers. Estimate the time at which the motors cut out. What was the velocity of the rocket at that time? By drawing a *sketch* of the velocity-time graph, calculate the acceleration before and after the motors cut out.

Acceleration is the rate of change of velocity. It has size (or magnitude); it also has direction. If the acceleration of an object has the same direction as the velocity the velocity of the object increases – it accelerates. If the acceleration has exactly the opposite direction, the velocity of the object decreases – it decelerates. In either case the direction of the velocity is unchanged. However, if the acceleration is at an angle to the velocity, both the size and the direction of the velocity may be changed. To study what happens we need to know how to combine two or more vector quantities. Scalar quantities can be added together or subtracted, using ordinary arithmetical rules. For example, if a man is walking *along* the roof of a railway carriage with a speed of 2m/s, while the train is moving at 50m/s, the man's speed relative to the track is either 52m/s or 48m/s depending on whether he is walking in the same direction as the train is moving, or in the opposite direction. We may add or subtract the two speeds, since both are directed along the railway track and we do not need to be concerned about directions at angles to the track. But if the man walks *across* the roof in a direction that is at an angle to the line along which the train is moving, we must take directions into account. We cannot simply add or subtract the directions of the *velocities*. In fig. 10.3a, we were able to add the displacement OA to the displacement AB and obtain the displacement OB, even though OA and AB had different directions. By drawing a scale diagram such as this we were able to add two vector quantities. This is a fairly easy way to add vector quantities, and we shall make more use of this method later.

10.05 Equations of motion

To solve problems connected with motion it is generally easier to use plotted graphs or simple sketches of graphs. However, some people prefer to use mathematical equations. We can deduce these from the velocity-time graph, fig. 10.9. This shows the velocity of an object which is travelling at velocity u, and accelerates *uniformly* to velocity v in a period t. Its acceleration a, is represented by the gradient of the curve, which numerically is AB/BC.

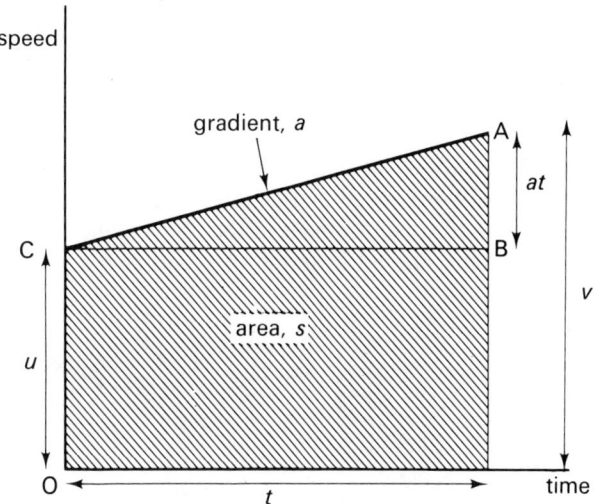

Fig. 10.9 A velocity-time graph illustrating the equations of motion.

Since BC = t, we can write:

$a = $ AB/t or AB $= at$

From this result we can obtain the first useful equation:

$v = u + at$ (the velocity-time equation).

If the object had started from rest we would have put $u = 0$, so that the equation would simplify to $v = at$.

The displacement (distance travelled in a straight line) of the object is represented by the area under the curve. This is the sum of the area of the rectangle ($= ut$) and the area of the triangle ($= \frac{1}{2} \times at \times t = \frac{1}{2}at^2$). From this we obtain the second useful equation:

$s = ut + \frac{1}{2}at^2$ (the displacement-time equation).

Another way of looking at the area under the curve is to think of it as a trapezium; the area of this is given by (base × average height):

$s = \dfrac{(u + v)\,t}{2}$ (alternative displacement-time equation).

The velocity-time equation can be re-written:

$t = \dfrac{v - u}{a}$

The alternative displacement-time equation can be re-written:

$t = \dfrac{2s}{u + v}$

99

Equating the right-hand sides of these two expressions for t gives:

$$\frac{v - u}{a} = \frac{2s}{u + v}$$

or $(v - u)(v + u) = 2as$.

Simplifying $v^2 - u^2 = 2as$.

Rearranging terms gives $v^2 = u^2 + 2as$ (the velocity-displacement equation).

Depending on what information you are given and what quantities you are asked to calculate you can use one or more of the above equations for solving problems on motion. The word speed can be substituted for velocity and distance for displacement in the above equations to solve problems quoted in terms of speed and distance.

Remember that the equations apply only to *uniformly* accelerated motion. If acceleration is not uniform, as in question 4, page 98, the equations cannot be used. To practise the use of these equations, try re-working some of the questions on page 98.

You have learned a lot about speed and velocity and how to use graphs or equations to solve problems about motion. Now it is time to put some of this knowledge to practical use. Try the following investigations and some of the projects at the end of this chapter.

Instructions: investigating the motion of a trolley
1 Place a trolley at the top of a slope and attach about 2 m of tickertape to it. Thread the end of this tape through the ticker timer and hold the trolley, ready to start (fig. 10.10a).

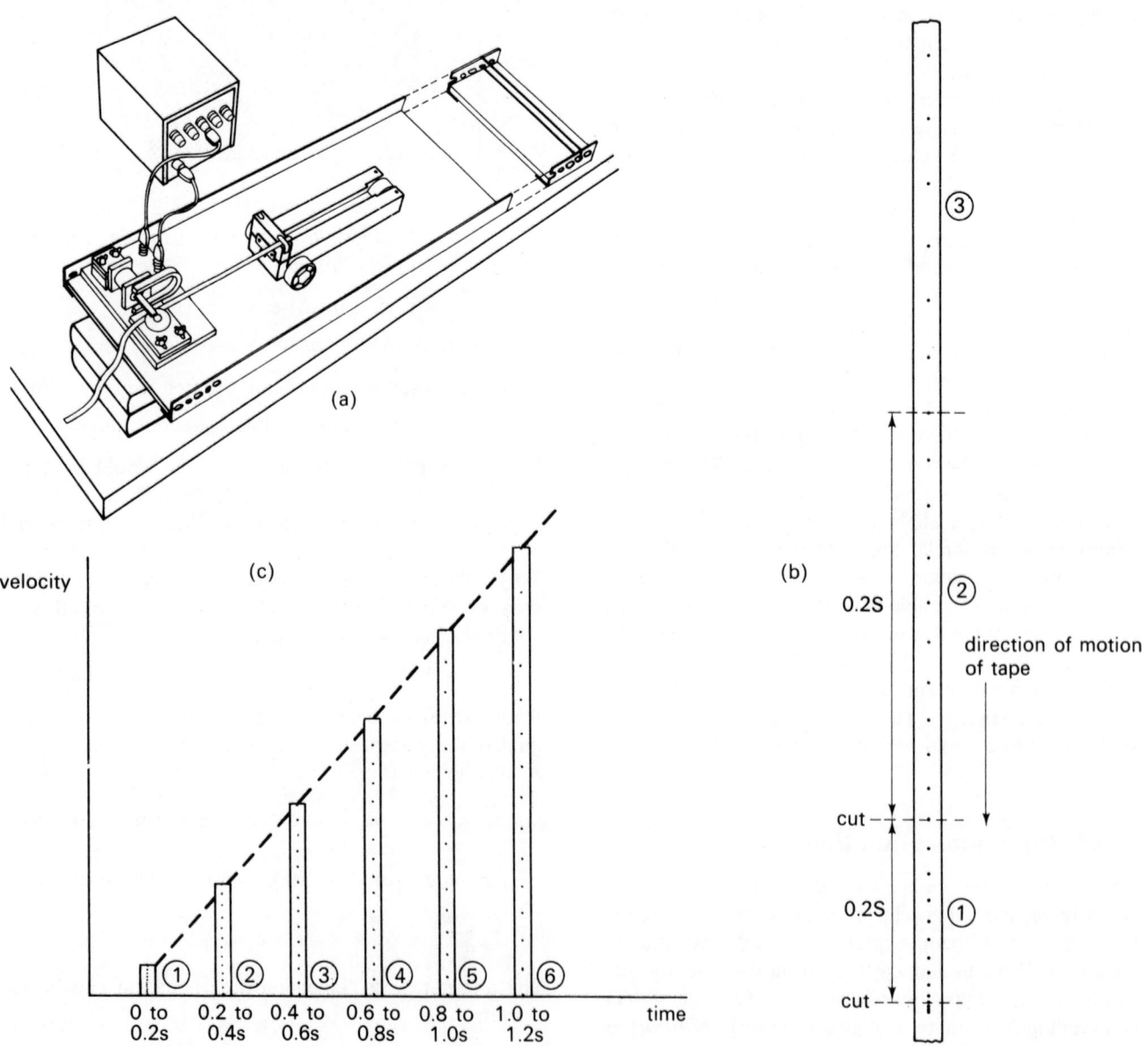

Fig. 10.10   Investigating the motion of a trolley.

100

2 Switch on the timer; let go of the trolley. When it reaches the bottom of the slope stop it and switch off the timer.

3 If the frequency of the timer is 50 Hz, the displacement (distance) between one dot and the next dot shows how far the trolley has travelled in 0.02 s. Cut the tape across the dots, every tenth dot, as in fig. 10.10b. Each piece of tape shows the displacement (distance travelled) in 0.2 s. Stick these strips, in order, on a sheet of paper to make a velocity-time graph. (fig. 10.10c). You may ignore the beginning and end of the tape, because at the beginning the dots are too crowded together and at the end there may be irregularities due to your stopping the trolley.

4 Measure the length, l, of each piece of tape and record the results in a table:

Piece no.	Length l in mm	Average Velocity u in m/s
1 2 3		

5 Calculate the average velocity corresponding to each piece of tape.

6 Calculate the acceleration of the trolley.

7 Repeat steps 1 to 6, but first make the slope steeper. What effect does this have on the acceleration of the trolley?

8 Repeat steps 1 to 6, but begin with the trolley at the bottom of the slope. Give it a push to start it moving up the slope. Find the deceleration of the trolley.

9 Make the slope exactly horizontal; push the trolley to start it and find its deceleration. What causes it to decelerate? Supposing that it does not come to rest before it reaches the end of the runway, calculate how far it would travel before stopping, if the runway was long enough.

Instructions: investigating uniform motion of a trolley

1 Arrange the trolley as before but on a slight downward slope.

2 Run the trolley several times and adjust the angle of slope until the trolley runs with zero acceleration, once it has been given a slight push to start it moving. What forces are acting on the trolley to make its acceleration zero?

10.10 Mass and momentum

10.11 Mass

The mass of a body is the quantity of matter it contains. We can roughly compare the mass of two objects if we think about what happens when we push them. Imagine a small, smooth rubber ball on a smooth table. A gentle push with a finger sets it rolling across the table at high velocity. Now imagine a large steel ball about half a metre in diameter resting on a hard smooth floor; you are asked to move it. It takes as much effort as your body can produce to start it moving, and even then it moves only slowly. The rubber ball offers little resistance when we try to move it. We say that it has low *inertia*. This is because it has low mass. The steel ball offers very great resistance. It has high inertia. This is because it has much mass. The same effect is noticed if the balls are moving and we try to stop them. It is very difficult to stop the steel ball when it is rolling because of its great inertia; this is because of its great mass.

The SI unit of mass is the kilogram. This is the mass of a block of a special alloy of platinum and iridium that is kept near Paris, France. This block is called the international prototype kilogram. All objects that have a mass of one kilogram contain the same quantity of matter as this block.

In everyday life, for convenience we sometimes use the kilogram as a unit of weight, but in physics we must remember that mass and weight are not the same thing. The difference is important.

Weight is a force, the force by which the Earth attracts an object, pulling it downwards toward the Earth's centre. Thus weight has *direction*: it is a *vector* quantity. The weight of an object depends on where the object is. Imagine a block of gold with weight 20 N. Although we do not normally say so, we should remember that this force has direction: towards the centre of the Earth. If the block is in Malaya and is then shipped to Brazil, the size of the force may not change but its direction changes, for Malaya and Brazil are almost on opposite sides of the Earth. In addition the force of gravity varies slightly at different parts of the Earth's surface as the result of local variations in the density of the crust; this means that the size of the force varies too as we transport the block of gold to different countries. If we take it to the Moon, its weight becomes much less. The gravitational attraction of the Moon is much less than that of the Earth. On the Moon the same block of gold weighs only about 3.3 N.

The weight of the block varies both in size and direction, depending on where it is, but, wherever it is, its mass remains constant. At any place on Earth or on the Moon or anywhere else in Space, the block contains a fixed amount of gold.

On the Moon we find it easy to lift a heavy rock that we could not lift when on Earth; you may have seen films of astronauts performing such feats. However if we try to roll the rock along a smooth surface, the effort required to start it rolling is *exactly* the same on the Moon as it is on Earth. We have to overcome the same amount of *inertia*. Inertia, and the effort required to overcome it, depends on mass, not on weight. This effect produces strange results. A man who weighs 800 N on Earth weighs only about 133 N on the Moon. You could easily lift him on the Moon, but if you and he were running along on the Moon and bumped into each other, you would be bruised by the impact just as much as if you were on Earth. To you he would feel just as 'heavy'. We should say that he feels just as *massive* on the Moon as he does on the Earth.

It has been mentioned that weight is a force and is therefore a vector quantity. By contrast, mass is a scalar quantity. To state the mass of an object we need say only how many kilograms of matter it consists of. We must next see how we can compare two bodies to find how much matter they each contain. One way was suggested at the beginning of this section. We can compare the masses of the two balls by giving each a push and seeing how quickly they roll. The more massive ball rolls more slowly than the less massive ball. Let us try to be more precise about this. Let us see how fast a trolley goes when we 'give it a push'.

Instructions: finding the velocity of a trolley, when pushed

1 You need a trolley that has a built-in spring-loaded impulse rod. You push the rod in against the force of the spring and secure it in position by a release pin. When the pin is struck, the rod is released and pushes against a heavy block placed on the track. In this way the trolley is 'given a push' of fixed amount and of very short duration. If you do not have this type of trolley, you may have a special catapult for delivering a short sharp force to the trolley, or you may be able to make a catapult, using elastic.

2 First adjust the slope of the runway so that when pushed the trolley runs at uniform velocity.

3 Place the trolley at the top of the slope, attach tape and thread it through the ticker-timer and push in the impulse rod to give its weakest push.

4 Start the timer, pull out the release pin.

5 The marks on the tape show a rapid acceleration as the impulse rod operates. From then on, velocity is uniform. Measure the uniform velocity.

6 Stack another trolley on top of the original trolley. What is the mass of the two trolleys together, compared with the mass of the original trolley?

7 Readjust the slope of the runway so that the stacked trolleys move with uniform velocity.

8 Repeat steps 3 to 5 for the double trolley.

9 Stack a third trolley on top of the other two.

10 Readjust the slope to give uniform velocity.

11 Repeat steps 3 to 5 for the treble trolley.

12 Record your results in a table.

What do you notice about the numbers in the third column?

Instead of stacking trolleys, you may be given two sheets of steel, each of which has the same mass as a trolley. These can be placed on top of the trolley at steps 6 and 9.

Mass (using a 'trolley' as the unit of mass)	Velocity v in m/s	mass $\times$ velocity
1 2 3		

Questions

1 If you had a trolley of mass 1.5 times that of the unit trolley, what would its velocity be when pushed by the impulse rod?

2 How could you use the apparatus of question 1 to measure out a quantity of rice which had a *mass* of exactly 1 kg?

3 How would you use this apparatus to measure out exactly 1 kg (mass) of rice if you were on the Moon?

10.12 Momentum

The quantity 'mass $\times$ velocity', which you calculated in the previous experiment is called *momentum*. The single, double and treble trolleys were all given the same amount of 'push'. They all acquired the same amount of momentum. If two objects are given the same momentum, and we measure their velocities, we can find their relative masses. This is one way of measuring mass.

If the mass of a body is measured in kilograms and its velocity in metres per second in a known direction, the unit of momentum is kilogram metre per second (kg m/s). For example, a man of *mass* 80 kg is walking with velocity 2 m/s north. His momentum is therefore 160 kg m/s north. Momentum is another example of a vector quantity.

Questions

1 Calculate the momentum of a tennis ball, mass 56 g, travelling at 5 m/s south.

2 Calculate the momentum of a bicycle and its rider, total mass 100 kg, travelling at 6 m/s east. Is this greater or less than the momentum of a large truck, mass 10 000 kg travelling at 5 cm/s east?

3 A train, mass 100 000 kg is travelling along a straight track running west. It accelerates at 2 m/s² for 20 s. By how much does its momentum increase?

4 A cricket ball, mass 0.1 kg is rolling along a smooth grassy lawn in a straight line south. It rolls through the grass for a distance of 40 m and comes to rest after 10 s. Calculate how much momentum it has lost while it was rolling.

5 A piece of rock has zero momentum. What can you say about this piece of rock?

10.13 Impulse

Instructions: finding the effects of different amounts of pushing force

1 Set up a trolley as before with its impulse rod in position to give the weakest push.

2 Measure the velocity it acquires and calculate its momentum. You can use the 'trolley' as a unit of mass, as you did before.

3 Repeat, with the rod set to give its next strongest push.

4 Repeat, with the rod set to give the strongest possible push.

5 Explain the results you obtain.

In this investigation the trolley is accelerated very rapidly by a strong force for a very short time. We did not measure either the force or the time, but we can be sure that the force at step 3 was stronger than the force at step 2 and probably acted for slightly longer. At step 4 the force was strongest of all and probably acted for longer than at the other two stages. Whatever the size and duration of the force, we can measure its *effect* by finding how much momentum it produced in the trolley.

If the battery of an automobile is 'flat', a crowd of helpers can push the vehicle to help make it start. Gradually it gains momentum; when it has gained enough momentum, the driver can let in the clutch and the engine may start. The more helpers, the sooner the vehicle gains enough momentum. If there are fewer helpers, they must push for longer. This example from everyday life shows that the gain in momentum is related to the amount of force on the car *and* the length of time for which it is applied.

The same applies for loss of momentum. If you apply the brakes of a bicycle gently, they may take many seconds to reduce its momentum to zero. If you apply the brakes strongly, the momentum is reduced in one or two seconds. The momentum of a sailing boat can be increased by a weak force acting for a long time, as when a gentle wind blows on the sails for several minutes. If the boat then crashes on to a reef, and it is brought to rest there, its momentum is lost in a fraction of a second. The time is very short and the force is therefore very great; it may be so great that the boat is seriously damaged.

In all of the examples above, and in the experiment, we say that the change in momentum has been caused by *impulse*. We define impulse as the product of the force and the time for which it operates. In symbols,

$$\text{impulse} = Ft.$$

For the present we will not give the units of impulse or say exactly what we mean by force. Note that impulse is a vector quantity. In the next experiment we try another way of delivering impulse to the trolley.

Instructions: investigating impulse and acceleration

1 Use a trolley with ticker tape, as before, and adjust the slope of the runway to give uniform velocity.

2 The force to accelerate the trolley is provided by an elastic cord. To use this attach one end to the trolley (fig. 10.11) and pull at the other end. Keep your finger just level with the front end of the trolley as it moves. In this way the force applied to the trolley is constant. Practise this a few times before running the trolley with tape.

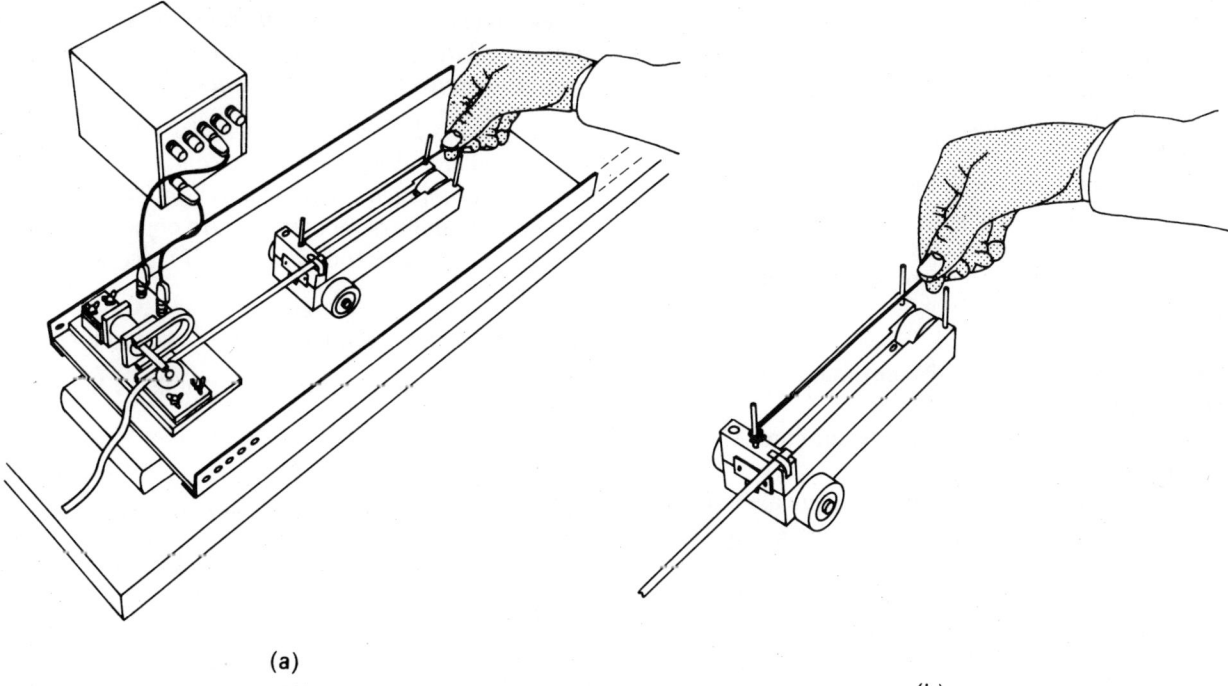

(a)

(b)

Fig. 10.11   Investigating impulse and acceleration.

3 Start the timer, pull on the elastic cord and draw the trolley along the runway.
4 Calculate the acceleration of the trolley. Is this acceleration uniform?
5 Repeat, using two cords instead of one. Now the force on the trolley is double. What happens to the acceleration?
6 Repeat, using three cords. Write all your results in a table. Make a rule to describe the relation between force and acceleration.

Consider an object (for example, the trolley) of mass, m, moving at velocity, u, in a certain direction (along the runway). As all the vectors in this description are in the same direction (along the runway), we need not bother to state the direction every time we mention a vector quantity. An impulse Ft acts on this object and its velocity changes to v.

To begin with the momentum of the object is mu.
To finish with the momentum of the object is mv.
Its change in momentum is $(mv - mu)$ or $m(v - u)$.

This change takes time, t, the time for which the impulse acts.

During any second of this period the change in momentum per second is $\dfrac{m(v - u)}{t}$.

For a period of one second, $t = 1$, so the impulse delivered during that time is F.

In the expression for change of momentum $\dfrac{(v - u)}{t}$

is also the rate of change of velocity, which we call acceleration, a.

Therefore the expression simplifies to ma. We can therefore say that:

if a force, F, acts on an object the rate of change of momentum of the object varies with the product of its mass and acceleration, ma.

This statement agrees with the results of the previous investigation in which you had a trolley of constant mass and applied various forces. To check this statement we could experiment with trolleys of various mass, using a constant force.

Instructions: accelerating different masses with constant force
1 Accelerate one trolley, using one elastic cord. Measure the acceleration.
2 Repeat with a second trolley (or equivalent mass) stacked on the first trolley. Use only one elastic cord. Measure the acceleration.
3 Repeat with a third trolley (or equivalent mass) stacked on the first trolley. Use only one elastic cord. Measure the acceleration.
4 Do the results confirm the relation between mass and acceleration as stated above?

10.20 Force

When a force, F, is applied to an object mass, m, the mass experiences acceleration, a, and rate of change of momentum varies with the force; we say ma varies with F.

Expressed mathematically, $ma \propto F$.

If we make use of a constant factor, k, we can re-write this as an equation,

$$ma = kF.$$

To make things easier, we choose our units so that we make $k = 1$. Then we can leave it out of the equation for good. We define the unit of force so that 1 unit of force causes an object of mass 1 kg to accelerate 1 m/s². Then $k = 1$, and the equation is:

$$ma = F.$$

If m is expressed in kg and a is expressed in m/s², then F is expressed in kg m/s² (said: kilogram metres per second squared).

The unit for force sounds rather complicated, but this is to be expected, since it is obtained by multiplying a mass by an acceleration. It is a derived SI unit and a vector. Force is such an important quantity in physics that we have given the unit a special name and symbol. This has the advantage of making it unnecessary to write kg m/s² and to say: 'kilogram metres per second squared' every time we measure a force. The unit of force is called the *newton*, and its symbol is N. To sum up we can define the newton like this:

1 newton (N) is the force which, if applied to a body of mass one kilogram, produces in that body an acceleration of one metre per second squared.

If impulse is defined as the product of force and time, Ft, the unit of impulse is *newton second*, (Ns), or in base units, *kilogram metres per second* (kg m/s). Finally, let us discover the units of momentum. An object mass, m, has velocity, u. It is acted on by impulse, Ft, and is accelerated to velocity, v.

Its increase in momentum is $mv - mu$ or $m(v - u)$.

The velocity-time equation (p. 99) tells us that

$$v = u + at$$
(or $v - u = at$).

So the increase in momentum of the object is

$$m(v - u) = mat.$$

But $ma = F$, so $m(v - u) = Ft$.

Increase of momentum is equal to Ft, or impulse.

Thus momentum and impulse have the same units, Ns or kg m/s. If you can calculate the impulse applied to an object this immediately tells you its change in momentum. Conversely, if you know the change in momentum of an object you immediately know what impulse has been applied to it. Then, if you know the length of time of the impulse you can quickly find its force. Try some of these examples.

Questions

1 In question 3, page 102, what is the change of momentum of the train? What impulse is needed to change its momentum by this amount? In what way is this impulse applied to the train? What is the size of the force?

2 In question 4, page 102, what is the change of momentum of the ball? What impulse is needed to change the momentum by this amount? What force is exerted on the ball by the blades of grass?

3 A rocket motor is turned on for 5 s while the rocket is on a vertical flight path. The velocity of the rocket increases from 120 m/s to 150 m/s. The mass of the rocket is 100 kg. Calculate the impulse and the force developed by the motor.

4 A boy pushes with a force of 2 N on a model automobile of mass 50 g. This has low-friction bearings and runs on a straight plastic track; friction can be ignored. The automobile is accelerated from rest to a velocity of 5 m/s. What impulse is required? For how long did the boy push the model?

5 At the end of the track (question 4) is a pile of sand into which the model crashes and is brought instantly to a halt. What is the impulse of the sand on the model? What is the impulse of the model on the sand? Can we calculate the force of the model on the sand?

10.21 Newton's laws of motion

In this chapter we have made discoveries and derived equations which are partly summarised by three laws first put forward by the mathematician Sir Isaac Newton in the seventeenth century. Stated simply, his laws are as follows.

Law 1 *The velocity of a body will not change unless a force acts on it.*

Law 2 *The rate of change of momentum of a body is proportional to the force that acts on it.*

Law 3 *If one body (A) exerts a force on another body (B), the force exerted on B by A and the force exerted on A by B are of exactly the same size and act along exactly the same line, but are in exactly opposite directions.*

Law 1 tells us that if $F = 0$, $a = 0$, which we know from our equation, $F = ma$. If $a = 0$, the velocity of the body does not change. If it is at rest, it remains at rest. If it is moving with velocity, u, in a certain direction, it continues to move with velocity u in the same direction. Neither the size nor the direction of its velocity changes. Law 2 is a statement in words of the meaning of the equation $F = ma$. Law 3 is the subject of the next section.

10.30 Collisions

Instructions: investigating collisions between two trolleys

1 Arrange the runway so that velocity is uniform. Place one trolley (A) at the top of the runway, and a second trolley (B) half-way down the runway. Line up the trolleys so that when A runs down it collides with B. Place a lump of Plasticine on the end of A so that when it collides with B the two trolleys stick together and move along joined together as one. Some types of trolley have a special cork and sharp spike to make a firm coupling between trolleys.

2 Fix tape to trolley A only.

3 Start the timer and push A so that it runs along the runway and strikes B.

4 Calculate the velocity of A before it strikes B.

5 Calculate the velocity of A and B together, after they have collided.

6 Repeat the investigation several times with extra trolleys (or equivalent masses) stacked on A or B or both, as indicated in the table at the bottom of the page.

Use the trolley as the mass unit to make calculations easier.

What is the momentum of trolley B before collision? What can you say about the total momentum (of A plus B) *before* collision when compared with the total momentum (of A joined to B) *after* collision?

Instructions: investigating elastic collisions

1 In this investigation the trolleys do not remain joined

Mass of A	Mass of A+B	Velocity of A, before collision	Velocity of A+B, after collision	Momentum of A, before collision	Momentum of A+B after collision
1	2				
1	3				
2	3				
2	4				

Mass of A	Mass of B	Before collision		After collision				
		Velocity of A	Momentum of A	Velocity of A	Velocity of B	Momentum of A	Momentum of B	Total momentum of A and B
1	1							
1	2							
2	1							
2	2							

after collision. A spring is placed on trolley A so that it strikes trolley B as they collide. Some types of trolley have a metal nose-piece which strikes against a rubber band on the other trolley.

2 Place trolley A at the upper end of the runway, and trolley B half-way along. Attach tape to both trolleys. The tapes may both be threaded through the ticker-timer. Place two carbon discs between the tapes, one disc with carbon side facing upwards and the other with carbon side facing downwards.

3 Use trolleys loaded with various masses (stacked trolleys or equivalent masses) and complete the details in the table above.

What is the momentum of B before collision? What can you say about the total momentum before collision when compared with the total momentum after collision?

Questions

1 In the investigation above the momentum of trolley A was shared between both trolleys as they collided. To transfer some of the momentum of A to B, and to accelerate B, some force must have been acting on B (Newton's first law). We do not know how long the collision lasted; it was certainly only a fraction of a second. Although we cannot calculate the force we can calculate the impulse. What was the impulse of A on B in each of your trials?

2 After collision the velocity of A was reduced. Some force must have acted on it to decelerate it. This must have been the force of B acting on A during collision. Calculate the impulse in each of your trials.

3 Set out the results of your calculations in a table:

Mass of A	Mass of B	Impulse B on A	Impulse A on B
1	1		
1	2		
2	1		
2	2		

What can you say about the impulses at each trial?

The impulses acted for the period during which the trolleys were in collision, so t was the same for both. We have shown that Ft was the same for both in size, but opposite in direction. This means that F must have been the same size for both, but opposite in direction. The trolleys illustrate the working of Newton's third law (see 10.21).

These experiments on collisions lead us to an important law called the *principle of the conservation of momentum*. This states that:

When two bodies interact, the total momentum of the bodies in any given direction remains unaltered. Momentum can neither be created nor destroyed.

Yet when an automobile crashes into a tree, or a person jumps down off a wall and hits the ground, or when a trolley hits a heavy block at the end of the runway, it looks as if momentum is being destroyed. If it cannot be destroyed, where does it go? To discover the answer to this problem we must remember one fact that we have so far ignored. The tree, the ground, the fixed block are all part of the surface of the Earth. When the trolley collides with the block it is, in effect, colliding with the Earth, for the block is firmly fixed to the Earth. The Earth is moving through Space, so has velocity and momentum. When the trolley strikes the block, the block exerts a force on the trolley, reducing its momentum to zero. During the same time the trolley exerts an equal but opposite force on the block. This force is transmitted to the Earth. The momentum of the Earth in the direction in which the trolley was moving is increased by an amount *exactly equal to* the momentum lost by the trolley. Momentum is conserved.

If the trolley has mass 0.75 kg and velocity 2 m/s before collision, its momentum is 1.5 Ns. The mass of the Earth is approximately 6 million million million million kilograms, so to increase its momentum by only 1.5 Ns means an exceedingly small increase in velocity: about 0.8 mm per hundred million million years! This explains why the momentum of the trolley appears to have been destroyed. The momentum has *not* been destroyed, it simply has no measurable effect after the collision with the massive Earth.

Questions

1 When you catch a hard ball (as in cricket) it is less painful to your hands if you swing your hands in the direction in which the ball was moving, as soon as it touches your hands. Explain this.

2 If you jump down off a wall, it eases the shock if you let your knees bend as your feet strike the ground. Explain.

3 A hammer has a heavy head. What is the advantage of this? Why is it much easier to hammer in a nail if the wood is supported on a firm surface?

4 An automobile has accidentally been driven off the road. To pull it back, a tow-rope is connected from the automobile to a truck. Slowly the truck pulls on the rope and the automobile is pulled slowly back on to the road. Next, the automobile is to be towed to the nearest garage, using the same rope. The rope hangs loose between the truck and the automobile. The brakes of the automobile are off. The truck driver is in a hurry; he accelerates the truck rapidly. The tow rope breaks as soon as it becomes tight. Explain why it breaks now, though it did not break before.

5 A girl, mass 60kg, is standing on roller-skates on a smooth level surface. She is holding a rock, mass 5kg. She throws the rock forwards; it travels through the air with a horizontal velocity of 1.2m/s. Explain what happens to the girl. If you cannot answer this question, stand on skates and throw a rock (beware of onlookers). Then try again to answer the question.

6 A tennis ball, mass 56g is travelling horizontally at 25m/s. It is hit by a tennis racket, and moves away in the opposite direction at 30m/s. What is the impulse of the racket on the ball? What is the impulse of the ball on the racket?

7 A pistol, mass 1.5kg, fires a bullet mass 2g horizontally at 300m/s. Find the velocity with which the pistol begins to recoil. Find the force needed to stop the recoil of the pistol in 1cm, assuming that the hand holding the pistol exerts a uniform force on it.

10.40 Using the principle of conservation of momentum

Questions 5, 6 and 7 above introduced a new aspect of the principle. The idea behind this can be checked by experiment.

Instructions: exploding trolleys

1 Place two trolleys on the runway with the impulse rod of one set to push against the other (fig. 10.12).

2 Fix tape to both trolleys and run this to two ticker-timers.

3 Start the timers, and release the impulse rod.

4 Calculate the velocity and the momentum of each trolley.

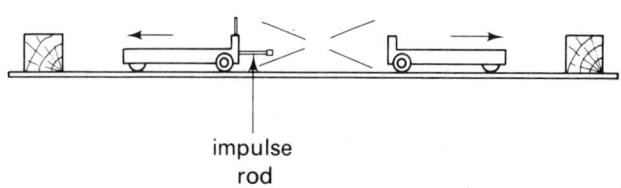

impulse rod

Fig. 10.12   Exploding trolleys.

5 Repeat with one or both of the trolleys loaded with stacked trolleys or equivalent masses. Describe what will happen if one trolley is very heavy (say, equivalent in mass to 100 trolleys) and the other is light (a single trolley).

6 What general conclusion do you obtain from this experiment?

Instructions: investigating recoil

1 Mount a catapult on a trolley and hold it in position with a piece of thread (fig. 10.13).

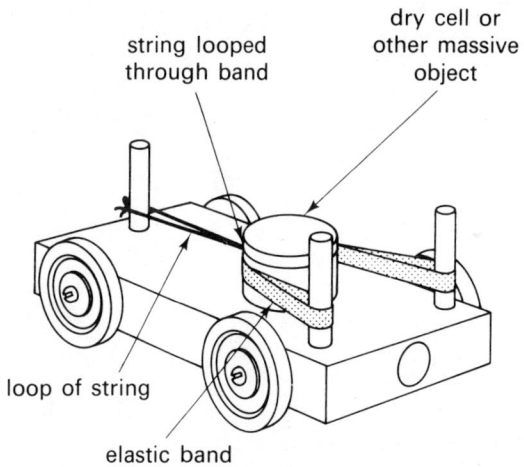

dry cell or other massive object

string looped through band

loop of string

elastic band

Fig. 10.13   Investigating recoil.

2 Put a heavy object, such as a dry cell, in the catapult.

3 Attach tape to the trolley and thread this through the timer.

4 Start the timer. Set light to the thread. What happens to the trolley as the object is thrown from it? Measure the velocity of the trolley.

5 Repeat, using a heavier object.

6 Repeat, using a stronger catapult.

7 What must you do to make the catapult throw the object a great distance? What must you do to make the trolley recoil with high velocity?

The firing of a gun is one way in which we make use of the principle. The explosion of the charge in the barrel forces the bullet and gun to move apart. Momentum must be conserved. The total momentum must remain zero, as it was before the gun was fired. The bullet has much less mass than the gun; it moves forward at high velocity; the gun moves backwards or recoils at low velocity. In a rocket, the fuel is burned and converted to a hot expanding gas. The gas is directed out at the rear of the rocket. It has low mass but very high velocity, so its momentum is high. To conserve momentum, the body of the rocket must move forward. Note that the rocket does not depend upon 'pushing against air' or pushing against the Earth. The rocket pushes against its exhaust gases and the gases push against the rocket, according to Newton's third law. Thus a rocket works just as well in a vacuum, better in fact, for there is no air resistance for the body of the rocket to overcome. This is why rockets are used for Space travel.

The jet engine of an aeroplane works on a similar principle. In this a jet of hot expanding air passes out from the rear of the engine and to conserve momentum the engine, with the aeroplane attached to it, moves forward. The details of the working of a jet engine are given in 28.20.

Projects
1 Measure the velocity of a cloud. Invent your own method for doing this.
2 Measure the acceleration of a bicycle. See who in your class can accelerate a bicycle at the greatest rate.
3 Measure the impulse needed to decelerate a bicycle from maximum velocity to rest.
4 Find out as much as you can about the methods used nowadays to reduce the dangerous effects of the impulse when an automobile collides with a stationary object or another vehicle.
5 Which animal living in your area runs at the greatest velocity? Try to measure this velocity. Find the maximum acceleration of this animal.

Fig. 10.14   The moment of impact during an experiment in which a car was in collision with a concrete block.

11 Investigating forces in action

11.00 Two or more forces acting together

Forces are vectors and if we are to find out what happens when two or more of them act together we must first know how vectors may be added.

11.01 The addition of vectors

The rules outlined in this section apply to all kinds of vector quantities: displacements; velocities; forces; impulses. Remember that we can add only vectors of the same kind. We can add a velocity to a velocity, but we *can-*

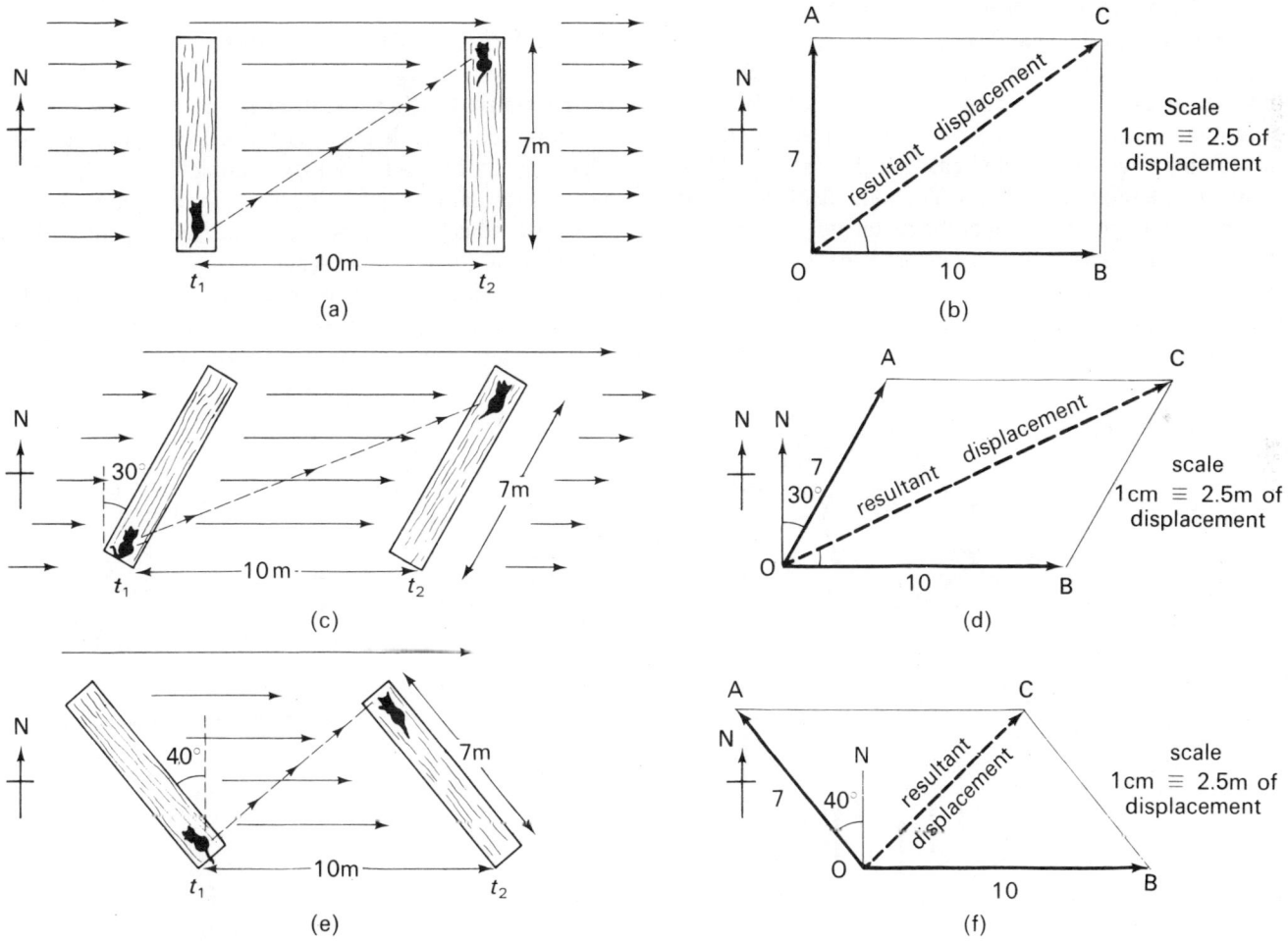

Fig. 11.1 Addition of vectors.

not add a displacement to an impulse. Remember also that both vectors to be added must be expressed in the same units. If one velocity is in metres per second and the other is in kilometres per hour, we must convert one to bring it to the same units as the other before we attempt to add them.

As an example of vector addition we will add two displacements. Imagine a log floating downstream; the length of the log is perpendicular to the direction of the stream. An overhead view is shown in fig. 11.1 a. At time, t_1, the log is at the position shown, and at a slightly later time, t_2, it has been carried further east. The displacement of the log is 10m east. At t_1 a cat is at the south end of the log. It walks along the log and reaches the north end at t_2. The displacement of the cat *from the south end* is 7m north. This is the second vector, to be added to the first. The displacement of the cat, as seen by somebody looking down from above, is the sum of the vector of the log's displacement downstream, and the vector of the cat's displacement along the log. To the onlooker the cat has moved along the path indicated by the dashed line. To find the exact displacement we draw a vector diagram (fig. 11.1b). The vectors are represented by arrows. The direction of each arrow indicates the direction of the vector. The length of each arrow represents the size of the vector, to some convenient scale. Starting from point O, draw the vector of the cat's displacement along the log; this is an arrow OA, directed north and 7cm long. From O draw the vector of the log's displacement downstream; this is an arrow OB directed east and 10cm long. Then complete the rectangle OACB. The total displacement is the vector OC, the diagonal of the rectangle. We call this the *resultant displacement*. By measurement we find that the length of OC is 12.2cm and angle COB is 35°. This tells us that the resultant displacement of the cat is 12.2m 35° north of east.

We can apply the same method to vectors that are not perpendicular. In fig. 11.1c the log floats downstream at an angle 30° east of north. The vector of the cat's displacement along the log is 7m 30° east of north. The vector of the log's displacement downstream is 10m east, as before. The vector diagram for adding these is fig. 11.1d. The diagonal OC of parallelogram OACB is the resultant displacement. Measurement tells us that the resultant displacement of the cat is 14.8m 24° 2′ north of east.

Questions
1 What is the resultant displacement of the cat when the angle of the log is as shown in fig. 11.1e and f?
2 If, in fig. 11.1c, the cat had walked more slowly and had walked only half-way along the log by time t_2, what would its displacement have been?
3 Use the same method to add these velocities. An aeroplane is flying west with velocity 60m/s relative to the air. The air is moving north-west with velocity 15m/s, relative to the ground. What is the velocity of the aeroplane relative to the ground?

11.02 Equilibrium of forces

Instructions: finding resultant force
1 You need a wooden board with pegs or nails around its edge (fig. 11.2), and three dynamometers. A dynamometer is similar to a spring balance. Its scale is marked in newtons, indicating the force that is extending the spring. (If you do not have dynamometers, you can use ordinary spring balances and work in force units of gram-force (gf). Spring balances are marked so that they give a correct reading when hung vertically, allowing for the weight of

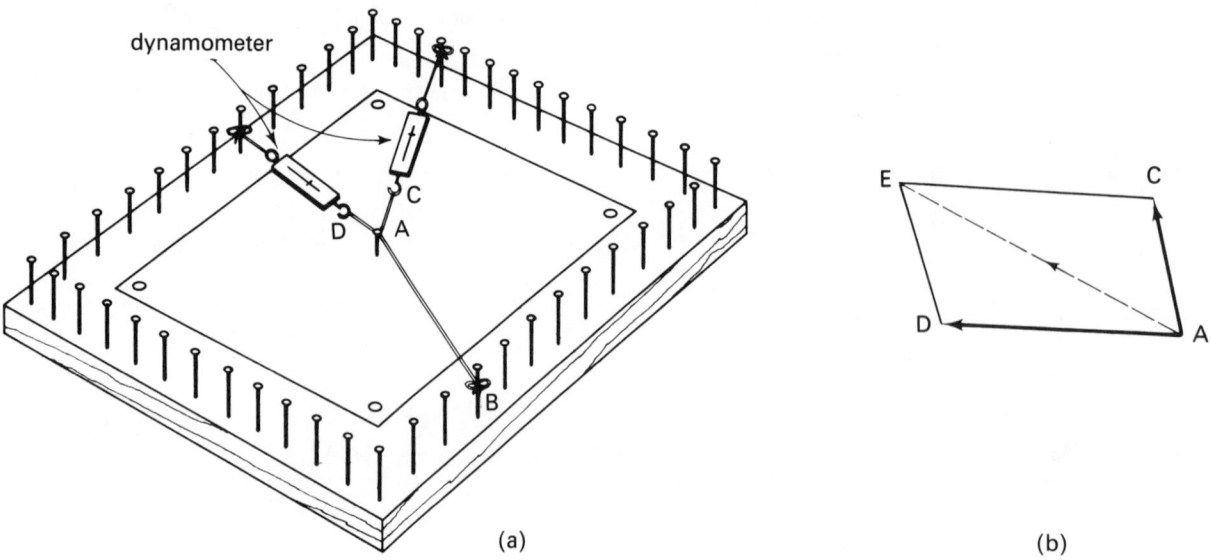

(a)   (b)

Fig. 11.2 Finding resultant force.

the mechanisms and hook; when used horizontally, as in this experiment, there is a slight error in their reading).

2 Tie two dynamometers to any two pegs. Tie threads to their hooks; join the threads by a knot (A).

3 Tie a third thread to the knot and run this to a third peg (B). Tighten this thread before you tie it to B, so that all three threads are tight and the dynamometers are not touching the board below. The dynamometer readings should all be greater than 10N (1 000gf); if not, tighten thread AB more.

4 Make a dot on the paper, directly below knot A.

5 From the dot draw lines below threads AC and AD (*not* AB).

6 Read the two dynamometers. Two forces acting on the knot are known. These are the tensions in the dynamometers, acting along AC and AD. We will calculate their resultant force.

7 Remove the dynamometers and threads, but keep them ready for step 11. *Do not remove the paper from the board.*

8 Mark off distances along lines AC and AD in proportion to the size of the forces (according to the dynamometer readings). Choose a scale for this so that you get a reasonably large diagram. Complete the parallelogram. Draw its diagonal. This represents the resultant force.

9 Measure the diagonal. Record its size and direction. Calculate the size of the resultant force.

10 When the forces were measured the knot A was at rest. We say it was *in equilibrium*. This means that the total force acting on it is zero. If it were not zero, the knot would accelerate in some direction or other. The only way for the total force on A to be zero is for the resultant of the two forces from the dynamometer threads to be exactly balanced by the force in thread AB. The tension in AB should be exactly equal in size to the resultant force, and exactly opposite in direction.

11 Reassemble the apparatus, using three dynamometers, the third one being in thread AB. Adjust the tension in AB until the knot is above A and the threads AC and AD lie along their lines as marked on the paper.

12 Draw a line below AB. Note the tension in its dynamometer.

13 Remove the threads and dynamometers. Measure out the vector along AB, according to the dynamometer reading. Is the vector AB equal in size but opposite in direction to the resultant vector of AC and AD? Can you think of the reason for an error?

14 Repeat two or three times, with the threads tied to different pegs, and with different tensions.

The measurements confirm that forces may be added by using the parallelogram method we used for other vectors. A diagram such as fig. 11.2b is sometimes called a *parallelogram of forces*. We can use the parallelogram of forces for solving many kinds of problem.

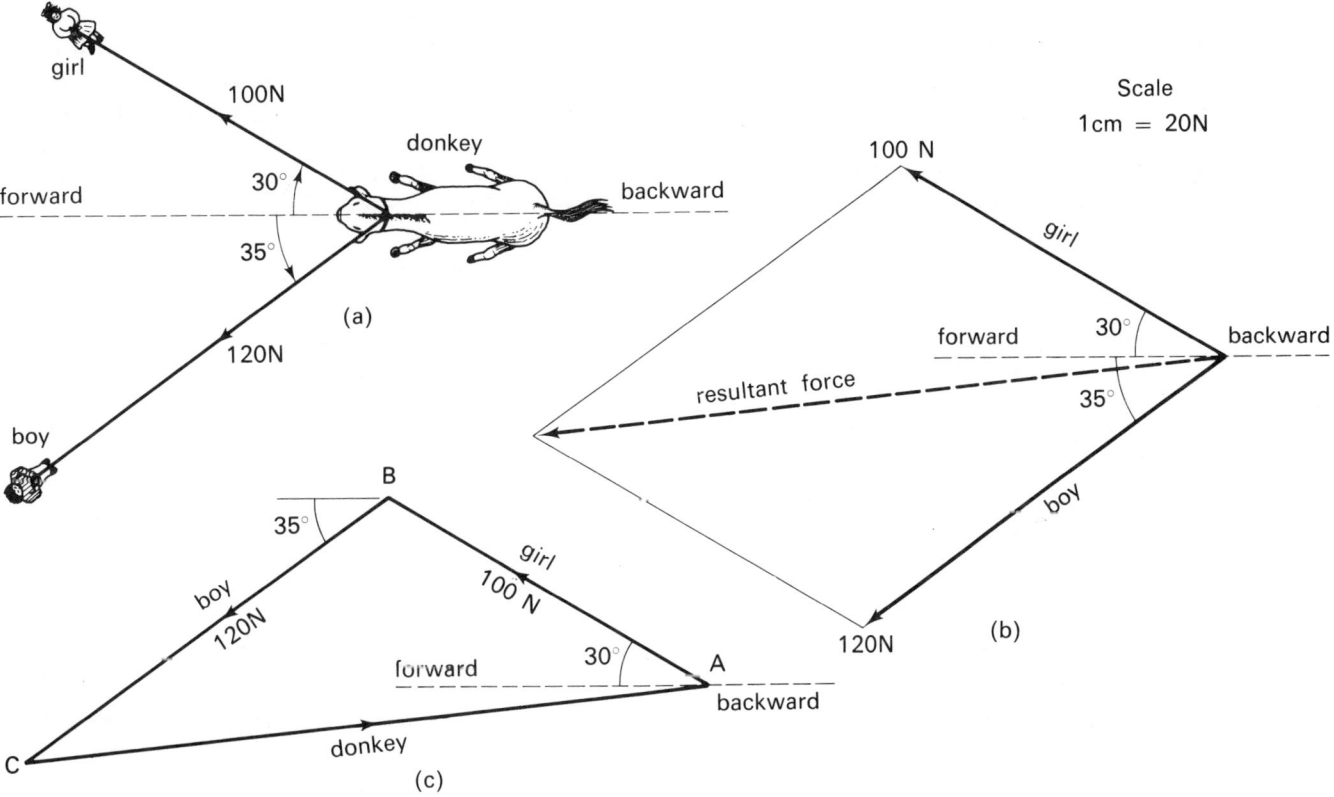

Fig. 11.3   Three forces in equilibrium.

Questions

Suppose a donkey is being pulled by a girl and a boy, but it refuses to move (fig. 11.3a). It grips the ground with its hooves, but the soil is loose so it may slide if pulled strongly. What happens to the donkey? Solve the problem in stages:

1 Draw the parallelogram of forces to scale (it is not to scale in fig. 11.3b). What is the resultant force produced by the children?

2 If the greatest force by which the donkey can resist the resultant force is 200N, what happens to the donkey?

3 If the greatest force by which the donkey can resist the resultant force is 190N, what happens to the donkey then?

4 If the donkey can exert 200N, what can the children do to increase their resultant force and so move the donkey?

If we have three forces (as in the previous example) and we know they are in equilibrium, we can solve problems about them by using the *triangle of forces*. The triangle of forces with the donkey in equilibrium (not moving) is shown in fig. 11.3c. This was drawn as follows. Start at A and draw AB to represent the vector of the pull of the girl; the direction and length of the line indicate the direction and size of her pull. From the *far end* of AB draw BC to represent the boy's pull. Now complete the triangle by drawing CA. The direction of the line, *from C to A*, represents the direction of the pull *of the donkey*. Its length represents the size of the donkey's pull. Measurement shows this to be 190N, the value for which the three forces are just in equilibrium. Note that the arrow-heads run around the triangle and that the triangle gives you the third force (the donkey's pull) not the resultant of the other two forces.

The triangle of forces applies only to three forces in equilibrium. There is no movement, the donkey is successfully resisting the children. In the example above we knew the size and direction of two forces, and found the size and direction of the third. We can use the triangle of forces in other ways too: for example we may know all three directions but only one size; we can then calculate the sizes of the remaining forces. Use the triangle of forces to solve these questions.

Questions

1 Two points 3m apart on the same horizontal level are joined by a string 4m long (fig. 11.4a). An object weighing 50N is hung from the string at a point 1m from one end of the string. Find the tension in each section of the string.

2 Three strings are knotted together at one point. Objects weighing 35N, 40N and 45N are tied to the loose ends of the strings. The strings with the 35N and 45N objects attached are looped over frictionless pulleys, so that the objects hang downwards (fig. 11.4b). The third object hangs down from the knot. Find the angles between the strings. If the pulleys are moved closer together, in what way will this alter the angles between the strings?

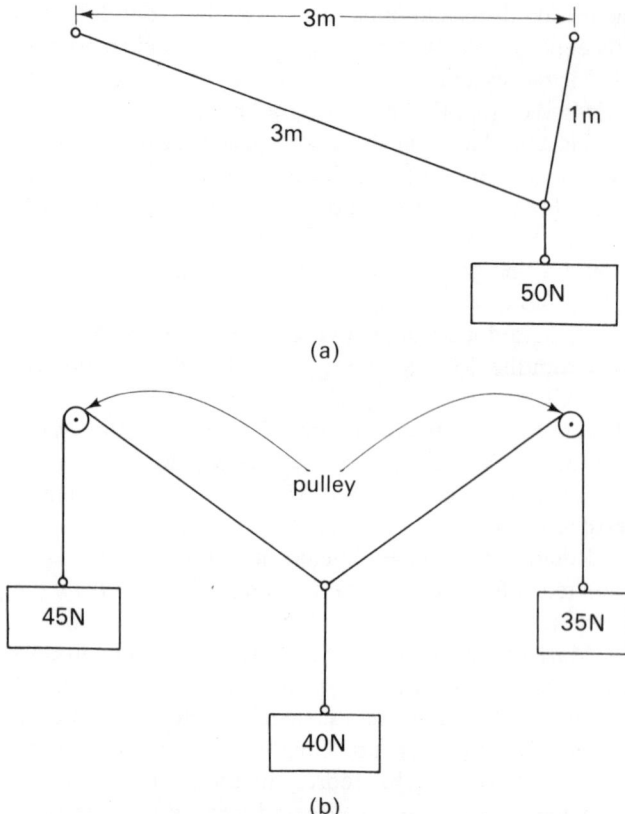

Fig. 11.4   More examples of forces in equilibrium (see questions – not drawn to scale).

Instructions: testing the triangle of forces with three dynamometers, one in each thread

1 Use the same apparatus as before. Cover the scales of two of the dynamometers so that you cannot see the pointers, taking care not to interfere with the movement of the pointers along the scales. You do not need a sheet of paper on the board.

2 Attach the threads to three pegs. Adjust and tie them so that all threads are tight.

3 Use a protractor to measure the angles between the threads.

4 Read the force in the one exposed dynamometer.

5 Draw a triangle of forces, given three angles and the length of one side. Find the lengths of the other two sides. These represent the forces in the hidden dynamometers.

6 Remove the covering from their scales to check the correctness of your calculation.

11.03 Forces which change the direction of motion of an object

If an object is in motion and a force acts on it, one of three things may happen.

a) It may accelerate without changing direction; this happens when the force acts along its line of motion and in the same direction.

112

b) It may decelerate without changing direction; this happens when the force acts along its line of motion but in the opposite direction.

c) It may change direction; this happens whenever the force acts at an angle to its line of motion.

In the third case the object may accelerate or decelerate in the direction of its original motion, depending on the angle of the applied force. The only kind of exception to case (c) is when the object is not able to change direction, for example a railway truck on a track, unless perhaps the force is strong enough to de-rail it.

Imagine a block of wood carried along in a stream at velocity 5m/s east (fig. 11.5a). A wind blows across a short section of the stream at an angle of 50° to the direction of flow of the stream. The force of the wind on the block is 2N; the mass of the block is 4kg. Ignoring water resistance, the effect of the wind on the block is to accelerate the block (Newton's second law) *in the direction of the wind*. Using the equation $F = ma$ (page 104), we calculate that $a = 0.5$m/s. If the block is exposed to the wind for 6s, it acquires a velocity of 3m/s in the direction of the wind. It now has *two* velocities: 5m/s east due to the stream, and 3m/s 50° north of east, due to the wind. To find the resultant velocity we add these vectors by drawing

a parallelogram, to scale (fig. 11.5b). Measurement tells us that the resultant velocity of the block is 7.3m/s 18° north of east. If there were no force to oppose it the block would continue to move with its new velocity, drifting across the stream until it hit the northern bank.

11.10 Gravitational forces

Gravitation is a fundamental property of matter. We do not know its cause but because we live on a massive body (the Earth) it plays an important part in our daily lives. We have studied its effects in great detail.

Gravitational force is an interaction between two pieces of matter. If we have two objects, each one attracts the other and, if they are free to move they will accelerate towards each other until they collide. In this chapter we are mainly concerned with the attraction between one body, the Earth, and another body, which might be any object on the Earth. If we hold an object and then let it go, it falls towards the centre of the Earth because of the gravitational force. Attraction operates in both directions, so the Earth moves upwards towards the object, but its movement is so slight that it cannot be detected. We will ignore it.

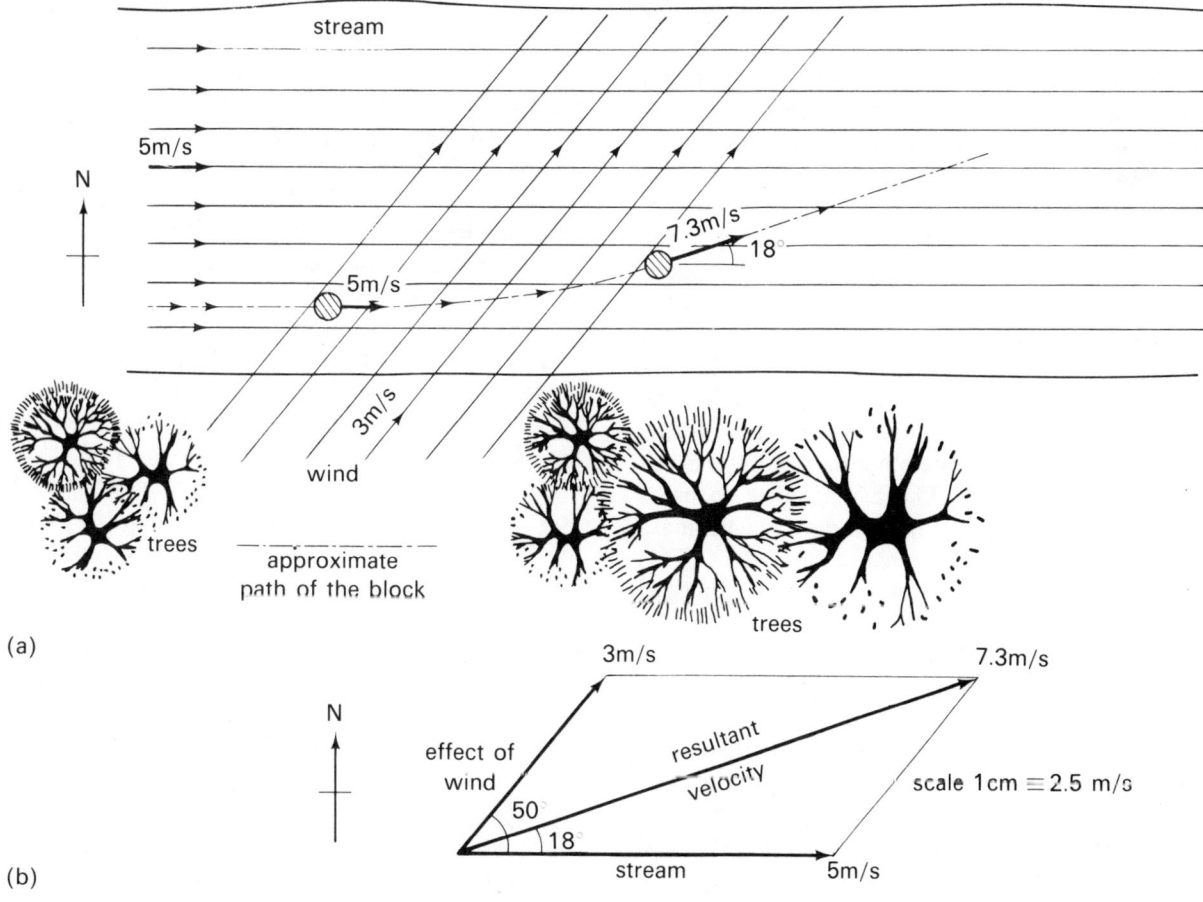

Fig. 11.5 *A force changes the direction of motion of an object.*

113

The direction of the gravitational force of the Earth on an object is, by definition, downward. The size of the force depends on:

(a) the mass of the Earth, which is great and constant;
(b) the mass of the object;
(c) the distance between the object and the centre of the Earth.

All objects at or near the Earth's surface are more or less the same distance from the centre, so we can usually ignore this third point. When we journey out into Space we must allow for it. The gravitational force of the Earth on an object is called its *weight*.

11.11 Weight and weighing

Since weight is a force it should be expressed in the SI unit of force, namely, the newton. You may have noticed that this was done in the questions above. In everyday life we sometimes, for convenience, use a gravitational unit, the kilogram-force kgf.

Suppose we wish to buy 1 kg of beans from a stall-keeper. We go to the stall and the man weighs out 1 kg of beans. Though we say he is 'weighing', we do not really care what force the Earth exerts on these beans. Our main interest is how many beans we will get: the quantity of beans; the quantity of matter. What interests us is the *mass* of the beans sold to us, not their *weight*. So 'weighing' is a misleading word. Unfortunately there is no common word such as 'massing', that might be a more accurate description of what the stall-keeper does.

He measures out the required mass of beans in one of two ways.
(a) *He uses a beam balance, or scales.* He puts a 'weight' on one scale pan and piles beans on the other until the beam balances. The 'weight' has a mass of 1 kg. He is not able to use the international prototype kilogram. Instead he uses a block of iron or brass which has more or less the same mass. The Earth exerts a force of 1 kgf on this 'weight'. He adds beans to the other pan until the Earth attracts the pile with a force of 1 kgf. When the two forces are equal the beam balances. Then we know that the mass of the beans is the same as the mass of the 'weight'. This method depends on the fact that bodies with equal masses are attracted to the Earth by equal forces.
(b) *He uses a spring balance.* When a force pulls on a spring, the spring is extended. The amount by which it is extended varies with the force applied. A spring balance must first be calibrated; we hang various masses from it and mark the scale to correspond. Calibration is usually done in the factory where the balance is made. When the stall-keeper uses the balance, he hangs a bag on the hook and puts beans in the bag until the pointer comes to the 1 kg mark. Then he knows that the bag of beans is extending the spring by exactly the same amount as a mass of 1 kg. When calibrated the force on the spring was 1 kgf (10 N), and the same force is acting now that the bag of beans is on the hook.

The beam balance is used to compare two forces (or masses) and to adjust one of them until they are equal. It does not measure forces: you do the measuring when you add up the 'weights' that you have placed in the pan. A calibrated spring balance is a force-measuring device, and gives a direct reading of the force applied to it.

In the experiments already carried out in this chapter you used spring balances calibrated in newtons. If we hang a 1 kg 'weight' from a dynamometer it reads about 9.8 N. Thus the force on a mass of 1 kg is approximately 9.8 N, when the mass is at the Earth's surface. You can use this factor, 9.8, to convert force from the *SI force unit* (N) into the *gravitational force unit* (kgf). In calculations in this book you may as a rough approximation use a conversion factor that makes the calculation very much easier: **1 kgf = 10 N.**

If we were to go to the Moon to buy beans, we would find things very different. At that distance from Earth, the gravitational attraction of the Earth is too small to have any effect. The Moon's gravitational attraction is important, but it is only one sixth of the force we experience on Earth. If the stall-keeper used a beam balance on the Moon his 1 kg 'weight' would be attracted to the centre of the Moon with a force of only about 1.7 N. The force by which the beans are attracted is also only one sixth of that experienced on Earth. He would pile beans on the other pan until the quantity of beans was attracted by a force of 1.7 N. This would give us the same mass of beans as we would get on Earth. If he used a spring balance, the force on 1 kg (mass) of beans is only 1.7 N and the spring balance would be extended only one sixth of the way to the 10 N mark on the scale. To make the balance read 10 N, you would need six bags full of beans!

If we measure out a mass of 1 kg of beans on Earth and then take them to the Moon, their mass is unchanged. The mass of an object is constant, wherever it is taken. But on the Moon, this quantity of beans has a weight of only 1.7 N. If we held the bag still in our hands and felt the downward force on it, we could tell that the force was much less than on Earth. Yet if we threw the bag of beans horizontally, accelerating it by applying a horizontal force to it, the force we would exert is exactly the same as on Earth. For a given acceleration, a, and a mass of beans, m, the force required is given by the equation $F = ma$ (10.20). F depends on *mass*, not on weight, and mass on the Moon is exactly the same as mass on Earth.

11.12 Falling

Like any other force, the force due to gravity produces acceleration when it acts on a body. This is what makes an object fall.

Instructions: measuring the acceleration due to Earth's gravity

1 Fix a ticker-timer so that the tape passes through it vertically. Support the timer about 2m above the floor.
2 Attach a small laboratory weight to one end of the tape. Thread the tape through the timer so that the weight hangs just below the timer.
3 Coil the rest of the tape loosely so that it can pass freely through the timer when the weight falls.
4 Switch on the timer. Let the weight fall, so that it pulls the tape through the timer.
5 Measure the acceleration of the falling weight (see chapter 10). What value do you obtain?
6 Repeat the experiment, using objects of different mass and different materials. What do you notice about the results?

The fact that falling bodies accelerate at the same rate was first demonstrated by Galileo Galilei, a sixteenth century physicist. In the Italian city of Pisa is a tower that is tilted instead of standing upright. It is said that Galileo let three iron balls of different masses drop from the top of this tower. They all hit the ground at exactly the same instant.

Not all objects accelerate as rapidly as metal balls. A feather or a piece of paper falls slowly, owing to air resistance. If we place a feather and a metal ball in a long tube and then suck all the air from the tube, both fall down inside the tube at exactly the same rate. The feather experiences no air resistance and falls rapidly to the bottom of the tube, as fast as the metal ball.

Even a rounded metal ball experiences some air resistance. If dropped from a great height it accelerates at 9.8m/s² at first. As its velocity increases, air resistance increases too. The rate of acceleration becomes less as the downward force of gravity becomes more and more opposed by the upward force of air resistance. At high velocity these forces become equal in size. Then there is no resultant force on the object and acceleration is zero. It has reached its maximum velocity, called its *terminal velocity*. The terminal velocity of an object falling through air depends on the density and shape of the object. For a well streamlined dense object such as a bomb or bullet the terminal velocity is about 400m/s. For a falling man it is about 50m/s, but if he is on a parachute this is reduced to about 6m/s.

The people to whom Galileo demonstrated the falling masses found it hard to believe that the masses would hit the ground at the same time. They could feel that the larger masses had stronger forces pulling them down. Perhaps you too find this hard to believe, even though you have proved it by practical trials. The gravitational force on a body (its weight) is proportional to its mass. In this way gravitational force is unusual. Frictional force and magnetic force, for example, do not depend simply on the mass of the body. For an accelerating body, $F = ma$; or by rearranging terms $a = \dfrac{F}{m}$. If we have two bodies with one having twice the mass of the other, the force of gravity acting on the one is twice as great as that acting on the other. If we double the mass we double the force and so on in proportion. Whatever the value of m, the ratio F/m remains constant. Therefore the acceleration, a, is always the same. At the Earth's surface it is always 9.8m/s², the value found by your experiment.

This particular value occurs so often in physics that we give it a special symbol, g.

If a body has mass m (m in kg),
the force of gravity upon it is expressed as $F = mg$ (F in N).
In other words its *weight* is mg (mg in N).
But we can also express its weight as m (m in kgf).
This is how we obtain the relation
1 kgf = 9.8N;
the factor 9.8 is the numerical value of g.

In the questions below take the value of g as 10m/s² to make calculations easier. Ignore effects of air resistance, if any.

Questions
1 A girl throws a ball vertically upwards. It leaves her hand with velocity 20m/s upwards. Calculate the time it takes to reach maximum height. What height does it reach? How long does it take to return to her hand again? What is its velocity as it returns to her hand?
2 A small metal ball is dropped from a window in a building. It takes 5s to reach the ground. From what height was it dropped?
3 If the mass of the ball is 0.1kg (question 2), what is its impulse when it hits the ground?
4 A coconut falls from the top of a tree 24.2m high. A boy standing beneath the tree sees the nut begin to fall and walks quickly away. If he accelerates at 4m/s² away from the tree, how far from the tree will he be when the coconut hits the ground?
5 A boy throws a ball horizontally with velocity 20m/s east. If the boy is standing on level ground, and his hand is 1.8m above ground when the ball leaves it, how long does the ball take to reach the ground?
6 How far will the ball (question 5) have travelled in a horizontal direction at the moment it touches the ground?

11.20 Orbits

In fig. 11.6 a bullet is fired horizontally from a gun. A horizontal force, F, acts on it until it reaches the end of the barrel. It leaves the gun with horizontal velocity, v. It continues to move with the same horizontal velocity (we ignore air resistance). While it is in the barrel its weight is supported by the barrel. As soon as it leaves the barrel, it

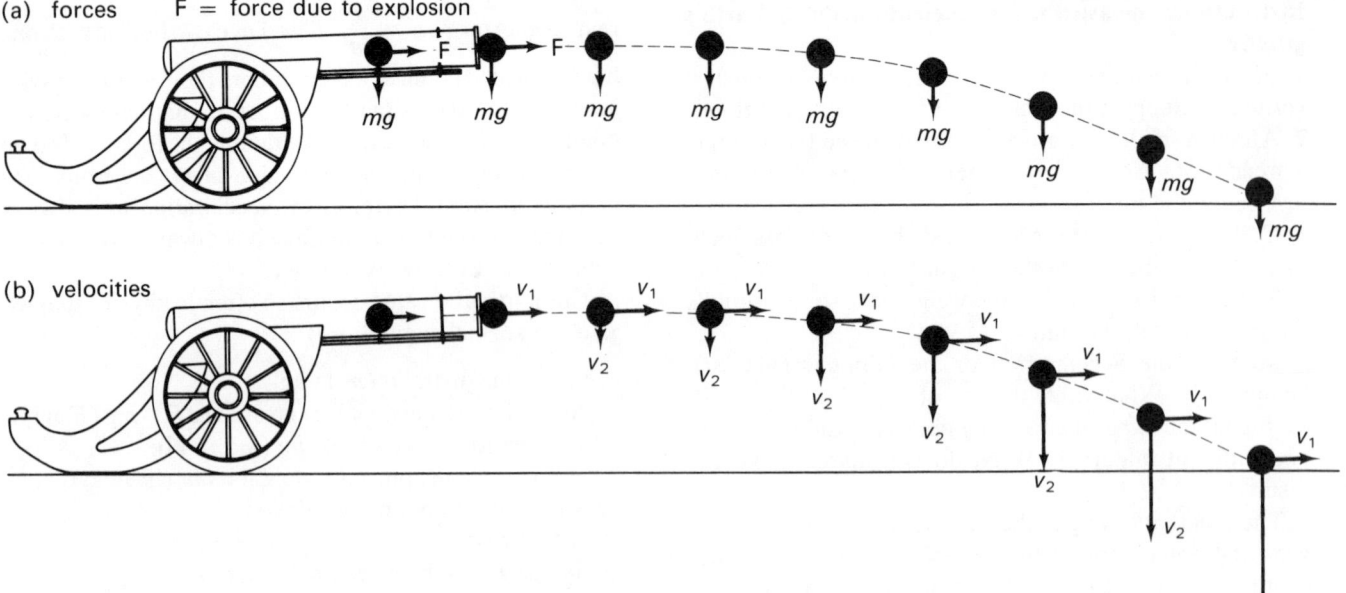

(a) forces   F = force due to explosion

(b) velocities

Fig. 11.6   *The path of a bullet.*

experiences downward acceleration due to its weight. The downward force is *mg*. The *time* taken for the bullet to reach the ground depends on the height of the barrel above ground, for, if we are thinking of vertical distance, only the vertical part of its motion and only the vertical downward acceleration are concerned. The time taken to reach the ground does not therefore depend on its horizontal velocity. This may be difficult to believe, so try this experiment.

Instructions: investigating the effects of horizontal velocity on the time taken for a falling object to reach the ground
1 The equipment is a special lever with three notches in it (fig. 11.7). It is called a disc-throwing catapult.
2 Place the catapult at the edge of the bench.
3 Pull the lever back as far as it will go and place a disc in each notch.
4 Allow the lever to flick forward, throwing the discs over the edge of the bench. Watch and listen to see where and when they hit the floor.

Questions
1 Which disc hit the floor first? Which hit the floor last?
2 Which disc had the greatest horizontal velocity as it left the bench?
3 Which disc travelled furthest before hitting the floor?
4 Which disc had the least horizontal velocity as it left the bench?
5 Which disc travelled the least distance before hitting the floor?
6 If we want to increase the horizontal distance that a gun can fire a bullet, what must we do?

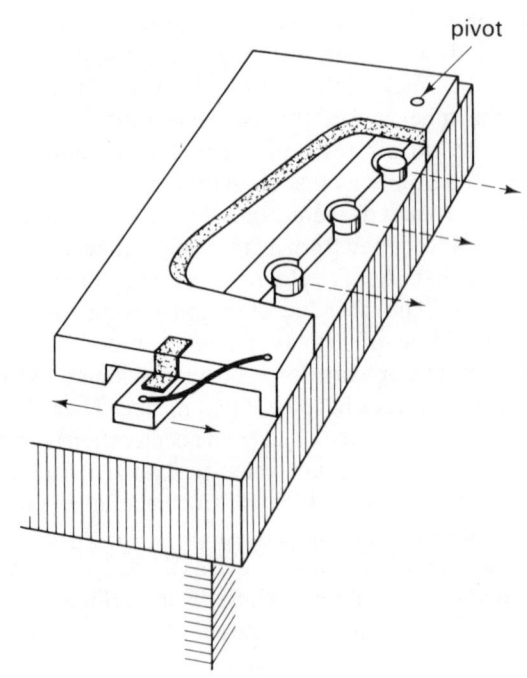

Fig. 11.7   *The disc-throwing catapult.*

We will not consider the effect of tilting the barrel of the gun upwards, but let us calculate the effect of firing the bullet with a very high horizontal velocity, say 400m/s. If fired from a height of 1.8m, on level ground, the bullet takes 0.6s to reach the ground, just the same time as the ball thrown by the boy in question 5 on page 115.

116

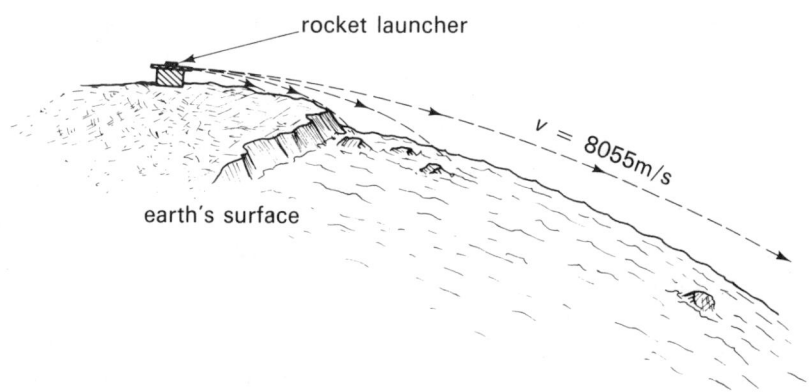

rocket launcher

v = 8055m/s

earth's surface

Fig. 11.8 Putting a rocket into orbit.

During 0.6 s it travels 240 m horizontally and then hits the ground. To obtain greater velocity we could fire a rocket; if we fired it at 1 000 m/s it would travel 600 m. In calculating this we are not allowing for any 'lift' caused by the fins of the rocket, which would delay its descent. Let us assume that such effects are absent for, in a moment, we shall be thinking about rockets travelling in Space, where there is no atmosphere. If the rocket is fired at 8 000 m/s it would travel 4 800 m in 0.6 s, but would it then hit the ground? When we think of these high velocities and great horizontal distances, we must remember that the surface of the Earth is not flat. It is curved. As the rocket falls to the ground, the ground is curving away beneath it (fig. 11.8). The rocket never reaches the ground; it is still 1.8 m above the surface. At this velocity the rocket circles the Earth. It is in orbit.

The example just quoted makes assumptions about lack of air resistance and lack of atmosphere, and assumes that there are no mountains to get in the way of the rocket as it circles the Earth. Yet the general conclusions are true. Suppose we fire the rocket vertically upwards, so that it goes quickly through the atmosphere. Near the Earth, air resistance is great, but this part of the journey is soon over.

When it is above the atmosphere (and well above the mountains), we turn it and accelerate it horizontally. If we accelerate it to a velocity of about 8 055 m/s, it will circle the Earth indefinitely. This is the normal procedure for launching rockets and the satellites that rockets carry into Space. Let us look more closely at the forces involved.

Instructions: investigating the forces acting on an orbiting object

1 Tie a large cork on one end of a piece of thin elastic, about 0.5 m long.

2 Go outdoors. Whirl the cork around on the elastic, slowly at first, then more quickly. What forces can you feel? What forces can you notice?

3 Suddenly let go of the elastic. What happens to the cork? How do you explain this?

Instructions: measuring the forces acting on an orbiting object

1 The apparatus is shown in fig. 11.9. The string is about 1.5 m long. Tie a large cork to one end, then thread the string through the handle and thread about 10 washers on to it. The paper clip prevents the washers from coming off. Place the crocodile clip (or another paper clip) just below where the string leaves the bottom of the handle. This is to act as marker of the position of the string.

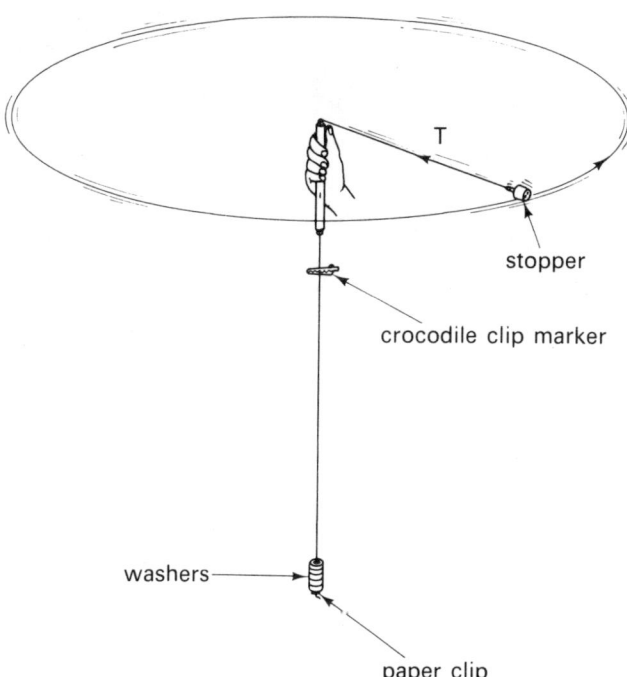

T

stopper

crocodile clip marker

washers

paper clip

Fig. 11.9 Measuring the forces acting on an orbiting object.

2 Whirl the cork around at such a speed that the marker just remains about 1 cm below the handle. Try not to let your hand move in a larger circle than necessary.

3 Keep whirling steadily while your partner counts how many orbits the cork makes in one minute.

117

4 Measure the radius of the orbit by measuring the length of string between the top of the handle and the cork. Allow a little for the circling motion of your hand. Then calculate the circumference of the orbit. How far does the cork travel in one minute? How far does it travel in one second? This is its speed around the orbit.

5 Weigh the clips and washers. Their weight is equal to the centripetal force that keeps the cork in orbit. Calculate this force in newtons. Record results in a table.

6 Repeat two or three times with a different number of washers on the string. What can you say about the relation between centripetal force and the speed of the cork?

7 Put 10 washers on the thread, but move the marker so that the radius of the orbit is reduced by about 20-30cm. Whirl the cork again then add or subtract washers until the speed of rotation is as near as you can get it to the speed you obtained at step 3. What can you say about the relation between centripetal force and the radius of the orbit?

8 It can be shown that for an object in circular orbit the centripetal force, F, is given by $F = \dfrac{mv^2}{r}$ where m is the mass of the object (the cork), v is its velocity (equivalent to its speed around the circle), and r is the radius of the orbit. Find the mass of the cork and for a few sets of your results see if this equation appears to be true. If the force is as stated above, what is the acceleration of the object towards the centre of the orbit?

Imagine an object being whirled on a string in a circular path (fig. 11.10). At one instant it is at point A, and its velocity is V_A, as indicated by the arrow. This velocity is at a tangent to the circle. If, at that instant the string breaks, the object continues to move with unchanging velocity. It travels along a path at a tangent to the circle, to A′ and beyond, still moving with velocity V_A, in the same direction it had at A. If the string does not break, the tension in the string causes the centripetal force F, acting towards the centre of the circle. The situation is similar to that of fig. 11.5. An object moving with velocity V_A experiences a force, F, acting at an angle to the direction of its motion. The effect of F is to accelerate the object towards the centre of the circle, giving it a velocity v_A towards the centre of the circle. The sum of the two velocities, by vector addition is the resultant velocity V_B. The path of the object is changed. Instead of moving to A′ it moves to B. At B, the same situation is found. Its velocity is V_B, and if the string broke it would fly off to B′ and beyond. If the string does not break, it is accelerated *towards the centre*

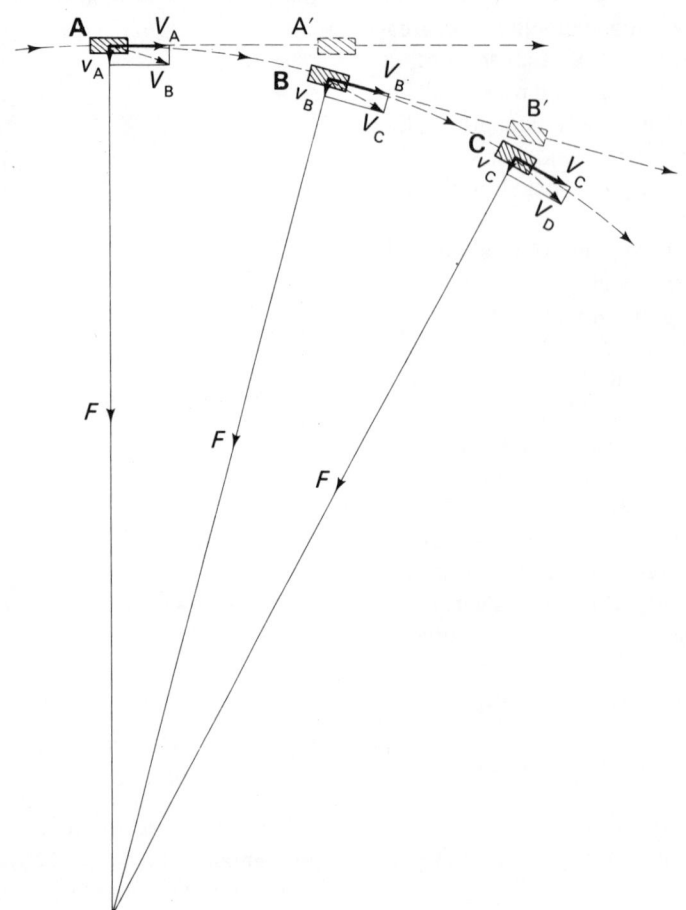

Fig. 11.10   The forces acting on an object in orbit, and the velocities.

by force F acquiring velocity v_B. So it goes to C, and continues round the circle. The diagram shows the object moving in 'jerks' or steps round the circle, but of course it really moves smoothly. Force F acts continuously, the object accelerates continuously towards the centre and the direction of its velocity changes continuously at constant rate as it moves round the circular orbit. In fig. 11.10 the force has been shown as the tension in a piece of string. In the example of a satellite in orbit around Earth, this is replaced by the force of gravity acting on the satellite and accelerating it towards the Earth. Instead of writing F, we could write mg (10.20), where m is the mass of the object and g is the acceleration due to gravity. For objects close to the Earth the value of g is approximately $9.8\,\text{m/s}^2$, but it decreases the further we go away from the Earth.

Questions

1 Where does centripetal force operate when a satellite circles round the Earth?

2 If we fill a bucket with water and attach it to a piece of string and then whirl it around, we can whirl it in a vertical circle. Though the bucket is upside down, the water does not spill from the bucket. Explain.

3 What would happen if the rocket shown in fig. 11.8 was fired with a horizontal velocity of $9\,000\,\text{m/s}$?

4 To put a satellite in orbit it must be accelerated to a certain velocity. Does this velocity depend on the mass of the satellite?

11.30 Satellites

Given that the centripetal force on an object in orbit is

$$F = \frac{mv^2}{r} \quad (F \text{ in N}),$$

we can perform some interesting calculations on the orbits of satellites. In this expression, v is the tangential velocity and, if T is the time taken to make one orbit, we can calculate that $v = \frac{2\pi r}{T}$.

Thus
$$F = \frac{m\left(\dfrac{2\pi r}{T}\right)^2}{r} = \frac{m\left(\dfrac{4\pi^2 r^2}{T^2}\right)}{r}$$

$$= \frac{4\pi^2 m r^2}{T^2 r}$$

$$= \frac{4\pi^2 m r}{T^2}$$

We know also that $F = mg$, so

$$mg = \frac{4\pi^2 m r}{T}$$

Rearranging terms, $\quad T^2 = 4\pi^2 \times \dfrac{r}{g}$

Taking square roots, $\quad T = 2\pi\sqrt{\dfrac{r}{g}} \;\; (T \text{ in s})$

If we are concerned with satellites close to the Earth, we may take the value of g to be $9.8\,\text{m/s}^2$. Then we find that $\dfrac{\pi}{\sqrt{g}}$ is very close to 1, which simplifies the expression a lot, giving us

$$T = 2\sqrt{r}.$$

The satellite *Cosmos I* had a nearly circular orbit, in which it was $174\,\text{km}$ above the surface of the Earth. The radius of the Earth is $6\,370\,\text{km}$, so the radius of the orbit of *Cosmos I* was $6\,544\,\text{km}$, or $6\,544\,000\,\text{m}$. Applying the equation we find,

$$T = 2\sqrt{6\,544\,000} = 2 \times 2\,558 = 5\,115 \;(T \text{ in s}) \text{ or}$$
$$T = 5\,115/60 = 85.3 \;(T \text{ in minutes}).$$

This approximation agrees well with the actual period of this satellite, which was $87.9\,\text{min}$. For all satellites that circle a few hundred kilometres above the Earth the calculations give similar results. You will note that in the calculation the only quantities required are r and g. The mass of the satellite does not come into the equation. Close to Earth g is reasonably constant and the actual altitude of the satellite above the Earth's *surface* makes little difference since the Earth's radius is much larger in comparison. Thus most satellites close to Earth have an orbital radius between about $6\,500\,\text{km}$ and $7\,500\,\text{km}$, giving them an orbital period of about $90\,\text{min}$, or $1\tfrac{1}{2}\,\text{h}$.

For satellites that are not close to the Earth we have to allow for the fact that g decreases as the distance from the Earth increases. The rule is as follows.

On Earth's surface we are $6\,370\,\text{km}$ from Earth's centre, and $g = 9.8\,\text{m/s}^2$.

At twice this distance *from Earth's centre,*

$$g = \frac{9.8}{2^2}\,\text{m/s}^2$$

At three times this distance from Earth's centre,

$$g = \frac{9.8}{3^2}\,\text{m/s}^2$$

At n times the Earth's radius $(6\,370\,\text{km})$ from the Earth's centre, $g = \dfrac{9.8}{n^2}\,\text{m/s}^2$

To calculate g at any distance from Earth's centre, we must first calculate how many times this distance is bigger than $6\,370\,\text{km}$, the Earth's radius. Satellite *Midas 7* was $3\,760\,\text{km}$ above the Earth's surface, making its orbital radius $10\,130\,\text{km}$. If we divide $10\,130$ by $6\,370$, we get 1.59, so we can say that its orbital radius is about 1.6 times the radius of the Earth, or $n = 1.6$. At this distance, $g = 9.8/1.6^2 = 9.8/2.56 = 3.83\,\text{m/s}^2$. Though we are still fairly close to the Earth, g has been considerably reduced by the distance. To calculate T, we must go back to the complete equation on page 119. Remember that for this, r must be expressed in m, not km.

$$T = 2\pi\sqrt{\frac{r}{g}}$$

$$= 2\pi\sqrt{\frac{10\,130\,000}{3.83}}$$

$$= 2\pi\sqrt{2\,645\,000}$$

$$= 2\pi \times 1\,626$$

$$= 10\,216 \ (T \text{ in s})$$
or 170 (T in min).

This result agrees well with the observed period, which was 168 min.

Questions

1 Calculate the period of our one natural satellite, the Moon. The radius of its orbit around the Earth is 400 000 000 m. You can consider that this is 60 times the radius of Earth.

2 Deimos is a satellite of the planet Mars and it has an orbit of 23 460 km radius. At this distance the acceleration due to the gravity of Mars is 0.078 m/s². Calculate the period of Deimos.

3 Calculate the radius of the orbit of an Earth satellite which has a period of exactly one day. In this orbit, $g = 0.225$ m/s². Do you know of any such satellites? What are they used for?

4 The satellite *Meteor 2-02* was launched from Plesetsk in January 1977 with an almost circular orbit, height 900 km above the Earth's surface. It is expected to remain in orbit for about 500 years. Calculate the period of its orbit. Why will it eventually return to Earth?

Projects

1 Make a display of charts and models of Earth satellites, showing their orbits, their heights above Earth, their periods, and the jobs that the satellites do.

2 Read as much as you can about the condition of 'weightlessness' that astronauts experience in Space. What are its effects? How can they be overcome?

Fig. 11.11 These astronauts are experiencing weightlessness during their training. This will help them to cope with it when they are in Space.

12 Investigating turning forces

When two forces act on a body, but do not act along the same line, they may cause the body to turn or rotate. We have seen examples of this in the work on levers (2.00). If the forces due to the weights or spring balances do not balance exactly, the lever turns. In this section we find out more about this.

12.00 Moments

The moment of a force about a point is the product of the size of the force and the perpendicular distance between the point and the line along which the force acts. In fig. 12.1, the moment, M, of the force about point P is given by

$M = Fx$ (M in Nm).

The SI unit of moment is the newton metre. Since force is a vector quantity, moment too has direction, which must be stated. It is usual to describe the direction of the turning force as clockwise or anti-clockwise. In fig. 12.1, the moment of the force is Fx anti-clockwise about P.

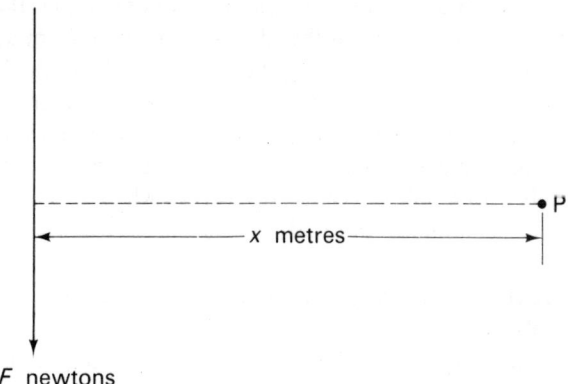

Fig. 12.1 The moment of a force about point P.

It is possible for two or more forces acting on a body to be balanced. In fig. 12.2a we see the forces acting on a lever as in trial 3 of fig. 2.2 on page 13. As the lever is horizontal and the weights are vertical forces, the weights act in a direction perpendicular to the lever. Therefore, to find the distance of a force from any point on the bar we simply measure the distance along the bar.

Questions

1 What is the moment of force L about the pivot P?

2 What is the moment of force E about the pivot P?

3 What is the sum of these moments? (You can make up your own rule for adding moments).

4 What is the moment of L about the end A of the lever?

5 What is the moment of E about the end A of the lever?

6 What is the sum of these two moments about A?

7 What would happen to the lever if the pivot were moved from P to A?

8 Point B is 0.1 m to the left of P. What are the moments of L and E about B? What is their sum?

9 What would happen to the lever if the pivot were moved from P to B?

10 Make up a rule about levers that are in equilibrium (balanced and not turning).

11 The figure shows two forces (L and E) acting downwards on the lever. Why does the lever not accelerate downwards?

12 Taking into account the upward force at P, what is the total moment of all forces, taken about A? What is their total moment about B?

With the pivot at P as shown in fig. 12.2a, and the *three* forces acting on it, the lever is motionless. The total downward force (2N + 4N) equals in size the total upward force (6N at the pivot). The total moment of these forces about the pivot is zero. The lever does not turn. We could get the same effect if we replaced the two downward forces by a single downward resultant force, size 6N, acting at the pivot (fig. 12.2b). The lever would be motionless, as before, because the resultant of the two downward forces is equal in size but opposite in direction

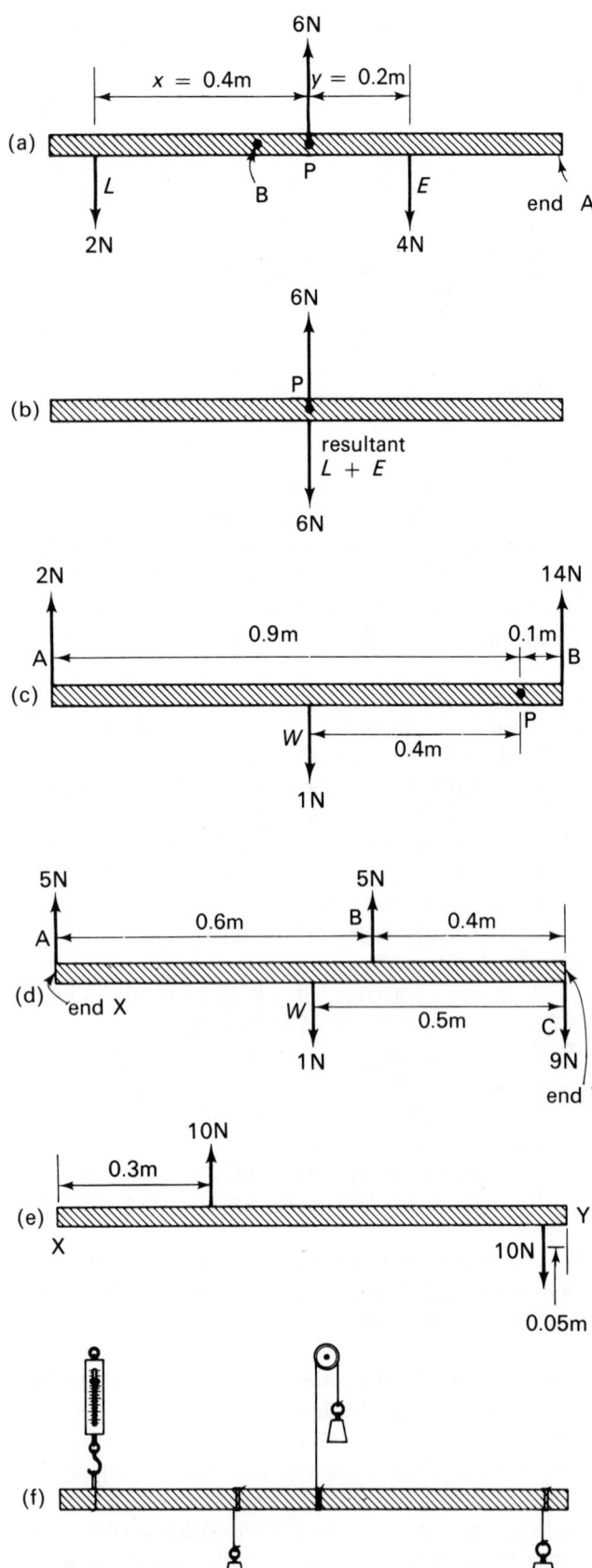

Fig. 12.2   Forces acting on levers (see text).

to the upward force. In replacing the two downward forces by a resultant force we have used rules which can be used for any number of *parallel* forces acting on a body.

1 The size of the resultant force is the sum of the sizes of the forces it replaces.

2 The moment of the resultant force about any point is the sum of the moments of the forces it replaces.

In this example, rule 1 tells us that the resultant force must be 6N. To apply rule 2 it is convenient to take moments about the pivot. The moment of the two downward forces about P is zero; therefore the moment of the resultant force about P must also be zero; therefore the resultant force must act *through* P. Try calculating moments for the lever shown in fig. 12.2c. The weight of the lever can be thought of as a force 1 N acting at point W.

Questions

1 Is the lever in equilibrium? Calculate moments to find out.

2 What single upward resultant force can replace the two upward forces A and B?

3 What is the force acting on the lever at the pivot?

4 What downward resultant force can replace the two downward forces acting on the lever?

5 What can be said about the resultant upward force and the resultant downward force?

6 In fig. 12.2d we see a lever which has no pivot and which is acted on by four forces, one of which is its weight. Let us find out what happens to this lever. Does the lever as a whole move either upwards or downwards?

7 What is the resultant of the two upward forces A and B?

8 What is the resultant of the downward forces C and W?

9 What can be said about the resultant upward force and the resultant downward force? What do you think happens to the lever?

If we have a body acted on by parallel forces we can find a resultant upward force and a resultant downward force. If these both act along the same line and are equal they cancel each other. Then the object is in equilibrium and does not move (fig. 12.2b). If they act along the same line but are unequal they partly cancel each other. We are left with a single force which makes the body move as a whole, either up or down. If they do not act along the same line we are left with a pair of forces (fig. 12.2e) which is called a *couple*. This makes the body rotate.

Instructions: investigating forces on a lever in equilibrium

1 You need a long light lever, such as a metre rule, several weights of different sizes and one or two dynamometers.

2 Arrange a lever (with no pivot) as in fig. 12.2f. Hang weights on the lever to produce downward forces. Use dynamometers to produce upward forces. If you have no dynamometers you can produce upward forces by carrying threads over pulleys and hanging weights from them. The arrangement and weights you use are your own choice. Adjust the size and position of the weights until the lever is in equilibrium.

3 Draw a simple diagram, like those in fig. 12.2 to show the lever and the forces acting on it. Remember to include the force due to the weight of the lever; think of this as acting at the centre of the lever.

4 Calculate the total upward force.

5 Calculate the total downward force. What can you say about the total upward force and total downward force?

6 Calculate the total moment of forces that act clockwise; choose any convenient point about which to calculate moments, for example, one end of the lever.

7 Calculate the total moment of forces that act anti-clockwise about the same point. What can you say about the total clockwise and anti-clockwise moments?

8 Find the resultant of all upward forces.

9 Find the resultant of all downward forces. What can you say about these two resultants?

10 *Imagine* that *one* of the forces is to be taken away. This would alter the position of the line of action of one of the resultant forces. Calculate what effect this would have. Then remove this force, and see what happens to the lever. Was your prediction correct?

11 Repeat for one or two other arrangements of forces.

12.10 Centre of gravity

Instructions: investigations with hanging shapes

1 Cut out some shapes from thick cardboard. You can cut some irregular shapes like the one in fig. 12.3, and some regular shapes such as a square, a rectangle, some triangles, a circle, an L-shape and so on. The pieces should be roughly 10-20cm across.

2 Make a small hole near one edge of one of the shapes and pin it to a support with a horizontally placed needle, so that the plane of the card is vertical and it hangs and swings freely.

3 Hang a small weight (a plumb bob) from a thread and attach this to the needle (fig. 12.3a).

4 Without disturbing the position of the card or thread, mark the lower edge of the card just behind the thread.

5 Remove the card and thread from the needle. Draw a straight line, using a ruler, from the hole to the mark. This indicates the vertical direction when the card is hanging from the hole.

6 Repeat steps 2 to 5, using another hole made at some other position near the edge of the card (fig. 12.3b).

7 Repeat steps 2 to 5, using a third hole. What do you notice about the three lines you have drawn? If the shape of the card is regular, what can you say about the meeting-point of these lines? Have you found any shape for which the meeting-point is not on the card?

8 Try balancing the card on your finger-tip, so that it balances horizontally. What do you notice about the point on which it balances? Are there any cards which it is not possible to balance on a finger-tip? How could you balance these if you were given a small piece of wire and some sticky tape?

The place where the lines meet is called the *centre of gravity* of the card. To be precise, the centre of gravity is not at the surface of the card, but exactly half-way between the two surfaces, below the point on the surface where the lines cross.

When the card hangs from a hole it acts as a lever, and the needle is the pivot. The pivot supports the card, exerting an upward force, F_1, on it. Every particle of the card is attracted downwards by gravity. Figure 12.4a shows the downward forces on the particles (or a few of them, at least). Each force has a moment about the pivot, either clockwise or anti-clockwise. These millions of

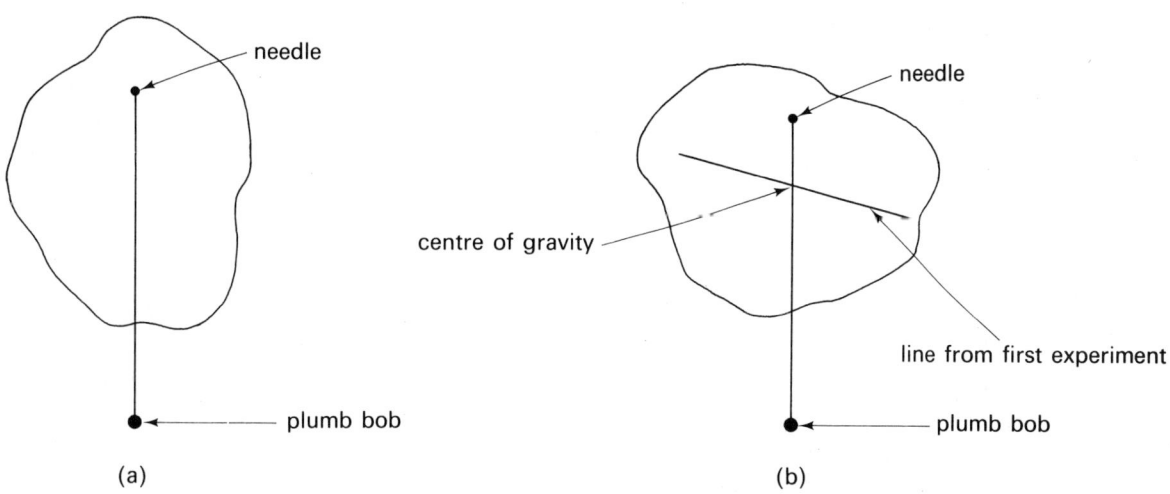

(a)   (b)

Fig. 12.3   Locating the centre of gravity.

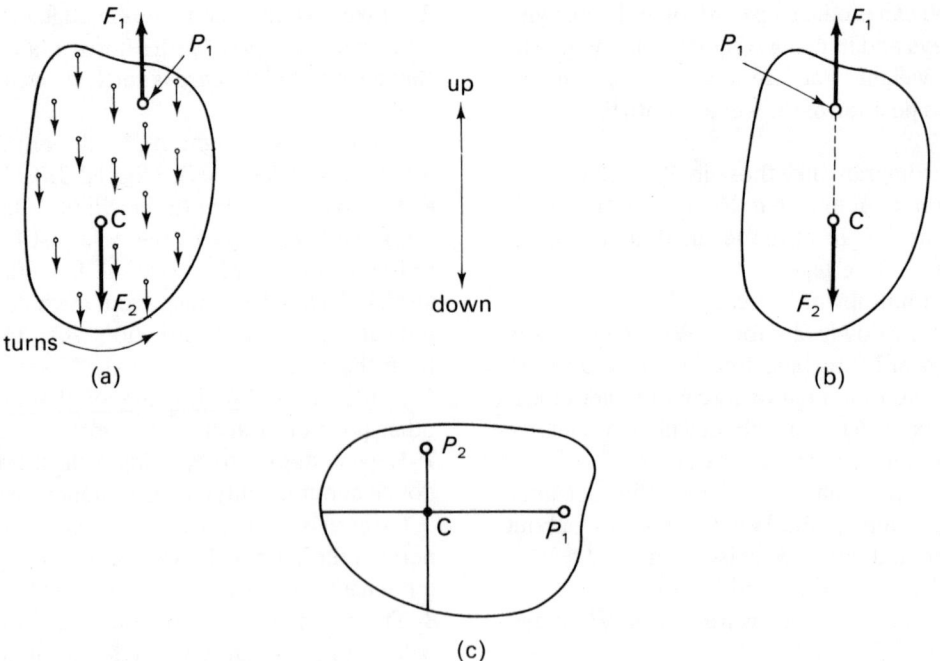

Fig. 12.4 *Explanation of the method of locating the centre of gravity.*

forces, if summed and replaced by a resultant force, can be represented by a single force, F_2, equal in size to the weight of the card and acting at some point in the card. The point at which F_2 acts is the centre of gravity, C. In fig. 12.4a forces F_1 and F_2 form a couple. The card rotates in the direction shown. In fig. 12.4b, they act along the same line; they are equal and cancel; the card is in equilibrium. In this position the centre of gravity is vertically below the pivot. In the experiment we draw a line vertically down from the pivot; then we know that C is *somewhere* along this line. If we then support the card from another hole (fig. 12.4c) we can draw another line because the same reasoning applies; C must be *somewhere* along this line too. The only place where C can be, is where the lines cross. This is how we locate the centre of gravity of a flat card.

After locating the centre of gravity, we find we can balance the card on a finger-tip placed below that point. Often you will have found that the centre of gravity is on an axis of symmetry of a regular shape. For a lever which has uniform width and thickness the centre of gravity is half-way along it; at the 50 cm mark on a metre rule. This was what we assumed in the calculations made on page 122.

Instructions: making a microbalance

1 Make the beam from a drinking-straw or cut a piece of grass-cane or thin bamboo, about 20 cm long. Shape one end as shown in fig. 12.5 to make a flat scoop.

2 Place a small nail or screw in the other end. A little glue or modelling clay will help it stay in place if it is not a tight fit.

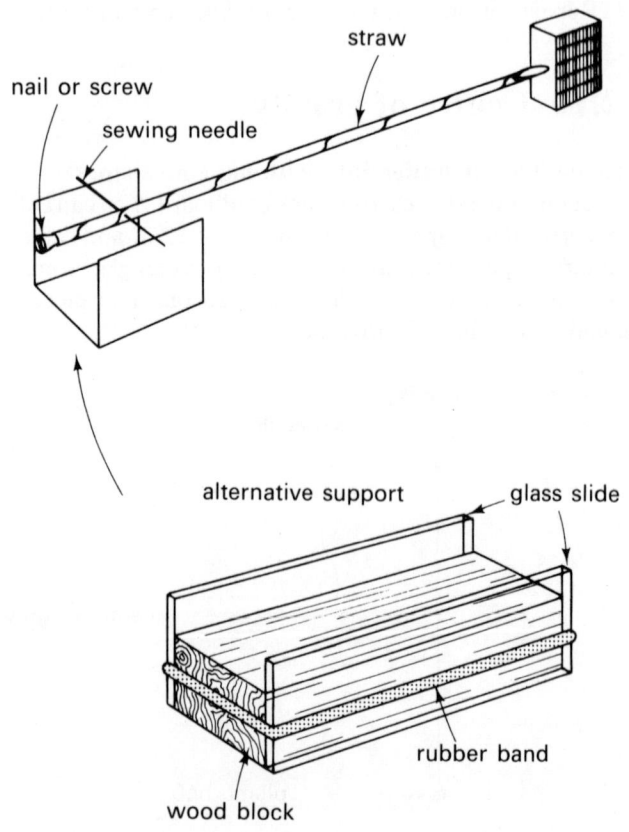

Fig. 12.5 *A microbalance.*

3 Balance the beam on your finger to find the approximate position of its centre of gravity. Then balance it on a sharp edge to find the position more accurately.

4 Push a needle through the beam just at the centre of gravity, but a little nearer the top side of the beam than the lower side. To support the beam you can use a piece of bent aluminium sheet, as shown, or fix two thin sheets of glass (microscope slides) to a small wood block using rubber bands.

5 Make a scale by sticking a piece of graph paper on to one side of a wood block.

6 When placed on the support the beam should come to rest with the scoop level with the top end of the scale. If it does not, pull the nail a little way out of the beam. Explain why this adjustment makes the scoop end rise higher.

7 To calibrate the balance, use 'weights' made by cutting graph paper into squares, each 2 mm × 2 mm. With the scoop empty, mark on the scale the level of the end of the scoop. Label this '0'. Then put one paper square on the scoop; mark the scale and write '1' beside the mark. Continue in this way using two, three or more paper squares, until the beam nearly touches the bench. You now have a balance that is calibrated in your own units of weight: 'squares'.

8 Use the balance to weigh all kinds of tiny objects such as grains of sand, small insects such as a mosquito or ant, a short piece of hair, a tiny seed and so on. Each time place the object on the scoop. Note the position of the end of the scoop when the beam comes to rest. Record weights to the nearest tenth of a 'square'.

9 What is the weight of a 'square' in grams? Work out your own method of answering this.

10 Convert all your weights in 'squares' into weights in grams.

12.20 Equilibrium

We have often used this term to mean a state in which an object remains level; it has no tendency to turn in any particular direction, though it may swing slightly. There are different kinds of equilibrium. These are best illustrated by looking at a cone resting on a flat level surface (fig. 12.6).

If the cone is resting on its base, its equilibrium is *stable*. When the cone is tilted (dashed lines) the weight of the cone, *W*, acting at the centre of gravity has an anti-clockwise moment about the point on the rim on which it is tilted. The effect of this is to make the cone return to its former position if it is let go. Provided that the cone is not tilted too far, it always returns to rest on its base. The same applies to a balanced beam, like the beam of the micro-balance. If pushed down at either end and then released, *it returns to its former position*. This is what we mean when we say that its equilibrium is stable.

If the cone is resting on its side, it can be rolled round and its weight is always above the part of its surface that it is resting on. Rolling the cone makes no difference to the way it is balanced. It will stay in any position to which it is rolled. This is called *neutral equilibrium*. The same kind of equilibrium is shown by a cylinder resting on its side or a pulley-wheel on a pivot. Whatever position it is turned to, *it stays in that position*. This is what is meant by neutral equilibrium.

If the cone is resting on its point, even a slight disturbance means that the centre of gravity is no longer directly above the point. If the cone in the figure is pushed a little way to the right, the force *W* makes it turn even further to the right; it topples over. This is called *unstable equilibrium*. If it is disturbed even slightly, it never returns to its original position but moves to some *very different position*.

Questions

1 You are given a pencil. In what positions can you place it to demonstrate all three types of equilibrium?

2 What type of equilibrium is shown by a sphere resting on a flat surface (a marble on a level table)? Explain.

3 What kind of equilibrium is shown by a marble resting on an inverted saucer? Explain.

4 What kind of equilibrium is shown by a pendulum? Explain.

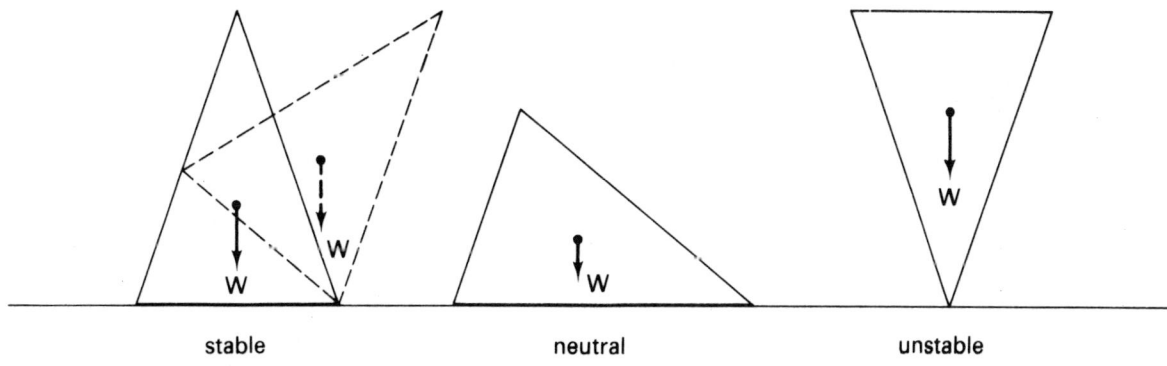

Fig. 12.6 *Three types of equilibrium, illustrated by a circular cone.*

Projects

1 What kind of equilibrium is shown by the toy illustrated in fig. 12.7? Explain. Do you know of any other toys that make use of this idea? Perhaps you could make some and display them in the laboratory.

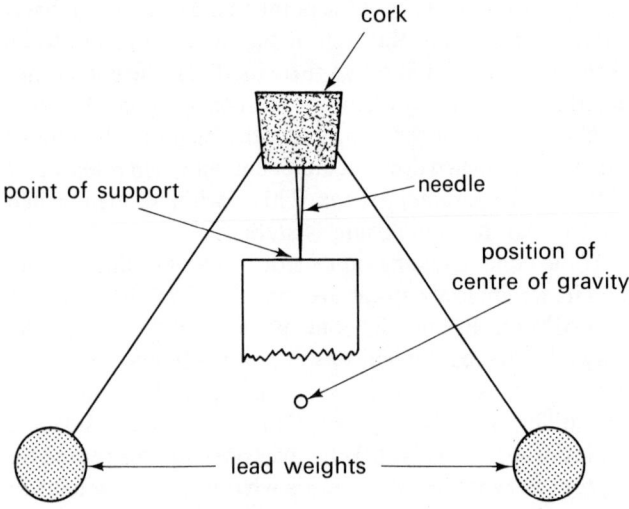

Fig. 12.7 A balancing toy – see project 1.

2 Try to increase the sensitivity of your microbalance. Think of ways that you could do this, then try them to see if they work.

13 Investigating force and pressure

A block of copper, size $5\,\text{cm} \times 5\,\text{cm} \times 10\,\text{cm}$, has volume $250\,\text{cm}^3$ ($0.000\,25\,\text{m}^3$). Its mass is $2.25\,\text{kg}$. We can calculate its density as follows:

$$\rho = \frac{\text{number of kilograms of mass}}{\text{number of cubic metres of volume}} = \frac{2.25}{0.00025}$$

$$= 9\,000 \ (\rho \text{ in kg/m}^3)$$

It is emphasised that we are dividing a *number* by a *number*. In particular the quantity of material in the block is represented by its *mass*, not by its weight. The density of a material is a property of that material and should not depend on where the density is measured. Mass is a property of the block and does not vary if we take the block to different places, so we must use mass when defining density. However, for practical purposes we can 'weigh' the block and use what we *call* its weight for our calculations of density, provided that we do all the 'weighing' at or near Earth's surface.

If the block is resting on a flat level surface, it presses on the surface because of its weight. If its mass is $2.25\,\text{kg}$, and if it is on Earth's surface where $g = 10\,\text{m/s}^2$, the total downward force on the block is given by:

$$F = mg = 2.25 \times 10 = 22.5 \ (F \text{ in N})$$

If the block is resting on one of its faces measuring $5\,\text{cm} \times 10\,\text{cm}$, the area of contact with the surface below is $50\,\text{cm}^2$, or $0.005\,\text{m}^2$. The pressure on the surface is:

$$P = \frac{\text{number of newtons of force}}{\text{number of square metres of area}} = \frac{22.5}{0.005}$$

$$= 4\,500 \ (P \text{ in N/m}^2).$$

Since the unit 'newton per square metre' is called the 'pascal' (p.10), the pressure calculated above can therefore be expressed as $4\,500\,\text{Pa}$.

13.00 Fluid pressure

If you take a can and punch three holes in it of equal size and fill it with water, the water runs out as shown in fig.

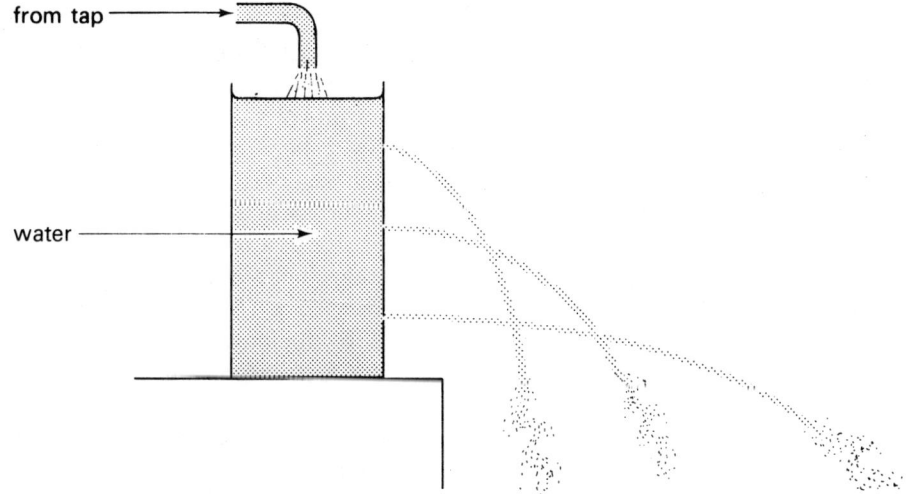

Fig. 13.1 Demonstration of water pressure.

13.1. The explanation of this is that pressure in the water increases with depth. Let us find the exact relation between pressure and depth, and also see what other factors affect pressure in liquids.

Imagine a container filled with a liquid, density ρ. At distance h below the surface is an object. The upper surface of this object has area A (fig. 13.2). The shape of the surface does not matter. The surface is pressed on by the weight of the liquid about it. Think of this liquid as a column with area of cross-section A, and height h.

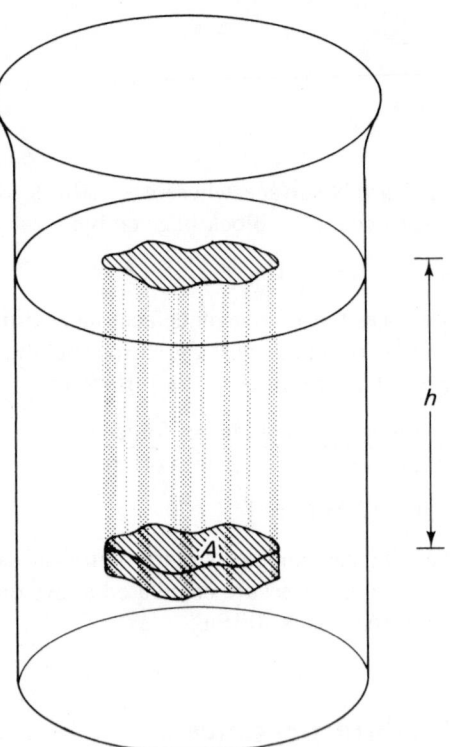

Fig. 13.2   Factors affecting water pressure (see text).

The volume of this column is:

$V = Ah$ (V in m³).

The mass of this column is:

$m = \rho V = \rho Ah$ (m in kg).

The force of gravity on this mass of liquid (in other words, its weight) is:

$F = mg = \rho Ahg$ (F in N).

The pressure on the surface due to this weight is:

$$P = \frac{F}{A} = \frac{\rho Agh}{A} = \rho gh \text{ (}P\text{ in Pa or N/m}^2\text{).}$$

The area has cancelled out of the calculation. The pressure in a liquid varies with the depth and the density (and also with g). Remember that in measuring h this is the *vertical* distance below the surface.

Instructions: finding the direction of pressure in a liquid

1 Use a metal can or plastic container that has been badly crumpled so that its sides are very bent. Make holes of equal size in its sides, so that there are several holes at the same level, but in surfaces that face in different directions (fig. 13.3a). Instead you can use a plastic bag (fig. 13.3b).

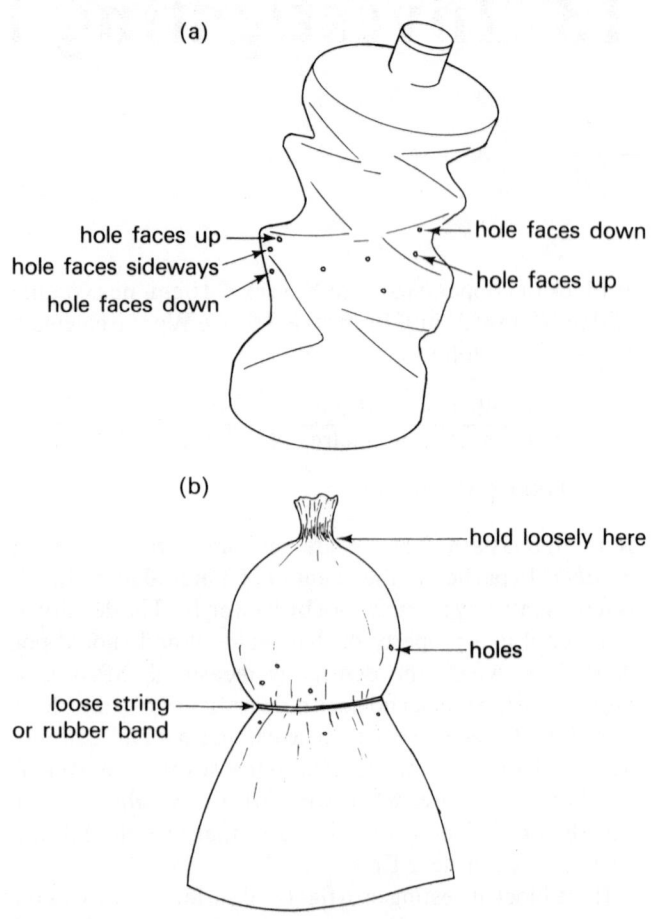

Fig. 13.3   The direction of pressure in a liquid.

2 Fill the can (or bag) with water. Watch the water run out. In which direction is pressure greatest: upwards, downwards or sideways?

3 Look closely at the jets of water just as they come through the holes. In what direction is the pressure on the inside surface of the can at each of these holes? The results do not give us a proof that pressure is *exactly* equal in all directions, but this can be proved by calculations similar to those above. If you think of the object in fig. 13.2 as a thin horizontal layer of water, the downward force on this layer is $F = \rho Agh$. This force should make the layer of water accelerate downwards, but it does not. There must be an equal force, acting upwards. Therefore the upward pressure must be exactly equal to the downward pressure.

The taps of a house are usually supplied from a water tank on the roof, or in a roof-loft. The reason for this is that

128

pressure is created by placing the tank high above the outlet, the tap. At the tap the water pressure depends on the vertical distance between the end of the tap and the surface of the water in the roof-tank. The greater the vertical height the greater the pressure and the more quickly the water flows from the tap. You may have noticed that the pressure at a tap in an upstairs room is less than the pressure at a tap in a downstairs room, in the same building. If the house catches fire it is not possible to connect a hose to the tap and use this to spray water on the roof. The water could not rise to a level higher than that from which it came, for the pressure is not great enough. Instead a pump must be used. The water supply to a town usually comes from a reservoir on a hill or tall tower. Unless the water comes from a dam or mountain stream at higher level it has to be pumped up to this reservoir. If the reservoir is high above the houses this creates enough pressure to force the water into the roof-tanks of the houses below. It may also create enough pressure to allow fire hoses to be used effectively when connected directly to the town main water supply.

13.01 The syphon

A syphon is a useful way of transferring liquids from one container to another (fig. 13.4). The tube is first filled with liquid; then it is placed with one end in each container. Liquid then flows through the tube from the container in which the liquid is at higher level. Since pressure in a liquid depends on depth, the pressure is equal at all points at a given depth in a body of liquid. At all these points, depth is h and pressure is ρgh. The pressure at the surface in container A could be zero. (It might be more, if there is an atmosphere pressing there, but a syphon works in a vacuum, so we need not take atmospheric pressure into account.) Let us call it P_0. Since pressure is equal at all points on the same level, it is P_0 also at points A and B

inside the syphon tube. Point C is a depth $(h_1 + h_2)$ below point B, so the pressure at the end of the syphon tube is P_1.

therefore $P_1 = P_0 + \rho g(h_1 + h_2)$

If we consider the liquid in container B, the pressure at its surface is zero, since it is exposed to the same atmosphere, if any, as the liquid in container A. Therefore the pressure in the liquid at the end of the syphon is

$P_2 = P_0 + \rho g h_2.$

The pressure difference at the end of the tube is given by

$$P_1 - P_2 = [P_0 + \rho g(h_1 + h_2)] - (P_0 + \rho g h_2)$$
$$= \rho g h_1$$

This pressure difference, acting outwards at the end of the tube, causes liquid to flow into container B. It continues to flow until h_1 becomes zero, because then the pressure difference becomes zero, as the equation shows. This happens when the liquid levels in the containers are equal. Then flow stops. Thus the syphon illustrates a rule which is often stated: liquids find their own level.

If the syphon tube does not reach down below the liquid surface in container B, its operation is not affected. The pressure at the end of the syphon tube is

$P_1 = P_0 + \rho g h_3.$

The pressure (if any) acting inwards on this liquid surface is P_0, so the pressure difference is

$P_1 - P_0 = (P_0 + \rho g h_3) - P_0 = \rho g h_3.$

This pressure difference, acting outwards at the end of the tube causes liquid to flow out of the tube and fall into the container. It continues to flow until h_3 becomes zero, because then the pressure difference becomes zero. This happens when the liquid in container A is level with the end of the syphon tube. Then flow stops.

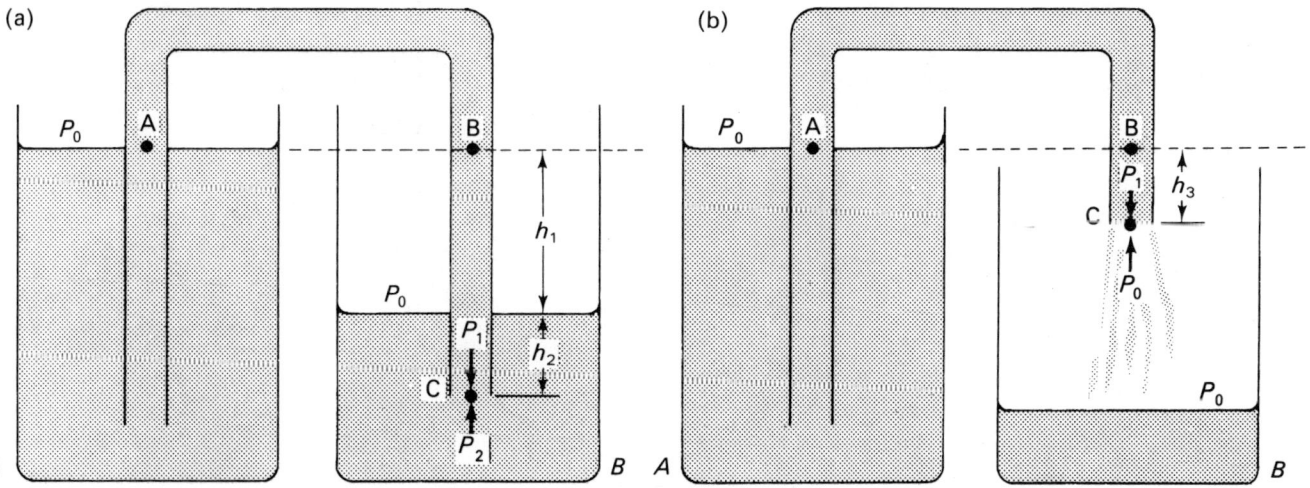

Fig. 13.4   The action of a syphon (see text).

13.10 Pascal's principle

Instructions: transmitting pressure by water

1 Connect two plastic disposable syringes together by a short length of rubber or plastic tubing. One syringe (A) should have large diameter (capacity about 20cm³ or more), the other (B) should be of small diameter (capacity 1cm³).

2 Support the syringes vertically in two clamps, as in fig. 13.5.

3 Fill them and the tube with water, so that the pistons are about half-way along each syringe.

4 Press on one piston. What happens to the other?

5 Balance a 20N weight on the wider piston (A), keeping your finger pressed down on piston B to stop it from rising. If the weight will not balance securely glue a sheet of card to the end of the piston, to make it wider.

6 Place weights on piston B so that the piston just does not rise. You may need to glue a thin card to this piston, too, if the end of the piston is not wide enough to hold your weights securely.

7 Find out what weight must be placed on piston B to make piston A (with the 20N weight) rise.

8 How far up does piston A rise when piston B is pushed 5cm downwards? This is a model of a hydraulic press or jack. It depends on *Pascal's principle* which states that *if a fluid completely fills a vessel, and if pressure is applied to any part of the surface of the fluid, the pressure is transmitted equally to the whole of the fluid*. Pascal's principle applies both to liquids and to gases, but liquids do not decrease in volume when pressure is applied to them. This is why liquids are used more often than gases. In the hydraulic jack there are two cylinders (fig. 13.6) the pump cylinder and the ram cylinder. Also, but not shown in the diagram, there are valves so that liquid pumped from the pump cylinder to the ram cylinder can be held there until a release valve is opened.

Questions

1 Suppose the area, A, of the pump piston is $0.01\,m^2$ and a force F of 100N is applied. What pressure is applied to the fluid in the jack (usually a special kind of oil)?

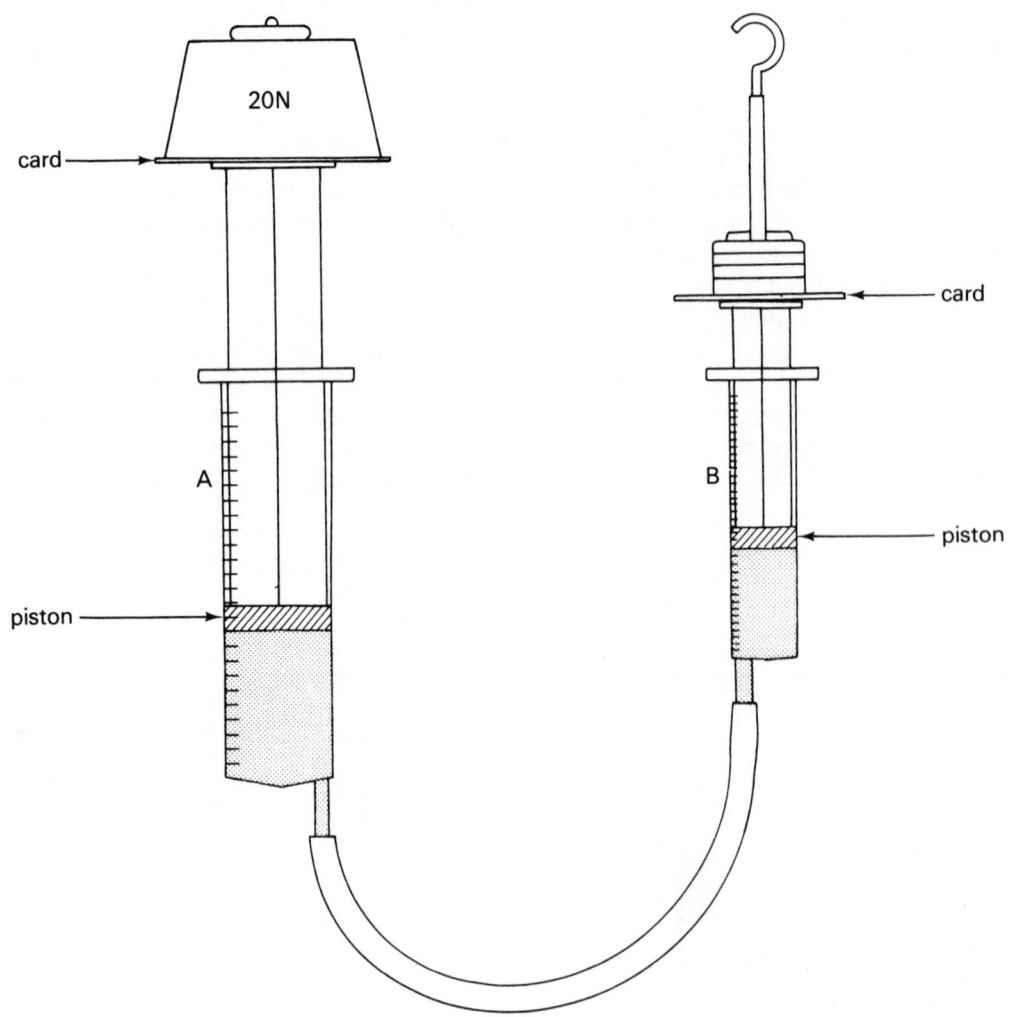

Fig. 13.5 Transmitting pressure by water.

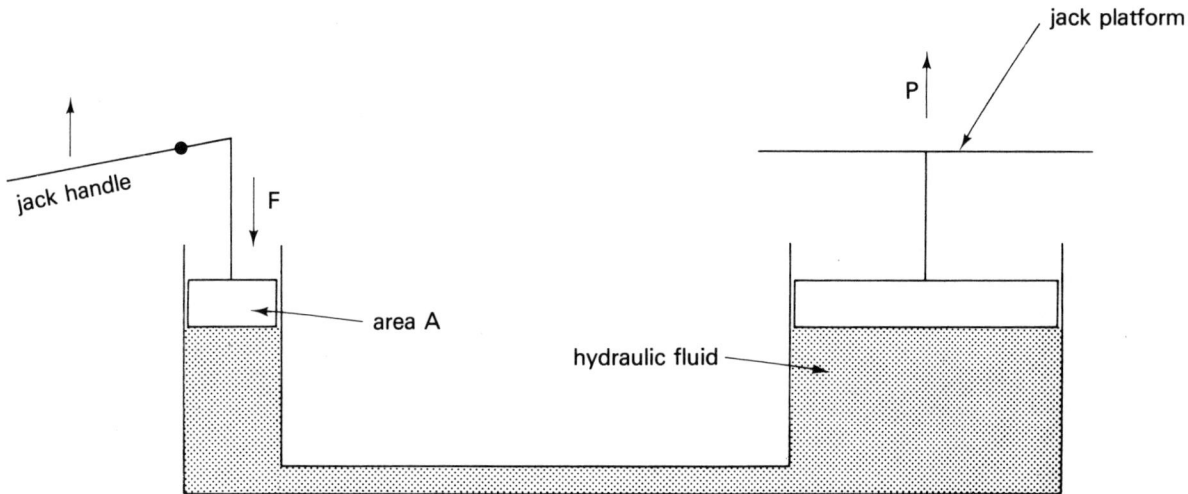

Fig. 13.6   Principle of a hydraulic jack.

2 If this same pressure is exerted over the whole surface of the ram piston, area 4m², what is the upward force *P* on the ram piston?

As with a lever, a small force applied over a long working distance can be used to produce a very great force acting over a relatively short distance. The advantage of the hydraulic jack is that if the ram piston is made large and strong enough enormous forces can be produced. In industry hydraulic presses are used for many purposes, including forming automobile bodies from sheet metal. In addition, the transmission of pressure through a fluid-filled pipe is often a very convenient way of transferring a force from one place to another. In most automobiles, fluid pressure is used to operate the brakes. The brake pedal is connected to a piston in an oil-filled cylinder (fig. 13.7). This is connected by tubes to the brakes on the wheels. When the pedal is pressed, the increased pressure is transmitted to cylinder B where it pushes outwards on

two pistons, which push outwards on two brake shoes (S). These press against the brake drum to which the wheel is bolted, exerting frictional force on the wheel, to prevent it from rotating. The advantage of hydraulic brakes depends on Pascal's principle that pressure is transmitted *equally*. The pressure of the driver's foot is applied equally to all brakes, braking force operates equally on all wheels of the vehicle, and there is much less risk of skidding.

13.20   Units for measuring at atmospheric pressure

The standard atmospheric pressure corresponds to a height of 760mm in a mercury barometer (2.31). We can use the equation from page 128 to calculate the pressure at the bottom of a column of mercury 760mm high. For mercury, $\rho = 13\,600\,\text{kg/m}^3$, so if the column is 0.76m high, and $g = 9.8\,\text{m/s}^2$, the standard atmospheric pressure

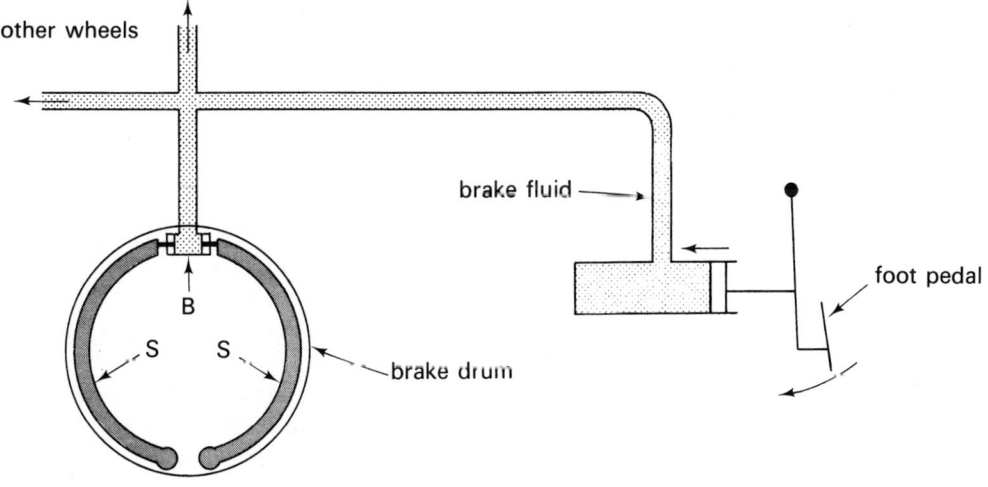

Fig. 13.7   Hydraulic brake system of an automobile.

is $P = \rho gh = 101\,000$ (approximately) (P in Pa or N/m².) This figure is near enough to 100 000 Pa for most practical purposes.

Weather-men use this quantity as a convenient unit for atmospheric pressure. Their unit is the *bar*, which is equivalent to 100 000 Pa (or N/m²). They subdivide this into *millibars*. These are the units generally marked on weather maps. Where you see '1 000 mbar', this means a pressure of 1 bar, or 100 000 Pa.

Questions

1 One tube of a manometer (fig. 2.12) is open to atmospheric pressure of 96 000 Pa. The water level in this tube is 50 mm higher than the water level in the other tube, which is connected to a closed container. What is the pressure in the container? (Take $\rho = 1\,000$ kg/m³ and $g = 10$ m/s².)

2 The pressure in a closed container is 150 000 Pa. Atmospheric pressure is 100 000 Pa. What difference in height will there be between liquid levels in the tubes of a manometer if the manometer is filled with (a) water, and (b) mercury?

3 If we wish to make a barometer, using water instead of mercury, what is the expected height of the column at standard atmospheric pressure?

13.30 Archimedes' principle

We have used this principle before (4.22); now we are in a position to prove it. Imagine a vessel containing a liquid (fig. 13.8). In this we imagine a particular region of the liquid, shaded in the figure, though we cannot really see it as a distinct region. It has cross-sectional area A, and height h.

The volume of the liquid $V = Ah$ (V in m³).

If the density of the liquid is ρ, its mass is:

$m = V\rho = Ah\rho$ (m in kg).

The weight of the liquid is:

$W = mg = Ah\rho g$ (W in N).

If the pressure on the upper surface of the region of liquid is P_1, the total force acting downward on the region is:

$F_1 = P_1A$ (F_1 in N).

Similarly, if the pressure on the lower surface of the region of liquid is P_2, the total force acting upward on the region is:

$F_2 = P_2A$ (F_2 in N).

The upthrust on the region of liquid is:

$U = F_2 - F_1 = (P_2 - P_1)A$ (U in N).

According to the equation relating pressure and depth,

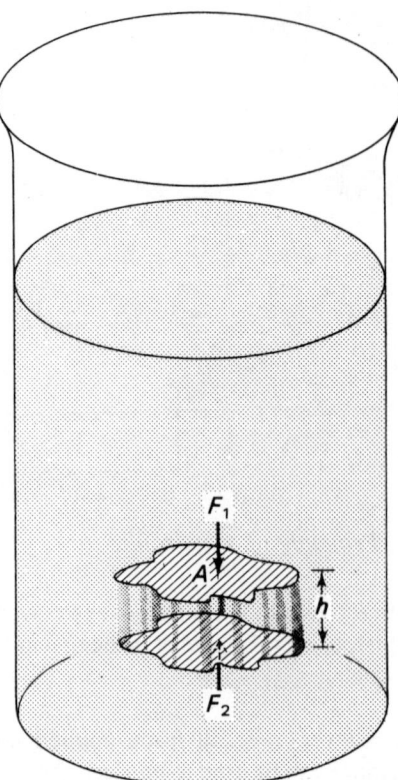

Fig. 13.8 Forces on a region of liquid (see text).

$P_2 = P_1 + \rho gh$ (P in Pa),

where ρgh is the difference of pressure between the upper and lower surfaces of the region. From this equation we can see that:

$P_2 - P_1 = \rho gh$ (P in Pa),

and $U = \rho ghA$ (U in N).

Referring back to the earlier equation for the weight of the liquid region we now find that:

$U = W$ (U and W in N).

In words, the upthrust on the region of liquid is equal to its weight. We have considered the case of a region of liquid, but we could also imagine the same region filled with any other material, such as iron or wood. Whatever material fills the region, the upthrust is unaffected. Always, it equals the weight of the liquid which would fill the region: a volume of liquid equal in volume to the region or equal in volume to any solid object filling the region. In other words, *the upthrust is equal to the weight of liquid displaced by the object*. This is *Archimedes' principle*.

Instructions: making a Cartesian diver

1 The apparatus is shown in fig. 13.9. For a bottle use one made from thin transparent plastic. The kind used for cooking-oil, fruit drinks or shampoo is suitable. It must have a secure water-tight cap.

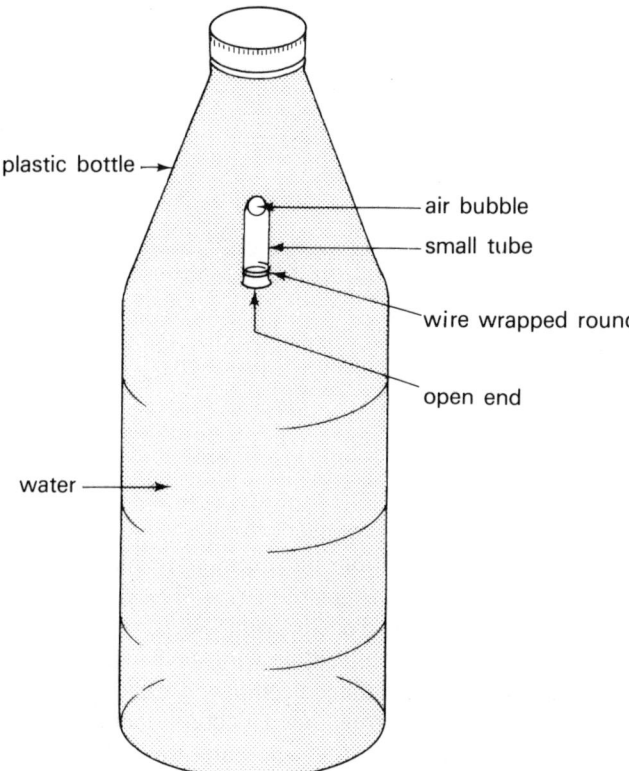

plastic bottle

air bubble

small tube

wire wrapped round

open end

water

Fig. 13.9   A cartesian diver.

2 The diver is made from any small tube of glass or plastic open at one end. Use a small test-tube, the barrel of a small syringe (plug the outlet with Plasticine), or the cap from a ball-point pen. Wrap wire around the open end so that when a *little* air is in the diver, it floats vertically, open end downwards.

3 Adjust the amount of air in the diver by floating it first in a bowl of water. When the correct amount is present, the diver floats on the surface but when given a slight push, it sinks and takes several seconds to come back to the surface again.

4 With this amount of air inside it, transfer the diver to the bottle, which is already full of water. Fill the bottle to the top, and screw on the cap so that there is no air above the water in the bottle.

5 Squeeze on the outside of the bottle; watch what happens to the diver.

6 Release the pressure on the bottle; watch what happens to the diver.

7 This makes an amusing toy for a younger member of your family. For you, the problem is to explain how it works.

Projects

1 Find out as much as you can about lift pumps and force pumps. Obtain models, if possible, to see how they work. Using your knowledge of pressure, explain their action.

2 Work out your own method for measuring the pressure of the mains water supply. Compare your result with the pressure estimated, knowing the height above your school of the town reservoir or the school's storage tanks.

3 Find out what you can about deep sea diving vessels and how they are made to stand up to the enormous pressures at depth.

4 Find out what you can about the effects of water pressure on divers who use Scuba equipment. If you can obtain a set of this equipment, study it to see how it works, and ask an expert to explain how to use it correctly.

Fig. 13.10   Deep-sea diving vessel. This one can hold two divers.

14 Investigating electromagnetic force

Instructions: demonstrating electromagnetic force

1 Fix two narrow strips of aluminium foil side by side, but not touching (fig. 14.1). It is better if the apparatus is surrounded by a transparent screen to prevent the strips from being blown about by draughts.

2 Connect the strips as shown in the figure, so that current flows upwards in both strips.

3 Switch on and off very briefly, observing the strips closely.

4 Reverse the current by exchanging the wires at the terminals of the battery. Switch on and off briefly, closely observing the strips as the current flows downwards in both of them.

5 Reconnect the strips so that the current will flow upwards in one and downwards in the other. Switch on and off, observing closely.

6 Reverse the battery connections and repeat step 5. Taking all your observations into account, make up a rule to describe the behaviour of the strips. The results are what can be expected from previous work on electromagnetism (9.20). A current flowing in a wire has a magnetic effect; it can deflect a compass needle. Here we see that one electric current can deflect *another current*. To be more precise, *each* of the currents deflects the other.

Instructions: finding the relation between force and current

1 If you already have a current balance; use it, following the manufacturer's instructions. If not, you can make a simple current balance, basing the design on the microbalance, fig. 12.5.

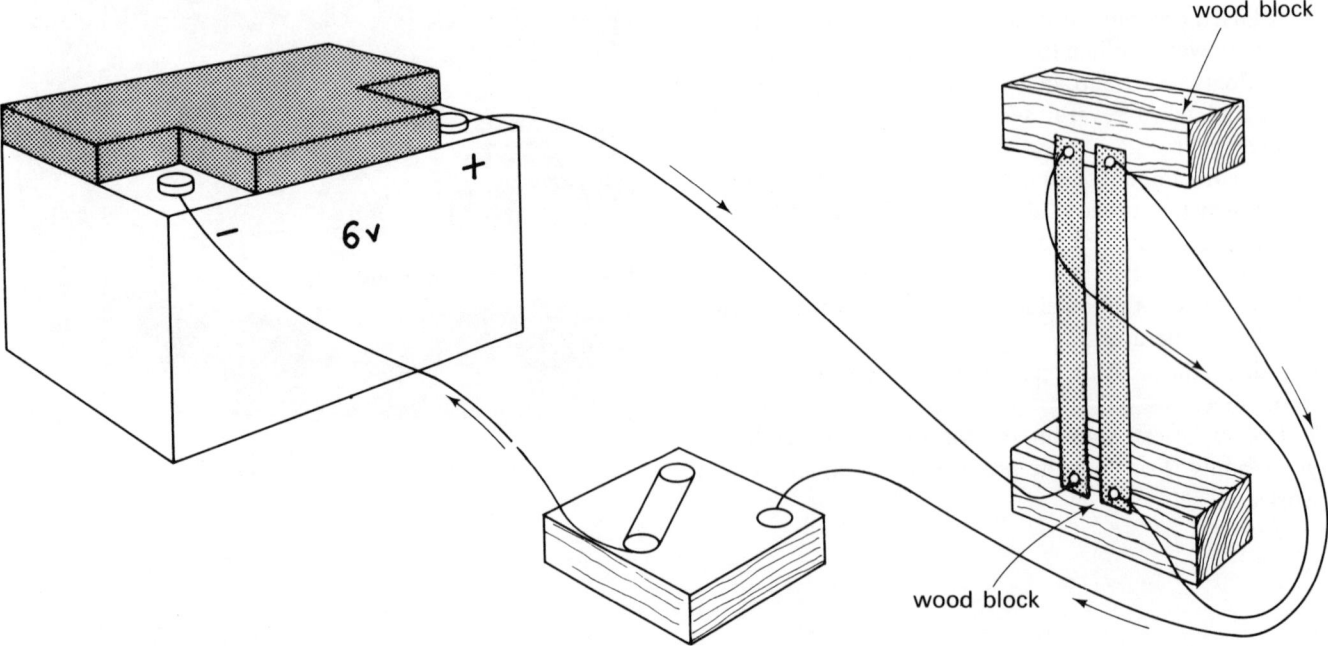

Fig. 14.1 Demonstrating electromagnetic force.

2 Use a drinking straw or a piece of grass cane or thin bamboo, about 20 cm long. Cut one end to make a pointer.

3 To the other end fix a small strong bar magnet (Alcomax, etc) so that its poles point up and down. Hold it in place with sticky tape.

4 Find the balance-point by balancing the beam across a sharp edge, such as a ruler. Then push the needle through the beam about 1 mm from this point, towards the pointer end. When the beam is rested on the support it should rest with the magnet end down and the pointer end up.

5 Make a rider from copper wire. This should be heavy enough to make the beam take up a horizontal position when the rider is about 4 cm from the needle. Mark the position of the rider.

6 Make a coil by winding about 100 turns of wire around a short cylindrical tube. The tube can be cut from card or from a plastic bottle. When in position the coil should fit around the magnet, so that the magnet is near the centre of the coil, but the coil must not prevent the beam from swinging.

7 With the beam at rest in a horizontal position, mark the scale to show this rest position.

8 Connect the coil in a circuit, with one lamp, a switch and a 2V accumulator cell, as shown in the circuit diagram.

9 Switch on. The magnetic field in the coil should attract the magnet further into the coil. The pointer should rise. If it falls, reverse the connections to the cell. If the effect is very slight, connect an additional cell or cells, or increase the number of turns in the coil.

10 Switch on. The pointer rises. Move the rider towards the pointer end of the beam. Find the position at which the additional moment produced by shifting the rider makes the beam again come to rest with its pointer level with the

mark. If the beam rests with pointer high, even when the rider is at the pointer end, the magnetic force is too strong to be measured. What can you do to measure it?

11 When the beam is balanced, measure the distance between the position of the rider and its original position. What extra moment has been produced by shifting the rider? What extra moment has been produced by the electromagnetic force between coil and magnet?

12 Connect a second lamp into the circuit (position A) so that the current flows through each lamp in turn. Switch on; what do you notice about the brightness of the lamps? What can you say about the current flowing in the circuit now?

13 With the current flowing, adjust the position of the rider until the beam is balanced again. What is the extra moment now? What extra moment is being produced now by the coil and magnet? Is this greater or less than that at step 11?

14 Repeat steps 12 and 13 with a third lamp connected in the circuit at position B.

15 Make a rule to relate the amount of electric current with the amount of electromagnetic force.

The current balance of fig. 14.2 measures the force between a magnet and the current in a coil. It has the advantage that the magnet is very strong, so even a small current produces an easily measurable force. The great disadvantage is that we do not know the strength of the field of the magnet; with a different magnet we might get bigger or smaller forces; also the strength of the magnet changes in time, especially if the magnet becomes suddenly demagnetised, so we can never really compare measurements made with the same magnet at different times. To get over this problem a reliable current balance makes use of the force between *two wires* or coils carrying the *same*

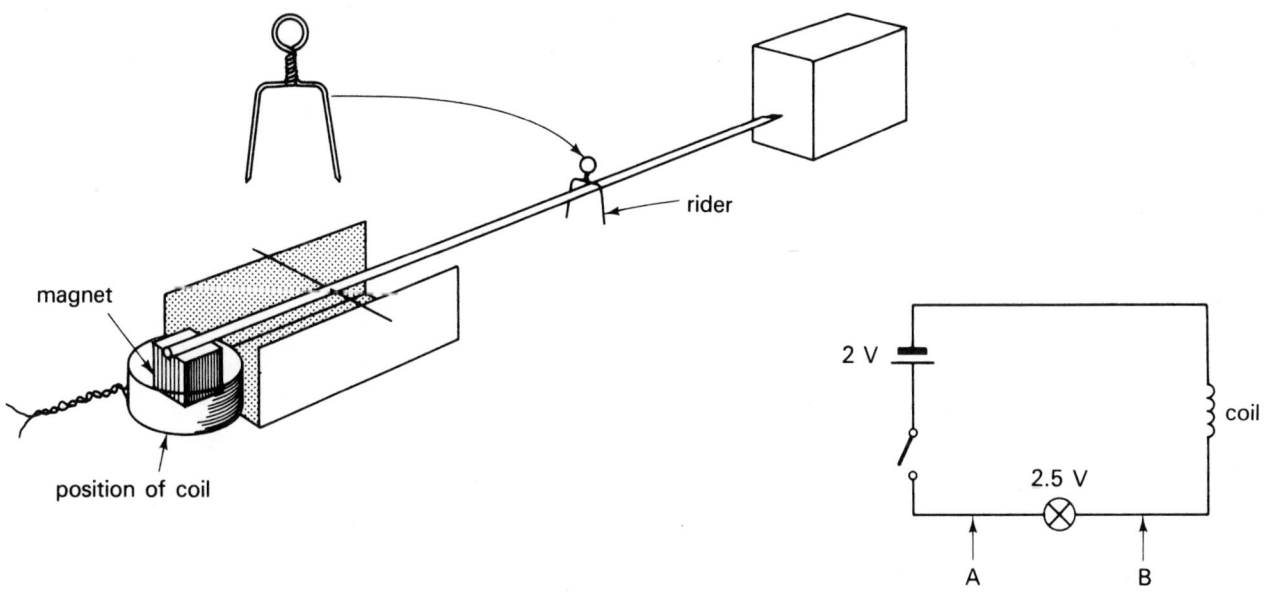

Fig. 14.2 Making a simple current balance.

135

current. Then we can measure the effect of the current *on itself*, and no magnets are required. In the experiment with two strips of aluminium carrying the same current we could *see* the effects of the force, but not measure it. The force between two wires is small and very sensitive balance mechanisms are needed to measure it (fig. 14.3). Current balances usually have carefully made coils with a very large number of turns, to increase the size of the force. When the balance is used the force is measured in newtons. This gives us a method of relating electric current to one of the SI units.

The SI unit of electric current is the *ampere*, often called an 'amp' for short. Its symbol is A. As with other SI units we have multiples, such as kiloampere (kA), and submultiples, such as milliampere (mA). The unit is named after André Ampere who made many of the early discoveries about electricity. We will not go into details of the way the ampere is defined. In brief, we can calculate what force is produced by a current of 1A flowing in a current balance with coils of given size and shape. We adjust the current until this force is registered. Then we know that the current flowing is exactly 1A. This same current can be passed through some other current measuring instrument (26.21) and its scale can be marked '1A'.

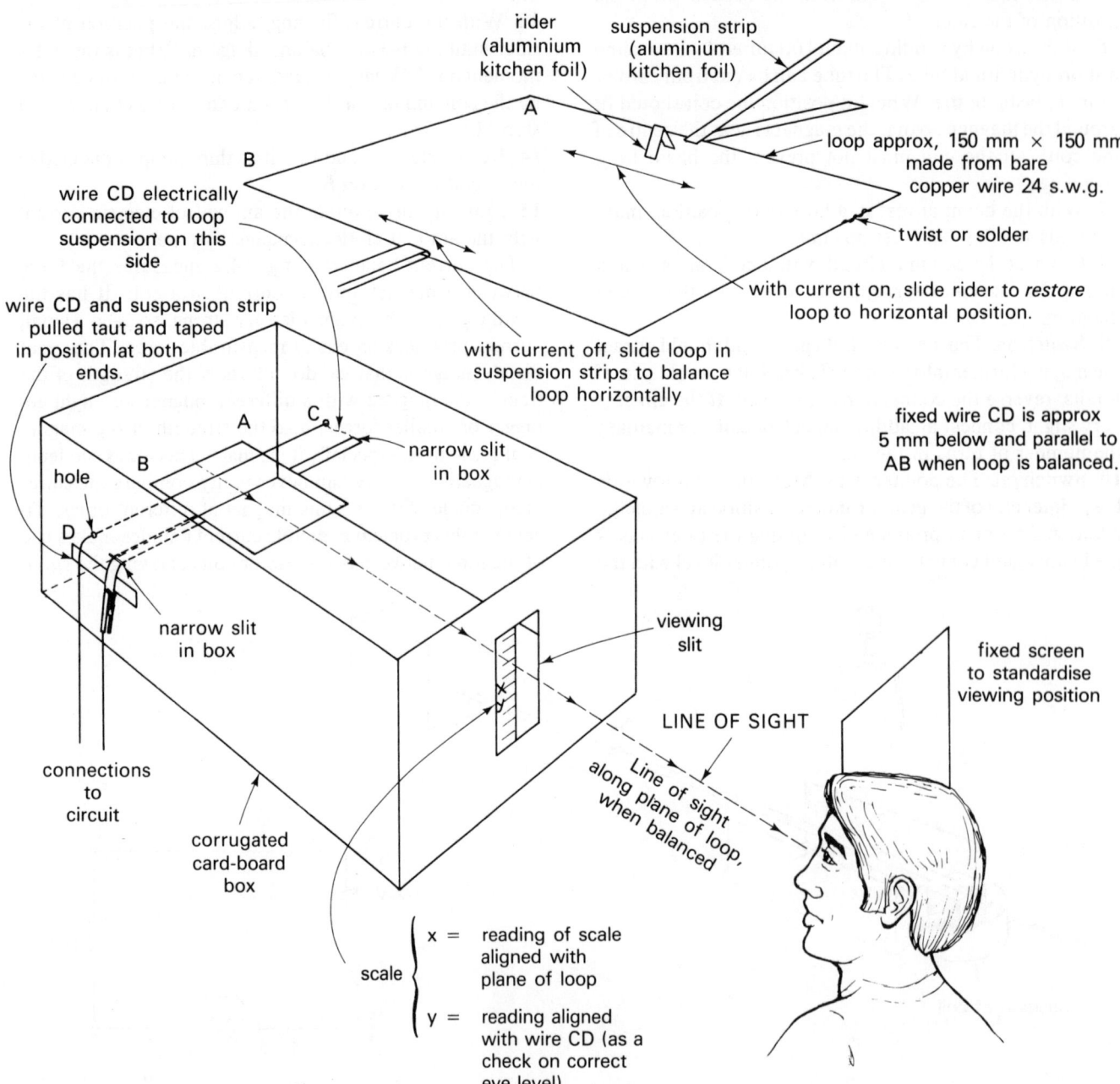

rider
(aluminium kitchen foil)

suspension strip
(aluminium kitchen foil)

loop approx, 150 mm × 150 mm made from bare copper wire 24 s.w.g.

twist or solder

wire CD electrically connected to loop suspension on this side

wire CD and suspension strip pulled taut and taped in position at both ends.

with current on, slide rider to *restore* loop to horizontal position.

with current off, slide loop in suspension strips to balance loop horizontally

fixed wire CD is approx 5 mm below and parallel to AB when loop is balanced.

hole

narrow slit in box

narrow slit in box

viewing slit

fixed screen to standardise viewing position

LINE OF SIGHT

Line of sight along plane of loop, when balanced

connections to circuit

corrugated card-board box

scale

x = reading of scale aligned with plane of loop

y = reading aligned with wire CD (as a check on correct eye level)

Fig. 14.3 A home-made current balance with parallel conducting wires.

Electric current is a *quantity of electricity* flowing past a point in a circuit *in a given time*. In many ways it is like a current of water. To measure the amount of water current flowing in a pipe we can let the water flow out of the pipe and into a bucket for a measured period of time. Then we could calculate the water current as:

$$\frac{\text{number of kg of water collected in bucket}}{\text{number of seconds during which water is collected}}$$

In the same way we can say that an electric current is

$$I = \frac{\text{number of units of electric charge passing a point in a circuit}}{\text{number of seconds during which the current passes}}$$

Suppose that you have an electric circuit. A current balance or some other type of current meter is part of this circuit. It measures the current and tells you it is 1 ampere. You switch on the current for exactly one second. During that time the quantity of electric charge passing any point in the circuit is 1 *coulomb*. This is how we relate the unit of electric charge to the unit of current. In brief we say: '*The coulomb is the quantity of electric charge which passes any point in a circuit when a current of one ampere flows for one second.*' The coulomb, symbol C, is named after the physicist, Charles Coulomb. If a current I flows for time t, the quantity of charge flowing past a point in the circuit is $Q = It$ (Q in C).

Questions

1 An electric lamp is switched on in a room for 3 h. The current passing through the lamp is 0.4 A. What quantity of electricity passes in this time?

2 A flashlamp operates at full brightness when a current of 0.06 A is passing. The battery can deliver this current for 27 h. What total quantity of electric charge can the battery supply?

3 A small portable radio uses a current of 10 mA. What quantity of electricity passes through the set while switched on to listen to a news broadcast lasting 5 min?

Project

Build a simple current balance like that shown in fig. 14.3. Use it to demonstrate that the force on the wire AB varies as the current that passes through it.

15 Work, energy and power

15.00 Work

The word 'work' has many meanings; it can mean some kind of job, what you do in school, an artist's painting or a musical composition. In physics it has a very special and very exact meaning: *if a force is applied to an object and the point at which the force is applied moves in the direction of the force, work has been done.* If you push against a door, and it opens as a result of your push, you have done work. If the wind blows and a leaf moves, the wind has done work. If you grasp a weight and lift it from the ground, you have done work. A force has produced motion. But if you push against the door and it does not move, you have done no work, no matter how hard you have pushed. If you have tried to lift the weight but it is too heavy for you to be able to lift it even a fraction of a millimetre, you have done no work.

Work is done only if there is a force *and motion*. The amount of work done is measured by the product of the force and the distance moved in the direction of the force. If you pull *upwards* on a rock with a force of 50N and lift the rock a distance of 1.5m *upwards*, the amount of work you have done is:

W = number of newtons of force $\times$ number of metres moved in direction of force = $50 \times 1.5 = 75$ Nm.

The derived unit 'newton metre' is used so often in physics, for work can be done in so many different ways, that it has a special name, the *joule*. The symbol for joule is J and it is named after the physicist, James Joule, who first put forward the idea of conservation of energy. We will discuss this subject in the next chapter. We can therefore define the joule as follows:

One joule of work is done when the point at which a force of one newton is applied moves one metre in the direction of the force.

To do a joule of work, lift an object weighing 1N a vertical distance of 1 metre. You could use a 100g weight as your object. As with other SI units, we can have multiples and submultiples of the joule. As you can see, 1J represents rather a small amount of work, so we most often use the multiples, the kilojoule (kJ) and the megajoule (MJ). The latter is 1 million joules.

Questions

1 A girl pushes a block of wood across a table, using her finger. She presses against the block with a force of 5N. The block moves 0.5m in the direction of the force. How much work has she done?

2 A man weighing 700N climbs a ladder, raising his body through a vertical distance of 3m. How much work has he done?

3 A boy throws a ball of mass 0.1kg. It leaves his hand at the end of 0.5s with a velocity of 2m/s. How much work has he done?

4 If the boy in question 3 threw the same ball, with a velocity of 2m/s, but took 0.2s to throw it, how much work would he have done?

15.01 Work done by machines

You studied some simple machines in 2.10 and 2.20 and discovered some relations between load and effort. Now we will look at these in more detail and see how much *work* is done by the machine. The experiment which follows deals with one particular system of pulleys but you can adapt the instructions to other pulley systems and to various types of levers. If you are working in a class it would be a good idea for different groups to use different kinds of machines, either levers or pulleys. Then you can see that the conclusions to be reached from the experiment apply to machines generally.

Instructions: investigating machines and work

1 Set up a pulley system, as in fig. 2.6, type 7. Attach a weight of 10N to the hook on the lower pulley block.

2 Place a ruler vertically beside the load; raise the load

Load	Effort	Mechanical advantage i.e. L/E	Efficiency i.e. $M.A./V.R. \times 100\%$

a measured distance, say 10cm. While you do this, measure the distance moved by your hand (effort) in pulling on the free end of the cord. The easiest way to do this is to tie a small knot in the cord, place a ruler along the cord and see how far the knot is moved. Record the distances moved by load and effort.

3 Calculate the *velocity ratio* for the pulley system.

$$\frac{\text{Velocity}}{\text{ratio}} = \frac{\text{distance moved by the effort}}{\text{distance moved by the load}} \quad \text{in the same time.}$$

4 With various loads on the hook, measure the effort required to just raise them. Record results in newtons, in a table as above.

You can measure effort either by adding weights to the free end of the cord or by attaching a dynamometer and pulling by hand.

5 The *mechanical advantage* of the pulley system is 4, in theory, but in practice the weight of the lower pulley block and of the cord are an additional load. Also, extra effort is required to overcome the friction in the bearings of the pulleys. The practical *M.A.* is therefore less than 4. Calculate *M.A.* for each pair of values of load and effort and write the results in the table.

6 Plot a graph of *M.A.* (on the vertical axis) against load (on the horizontal axis).

7 The efficiency of a machine is defined as a percentage:

$$\text{Efficiency} = \frac{\text{Useful work done by the machine}}{\text{Work put into the machine}} \times 100\%$$

$$\frac{\text{Useful work done by}}{\text{machine}} = \text{load} \times \frac{\text{distance load}}{\text{is lifted}}$$

$$\frac{\text{Work put into the}}{\text{machine}} = \text{effort} \times \frac{\text{distance moved}}{\text{by effort}}$$

$$\text{Efficiency} = \frac{\text{load} \times \text{distance load is lifted}}{\text{effort} \times \text{distance moved by effort}} \times 100\%$$

$$= M.A. \times \frac{1}{\text{velocity ratio}} \times 100\%$$

$$= \frac{M.A.}{V.R.} \times 100\%$$

The last expression is a convenient way of calculating efficiency without the need to calculate the actual amounts of work required for moving the loads a given distance. Calculate efficiency for each line of the table and enter this in the fourth column of the table.

8 Plot a graph of efficiency (on the vertical axis) against load (on the horizontal axis).

Questions

1 What are the units of velocity ratio?

2 What is the velocity ratio for the pulley system of fig. 2.6, type 7?

3 Describe and explain the shape of the *M.A.* – load graph.

4 Is it possible with this pulley system for the value of the *M.A.* on the graph to rise beyond 4, if the load is made sufficiently large?

5 What are the units of efficiency?

6 Describe and explain the shape of the efficiency-load graph.

7 Can efficiency ever be greater than 100%?

Instructions: investigating the efficiency of tools and other machines

The procedure of the previous experiment can be applied to tools and other simple machines in everyday use. Examples include pliers, screwdriver, spanner, carpenter's brace and bit, wheelbarrow, engineer's vice, automobile lifting-jack, inclined plane or wedge (fig. 15.1), crowbar, bicycle, windlass and any other simple machines used locally. You will need sometime to design your own methods for measuring load and effort. The general procedure is as follows:

1 Measure distances moved in the same time by load and effort; calculate velocity ratio.

2 With various loads, measure the effort required. It may not be easy to arrange for many different loads, but try to use more than one, so that you can get an idea how efficiency varies with load (*if* it does).

3 For each load, calculate mechanical advantage and efficiency.

4 Record your results in a table. If you have several values of load, plot graphs so that you can see how *M.A.* and efficiency vary with load.

5 Calculate the theoretical *M.A.*, ignoring effects of friction, weight of parts of the machine etc.

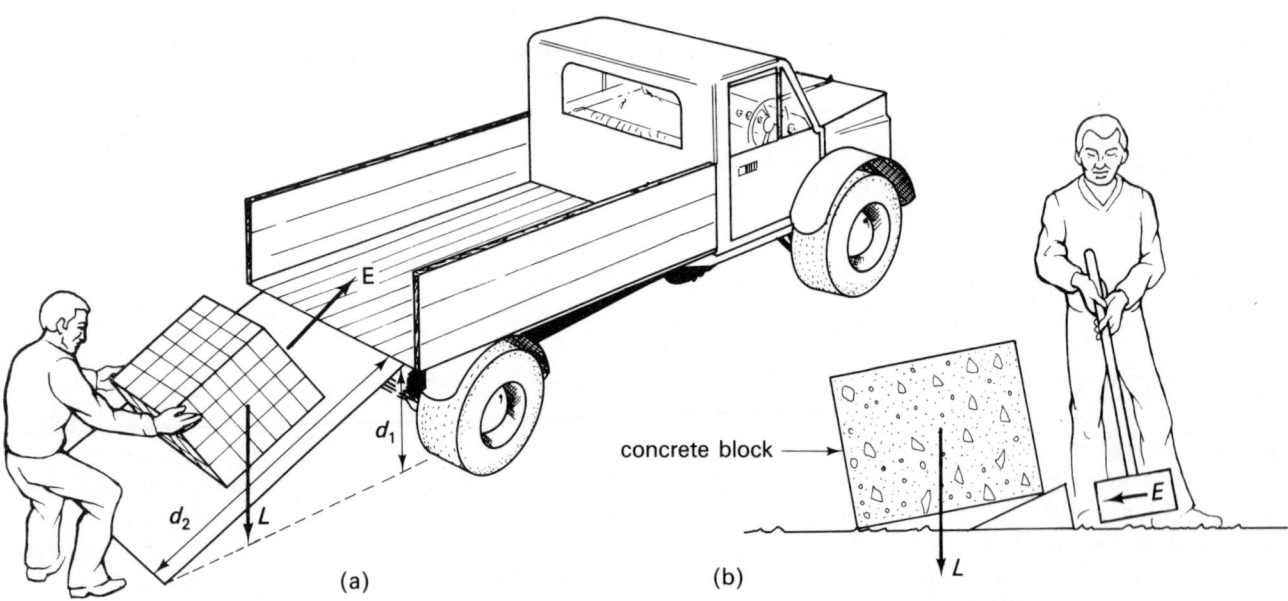

Fig. 15.1 (a) Inclined plane: Which distance is moved by the load? Which distance is moved by the effort? (b) wedge.

Questions
1 Which of the machines listed above has the highest mechanical advantage?
2 Some, like the inclined plane, have a low mechanical advantage. What is the reason for using these?
3 Which of the machines listed above has the highest efficiency?
4 What is the reason for four-speed or five-speed gears on a bicycle?

15.10 Energy

Energy is defined as the capacity for doing work. If you wind up a clockwork motor, the effort you make in turning the winding-knob makes the spring become more tightly coiled. When the motor is wound, with its brake on, the motor is not doing any work. It has the *capacity* to do work for, if you release the brake, the spring will start to uncoil and drive the wheels of the motor. It can be connected to a pulley system and can be used to lift a load: to do work. While the motor was fully wound with the brake on, the spring was storing energy; it had the capacity to do work. The amount of energy it stored depended on the amount of work you did in winding it (less any losses through friction). The amount of work that it could do while unwinding depends on how much energy it stored. Ignoring losses due to friction during winding and unwinding, the amount of work done by the motor should be equal to the amount of work you put into it. Before winding, the motor had no energy; it could do no work. In between winding and unwinding, the stored energy of the motor was high. After it had unwound, it could do no work; it had lost all the energy you had given it. Since work and energy are related in this direct way, they are expressed in the same unit, the joule.

Energy exists in many forms. In the example above the energy is *mechanical energy*. Your experiments in 9.20 show that an electric current can cause a magnet to move: a magnetic force acts and the magnet moves in the direction of the force. Work is done. If electricity has the capacity to do work we can say that electricity is a form of energy: *electrical energy*. Heat can do work, as when the expanding steam in a steam engine or the expanding gases in an automobile engine move the piston; so heat is a form of energy: *heat energy* or *thermal energy*. These and many other forms of energy are of great importance in the world today. In chapter 16 we will study some of the many forms of energy and how one form can be converted into another. For the present we will return to the study of mechanical energy.

A moving object can strike another object, and the impulse of the blow can cause the second object to begin to move. The first object has done work on the second by making it move. This means that the first object has energy, the capacity to do work. This energy results from the fact that *it is moving*. This type of mechanical energy which is the result of the motion of an object is called *kinetic energy*.

15.11 Kinetic energy

An object of mass m is moving with a velocity v. We wish to know how much energy it has. The amount of energy is equal to the amount of work done in giving the object its velocity. Suppose that the object was accelerated by a force F, acting while the object moved along a distance s. The initial velocity of the object was zero, so $u = 0$.

The equation of motion (10.05) gives us the relation

$$v^2 = u^2 + 2as$$

Since $u = 0$, we can write

$$v^2 = 2as \text{ or } a = \frac{v^2}{2s}$$

To accelerate the object, a force F is required such that

$$F = ma$$

$$\therefore F = m\frac{v^2}{2s} \text{ } (F \text{ in N})$$

The point of application of this force moves over a distance s, so the work, W, done by this force is given by

$$W = Fs = m\frac{v^2}{2s}s \text{ } (W \text{ in joules})$$

Work done, W, is equal to the kinetic energy E_k of the object.
Cancelling s and rearranging the expression gives

$$E_k = \tfrac{1}{2}mv^2 \text{ } (E_k \text{ in joules})$$

Questions

1 A train, mass 1 000 000 kg, travels at 90 km/h. What is its kinetic energy?
2 A metal ball, mass 2 kg, falls from a height 5 m. What is its kinetic energy just as it reaches the ground?

In question 1 above we can see that a heavy object travelling at high velocity can have enormous kinetic energy. If the velocity is doubled, the energy is increased four-fold. This point is of interest to people concerned with fuel consumption of locomotives and automobiles. The energy required to accelerate a vehicle from rest is gained by burning fuel of some kind. Since E_k is proportional to the *square* of the velocity, it takes four times as much fuel to accelerate an automobile from rest to 20 m/s as it takes to accelerate it from rest to 10 m/s. To accelerate it from rest to 30 m/s takes nine times as much fuel.

To reduce velocity, kinetic energy must be removed from the vehicle. Usually this is done by applying brakes. Then, through friction, kinetic energy is converted to thermal energy and the velocity of the vehicle decreases. If the vehicle is brought to rest by crashing into a tree, the kinetic energy is used to change the shape of the tree and the vehicle. A vehicle that is travelling at 20 m/s has four times as much energy as one travelling at 10 m/s, which explains why damage to a vehicle which crashes at high speed is so much more serious than damage to one which crashes at low speed.

Once a vehicle has accelerated, it has gained kinetic energy. This energy is still there, even though the vehicle changes direction. The energy depends on the speed of the vehicle, but not its direction of movement. Energy is a scalar quantity.

In question 2 above we saw that the metal ball has an energy of 100 J on reaching the ground. How did it acquire this energy? It accelerated under the force of gravity and by this means gained energy during its fall from a period of rest. Does this mean that energy can be created from nothing? To answer this we will look at another type of mechanical energy.

15.12 Potential energy

We will continue with the example of the metal ball. If we begin with a ball resting on the ground, we might say it has no energy. If we lift it a vertical distance of 3 m, we have to do work *against* the force of gravity. The force required is 20 N*, and the distance moved in the direction of the force is 5 m, so the work done in lifting the ball is $W = 5 \times 20 = 100$ J. You can see that this amount is the same as the kinetic energy of the ball just as it reaches the floor. What energy did the ball have when it was 5 m above the floor, being held motionless, ready to be dropped?

At 5 m above the ground the ball had no motion, so it had no kinetic energy, but it *can* be thought of as having energy because of its *position*, 5 m above the ground. We call this *potential energy*. The changes of energy of the ball are summarised in fig. 15.2. At (a) the ball is on the ground and has zero energy. At (b) it is being lifted and is being given potential energy. At (c) it has no kinetic energy, and its potential energy is 100 J. It is then allowed to fall. Half-way down, at (d), we can find from the equation of motion or a velocity-time graph that its velocity is 7.071 m/s. Therefore

$$E_k = \tfrac{1}{2} \times 2 \times 7.071^2 = 50 \text{ } (E_k \text{ in J}).$$

* You may object that, if the downward force on the ball is 20 N due to gravity, and if we push up on the ball with a force of exactly 20 N the forces will exactly balance and the ball will not rise. This is true. But if we exert a little extra upward force at the beginning of the journey, say a total of 21 N upward, the ball accelerates upwards. Then, when the journey is half complete (at 2.5 m), we can reduce our upward force to 19 N. The ball continues to travel upwards but gradually decelerates, since the total force on it is now 1 N downward. It comes to rest at exactly 5 m above ground. In this case we have raised the ball 5 m by applying 21 N for 2.5 m and 19 N for 2.5 m. Total work done $= (21 \times 2.5) + (19 \times 2.5) = 52.5 + 47.5 = 100$ J. The same result as above. In short, in examples of this kind a little extra force applied at the beginning to accelerate the object is exactly compensated for by a reduction of force later in the journey to allow the object to decelerate to rest. We can therefore ignore the problem of accelerating and decelerating the object and simply consider the magnitude of the force against which we work.

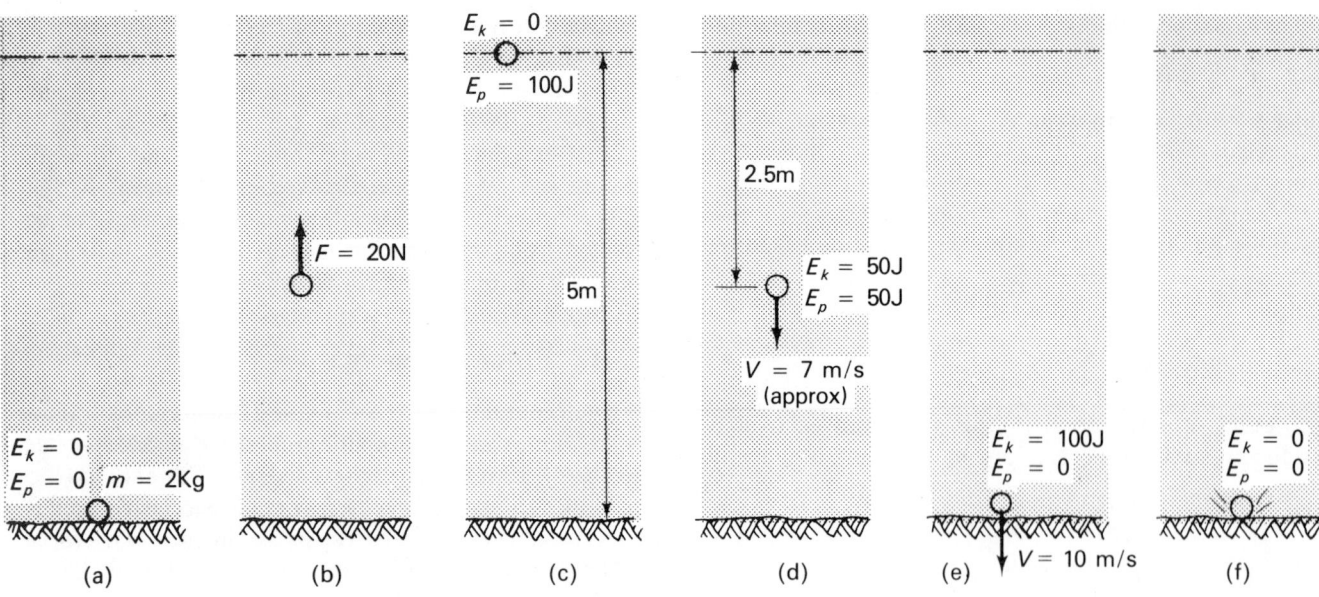

Fig. 15.2   Kinetic energy and potential energy (see text).

Its potential energy is the amount of work needed to lift it from the ground to half-way position; $W = 2.5 \times 20 = 50J$. As it has fallen it has lost potential energy and gained kinetic energy. Note that the total mechanical energy, $E_k + E_p$, remains unchanged at 100J. As it reaches the ground, at (e), all its potential energy is gone and it has gained a total of 100J kinetic energy. Again $E_k + E_p = 100J$.

We can state this as a general rule that for a moving body

$E_k + E_p =$ **constant**.

At (f) we see the ball immediately it has hit the ground. It has now lost all its kinetic energy, since its velocity is now zero. The ball now has no energy and is in the state in which it began at (a). The kinetic energy has been converted into other forms of energy: perhaps some energy to deform the ground; some to make sound; some to heat the ball and ground (thermal energy). This energy is that which was given to the ball when it was lifted. Like all energy most of it eventually becomes converted into heat.

Figure 15.3 shows how a pendulum is continually converting potential energy to kinetic energy and back again, with very little loss. At A the bob is at the top of its swing to the right. It is motionless for an instant, so $E_k = 0$. If the weight of the bob is 10N, this is the force that had to be overcome for it to rise the distance h above its lowest position. Note that h is the *vertical* distance as the force of gravity is the only force that has to be overcome to raise the bob, and gravity acts vertically downwards. Its E_p is therefore $10h$ J. At its lowest point, B, it has lost potential energy, and its kinetic energy must now equal the potential energy lost, so $E_k = 10h$ J. As it swings up to the left, it loses velocity and therefore kinetic energy and at C it has $E_k = 0$ and $E_p = 10h$ J again.

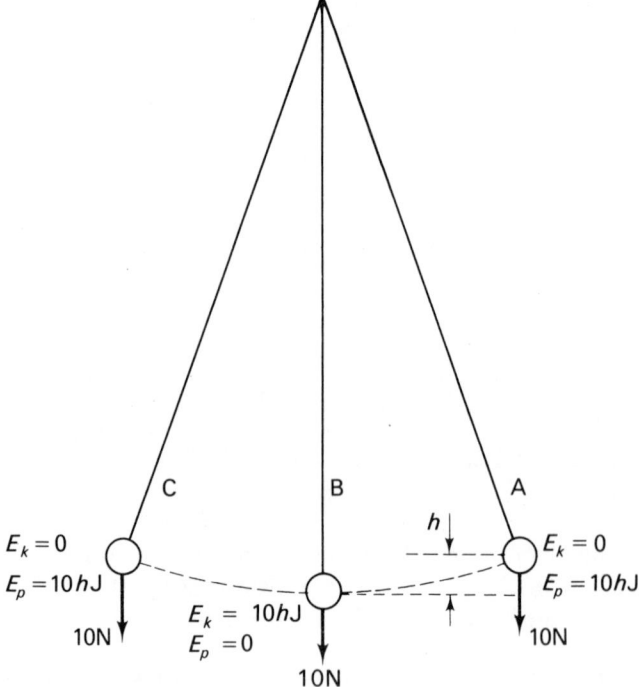

Fig. 15.3   Kinetic energy and potential energy of a pendulum.

Questions

1 It was stated above that there is very little loss of energy in a swinging pendulum. What losses are there?

2 Where is the bob when its velocity is greatest?

3 For the pendulum shown, with a bob weighing 10N, and a vertical height of swing of 0.1m, what is the maximum velocity of the bob?

Another type of potential energy can arise from the *state* that an object is in. At the beginning of this chapter

142

we referred to winding a clockwork motor. When the spring is tightly coiled it has no motion, so has no kinetic energy, but it is capable of motion if released. When the spring is coiled and the brake is on, it has potential energy. We call this *elastic potential energy*. It gained this when it was wound, and it can release it and convert it to mechanical energy when the brake is taken off. Here the energy may first appear as a *rotation* of the parts of the motor. This is *rotational energy*, another form of kinetic energy. When a catapult is used the operator pulls back the elastic. He does work on the elastic and it stretches. It gains elastic potential energy due to its stretched state. When it is released, the energy is converted into the kinetic energy of the moving elastic and the moving catapulted stone. We must therefore define potential energy as energy due to the *position or state* of an object.

We can state when we consider an object to have zero kinetic energy; this is when it is not moving. If you think about the Earth in space you soon realize that a pencil which is resting on your desk has a high velocity. The Earth is spinning on its axis and it is in orbit around the Sun. Its speed around its orbit is about 30 km/s. Someone out in Space might consider that your pencil has an exceedingly high kinetic energy. For most purposes we forget about these velocities and think of motion *relative to the Earth*. We choose to consider that objects which are at zero velocity relative to the Earth have zero kinetic energy. In the discussions which we have had on potential energy, we have taken zero potential energy as the energy possessed by the object in its lowest position, that is to say when resting on the ground. This again is just a convenient way of thinking of things. In the pendulum example we did not use the ground as the level for zero potential energy. Instead we used the lowest position of the bob during its swing. It is in order for us to choose any level (or state) as zero and work from that, provided we remember which level (or state) we have chosen as zero and state our results *relative to* this level.

15.13 Electrical potential energy

First we will look at an example of mechanical potential energy, similar to the example of fig. 15.2. In fig. 15.4a a metal ball, mass 1 kg, is lifted 5 m above ground level. Do not look at fig. 15.4b yet. For convenience, consider that the ball has zero potential energy when it is on the ground. The ball is a *quantity of matter* (1 kg); it is in a gravitational field; it is acted on in a downward direction by gravitational force. We lift the ball upwards a distance 5 m. At all stages on its journey a force 10 N acts downwards on it. We do work against this force. When we have raised the ball 5 m, the work done is 50 J (page 142). We say that $E_p = 50$ J. If the mass of the ball is 2 kg, work done is 100 J, and so on, in proportion. In a given gravitational field, the work done depends on the mass of the ball and the distance through which it is moved. We can say that, for mass m, distance s and field g, work done is equivalent to potential energy and

$$E_p = smg \quad (E_p \text{ in J}).$$

If we think of different balls of different masses, all lifted the *same distance*, to the same level above ground, 5 m, we find that $E_p = 5 \times m \times 10 = 50m$ J. Potential energy is 50 joules *per kilogram*. This figure is a property of that particular level above ground. No matter what object is raised to that level, if we take the number of kilograms of its mass and multiply by 50, we get its potential energy. Each kilogram of it acquires 50 J in being raised to a height of 5 m. This is what we mean when we say that the gravitational potential of that level is 50 J/kg.

Figure 15.4b is similar but represents an electrical field. Electrical fields and forces are similar in many ways to the magnetic fields and forces you studied in chapter 8. In this diagram there is no gravity and no mass; the plane of the diagram is not necessarily vertical, it could lie in any direction. The 'object' in the figure is a *quantity of electric charge* (1 coulomb); it is in an electrical field; it is acted on by an electrical force towards the bottom of the figure where, for convenience, we consider its electrical potential to be zero. We move the charge towards the top of the diagram, a distance 5 m. The strength of the field is such that at all stages on its journey a force of 10 N acts on the charge, tending to make it return to the region of zero potential. We do work against this force. When we have moved the force 5 m, the work done is 50 J. We can say that its *electrical potential energy*, E_p, is 50 J. If we release the charge it is forced back to the region of zero potential energy, losing potential energy as it goes. A charge 1 C experiences in this field a force 10 N, a charge 2 C experiences a force 20 N, and so on in proportion. So, as in the gravitational example, we can say that when we move a charge from zero potential to a point 5 m away (in *this* field), the work done is 50 joules *per coulomb*. This is the potential energy acquired by every coulomb of charge moved to this position in this field. The quantity 50 J/C is a property of this region of the field. We call it the *electrical potential* of the region.

Other regions have different electrical potentials. In the figure, regions with potential 30 J/C and 45 J/C are marked. If we take a charge of 1 C from the 30 J/C region to the 45 J/C region we must do 15 J work, and the charge gains 15 J potential energy. Moving in the reverse direction it loses 15 J potential energy. The *potential difference* between these regions is 15 J/C. The potential difference between zero and a region 5 m away is 50 J/C. Since we use the term 'potential difference' quite often we shorten it to '*p.d.*'. Electrical potential and p.d. can be expressed in joules per coulomb, but because this unit is often used it is given a special name. The joule per coulomb is called the *volt*, after the physicist who did much of the pioneer work

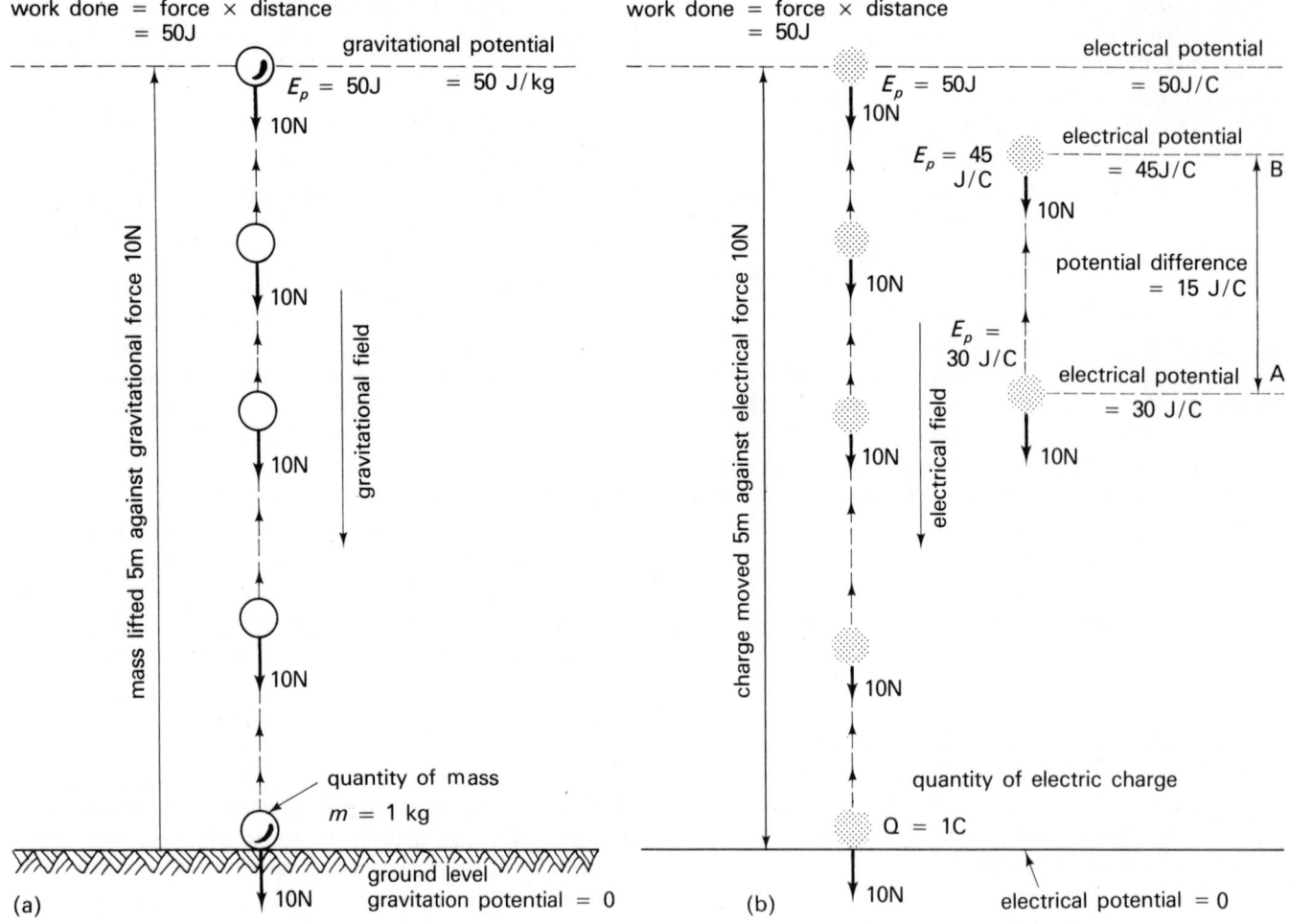

Fig. 15.4 Gravitational and electrical potential (see text).

in electricity, Alessandro Volta. Its SI symbol is V. Therefore we can say that the electric potential at the top of fig. 15.4b is 50 volts. The p.d. between levels A and B is 15 V. If a charge, Q, is moved against an electric field and the p.d. between the beginning and end of its journey is V its gain in electrical potential energy is

$$E_p = VQ \quad (E_p \text{ in J}).$$

Conversely, if the charge moves under the action of the field, its loss of potential energy is VQ.

In chapter 9 we measure the e.m.f. of a cell in volts. We can apply the idea of electrical potential to such a cell. Figure 15.5a shows a zinc-carbon cell connected to a lamp. Current flows from the positive electrode, through the lamp, to the negative electrode. It also flows through the electrolyte and completes the circuit. In fig. 15.5b the circuit has been straightened out. We can take any point in the circuit as zero potential; we will take the zinc electrode. The current is made to flow in the circuit, carrying electrical charge around the circuit, because of the electrical field that exists in the circuit.

If we tried to move a charge from the zinc electrode,

through the wires and lamp to the carbon electrode and, finally, through the electrolyte and back to the zinc electrode, we would do work *against* the electric field. We would be moving the charge *against* the direction in which the current normally flows. Just as we have to do work to run around a race-track and return to our starting point, so we must do work in carrying a charge backwards around a circuit and back to its starting point. Over the whole journey there are electrical forces opposing the motion of the charge; forces of different size operate at different points in the circuit and are not shown on the diagram. Measurements show that the work done to move a charge of 1 coulomb around this circuit is 1.1 J. To move a charge of 2 C requires 2.2 J, and so on, in proportion. If we take a charge of 1 C once round the circuit it gains the work we have done and its potential energy increases by 1.1 J. Thus when a charge of 1 C flows through this circuit freely under the action of the electric field, it loses 1.1 J of potential energy during one complete circuit.

Where does the potential electrical energy go? Most of it is converted to internal energy (heat, chapter 18) as the current passes through the wires. The wires become

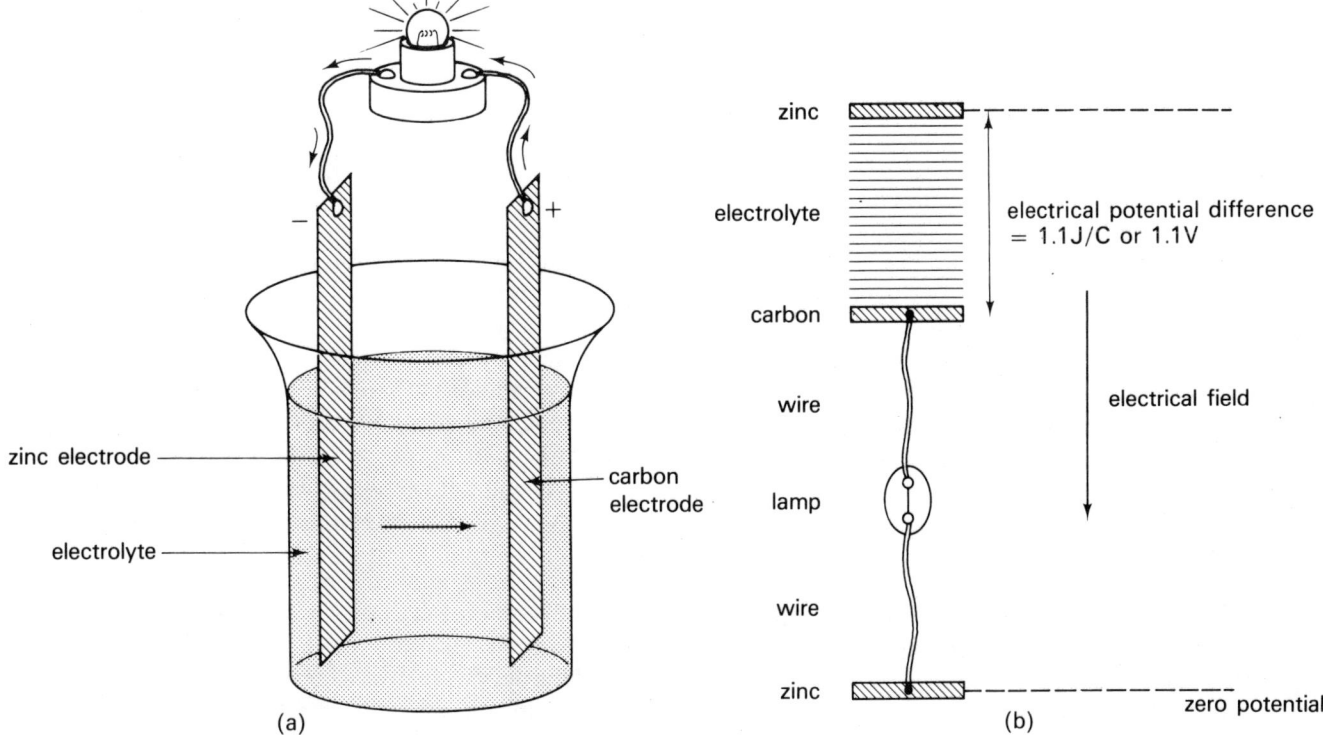

Fig. 15.5 Electrical potential in a circuit containing a cell.

slightly heated; the thin wire of the lamp gets so hot that it glows and gives off another form of energy: light energy.

Where does the electrical energy come from? The electrical field is the result of the particular arrangement of chemical substances of the cell: the electrodes and electrolyte. This results in chemical energy and the cell is a device for *converting* it into electrical energy. It creates an *electromotive force* (e.m.f.) to drive the charge around the circuit. In other words, it maintains the electric field, providing the energy to move charges along in the field.

Even if there are no wires or lamp connected to the cell, the electric field still exists between the terminals of the cell. There is a *potential difference* between them; if a current can flow, it will.

15.20 Power

Power is the rate of doing work. If a boy lifts a book, weight 2N, from a table and puts it on a shelf 1.5m above the table, he does 3J of work. No matter how quickly or how slowly he does this job, the total amount of work done is 3J. If he lifts the book quickly, taking 2s to lift it to the shelf, he has done 3J of work in 2s. This is 1.5J *per second*. His rate of working, or *power* is 1.5J/s. If he does the same job, but perhaps reads a few pages of the book before putting it on the shelf, he may take 100s to do the job. The total amount of work done is still 3J, but his rate of working is now only 0.03J/s. His power is 0.03J/s.

Instructions: measuring your personal power
1 What is your weight? Did you express this in newtons?
2 Run up a flight of steps as quickly as you can. Ask your partner to measure how long you take from bottom to top.
3 What is the vertical height between the bottom and top of the steps?
4 Calculate how much work you did in raising your body through this vertical distance.
5 What is your power?

The SI unit of power is joule per second but, since it is so often used, it has a special name, the *watt*, symbol W. The watt is named after the engineer James Watt, the inventor of many early steam engines.

Questions
1 A light truck, mass 2 000kg accelerates from rest to a velocity 36km/h in 25s. What is the acceleration of the truck?
2 What is the useful force produced by its engine?
3 If the truck is travelling at 36km/h, what power is developed by its engine?
4 Some men are carrying a box weighing 1 000N up a stairway to a landing 5m above ground level. What potential energy has the box gained by being lifted up the stairs and placed on the landing? What work is done by the men in raising the box? If the men carry the box upstairs quickly without resting and it takes them 200s, what is their power? If they rest several times during the job, and it takes them 10min to raise the box, what is their power? If

the box accidentally falls vertically downwards from the edge of the landing, what is its kinetic energy as it reaches ground level?

5 A locomotive is towing a railway truck at uniform velocity 5m/s. The tension in the coupling is 450N. How much work is done by the locomotive in towing the truck 20km? What is the useful power expended by the locomotive in towing the truck?

The questions above are about mechanical work and mechanical power. The idea of power can be applied to any kind of work. Since 'doing work' is really a matter of converting one kind of energy into another (muscular energy into kinetic energy of a cricket-ball, electrical energy into light energy from an electric lamp, and so on), another way of defining power is to say:

Power is the **rate of conversion of energy** *from one form into another form.*

When a sound arrives at your ear it makes your eardrum vibrate. The kinetic energy of vibrating air is converted into kinetic energy of the moving parts of your ear. The power is very small, only a tiny fraction of a watt. At the other extreme the rate of conversion of chemical energy (from fuel) into electricity at a power station may be measured in kilowatts or even megawatts. The energy radiated by the Sun is 100 000 000 000 000 000 000 000 watts. This really is a tremendous rate of energy conversion.

Let us go into the topic of electrical power in more detail.

15.21 Electrical power

If you examine a household lamp, you generally find its wattage marked on the glass. Lamps are commonly sold in various wattages: 25W, 60W, 100W and 150W. The higher the wattage, the greater the rate of conversion of electrical energy to light energy, and the brighter the lamp. The greater the heat production, too. You may also find the correct mains voltage marked on the glass: this may be 110V, 240V or other voltages.

Suppose you have a lamp marked: 100W, 250V. You plug it into a 250V supply. The power of the lamp is 100W, meaning that every second it converts 100J of electrical energy into light and heat. The potential difference between the terminals of the lamp is 250V, or 250J per coulomb. This means that for every coulomb of electric charge passing through the lamp, 250 joules of potential energy are lost: they appear as light and heat. If the lamp converts 100J in 1 second, it requires 2.5s to convert 250J. Thus the electric charge is passing through the lamp

at the rate of 1 coulomb every 2.5s, that is to say, 1/2.5 coulomb per second, or 0.4C/s. The amount of charge passing per second is what we call the current: 0.4C/s means 0.4 amperes. To sum up, if a 100W lamp is rated to operate on a 250V supply, the current through the lamp when operating on this supply is 0.4A.

Questions

1 A flashlamp rated to operate from a 6V battery operates at 0.06A. How much electrical charge flows through the lamp in 1 second?
2 How many joules of electrical energy are converted while this amount of charge flows through the lamp, given that the potential difference between the terminals of the lamp is 6V?
3 If this amount of energy is converted to light and heat each second, what is the wattage of the lamp?

Given voltage and current, we have been able to calculate wattage. Let us see how to simplify the stages of the calculation.

Consider a device (such as a lamp) which operates with a potential difference V between its terminals, and carries current I.

The amount of charge flowing through the device in time, t, is

$$Q = It \ (Q \text{ in C}).$$

This charge flows through a potential difference V, so the potential energy lost and converted to light (or heat etc, depending on what sort of device it is) is

$$E_p = VQ = VIt \ (E_p \text{ in J}).$$

The power, or rate of energy conversion, is

$$P = \frac{E_p}{t} = \frac{VIt}{t} = VI \ (P \text{ in W}).$$

This gives us the relation:

$$P = VI \ (P \text{ in W})$$

This relation is one that is frequently used for calculating the power of electrical devices and has the advantage that we do not need to calculate the numbers of joules or coulombs involved in the energy conversion.

Questions

1 The current flowing through the motor of a model racing car is 700mA; the p.d. between its terminals is 6V. What is the rate of conversion of electrical energy in the motor?
2 The element of an electric kettle is rated at 3kW, when connected to main supply at 240V. What current flows through the element?

16 Using energy

16.00 Types of energy

Several types of energy have been mentioned in chapter 15. The list below summarises the main types.

(a) *Mechanical:* either *kinetic* (motion) or *potential* (gravitational or elastic). Elastic potential energy is the storage of energy in springs etc.
(b) *Radiant:* includes light, radio waves, ultra-violet radiation, X-rays, infra-red radiation.
(c) *Chemical:* includes fuels such as oil, wood, coal; explosives; electrical cells; food.
(d) *Electrical*
(e) *Nuclear:* the production of energy from matter (chapter 33).
(f) *Internal energy:* the kinetic and potential energy of the molecules of which an object is composed (chapter 18), sometimes inaccurately called 'heat energy'.

16.10 Conservation of energy

The law of conservation of energy states:
Energy can neither be created nor destroyed. This law was believed to be true for many years, until early in the twentieth century. Then Einstein, in his Theory of Relativity, put forward the idea that matter could be converted into energy and energy into matter. This new idea meant that the law of conservation of energy and the law of conservation of matter could no longer be believed. Later experiments showed that Einstein was right. We are now able to build nuclear power stations in which matter is converted into energy; we call this *nuclear energy*. As we shall see in chapter 33, processes of a similar kind are responsible for the production of nuclear energy in the Sun and other stars, and by atomic bombs. Though the laws of conservation of energy and conservation of matter have now been shown to be untrue, they still apply to most aspects of physics and of everyday life. So we still find

them useful. It is only under certain special conditions (eg in the centre of the Sun, or in an atomic pile) that they are untrue and that we need to apply different laws. The next experiment tests the law of conservation of energy.

Instructions: testing the conservation of energy
1 Set up two dynamics trolleys for inelastic collision. Trolley A is to be connected to a ticker-tape timer. The trolleys may be loaded to have unequal masses.
2 One trolley (B) is stationary. Accelerate the other trolley (A) so that it collides with B and the two remain joined together and move on together for a distance.
3 Measure the tape and calculate the following velocities:

u_A, the velocity of trolley A before collision;
v_A, the velocity of trolley A after collision.
For trolley B we know $u_B = 0$ and $v_B = v_A$.

4 The total *momentum* before collision is
$m_A u_A + m_B u_B = m_A u_A$ (since $u_B = 0$).
Calculate this.
5 The total momentum after collision is $(m_A + m_B)v_A$.
Calculate this.
6 The total kinetic energy before collision is
$\frac{1}{2} m_A u_A^2 + \frac{1}{2} m_B u_B^2 = \frac{1}{2} m_A u_A^2$ (since $u_B = 0$).
Calculate this.
7 The total kinetic energy after collision is
$\frac{1}{2}(m_A + m_B)v_A^2$. Calculate this.

Questions
1 What is the relation between momentum before and after collision? Explain.
2 What is the relation between kinetic energy before and after collision? Explain.

16.20 Conversion of energy

In the previous investigation kinetic energy was converted to sound and internal energy. The conversion of energy

Examples of man-made energy convertors are:

	Converts energy	
	from	**to**
Electric lamp	electricity	light and heat
Candle	chemical energy	light and heat
Electric cell	chemical energy	electricity
Edge of matchbox	kinetic energy	heat
Musical instrument	kinetic energy	sound vibrations
Electric motor	electricity	kinetic energy
Dynamo	kinetic energy	electricity
Photocell	light	electricity

We can also add some natural energy converters:

Muscle	chemical	kinetic energy or potential energy
Leaf	light	chemical
Oil, coal	chemical	light, heat

You can probably think of several more examples.

from one type to another happens frequently and we have invented many devices for making these conversions. Examples are listed above.

Questions
1 What energy conversions are involved when we use matches to light a wood fire? Where did the energy of the fire come from? List the stages and conversions.
2 What energy conversions are involved when we generate electricity by a hydro-electric generating station? Where does the energy come from?

Instructions: some experiments on energy conversion
It is interesting to study some simple systems in which energy is converted. We can measure how much energy is put in and how much is got out of the system. We can also try to account for energy that seems to be lost during the conversion. There are many systems that can be studied. Below are a few examples for you to try, but you should also try investigations on systems of your own design, using any apparatus that you have available in the laboratory or at home. Discussion of some of the results appears in the Answers and Discussion section (to be looked at *after* you have tried the experiments).
1 Fix up a pendulum and arrange some graph paper behind it so that you can judge the vertical height of the bob above its lowest position (fig. 15.3). Set the pendulum swinging and measure h. Leave it swinging for a measured length of time, say 10min. Then measure h again. Calculate E_p for the bob when it is at its highest position

(a) immediately it was set swinging and (b) 10min later. How much energy has it lost? What has happened to this energy? What is the rate of loss of energy in watts? Note: another way to measure h is to place a protractor behind the support of the pendulum length d and measure θ the angle of swing (to *one* side). By calculation you can then obtain h, for $h = d(1-\cos\theta)$.
2 Use a small electric motor which has a long spindle. Connect it to a battery, switch, voltmeter and ammeter as shown in fig. 16.1a. A length of thin cord is attached to the motor spindle; the cord hangs down over the edge of the table and a weight (try a 20g mass) is hung on this. You may need to experiment to find a weight which the motor can lift as it winds the thread on its spindle. It should lift the weight from floor to table in about 5s. Find the exact time taken for the motor to lift the weight a measured distance. What is the downward force on the weight? What work is done on the weight in lifting it? What is the lifting power of the motor? This tells you the power output from the device. This must be compared with the electrical power used to drive the motor. While the motor is lifting the load, read the current and voltage. Calculate the power input to the motor. Compare input power and output power, and account for the difference.
3 Use the same motor, weight and thread, ammeter and voltmeter but disconnect the battery and replace it with a low value resistor (about 100ohms). Let the weight fall, driving the motor; the motor is now being used as a dynamo. Time the fall for a measured distance and calculate the power input. Read the voltmeter and ammeter

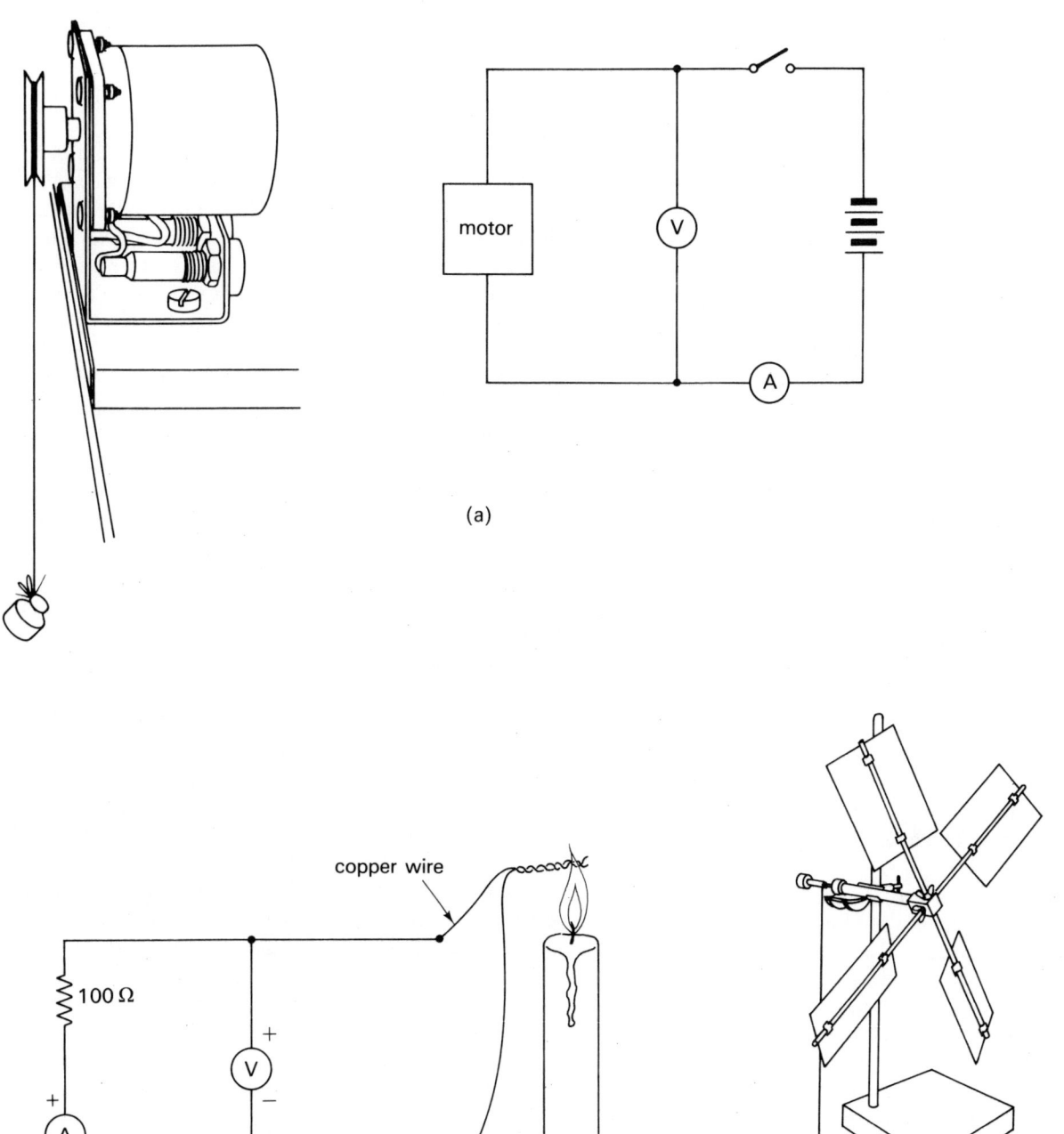

Fig. 16.1 Energy conversions.

and calculate the power output. Compare input and output power, and account for any difference.

4 Wind up a clockwork automobile or similar toy, measuring the work you do in winding it. To measure the force required to turn the winding knob you may need to attach a long lever to the knob and use a dynamometer. Calculation of 'force times distance' gives work done.

Since force required may vary as winding proceeds, measure it several times during winding and take an average. Calculate the total energy stored in the wound spring. Place the toy on a level smooth surface and let it run until the motor is unwound. Measure the time taken for it to run down and the distance travelled. Find the mass of the toy.

If the amount of energy stored in the spring had *all* been available to accelerate the toy, and had been transmitted to the wheels at a steady rate during the whole time of its journey, how far would it have travelled? Compare this with the distance actually travelled.

You can try other variations of this experiment. For example, use a flying model plane, measure the work done in winding the elastic motor and calculate the total energy stored. Allow the plane to fly from the hand, measure the distance travelled and time taken before the motor runs down. Another variation is to use a battery-powered toy and measure the current and voltage to find the power of the motor. The amount of electrical energy supplied during a timed run can then be calculated.

5 Estimate your expenditure of energy during a day. Keep a record for a period of 24 hours of how much time you spend in these main types of activity:

	Energy required kJ/h
Resting or sleeping	250
Sitting	375
Standing still	420
Light activities such as slow walking, working in laboratory, household duties	630
Medium activities such as playing table-tennis, physical exercises, fast walking	1 250
Vigorous activities such as playing football, dancing, cycling, climbing, running	2 000

These values are very rough and you will have to use the descriptions to estimate how much energy you are using for other activities. When you have worked out your total for the day, calculate your average expenditure of energy per second. This is your average power, in watts. You can also measure your expenditure more accurately for an activity such as running upstairs. If you know your weight, the height you climb and the time you take, you can calculate your rate of working.

6 Make a thermocouple by twisting together a piece of copper wire and a piece of constantan wire at their ends. It is best to solder the joint to make it secure. Connect the thermocouple to a voltmeter and ammeter (fig. 16.1b). The power developed is small, so use a voltmeter capable of measuring 0.1 V or less, and an ammeter measuring 50μA (microamperes, millionths of an ampere). A resistor of about 100 ohms completes the circuit. The constantan wire should be connected to the *negative* terminal of the meters. Light a wax candle or small oil lamp and hold the thermocouple in the flame. The heat energy from the flame is converted into electrical energy by the thermo-couple. After a few minutes, the voltage and current reach fairly steady values, which can be recorded. Calculate the power output of the thermocouple.

The power input is not easy to calculate but can be estimated. A fair estimate is that 1 g of wax when burnt, releases chemical energy which is converted to heat and light; the amount of heat energy is about 35 kJ. Let the candle burn for an hour or so and see what weight of wax burns in that time. This gives you an estimate of the power of the candle. Compare this with the power of the thermocouple. Account for the difference.

When you have finished with the thermocouple, place it on a hard surface, such as a large stone, and hammer it vigorously for a few seconds. What energy conversions are happening now?

7 Make a windmill (fig. 16.1c) with a weight attached to its shaft. The angle of the blades should be adjustable. With the blades set at various angles, let the weight fall and turn the mill. Calculate the power of the mill by timing the fall of the weight over a measured distance. How does this vary with the angle of the blades? Also put the windmill outdoors when the wind is blowing. Measure the rate at which the weight is raised and calculate the power input to the blades, assuming friction can be ignored.

8 Connect a photoelectric cell to a voltmeter and ammeter The circuit is the same as in fig. 16.1b, with the photo-electric cell in place of the thermocouple. Place the equipment outdoors when the Sun is shining, preferably around mid-day with the Sun nearly overhead. The exact amount of light energy reaching the ground depends on the angle of the Sun and whether the sky is clear or cloudy. With a clear sky, take the energy of sunlight as 80 mW/cm^2. If it is cloudy, it might be 40 mW/cm^2 or less. Measure the area of the photoelectric cell and calculate the power of sunlight falling on it. This is the power input to the cell. Measure the voltage and current delivered by the cell. Calculate its power output. What is the efficiency of the photoelectric cell? Figure 16.2 shows an array of photoelectric cells being used to convert sunlight into electricity to provide power for the satellite.

Small electric motors are made with low-friction bearings that can be driven by the output from a photoelectric cell. You could attach a small weight to the spindle of such a motor and measure the time it takes to raise the weight a given distance. Then you can calculate the power output of the motor and the efficiency of the cell-motor combination.

The above are but a few examples of conversion of energy. It is not difficult to devise equally interesting examples from other equipment you have available. You may need to make some assumptions when working out the result. You may need to estimate a few of the quantities rather roughly. However, this does not matter very much, provided you understand the relations between force,

Fig. 16.2 A satellite, showing panels of solar cells.

work, energy and power that were the subject of chapter 15. In all the systems you will find one clear result: power output is generally much less than power input. In other words, efficiency is low. The systems you have used are cheap and simple, so we might not expect them to be very efficient. Their bearings are probably simply made so that friction is high and a large proportion of the energy is converted to internal energy. Most conversions of energy are inefficient and in most cases much of the energy is converted to internal energy. In this form, the energy is still there, so the law of conservation of energy has been obeyed, but the energy is not available for doing useful work. We can no longer use it to drive vehicles, generate electric power or do any of the things for which we need energy. One of the tasks of engineers and physicists is to devise efficient ways of converting and using energy.

16.30 Conversion of other forms of energy to internal energy

In many of the examples in the previous sections we have seen that when energy is converted from one form into another form, much of it is converted into internal energy. This appears as a rise in temperature. Sometimes we make use of the temperature rise, for example in electric heaters, to give light in electric lamps, and in striking matches. Sometimes the temperature rise represents wasted energy. Good examples of this are the heating of bearings of a machine and the heat produced by an electric lamp. It is interesting to find out just how great a rise in temperature is produced by conversion of other forms of energy into internal energy.

Instructions: measuring the conversion of mechanical energy into internal energy

1 Weigh out a quantity of lead pellets (lead shot); the amount can be between 0.5 kg and 1 kg and you should record the mass to the nearest 10 g. Call this mass of lead shot m.

2 Put the shot in a stout cardboard tube, about 1 m long and about 4 cm in diameter. A cardboard mailing-tube is suitable. Put a cork or rubber stopper in each end. Preferably one stopper should have a hole wide enough to let a thermometer slip easily through it, and this hole should be closed with a small stopper. Measure the exact length l, between the stoppers when they are pushed firmly in the ends of the tube.

3 Hold the tube vertically. Insert the bulb of the thermometer among the pellets which are at the lower end of the tube. After a few minutes record their temperature, θ_1. *Remove the thermometer* and close the tube.

4 Turn the tube upside-down quickly so that the end containing the pellets is uppermost. The pellets fall straight down to the lower end of the tube. They gain potential energy as you turn the tube. They lose this potential energy and gain kinetic energy as they fall. How much kinetic energy do the pellets have as they reach the bottom of the tube? As they hit the stopper at the lower end of the tube, they lose this kinetic energy. Can you explain what happens to it?

5 Turn the tube a total of 100 times so that the pellets fall from top to bottom a total of 100 times. Work as quickly as possible. Explain why this needs to be done quickly. What is the total amount of kinetic energy lost by the pellets?

6 Measure the temperature of the pellets, θ_2. How great is the rise in temperature?

7 Assuming that all the kinetic energy lost by the pellets has been converted into internal energy of the pellets, how much internal energy have they gained during the experiment?

8 This amount of energy raised the temperature of the pellets by $\theta = \theta_2 - \theta_1$ degrees. Calculate how many joules of internal energy would raise the temperature of the pellets by one degree. We call this amount the *heat capacity* of the pellets. It depends on how many pellets you used (their total *mass*) and the *material* from which they are made (in this case, lead).

9 If you had 1 kg of lead pellets, how many joules of internal energy would be needed to raise the temperature by one degree?

16.31 Specific heat capacity

In the previous investigation we found that a definite amount of heat is needed to raise the temperature of 1 kg of lead by one degree. We measured the change of temperature on the Celsius scale (see 3.10). However, the degree Celsius is not used as a unit of temperature change in SI, so we cannot use the degree Celsius when we define heat capacity or specific heat capacity. We must use the true SI unit instead. This is the kelvin, symbol K, named after the distinguished nineteenth century physicist, Lord Kelvin. We will find out more about the kelvin scale (sometimes known as the absolute scale) in 18.33, but for the present it is enough to note that a temperature change of $\theta_2 - \theta_1$ on the Celsius scale is equivalent to a temperature change of $\theta_2 - \theta_1$ on the kelvin scale. A rise in temperature of 5 degrees on the Celsius scale is equivalent to a rise of 5 K.

We measure our temperatures using a Centigrade thermometer but we express the *change in temperature* in K rather than °C.

We can therefore say that we require a definite amount of energy to raise the temperature of 1 kg of lead by 1 K. This amount of energy is the *specific heat capacity* of lead. The exact amount required is a property of the material itself. You could repeat the measurement using brass or iron rivets instead of lead shot and you would obtain a different result. For lead the specific heat capacity, c, has the value 130 J/kg K.

This means that to raise the temperature of 1 kg of lead through 1 K, we have to supply 130 J internal energy. The internal energy can be increased by mechanical means: by dropping from a height, by rubbing (friction) or in various other ways. In the next experiment, we increase the internal energy of the material by heating it with an electric heater. This gives us a very accurate method of measuring specific heat capacity because the amount of energy supplied can be measured accurately.

Instructions: measuring specific heat capacity by the conversion of electrical energy into internal energy

1 The apparatus is shown in fig. 16.3. The block, made from aluminium, brass, copper or iron, has a mass of 1 kg. This simplifies calculations. The aim of the experiment is to find how much energy is needed to raise the temperature of the block by 1 degree.

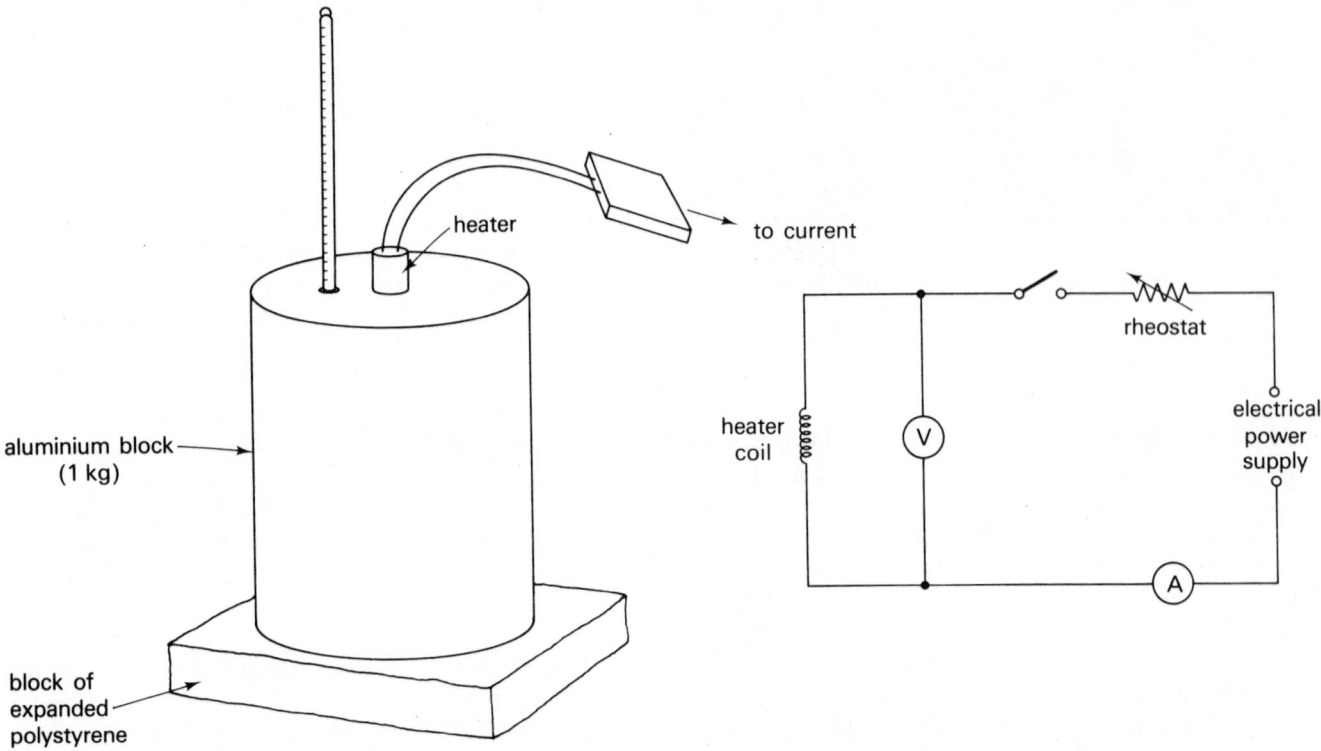

Fig. 16.3 Measuring specific heat capacity by an electrical method. The expanded polystyrene or felt cover for the aluminium block is not shown. The voltage of the power supply depends on the rating of the heater.

2 The amount of energy supplied is measured by supplying a steady current, for a known period of time, to the heater. Before placing the heater in the central hole in the block, switch on the current and adjust the rheostat so that a current of 1 A flows in the circuit. Switch off and allow the heater to cool.

3 Place the heater in the central hole of the block, and a thermometer in the other hole. Surround the block with a layer of expanded polystyrene or thick felt, to prevent undue loss of heat. Read the temperature of the block, θ_1.

4 Switch on the current and begin timing, using a stop-clock.

5 Read the current, I, and the p.d., V, across the terminals of the heater. If necessary adjust the rheostat to keep the value of I constant.

6 When the temperature of the block has risen by about 10 degrees, note the time t, and the exact temperature, θ_2.

7 Calculate the rate of conversion of electrical energy in the heater.

8 Calculate the total amount of electrical energy converted during the experiment.

9 Assuming that all this energy went to increasing the internal energy of the block, how much energy would be needed to increase the temperature of the block by 1 degree? Since the block has mass 1 kg, this amount is the specific heat capacity *of the material* of the block.

10 The assumption made at step 9 is not entirely justified. What other parts of the apparatus received internal energy from the heater? May this account for inaccuracies in your final result?

Instructions: measuring the specific heat capacity of a liquid

1 The apparatus is similar to that used above except that instead of the block of solid metal you use a copper or aluminium can or calorimeter to hold a measured quantity of the liquid. A thick jacket and lid of expanded polystyrene or felt surrounds the calorimeter to minimise the loss of heat.

2 Weigh the calorimeter empty, m_1. How much heat is needed to raise its temperature 1 K?

3 Add water to the calorimeter so that it is two-thirds to three-quarters full. Weigh the calorimeter and liquid, m_2.

What is the mass of the liquid, m? For a first trial the best liquid to use is water, at room temperature.

4 Before putting the heater in the liquid, switch on and adjust the rheostat so that the current through the heater is about 1 A. Let the heater cool before putting it in the liquid.

5 Put the heater in the liquid. After a few minutes measure the temperature of the liquid, θ_1.

6 Switch on and begin timing, using a stop-clock.

7 Read the current, I, and the p.d., V, across the terminals of the heater. If necessary, adjust the rheostat to keep the value of I constant.

8 When the temperature of the liquid has risen by about 10 degrees note the time, t, and the exact temperature, θ_2.

9 Calculate the rate of conversion of electrical energy in the heater.

10 Calculate the total amount of electrical energy converted during the experiment.

11 Assuming that all this energy went to increasing the internal energy of the liquid *and the calorimeter*, how much energy went to the liquid?

12 Given U_1, the energy which went to the liquid, how much energy is needed to increase the temperature of the liquid by 1 degree? How much energy is needed to increase the temperature of 1 kg of this liquid by 1 degree? This is the specific heat capacity of the liquid.

The specific heat capacity of water is exceptionally high. This means that it can absorb or give out relatively large quantities of heat with relatively small change in its temperature. This explains why the temperature of the ocean and of lakes remains within a narrow range during hot and cold seasons, while at the same time the soil temperatures of the same geographical region rise high and fall low. The high specific heat capacity also makes water a good cooling liquid. It can absorb large quantities of heat from the engine of an automobile and carry this to the radiator where it is lost to the atmosphere. For similar reasons it is used in central heating systems to carry heat from the furnace to the rooms. If its specific heat capacity was lower, it would get much hotter and probably boil, yet would not convey as much heat. Water is also commonly used to reduce the temperature of certain types of fire, so acting as a fire extinguisher by absorbing large quantities of heat from the burning materials.

PART E   INVESTIGATING MATTER

In physics we often use numbers that are very large and numbers that are very small. In this chapter we consider objects as small as 0.0000000001 m; in a later chapter we discuss a type of radiation with a frequency of 10 000 000 000 000 000 000 000 Hz. There are too many zeros; they lead to confusion and take up a lot of space. To avoid such awkward numbers there are two things you can do:

1 Use SI prefixes: You have used several of these already. Below is a summary list of the most commonly-used ones.

An exception to the system is the kilogram, which is the *base* unit. The prefix 'centi' is not much used, the only common use for it being in centimetre, which is a very convenient length for many measurements in the laboratory and in everyday life.

2 Use standard form: A number written in standard form has a single digit (never zero) before the decimal point. Following the number is a multiplying factor in the form of the number 10 and an exponent. For example: 7.45×10^6; 3.01×10^2; and 5.7923×10^{-5} are all in standard form.

The way to handle SI prefixes and to work with numbers written in standard form will be illustrated in the many calculations which appear in the chapters which follow.

	Multiple or sub-multiple	Multiplying factor	Prefix	Symbol	Example
Multiples of a base unit or derived unit	million times unit	$\times 10^6$	mega	M	MW, megawatt
	thousand times unit	$\times 10^3$	kilo	k	kV, kilovolt
Submultiples of a base unit or derived unit	hundredth of unit	$\times 10^{-2}$	centi	c	cm, centimetre
	thousandth of unit	$\times 10^{-3}$	milli	m	mA, milliampere
	millionth of unit	$\times 10^{-6}$	micro	μ	μm, micrometre
	thousand-millionth of unit	$\times 10^{-9}$	nano	n	ng, nanogram
	million-millionth of unit	$\times 10^{-12}$	pico	p	pF, picofarad

17 The structure of matter

17.00 Molecules

For many hundreds of years, men who have studied science have believed that all matter is composed of tiny particles. During the last century scientists began to prove that this idea is true. During this century we have learned a lot about the structure of matter and the particles of which it is made. We have been able to study their sizes and shapes and find out much about the forces that exist between them.

The smallest particle of a substance which can exist on its own is called a *molecule*. We can speak of a molecule of water, a molecule of sugar, a molecule of sodium chloride. A single molecule of these substances has the normal chemical properties that would be shown by a whole test-tube-full of these substances, though we would find it very hard to do experiments with single molecules! If we could cut these molecules into pieces they would no longer be the same *substance*. The molecule of water is composed of three smaller particles called *atoms*: two atoms of hydrogen and one of oxygen. If we break a water molecule we get two different substances and neither of them is water. The same happens if we break sugar; we get carbon, hydrogen and oxygen. If we break sodium chloride (a harmless substance which we use in cooking) we get sodium (a metal which explodes violently when in contact with water) and chlorine (a highly poisonous gas). Splitting a molecule produces big changes!

Most molecules consist of two or more different kinds of atom, though some consist of two or more atoms of the same kind, and a few consist of just a single atom. The atoms are the smallest particles in which an *element* can exist. Atoms too are found to be made up of even smaller particles. We shall study atomic structure in chapter 20.

In chemistry we are concerned with what atoms are combined together in the different kinds of molecules and how these molecules react with each other. In physics we are less concerned with the chemical properties of different kinds of molecule and more interested in the properties possessed by all molecules, independently of their chemical composition. For this purpose it is easiest if we *think* of the molecules as being very small solid spheres, all looking very much alike. This makes it easy to draw the diagrams and understand the explanations. In many ways molecules behave *as if they are* small solid spheres, so this simple idea is useful. However from time to time we must remember that they are not really solid and not really spherical.

17.01 Evidence for molecules

One of the first pieces of evidence of the existence of molecules was the discovery made by Robert Brown in the early nineteenth century, that has since been called Brownian Movement.

Instructions: investigating Brownian movement (1)
1 Put a few drops of Indian ink in water so that a pale grey mixture is produced.
2 Place a drop of this on a microscope slide; put a cover-glass on top.
3 Examine beneath a microscope, focussing on the fine black particles of carbon that are suspended in the water. Adjust the illumination to obtain the clearest view of the particles. Often it is best to illuminate the slide from one side rather than from below. What do you see?

Instructions: investigating Brownian movement (2)
1 If you have a 'smoke cell' apparatus (fig. 17.1), this can be used for demonstrating Brownian movement in air. The apparatus consists of a small glass cell or smoke chamber (a short length of glass tube) and a small lamp and lens (a glass rod) which produces a sharply-defined beam of light that illuminates the tube from the side.
2 Fill the cell with smoke from a piece of smouldering string.
3 Place the cell in its clip, and put the apparatus on the stage of a microscope.

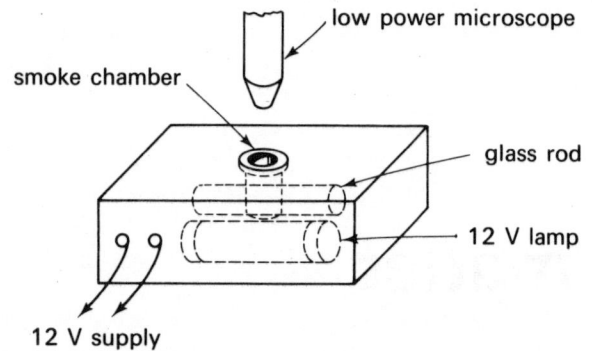

Fig. 17.1 *Smoke cell apparatus for demonstrating Brownian Movement.*

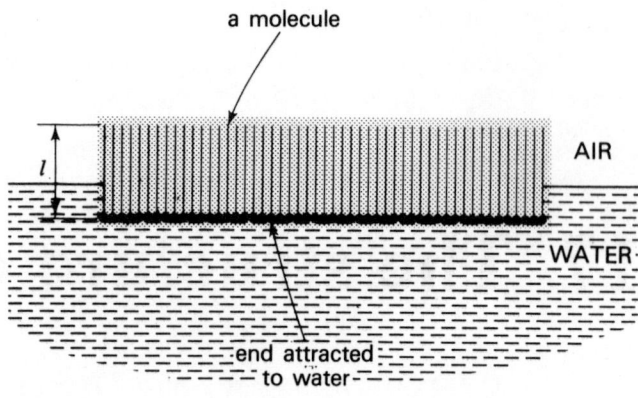

Fig. 17.2 *Diagram of molecules of oleic acid, on water.*

4 Switch on the lamp. Focus the microscope until you can see tiny white points of light. These are smoke particles. They are too small to be seen clearly if lit by normal microscope illumination from below. The light from the side is very bright and is scattered by the smoke particles; each particle is very small but reflects a lot of light, so can be seen.

Questions

1 Describe the motion of the carbon particles.
2 Can you explain why they move as they do?
3 Describe the motion of the smoke particles. Can you explain why they move as they do?
4 Can you account for any difference in the results of the two experiments?

17.02 The size of molecules

Although it is usually convenient to think of molecules as being spherical, the next experiment is done with molecules which have a long narrow shape, like thin string. These are the molecules of a fatty substance called oleic acid. The opposite ends of the molecules have different properties: one end is very strongly attracted by water and the other end is repelled. If this substance is floated on water, the molecules can form a thin film just one molecule thick. In this film the molecules lie side by side parallel to each other, their water-attracted ends down and the other ends pointing straight up (fig. 17.2). If we could measure the thickness of the film, we could know the length of the molecule.

Instructions: measuring molecules; the oil-film method

1 Make up a 0.5% solution of octadecenoic acid in methanol. You need only a small quantity of solution, just a few drops, but a larger quantity may be made up and kept for other people to use later. 1 cm³ acid made up to 200 cm³ solution is suitable.
2 Weigh an empty beaker.
3 Use a fine dropping pipette and quickly run exactly

100 drops into a small beaker. Weigh the beaker containing the solution. Calculate the mass of 100 drops.
4 Calculate the mass of one drop.
5 The density of methanol is $791\,\text{kg/m}^3$; calculate the volume of 1 drop. If the solution is 0.5% by volume, calculate V, the volume of acid in the drop.
6 Put water in a large plastic dish; dust the surface thinly with lycopodium powder.
7 Allow exactly *one* drop of the solution to fall on to the surface of the water from a height of 1 or 2 mm, no further. The drop spreads quickly over the water surface and the methanol partly evaporates and partly mixes with the water, leaving a circular film of octadecenoic acid. This pushes aside the floating particles of lycopodium powder, making a clear area (fig. 17.3), roughly circular, covered with acid.

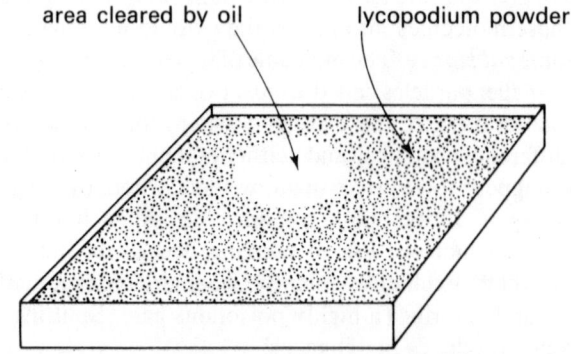

Fig. 17.3 *The oil film, ready for measuring.*

8 Measure the diameter of this area.
9 Repeat several times and calculate the average diameter. Then calculate the average radius, r.
10 The volume of the film is $\pi r^2 l$, where r is its radius and l its thickness. This is equal to V, the volume of acid in one drop of solution.

If $V = \pi r^2 l$, we can calculate l from $l = \dfrac{V}{\pi r^2}$

Questions
1 What value did you obtain for *l*?
2 What possible sources of error could occur in this method?

17.03 Intermolecular forces

It has already been mentioned that molecules are in motion. From this we can deduce that there are forces which cause this motion. One physical force is gravity by which molecules attract each other, tending to move towards each other. But the force of gravity is a weak force. We notice it only when we have an extremely large *group* of molecules (such as the millions upon millions of molecules which make up the Earth) which *together* produce a force strong enough to attract other *groups* of molecules (such as those making up a coconut). The gravitational attraction between two molecules is so small that we can forget about it.

On the molecular scale, other forces are of much greater importance. They are much stronger than gravity, but they act over a much shorter distance. They are electromagnetic forces. They exert an attractive force, causing two adjacent (neighbouring) molecules to be attracted to one another. They move towards each other, but do not come together completely for, when they get close, a repulsive force comes into operation.

We can imagine the force operating as in fig. 17.4a. Molecules which are relatively distant have an attractive force between them (dashed line with arrow). They respond to this force and move towards each other; then as they approach each other closely, a strong repulsive force (thick line and arrows) comes into play forcing them apart.

They then move apart again, gaining velocity until they are decelerated by the attractive force. Molecules at a certain distance apart (wavy line) are subject to equal attractive and repulsive forces and, if they are not already in motion, neither come together nor move apart. Molecules which are *more* than this distance apart have potential energy due to the attractive force; molecules which are *less* than this distance apart have potential energy due to the repulsive force. We can think of the molecules tending to become arranged so that they are all roughly the same distance apart (fig. 17.4b), vibrating to and fro in various directions around their relatively fixed positions. They have kinetic energy. If they depart from their position they get nearer to some of the adjacent molecules and further from others. Intermolecular forces increase their potential energy and act to bring them back into their equilibrium position. It is almost as if the molecules were joined together by spiral springs, which produce force when extended and when compressed. You may have seen a model consisting of polystyrene spheres joined by wire springs which is intended to represent intermolecular forces. If you have one available, try shaking one molecule slightly, by hand. Note how the extra kinetic energy is transmitted along the springs to the other spheres. We will return to this topic later (chapter 18).

17.10 States of matter

Matter can exist in three states, solid, liquid and gas. In these three states the intermolecular forces are operating in different ways, producing the characteristic properties of each state. We will consider the states in turn.

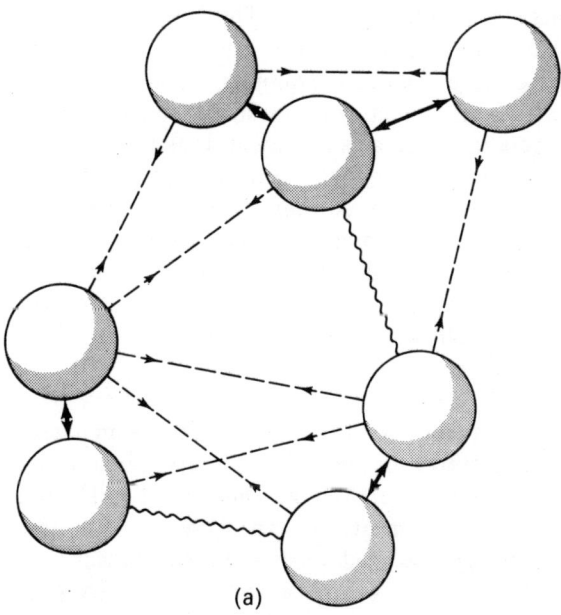

(a)

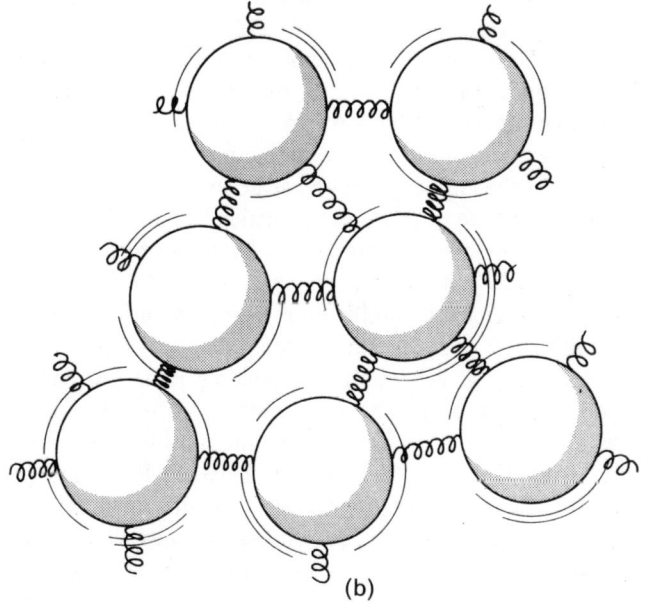

(b)

Fig. 17.4   Inter-molecular forces (see text).

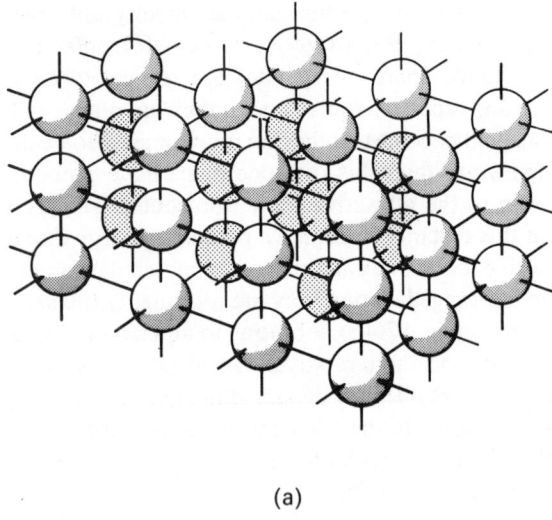

(a)

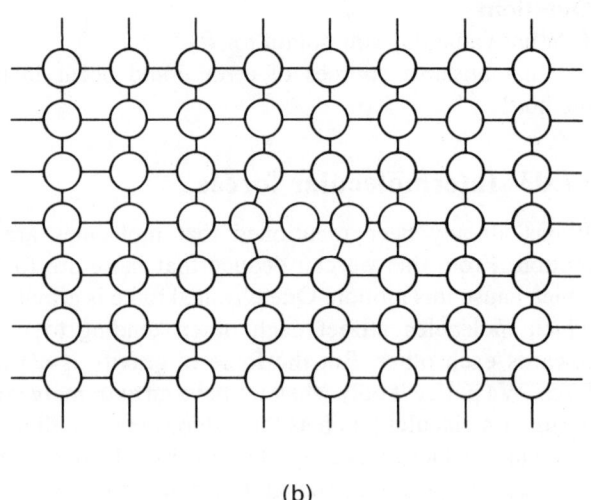

(b)

Fig. 17.5   Cyrstal lattices (see text).

17.11 Solids

Many solid substances are obtainable in the form of crystals. You have probably 'grown' crystals in chemistry lessons and know that they have a definite shape. This shape is a characteristic of the substance of which the crystal is composed. We interpret this as showing that the molecules are arranged in a regular three-dimensional *lattice* (fig. 17.5 a). Many types of lattice are possible and the type depends on the kind of molecule or molecules making up the lattice. In this way the regular geometrical shape of the crystal as a whole is obtained. Under stress the crystal tends to split along the planes of the lattice, producing smaller crystals of the same general shape. The regularity of the lattice may be broken at certain points where, for some reason, the rows and columns of the lattice do not quite fit together evenly (fig. 17.5b). Throughout a regular lattice the molecules are spaced so that their intermolecular forces are in equilibrium. If the lattice has irregularities, intermolecular forces produce tensions at these points and the structure is weaker. The number and type of these irregularities has an important effect on the mechanical properties of the substance. Most substances do not exist as single crystals. As they are formed, either by crystallization from a solution or by crystallization from the liquid state, small crystals form and gradually grow, touch, and become joined to each other. The mass of substance consists of many small crystals; each has its own regular lattice, but there are great irregularities at the surfaces between these small crystals. This too has large effects on the mechanical properties of the material. In solid substances which are not pure, a mixture of crystals of different kinds may be present. Examination under a lens of such substances (for example, rocks) shows the complex crystalline structure.

Each molecule has its fixed equilibrium position in the crystal lattice. Molecules can vibrate slightly from their equilibrium positions, but intermolecular forces are strong enough to prevent the molecules from moving out of position to some other position. As a result, solids have a distinctive pair of properties:

(a) They have an almost *constant volume*; any attempt to push the molecules closer together or to pull them further apart is opposed by intermolecular forces.

(b) They have a *constant shape*; any attempt to push the lattice into a different shape is opposed by intermolecular forces.

Solids are incompressible, but it is possible to change their size slightly. One way of doing this is to heat the solid, making it expand (18.30). This produces only a small change in size (6.20). We can apply force from outside to change shape; for example, to bend a piece of copper wire. The wire has constant shape until we bend it; then the bending force acts mainly at weak points in the lattice and between crystals, producing a permanent change of shape. Some substances can be bent only slightly; for example, if force is applied to a rod of carbon, the lattice gives way suddenly and the rod snaps into pieces. In the next experiment we see some of the effects of force on a length of wire and we relate this to inter-molecular forces.

Instructions: investigating the effect of force on a wire
1 Figure 17.6 shows the apparatus. The comparison wire and test wire (both taken from the same reel of wire) hang side by side from a firm support. The wires are between 2 and 3 metres long. The fixed weight on the comparison wire is to hold it straight. As weights are added to the test wire, changes in its length are measured by reading the scale and vernier.

158

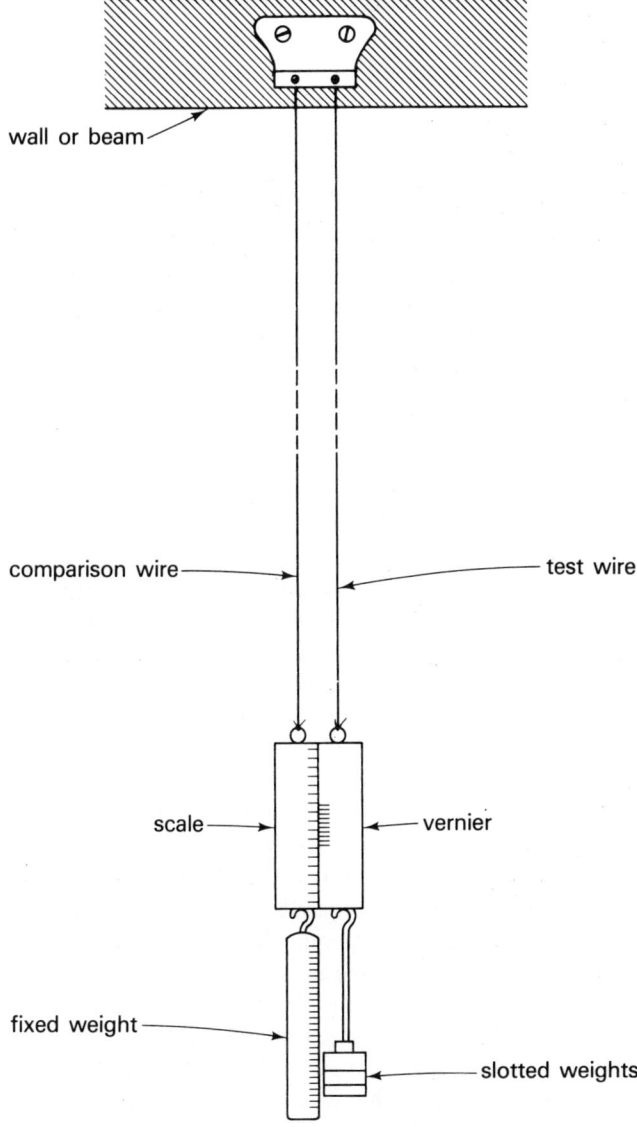

wall or beam

comparison wire

test wire

scale

vernier

fixed weight

slotted weights

Fig. 17.6 Apparatus for stretching a wire.

2 Hang a small load (5N) from the test wire. After making sure that all the bends and kinks in the wire have been pulled straight, read the scale and vernier.

3 Add weights in steps of 2.5N. Read the scales at each step.

4 When the total load is 10N, take the readings, then *remove* weights in steps of 2.5N. Read the scales at each step, down to and including zero.

5 Start again adding weights by steps, from zero up to 50N or more. Read the scales at each step. When loading is heavy, take care not to drop the slotted weights on to the pile. Place them on top as gently as possible.

6 When loading is heavy, the addition of 2.5N makes the wire extend a lot. At this stage, add the weights in steps of 1N.

7 Be ready for the wire to break at maximum load. Place a pad of soft material below to catch the weights when they

fall. Take care that the snapped ends of wire cannot get near your eyes.

8 Plot a graph to show the relation between load and extension. The exact results you obtain depend on the metal and thickness of the wire. The curve should have a shape similar to that of the curve in fig. 17.7.

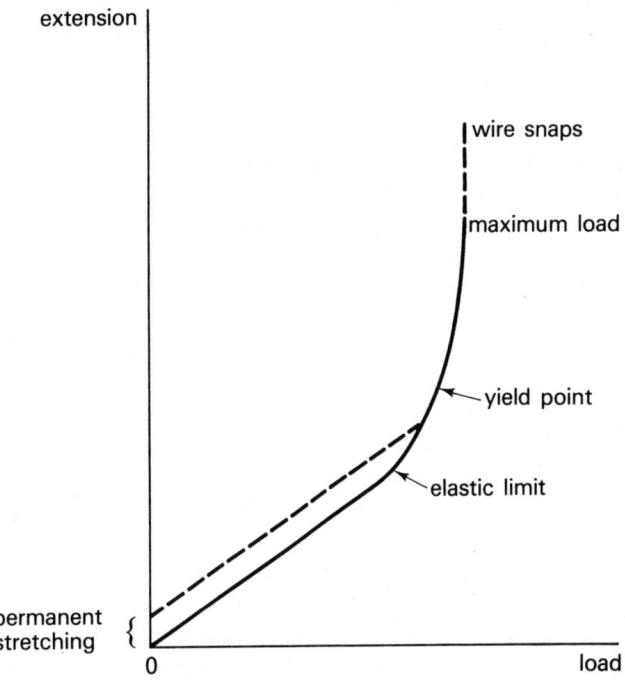

Fig. 17.7 The effect of loading a copper wire.

Questions
1 What happened to the length of the wire at step 4?
2 Make up a rule to describe the relation between load and extension, for a wire below its elastic limit.
3 Explain the behaviour of the wire, using ideas about intermolecular forces.
4 What is the value of the elastic limit for your piece of wire?
5 What is the relation between load and extension for a wire above its yield point?
6 Describe what happens when maximum load is reached and the wire is just going to snap.

Below the elastic limit the wire has elastic properties. The wire obeys *Hooke's law* which states:

If a material is acted on by an applied force tending to change its shape, the amount of change of shape varies as the size of the applied force.

In a stretched wire, the change of shape is an increase in length and a decrease in thickness. Hooke's law applies to other changes of shape: stretching springs (the spring balance and dynamometer make use of Hooke's law); bending beams; coiling springs (as in a clockwork motor). The volume of a stretching wire does not change; as it gets longer it becomes narrower; in the end it becomes so narrow that it easily snaps.

17.12 Liquids

Molecules in a liquid interact with each other by intermolecular forces. They keep to roughly the same distances as molecules in solids. Their kinetic energy is greater than that of molecules in solids. As a result, molecules in liquids do not stay permanently in a fixed position in a lattice. They do tend to remain together in small groups, but there is no lattice extending through the whole of a quantity of liquid, as there is in a crystal. The groups of molecules are continually changing their positions. In addition, molecules frequently leave one group and join another. This type of interaction between molecules gives a distinctive pair of properties to liquids:

(a) They have almost *constant volume*. This property they share with solids, for the distances between molecules are almost the same as in solids. As a result of this, the intermolecular forces are about the same and they resist compression or expansion just as readily.

(b) They are *variable in shape*. Groups of molecules can flow past one another. The liquid as a whole *flows*; it takes the shape of the lower portion of its container.

These properties explain Pascal's principle (13.10). The incompressibility of liquids is made use of in pressure-transmitting machines, such as the hydraulic press and hydraulic brake systems.

Instructions: investigating diffusion in a liquid

1 Fill a large beaker or glass jar (about 2 litres) with water and put it in a place where it can be left for several days without being disturbed. There should be no vibration in its support and it must never be in direct sunlight.

2 Take a piece of glass tubing about 6mm wide and long enough to reach to the bottom of the beaker. Place it in the beaker, resting on the bottom.

3 Drop some small crystals of potassium manganate (VII) down the tube. They sink and rest on the bottom of the beaker.

4 Place your finger over the upper end of the tube and carefully withdraw it from the water. With your finger over the tube, most of the coloured solution inside the tube will be removed. The aim is to leave the crystals in the water, with the water above them free from colour.

5 Place a lid or piece of card over the beaker, to keep out draughts.

6 Observe the beaker daily for at least a week. Record what you see. Describe what happens. Account for the changes you notice.

7 When the solution is the same even colour throughout, does the movement of the molecules stop? Why must this investigation not be done where sunlight can fall on the beaker?

Instructions: supporting a steel needle on water

The density of steel is 7 800kg/m³; the density of water is 1 000kg/m³. According to Archimedes' principle steel cannot float on water. Perhaps it can be supported in another way.

1 Fill a dish with water.

2 Cut out a piece of thin paper tissue a little longer than a sewing needle and about 2cm wide.

3 Place the needle centrally on the tissue and gently place both on the water surface.

4 The tissue should become saturated with water and slowly sink, leaving the needle supported on the water surface.

5 Examine the water surface around the needle. What can you see?

6 Add *one drop* of liquid detergent to the water. What happens?

We have demonstrated a property of liquids called *surface tension*. There is *no* elastic film over the surface of the water, but the water behaves *as if* the film were there. We can explain this by referring to our ideas on intermolecular forces. In the liquid all molecules are being attracted by their neighbours and are vibrating about their equilibrium positions (fig. 17.8a). On average, the forces acting on a molecule from one side are balanced by forces acting on it from the other side. At the surface, forces acting on a molecule from below are *not* balanced by a force acting on it from above. On average, there is an unbalanced downward force. It is better to call this an *inward* force, acting perpendicular to the surface. The effect of this is to make the molecules move to make the surface as small as possible. A small drop of liquid (fig. 17.8b) is pulled into a spherical shape, for a sphere has the smallest surface area for a given volume. Thus the forces acting on molecules at the surface cause them to move in a way that is just the same as if the surface is covered by a thin elastic skin. This surface tension can support the weight of a light object resting on the surface, provided that the water does not *wet* the object, and provided that the object is lowered so slowly that its impulse on the surface does not carry it through the surface layers. However, if the object is wetted by the water, there is attraction between the molecules of water and the molecules of the needle, then the needle is attracted *into* the surface and sinks. Normally a needle is covered with a thin film of grease from our fingers and this repels water, so the needle is not wetted. But if a drop of detergent is added to the water, this is highly active and allows the water to wet the needle. Immediately, the needle sinks.

Clean glass is wetted by water and this accounts for the way in which the water surface curves upwards at the edge in a beaker, measuring cylinder or pipette (fig. 17.8c). Two forces are involved:

1 *cohesion*, attractive intermolecular forces between water molecules;

liquid surface.

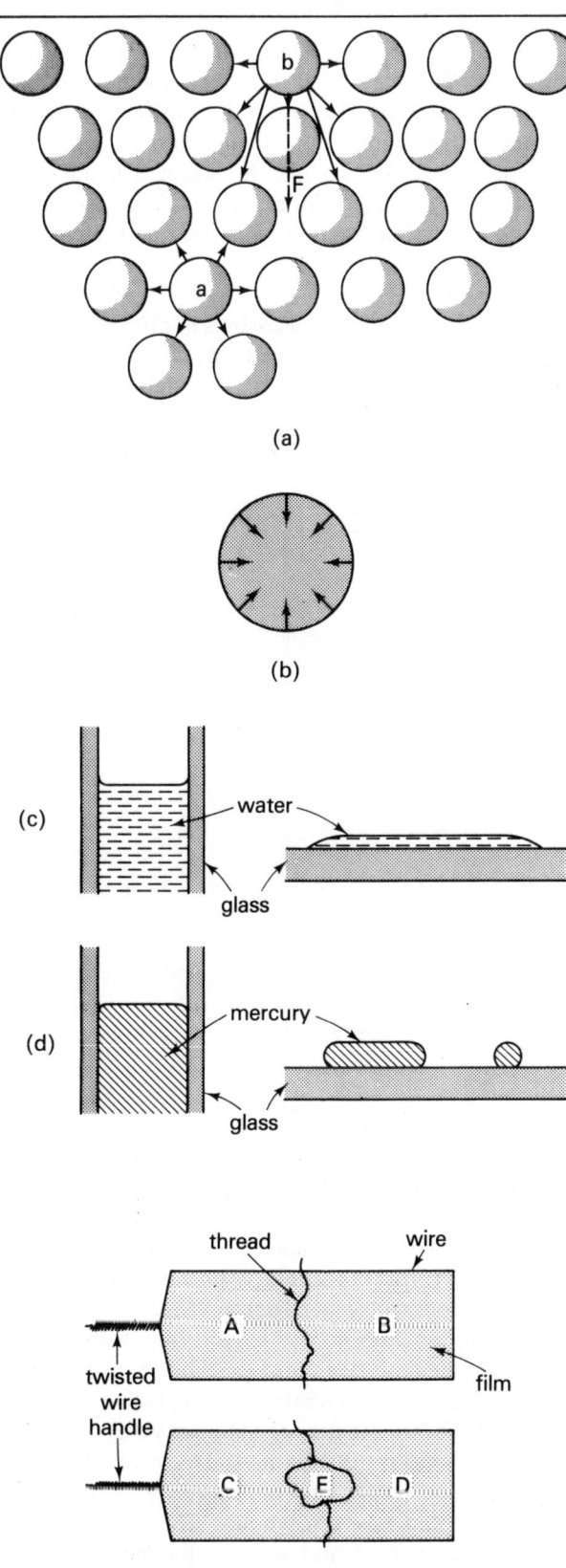

(a)

(b)

(c)

(d)

thread   wire

twisted
wire
handle

film

(e)

Fig. 17.8   The effects of surface tension.

2 *adhesion*, attractive intermolecular forces between water molecules and molecules in the glass.

Adhesion is stronger than cohesion and the result is that the water film slopes upwards to the glass. For the same reason a drop of water spreads over a horizontal glass surface. By contrast the cohesive force of mercury is *greater* than its adhesive force to glass. In other words, mercury does not wet glass. The result is shown in fig. 17.8d.

If a glass tube is very narrow, the effects of these forces exceed the effect of gravity and liquid pressure, which would normally tend to make the liquid levels inside and outside the tube equal (13.01). Water rises in fine glass tubes, the finer the tube the further it rises. In a fine glass tube with a bore of half a millimetre or less, rises of several centimetres are obtained. This effect is known as *capillarity*. Can you think of examples of capillarity in everyday life?

Instructions: investigating further examples of surface tension

1 Make some small wire frames, with cotton thread attached, as in fig. 17.8e.
2 Dip them into a dilute solution of liquid detergent, so that a film is formed where shown, with the thread lying *in* the film.
3 Burst the film A; dip the frame in the solution again to remake the film. Then burst the film B. What happens to the thread?
4 Burst the films C, D and E, in turn. What happens to the thread?
5 Make a rule to describe what happens when any *one* of the films is burst.

What is the significance of the threads being pulled into arcs of *circles*? When an object falls through a liquid or when a liquid flows past an object, or through a pipe, the forces of adhesion and cohesion act to reduce the rate of fall or the rate of flow. In the case of a falling object, the shape of the liquid is being rapidly changed both in front of the object and behind it (fig. 17.9). In front, a gap must be made into which the object can fall; behind the object the gap is closed. These changes of shape mean that the liquid must be parted and must return. Attractive forces between molecules must be overcome. Work must be done, and energy is needed for this. We can describe the fact that the liquid opposes any attempt to change its shape by saying that it is *viscous*.

At the sides of the object another effect is seen. The liquid closest to the surface of the object may show adhesion with it; there is often a surface film of liquid which is carried along with the object. You can often notice a similar effect if you are travelling at high speed in an automobile. Though you are travelling at 100km/h or more, and the air is rushing past, the air closest to the surface of the automobile is almost still. Loose dust and small insects are *not* blown off the surface of the body, in

161

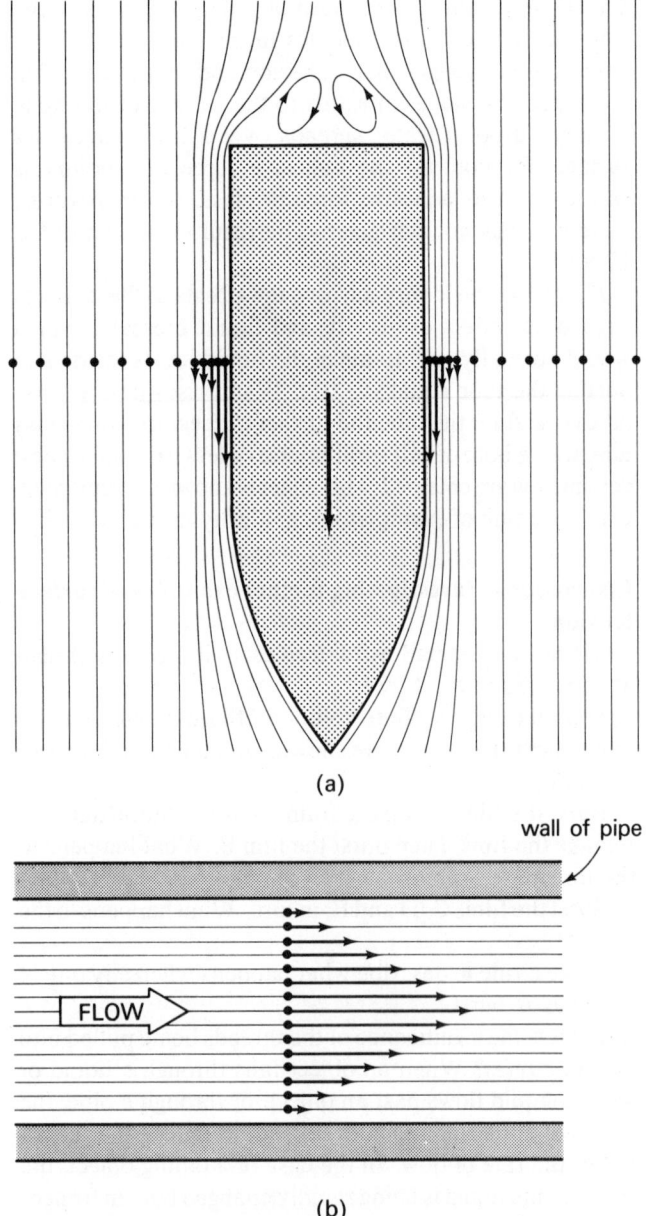

(a)

wall of pipe

FLOW

(b)

Fig. 17.9   The effects of viscosity.

The same situation occurs when liquid is flowing through a pipe (fig. 17.9b). The liquid flows with greatest velocity in the centre and with almost zero velocity close to the walls of the pipe. As these layers of liquid flow past one another, attractive forces develop between molecules in adjacent layers, but almost immediately the molecules are carried away from one another again. The attractive forces resist the relative motion of the layers. This effect produces an internal friction, which is another effect of viscosity.

Instructions: comparing viscosities of different liquids
1 Use a long glass tube, diameter 4cm or more, and at least 60cm long. Stopper the tube at one end and support it vertically. Mark the tube at two points, which can conveniently be 50cm apart, or more, if the tube is long enough (fig. 17.10).

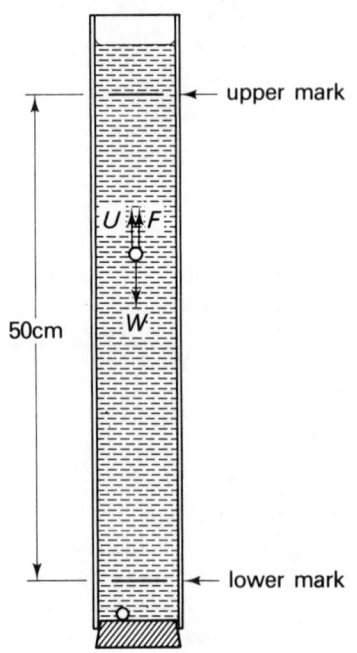

upper mark

U F

W

50cm

lower mark

Fig. 17.10   Comparing viscosities.

spite of the 100km/h wind that you feel if you put your hand out of the window. In a fast mountain stream, very small animals are able to crawl on the surface of rocks. Even in a fast stream the water within about 4mm of the rock surface has very low velocity; small animals can remain there, yet larger animals would be torn away from the rocks and carried downstream by the force of the water. Close to the falling object (fig. 17.9a) the liquid has almost the same velocity as the object. Further away the velocity is less and at a distance the liquid is still. We can think of the liquid around the object consisting of layers each moving at a different velocity; the arrows in the figure represent the relative velocities at different distances from the object.

2 Fill the tube with the liquid to be tested, for example: castor oil; glycerol (propane–1,2,3 –triol) 100%, and in various dilutions with water, such as 95%, 90% and 80%; golden syrup or concentrated sugar syrup; water.
3 Measure the time taken for a steel ball to fall from the upper mark to the lower mark. Use steel balls about 5mm in diameter: bearing balls are suitable. After they have fallen to the bottom, the balls may be recovered by using a strong magnet placed against the outside of the tube. Time several falls and take an average. The ball should be released just at the liquid surface. By the time it has reached the upper mark it should already have reached its terminal velocity (11.12). The velocity which you are measuring is thus its terminal velocity.

It can be shown that the terminal velocity is inversely proportional to the viscosity of the solution. For example if liquid A has *twice* the viscosity of liquid B the terminal velocity of a body falling in liquid A is half the terminal velocity of the same body falling in liquid B.
Set out your results in a table:

Liquid	Terminal velocity v in m/s	Relative viscosity ($1/v$)

Note that the 'relative viscosity' you obtain in this experiment depends also on the size of the ball and the material from which it is made; this is just a way of comparing the results *you* get with the different liquids you have tried. Which liquid was the most viscous? Which liquid was the least viscous?

In fig. 17.10, three forces are shown acting on the ball: W, its weight; U the upthrust (page 132) and F the viscous force opposing its motion through the liquid. What does W depend on? What does U depend on?

Viscosity depends on temperature. Liquids become less viscous as their temperature increases. You could test this effect using the method of the previous experiment. A good lubricating oil should have roughly the same viscosity over a wide range of temperature. If the engine is cold, the oil must be fluid enough to allow the parts to move easily. Yet when the engine has been running for some time and is very hot, the oil must still retain the right amount of viscosity. Lubricating oils are specially prepared to maintain suitable viscosity over the normal operating temperature range of the engine. Even so, unless special oils are used, problems arise when vehicles and other machines operate in arctic conditions.

One class of liquids is so viscous at ordinary temperatures that for all practical purposes we can consider them to be solids. These include pitch and glass. Pitch makes a hard road surface, which is apparently solid. But when the Sun has been shining on it during the hot season, the viscosity of the pitch decreases gradually, and it begins to show its true liquid properties. We are used to thinking of glass as a solid, but really it is a highly viscous liquid. It does not have the crystalline properties of a true solid. As glass is heated it *gradually* softens. This is *not* melting; it is simply becoming less viscous (more 'runny'). If it were melting, it would be solid at first and then quite suddenly change to a runny liquid, like ice does when it melts. If a piece of glass tubing is heated, it gradually softens and can then be bent into some required shape. Then it is left to cool. It becomes more viscous again and retains its new shape.

17.13 Gases

In gases the molecules are much farther apart than in solids or liquids. This explains why gases have low density. Since intermolecular forces act over a very short range, the molecules of a gas do not attract or repel one another, except when they happen to pass close by. The molecules are moving relatively faster than in solids or liquids, so that attractions and repulsions occur for only short periods, in passing, and no permanent groups of molecules are formed. Each molecule moves singly. As a result of this we have two distinctive properties of gases (compare 17.11 and 17.12):

(a) They have *variable volume*. Since the molecules are far apart, with space between them, it is possible to compress or expand the gas without any opposition from intermolecular forces.

(b) They are variable in shape. Molecules are free to move independently and can travel to any part of the container they are in. The gas takes the shape of its container and always fills it completely.

Instructions: investigating diffusion in a gas
1 Mix a few cubic centimetres of water with an equal volume of concentrated nitric acid.
2 Place some copper filings (turnings) in the bottom of a gas jar; pour the diluted nitric acid on the turnings. The brown gas nitrogen dioxide is given off.
3 When the action has stopped and the gas has cooled to room temperature, the jar is full of nitrogen dioxide. Invert a second jar (containing air) on top of the first jar.
4 Place a piece of white paper behind the jars to make it easier to see the diffusion of the gases. How long does the nitrogen dioxide take to diffuse through the air? In what way does the diffusion of this gas differ from the diffusion of potassium manganate(vII) in water?

Imagine a container which has a gas in it (fig. 17.11). In reality there would be far more molecules in it than shown in the figure, but this does not affect the description. The molecules are all moving quickly, in all directions, as indicated by the arrows. Occasionally they collide with one another, or, to be more accurate, come so close together that they interact, being attracted together at first

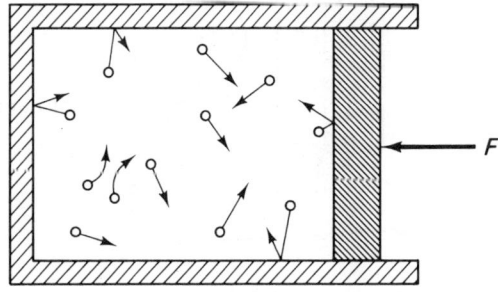

Fig. 17.11   Gas in a closed container (see text).

and then repelled violently when they get *very* close. The same happens when they reach the walls of the container. All the time, the walls of the container are being hit by molecules. Each time a molecule hits the wall it exerts an *impulse* on it. Over any given area of wall, hundreds of molecules are hitting it all the time; the total effect of their impulses is what we call *pressure*. In the figure, one wall is moveable. The impulse of the molecules hitting this wall would gradually accelerate the wall, making it move to the right. But as you can see, there is a steady force, *F*, acting on the wall from the outside. This force just balances the force due to the molecules hitting the wall. This external force might be caused by the impulse of molecules of the air outside hitting the wall, or it might be a force applied by hand, or a combination of both. What happens if the external force on the wall is increased? Before answering this question, try the next two experiments.

Instructions: investigating the effect of increasing the pressure on a gas

Solids and liquids are incompressible; now we find out what happens if we increase the pressure on a quantity of gas.

1 The apparatus (fig. 17.12) consists of a vertical tube containing air. It is connected to a reservoir containing oil. The interior of the reservoir is connected to a foot-pump. A Bourdon gauge indicates the pressure inside the apparatus.

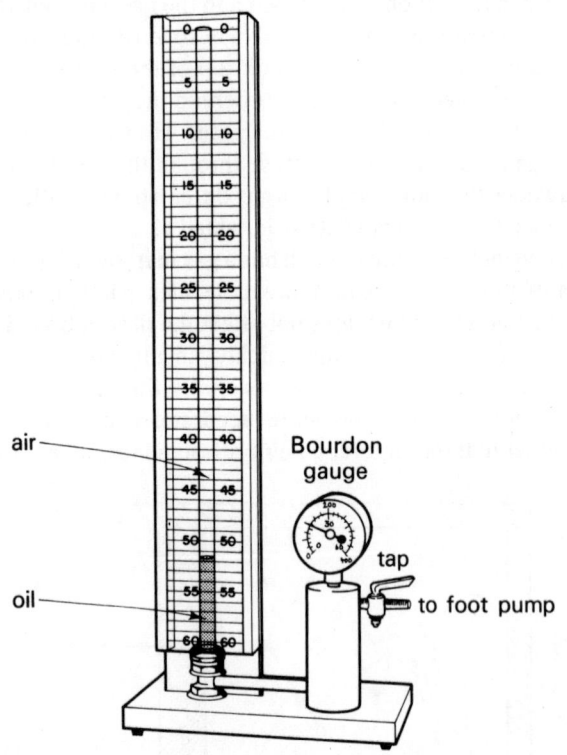
air
Bourdon gauge
oil
tap
to foot pump

Fig. 17.12 Apparatus for investigating the relation between pressure and volume of a gas.

2 With atmospheric pressure in the apparatus (tap open) measure the volume, *V*, of the air, using the scale behind the tube. Record the air pressure, *p*. Enter the results in a table:

Pressure p in Pa	Volume V in cm³	Product $p \times V$	$1/p$

3 Connect the foot-pump and, with tap open, pump gently until the oil has risen about 5 cm up the tube. Close the tap. Read the exact values of *V* and *p*.

4 Repeat four or five times more, for increasing pressures.

5 For each pair of values of *p* and *V*, calculate the product *pV* and enter this in the third column of the table. What do you notice about *pV*?

6 Calculate $1/p$, and enter it in the table. Plot a graph of *V* (vertical axis) against $1/p$ (horizontal axis.). Comment on the graph.

Instructions: further investigations of the effect of pressure on the volume of a gas

1 Fix a large disposable plastic syringe (50 cm³ or larger) vertically in a clamp (outlet downward). Block the outlet with Plasticine when the piston is pulled almost fully out.

2 Add weights to the top of the piston to increase the pressure on the air inside.

3 Knowing the area of the piston, calculate the extra pressure due to the weight; add to this the atmospheric pressure to obtain total pressure, *p*. Measure *V*.

4 Repeat this several times and make a table of *p* and *V* as in the previous experiment. Analyse the results as in steps 5 and 6 above.

5 Invert the syringe. Begin with 20 cm³ of air in it at atmospheric pressure. Hang weights from the end of the piston to *decrease* the pressure inside. Investigate the relation between *p* and *V*, as before. Does the same rule apply for pressures less than atmospheric pressure?

The rule connecting pressure and volume was first discovered by Robert Boyle in the seventeenth century. It is now known as *Boyle's law*, and may be stated:

The volume of a quantity of gas is inversely proportional to the pressure, provided that the temperature is constant.

Returning to fig. 17.11 we can imagine the moveable wall being pushed inwards by increasing the value of *F*. This is what we did, when, by increasing the pressure in the oil reservoir, we pushed more oil into the glass tube.

The molecules of gas inside the container (the glass tube) are forced closer together, but this does not really affect their intermolecular attractions to any measurable extent. But now that the volume inside is smaller, molecules hit the moveable wall (and the other walls too) more often. There are more impulses *per second*: the outward pressure on the walls increases. If we double the external force on the moveable wall, the wall moves inwards until the inward force is balanced by the now increased pressure of the gas molecules; this occurs when the volume occupied by the gas has been reduced to half its former size. Similarly, as we increase the pressure in the oil reservoir, oil passes into and up the tube until the increased pressure on the surface of the oil (caused by the increased rate of molecular collisions between the air molecules and oil surface) exactly equals the new higher pressure.

18 Matter and heat

In this chapter we use our knowledge of the molecular structure of matter to explain how matter behaves when heated. We explain some of the results from earlier chapters and investigate some more effects of heating.

18.00 Vibrating molecules

In a solid the molecules each have an equilibrium position in the lattice of the crystal. They vibrate slightly about this position, under the action of attractive and repulsive intermolecular forces (17.03). When a molecule is at its equilibrium position it has maximum kinetic energy. It is moving towards some of its neighbours and away from others. When it gets very near to a neighbour it is repelled; when it gets far from a neighbour it is attracted; both effects make it lose velocity as it moves away from its equilibrium position. Its velocity decreases until it comes to rest. The molecule then has no E_k (fig. 18.1a): all its energy is potential energy, because it is 'too close' to some neighbours and 'too far' from others. We can say that E_p is a maximum. The situation is rather like that of a pendulum

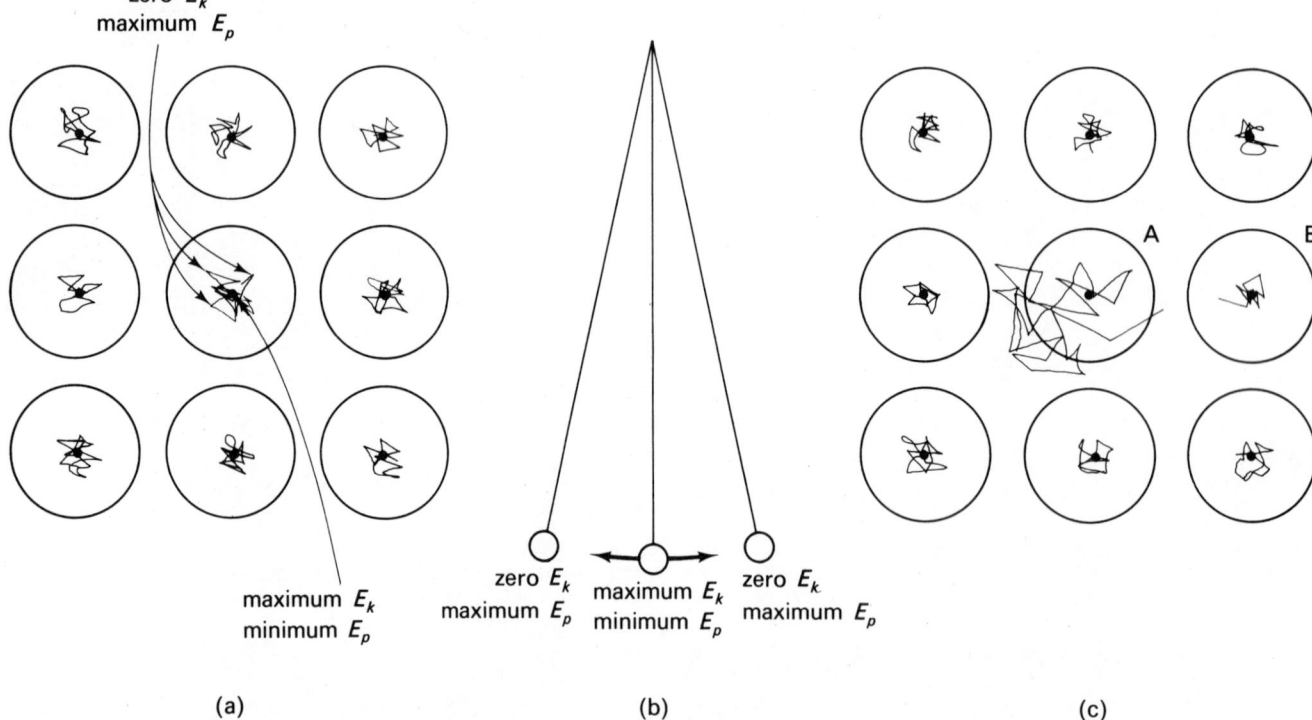

Fig. 18.1 *Vibrating molecules (a) the circles represent the molecules in the lattice of a solid; dots indicate the positions of molecules in equilibrium: the zig-zag lines represent the motion of the centre of each molecule as it vibrates; (b) vibration of a pendulum; (c) molecule A has more energy than its neighbours.*

(fig. 18.1 b) vibrating to and fro; at the ends of its path it has no E_k and maximum E_p; at its lowest point (its equilibrium position) it has maximum E_k and minimum E_p (or zero E_p if we take this point as the zero reference level for E_p). The difference between the pendulum and the molecule is that the pendulum bob moves along an arc of a circle, while the molecule vibrates in three dimensions under the continually changing forces produced by its vibrating neighbours. The vibrations of a pendulum gradually become smaller owing to loss of energy brought about by air resistance and friction. Molecules have no forces to reduce their energy. Energy lost by one molecule is gained by its neighbours, as explained in the next section. Therefore the molecules continue to vibrate for ever.

18.01 Transfer of energy between molecules

If, for any reason, one molecule has more energy than its neighbours, this extra energy is soon shared with these neighbours. In fig. 18.1 c the central molecule A has extra energy. Its neighbours all have less energy than A. The figure shows molecule A approaching molecule B. Because of its extra energy, A has moved further than the other molecules from its equilibrium position. Its velocity when it passes through its equilibrium position is greater because of its extra kinetic energy. Therefore it travels a greater distance before being brought to rest by intermolecular forces. A and B are much closer together than molecules normally are; the result is a *very strong* repulsive force between them. Molecule B moves off with extra energy due to this extra strong repulsion. This gain in energy is balanced by a loss in E_p for A, for energy must be conserved in this interaction. Some of A's energy has been transferred to B. A goes close to all of its neighbours several times as it vibrates. A little of its energy is lost to each of them every time. In turn these neighbours, vibrating a little more strongly because of the extra energy they have gained from A, pass some of *their* extra energy on to *their* neighbours. The extra energy from A gradually spreads until it is shared by all the molecules of the lattice. Now all are vibrating a little more vigorously than they were before. In short, we can say that if some molecules have more energy than others, there is a tendency for this energy to be transferred until the energy is shared more or less equally by all.

This exchange of energy goes on all the time. At any instant most molecules have about the same energy, though some have a little more than average and some a little less. Some gain, some lose, but the total amount of energy available is shared fairly evenly among all the molecules.

18.02 Giving energy to molecules

Energy can be given to molecules in two different ways.
(a) *By doing WORK*. If an impulse Ft acts on a solid body (fig. 18.2a) it affects *all* its molecules. The body moves from rest with velocity v. Every molecule in the body is still vibrating around its equilibrium position with

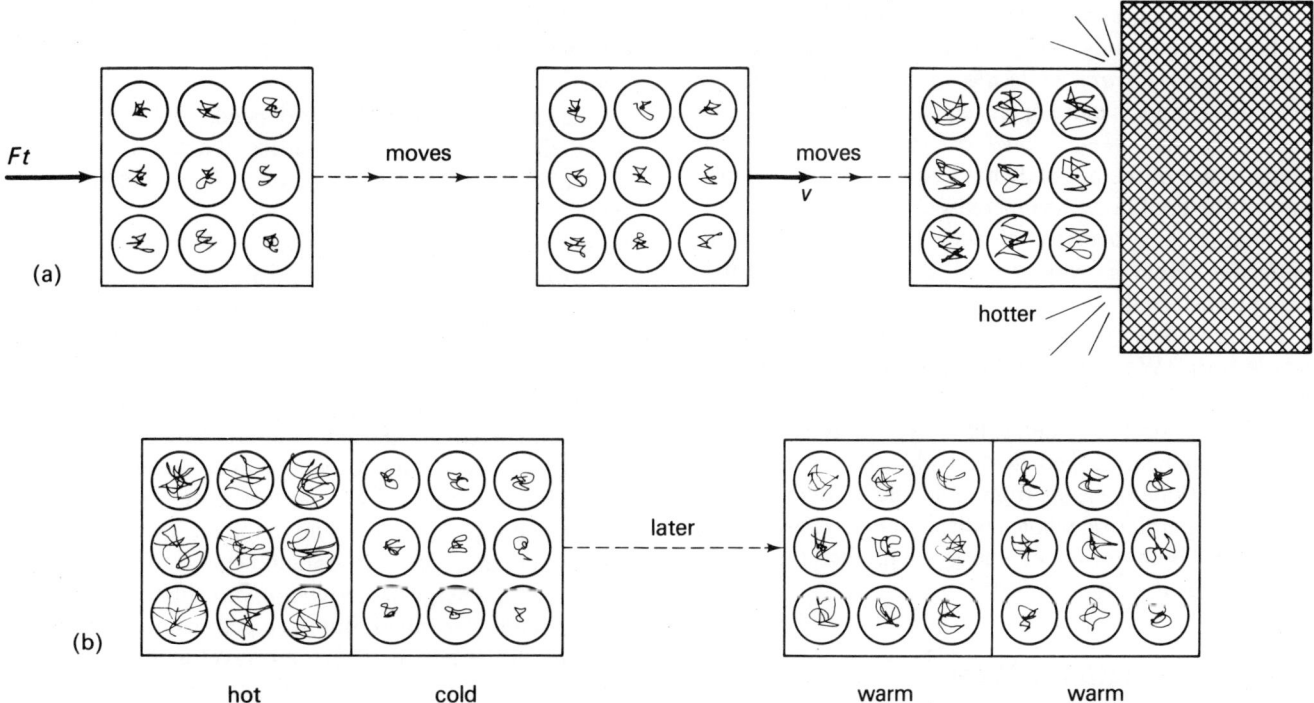

Fig. 18.2   *Transferring energy to molecules by (a) doing work, (b) heating.*

changing velocity but added to this velocity each molecule has the velocity $\dot{v}$ in the direction taken by the body. It is like a railway compartment full of mosquitoes buzzing around while the train travels along the track. In this way the energy of each molecule has been increased by $\frac{1}{2}mv^2$. If the body hits a fixed stationary object, it is brought to rest rapidly. Some of the energy becomes noise (vibrations of air molecules), some is transferred to the molecules of the stationary object, some of it appears as increased vibration of the molecules of the object itself. You can imagine that the sudden jolt will set all the molecules vibrating very vigorously. They no longer have the extra energy due to $\frac{1}{2}mv^2$, but each has extra energy (not as much as $\frac{1}{2}mv^2$) due to their greater kinetic energy of vibration.

This is what we mean when we say that the *internal energy* of a body has increased. Internal energy is the energy belonging to the molecules of which the body is composed. If you were to measure the temperature of the body after impact you would find that its temperature is greater than before impact. Increased internal energy means increased temperature. When we measure the temperature of a body we are measuring the average energy of its molecules. This leads us on to the second way in which energy can be given to molecules.

(b) *By HEATING.* If, instead of applying an impulse to the body, we put a hotter body in contact with it, we get the situation shown in fig. 18.2b. In the hot body the molecules are vibrating strongly, they each have more energy than the molecules of the cool body. If this is so, there is transfer of energy between *molecules* near to the contact surfaces, as explained in section 18.01. In effect, we have a single body, hot at one end and cool at the other. Energy passes *from molecule to molecule*, until the available energy is more or less equally shared by all. The cool body has been *heated* by the warm one. The temperature of the hot body has decreased; the temperature of the cool body has increased.

Heating is a *transfer* of energy from *molecule* to *molecule*. 'Heat' is what we call this energy when it is being transferred.

Doing work is a transfer of energy by applying force to a *whole body*. 'Work' is what we call this energy when it is being transferred.

Compare the two statements above carefully and you will see that 'heating' and 'doing work' are similar actions, except that one operates from molecule to molecule and the other operates on whole bodies. Similarly, 'work' and 'heat' are both forms of energy and, therefore, both measurable in joules (16.30).

18.03 Conduction of heat

Conduction of heat has already been explained in fig. 18.2, but we consider it in a little more detail in fig. 18.3.

There we have a body which is hotter at one end than at the other. By transfer of energy (heat) from molecule to molecule, the available energy is gradually shared among all molecules. If we think of this in terms of temperature, we would say that the body has reached the same temperature throughout. Heat always behaves in this way, so that we can make a general statement:

If there is a difference of temperature between two points, energy (heat) flows from the point with the higher temperature to the point with the lower temperature.

Since the mechanism of transfer of energy is from molecule to molecule and depends on intermolecular forces, the rate of transfer is affected by the types of molecules, their average distance and the pattern of the lattice, if there is one. Though we have referred throughout this discussion to solids, similar interactions occur between molecules in liquids and in gases, so that heat can be conducted in these too. Heat can also be conducted from a solid to a liquid or gas. For example, if the solid in fig. 18.3a is surrounded by air, the molecules of air interact with the molecules at the surface of the solid; they gain energy in this interaction. Some of the heat from the solid passes to the air and the object gradually cools. We could imagine the opposite situation in which the object was surrounded by a very hot gas, then it would gain heat from the gas. We can also imagine what happens if you put your finger on a warm object. The molecules in your skin gain energy from the molecules of the object; this energy passes from molecule to molecule in your skin. Eventually it passes to molecules inside the heat-sensitive nerve cells in your skin. Their increased energy is detected and this triggers off a nerve message to your brain: the object 'feels hot' to you.

The rate of transfer of heat depends on the molecules and their arrangement in the lattice (if a solid). In metals, an additional mechanism operates. The molecules of a metal crystal are in fact single atoms of the metal. The atoms of metals, like all atoms (chapter 20), have electrons circling around them. In metals, some of these electrons are less tightly bound to the atom. They are free to wander around in the space between the atoms. We can think of a metal as a lattice with fixed atoms but between them is a cloud of freely-wandering electrons, none of them attached to any particular atom (fig. 18.3b). Moving electrons possess kinetic energy and, if they are given extra kinetic energy, they move even faster. If one part of a metallic body is hotter than another, the atoms *and electrons* in that region have greater energy. The atoms can transfer their energy *to their neighbours* in the way that we have already described. But the electrons are not fixed in position in the lattice. They can travel quickly from one part of the object to another. In the hotter region, they pick up energy from the electrons and atoms of that region; in the cooler region, they give up energy to the atoms and

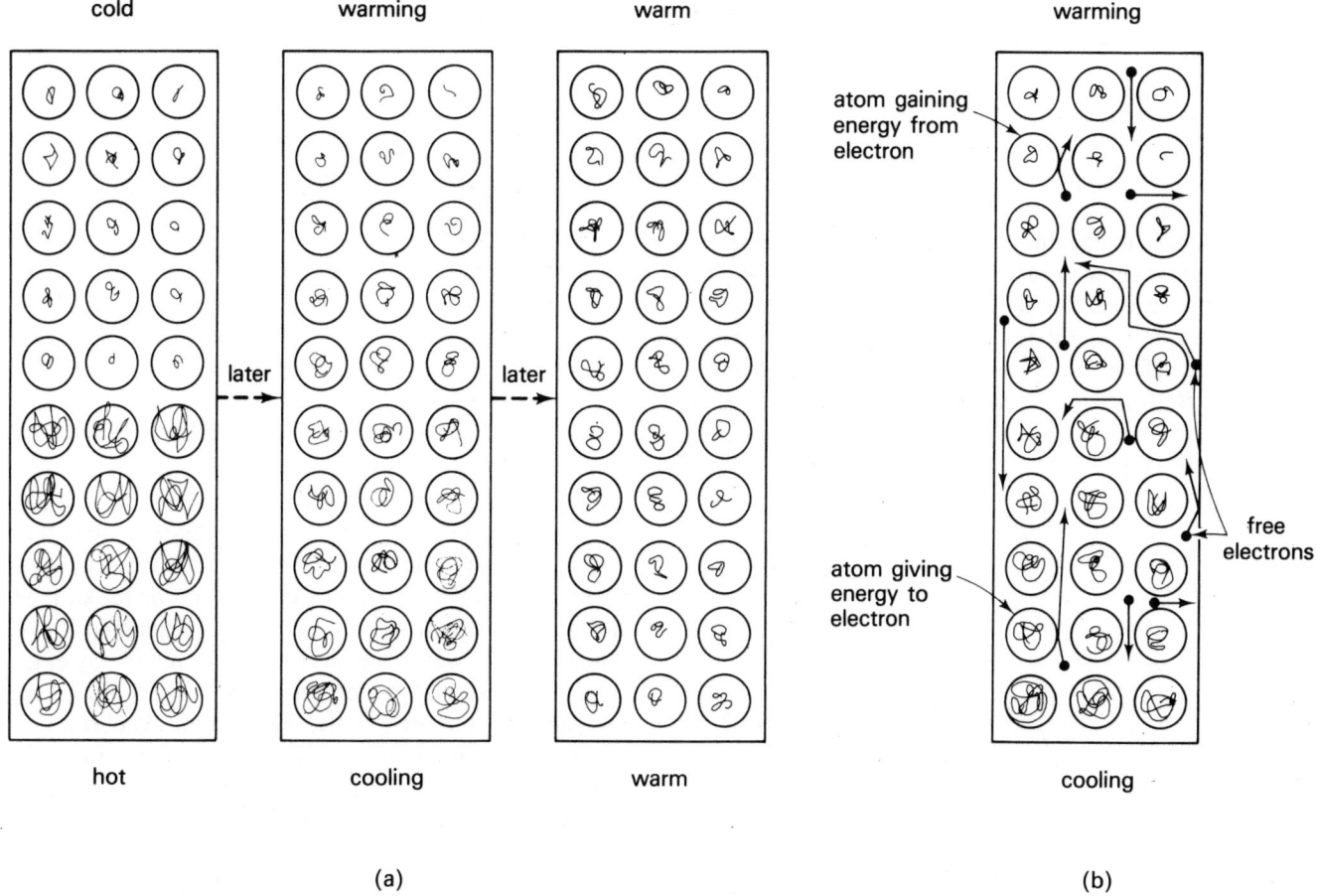

cold　　　　　warming　　　　　warm　　　　　　　　　warming

atom gaining
energy from
electron

free
electrons

atom giving
energy to
electron

hot　　　　　cooling　　　　　warm　　　　　　　　　cooling

(a)　　　　　　　　　　　　　　　　　　(b)

Fig. 18.3　Conduction of heat (a) from molecule to molecule; (b) by free electrons, as in metals.

electrons there. This cloud of *moving* electrons quickly transfers heat from hot regions to cooler ones. The rate of transfer by this means is much faster than the rate of transfer from atom to atom fixed in the lattice. This is why metals are much better conductors of heat than other types of matter are.

18.04 Change of state

The three states of matter were described in chapter 17. Let us see what happens when matter changes from one state to another.

Instructions: relating temperature to change of state
1 Place some crushed ice in a beaker. Measure its temperature.
2 Leave the beaker on the bench and measure the temperature of the ice every 5 minutes. Keep a record of the temperatures and note when the ice begins to melt.
3 When all the ice has melted, place the beaker over a small flame. Heat steadily and record the temperature every minute.
4 Continue heating and recording temperature until the water has been boiling for 5 minutes.

Questions
1 What happened to the temperature of the ice before it began to melt?
2 What happened to the temperature of the ice-water mixture while the ice was melting?
3 What happened to the temperature of the water while it was being heated by the flame?
4 What would happen to the water if it was kept over the flame for an hour or more?
5 What would happen to the water if (after boiling for 5 minutes) it was put on the bench to cool, and then left for several days?

During the first stage of this investigation the molecules of the beaker and the air of the room are at a higher temperature than the molecules of water in the ice. Heat will be transferred from the beaker and air to the ice. The molecules of water in the ice gradually acquire more and more energy. The temperature of the ice increases.

As the water molecules (in ice) acquire more energy, they move more rapidly. When their energy reaches a certain level (represented by 0°C), they are moving so fast that intermolecular forces are not strong enough to convert their kinetic energy to potential energy and hold

169

the molecules vibrating about fixed equilibrium positions. In other words, the molecules are able to escape from their fixed positions in the lattice. The regular arrangement of the crystal is destroyed. Now the molecules are in roughly-defined groups (17.12), but many are able to leave one group and join another. The solid melts; the ice turns into water.

While this is happening, all the energy entering the ice from outside is being used to overcome the intermolecular forces which previously held the molecules in a regular lattice. The molecules gain E_p, but their E_k remains about the same as before; the effect of this is that the *temperature does not rise while the ice is melting* (fig. 18.4). Heat is entering the ice but there is no increase of temperature. The heat is being used to turn ice into water: *a change of state*.

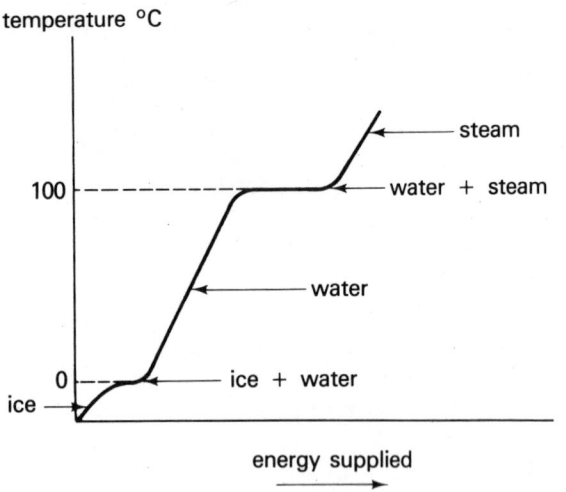

Fig. 18.4 *Temperature changes as ice is heated and changes state to liquid and to vapour.*

During the second stage of the investigation the water molecules gain more kinetic energy, this time from the flame. As their temperature increases their average velocity increases and the intermolecular forces are less able to hold them in groups. Occasionally a molecule near the surface may be approached from below by two or more molecules. If this happens the repulsion can accelerate the molecule so strongly that it accelerates away from the surface (fig. 18.5). Then it might fall back in, attracted by the water molecules below, or perhaps because it is hit by a molecule in the air above. A lot of molecules escape from the surface and move away among the molecules of the air. We say that these molecules have *evaporated*.

When we say that water has a temperature of, say, 50°C, we mean that *on average* the molecules have kinetic energy equivalent to a temperature of 50°C. This does not mean that all molecules have exactly the same E_k. Some have more than average, others less. The molecules approach one another continually, transferring

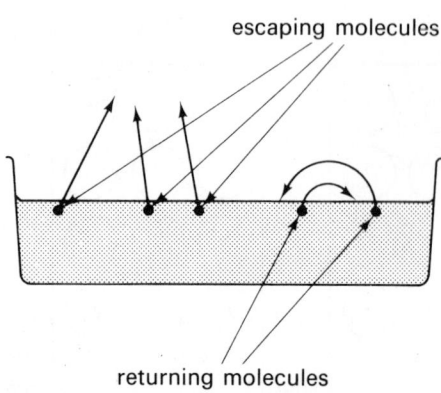

Fig. 18.5 *Evaporation from the surface of a liquid.*

energy. Occasionally, *by chance*, a molecule may receive energy from two or more molecules, giving it far more than average energy. If it is deep in the liquid it will soon pass this extra energy to other molecules. If it is on the surface, this extra energy makes it evaporate, as explained above. The hotter the water, the higher the average energy of the molecules and the greater chance of a molecule getting enough energy to evaporate. The rate of evaporation is greater if temperature is higher. We know this from everyday life; pools of water evaporate quickly in the hot season, but may remain for weeks in the cool season.

When water reaches 100°C, the average energy of the molecules is enough to allow them to evaporate; evaporation does not now depend on the molecule gaining a lot more energy than the others. Molecules at the surface are able to leave easily. Molecules deep in the liquid can now move away from each other, forming bubbles of vapour (steam) in the liquid. The bubbles are less dense than the liquid; they rise to the surface and the vapour escapes to the air above. The water is *boiling*.

As water boils it *changes state* from *liquid* to *vapour*. We use the word 'vapour' to remind ourselves that though the water now has the properties of a gas and, in fact, *is* a gas, it is a liquid at ordinary room temperatures. We usually do not call a substance a gas unless it is in the gas state at ordinary temperatures. Thus we speak of water vapour, sodium vapour and mercury vapour but oxygen gas and hydrogen gas.

18.05 Latent heat

While the water boils its temperature remains at 100°C (fig. 18.4). The energy being supplied by the flame is being carried away as the kinetic energy of escaping molecules. The energy of the flame is being used to *change the state* of the water, *not to raise its temperature*. The amounts of energy used in changing state (from solid to liquid and from liquid to vapour) are relatively high. They can be measured, as in the experiments following. Since a change of state means that heat is supplied but temperature does not rise, the heat used in changing state has been called *'hidden heat'*. The word actually used is *'latent heat'*, which means the same thing. If we measure the latent heat required to change the state of 1 kg of a substance, we call this the *specific latent heat*.

Instructions: measuring the specific latent heat of fusion (melting) of ice

1 The method and apparatus are the same as those given in chapter 16 for measuring the specific heat capacity of a liquid (page 153), except for the differences noted below.

2 Find the mass of the calorimeter empty, m_1.

3 Place dry melting ice in the calorimeter; the ice should be dried in a cloth immediately before use.

4 Switch on the current and begin timing; read I and V. Stir the ice as the heater melts it. As the level falls, add more dry melting ice.

5 When the calorimeter is full, allow all the ice to melt. As it just finishes melting, switch off and record the time.

6 Find the mass of the calorimeter and melted ice, m_2. Calculate the mass of ice melted, m.

7 Calculate the total electrical energy converted into latent heat.

8 If this amount of energy is required to melt ice mass m, what amount is required to melt 1 kg of ice? This is the specific latent heat of fusion of ice, l. Calculate l. What are its units?

Instructions: measuring the specific latent heat of vaporization of water

(This is often called the specific latent heat of steam.)

1 The method and apparatus are the same as in the previous experiment, except that the calorimeter is filled with water and the heater is switched on. When the water begins to boil, begin timing. After 10-20 min, switch off and, by 'weighing', find the mass of water which has been boiled away.

2 Given I, V, m, and t, calculate l, the specific latent heat of vaporization of water.

Latent heat is given out again when a substance changes from the vapour state to liquid state, or from the liquid state to the solid state. We can use this fact, as below.

Instructions: measuring the melting point of paraffin wax

1 Place solid paraffin wax in a beaker. Heat it gently to melt the wax, but as soon as it is melted, do not heat it further. Wax can become very hot and you might make it so hot that your thermometer would be damaged.

2 Place a thermometer in the wax. Watch it closely as you do this. If the thread seems to be rising rapidly to over 100°C, take the thermometer out again, very quickly. Wait for the wax to cool. Then support the thermometer so that its bulb is centred in the wax. Allow the wax to cool and record its temperature every 5 min.

Question

The temperature drops steadily at first. Then it remains steady. Explain why it remains steady. What happens to the wax during this time?

These results show that the heat used in changing state is hidden but not lost. It can be recovered. If a jet of steam strikes a cool surface it loses heat and becomes water again. We say that it *condenses*. As it does so the latent heat is released and the surface it strikes becomes very hot.

It is interesting to note that to heat 1 kg of water from 0°C to 100°C requires 420 000 J. Then to turn that boiling water into steam (at the same temperature) requires 2 260 000 J. This relatively large amount of heat is released when the steam condenses. This explains why steam is so much more effective than hot water as a conveyor of heat. Steam is carried in pipes in many kinds of engine and industrial equipment. In the coffee-bar, a steam jet is often used to bring a cup of coffee or bowl of soup almost to the boil. The attendant bubbles steam through the cool liquid. The steam condenses (making a loud noise, the sound of collapsing bubbles), and gives off its large amount of latent heat to the coffee. This gets hotter. The amount of water produced by condensing steam is so small that the coffee is not made weak by this process.

Though vapours such as steam can hold much energy in the form of latent heat, they are poor conductors of heat. This is because the molecules are far apart. They meet one another rarely, so there is less opportunity for molecules to transfer energy in the way described in 18.01. This limits heat conduction. In gases the main method of heat transfer is by convection, which involves a mass flow of highly energetic molecules from one region to another.

18.10 Evaporation and boiling

Imagine a bottle into which a quantity of liquid is placed. The bottle is closed with a stopper. Molecules evaporate from the liquid into the space above the liquid. They move around, between the molecules of air gases in the bottle. They hit the walls of the bottle. There is pressure on the walls caused by the air; this is air pressure. There is pressure on the walls caused by the molecules evaporated from the liquid; this is *vapour pressure*.

As evaporation continues, the number of vapour molecules increases. After a period of time the rate of evaporation is equalled by the rate of return of molecules from the vapour back to the liquid. There is equilibrium between the liquid and the vapour. This is an equilibrium *with movement*; molecules are evaporating still, and molecules are returning; such an equilibrium is called a *dynamic equilibrium*. When this stage is reached, the number of vapour molecules in the space above the liquid is at its maximum. The vapour is now a *saturated vapour*. The pressure of the vapour on the walls of the bottle is also at a maximum. This is the *saturated vapour pressure*.

If temperature increases, the molecules of liquid gain kinetic energy. They are more easily able to evaporate. The number of molecules above the liquid increases; eventually a new equilibrium is reached, at a higher saturated vapour pressure. As temperature increases, so does saturated vapour pressure. When the liquid reaches its boiling point its saturated vapour pressure is equal to the air pressure. Then molecules freely leave the liquid surface and the liquid boils. At high altitudes, where atmospheric pressure is low, the saturated vapour pressure of a liquid becomes equal to atmospheric pressure at a low temperature, for example, water boils at temperatures lower than 100°C. This explains why liquids boil at lower temperatures if pressure is low (3.40).

Boiling happens at the temperature at which the saturated vapour pressure of a liquid equals atmospheric pressure. Evaporation can happen at any temperature lower than this. It can happen at temperatures lower than the melting point. Evaporation from the solid state is generally slow, for it is not often that a molecule acquires enough energy to escape from the lattice. A lump of camphor, or a crystal of iodine gradually 'disappears' if left on an open dish. These substances evaporate readily from the solid state. Tungsten evaporates from the hot filament of an electric lamp and condenses on the inside of the glass bulb, causing the blackening of the glass that we see in old lamps.

18.20 Evaporation and cooling

Instructions: investigating cooling by evaporation
1 Pour a little water on a block of wood and stand a small beaker in this pool of water (fig. 18.6).

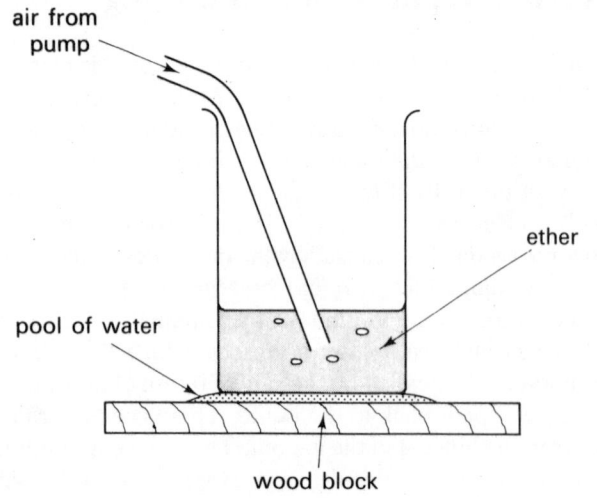

Fig. 18.6 Cooling by evaporation.

172

2 Put a small amount of ether in the beaker.
3 Bubble air through the ether. What appears on the outside of the beaker?
4 After a few minutes, try lifting the beaker. What has happened to the water?

The boiling point of ether (diethyl ether, ethyoxyethane) is 34.5°C. At room temperature it is near to its boiling point; its saturation vapour pressure is high. By bubbling air through it we help make the rate of evaporation as high as possible. When molecules escape from a liquid, those with the greatest amounts of energy escape most easily. Those which are left behind are the least energetic ones. This means that as molecules evaporate the average energy of molecules in the liquid is reduced. To say that the average energy becomes less is the same thing as saying that the temperature becomes lower. In the experiment the temperature of the ether becomes so low that it removes heat from the pool of water, which freezes.

We notice the same effect on a windy day after swimming. The wind helps the water evaporate from the surface of the skin. The temperature of the water is lowered; heat is removed from our skin; we feel cool. On a still day the effect is less; as water evaporates the air around the skin becomes saturated with water vapour; the rate of evaporation is much reduced, and so is the cooling effect. The cooling effect is also reduced if the air is humid and already saturated with water vapour.

In some countries water is stored in porous containers, made of unglazed baked clay. A small amount of water soaks through the pot to the outer surface. There it evaporates. This cools the water inside, so making it pleasant to drink. The goat-skin bag works on the same principle.

One of the ways the human body keeps cool is by producing sweat from sweat glands in the skin. The evaporation of this sweat keeps us cool. In humid conditions the sweat cannot evaporate but simply runs off the skin in drops; there is no cooling effect. In hot climates people use electric fans indoors to increase evaporation of sweat, so keeping the body cool. Note that the fan does *not* cool the air of the room, it works only by cooling the moist skin. In desert areas, where humidity is nearly always low, a good way of keeping the house cool is to use a 'desert cooler'. This consists of a large electric fan mounted in an enclosure on an outer wall of the house. The fan blows large quantities of air through a special pad made of straw, which is automatically kept wet with water. The water evaporates readily in the dry hot air, and a stream of cool moist air is produced which is blown into the house. This device is much cheaper to buy and to operate than an air-conditioner but unfortunately is only of practical use in areas where the atmosphere is very dry for most of the year.

The cooling effect is put to scientific use in the wet-and-dry-bulb thermometer. This consists of two identical thermometers, one of which has its bulb surrounded by a tube of open-mesh cotton fabric. The fabric dips into a small bottle containing water. Water soaks into the fabric and keeps the bulb moist. Water evaporating from the fabric is replaced by more drawn up from the bottle by the wick action of the fabric. If the air is humid, evaporation is little or none, and there is little or no cooling effect. The two thermometers indicate the same temperature or very close temperatures. The drier the atmosphere, the greater the evaporation, the greater the cooling effect, and the bigger the difference between the readings of the thermometers. Special tables have been prepared by means of which the temperatures shown by the thermometers can be interpreted to give the *relative humidity* of the atmosphere. This is a measure of the saturation of the atmosphere with water vapour. This knowledge is important for making weather forecasts and, for example, deciding if conditions in a store-room are suitable for the storage of products which are badly affected by a damp atmosphere.

The principle of the refrigerator is illustrated in fig. 18.7. The air conditioner works in a similar way. The refrigerating system contains a special liquid, or refrigerant with boiling point around 30°C. This enters the coil in the ice-making compartment through a valve. The pressure in this coil is low, and the refrigerant evaporates very rapidly, or may boil. The latent heat of vaporization comes from the liquid itself, which is cooled, as was the ether in the experiment on page 172. A pump removes the vapour, pumping it to another set of coils on the outside of the refrigerator box, usually at the rear. The action of the pump keeps the pressure low in the cooling coil and high in the external coil. In the external coil the high pressure makes the vapour condense to the liquid state, giving out its latent heat. This heat passes out to the atmosphere. If you place your hand on the external coils of a refrigerator, you can feel that they are slightly warm. The liquid refrigerant is then pumped on through the valve, back to the cooling coil again. Thus in the refrigerator we have a method of transporting heat from one place to another. The rate at which this occurs is controlled by some kind of thermostat switch (fig. 6.4) inside the refrigerator, which switches the pump on or off as required.

18.30 Expansion

Imagine the vibrating molecules of a solid substance (fig. 18.1a). If the temperature increases, they vibrate more strongly. They move faster and they move further from their equilibrium positions. Molecules come closer to their neighbours and are repelled more strongly. To make room for this extra activity the average distance between molecules must increase: the solid *expands*.

The same happens in liquids. Expansion of solids and liquids is the result of intermolecular forces. These forces are very strong; they are the forces which hold matter together. Forces of this size are strong enough to pull matter apart too. It is not surprising that thermal expansion can produce forces great enough to buckle steel girders and shatter huge rocks.

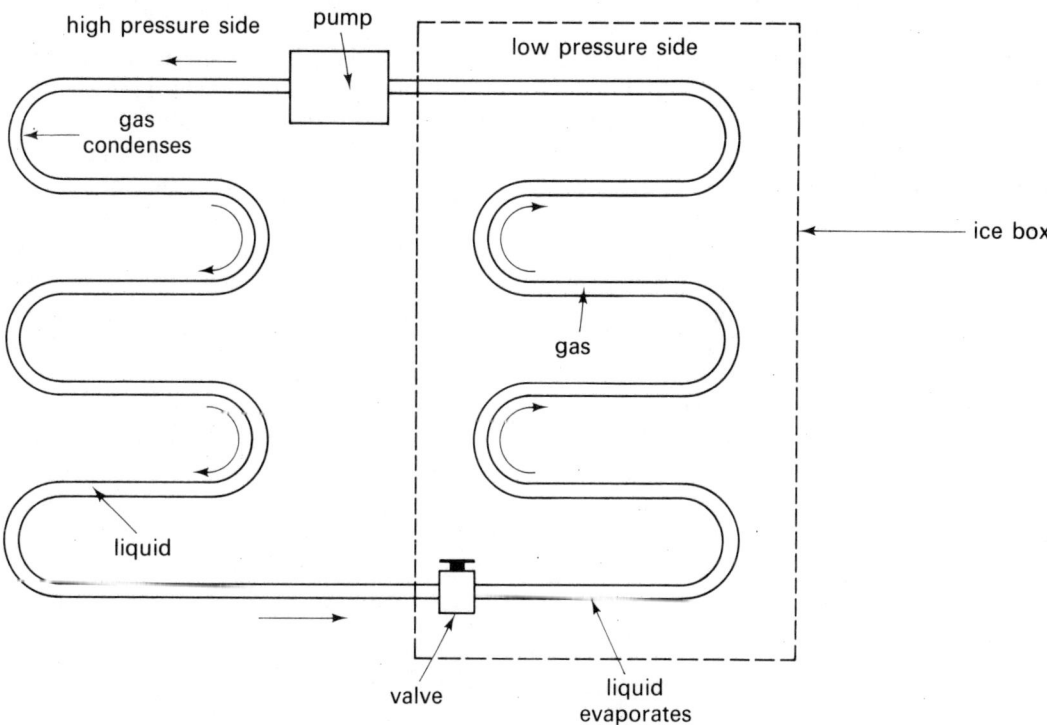

Fig. 18.7 Diagram to show the working of a refrigerator (see text).

Expansion leads to increase in volume, but the quantity of matter and its mass remain unchanged. Therefore expansion leads to reduction in density. If a quantity of liquid is heated from below, the lower regions are hotter than the upper regions. They are therefore less dense. The heated liquid rises to float on top of the cooler liquid. The cooler liquid sinks below the heated liquid. This creates the circulating currents which we call convection (6.43). Convection is a mass movement of hotter molecules (with greater kinetic energy) from the lower heated region of the liquid and their replacement by cooler (less energetic) molecules from the cooler region. If the liquid is heated from above, convection does not occur; the less dense heated water is already above the denser cooler water.

During the hot season the Sun shines on tropical seas and lakes; the surface layers of water are heated. There is no convection, so these layers do not mix with the cooler layers below. The water below remains very cold. In the cooler season the surface layers lose heat and then may be colder than the lower layers. Then convection may occur, stirring the waters of the lakes and bringing dissolved nutrients to the surface. This stirring is very important since it supplies essential nutrients to plants and animals living in the surface layers.

18.31 Expansion of gases

When a solid or liquid expands or contracts the forces are great. We can do little to prevent the change in size. Gases are compressible and expansible; by altering the pressure on a quantity of gas we can make it almost any size we want, whatever temperature it is at. We must therefore think about pressure when we are investigating the effects of temperature on the volume of gases.

The air thermometer (fig. 3.2) shows that gases expand a lot when heated. In the air thermometer the pressure in the thermometer remains constant at atmospheric pressure, plus a very slight increase due to the water-drop in the tube. As temperature is increased, the molecules in the thermometer travel with increasing velocity. Their impulse is greater as they hit the walls of the thermometer or the lower end of the water-drop. This means increased pressure but, as pressure tends to increase, the water-drop is pushed outwards. It moves outwards until the pressure outwards once again equals pressure inwards. This increases the volume of space available to the molecules. They become more widely spaced. In a given period of time *fewer molecules* hit any given area of wall although the molecules have greater energy. In other words, the pressure remains unchanged.

18.32 Expansion of a gas at constant pressure

If we have a quantity of gas at 0°C and heat it, we can measure its expansion. If we arrange for its pressure to remain constant during this operation we can calculate a quantity called the *coefficient of expansion at constant pressure* α.

$$\alpha = \frac{\text{increase in volume (m}^3)}{\text{volume (m}^3) \text{ at } 0°C \times \text{ rise in temperature (deg. C)}}$$

Instructions: measuring the expansion of a gas at constant pressure

1 Use a plastic disposable syringe, capacity 25 cm³ or more. Set the barrel to about half the total capacity of the syringe. Seal the nozzle with Plasticine or with a short piece of tubing and a clip.

2 Put the syringe in a water bath containing a mixture of water and melting ice and leave it there for a few minutes.

3 Ease the plunger gently back and forth a few millimetres; you will be able to *feel* when the air inside is at atmospheric pressure. Then read the volume of air in the syringe.

4 Put the syringe into water baths at other temperatures and repeat steps 2 and 3. Suitable temperatures are 25°C, 50°C, 75°C and 100°C.

5 Plot a graph to show the relation between volume (vertical axis) and temperature (horizontal axis). This graph should be set out so that the temperature scale reads from −300°C to +100°C. Your plotted points will be only on the right-hand third of the paper. Draw the best straight line through your points.

6 From the graph read the increase in volume when the air is heated through 100°C. Calculate α.

7 Place a ruler along the line of the graph. Continue the line toward the left to where it cuts the horizontal axis. This is the temperature at which the gas would have zero volume. Read this temperature.

Measurements show that all gases expand by the same amount when heated; the value of α is 0.00366 per degree Celsius (or per kelvin). Your graph probably looked like fig. 18.8, in which the thick line is the part actually measured. Other measurements show that gases do contract as they are cooled, so we are justified in continuing the line towards the left. Eventually the cooled gas turns to liquid and then to solid so that its volume does not actually decrease to zero. Nevertheless, the idea of 'zero volume' is useful; we can imagine a gas that does not liquify or solidify; it stays as a gas until its molecules lose all their energy and are so close together that their volume can be ignored. This would eventually happen at about −273°C.

In fig. 18.8 the gradient of the thick line is:

$$\frac{\text{change in volume}}{\text{increase in temperature}}$$

The gradient of the thin line is $\dfrac{\text{volume at } 0°C}{273}$.

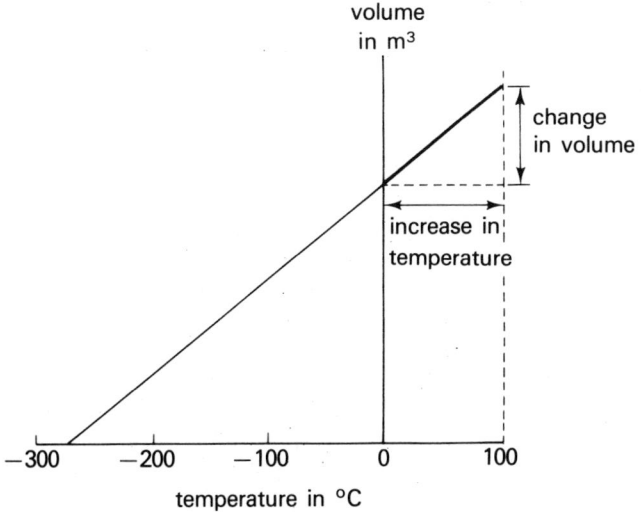

volume
in m³

change
in volume

increase in
temperature

-300 -200 -100 0 100

temperature in °C

Fig. 18.8 Expansion of a gas at constant pressure.

These gradients are equal, being gradients of parts of the same straight line, so:

$$\frac{\text{change in volume}}{\text{increase in temperature}} = \frac{\text{volume at } 0°C}{273}.$$

Rearranging gives

$$\frac{\text{change in volume}}{\text{increase in temperature} \times \text{volume at } 0°C} = \frac{1}{273}.$$

The left-hand side of this equation corresponds with the definition of α so

$$\alpha = 1/273 \text{ per degree C}$$

This is the same for all gases and this is why it was possible to use a mixture of gases (air) in the experiment. This relation is expressed as *Charles' law* (18.33).

18.33 The effect of temperature on pressure

In this section we find out what happens to the pressure of a heated gas, if we keep its volume constant.

Instructions: measuring the pressure coefficient of air at constant volume

1 Immerse a glass or plastic flask or bottle, capacity 200 to 400cm³, in a water bath. Seal the flask with a stopper with a tube passing through it, connected by rubber tubing to a Bourdon gauge. The connecting tube must be as short as possible.

2 Place iced water in the bath, and have the air in the flask at atmospheric pressure. Record its exact pressure.

3 Heat the water in the bath, stirring occasionally, and record the air pressure and water temperature at a number of temperature points, say every 20 degrees from 0°C to 100°C. Remove the flame and stir well just before you take each set of measurements.

4 Plot a graph of pressure (vertical axis) against temperature (horizontal axis). As in the previous experiment set out the graph with the temperature scale ranging from −300°C to +100°C. Draw the best straight line through the points.

5 The pressure coefficient, β, is calculated as follows:

$$\beta = \frac{\text{change in pressure (Pa)}}{\text{pressure (Pa) at } 0°C \times \text{rise in temperature (deg. C)}}$$

Calculate the value of β. What does this value remind you of?

6 Continue the line toward the left to where it cuts the horizontal axis. At what temperature does the line cut this axis?

7 A gas at −273°C would have zero pressure. What does this tell you about the kinetic energy of its molecules at this temperature?

Measurements made in the two last experiments both led to the temperature −273°C. This seems to be a temperature below which it is impossible to cool the gas. At this temperature all molecules are still; they have no kinetic energy. This temperature is called *absolute zero*. The SI temperature scale is based on this. The SI unit of temperature is the *kelvin*, symbol **K**. We do not need to say or write 'degrees' or use the symbol '°'. The kelvin scale begins at absolute zero (0K). This is its lower fixed point (3.11). Increases of temperature are the same size as those on the Celsius scale. The upper fixed point is the melting point of ice. It has the value 273K. Consequently, the boiling point of water is 373K. To convert temperatures in °C to temperatures in K, we simply add 273.

Although degrees Celsius are not really part of SI, we continue to use this scale for convenience in everyday life and often also in the laboratory. Most thermometers are marked in °C, not in K. But for defining physical quantities we must always use the correct SI base units. The kelvin is one of these, like the metre, kilogram and second. It does not depend on any other SI units. Since a kelvin and a degree Celsius are the same size, the figure obtained is not affected by the change of unit. Similarly the unit for α and β is 'per kelvin' (symbol, /K) and not 'per degree Celsius'.

We can now restate *Charles' law:*

The volume of a quantity of gas at constant pressure increases by 1/273 of its volume at 273K for each kelvin rise in temperature.

Similarly the pressure law can be stated as follows:

The pressure of a quantity of gas at constant volume increases by 1/273 of its pressure at 273K for each kelvin rise in temperature.

In the last experiment you could have marked the scale of the Bourdon gauge with the temperature of the water bath, for pressure varies as temperature. You would then have a thermometer of the type known as a *constant volume gas thermometer.* This type of thermometer has a very wide range, it is very sensitive and is very accurate. By using it with other gases we can cover a range of temperature from 3K to 1 750K. To use it at low temperatures we fill the thermometer with hydrogen or helium, which remain in the gas state at temperatures close to absolute zero. The disadvantage of this kind of thermometer is its large size and the fact that accurate types are very expensive. Their chief use is for accurately calibrating other types of thermometer.

18.34 Combining the gas laws

We have investigated three gas laws:

Boyle's law $\qquad pV = \text{constant}$ $\qquad$ (if T is constant)

Charles's law $\qquad \dfrac{V}{T} = \text{constant} = \alpha$ (if p is constant)

Pressure law $\qquad \dfrac{p}{T} = \text{constant} = \beta$ (if V is constant)

In these equations, p is pressure, V is volume and T is temperature (in kelvin). The value of constants α and β is 1/273. The value of the Boyle's law constant depends on how much gas was enclosed and what its pressure was when enclosed. These three laws can be expressed in one equation:

$$\frac{pV}{T} = \textit{constant.}$$

You can see that by making any one of the three variable quantities pressure, temperature or volume, constant, you obtain the equation of the law which relates the other two quantities.

Suppose we have a quantity of gas in a container, at pressure p_1, volume V_1 and temperature T_1. If we keep the gas sealed in its container but alter one, or two, or all three of pressure, volume and temperature, the equation must still be true. Suppose we alter the pressure to p_2 and alter the volume to V_2 and alter the temperature to T_2, the value of pV/T must remain constant. We get this equation:

$$\frac{p_1 V_1}{T_1} = \frac{p_2 V_2}{T_2}$$

This is a very useful equation. For example, suppose we have a cylinder volume $0.05\,m^3$, containing gas at pressure $2 \times 10^5\,Pa$ (or N/m^2) and temperature $15°C$. If we release this gas into a larger container so that its volume becomes $1.25\,m^3$, and at the same time increase its temperature to $25°C$, what is its final pressure? To solve this type of problem, set out the data in two columns, taking care to use the correct units:

In the cylinder $\qquad$ **In the larger container**

$p_1 = 2 \times 10^5$	$p_2 = ?$	(p in Pa or N/m^2)
$V_1 = 0.05$	$V_2 = 1.25$	(V in m^3)
$T_1 = 288$	$T_2 = 298$	(T in K [*not* $°C$])

Rewrite the equation in the form which leaves the unknown quantity on the left:

$$p_2 = \frac{p_1 V_1 T_2}{V_2 T_1}$$

$$= \frac{2 \times 10^5 \times 0.05 \times 298}{1.25 \times 288}$$

$$= 8\,278 \quad (p \text{ in Pa or } N/m^2)$$

or, in standard form, $p_2 = 8.278 \times 10^3\,Pa$ (or N/m^2).

Questions

1 On a warm day, when the temperature is $27°C$, and atmospheric pressure is $99\,000\,Pa$ (or N/m^2) a bottle of volume $250\,cm^3$ is sealed. It contains only air. The bottle is then placed in a refrigerator, and cools to $3°C$. What is the pressure inside the bottle?

2 A fixed quantity of hydrogen in a balloon has volume $10\,m^3$ at ground level, where the pressure is $10^5\,Pa$ (or N/m^2) and temperature is $25°C$. The balloon rises to the upper atmosphere where the pressure is $0.5 \times 10^5\,Pa$ (or N/m^2) and the temperature is $-20°C$. Assuming that the balloon is large enough to allow for expansion (if any) without causing additional pressure on the contained gas, what is the volume of the gas?

19 Matter and electricity

19.00 The structure of atoms

Before we can understand how matter can carry electricity we must know more about the structure of the atoms from which matter is made. It was said on page 155 that matter is made of molecules and that molecules are made of one or more atoms. If we break a molecule into its separate atoms, we no longer have the substance we began with. We have separate atoms of one or more different elements. Atoms too can be broken to pieces (chapter 20) and we find that they are made up from three kinds of particle:

(a) *electrons,* very light particles with a negative electric charge;

(b) *protons,* particles which are about 1 800 times more massive than electrons; they have a positive electric charge which is equal in size but opposite in sign to the charge on the electron;

(c) *neutrons,* particles with about the same mass as protons, and with no electric charge.

Modern physics tells us that the structure of atoms is more complex than was once thought. In addition to the three types of particles listed above, there are several other types which occur under special conditions. These and protons and neutrons are now thought to be made up from other fundamental particles with properties that are not yet fully understood. The search for new understanding about the structure of atoms is still going on and much remains to be discovered. Fortunately, the simple ideas outlined above are enough to allow us to explain as much as we need to know about electricity.

The atoms of all the known elements can be built from electrons, protons and neutrons. The mass of an atom is concentrated in its *nucleus* (fig. 19.1). This consists of a number of protons and neutrons, the numbers depending on the element. The nuclear particles, protons and neutrons, are called nucleons. Thus the nucleus of the carbon atom contains twelve nucleons consisting of six protons and six neutrons. The total number of nucleons

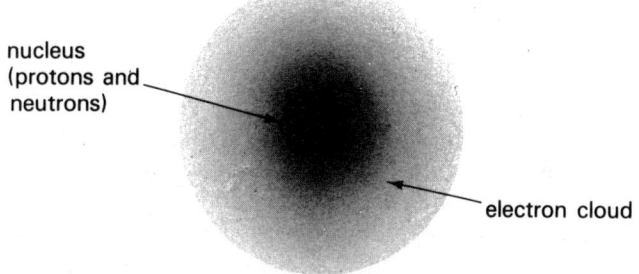

Fig. 19.1 *The structure of an atom.*

(protons plus neutrons) is called the *nucleon number, A* (this was formerly called the mass number). It gives us a measure of how great the mass of an atom is. Note that there are no electrons in the nucleus, only protons and neutrons occur there. Since in the nucleus only the protons have an electric charge, the nucleus is positively charged. Its total charge depends on the number of protons present. The number of protons in the nucleus is called the *proton number, Z* (formerly called the atomic number). It gives us a measure of the electrical charge on the nucleus.

Around the nucleus is a cloud of electrons. This cloud has negligible mass. The number of electrons in the cloud is equal to the number of protons in the nucleus, so the total negative charge of the electrons balances the total positive charge of the protons, and the atom *as a whole* is electrically neutral.

Some of the evidence for these ideas on the structure of the atom will be given in chapters 20 and 23.

19.10 Carriers of electricity

To produce an electric current we need to move electric charge from place to place (15.13). To move the charge we must have *charge carriers*. Protons and electrons both have electric charge built in to their structure, so they could be good charge carriers. Protons are part of the atomic nucleus and, in solids, this is fixed as part of the lattice. The protons are fixed in position and therefore are not suitable for carrying charge from place to place. Electrons are located in the outer part of the atom; they are relatively easy to remove from the atom. In metals they are able to leave the atom and move freely within the lattice (18.03). They could be very good charge carriers. If we can remove electrons from atoms, we can use them to carry charge. Let us see one easy way in which this can be done.

19.11 The behaviour of electric charges

IMPORTANT The investigations in this section do not work well in damp conditions. It is best to try these investigations only during dry seasons.

Instructions: removing electrons from atoms

1 Cut a sheet of polythene (from a plastic bag) as shown in fig. 19.2a.
2 Place it on the bench, hold it at one end and rub lightly but briskly with a *dry* soft cloth. Rub it in one direction only to avoid tearing the fringe of strips, as in fig. 19.2b.
3 Hold the sheet above the bench with the strips dangling (fig. 19.2c). What happens to the strips?
4 Repeat the rubbing; hold the sheet near the wall of the room. What happens?
5 Let the strips touch the wall, then remove the strips from the wall. How do they behave now?
6 Repeat the rubbing; hold the sheet above the bench; bring one of your fingers towards one of the strips. What happens?
7 Repeat the rubbing at the same time as your partner rubs on another sheet of strips. Hold the sheets above the bench and gradually bring them closer together. What happens?

Questions

1 If the strips do not hang vertically downwards under the force of gravity, what must be acting on the strips?
2 There was something different about the force in steps 4 and 6. What was different?
3 Explain what was happening in step 7. Does this behaviour suggest a rule?
4 Can you say what kind of electric charge has been given to the sheet?
5 In what previous chapter have you found a similar rule?

Instructions: investigating positive and negative charge

1 Place two strips of polythene film, about $3 \text{cm} \times 20 \text{cm}$, side by side on the bench. Charge them at the same time, by rubbing with a dry soft cloth (fig. 19.3a).
2 Hold them above the bench, one in each hand (fig. 19.3b) and slowly bring them closer together. What happens to their lower ends? How do you explain this?
3 Repeat steps 1 and 2, using two strips of acetate sheet (you can use exposed 35 mm photographic film if it is not curled). Answer the same two questions.
4 Repeat steps 1 and 2, using one strip of polythene and one strip of acetate. Answer the same two questions. What do you deduce about the charge on the polythene and that on the acetate? Think of another rule that is demonstrated by these results.

In some way that is not properly understood, rubbing has removed electrons from some of the atoms in the acetate film. This leaves some of the atoms with their positive nuclear charge unbalanced. The positive charge on the acetate is thus due to the protons of its atoms. The protons are not mobile; they cannot act as charge *carriers*; but we can use protons as a way of *storing* positive charge on the acetate film. The polythene sheet receives extra electrons when rubbed; this is a way of storing negative electric charge. In the act of rubbing, electrons can be moved, since they are mobile *charge carriers*. Once the

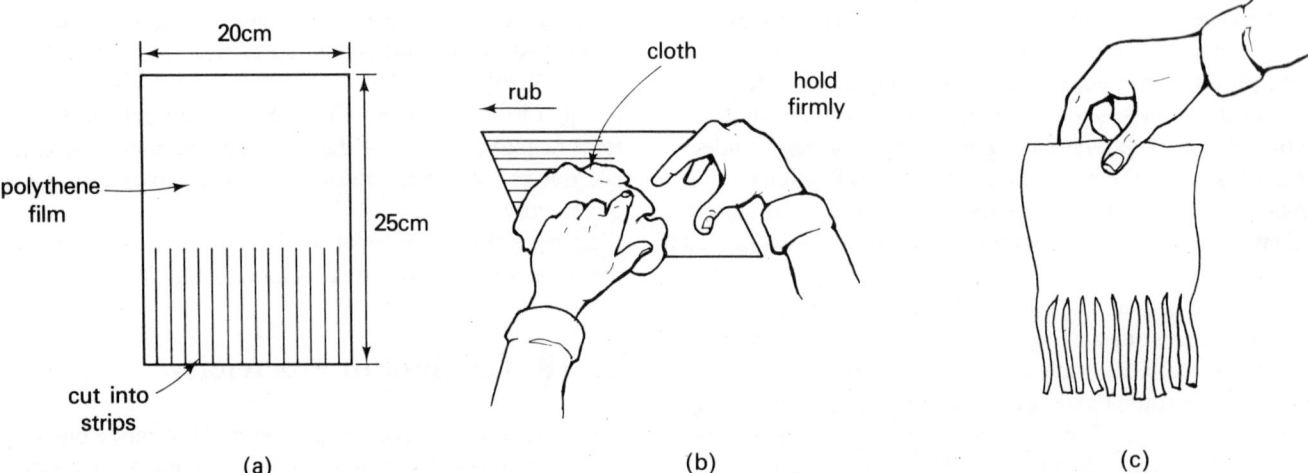

Fig. 19.2 Charging a sheet of polythene film.

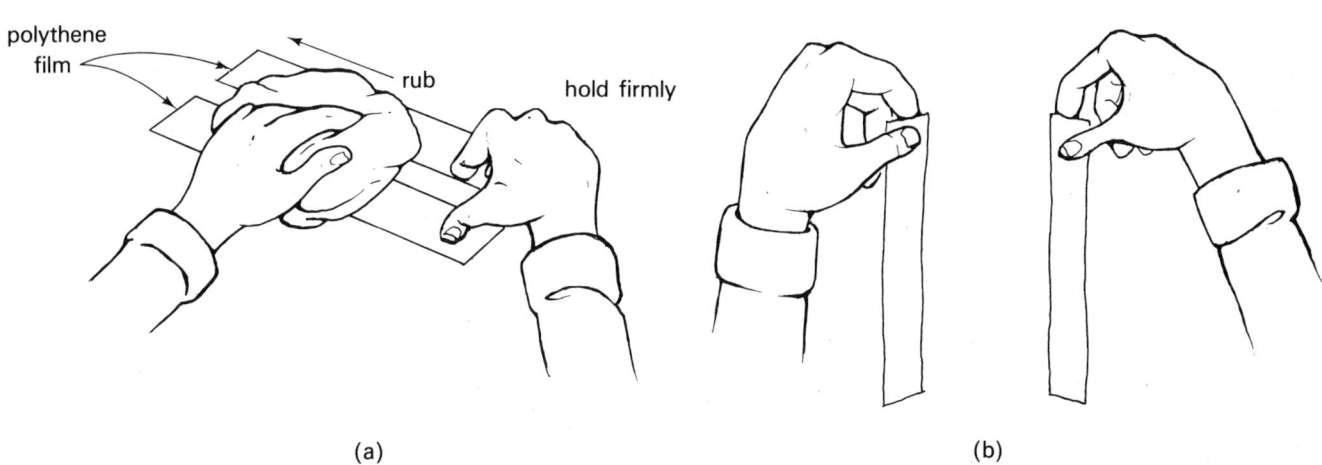

Fig. 19.3 Giving two strips of polythene film the same electric charge.

rubbing is completed, the sheets remain charged because no further transfer of electrons is possible. By rubbing non-conducting materials we are able to charge them positively or negatively. We have done work and we have created an *electric field*.

The field on a polythene strip can be represented in the form of a diagram (fig. 19.4). In this, the field is represented by lines with arrows (though remember that this is a *diagram* only, that no such lines exist). The lines tell us in which direction a small positive charge would move, if it was placed in the field. Wherever such a charge is placed in fig. 19.4 it moves toward the strip; it is attracted. The electric forces acting on the charge make it move toward the polythene strip in the direction of the arrow. These lines are *lines of electric force*. If we have a

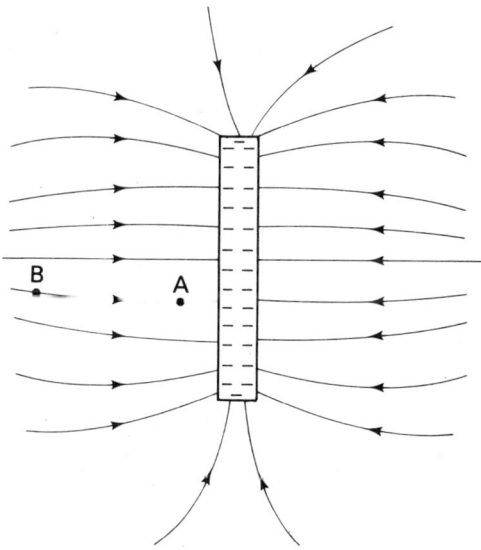

Fig. 19.4   The electric field around a charged rod. The thin lines show the direction moved by a positive charge free to move in the field.

small positive charge at point A and move it to point B we would have *to do work* against the force. In moving the charge from A to B it gains potential energy. If we released it, it accelerates toward A again, gaining kinetic energy. When it is at B it has greater potential energy than at A. We can say that the electric potential at B is higher than that at A. There is a *potential difference* (p.d.) between point A and point B.

This idea was mentioned earlier (15.13) and we saw that the unit for measuring the potential difference between two such points as A and B is the *volt*. When we finished rubbing the polythene and pulled the cloth away from it, we were moving positively charged atoms (on the cloth) away from the polythene strip. We were doing work. Though you did not *feel* the difference, you needed to do extra work to separate the cloth from the strip. This extra work has appeared as a potential difference between the strip and the objects around it.

To study electric fields and their effects in more detail we need an instrument which we can use to detect charges and to identify them as positive or negative. A simple instrument that will do this is the *gold-leaf electroscope* (fig. 19.5a).

Instructions: using the gold-leaf electroscope
1 Touch the plate with your finger to discharge it.
2 Charge a polythene rod or plastic ruler by rubbing it with a soft cloth.
3 Bring it close to the plate. What happens?
4 Remove the rod. What happens?
5 In steps 3 and 4 the effect was only temporary. To charge the electroscope permanently (or semi-permanently, for it will gradually discharge through leakage), *rub* an uncharged rod or strip across the plate several times. In effect you are rubbing the strip with the plate, just as previously you rubbed it with a cloth. What charge should this procedure leave on the electroscope?

6 Discharge an acetate strip by passing it quickly through a bunsen flame. Then draw it firmly across the plate several times. What happens during this procedure? Is the charge permanent?

7 Bring a charged polythene strip towards the plate, start with the strip about 50cm away from the plate and bring it slowly towards the plate. Observe what happens.

8 Bring your hand close to the plate without touching the plate. What happens?

9 Bring a charged acetate strip towards the plate (slowly from 50cm). What happens?

10 You have been using an electroscope charged negatively by rubbing acetate on the plate. Repeat, using an electroscope charged positively by rubbing a polythene strip on the plate. Complete a table, as below, to show whether the divergence of the leaf *increases* or *decreases*.

Questions

1 Why does a decrease of divergence not *prove* that an object is charged?

2 How can you prove that an object is positively charged?

3 How can you prove that an object is negatively charged?

4 How can you estimate the amount of the charge?

The behaviour of the electroscope can be explained by thinking about electric fields (fig. 19.5). In (a) we see what happens when a negatively charged rod is brought to an uncharged electroscope. Electrons are repelled from the plate because of the field around the rod. Remember that electrons are negatively charged, so they move in the direction *opposite* to that indicated by the arrows of the lines of electric force. Remember too that the charged rod is a non-conductor; electrons on that rod cannot move. By contrast, the plate, stem and leaves are conducting metal and electrons can move. They move down into the leaves; these both become negatively charged owing to the arrival of extra electrons from the plate; they diverge. The field continues across the gap between the leaves and the case,

which is earthed. The earth connection can be made by wire, but even without this connection there is usually good enough contact through the bench to ground to allow electrons to flow to earth, leaving the inside of the case with a positive charge. Thus the leaves diverge because they repel each other *and* because the gold leaf is attracted towards the case.

In (b) a positively charged electroscope has its leaves diverging. They repel each other *and* they are attracted towards the case because of the field between themselves and the earthed case.

In (c) a positively charged rod has been brought near, attracting electrons towards the plate, so cancelling the positive charges there. At the same time it creates an increased positive charge in the leaves; they diverge even further under the strong field created. Note the flow of electrons from earth to case, influenced by this field.

In (d) the reverse happens: electrons flow down to the leaves and also from the case to the earth. The earth can act as a reservoir, able to provide or receive electrons without any appreciable change in its potential. This is a good practical reason for considering the earth to be at *zero* potential for most measurements; it makes a very convenient reference point.

Instructions: charging by induction

1 Place two polystyrene tiles on the bench to act as insulators. On each, place a metal can; the cans should be touching (fig. 19.6a).

2 Bring a negatively charged polystyrene rod or strip close to one of the cans. What do you expect to happen to the electrons in the metal?

3 While the rod is still in position, slide the tiles apart to separate the cans.

4 Check your prediction by bringing the cans towards charged electroscopes. It is essential not to touch the cans during this operation; carry them on the polystyrene tiles. Keep your hands well away from the electroscope plate, as they can influence the electroscope leaves.

Plate rubbed with	Charge on electroscope	Charge brought near to plate	Effect on divergence
Acetate	negative	positive	
		negative	
		uncharged (hand)	
Polythene	positive	positive	
		negative	
		uncharged (hand)	

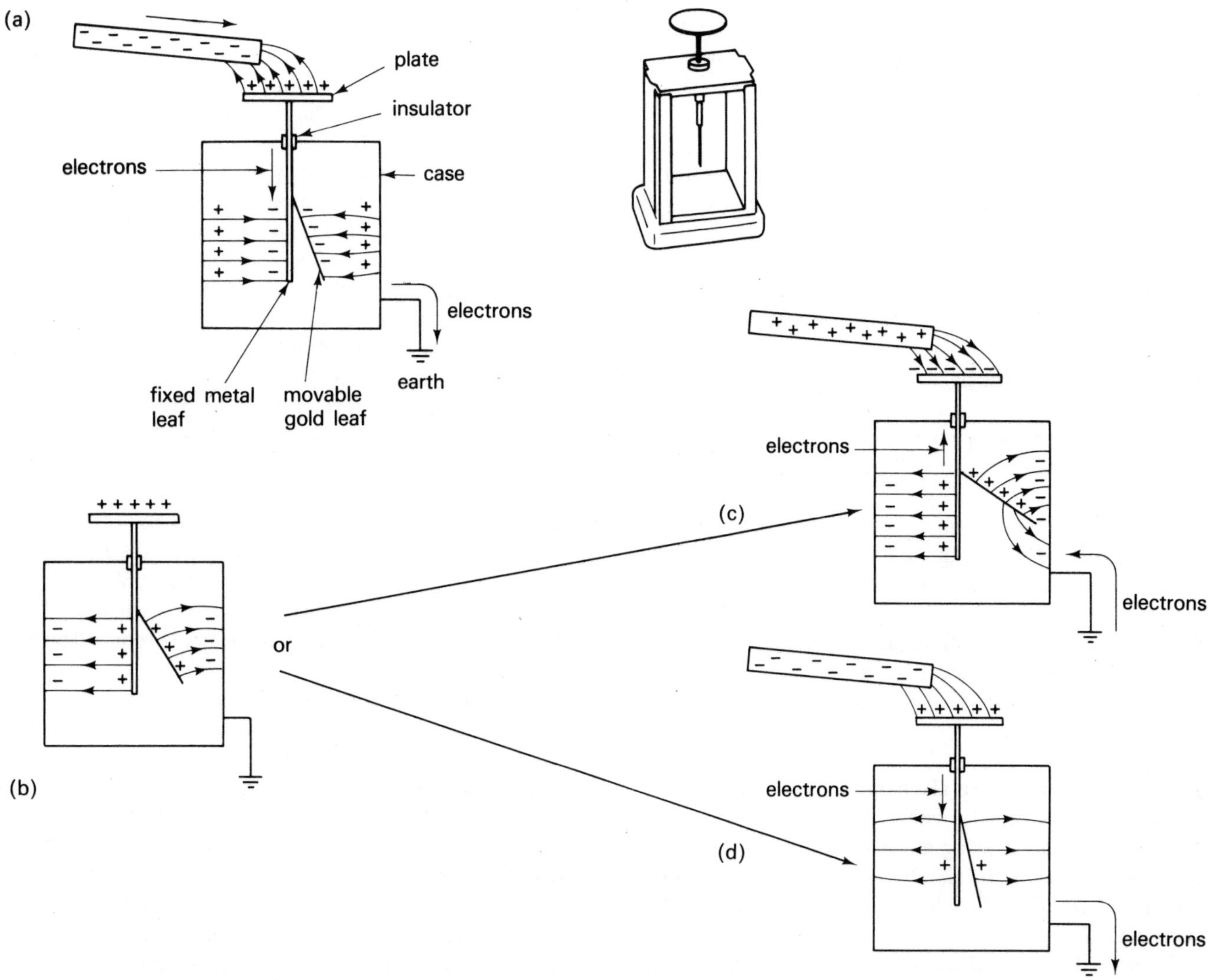

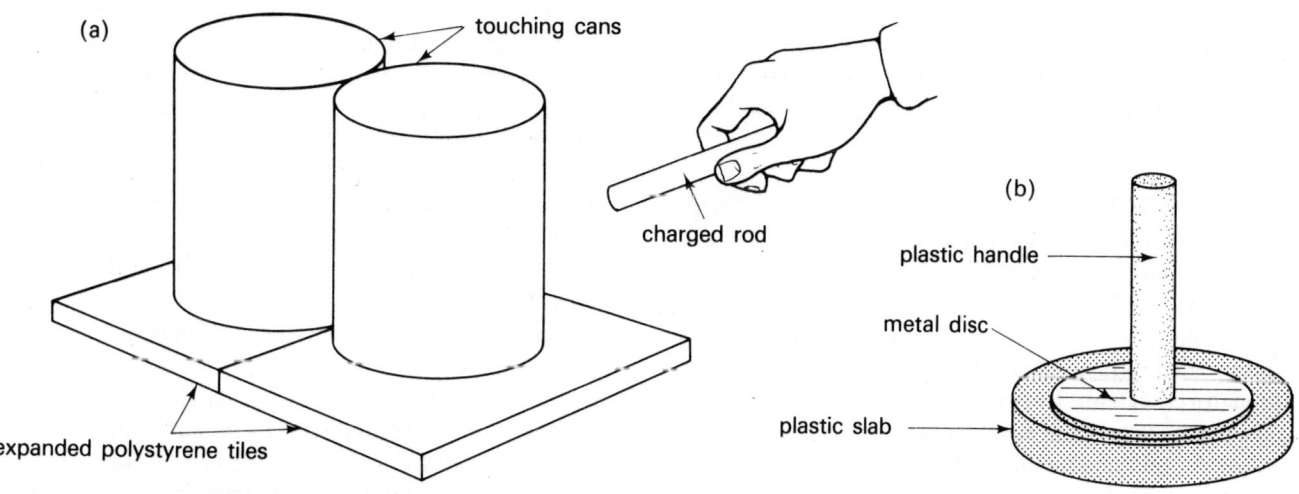

(a)

plate

insulator

electrons

case

electrons

earth

fixed metal leaf

movable gold leaf

(b)

or

(c)

electrons

electrons

(d)

electrons

electrons

Fig. 19.5   *The gold-leaf electroscope (a) illustrating the effect of bringing a negatively charged rod towards the plate of an uncharged electroscope. The thin lines represent the electric field between the rod and the plate and between the leaves and the case. (b), (c), (d) illustrating the effect of bringing charged rods towards the plate of a charged electroscope.*

(a)

touching cans

charged rod

(b)

plastic handle

metal disc

plastic slab

expanded polystyrene tiles

Fig. 19.6   *(a) Charging a can by induction. (b) An electrophorus.*

5 The two rules about electric charge have been rather like two of the rules for magnets. This investigation demonstrates one difference between magnets and charged bodies. What is this difference?

Instructions: another way of charging by induction

1 Use *one* can on a polystyrene tile. Bring a charged rod close to it.
2 Touch the can with your finger while the rod is still close by. Move your finger away.
3 Now remove the rod. What charge, if any, do you think will be on the can?
4 Check your answer by using a charged electroscope.

Instructions: charging an electroscope by induction

1 Discharge the electroscope by touching the plate with your finger.
2 Bring a negatively charged rod close to the plate.
3 While the rod is still there, touch the plate with your finger. Move your finger away. Note the position of the leaves.
4 Now remove the rod. Note the behaviour of the leaves. What charge is on the electroscope now?
5 Check the charge of the electroscope by bringing charged rods to it.

Instructions: using the electrophorus

1 Rub the slab briskly with a soft cloth to charge it. Identify the charge, using a charged electroscope.
2 Place the disc on the slab (fig. 19.6b), holding it by the handle *all the time*. Touch the upper surface of the disc with your finger to earth it.
3 Remove the disc and test it for charge. Bring the edge of the disc close to a knuckle. Can you detect a spark? The disc is now discharged; check this.
4 Repeat steps 2 and 3 several times (*not* step 1). Is it possible to recharge the disc many times?

Question

Explain the working of the electrophorus, using your knowledge of induction of electric charge. How is it possible that the disc can be recharged many times without the need to recharge the slab?

We have made much use of the electron as a *charge carrier*. We have placed it, immobile, on non-conductors, so creating an electric field. Note that we have *not created electricity*. We have merely redistributed electrons, and we have had *to do work* to redistribute them. We have also caused electrons to flow from one place to another, through a conductor. For example, electrons flow from the plate to the leaves of an electroscope through the stem (fig. 19.5a). In this and similar examples the electrons carry charge from place to place. We have an *electric current*.

Let us examine this current more closely. In fig. 19.7a we see a strip of metal and the dots represent the outer electrons of the atoms, the ones that are free to move around (18.03). These dots have arrows to represent the motion of the electrons. The electrons are responsible for carrying heat. In (a) they are more or less evenly distributed, so their charge is balanced by the positive charge of the nuclei of the atoms in the crystal lattice. In (b) a charged polythene rod has been brought near to one end of the metal strip. The dots on the rod represent electrons. There are no arrows as the electrons are fixed in position; the polythene is a non-conductor. The electric field lines show the direction in which a positive charge would move if free to do so. Any *negative* charges would move in the opposite direction to the arrows on the field lines. The electrons in the metal are free to move; they move towards the far end of the metal strip. There is an electron flow or electric current. To put it in other words, the electrons are repelled towards the far end of the strip. They do not go into one tightly packed group at the far end. Being all charged alike, they tend to *repel each other* too, so there is a limit to how close together they can be. This partly depends on the strength of the field produced by the polythene rod.

When the polythene strip is brought near there is an almost instantaneous flow of electrons. This leaves the atoms at the top end of the strip deficient in electrons; they now have an overall positive charge. The excess of electrons at the far end creates an overall negative charge there. From this point on, though the electrons still have their own kinetic movements, the distribution of electrons *as a whole* remains unchanged; there is equilibrium.

What would happen if there was a continual supply of electrons at the top end of the strip and there was some way for the electrons to leave at the bottom? We could arrange this by connecting the strip to a cell as in (c). The negative electrode of the cell is a source of electrons. The positive electrode of the cell is a receiver of electrons. The electric lines of force representing the field between the electrodes run through the wires (drawn too narrow to show them) and through the metal strip. Electrons flow from the negative electrode, through the wire, enter the strip, pass through it, pass along the other wire to the positive electrode. We have a *continuous* electric current. Negative charge is carried by electrons from the negative electrode to the positive electrode. Though not shown in the diagram, the wires from the cell carry electrons in exactly the same way as the metal strip, for they too are made from metal.

This is the way in which we most often carry electricity, for metals are the most commonly used electrical conductors. It is because of their large supply of free electrons, not bound tightly to the atoms, that metals are such good conductors of electricity, and also of heat. In (c) there is an arrow to indicate the direction of electron flow. This is the direction of the negative electric charges. You

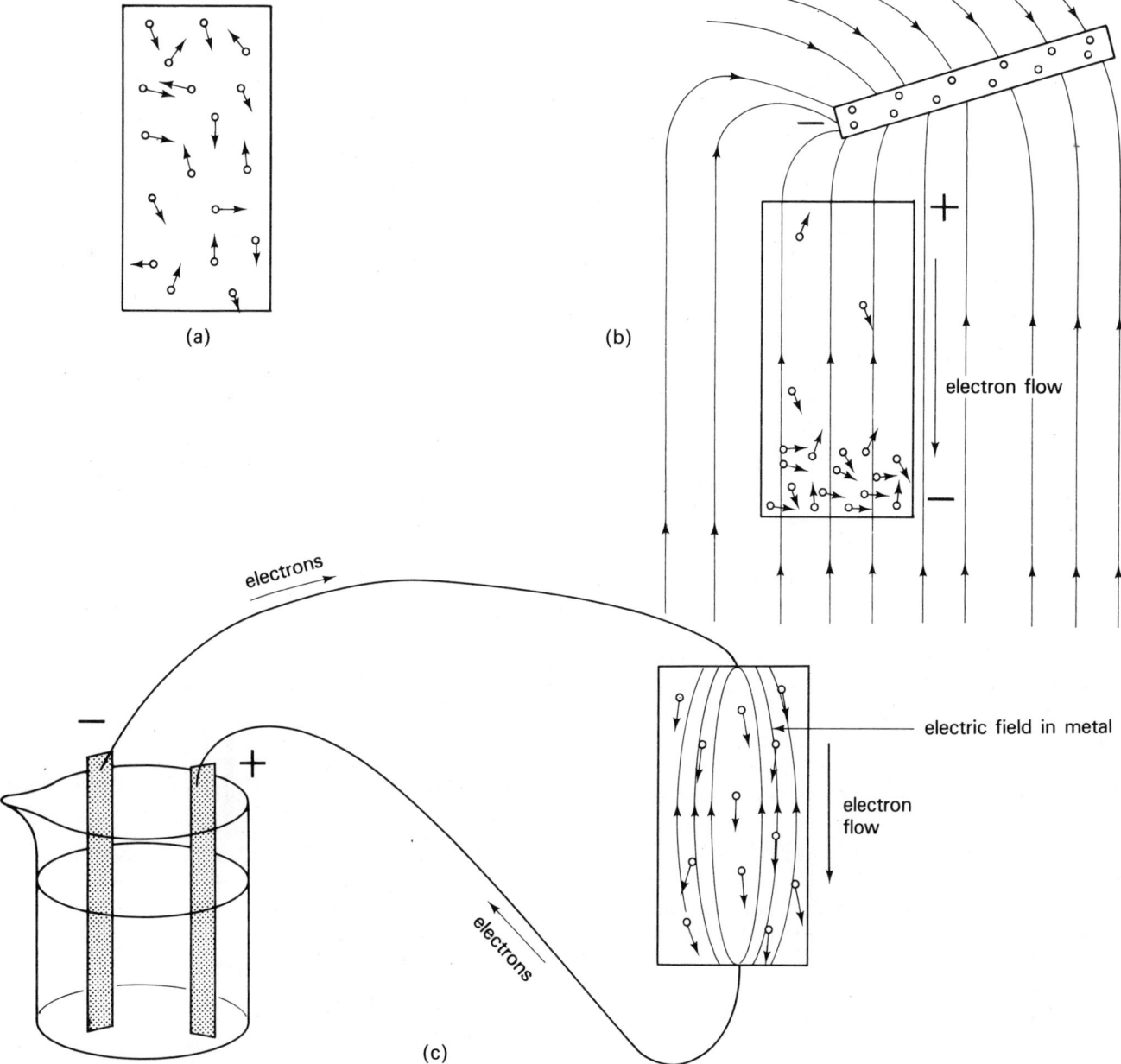

Fig. 19.7 *Electric currents in a bar of metal; (a) the dots represent free electrons in the metal, the arrows indicating their movement; the atoms of the lattice are not shown: (b) the movement of the electrons when a charged rod is brought near one end of the metal bar; the thin lines represent the electric field of the rod: (c) the movement of electrons in the electric field produced by connecting the metal bar to a cell.*

may have noticed that this is the *opposite direction* to that in which we normally imagine the current to be flowing. In most of our diagrams and descriptions we imagine that current flows from positive to negative. This idea is used as the basis of many of the rules covering electricity and electromagnetism: we use it for the swimming rule and the right-hand grip rule (9.20) and many others. This idea of current flowing from positive to negative is called *conventional current*. We shall continue to use it, since most people accept it as the most useful convention but when we are thinking of exactly how the current is carried, we

must remember that *when the current carriers are electrons*, they move in the opposite direction to the conventional current.

19.12 Conduction in electrolytes

Instructions: investigating the effect of passing a current through sodium chloride solution
1 Put two carbon electrodes in a beaker containing a weak solution of sodium chloride.

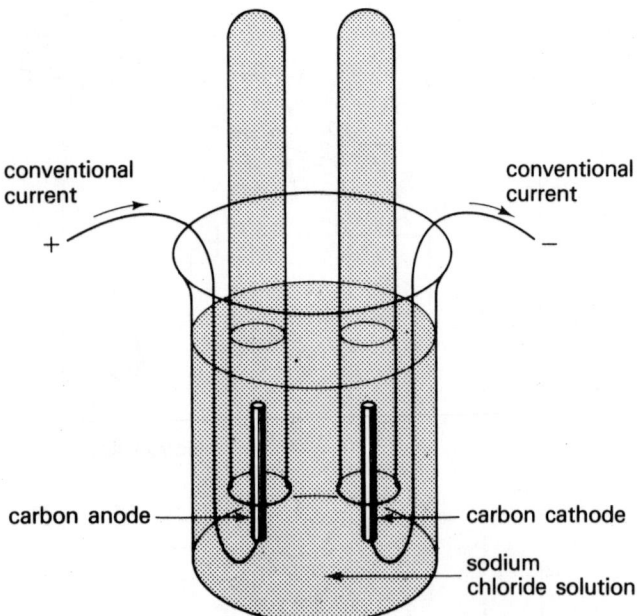

Fig. 19.8 *Passing a current through a solution of sodium chloride.*

2 Invert two test-tubes over the electrodes, so that both are full of solution (fig. 19.8). The electrodes are both made of carbon, so we need names to tell one from the other. The electrode connected to the positive supply is called an *anode*; the other is called a *cathode*.

3 Connect the electrodes to a battery or power supply of about 4.5 V.

4 Watch the electrodes; bubbles of gas form there and are collected in the tubes (fig. 19.9a).

5 When a few centimetres depth of gas has collected in each tube, switch off the electric current. In which tube has the greater volume of gas collected?

6 Remove the tube from the cathode without letting the gas escape. Bring the mouth of the tube close to a flame; open the tube. Record what happens. What gas is this?

7 Remove the tube from the anode without letting the gas escape. Pass a smouldering string or taper into the gas. Record what happens. What gas is this?

Questions

1 In a quantity of water, a portion of the water molecules splits into two parts. One part consists of an atom of hydrogen; the other consists of an atom of hydrogen and an atom of oxygen joined together. On splitting, the atom of hydrogen loses an electron: it is positively charged and we represent it by H^+. The electron is attached to the hydrogen-oxygen group, or hydroxyl group, making it negatively charged. We represent it as OH^-. These charged parts of a molecule are called *ions*. Using this information, try to explain how hydrogen and oxygen collected at the electrodes when a current was passed through the electrolyte.

2 Sodium chloride becomes ionized too. It forms Na^+ and Cl^- ions. In which direction do these ions move when the current is switched on?

Instructions: finding out the effect of passing a current through copper sulphate solution

1 You need a beaker, some copper sulphate solution and two electrodes made from sheet copper.

2 Weigh the electrodes and record their masses m_A (anode) and m_C (cathode).

3 Use a hydrometer to measure the density of the copper sulphate solution.

4 Set up the apparatus as in fig. 19.9b.

5 Switch on the current for 30 min. A current of about 0.5 A is best.

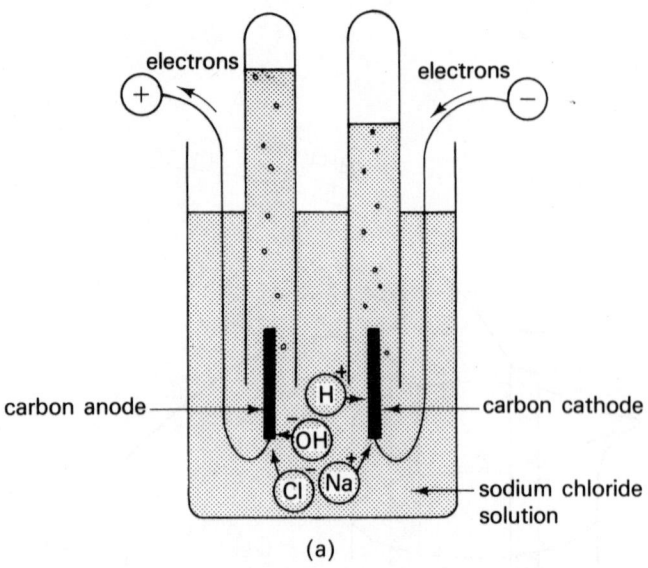

(a)

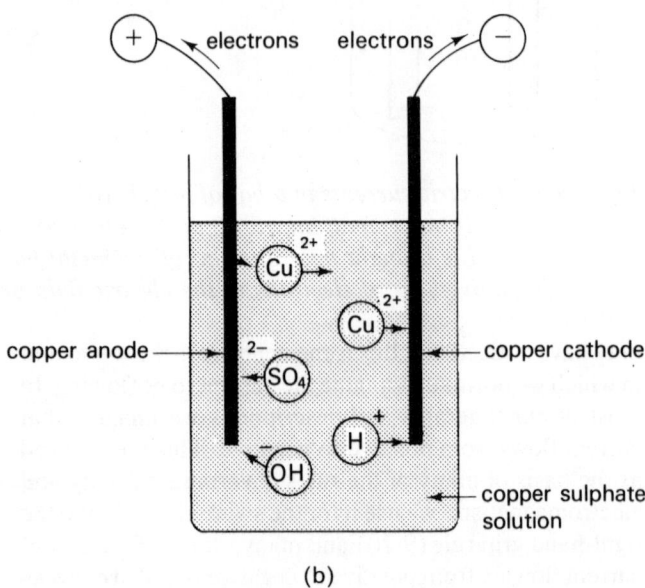

(b)

Fig. 19.9 *Electrolysis in (a) sodium chloride solution; (b) copper sulphate solution.*

6 While the current is flowing, look for gases appearing at the electrodes.

7 After the current has passed for 30min, remove the electrodes and note their appearance. What changes do you observe?

8 Weigh the electrodes; what changes do you observe in their masses?

9 Measure the density of the electrolyte; what change do you observe?

10 Given that copper sulphate ionizes to form Cu^{2+} and SO_4^{2-}, and that water ionizes as already described, try to explain what has happened.

Instructions: an experimental electrolysis of copper sulphate solution

1 In this experiment, use copper sulphate solution as electrolyte, and use *carbon rods* as electrodes. Before you begin the experiment, try to think out what will happen. Then perform the experiment to check your predictions.

2 Carry out the electrolysis; look for gases being produced at the electrodes; look for changes (if any) in the electrodes and in the electrolyte.

In these experiments, the *charge carriers* are ions. They are of two kinds:

(a) *negative ions,* pass to the anode and *give up* an electron (or two), which then is conducted along the anode, through the connecting wire to the cell;

(b) *positive ions;* pass to the cathode and *receive* an electron (or two) which has come from the cell by flowing through the connecting wire and cathode.

Electrons enter the electrolyte at the cathode, other electrons leave it at the anode. *In effect* the circuit is complete, owing to the action of the charge carriers of the ionized electrolyte. Ions which are not discharged at the electrodes do not take part.

Electrolysis has several industrial applications. It is used for the refining of impure copper. The anode is made from impure copper and the cathode from a thin sheet of pure copper. They are immersed in copper sulphate solution. Electrolysis transfers copper *but not the impurities* from the anode to the cathode. The cathode becomes coated with a thick layer of pure copper. Electroplating makes use of the same principle. A layer of one metal may be deposited on another metal. For example, cutlery made from brass may be plated with silver or gold. Steel or copper can be plated with chromium or cadmium to give protection against corrosion. Chromium-plated steel is frequently used for metal parts of bicycles, automobiles and household goods. We use it when we require a highly-polished appearance combined with resistance to corrosion.

Before leaving this subject we return to the related subject of electric cells. In the Leclanché cell (9.01) the electrolyte is ammonium chloride solution. At the negative electrode of the cell, zinc dissolves in the electrolyte. As it dissolves, it becomes positively charged, forming Zn^{2+} ions. For each ion formed, two electrons are left behind on the electrode. These flow away into the external circuit. Electrons arriving at the positive electrode (carbon) discharge the ammonium ion (NH_4^+) from the electrolyte. This then forms ammonia gas and hydrogen gas. The hydrogen gas would collect on the electrode and cause polarization but is removed by a chemical reaction with the manganese dioxide.

In the copper-zinc cell, with sodium chloride as an electrolyte, the supply of electrons at the negative electrode (zinc) is provided by the zinc dissolving in the electrolyte and ionising, as in the Leclanché cell. Electrons arriving at the positive electrode (copper) are neutralised by hydrogen ions produced from the water. The hydrogen gas collects in bubbles on the copper and acts as a poor conductor, so reducing the flow of charge through the cell. It also creates a back e.m.f. which opposes the e.m.f. of the cell. The cell becomes polarized and there is no way of overcoming this in this simple cell, apart from scraping the bubbles from the copper electrode.

The cell is a device for converting chemical energy into electrical energy. Normally zinc would dissolve and ionise, with the production of a quantity of heat. Because of the arrangement of electrodes to conduct away the electrons, the energy appears as electrical potential energy instead.

19.13 Conduction in semi-conductors

The two most frequently used semi-conductors are germanium and silicon. Crystals of these substances are *non-conducting* at low temperatures because all the electrons are within the electron cloud of the atom and the atoms are in fixed positions in the crystal lattice. Thus there are no charge carriers. As temperature is increased to around room temperature, some of the electrons are able to escape from the atoms. These escaped electrons can act as charge carriers and the material becomes *conducting*, though not as conductive as metals. When an electron escapes from its atom, it leaves the atom lacking an electron; there is a vacancy, due to the missing electron. Another electron could arrive and fill this vacancy. Usually we call this vacancy a *hole*.

If your head student leaves the school, this creates a vacant position. This vacancy could be filled by promoting the deputy head student to the position of head student. This leaves a vacancy for the position of deputy head student. This vacancy could be filled by promoting a prefect. The promotion of the prefect leaves a vacancy among the prefects. This vacancy could be filled by promoting one of the less senior students to the rank of prefect. In this procedure the individual students are promoted; they move *up* the scale of seniority. The vacancy moves *down*.

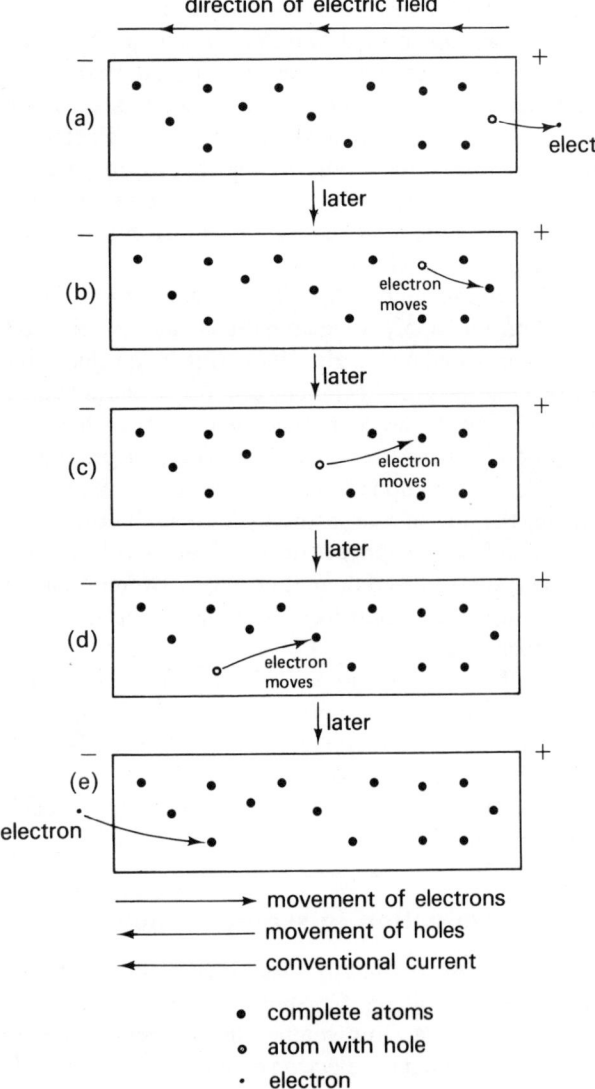

direction of electric field

(a)

later

(b)
electron moves

later

(c)
electron moves

later

(d)
electron moves

later

(e)
electron

——————→ movement of electrons
←—————— movement of holes
←—————— conventional current

• complete atoms
∘ atom with hole
· electron

Fig. 19.10   Conduction by holes in a semi-conductor.

The same kind of thing happens in a semi-conductor crystal *that is in an electric field* at room temperature. In fig. 19.10 we see a sequence of events in a bar of semi-conductor. Only a few of the atoms of the lattice are shown. In (a) an electron leaves the bar at one end (perhaps to flow along a wire), creating a hole. In (b) this hole is filled by an electron drifting along the bar under the influence of the electric field, in the circuit to which the bar is connected. In (c) this hole is filled by another electron from further along the bar. This continues along the bar until the final hole is filled by an electron entering the bar from a connecting wire. In the bar the electrons have moved for relatively short distances. The hole has moved the whole length of the bar, in the opposite direction. The arrows at the bottom of the figure show that electrons travel in the opposite direction in conventional current, as already explained. Holes travel in the same direction as conventional current.

Semi-conductors have two types of charge carrier: the electron and the hole. To improve the conduction of electric charge by semi-conductors it is usual to introduce a very low percentage of atoms of other elements into the germanium or silicon lattice. These are specially chosen to provide either extra electrons or extra holes. If the introduced element provides extra electrons, we call this an *n*-type semi-conductor (*n* for negative charge carrier). If the introduced element provides extra holes, we call this a *p*-type semi-conductor (*p* for positive charge carrier). The introduction of extra carriers of either type increases the conducting ability of the material and improves its properties. The uses to which these properties may be put will be described later.

19.20   Ohm's law

The meaning of the symbols used in the circuit diagrams is given in the table inside the back cover of the book. There is also a table explaining the resistor colour code, to help you identify any colour-coded resistors that you may be given to work with. The code is useful, but you need not learn it.

Instructions: investigating the relation between potential difference and current
1   Connect the circuit of fig. 19.11. The resistor, R, is a piece of 34 SWG constantan wire about 1 m long, wound into a loose coil, so that turns do not touch. The exact gauge and length do not matter; nichrome wire can be used instead. The cells shown are dry cells and can be conveniently held in a battery-holder, if the connections *between* the cells are accessible. Alternatively, 2 V accumulators can be used. The '1 A' beside the symbol for the ammeter means that the meter should read 1 ampere as the highest value on its scale: this way of indicating *full scale deflection* is used for ammeters and voltmeters on all diagrams.

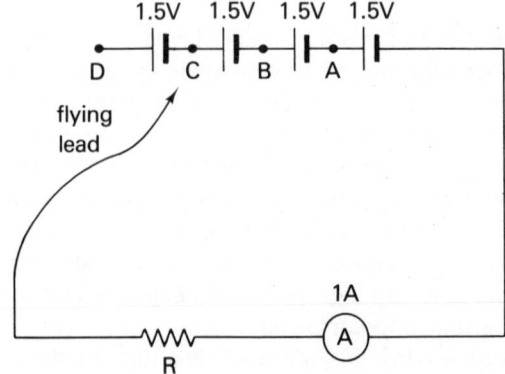

Fig. 19.11   Circuit for investigating the relation between p.d. and current.

2 The cells provide a range of potential differences, V, from 1.5 V to 6 V depending on where the flying lead is connected. Connect the lead to point A. Read the current, I. Record your results in a table:

Potential difference V in V	Current I in A	V/I
1.5 3.0 4.5 6.0		

3 Repeat for connections to B, C and D.

4 Calculate the ratio V/I. What can you say about the ratio V/I? Express this result as a rule, linking potential difference and current.

Experiments similar to the above were performed by Georg Ohm. His investigations led him to state that:

The current passing through a wire at constant temperature is proportional to the potential difference between its ends.

This has since been known as *Ohm's law*.

If you use different lengths of wire of different gauges and made from different metals or alloys, you find that Ohm's law is always true. Every time, you get a constant value for V/I; the value you get depends on the wire you are using. This value is a property of a particular piece of wire. We call it the *resistance* of the wire. Resistance could be given the units 'volts per ampere', because this expresses the idea of resistance. Resistance tells us what p.d. is required to make a given current flow through the wire. The resistance unit has a special name, the **ohm**. Its symbol is the Greek letter 'omega', written Ω. The ohm can be defined as a derived SI unit:

A wire has resistance 1 ohm when a p.d. 1 volt across its ends makes a current 1 ampere flow along it.

Numerically, this definition can be expressed as an equation:

Resistance, $R = \dfrac{V}{I}$ (R in Ω)

There are two other useful forms of this equation:

$V = IR$ and $I = \dfrac{V}{R}$

Instructions: investigating currents and potentials in circuits

1 Connect the circuit of fig. 19.12. The exact values of components are not important, but they should be reasonably close to the values shown in the figure.

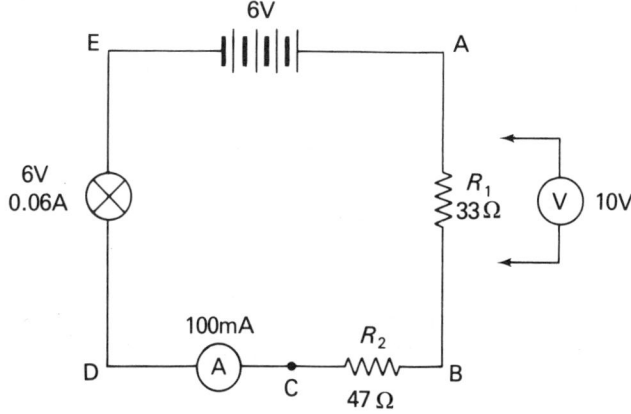

Fig. 19.12   Currents and potentials in circuits.

2 Use the voltmeter to measure the p.d. between various pairs of points in the circuit. Record these in a table, in which for example V_{AB} means the p.d. between points A and B.

p.d. measured	V in V
V_{AB} V_{BC} V_{CD} V_{DE} V_{AE}	

What do you notice about the total of V_{AB}, V_{BC}, V_{CD}, and V_{DE}?

3 Read the current, I. Then reconnect the circuit with the ammeter in various positions. Read the current each time and record the results:

Position of ammeter	I in A
as in fig. 19.12 at A at B at E	

What do you notice about the current at different points in the circuit?

4 Given your value V_{AB}, and your value of I, calculate the resistance of resistor, R_1.

5 Similarly, calculate the value of R_2, the ammeter and lamp.

6 A lamp with the ratings shown in the figure passes a current of 60 mA when the p.d. across its terminals is 6 V. What is its resistance, based on these values?

7 Explain why the resistance of the lamp calculated at step 5 does not agree with the value calculated at step 6.

8 Calculate the total resistance of the circuit (except for the resistance of the cells), using the equation $R = V_{AB}/I$. Does this differ from the sum of the resistances of the resistors, the lamp and the ammeter?

19.30 Resistance

Resistance is a property of a piece of wire, a carbon resistor, a lamp or some other component of an electric circuit. Here are some ways of measuring resistance.

Instructions: measuring resistance by the ammeter-voltmeter method
1 Connect a circuit as fig. 19.13. *R* is the unknown resistance of a resistor which could be a length of wire (such as narrow-gauge constantan wire), or a carbon resistor with unknown resistance value, or perhaps a carbon resistor that needs to have its resistance value checked. To begin with it is best to use an ammeter having a full scale deflection 1 A. If the current is too small to read accurately, change to an ammeter with f.s.d. 100 mA, 10 mA or 1 mA; change *in that order*.

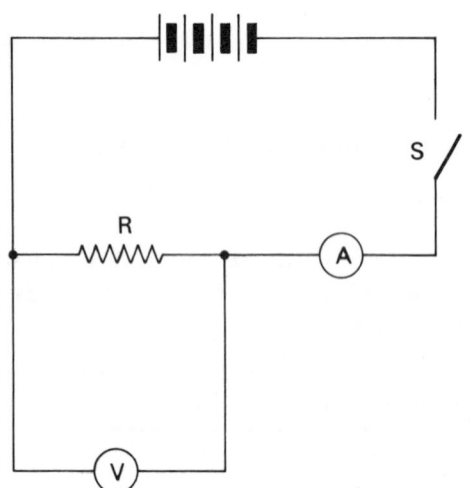

Fig. 19.13 Circuit for measuring resistance by the ammeter-voltmeter method.

2 Switch on; read p.d., *V*, across the resistor and current, *I*, flowing through the resistor. Calculate *R*, using the equation, $R = V/I$.
Note This circuit is suitable only for measuring low resistance (1 000 Ω, or less). The reason for this and an alternative circuit for higher resistances will be discussed in chapter 26.

Instructions: measuring resistance by substitution
1 Connect a circuit as fig. 19.14. R_x is the unknown resistor. R_s is a standard resistor, with variable resistance

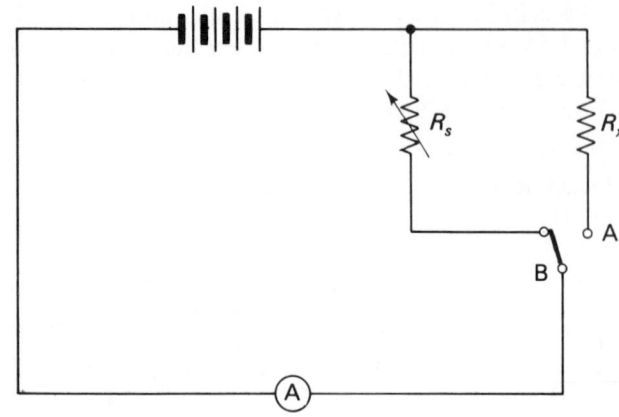

Fig. 19.14 Circuit for measuring resistance by the substitution method.

which is accurately known. For R_s use a resistance box; the value of its resistance can be altered by changing plugs or rotating knobs. A typical resistance box can be adjusted from 0 Ω to 9 999 Ω in steps of 1 Ω. Accuracy may be 1% or even 0.1%.
2 Begin with the switch in position A, so that current flows through R_x. Use a meter reading to f.s.d. 1 A. Change to a meter with smaller f.s.d. if currents are low. Measure I_x, the current flowing through R_x.
3 Set R_s to its highest value. Turn the switch to position B, so that current now flows through R_s.
4 Reduce the value of R_s by stages until the current has the same value as I_x. The p.d. is the same, so we know that $R_x = R_s$. The best way of reducing R_s is first to reduce the 'thousands of ohms', until you find that the current is just *higher* than I_x; then increase R_s by 1 000 Ω, to make the current less than I_x. Next reduce the 'hundreds' until again the current is higher than I_x; then increase it by 100 Ω, to make it less than I_x again. Similarly adjust the 'tens'. Finally adjust the 'units' to the nearest ohm. Can you think why the substitution method is not much use for measuring low resistances, 50 Ω or less?

A third method of measuring resistance is the *ohmmeter*. This is usually part of a multi-purpose test-meter (or multimeter). It has a special circuit inside, with its own battery. The unknown resistance is connected between the probe terminals of the ohmmeter; the resistance value is indicated by a pointer on the dial. This type of meter is very handy to use, but its accuracy is rather low.

Instructions: measuring the resistance of a length of wire
1 Use the ammeter-voltmeter method with a single cell (1.5 V or 2 V) and an ammeter with f.s.d. 2 A.
2 Measure the resistance of a piece of constantan or nichrome wire exactly 1 m long. The wire should be narrow gauge, about 34 SWG if available. Take a piece of wire just longer than 1 m and make connections to it

through crocodile clips placed exactly 1 m apart, measured along the wire. Take care that the wire does not coil and short-circuit parts of itself.

3 Reduce the distance between the clips, in steps of 10 cm. Measure the resistance at each step, down to 10 cm.

4 Plot a graph of resistance (vertical axis) against length (horizontal axis).

5 Make up a rule which gives the relation between length and resistance.

6 Suggest how you might make a variable resistor.

Instructions: making more measurements of the resistance of wire

1 Use the voltmeter-ammeter method.

2 Measure the resistance of wires all constantan or all nichrome and all 1 m long, but of different gauges. Suitable gauges are 20 SWG, 28 SWG and 34 SWG.

3 Measure the diameter of each gauge of wire, using a micrometer screw gauge. Measure the diameter at several places along each wire and at each place take two measurements across perpendicular diameters (in case the wire has oval section). Calculate the area of cross section, A of each wire.

4 What is the relation between A and R?

5 Lay two pieces of 34 SWG wire side by side and parallel (but not touching), and twist their ends together. The arrangement is like that of r_1 and r_2 in fig. 19.17. Connect the pair of roughly parallel wires so that the current divides, part flowing along one wire and part

flowing along the other wire, and rejoins at the other end. Measure the resistance of these parallel wires. What is their total area of cross-section? Do the results obtained with these wires agree with the relation found at step 4?

6 Repeat step 5 with three parallel wires. The arrangement is like that of the three resistors in fig. 19.17.

19.31 Series and parallel connections

Instructions: investigating cells in series and parallel

1 Figure 19.15a shows various numbers of cells connected *in series*. Current passes through one cell after another, the same current flowing through all cells and through the resistor.

2 Connect each circuit in turn. Use a voltmeter connected to A and B to measure the p.d. of the cells in series. Make up a rule for calculating the total p.d. of a number of identical cells wired in series.

3 Figure 19.15b shows various numbers of cells connected *in parallel*. The current through any one cell does not pass through the other cells. Connect each circuit in turn. Use an ammeter to measure the currents flowing through each cell (I_1, I_2, etc) and the current flowing through the resistor (I_r). What is the relation between I_r and the other currents?

4 Use a voltmeter connected to A and B to measure the p.d. of the cells in parallel.

5 Make up a rule for calculating the p.d. of a number of identical cells in parallel.

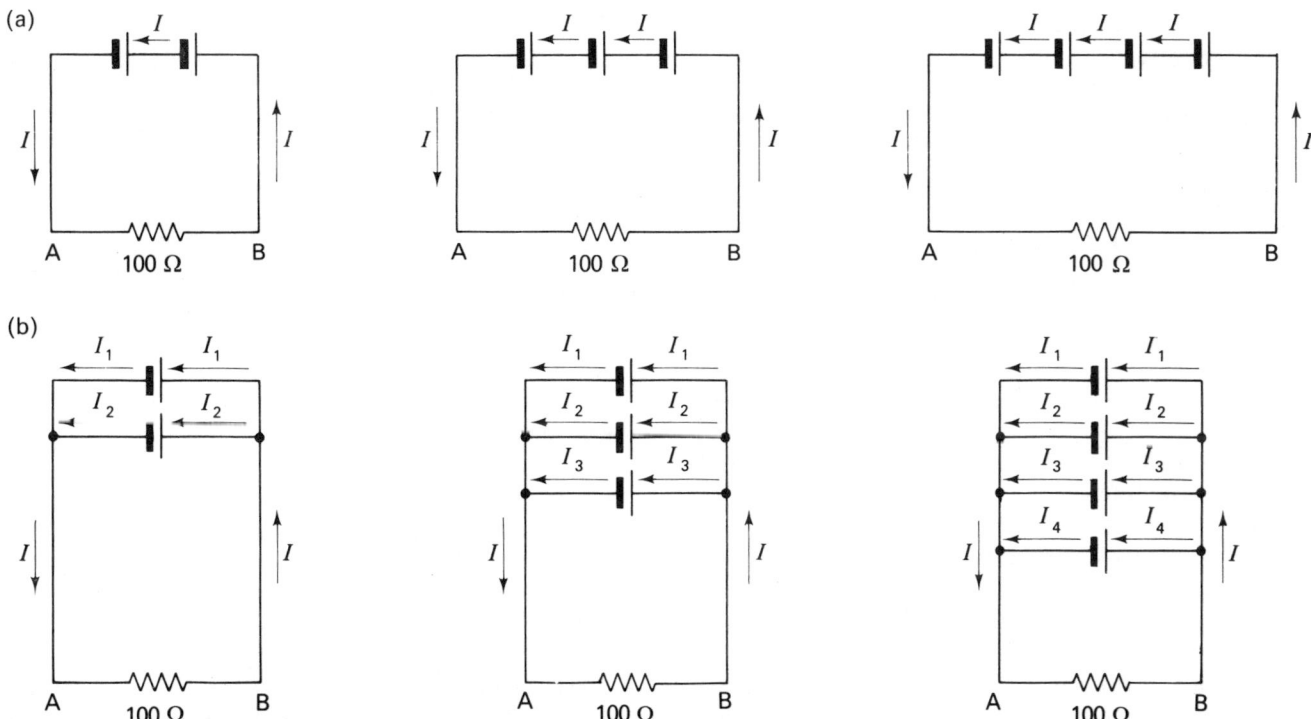

Fig. 19.15 (a) cells in series; (b) cells in parallel.

Instructions: finding out more about cells in series

1 Connect cells in series, as in fig. 19.15a, but use different types of cell. For example, use a dry cell, a nickel-iron cell and a lead accumulator.

2 Measure the p.d. across each cell. Measure the total p.d. across points A and B.

3 Make up a rule for calculating the total p.d. when different types of cell are connected in series. Suggest why it is unwise to connect different types of cell in parallel with each other.

Instructions: investigating resistances in series

1 Use the ammeter-voltmeter method or the substitution method to measure the total resistance of two or more resistors connected in series as in fig. 19.16. First measure the resistances of four or five different resistors, R_1 to R_5 singly. If you are using colour-coded carbon resistors, do not rely on the colour code for this experiment. Their true resistance may differ by 5% or 10% from the coded value; we wish to know the true value, so must measure it.

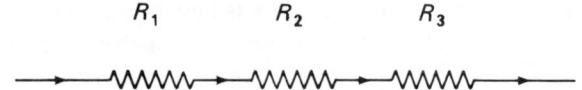

Fig. 19.16   Resistances in series.

2 Connect the resistors in series in twos, threes, fours, etc. Each time measure the total resistance of the series and record the details in a table:

Resistors tested	V in V	I in A	R in Ω
Tested singly R_1			
R_2			
R_3			
R_4			
R_5			
Two in series R_1 and R_2			
R_2 and R_5			
Three in series R_3, R_4 and R_5			
etc			

3 Make up a rule for calculating the total resistance of two or more resistors in series.

Instructions: investigating resistances in parallel

1 Use the ammeter-voltmeter method to measure the total resistance of two or more resistors in parallel (fig. 19.17).

2 You can use the same resistors as you used in the investigation above but it is best not to use low value resistors for this experiment. Preferably, resistors should have values between 100 and 500 ohms.

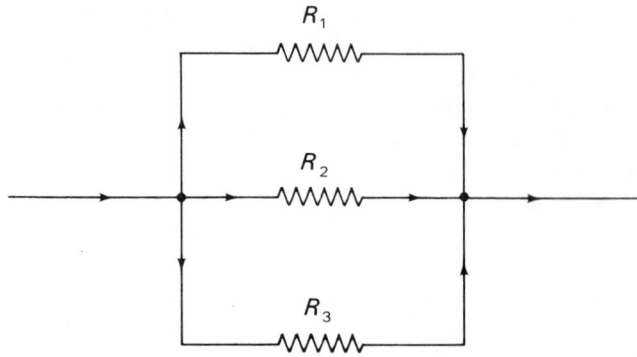

Fig. 19.17   Resistances in parallel.

3 Connect two or three resistors in parallel and measure their total resistance.

4 Repeat for several combinations of resistors.

5 Record your results in a table. When two or more resistors are in parallel, is their total resistance greater than or less than the value of the individual resistors? Can you find any rule for calculating the total? Do not worry if you cannot, for the next investigation will help solve the problem.

Instructions: investigating currents through resistors in parallel

1 Connect the circuit of fig. 19.18, using any two of the resistors from your previous experiment. This is really the circuit of the ammeter-voltmeter method, but the ammeter has been placed so as to measure the current through just one resistor.

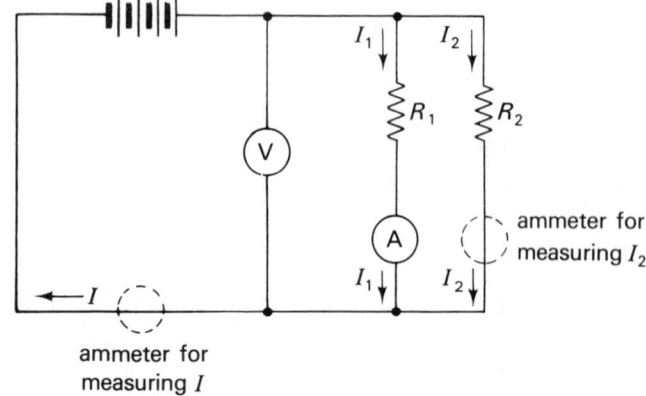

Fig. 19.18   The current through resistors in parallel.

2 Record the p.d., V, and current, I_1. Calculate R_1.

3 Put the ammeter in series with R_2, and measure I_2. Calculate R_2.

4 Restore the ammeter to its original position for the ammeter-voltmeter method (dotted circle) and measure the total current, I. Calculate total resistance R.

5 Do you agree that $I = I_1 + I_2$?

6 If $I = I_1 + I_2$, we can also write: $\dfrac{I}{V} = \dfrac{I_1}{V} + \dfrac{I_2}{V}$

What does $\dfrac{I}{V}$ equal? What does $\dfrac{I_1}{V}$ equal?

What does $\dfrac{I_2}{V}$ equal?

7 Make up a rule for combining two resistors in parallel.
8 Calculate I_1/I_2; calculate R_2/R_1. What do you notice about these two ratios?

The rules for resistors in series and parallel can be extended to any number of resistors. Suppose we have a number of resistors R_1, R_2, R_3 and so on in series (fig. 19.16). The same current, I flows through all. The p.d. across each resistor is V_1, V_2, V_3 and so on.

The total p.d. across the whole series is V, and

$$V = V_1 + V_2 + V_3 + V_4 + V_5 \ldots$$

We can divide all terms in this equation by any constant, such as I, giving

$$\frac{V}{I} = \frac{V_1}{I} + \frac{V_2}{I} + \frac{V_3}{I} + \frac{V_4}{I} + \frac{V_5}{I} \ldots$$

But these terms, in the form 'p.d./current', are measures of resistance, so

$$R = R_1 + R_2 + R_3 + R_4 + R_5 \ldots$$

This is the rule we have already discovered.

In fig. 19.17 the resistors are in parallel. They all have p.d. V across them. The total current I is the sum of the individual currents, and

$$I = I_1 + I_2 + I_3 + I_4 + I_5 \ldots$$

We can divide all terms by a constant such as V, giving

$$\frac{I}{V} = \frac{I_1}{V} + \frac{I_2}{V} + \frac{I_3}{V} + \frac{I_4}{V} + \frac{I_5}{V} \ldots$$

But there terms, in the form 'current/p.d.', are measures of 1/resistance, so

$$\frac{1}{R} = \frac{1}{R_1} + \frac{1}{R_2} + \frac{1}{R_3} + \frac{1}{R_4} + \frac{1}{R_5} \ldots$$

If there are only two resistors in parallel, we can obtain another useful rule; for

$$V = I_1R_1 = I_2R_2$$

This gives us $\dfrac{I_1}{I_2} = \dfrac{R_2}{R_1}$

In words, *the ratio of the currents is the inverse of the ratio of the resistances.*
For example, if two resistors A and B are in parallel, and resistor A has *twice* the resistance of resistor B, the current through A is *half* that through B.

With two resistors in parallel,

$$\frac{1}{R} = \frac{1}{R_1} + \frac{1}{R_2} = \frac{R_2 + R_1}{R_1R_2}$$

This gives an equation which is easy to handle, for we find that

$$R = \frac{R_1R_2}{R_1 + R_2}$$

In words, *total* resistance $= \dfrac{product \text{ of resistances}}{sum \text{ of resistances}}$

Questions
1 Given three resistors, $56\,\Omega$, $100\,\Omega$ and $200\,\Omega$, what is their total resistance when connected (a) in series and (b) in parallel?
2 A $10\,\Omega$ resistor is connected in parallel with a $1\,000\,\Omega$ resistor. What is their total resistance?
3 Given three resistors, $1\,500\,\Omega$, $50\,\Omega$ and $100\,\Omega$, how should these be connected to give a total resistance $136\,\Omega$?
4 Two resistors are connected in parallel. The current flowing through the resistor of value $100\,\Omega$ is $5\,mA$. The current flowing through the other resistor is $2\,mA$. What is its resistance?
5 Resistors $10\,\Omega$, $20\,\Omega$, $25\,\Omega$, $12\,\Omega$, and $6\,\Omega$ are connected in parallel. What is their total resistance? If a p.d. of $2\,V$ is applied across the terminal of the parallel set of resistors, what is the current flowing through each resistor?

Resistors in series are often used to make a *potential divider* (fig. 19.19). We have already learnt that

$$V_{AC} = V_{AB} + V_{BC}.$$

By choosing the right values for R_1 and R_2 we can arrange to make V_{BC} any value from almost zero to almost V_{AC}. The two resistors divide the p.d. between A and B into two parts, V_{AB} and V_{BC}. If R_1 has twice the value of R_2, V_{AB} has twice the value of V_{BC}.

If $V_{AB} = 2 \times V_{BC}$,
$V_{AC} = 2V_{BC} + V_{BC} = 3V_{BC}$

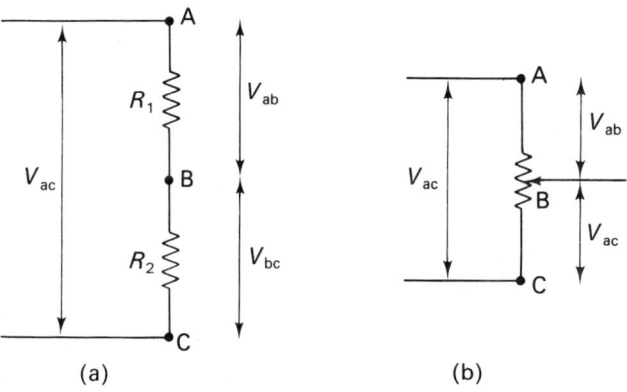

(a)   (b)

Fig. 19.19   Potential divider made from (a) two fixed resistors in series; (b) a variable resistor.

So if we apply a p.d. of 6 V across the two resistors, the p.d. across R_2 is 2 V. Whatever p.d. is applied across the two resistors, one third of it appears across R_2. The p.d. has been divided by three. By choosing other pairs of values for R_1 and R_2 we can obtain other fractions of V, the applied p.d.

The same effect can be obtained by using a variable resistor in place of the two resistors in series (fig. 19.19b). Point B is the sliding contact of the variable resistor. When B is close to end C of the resistor, V_{BC} is very low. As we slide the contact up towards A, V_{BC} increases and eventually equals V_{AC}.

The calculations above show that *potential* at B can vary in value between zero and V_{AC}. This assumes that we do not make an electrical connection to B and allow current to flow out through B to some other circuit. If we do this, potential at B will fall. However, if the current flowing out through B is very small compared with the current flowing through R_1 and R_2 it makes only a small difference to the value of V_{BC}.

19.40 Other relations between p.d. and current

Ohm's law applies to conduction in metals and alloys, and in some other instances. Your teacher may demonstrate to you the relationship between p.d. and current through an electrolyte, using copper sulphate solution and two copper electrodes. Calculation of V/I or plotting a graph of I against V shows that Ohm's law applies in electrolysis, provided the current is not too large. With a large current, the effects of electrolysis alter the resistance of the electrolyte close to the cathode and current falls. If the electrodes are platinum and the electrolyte is acidulated water, we find that no current passes through the electrolyte unless the applied p.d. is 1.7 V or greater. This is necessary to overcome the back e.m.f. of the system. So unless $V \geqslant 1.7$ V, Ohm's law does not apply. If $V > 1.7$ V, we obtain a straight-line graph indicating that $I = \dfrac{V - 1.7}{R}$.

This is a variation on Ohm's law.

Another demonstration can be performed to show conduction in a gas at low pressure. A neon tube is connected to a power supply which can give a p.d. up to 200 V. At first the current is extremely small, too small to be easily measured. When the p.d. exceeds 70 V or slightly more, a current begins to flow. This is due to electrons passing along in the gas. As the p.d. is increased the electrons travel with greater velocity and more kinetic energy. They hit the neon atoms and knock electrons from them. This creates additional charge carriers: more electrons and neon ions (positively charged). These carry an increased current; the resistance of the gas is effectively

decreased. As p.d. increases, the ratio V/I decreases. It does not remain constant and Ohm's law does not apply. The passage of electrons and ions excites the atoms of neon which glow with a reddish light. Note that although resistance decreases with increasing p.d., and Ohm's law is not true, we can still use the relation $R = V/I$ to calculate the *resistance* of the gas at any given p.d. When we say Ohm's law is not true we are simply saying that the resistance of the gas is *not constant* if p.d. varies.

19.41 The resistance of a semi-conductor diode

A semi-conductor diode consists of two pieces of semi-conducting material joined together: a piece of n-type and a piece of p-type (19.13). We can think of the diode as being like fig. 19.20a, though in practice its actual construction may be different from this.

Instructions: investigating the resistance of a semi-conductor diode

1 Connect the circuit shown in fig. 19.20b. The 1 k Ω variable resistor acts as a potential divider. It allows a variable p.d. to be applied; the small diagram below shows the correct way to connect a radio volume control potentiometer in this circuit. The 50 Ω resistor is to limit the current so that the diode is not damaged. Turn the variable resistor to its full extent anti-clockwise before switching on. This means that you begin with zero p.d.
2 Switch on, turn the variable resistor knob *slowly* until the voltmeter shows a p.d. of 0.1 V. Record the current, if any is measurable.
3 Increase p.d. in steps of 0.1 V, until the p.d. is 1.5 V, recording current at each step. DO NOT INCREASE p.d. ABOVE 1.5 VOLTS.
4 Plot a graph of current against p.d.
5 Reverse the diode connections; replace the voltmeter with one reading to 10 V; replace the milliammeter with a microammeter or a very sensitive galvanometer.
6 Switch on with the variable resistor turned to zero p.d. setting. Increase p.d. in steps of 1 V up to 6 V and record the current at each step.

Questions
1 In the first set of readings what was the resistance of the diode when the p.d. was 0.3 V?
2 What was the resistance when the p.d. was 1 V and when it was 1.5 V?
3 Does the diode conform to Ohm's law?
4 In the second set of readings, what can be said about the resistance of the diode?

The useful property of the diode is that it has a very low resistance to a current flowing from anode to cathode, but an extremely high resistance to a current flowing from

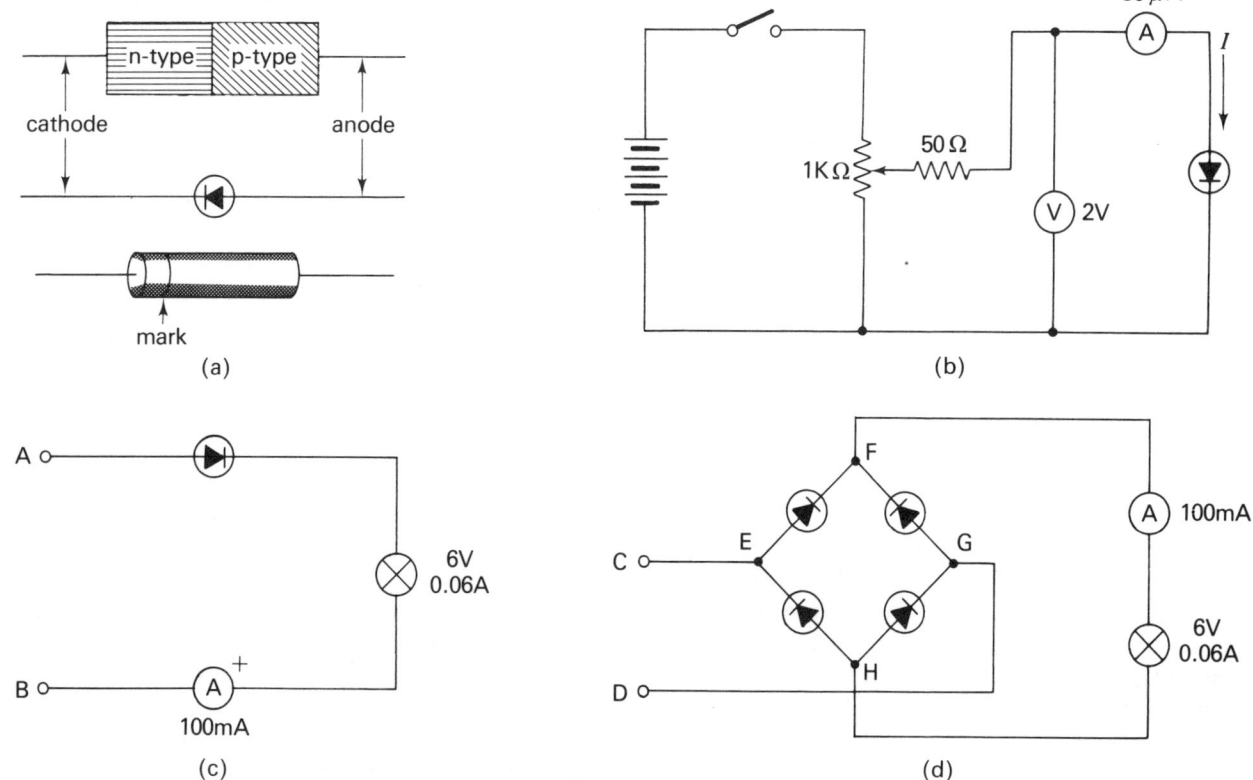

Fig. 19.20 (a) Semi-conductor diode, principle of construction, symbol, and external view of typical diode. (b) Circuit for investigating the resistance of a diode. (c) and (d) Rectifying circuits using diodes.

cathode to anode. In effect, a diode permits current to flow only in one direction.

Instructions: using the diode as a rectifier

1 Connect a diode, a lamp and an ammeter in series as shown in fig. 19.20c.

2 Connect a 6V battery to terminals A and B, first connect it with positive to A and negative to B, then with reversed connections.

3 Connect four diodes as shown in the fig. 19.20d to form a *bridge*; take care to join the cathodes and anodes in the correct positions.

4 Connect a 6V battery to terminals C and D, first with positive to C and negative to D, then with reversed connections.

Questions

1 With the single diode, why did the lamp light quite brightly with the battery in one position but not when it was connected in reverse?

2 With the bridge of four diodes, what happened to the bulb in the two battery positions? What happened to the direction of the current?

3 Try to work out which way the (conventional) current flowed when the positive terminal of the battery was connected to C. What was its path when connected to D?

This experiment shows that if we have a circuit in which the p.d. is reversing from time to time, current flows *in only one direction* through the lamp and ammeter; there is no reversal of p.d. in their part of the circuit. We say that the reversing or *alternating* p.d. has been *rectified*. The bridge is a better rectifier than the single diode, for the lamp lights whichever way the p.d. is applied to the circuit. Later we shall see how this rectifying property of the diode can be used.

19.50 Measuring temperature electrically

Instructions: investigating the effect of temperature on the resistance of a lamp

1 Use the voltmeter-ammeter method (fig. 19.13). *R* is a 6V or 12V lamp.

2 To begin with, connect only one cell in the circuit; record *I* and *V*. Does the lamp glow?

3 Increase the number of cells one at a time, to give increasing p.d. and increasing brilliance of the lamp. This means increasing temperature of the lamp filament. Do not exceed the rated voltage of the lamp.

4 Make a table of your results. For each pair of readings, calculate the resistance of the filament.

Questions

1 What effect does increasing temperature have on the resistance of the filament?

2 Lamps most often seem to 'burn out' when they are first switched on; they seldom fail when they have been running for some time. Can you explain why this might be, using your knowledge about temperature and resistance?

Instructions: investigating the effect of temperature on a thermistor

A thermistor is a piece of semi-conducting material of special kind, with wires connected to each end. Use one which has a resistance of around 100Ω at room temperature.

1 Use the voltmeter-ammeter method (fig. 19.13), with a single cell to supply the current at 1.5V p.d. The ammeter can be one which reads to 50mA, f.s.d.

2 Place the thermistor in various places to obtain a range of temperatures. You could use a freezing mixture, melting ice, room temperature, and water-baths at temperatures up to $100°C$. At each temperature, leave the thermistor in position for a few minutes before taking the reading. This gives it time to settle to the new temperature. At the time of measuring its resistance read the temperature, using a mercury thermometer. Record current, p.d. and temperature in a table.

3 Calculate resistance. Plot a graph of resistance against temperature.

4 Put the thermistor in other places: refrigerator, other rooms, outdoors. Do not use a mercury thermometer in these places, for there is no need. Calculate the resistance of the thermistor and use your graph to find the temperature which corresponds.

Instructions: investigating the effect of temperature on the resistance of a metal wire

If you have a coil of fine copper wire, you can design your own investigation on the lines of the one above to find how its resistance varies with temperature. Then plot a graph and use this for measuring temperatures.

Resistance depends on temperature. For a metal, resistance *increases* as temperature increases. For a semi-conductor, resistance *decreases* as temperature increases. In the metal conduction is by electrons and a cloud of free electrons exists ready to move among the molecules of the lattice whenever an electric field is applied. If the metal is hot, the molecules are vibrating more vigorously. Their motion makes it more difficult for the electrons to pass through the spaces between them. The electrons experience *more resistance* to their motion. The hotter the metal, the greater the resistance. In a semi-conductor there are few free electrons, if any. Heating *increases* the number of electrons which escape from the atoms in the lattice. Though the increased vibration of molecules might increase resistance by a certain amount, this is more than compensated for by the fact that there are far more charge carriers (electrons and holes) as temperature increases. The overall effect is that the extra charge carriers make it easier for electric charge to be carried in the semi-conductor. In both cases we have a physical property which depends on temperature. We can use this effect in making *resistance thermometers*.

The platinum resistance thermometer consists of a coil of platinum wire in a protective container. A special electrical circuit is used to measure its resistance very accurately, so temperatures can be accurately measured. It can measure temperatures over a range from 83K to 1 400K. Its chief disadvantage is that, since the coil is large, it and its container take a long time to reach the temperature of the surroundings, it cannot follow rapid changes in temperature. Also, while reaching the temperature of the surroundings, the thermometer may take too much heat from or give too much heat to its surroundings, so altering the temperature which it is measuring.

The thermistor is less accurate but it is cheap and can be made very small. A bead of semi-conductor material a millimetre across can be used to measure temperature. It quickly reaches the temperature of its surroundings, it takes little heat in and gives little out. It can be used to measure the temperature in very small spaces (for example *inside* the leaf of a plant) or of very small quantities of material. Thermistor thermometers are becoming used more and more widely and, since they operate through an electrical circuit, they can be used for automatically controlling all kinds of equipment: switching on heaters or cooling fans at selected temperatures, signalling temperature readings from satellites or space vehicles and many other applications.

Projects

1 Design and make your own ohmmeter. This will probably make use of an ammeter. The scale can be covered with paper and marked to read resistance, in ohms.

2 Investigate the effects of temperature on the resistance of a semi-conductor diode.

3 Find out as much as you can about the use of electrolysis in industry. Find out about electroplating and its many uses. Collect pictures, samples of materials and other items to make a display in the laboratory.

4 The light-dependent resistor is a semi-conductor device; its resistance depends on the amount of light falling on it. Connect an LDR in a resistance measuring circuit and investigate the relation between the amount of light falling on the LDR and its resistance. LDRs are sometimes called photoconductive cells or cadmium sulphide cells.

5 Try some electroplating. Try copper-plating on brass: a little sulphuric acid in the copper sulphate solution will improve things. Try nickel-plating on to copper using an anode of nickel foil, and a solution of nickel ammonium sulphate. Try zinc-plating on to copper using a zinc anode

and a strong zinc sulphate solution, to which a little sulphuric acid and a pinch of boric acid have been added. For good plating it is essential that electrodes or articles to be plated are cleaned with emery paper and thoroughly washed to remove grease; do not touch them when they are clean. Current should not be too high if a firm deposit is required. You could also try plating on to a paraffin wax model or impression; coat it with fine colloidal graphite (Aquadag or some types of grate polish) to make the surface electrically conductive. There is plenty of chance for experiment here.

20   Atoms

All matter is made from atoms and atoms are the stuff from which the Universe is made. Things that we can discover about the structure and behaviour of atoms here on Earth can help us understand happenings in the farthest parts of our galaxy.

20.00   Elements, nuclides and isotopes

Atoms are of many different kinds, with different properties. Chemists have investigated these properties and have discovered 89 different kinds of atom that occur naturally. To this list can now be added several more, made artificially by methods that we shall study later. Each of these different kinds has different *chemical* properties; we call them *elements*. From these elements we can build up all the different kinds of molecule that exist. The chemical study of elements shows that, if the elements are arranged in order of proton (atomic) number, Z, and set out in a special table (the Periodic Table), it is possible to group the elements into families according to which column of the table they are placed in.

Elements in the same family have many chemical properties in common. For example, we find that one family consists of the elements fluorine, chlorine, bromine, iodine and astatine. These are the halogens, elements that are all very poisonous; they or their compounds can be used for killing bacteria (except astatine which is very rare). Another family consists of helium, neon, argon, krypton, xenon and radon. All are gases and almost never combine with other substances to form compounds. Another group is lithium, sodium, potassium, rubidium, caesium and francium; these are the alkali metals. They are all silver-white metals and tend to react very violently with other substances. For example, a violent reaction occurs when sodium is placed on water. Study of these groupings has shown that members of a family are all alike in the number of electrons found in the outermost shell of their electron cloud. It is the electrons on the *outside*

which are most concerned when atoms react with one another and therefore which determine the *chemical* properties of the different atoms. The exact composition of the *nucleus* of the atom is of less importance.

20.01   Relative atomic mass

The relative atomic mass is not the same thing as the nucleon (mass) number, A. The nucleon (mass) number is the total number of neutrons and protons: it *must* be an integer. The relative atomic mass A_r (formerly atomic weight) is the mass of one atom of the element relative to the mass of one atom of carbon, taking the mass of the carbon atom to be 12. The relative atomic mass:

$$A_r = \frac{\text{mass of the atom}}{\frac{1}{12} \text{ mass of } {}^{12}_{6}\text{C atom}}$$

The relative atomic mass of carbon-12 is by definition $\dfrac{12}{\frac{1}{12} \text{ of } 12} = 12$ and is the same as its nucleon (mass) number. If we measure the relative atomic mass of other elements, we find that their relative atomic mass and their nucleon (mass) number are *not* exactly equal. For example, chlorine has a relative atomic mass of 35.5. This might suggest that the nucleus of its atom has 35 protons or neutrons, plus half a proton extra. The real explanation is that, in nature, chlorine exists in two forms. One form has relative atomic mass 35; about three-quarters of the chlorine atoms are of this form. The other form has relative atomic mass 37; about one-quarter of the atoms are of this form. When we measure the relative atomic mass of chlorine we are working with a *mixture* of the two forms. The result is an average relative atomic mass, 35.5. One form has 17 protons and 18 neutrons ($A = 35$ nucleons) and the other form has 17 protons and 20 neutrons ($A = 37$ nucleons). Since they both have 17 protons, they both have 17 electrons in the electron cloud. Therefore, they have identical chemical properties. The two forms of chlorine are the same element.

Investigation has shown that all elements exist in several forms. Some elements have many more than two forms. For example, mercury has seven forms. Their relative atomic masses are 196, 198, 199, 200, 201, 202 and 204, giving an average relative atomic mass 200.5. Each element, then, exists in a number of forms of differing relative atomic mass but with identical chemical properties. These different forms of each element occupy the *same place* in the Periodic Table. We call them *isotopes*, a word which means 'equal place'.

In the descriptions in this chapter we shall need to be able to make clear exactly which isotope we are referring to. If we speak of chlorine, do we mean the isotope with relative atomic mass 35, or the isotope with relative atomic mass 37? It makes no difference chemically, but an atomic physicist must know which isotope is meant. To specify an isotope, we write its relative atomic mass after its name: chlorine-35 *and* chlorine-37.

When describing atomic changes and reactions we use another system of symbols: $^{35}_{17}Cl$ *and* $^{37}_{17}Cl$.
The upper number is the nucleon (mass) number A; it tells us how many nucleons (protons + neutrons) are in the nucleus; it also tells us the mass of the atom relative to an atom of carbon-12. The lower number is the proton (atomic) number Z; it tells us how many protons are in the nucleus. From this number we know the charge on the nucleus. Isotopes of the same element have the same proton (atomic) number. It follows that all isotopes of chlorine have '17' as the lower number. Similarly, the seven isotopes of mercury are:
$^{196}_{80}Hg$; $^{198}_{80}Hg$; $^{199}_{80}Hg$; $^{200}_{80}Hg$; $^{201}_{80}Hg$; $^{202}_{80}Hg$; $^{204}_{80}Hg$. The number of neutrons ranges from 116 to 124. The number of protons is 80 in all the isotopes of mercury; the number of electrons is 80 in all cases; the chemical properties of all are identical.

20.02 The mass spectrometer

This is an apparatus which separates atoms according to their mass. The principle of it is illustrated in fig. 20.1. If a gas (for example, neon) is to be analysed, it is passed into the apparatus under very low pressure through a hole in an electrode, X. The entire apparatus is enclosed and its interior is at very low pressure, practically a vacuum. Between the electrodes X and Y the neon atoms are bombarded by electrons produced by an electron gun.

Electrons are knocked from the atoms, which thus become positively charged ions. The weak electric field between X and Y (with Y positive compared with X) causes the ions to move towards Y; some pass through the hole in Y. There is an exceedingly strong electric field between Y and electrode Z, with Z positive compared with Y. The ions are accelerated and pass through the hole in Z as a beam of ions moving at very high speed. They next pass between the poles of a powerful magnet. The magnetic field exerts a *sideways* force on the ions. This has the same effect that we have already noted in 11.03. Travelling at high speed, the ions are forced to change direction. The extent to which they change direction depends on the momentum of the atoms.

If a goat runs past you, you can catch it and make it change direction. You could not do this with an elephant or an automobile. The goat has relatively small mass, so its momentum is small (recall that momentum = mu); the elephant and automobile have large mass and high momentum, even if their velocity is the same as that of the goat. The same idea applies to the atoms. They all have the same velocity in the beam, as they have all been accelerated by the same electric field but, if some have greater mass than others, the atoms with greater mass will not be deflected as much as those with lesser mass. Neon

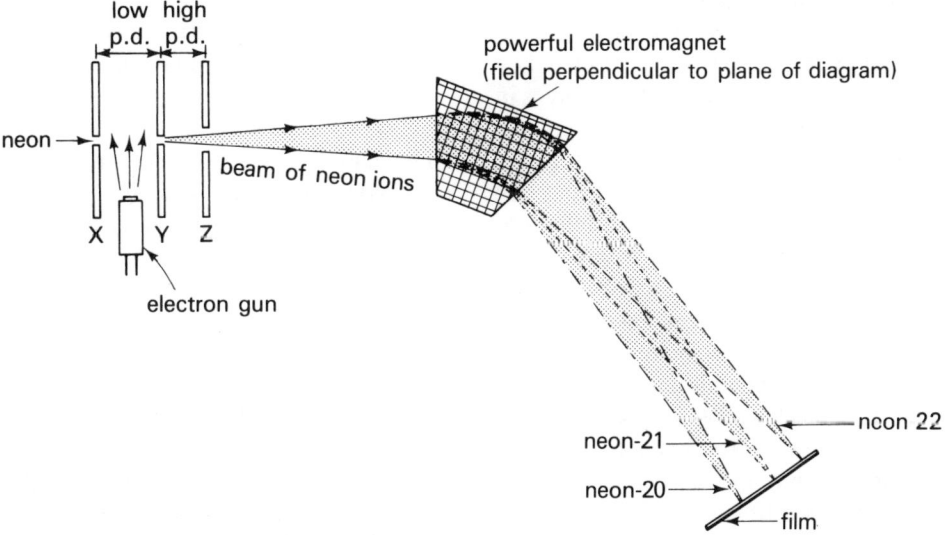

Fig. 20.1 Diagram to show the principle of the mass spectrometer.

197

has three naturally occurring isotopes. The commonest (about 90%) is $^{20}_{10}Ne$, the next is $^{22}_{10}Ne$, and the third is the rare isotope $^{21}_{10}Ne$. The mass spectrometer sorts out these three from the natural mixture, as shown in the figure. The most massive isotope, neon-22, is deflected least because it has the highest momentum. The least massive, neon-20, is deflected most. The magnet also acts as a lens to focus the beam to a sharp point on the photographic film. On arrival at the film, the ions affect the film when the film is developed, a dark spot is seen where each beam was focussed. These are three spots, corresponding to the three isotopes of neon, *in order corresponding to their relative atomic masses*. We can also compare the intensity of the spots; this gives us an idea of which isotope is the most plentiful and which the least. In natural neon, the spot for neon-20 is a very dark one, but that for neon-21 is very faint.

This instrument can be used with all kinds of elements, as well as with compounds and mixtures of substances, to find the masses of all the different kinds of atom present. We need a name for 'different kinds of atom', and this name is *nuclide*. Work with the spectrometer has shown that there are hundreds of nuclides. The nuclides can be classified according to proton (atomic) number; all nuclides with the *same proton (atomic) number* are isotopes of the *same element*.

20.10 Radioactivity

The nuclei of known nuclides contain up to a hundred or so protons with up to 150 or so neutrons. Hundreds of different numerical combinations of protons and neutrons are possible and hundreds are known to exist. But of all these nuclides, only a few are stable. These are ones in which the balance of forces within the nucleus allows the protons and neutrons to remain together permanently. Most of the nuclides are unstable. The ratio of protons to neutrons is unbalanced. After they have been formed they remain unchanged for a while but, sooner or later, forces in the nucleus break it apart. When this happens, fragments of nucleus are thrown out. The nuclide becomes a different nuclide. This may be stable, or it may not. If it is not stable, it too will break after a while and so the chain of breakdown continues until all the products are stable nuclides.

An example of this is the nuclide, uranium-238, which gradually breaks down, throwing off small pieces of the nucleus and changing into other nuclides, such as radium-226 and eventually becoming lead-204, which is a stable nuclide. This process is known as *radioactive decay*, for the process of breaking down of the nucleus leads to the production of radiation. Two of the types of radiation produced are the pieces of nucleus which have been thrown out. The more massive of these is called an *alpha*

particle. The last stage in the production of lead from uranium involves the emission of an alpha particle. The uranium breaks down by stages and the element polonium-210 is formed. Polonium-210 breaks down like this:

$$^{210}_{84}Po \rightarrow {}^{206}_{82}Pb + {}^{4}_{2}\alpha$$

Let us examine this equation to see what it means. The isotope of polonium represented here has a nucleus containing 84 protons and 126 (=210–84) neutrons. This combination is unstable, so cannot exist for ever. Eventually the nucleus throws out an α-particle which consists of two protons and two neutrons. Its nucleon (mass) number is 4, which means that it is a relatively heavy particle. The remaining nucleus has 82 protons and 124 (206–82) neutrons. This combination is stable and is an isotope of lead. This type of decay, which is called alpha-decay means a reduction of 4 in the nucleon (mass) number (since the particle contains 2 neutrons and 2 protons) and a reduction of 2 in the proton (atomic) number (since it contains 2 protons). You will note that the total nucleon (mass) number on the right-hand side of the equation equals that on the left; similarly for the total proton (atomic) number. We have lost or gained no protons or neutrons in the process of decay. The alpha-particle is a relatively large piece of nucleus. It is large enough to be a nucleus itself. If we look through the list of known nuclides, we find that helium has nucleon number 4 and proton number 2, so the α-particle is the same thing as the nucleus of a helium atom. The α-particle is thrown out of the polonium nucleus as a particle without electrons; it has a double positive charge due to its two protons. If it could acquire some electrons, it could become an atom of helium:

$$^{4}_{2}\alpha + 2 \text{ electrons} = {}^{4}_{2}He.$$

This idea was confirmed by an experiment performed by Rutherford and Royds. They placed some radon gas in a thin-walled tube surrounded by a thicker glass tube. Radon decays by emitting an alpha particle:

$$^{220}_{86}Rn \rightarrow {}^{216}_{84}Po + {}^{4}_{2}\alpha.$$

The nuclide produced is polonium. Note that this isotope of polonium is different from the isotope in the previous equation; they both have the same proton number ($Z=84$), so are the same element. The α-particles produced from the radon could pass through the thin walls of the inner container, but were retained inside the outer container. They gradually acquired electrons from their surroundings and became uncharged helium atoms. After a week, the helium which had collected in the outer container was passed into a narrow tube, in which were two electrodes. A high potential difference was applied to these electrodes, making the helium glow with a golden-coloured light. Analyses of the spectrum (5.23) confirmed that the gas was helium.

The second type of radiation is called a *beta-particle*. A nuclide which produces this is one of the isotopes of sodium:

$$^{24}_{11}\text{Na} \rightarrow {}^{24}_{12}\text{Mg} + {}^{0}_{-1}\text{e}$$

The last item in the equation represents the beta-particle. In fact this is an electron, with zero mass and a negative charge. Note once again that the sums of nucleon numbers and proton numbers are equal on both sides of the equation. In losing an electron the mass of the atom has not been significantly altered. But the *nucleus* has given off an electron, *increasing* its proton number from 11 to 12, so the product is magnesium. We have not thought of there being electrons in the nucleus, for we have said that only protons and neutrons are present there. The mechanism by which an electron can be emitted from the nucleus is complex and not fully understood. The end result is that an electron is emitted and one of the neutrons in the nucleus is replaced by a proton. *A* remains unaltered, *Z* is increased by 1.

Early experiments by Becquerel, Rutherford and Marie and Pierre Curie had shown that uranium, radium and some other radioactive substances gave off three types of radiation. They performed many experiments to discover the properties of these and the results of several of their experiments are summarised in one diagram as fig. 20.2. A small amount of radium is placed in a lead box. A beam of radiation leaves the open top of the box. If a strong magnetic field is applied, the rays are sorted out into the three types. The alpha-rays are *slightly* deflected one way; the beta-rays are *strongly* deflected the other way; the gamma-rays are not deflected. The direction of

deflection of the alpha rays shows that they are positively charged and the fact that their deflection is slight shows that they are relatively massive, they have relatively high momentum and tend to continue in the direction in which they were travelling before they entered the magnetic field. An alpha-ray is a beam of alpha-particles. The direction of deflection of the beta-rays shows that they are negatively charged and the ease with which they are deflected shows that they are relatively very light. A beta-ray is a beam of beta-particles. We can also notice another difference between alpha-rays and beta-rays. Alpha-rays are all deflected by about the same amount, showing that all particles have about the same velocity. Beta-rays are deflected by differing amounts, showing that the particles have differing velocities.

The gamma-rays are not deflected, showing that they are not charged. It was afterwards discovered that these are electromagnetic waves (21.40) of very short wavelength. They are a form of energy. We can think of them as a product of changes in the nucleus. When a nucleus decays, the new arrangement of protons and neutrons involves different forces between them and different energy levels within the nucleus. Energy is lost from the nucleus in the form of gamma-rays.

The diagram shows three different kinds of radiation apparently all coming from radium. This is because radium decays into radon which in turn decays, giving a series of other nuclides, each one decaying either by emission of an alpha- or beta-particle and often with the emission of gamma-rays at the same time. What the diagram shows as 'radium' is in fact a mixture of radium and the products of its decay; *between them* they produce all three kinds of radiation.

When the three types of radiation were first discovered their exact nature was unknown and they were given the names alpha, beta and gamma. Subsequent experiments, such as that by Rutherford and Royds showed the exact nature of the particles and gave further information about their properties. We will discuss their properties in more detail later.

20.20 Detecting radioactivity

20.21 Photographic emulsions

All three kinds of radiation affect a photographic emulsion. This property led to the discovery of radiation from uranium by Becquerel. He had placed crystals of a uranium salt on top of a photographic plate that was wrapped in black paper to prevent light from reaching it. When developed later, the plate showed darkening in the region close to where the crystals had been. This was the first indication that uranium gave off rays that were able to penetrate black paper.

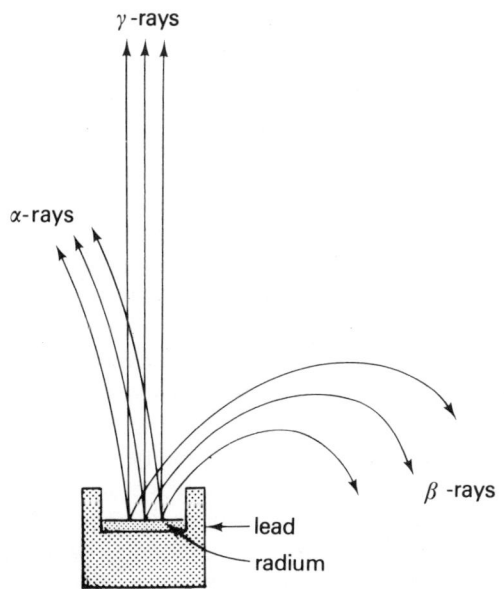

Fig. 20.2   Summary of the results of experiments by Rutherford and the Curies.

Nowadays, photographic film is used to detect small quantities of radioactive substances (20.42 and fig. 20.8). Workers at risk in places where radioactive substances are handled are given badges, which contain a small piece of unexposed photographic film. The badge is light-proof so, when the film is developed later, it *should be* clear. If, during the period the badge has been worn, the worker has been exposed to radiation, the film will show darkening when developed.

If this happens, it is a warning that something is wrong. Other personnel may be in danger of over-exposure to radiation. Steps must be taken to check safety precautions. The methods of handling and storing radioactive materials must be reviewed.

20.22 The spinthariscope and scintillation counter

The spinthariscope (fig. 20.3) was invented by Sir William Crookes. If there is a spinthariscope in your laboratory, look through the lens at the enlarged image of the screen. The screen is coated with zinc sulphide. If an α-particle is emitted by the radioactive source (usually a radium salt) and strikes the screen, it causes a small bright flash of light where it strikes. The flashes, or scintillations, occur irregularly over the surface of the screen, showing the chance way in which radioactive decay occurs. If the source is weak enough, the flashes can be counted. This instrument can be used for estimating the amount of radioactive substance present. Good results are obtained with α-radiation. Gamma-radiation has much less effect on zinc sulphide, but produces bright scintillations in a crystal of sodium iodide.

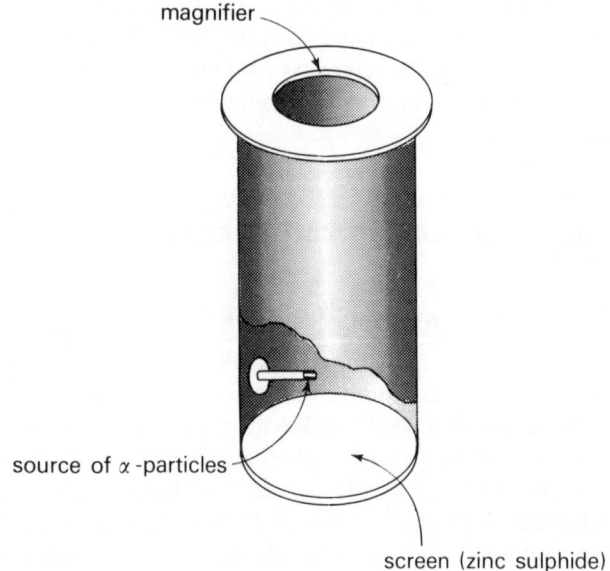

magnifier

source of α-particles

screen (zinc sulphide)

Fig. 20.3   Cut-away view of a spinthariscope.

In a scintillation counter, the flashes of light are detected electronically. Even the weak flashes caused by beta and gamma-radiation can be detected. The detector is connected to an electronic circuit which indicates the amount of radiation on a dial.

20.23 The electroscope

Instructions: discharging an electroscope with ions
1 Charge an electroscope by induction (page 182).
2 Hold a bunsen burner so that its flame heats the air *just above* the plate. The flame increases the energy of the molecules of the air above it. Electrons are knocked from some of the molecules, creating positive ions. The electrons become attached to other molecules, creating negative ions. Either the positive ions or the negative ions are attracted towards the plate, depending on whether the plate is negatively or positively charged. When the ions reach the plate, they discharge it. Discharging by ions is a good way of discharging non-conductors such as strips of polythene; it was used on page 180.
3 Re-charge the electroscope by induction.
4 Watch the behaviour of the leaves when your teacher holds a radioactive source above the plate. Explain what happens. Suggest how this effect could be used for measuring radiation.

The principle of the electroscope is used in the *ionisation chamber dosimeter*. It is a tiny electroscope in which the leaf is a very small fibre of quartz. This is examined through lenses. It can be charged to a known potential by an external power supply. Its position can be measured exactly against a scale in the instrument. Radioactive materials are placed in the chamber of the instrument. Ionisation causes the electroscope to discharge. As the fibre moves across the scale the amount of discharge can be timed and the rate of discharge calculated.

20.24 The cloud chamber

This makes use of ionisation in a quite different way. There is a limit to the amount of water vapour that the air can hold at any given temperature. If this amount is exceeded, the air is saturated. The excess vapour condenses into fine drops of water. This explains why mists appear beside rivers and lakes in the evening. During the day the air is warm and water evaporates into the air. In the evening the air cools. Its ability to hold vapour becomes less and less; eventually the air is saturated. Further cooling makes some of the water vapour condense into fine drops of water, forming a mist. Under certain conditions air can be more-than-saturated or *super-saturated* yet the mist does not form. This often happens high in the atmosphere on a clear day. Then as an aeroplane flies through the air it makes the vapour condense suddenly, leaving a 'vapour trail' where the aeroplane has been. This is not really vapour, but is condensed vapour, so is better called a cloud trail. Long after the aeroplane has gone, we can see its trail in the sky.

When a charged particle moves through air, it leaves a trail of ions behind it. Ions act as sites on which water drops can begin to condense. A charged particle moving through supersaturated air can be detected by the cloud trail it leaves behind. This explains how cloud chambers work. The type described in the experiment below uses condensing alcohol vapour instead of water vapour, but the idea is the same.

Instructions: using a cloud chamber

1 Unscrew the base of the chamber (fig. 20.4). Remove the sponge. USE FORCEPS to put some pieces of 'dry ice' (solid carbon dioxide) next to the partition. Replace the sponge and base. Dry ice *must* be handled with forceps, NEVER WITH FINGERS. Dry ice can be obtained from local suppliers or from a carbon dioxide cylinder with special attachment for collecting dry ice.
2 Saturate the felt ring in the lid of the chamber with alcohol (surgical spirit etc). Place the lid on the chamber; alcohol vapour diffuses through the chamber, until the air is saturated. Near the bottom of the chamber, where the air is cooled by the solid carbon dioxide, a slight mist is formed. Adjust the wedges until a layer of mist of even thickness is seen across the whole bottom of the chamber.
3 Adjust the illumination. It is best if this is performed in a darkened room. The light source should be a focussed horizontal beam passing through the chamber from the side and not shining directly on the black floor of the chamber. In this way, the maximum contrast is obtained and the trails will be clearly visible.
4 Place the source in position, so that it is not behind either of the screens. Rub the top of the chamber with a soft cloth to create an electrical clearing field. Look for cloud trails.
5 Repeat with the source behind the thinner of the two metal foils.
6 Repeat with the source behind the thicker of the two metal foils.

Questions

1 Describe the appearance of the trails at step 4. Did you see any trails which appeared curved or bent? What can you say about the speed of the particles? What can you say about the distance the particles travel in air?
2 Answer the same questions for the trails seen at step 5. Can you account for the differences?
3 Did you see trails at step 6? If not explain why.

Instructions: investigating the effect of a magnetic field

1 Set up the cloud chamber as in the previous instructions but, after placing the solid carbon dioxide beneath the

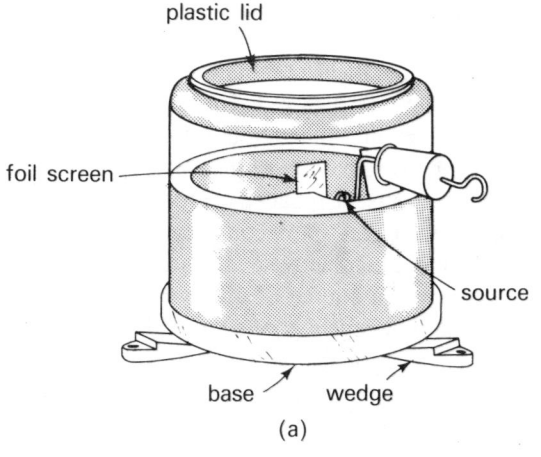
(a)

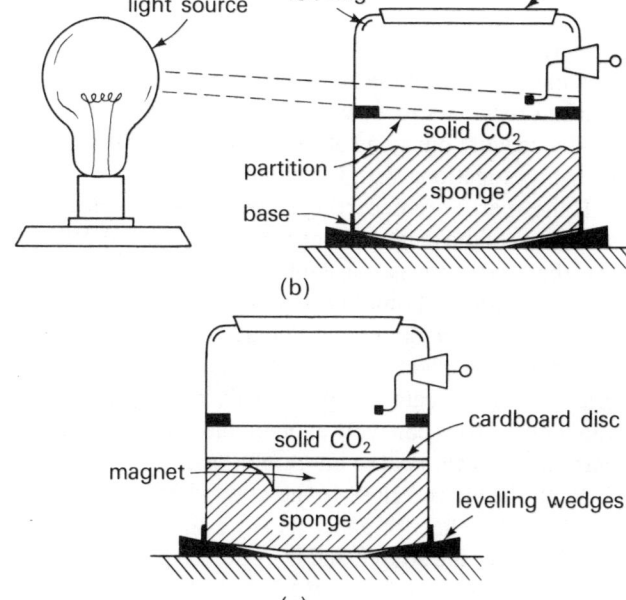
(b)

(c)

Fig. 20.4 Diffusion cloud chamber (a) the chamber, levelled and ready to display cloud trails; (b) section of the chamber; (c) section of the chamber showing a block magnet in position.

floor of the chamber, insert a cardboard disc and then a *block* magnet before replacing the sponge and screw cover (fig. 20.4c). A block magnet is specially made from a magnetic-ceramic compound. It has its poles at opposite surfaces instead of at its ends. The magnet can be placed either with its north pole or its south pole uppermost.
2 Repeat the previous instructions steps 1 to 5, and take special notice of curvature of the tracks.

Questions

1 What do you notice about the curvature of the tracks at step 4?
2 In what way is the curvature different at step 5? What does this indicate?

20.25 The spark counter

This consists of a metal gauze mounted a few millimetres above a metal plate. A high p.d. is applied between the two. If the p.d. is greater than about 4.5 kV, sparks occur frequently owing to breakdown of the electrical resistance of the air. The p.d. must be reduced to around 4.5 kV, when the sparking just stops. A spark can occur now only when the air is made more conductive by the production of ions. This can happen if a source of alpha-particles is near. When such a source is held above the grid, each particle that passes down through the grid and into the space between the grid and the plate causes ion-formation, leading to a spark. By counting the sparks, we can count the number of alpha-particles emitted by the source in a given time.

Questions

It is more helpful if the spark counter is demonstrated to you before you try to answer these questions, but you may be able to work out some of the answers from the results of your investigations with the cloud chamber.

1 If the sparks are occurring slowly enough, you can get some idea of the length of time *between* sparks. Would you say that this time is constant, that the sparks occur at regular intervals of time? Or would you say that this time is variable, that sparking is irregular?

2 What is the effect of raising the source further from the grid? Does the rate of sparking decrease gradually or suddenly? How can you explain the fact that above a certain distance, sparks are rarely heard?

3 With the source close to the grid, what is the effect of placing (a) a sheet of paper, or (b) a sheet of aluminium foil, between the source and the grid? Explain this effect.

Instructions: making a model of radioactive decay

As far as we know, no external influence can make a nucleus decay: heat, pressure, etc. have no effect. A nucleus may remain for years without decaying, or it may decay in less than 1 second from now. We have no means of telling when. It appears to be purely a matter of chance. In this model, radioactive atoms are represented by ten students who each toss coins to decide when they will 'decay'. The tossing of coins is a matter of chance. Our hypothesis is that radioactive decay is a matter of chance too. The experiment is to find out what happens when chance operates in this way.

1 Select ten students to represent ten radioactive atoms. Give each student five coins.

2 On the word 'throw', each student must spin all five coins. If *all five* fall 'heads' uppermost, this counts as a 'decay'. For other combinations of heads and tails, the atom has not decayed. If all five fall 'heads', the student takes no further part in the experiment, for an atom cannot decay twice. Students whose atoms have not decayed take part in second and third throws and so on, dropping out if they decay.

3 Record the results of successive throws in a table.

At any one throw the number of decaying atoms may be none, one or two. It is *possible* that three or more may decay, but this is *unlikely*; the chance of three students throwing 'five heads' is very low. In a quantity of radioactive atoms it is *possible* for a large portion of them to decay all at once, but it is *unlikely*.

Questions

1 Do the atoms decay at a regular rate, the same number at every throw?

2 Before you throw, is it possible to tell which atoms will decay at that throw?

3 What is the average rate of decay: the average number of decays per throw?

4 Would you find the same rate if you repeated the experiment?

5 If you continue the experiment, so that the five remaining atoms (students) decay, how many throws will you need? If you cannot answer this, find out by experiment.

6 What is the average rate of decay for the last five atoms?

20.26 The Geiger-Müller tube

This is another device that relies on ionisation. It consists of a hollow cylinder of conducting material (fig. 20.5 a). Running along the axis of the cylinder, and insulated from it, is a straight wire. The space inside the cylinder is filled with argon gas and bromine vapour at low pressure. One end of the cylinder is sealed with a very thin sheet of metal or mica. The tube is connected to a high-voltage power

Throw no.	Start	1 2 3 4 5 6 7 8	. . . until 5 atoms have decayed
Number of atoms decayed	0		5
Number of atoms remaining un-decayed	10		5

supply, usually about 400V, with the wire positive and the cylinder negative.

Questions

1 When a charged particle, such as an α-particle or β-particle enters the tube, what might it do to the gas molecules in the tube?

2 There is a high p.d. between wire and cylinder. What effect will this have on charged particles and ions?

3 What will fast-moving ions do to molecules of gas in the tube?

4 What is the effect of a large number of ions moving in the electric field in the tube?

The arrival of an alpha- or beta-particle in the tube brings about an electric discharge. Gamma rays are ionising and they too cause discharge. The discharge causes a brief change of p.d. in the tube, which can be detected by electronic circuits. The circuits are connected either to a *scaler* which counts the discharges and displays the total number of discharges or 'counts', or to a *rate-meter* which indicates 'counts per minute' on a dial. Counters often have a loudspeaker which clicks each time the tube discharges. If a strong radioactive source is near the tube, the speaker clicks thousands of times each second, producing a high-pitched screeching noise. When no radioactive source is near the tube, clicking is at a rate of about ten per minute. This *background* count is caused by very small quantities of naturally occurring radioactive materials near the tube. Naturally occurring potassium, for example, contains a small proportion of potassium-40, which emits β-particles. There may also be other radioactive atoms near by, the result of atomic explosions in other parts of the world. Radiation of various kinds also reaches us from distant parts of Space; it can easily pass through the roof and ceiling, adding to the background count of the laboratory.

The Geiger-Müller tube may be used to demonstrate the properties of radioactive substances. In fig. 20.5b, the tube is shown directed at a radioactive source. Between them is a sheet of absorbing material. The source may be one that emits mainly α-particles (americium-241, or plutonium-239), or β-particles (strontium-90), or γ rays (cobalt-60), or all three types of radiation (radium-226). Absorbing sheets consist of paper, aluminium of various thicknesses, and lead up to about 2 cm thick. The source is placed not more than 5 cm from the tube; then the rate of counting is measured first with no absorber and then with various absorbing sheets in position. Results show that absorption properties differ for different types of radiation.

α-particles: easily absorbed, even by a sheet of paper.

β-particles: absorbed completely by aluminium a few millimetres thick. The exact thickness required depends on the energy of the particles, and this depends on the nuclide.

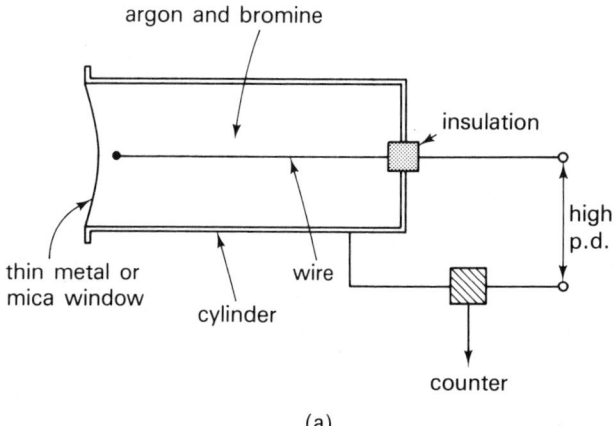

(a)

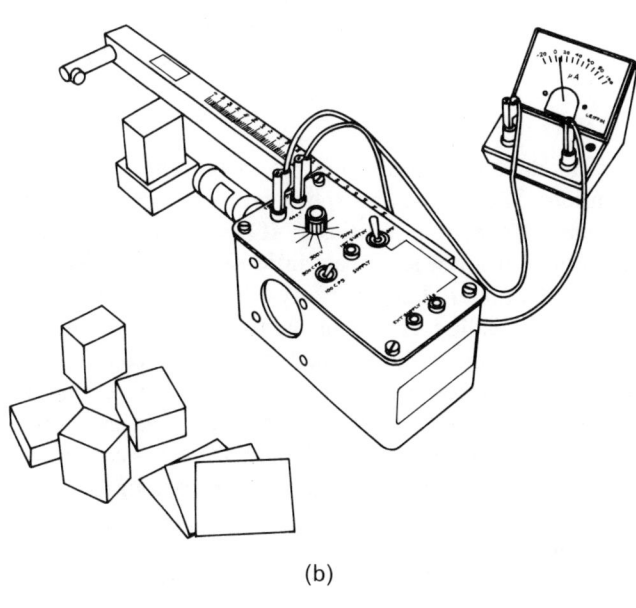

(b)

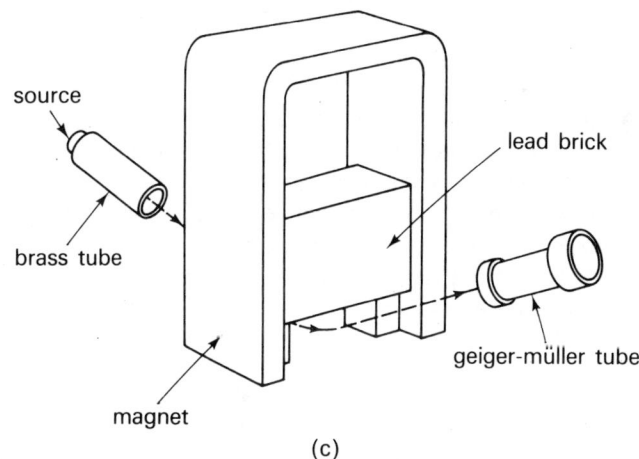

(c)

Fig. 20.5 Experiments with the Geiger-Müller tube: (a) section of the tube; (b) investigating the absorption of radiation by screens of various materials and thicknesses; (c) investigating the effects of a magnetic field on a radiation beam.

203

γ **-rays:** very penetrating. Paper and sheet aluminium have little effect. For complete absorption we must use lead, several millimetres thick. The most penetrating γ-rays need lead up to 10cm thick to absorb them completely.

In another demonstration, we can place the source at various distances from the tube, to find out how far the radiation can travel in air.

α **-particles:** the rate of counting decreases sharply when the distance is increased to more than about 5cm. These particles are easily decelerated by collision with molecules in the air, so have little penetrating power.

β **-particles:** can penetrate several metres in air. The rate of count decreases as the source is taken further from the tube; this is not due to absorption by the air, but because the tube is far from the source. The tube can catch only a very small fraction of the radiation when it is a long way away.

γ **-rays:** as for β-particles.

The demonstrations point to important safety precautions. We can use absorption to protect ourselves from α-particles and β-particles. If we keep radioactive materials in a bottle or can, little radiation can escape, unless the material emits γ-rays. For a γ-ray emitter a very thick-walled container made of lead is essential. Even then, some radiation may escape and people should not come close to the container except when necessary. In laboratories where γ-ray emitters are handled and stored, walls of lead blocks are built around the areas concerned.

Since distance is a safety-factor, forceps, long tongs and special remotely-controlled handling machines are used. The operator can handle the radioactive material, yet stay at a safe distance. It is important to remember that although α-particle emitters are normally quite safe when enclosed, they may severely damage the surface of the skin when an exposed source is handled. Also, if any radioactive substances are accidentally taken into the body, they can damage whatever part of the body they reach. Some of these substances, taken in small amounts at intervals can accumulate in large amounts in such regions as the bone marrow, where they may cause severe damage to health. It is especially important that food and drink should not be taken into areas where radioactive substances are handled.

Even the tiniest quantities of radioactive substance emit large amounts of radiation, given enough time. Think of the radiation coming night and day, year after year from the very small sources in your school. Particles far too small to be seen can produce damaging quantities of radiation. In laboratories a Geiger-Müller tube can be used to detect small quantities, and to check that any lost or spilt material is found and properly cleaned away.

The Geiger-Müller tube is used to demonstrate the effect of a magnetic field on radiation. In fig. 20.5c the brass tube is used to form a beam of β-particles. This is aimed downwards so that it does not enter the tube. A lead block prevents rays travelling direct from source to tube. If the magnet is correctly positioned the beam of β-particles can be bent so as to enter the tube. When the magnet is removed or its poles reversed, no count is recorded. If a γ-ray emitter is used as source, this effect cannot be shown, for these rays are not deflected by a magnetic field.

20.30 Half-life

For how long does a radioactive substance remain radioactive? When will your school need to buy a new source for use in the cloud chamber? We will answer these questions later; first try this model of decay.

Instructions: making a model of the rate of radioactive decay

1 The idea of this model is similar to that of the previous one, but we have more 'atoms' and they 'decay' more rapidly. Each 'atom' is represented by a die. You can use ordinary dice or cubes of wood painted or marked on one side to represent 'decay': a red dot will do. If you use ordinary dice, the 'six-dots' side means 'decay'. You need at least 100 dice or cubes, to be shared among members of the class, so that all the students have about the same number of dice.

2 On the word 'throw', each student must roll all the dice he has. Those which show 'six spots' (or the coloured side of a wooden cube) are said to have decayed. These must be handed in to the teacher or leader. They are 'decayed' and are not to be rolled again.

3 The teacher or leader records the results in a table:

Throw no.	Start	1 2 3 4 5 6 7 8	. . . until 90 or more atoms have decayed
Number of atoms decayed	0		90 (or more)
Number of atoms remaining undecayed	100		10 (or less)

4 Continue throwing as many times as necessary to reduce the number of atoms undecayed to ten or fewer. At each throw, decayed atoms are handed in for counting and recording.

5 Plot a graph of number of atoms remaining undecayed (vertical axis) against throw number (horizontal axis).

Questions

1 In this model, what is represented by 'throw number'?

2 How many atoms would you expect to decay at the first throw?

3 How many atoms would you expect to decay at the second throw?

4 After how many throws would you expect the number of atoms remaining undecayed to be half the original number?

5 After how many throws would you expect the number of atoms remaining undecayed to be a quarter of the original number?

The model shows what happens if the number of atoms that decay in a given period is a *fixed fraction* of the number present at the beginning of the period. In any given period we do not lose a fixed number, but a fixed fraction or percentage of the number. In this example we expect to lose about 1/6 of the atoms at each throw. At the next throw we lose 1/6 of those that remain, and so on. The expected behaviour of this model is graphed in fig. 20.6. Your graph was probably not quite as smooth as this curve, for this curve was the result of calculations based on the laws of chance not the result of a practical trial. The curve is steep at the beginning; there are many atoms, so a large number decay at each throw. As time passes the curve is less and less steep; the rate of decay becomes less and less; there are fewer and fewer atoms left to decay.

In this model each atom has a one in six chance of decaying at each throw. But before we throw, we cannot tell which atoms will decay and which will not. Taking the atoms as a group we can be fairly certain that about 1/6 of them will decay, giving us the curve of fig. 20.6a. If this is a true model of radioactive decay, we should be able to obtain a similar curve using a radioactive source.

Radon gas is a suitable source. This nuclide is itself formed by the radioactive decay of thorium. A quantity of thorium salt in a squeeze-bottle decays radioactively and the thorium becomes radon. This gas can be squeezed out from the bottle into the chamber of an ionisation chamber dosimeter (20.23). As the radon-220 decays, emitting α-particles, the quartz fibre is gradually discharged and moves across the scale. A graph of scale readings is given in fig. 20.6b. The rate of movement of the quartz fibre varies with the rate of decay. This means that the graph is

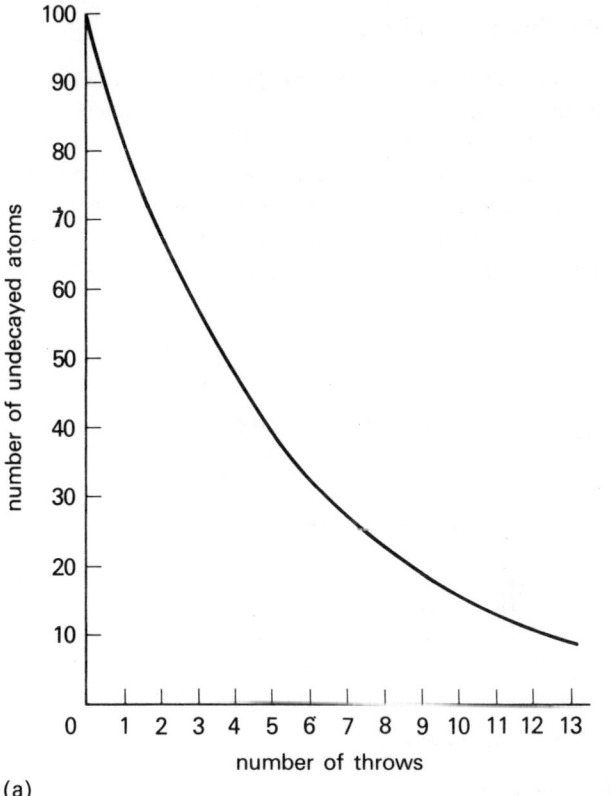

(a)

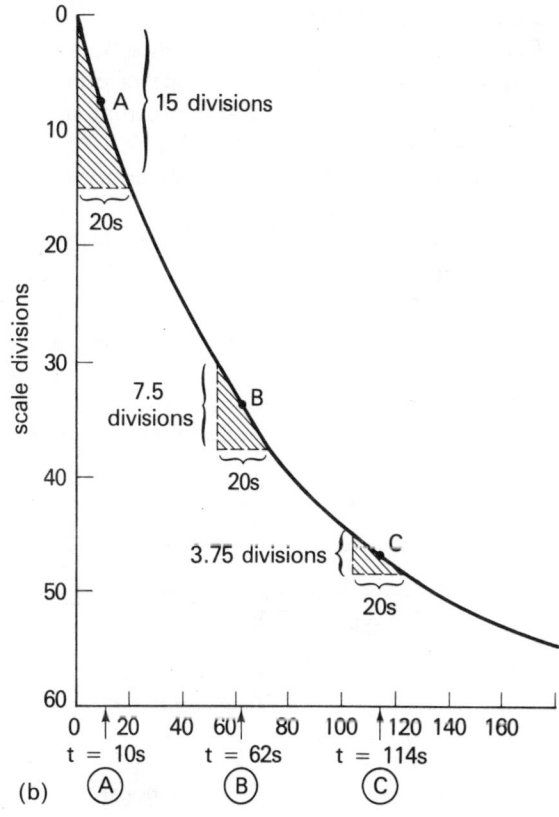

(b)

Fig. 20.6   Radioactive decay and half-life: (a) theoretical curve for the model of decay; (b) experimental curve for the decay of radon-220, as measured by an ionisation chamber dosimeter.

205

steep to begin with and gradually becomes less steep. From the start the fibre moves 15 scale divisions in 20s. The time half-way through this period is 10s from the start, represented by point A on the curve. Later the curve is less steep; the fibre moves more slowly; the rate of radioactive decay is less. At what time is it exactly half its original value? Looking along the curve, we find point B; here the rate of movement of the fibre is 7.5 divisions in 20s: half the original rate. Measurement on the curve shows that B is at $t = 62$s, 52s later than point A. Further along the curve, we find point C; here the rate of movement is 3.75 divisions per second, this is half the rate at B and quarter the rate at A. Point C is at $t = 114$s, 52s later than point B.

Thus the graph shows that the rate of radioactive decay is halved every 52s. The curve has the same shape and properties as the curve of our model. Our model of chance decay of atoms is a true one.

The curve for the decay of radon shows that whatever quantity of radon we begin with, 52s later half of it will have decayed and half of it will still be undecayed. We say its *half-life* is 52s. We can calculate that after 5 min, which is about 12 half-lives, the quantity will have halved itself 12 times, and only about 1/4 096 of the original quantity will remain. This is lucky for us, for it solves the problem of what to do with the radioactive gas when the investigation is over. If we leave it for 10min the amount left is so small that we can forget about it.

Radon was chosen for the investigation because it has a relatively short half-life. The half-life of other nuclides may be much longer. The half-life of strontium-90 is 27 *years*, and that of plutonium-239 is 2.44×10^4 years. This is why we cannot use the usual laboratory radioactive sources for demonstrating half-life.

In the natural world radioactive substances are rare. Those which were formed early in the history of the Solar System have had plenty of time to decay. Only those nuclides with very long half-lives occur naturally today. An example is uranium-238, with half-life 4.51×10^9 years. Potassium-40 (20.26) has half-life 1.26×10^9 years. Most of the radioactive nuclides are produced artificially. Some of these have long half-lives. This gives rise to one of the problems of working with radioactive materials: the disposal of materials that are no longer needed. The disposal of wastes from nuclear power stations is an especially serious problem. Many of the nuclides will remain highly radioactive for millions of years. We can do nothing to make them decay more quickly. They can be buried underground or dumped in deep oceans. When this is done, the containers used must be very resistant to damage and corrosion, so that radioactive materials do not escape during the long time that must pass before their radioactivity is reduced to a safe level.

20.40 Natural and artificial nuclides

For centuries men have wanted to be able to turn common metals into rare and costly ones such as gold. Now we are able to do this, though the quantities produced are far too small to be of any commercial value: often only a few atoms are produced.

20.41 Nuclear change

The first nuclear change was demonstrated by Rutherford. In his apparatus a source of α-particles was placed in a container filled with nitrogen (fig. 20.7). If the source was placed near the silver foil window, α-particles passed out and caused flashes on the screen. If the source was taken further from the screen, and if the nitrogen was replaced by oxygen, the flashes stopped. The range of the particles

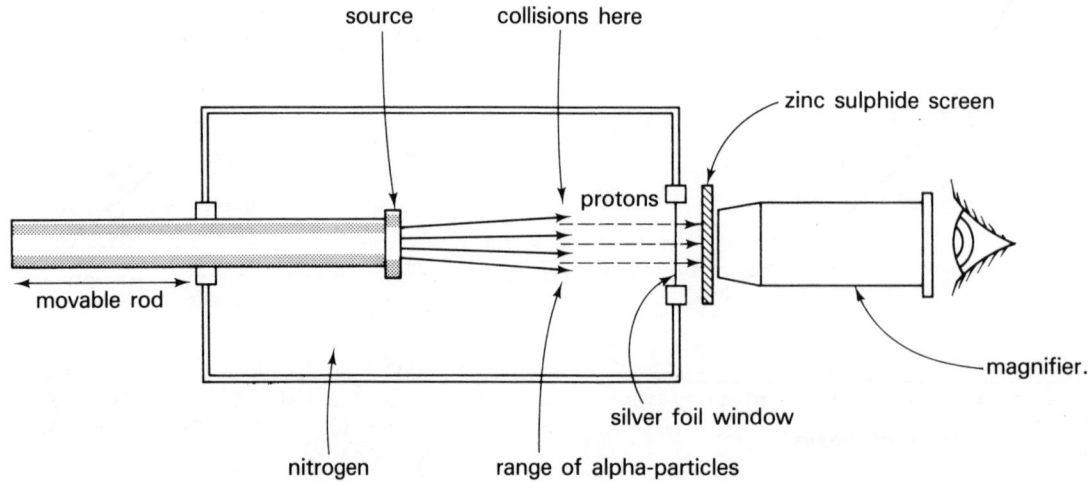

Fig. 20.7 Atomic change: simplified diagram of Rutherford's experimental production of oxygen-17 by the bombardment of atoms of nitrogen-14 with α-particles.

was not long enough for them to reach the screen. When nitrogen was put back into the tube, the flashes began again (the stage shown in the figure). Rutherford found that the flashes were caused by *protons* hitting the screen. He reasoned that the protons had been knocked out of the nuclei of nitrogen atoms when hit by α-particles. The mass of a proton is only one quarter that of an α-particle. By the principle of conservation of momentum we would expect that the proton would leave the nucleus with much greater velocity than the α-particle which had entered the nucleus. The higher velocity would give the proton a greater range, so that it could reach the screen and cause the zinc sulphide to flash. This experiment showed that nitrogen atoms were being changed into oxygen atoms:

$$^{14}_{7}\text{N} + {}^{4}_{2}\alpha \rightarrow {}^{17}_{8}\text{O} + {}^{1}_{1}\text{p}$$

The proton is represented by p. Oxygen-17 is a stable isotope, which does occur naturally but is rare. The common isotope is oxygen-16.

By long irradiation in a nuclear reactor (33.30) we are able to produce useful quantities of radioactive nuclides for special purposes. An example is cobalt-60, used as a source for experimental work and having many medical and industrial uses. It does not occur naturally, but can be made from cobalt-59, which is placed inside a nuclear reactor. There it is bombarded with neutrons:

$$^{59}_{27}\text{Co} + {}^{1}_{0}\text{n} \rightarrow {}^{60}_{27}\text{Co}$$

The neutron (n) enters the nucleus *and stays there*. Because of this extra neutron the cobalt-60 nucleus is not stable; sooner or later it decays. Its half-life is 5.23 years. When it decays it emits a β-particle (an electron) and high-energy γ-rays:

$$^{60}_{27}\text{Co} \rightarrow {}^{60}_{28}\text{Ni} + {}^{0}_{-1}\text{e}$$

The nickel-60 formed in this decay is stable.

20.42 Using radioactive nuclides

High-energy γ-rays from cobalt-60 have similar properties to X-rays. A cobalt-60 source can be used instead of an X-ray generator. In industry, the source is put on one side of a piece of forged metal and a photographic film is placed on the other. When the plate is developed, we see the equivalent of an X-ray picture of the metal. Bubbles and defects inside the metal can be clearly seen. In this application, cobalt-60 has the great advantage that it is portable and does not need any power supply, unlike the X-ray generator. It can be taken to construction sites and used on the spot. When not in use the source must be kept in a thick lead container. It must be handled with care; tongs are needed to allow the operator to work at some distance from the source.

In medicine, cobalt-60 sources are used for treating cancer. The γ-radiation penetrates the body, killing tumours and other cancerous tissues. The radiation also kills bacteria and is used for sterilising hospital equipment. Hypodermic needles, surgical blades and other items are sealed in packages and carried on a conveyor-belt past a cobalt-60 source in the packing machine. The radiation penetrates the package, killing all bacteria inside. The instrument inside the package is now sterile (has no live bacteria on it), and remains so until the package is opened, just before the instrument is to be used. This method of sterilising has the advantage that it does not spoil the properties of the metal of the needles or blades and does not corrode them. The other main method of sterilisation, heating to high temperature, does not have this advantage.

Sources of β-particles or γ-rays are used for measuring the thickness of plastic film during manufacture. As the film comes from the machine it passes over a source. A Geiger-Müller tube detects how much radiation penetrates the film. This depends on the thickness of the film. Readings from the Geiger-Müller tube automatically control the machine so that it produces film of the required thickness. This method is used in machines making the polythene film used for plastic food-bags. It is difficult to imagine a micrometer screw-gauge being successfully used for the same purpose! We can use a similar method for measuring the depth of material in a container. The source is below the container and the Geiger-Müller tube above. The amount of radiation reaching the tube depends on how deeply the container is filled with material. In a machine that is filling containers with materials in liquid or powder form, the tube can detect when the container has been filled to the correct depth. Containers that are not full enough, or too full, can be automatically rejected.

The sources used in the school laboratory are small, yet we know that they will continue to emit radiation at the rate of thousands of counts per second for millions of years. During this time there will be a relatively small fall in their activity. The number of radioactive atoms in even a small source is far larger than we can imagine. A tiny speck of radioactive material, too small to be seen, can emit measurable amounts of radiation. This creates problems in handling radioactive substances but has some useful applications. For example we can detect the extremely small quantities of metal worn from the piston of an engine. Part of the piston is made radioactive. If any part of this wears away when the engine is run, we can use a Geiger-Müller tube to find out how much has worn away and where it has gone. We can test the parts that the piston rubs against and we can test the oil in the engine to see how and why the wear occurs. This helps us to design an engine that will not wear out so rapidly. To get this information by running the engine until it *did* wear out would take a very long time, perhaps several months or years for each trial. In other types of test oil is mixed with a small amount of radioactive substance. This special oil is placed on one area of a bearing. When the machine is run for a few minutes, we can follow how the oil spreads to different parts of the bearing. This helps in the study and design of bearings. We use a similar method for detecting leaks in pipes. A small amount of radioactive substance is mixed with the fluid in the pipe. A Geiger-Müller tube is used to explore the outside of the pipe. We can detect leaks that are undetectable by other methods.

A radioactive isotope and a stable isotope of the same element have exactly the same chemical properties. In the body of an animal or plant they take part in chemical reactions in exactly the same way. To the animal or plant, there is no difference between them. For example, we ask a person to swallow a quantity of iodine compound in which a fraction of the iodine atoms are the radioactive isotope, iodine-131. The remainder are the natural isotope, iodine-127. The tissues of the body use both isotopes in exactly the same way. In the human body, iodine is taken to the thyroid gland in the neck, where it is used for making the iodine-containing substance thyroxine. This is concerned with regulating several activities of the body, including growth and development. After we have given a person iodine-131, we place a Geiger-Müller tube close to the throat. As the iodine reaches the thyroid gland, the radioactive isotope can be detected by the increased count; we can assume that the stable isotope, iodine-127, has also reached the thyroid gland. We are using the radioactive isotope as a *tracer*; It helps us trace the path of iodine as it is taken to the body, used in the thyroid gland and afterwards transferred as thyroxine to other parts of the body. This can be done without the need to perform surgery. In a similar way, salt containing sodium-24 is injected into the blood; its journey around the body is followed by Geiger-Müller tubes placed in contact with the skin. This method is used for looking for blockages in blood vessels. Naturally we do not wish to risk damage to a person's health by putting large quantities of radioactive substances inside the body. Only small quantities are needed in these techniques. We choose isotopes with a short half-life; after a few minutes or days they have decayed. There is then no further risk of damage.

Similar methods are used to trace the path of elements in plants. The phosphate ion plays an important part in plant nutrition. If we prepare phosphate of which the phosphorus atom of some molecules is phosphorus-32, we can use a Geiger-Müller tube to detect and measure it. We can find out exactly what happens to phosphate when it is taken from the soil and used by the plant. From the information we obtain from these experiments we learn the best techniques for using chemical fertilisers most efficiently. Another way of detecting the radioactive phosphate is illustrated by fig. 20.8. Some time after the plant has taken up phosphate containing phosphorus-32, we take parts of the plant and place them on a sheet of glass. Photographic film is left in contact for several days. Later the film is developed. The β-particles from the phosphorus-32 cause the film to blacken when developed. The blackening is greatest where the film is in contact with parts of the plant which contain the greatest amounts of

Fig. 20.8   *Autoradiograph of a sprig of privet.*
The darkest parts of the image indicate the highest concentrations of phosphorus.

phosphorus. We get a picture, an autoradiograph, showing which parts contain most phosphorus and which contain least. This technique can be used with thin slices of tissue taken from animals or plants. We can even tell to which cells or parts of a cell the radioactive atoms have gone. Much has been learned about the working of plants and animals by using methods such as these.

Alpha-radiation produces a much greater amount of ionisation in the air than beta or gamma radiation. Sources of α-particles can be placed in machines used for manufacturing fabrics. Normally the movement of the fabric through the machine gives an electric charge to the fabric, just as we charge a polythene rod and cloth (19.11) by rubbing them together. Charged fabric attracts dust particles from the air, making the fabric dirty. If the air has been ionised by α-radiation, the fabric is quickly discharged, it does not attract dust, and so stays clean.

20.43 Radioactive dating

Air contains carbon dioxide. In most carbon dioxide molecules the carbon is carbon-12, but a small proportion of molecules contain the radioactive isotope, carbon-14. This decays, emitting β-radiation. The amount of carbon-14 is renewed as quickly as it decays by the bombardment of nitrogen atoms by neutrons coming from outer Space. The reaction occurs in the upper atmosphere. As a result, the ratio of carbon-14 to carbon-12 remains constant. When a plant takes carbon dioxide from the atmosphere, a fixed fraction of this contains carbon-14. This and the carbon-12 both become built into cellulose, starch, proteins and other compounds in the plant. Wood, paper, cloth and other items made from plants all contain both isotopes of carbon. The ratio between them is at first the same as the ratio found in atmospheric carbon dioxide.

Imagine a piece of wood cut from a tree and used to make furniture. As time passes carbon-14 decays by emitting a β-particle, turning to nitrogen-14. The half-life for this decay is 5 570y. In 5 570y, half the carbon-14 decays; the ratio of carbon-14 to carbon-12 is half what it was to begin with, for carbon-12 is stable, but carbon-14 is not. As time passes the ratio of carbon-14 to carbon-12 gets steadily less and less. If we take a piece of this wood and test it, we can find out what this ratio is. From this we can calculate how long ago the wood was formed in the plant. This method is used for finding the age of ancient Egyptian furniture and papyrus and for finding the age of fossil plants and other items of plant matter that are several thousands of years old.

A similar method is used for dating rocks. From the time it is formed by solidifying from a melted state, or by the settling of a deposit, molecules are not able to enter or leave the crystal lattice of the rock. But atoms in the rock, if radioactive, may decay. If a rock contains uranium, we can test it to see how much of the uranium has decayed to a stable isotope of lead; then we can calculate how long ago the rock was formed. Potassium-40 is also used in measurements of this kind, for it decays to the stable isotope argon-40. The ratio of potassium-40 to argon-40 is used for dating younger rocks.

21 Waves

21.00 Features of waves

How many kinds of waves do you know? In how many different ways can you make waves? There are many answers. There are many kinds of wave motion in the Universe. In this chapter we begin by looking at several different kinds of waves to see what features they share. Then we shall know what we mean by wave motion.

21.01 Making waves

Instructions: making waves in a thread

1 Tie a small metal pendulum bob to a piece of thin thread which is about 2m long. Tie the bob to the thread about 15cm from one end.

2 Attach the thread to a support high above the floor, or you can stand on a chair and hold one end of the thread in your hand. Fix it or hold it by the shorter part of the thread, making a pendulum about 15cm long, with the longer part of the thread hanging down below the bob. It should hang straight down with its free end just resting on the floor.

3 Set the pendulum vibrating. Observe the shape taken by the hanging thread.

4 What can be seen travelling along the hanging thread?

5 In which direction are they travelling?

6 What is their frequency?

7 What is their amplitude?

8 Make a mark on the thread about half way along the hanging part. Use brightly coloured ink for this. As the pendulum vibrates, watch the mark. In which direction does the mark move?

9 Fix a small cork or piece of Plasticine to the thread, near the lower end (but *not* resting on the floor). Watch the cork as the pendulum swings. Does it have energy? What kind of energy? Where does it get its energy?

Instructions: making more waves

1 Other vibrating objects can be used to produce waves. We can use a weight vibrating up and down on the end of a spiral spring; we can use a tuning fork; we can use a springy wooden stick or hacksaw blade held in a clamp. You could try using these to make waves in a thread. Check to see if these waves have a definite *frequency* and a definite *amplitude*. Look to see if energy is transferred and how it is transferred.

2 Get a piece of plastic tubing or hose-pipe several metres long; lay it on a smooth horizontal surface. Give one end of the pipe a flick, making *one* wave. Then flick at a steady rate, making a *train of waves*. Try flicking horizontally or vertically. Measure how long a single wave takes to travel along the pipe. What is the *speed* of the wave? Does the speed depend on how hard you flick the pipe?

3 Repeat step 2 using a length of rope or a 'slinky' spiral spring.

21.02 Transmission of waves

Figure 21.1 shows the thread used in the instructions on this page with arrows to represent the velocity of points on the thread at one instant of time.

Questions

1 What can be said about *all* the velocity vectors?

2 In what direction are the waves moving?

3 Describe the changes in the displacement and velocity of any one point on the thread while the bob makes one complete swing.

4 Do all points on the thread go through the same cycle of changes?

5 In what ways are the parts of the thread at C and H alike?

6 Name any other pairs of points that have the same relation as C and H.

7 What can be said about the vertical distance between C and H, and between the other pairs of points listed in the answer to question 6?

8 How many complete wavelengths are shown in the figure?

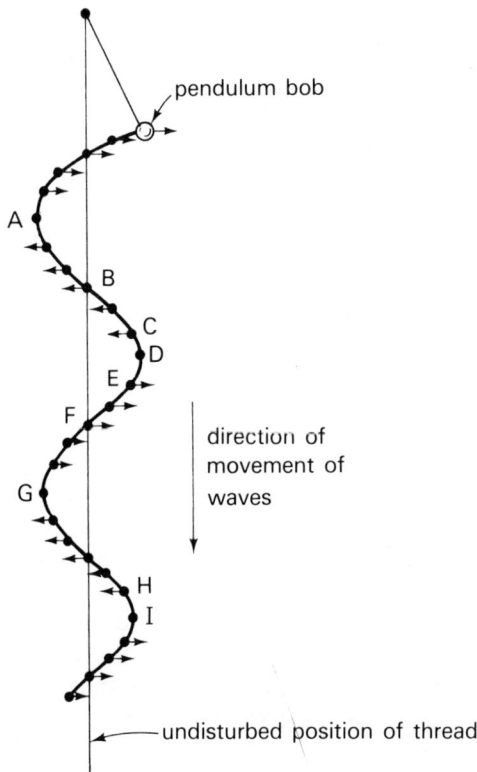

pendulum bob

A

B

C

D

E

F

G

direction of
movement of
waves

H

I

undisturbed position of thread

Fig. 21.1   Waves made by a vibrating pendulum.

The *ripple-tank* is a useful apparatus for studying waves. There are various designs of ripple-tank, but in all we have a shallow tray containing water. We make waves or ripples on the water surface and study their motion and behaviour. The bottom of the tray is transparent. Light shines through the water and is focussed to form an image of the water surface on a screen. We can see the waves as light and dark bands passing across the screen. Waves travel quickly on water, so most ripple-tanks have some way of making the waves *appear* to stand still, or move

very slowly. The lamp in the apparatus has a shutter in front of it, to make it flash very quickly, so quickly that you do not notice it is flashing. The effect is like that which you get with a cinema film or TV set when a cart with spoked wheels is seen running along a road. We know that its wheels are turning quickly and the spokes are moving quickly, but they appear to be still or moving very slowly, either forwards or backwards. The same effect may be seen in films of helicopters in flight. We call it the stroboscopic effect. Your teacher may be able to demonstrate this effect with a special stroboscopic lamp, which flashes very rapidly. If the lamp is made to flash at the same rate that a motor is turning, the motor appears to be still. Such lamps are used for studying the action of motors and other machines that operate at high speeds. The stroboscopic effect is used in the ripple-tank so that we can make the waves apparently stop or move slowly when we want to see exactly how they behave, in 'slow motion'.

The ripple-tank has a vibrator to produce waves. Usually it is electrically driven. If the vibrator is set vertically we get circular waves (fig. 21.2a). In the centre is the shadow of the vibrator, or source of waves. From this *point source*, waves spread out at the same speed in all directions, forming concentric circular waves. If the vibrator is a horizontal rod, the waves are partly circular and partly plane (fig. 21.2b). The bright rings and bands on the screen are the *crests* of the waves (fig. 21.3). Where the water surface is above its undisturbed level, it focusses the light, concentrating it on the screen in a bright band. The darker rings and bands correspond to the troughs, which scatter the light, leaving the screen dark. In fig. 21.2 the lines represent the dark bands on the screen, the troughs of the waves. All the water particles along a line are at the same stage in their motion; they all have the same displacement and velocity. Since they are all at the bottom of a trough, they have maximum displacement below the undisturbed level, and they have zero velocity.

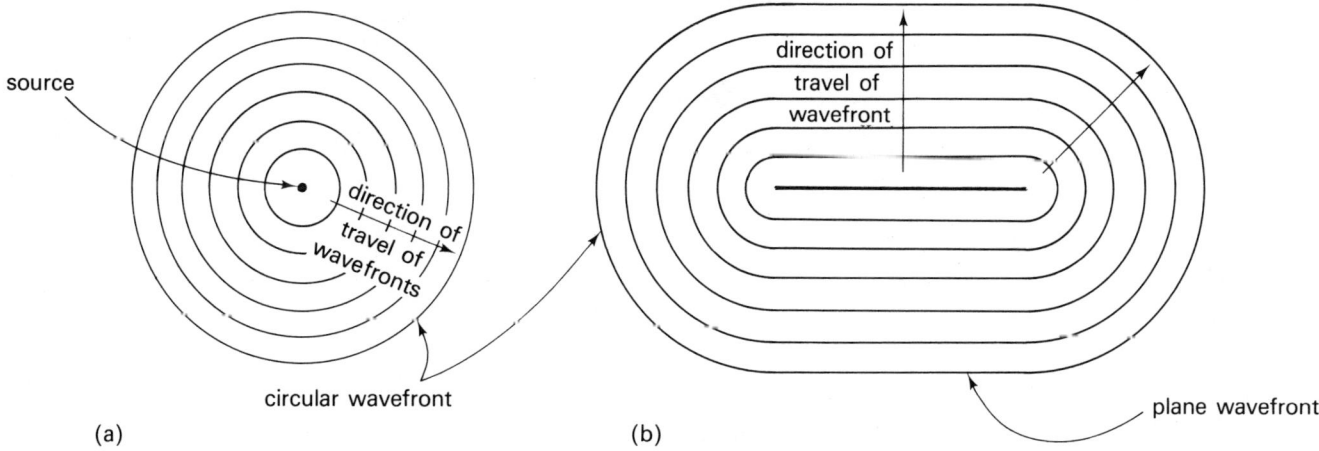

source

direction of
travel of
wavefronts

circular wavefront

(a)

direction of
travel of
wavefront

plane wavefront

(b)

Fig. 21.2   Waves produced in a ripple-tank: (a) by a point source; (b) by a horizontal rod.

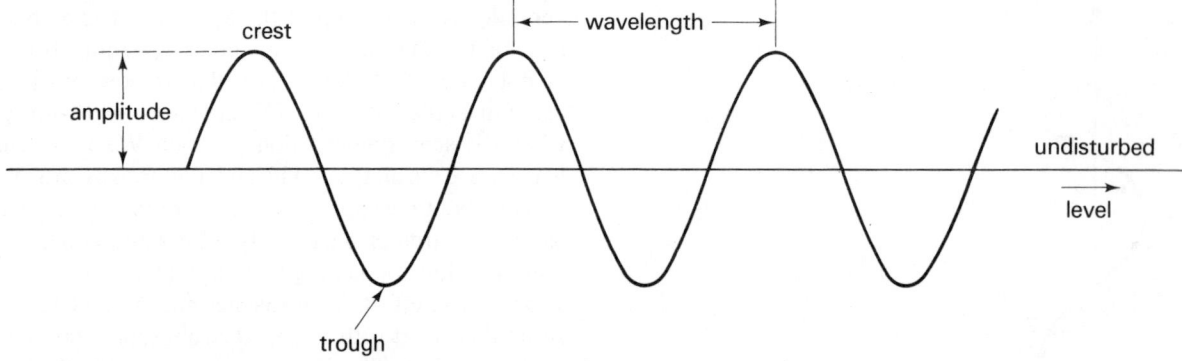

Fig. 21.3 The main terms used to describe waves.

Any line which runs through particles all at the same stage in their cycle is called a *wavefront*. Between the lines drawn in fig. 21.2, we could draw other lines representing any other stage in the cycle; these too are wavefronts.

Waves on water (fig. 21.3) are like the waves in thread (fig. 21.1) for in both, the particles move in a direction *perpendicular* to the direction of motion of the waves. The vibrator of the ripple-tank vibrates vertically, the molecules of water vibrate vertically, but the waves travel horizontally.

Instructions: investigating another type of wave motion
1 Make a toy telephone from two used food cans and a few metres of thread or thin string. Make a small hole in the centre of the bottom of each can. Pass the thread through and knot it to stop it coming out (fig. 21.4).
2 Two students stand sufficiently far apart so that the thread is pulled fairly tightly between the cans.
3 One student places the can to his ear. The other student speaks into the other can.

4 One student taps on the bottom of one can. The other student listens, with his ear to the other can. What way does the bottom of the can move when made to vibrate? What forces does the bottom of the can exert on the string? Is energy transferred from one can to the other?

There must be waves passing along the thread, but these must be different from the waves in the thread of the instructions on page 210, and in the hosepipe. We cannot see the waves in the string as they pass from one can to the other. The next experiment makes it easier to imagine what these waves are like.

Instructions: investigating waves in a slinky spring
1 In the instructions on page 210, step 3, you were asked to make waves in a slinky spring like those shown in fig. 21.5a. These are like the waves in the thread and in the ripple-tank. What can we say about the direction of movement of the particles and the direction of movement of the waves?

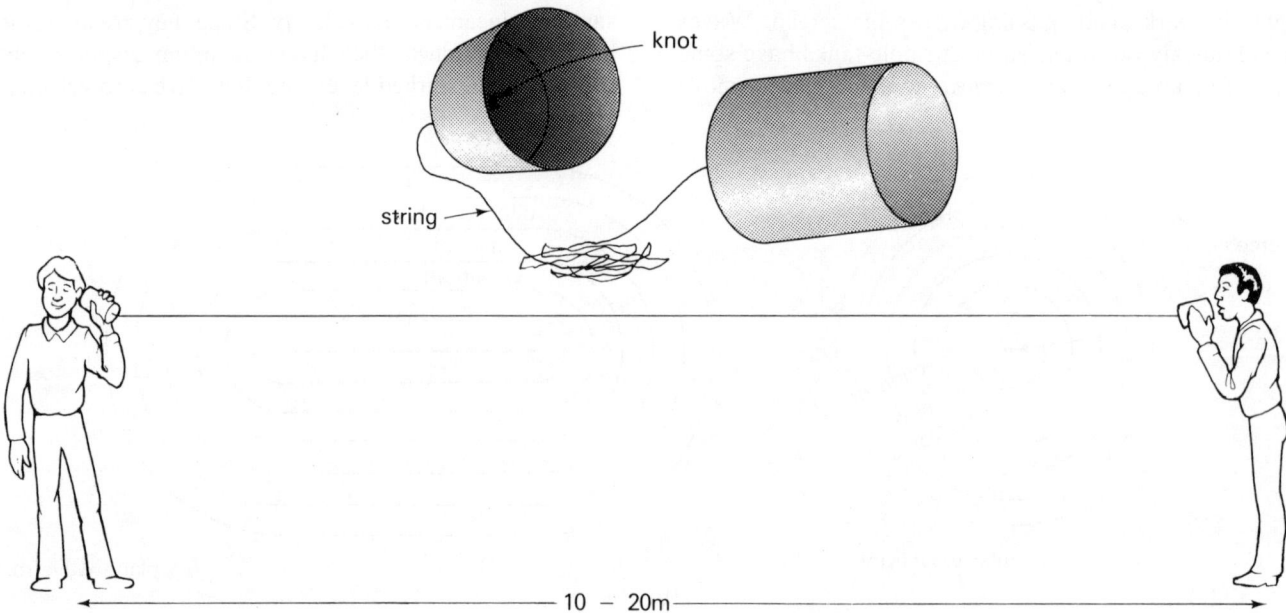

Fig. 21.4 Making a toy telephone.

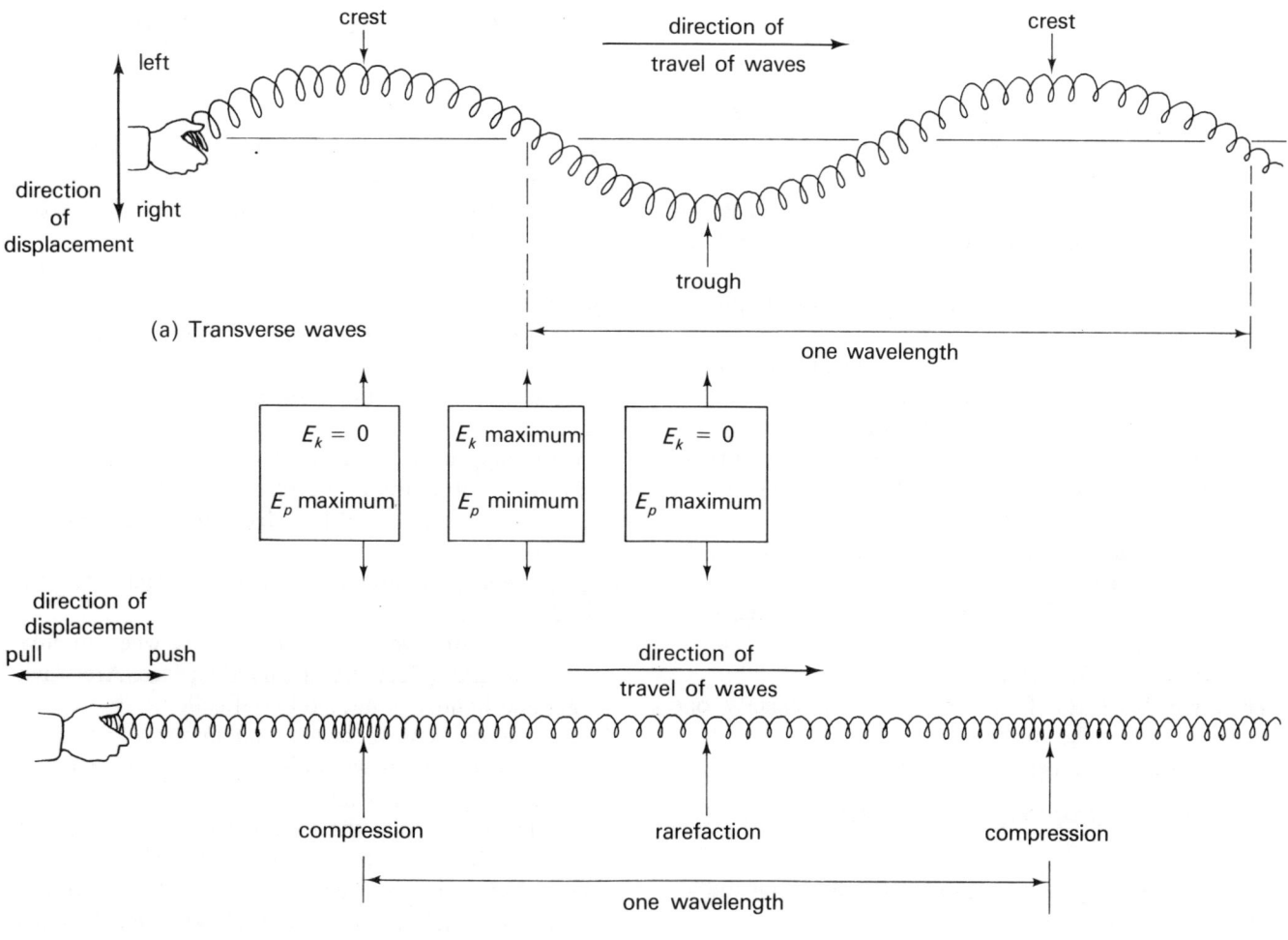

left

crest

direction of
travel of waves

crest

direction
of
displacement

right

trough

(a) Transverse waves

one wavelength

$E_k = 0$

E_p maximum

E_k maximum

E_p minimum

$E_k = 0$

E_p maximum

direction of
displacement
pull   push

direction of
travel of waves

compression

rarefaction

compression

one wavelength

(b) Longitudinal waves

Fig. 21.5 Waves in a slinky spring.

2 Make some transverse waves with the slinky spring, to remind yourself what they are like.

3 Put the slinky spring on a smooth surface; hold it at one end and make waves by pushing and pulling the end of the slinky spring, as in fig. 21.5b. Watch the waves moving along the spring.

4 Tie a piece of ribbon on to one turn of the spring about half way along, or fix a small piece of Plasticine there. Watch it as you make waves.

5 In which direction do the molecules of the spring move as waves pass along the spring? In which direction do the waves move?

If the particles and waves move in the same direction, we have longitudinal waves. The two types of waves seem to be very different, but they are alike in some ways. They both transfer energy. In both types the particles show the same alternation between E_p and E_k. In transverse waves the particles at the crests and troughs have maximum E_p (they are at their greatest distance from their undisturbed position), and zero E_k (zero velocity). In longitudinal waves, some parts of the spring are compressed more than

average (a compression) and other parts are compressed less than average; they are extended or stretched (a rarefaction). At the points of compression or rarefaction the spring has maximum E_p; it has the ability to return to its normal average compression. Between these points the spring is moving, being extended or compressed; it has maximum E_k. At the points of maximum compression and rarefaction it is not moving; it has zero E_k. In both types of wave motion, a particle alternates between a state of zero E_k and maximum E_p, and a state of maximum E_k and minimum E_p.

Comparing the two types of waves in fig. 21.5, we can find positions along the spring where two particles are in exactly the same part of their cycle. These two points are separated by one wavelength. In transverse waves we measure the wavelength from one crest to the next, from one trough to the next, or from any point to the next point at the same stage. In longitudinal waves we measure the wavelength from one compression to the next, from one rarefaction to the next, or from any point to the next point at the same stage.

213

Instructions: measuring the speed of waves

1 Use a slinky spring and stop-watch. You can use a plastic hosepipe instead of the slinky spring.

2 Lay the spring straight on a smooth surface. Make a single transverse wave and get your partner to measure the time it takes to travel a measured distance.

3 Calculate the speed of the waves. If the measured length is s, and the time taken is t, the speed is calculated as s/t.

4 Make waves of different amplitudes. Does amplitude affect speed?

5 Make waves of equal amplitude but different wavelength. Does wavelength affect speed?

6 Now that you know the speed of waves in your spring, calculate how many waves you must make per second so that exactly two waves fit into the length of the spring (it will then look like fig. 21.5a).

7 Try making waves at this frequency to see if your calculation is correct.

8 If you want to fit three waves into the spring, what must you do? See if you can do it.

9 Try to fit in four waves.

10 What must the frequency be to fit exactly one wavelength into the spring?

The results obtained above lead to an equation:

wave speed = frequency × wavelength
or, $c = f\lambda$

This equation applies to all types of wave motion and we shall use it often.

21.03 Detection of waves

Detection of waves is really the detection of the energy they are transferring. We can touch the thread and feel its motion. We can hear the sounds coming from the toy telephone. We can feel the vibration of one end of the slinky spring when someone vibrates the other.

When there is an earthquake, a great amount of potential energy of the Earth's crust is suddenly converted to kinetic energy. This energy is transferred through the crust by wave motion. Some waves are transverse: the ground rises and falls. Some waves are longitudinal: the ground shakes with a horizontal motion. The energy released at the centre of an earthquake can destroy buildings tens of kilometres away. If the earthquake is beneath the sea, the disturbance causes huge tidal waves, the tsunami. Those who have experienced the damage to buildings caused by earthquakes do not need to be reminded that wave motion is a method of transfer of energy. Wave motion in the sea represents an enormous store of energy; methods are being devised to convert this energy into a more useful form, such as electricity (see Project 1, at the end of this chapter). We shall return to the subject of wave detection later.

21.10 The behaviour of waves

21.11 Reflection

The easiest way to study reflection is to use a ripple-tank. A vibrating rod produces plane waves and a reflecting surface is placed at an angle to the wavefronts. The result is shown in fig. 21.6a.

Questions

1 What happens to the direction of the waves when they are reflected?

2 What happens to the wavelength when waves are reflected?

3 What happens to frequency when waves are reflected?

4 In fig. 21.6b, circular wavefronts are being reflected from a plane surface. What shape have the reflected wavefronts?

5 From what point do the reflected wavefronts appear to come?

6 In fig. 21.6c, we see what happens when circular wavefronts are reflected at a curved surface. Are wavelength and frequency altered by reflection?

7 A wavefront represents a certain amount of energy. If we could measure this we could state how much energy it represents, in joules per centimetre length of wavefront. In the figure we see that the reflected wavefronts become shorter and shorter as they move from the reflector toward point F. What must happen to the energy of the wavefront as the wavefront becomes narrower? What might happen to the amplitude of the waves as they approach F? What effect does this have on the appearance of waves as seen on the screen of a ripple-tank?

8 After the waves have passed through point F, the wavefronts appear to radiate from F. What happens to the amplitude of the waves beyond F?

9 If we put a vertical vibrating rod at F, we could make circular waves, which radiate outward from F. What will these waves look like after reflection from the curved reflector?

21.12 Interference

The demonstrations of reflection showed that waves can pass through each other and do not affect each other. You can see the same effect if you and your partner send longitudinal waves from opposite ends of a slinky spring. The waves meet near the centre of the spring, pass, and continue their separate journeys to the opposite ends of the spring. In fig. 21.7 we see what happens in a ripple-tank when there are two sources of waves. These are produced by a two-pronged vibrator, so both sets of wavefronts have the same frequency, wavelength and amplitude. We can see that each set spreads as a train of concentric circular wavefronts. Each set is unaffected by

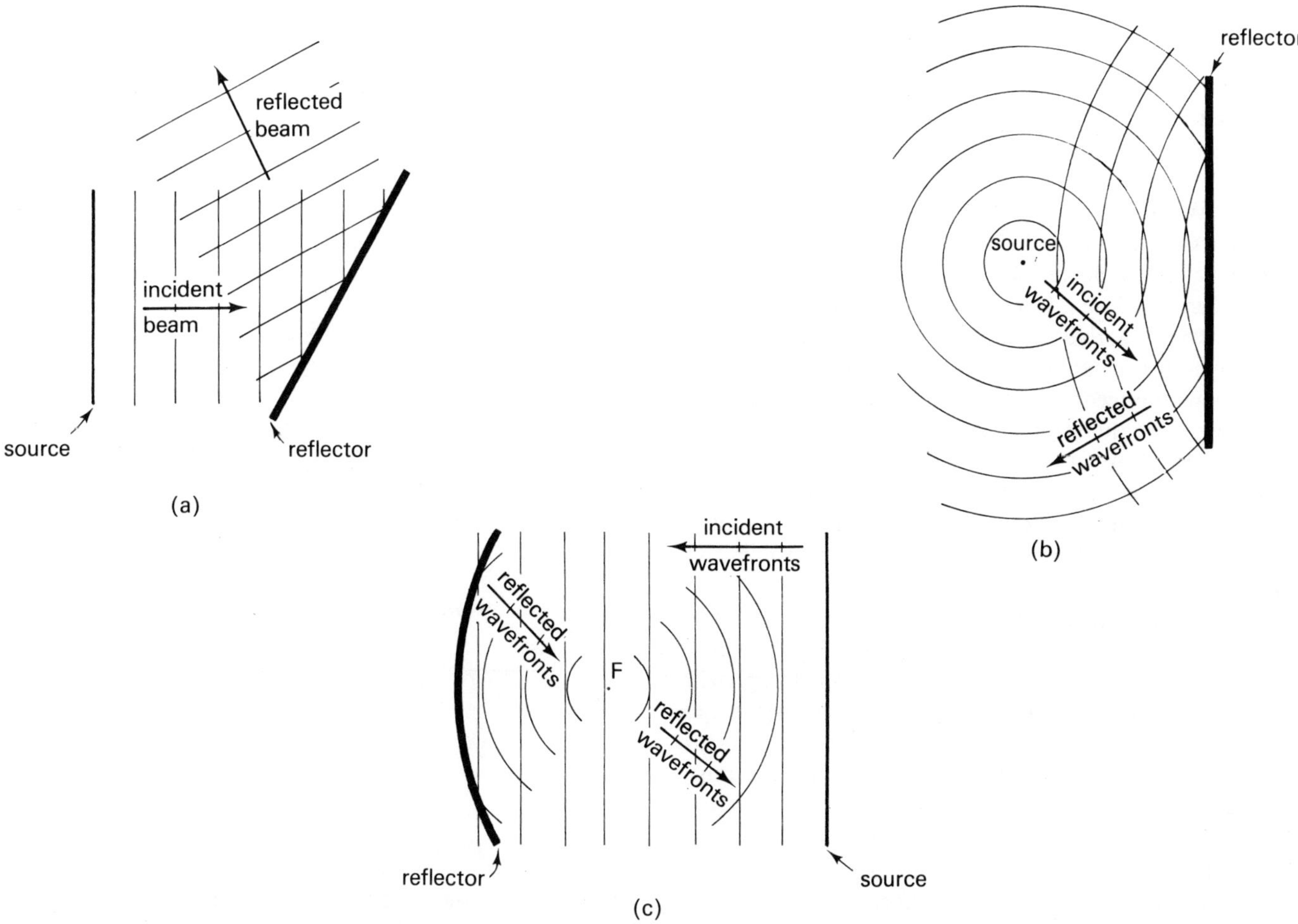

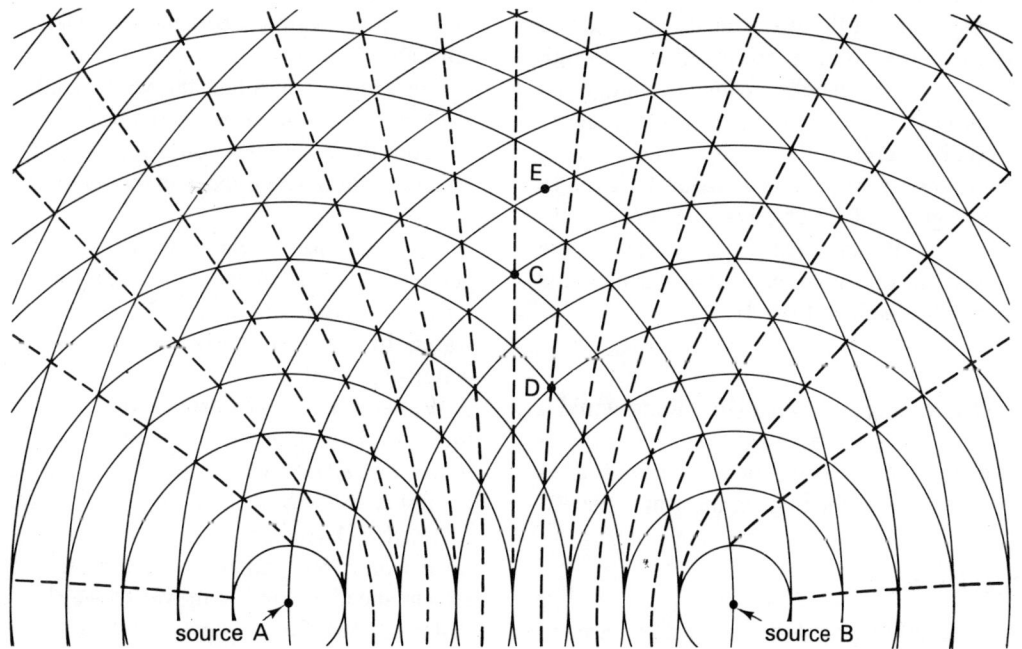

Fig. 21.6 The reflection of waves in a ripple-tank: (a) reflection of plane wavefronts at a plane surface; (b) reflection of circular wavefronts at a plane surface; (c) reflection of plane waves in a curved surface.

Fig. 21.7 Interference between two sets of circular wavefronts in a ripple-tank.

the other. At any point on the water surface the molecules are acted on by two forces, one force caused by the wave from one source, the other force caused by the wave from the other source. Let us see how these two forces combine together. In the figure, waves are spreading outwards in circles from A and from B. Point C is equidistant from A and B. When a crest from A reaches C, a crest from B reaches C at the same instant. The two crests combine making a very high crest at C. We say the waves interfere *constructively*. Later, when a trough from A arrives at C, a trough from B arrives at C at the same instant. The two troughs combine, making a very low trough at C. Again we have constructive interference; the two waves combine crest with crest and trough with trough, so that the water surface at C rises very high and sinks very low. If an insect was floating on the water at C, it would be shaken up and down very violently. The same effect is found at all points which are equal distances from A and B, as marked by the dashed line running through C.

Point D is further from A than from B, but the extra distance is exactly one wavelength. At point D there is constructive interference, crest with crest and trough with trough. The same occurs at all points which are one wavelength further from A than from B. These points are indicated by the dashed line drawn through D. The other dashed lines on the diagram show the other points of constructive interference. At all points along these lines the distance from A is a *whole number* of wavelengths more or less than the distance from B.

In the regions between the regions of constructive interference we find the opposite effect. For example, point E is half a wavelength further from A than from B. When a trough from A arrives at E, a crest arrives from B; the two cancel out, leaving the water at its undisturbed level. Later when a crest from A arrives at E, a trough arrives from B; the two cancel out, leaving the water at its undisturbed level. Here we have *destructive interference*: the two waves combine crest with trough and trough with crest, so that the water is always at its undisturbed level. In the diagram the regions of destructive interference are half-way between the dashed lines. In these regions the distance from A is an odd number of half-wavelengths more or less than the distance from B.

Questions
1 If an insect was floating at E, what conditions would it find?
2 If the water is calm at points on the dotted lines, does this mean that no forces are acting on the water there?

21.13 Refraction

A sheet of glass placed in a ripple-tank makes an area of water that is less deep than normal. If the layer of water above the sheet is thin enough, we get an interesting effect

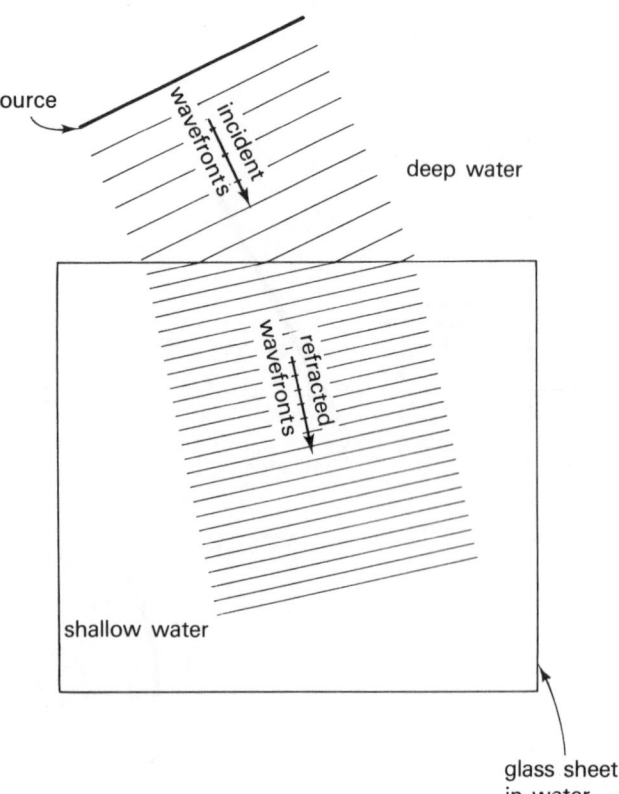

Fig. 21.8 *Refraction of plane wavefronts in a ripple-tank.*

(fig. 21.8). Incident wavefronts cross the boundary between deep water and shallow water. At the boundary they change direction; they are refracted.

Questions
1 Which is greater, the angle of incidence or the angle of refraction?
2 Does refraction affect the wavelength?
3 Does refraction affect frequency?
4 From the answers to questions 2 and 3, what can we conclude about wave speed in the region of shallow water?

The reduction of wave speed explains why refraction occurs. In fig. 21.9 we see a train of wavefronts being refracted at the boundary AD between deep and shallow water. In the deep water, $\lambda_1 = c_1/f$; in the shallow water, $\lambda_2 = c_2/f$.

Wavefront AB has just reached the boundary. During the period of time for one wave to pass ($t = 1/f$), the wavefront AB moves forward to the position at present occupied by wavefront CD. In this time end B moves forward to D a distance λ_1, with speed c_1, since it is still in deep water; but end A, in shallow water, moves with a slower speed, c_2, and so moves forward only as far as C, a distance λ_2. Points between A and B move forward intermediate distances and, as a result, the front AB moves forward to CD.

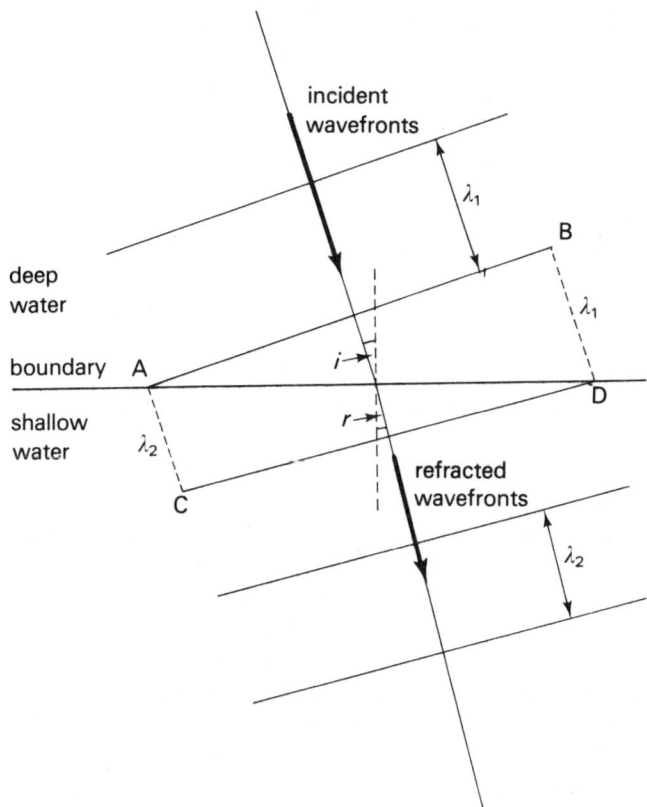

Fig. 21.9 Calculating the refractive index (see text).

In the triangle ABD, angle BAD = i.
$$\sin i = BD/AD = \lambda_1/AD$$

In the triangle DCA, angle CDA = r.
$$\sin r = CA/AD = \lambda_2/AD$$

We represent the amount of refraction by a ratio called the *refractive index, n*:

$$n = \frac{\sin i}{\sin r} = \frac{\lambda_1}{AD} \times \frac{AD}{\lambda_2} = \frac{\lambda_1}{\lambda_2}$$

But $\lambda_1 = c_1/f$ and $\lambda_2 = c_2/f$, so

$$n = \frac{\lambda_1}{\lambda_2} = \frac{c_1}{f} \times \frac{f}{c_2} = \frac{c_1}{c_2}$$

$$n = \frac{c_1}{c_2} = \frac{speed \ in \ deep \ water}{speed \ in \ shallow \ water}$$

If the depths of water are fixed, this ratio is a constant. The refractive index depends on the speeds in the two regions of water. If we know these we can always calculate how much refraction will occur at the boundary between them.

21.14 Diffraction

If a metal barrier is put in a ripple-tank, waves which strike it are reflected. Behind the barrier the water is calm; there is a *shadow* (fig. 21.10a). If you look closely at the

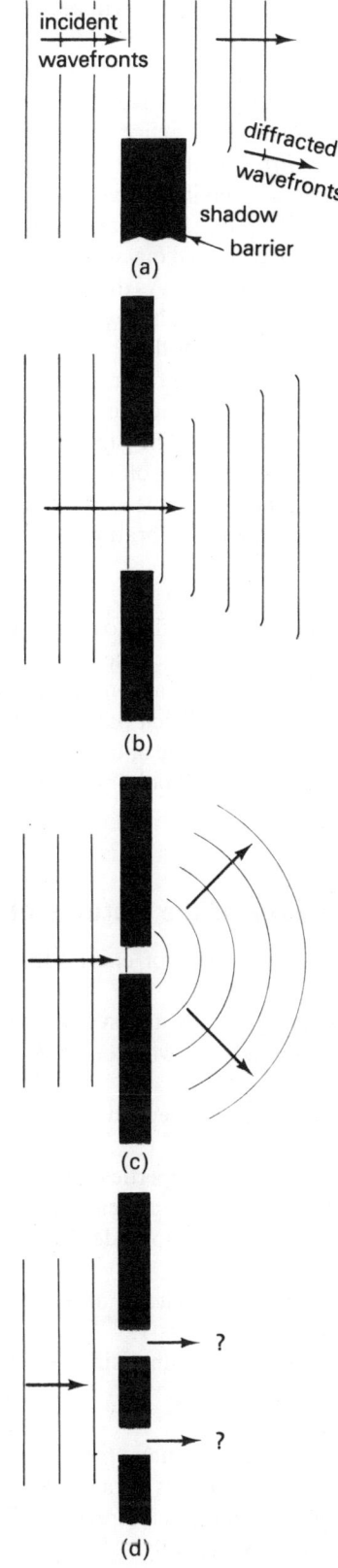

Fig. 21.10 Diffraction of wavefronts in a ripple-tank: (a) around a barrier; (b) through a wide gap; (c) through a narrow gap; (d) through two narrow gaps.

217

wavefronts on a ripple-tank you will see that the edge of the shadow is not exactly perpendicular to the wavefronts. The ends of the wavefronts turn 'round the corner' and spread slightly into the shadow region. This is a change of direction. A change of direction caused by passing a barrier is called *diffraction*.

Diffraction can be seen if we have a barrier with a wide opening in it (fig. 21.10b). After passing through the gap, the plane wave fronts spread out slightly at both ends. In fig. 21.10c the gap is narrow. As the wavefronts reach the gap they are plane. After they have passed through the gap they spread out widely. With a very narrow gap they become circular and radiate from the gap. These waves are weak, for only a small fraction of the wave energy is able to pass through the gap. The gap acts as a point source of waves. To an insect floating on the water to the right of the barrier, the effect is the same as if we replaced the gap by a vertical vibrating rod, vibrating rather gently.

Question

Suppose we have a barrier with two narrow gaps, as in fig. 21.10d. Plane wavefronts reach the barrier. The barrier is parallel to the fronts, so that a front reaches both gaps at exactly the same instant. What pattern of waves do you expect to see on the other side of the barrier? Try to work out the answer first. Then, if possible, use a ripple-tank to see if you are right.

21.15 Summary of the features of waves

Wave motion is a transfer of energy from a vibrating source.

Waves can be of two kinds, transverse or longitudinal.

Waves have amplitude, a, wavelength, λ, speed, c, and frequency, f.

Frequency and amplitude depend on the frequency and energy of the source.

Speed depends on the properties of the substance or medium in which the waves are carried.

Speed is not affected by amplitude, frequency or wavelength.

The equation $c = f\lambda$ applies to all kinds of waves.

Waves can be reflected: $i = r$.

Waves can be refracted: $\sin i / \sin r = n$ (a constant, called the *refractive index*).

Waves can be diffracted.

Waves from two or more sources (or from a single source and its reflected images) pass through each other without loss of energy or change of direction. They interfere constructively or destructively.

Now we will study sound and light, making use of what we already know about them, to see if they have the features listed above. If they have, we can assume that they are a type of wave motion.

21.20 The nature of sound

21.21 Information from previous investigations

This is what we have already discovered about sound in chapter 7:

It is produced by vibrating sources.
It can be reflected.
Its speed is about 346 m/s, but varies in different materials.
It involves transfer of energy.
Loudness depends on amplitude.
Pitch depends on frequency.
It probably consists of longitudinal waves (page 212).

Questions

1 If the velocity of sound in air is 340 m/s, and assuming that we can use the equation discovered in our study of waves, what is the wavelength of (a) a high-pitched whistling sound, frequency 10 kHz, such as might be produced by a jet engine, (b) the middle C note, frequency 256 Hz, and (c) the hum, frequency 50 Hz, often heard coming from electrical equipment operating from the mains supply.

2 Refer to fig. 21.9, but think of the waves as sound waves in air. *If* sound can be refracted, we can predict how it *ought* to behave. In questions 2 to 6, we will work out what ought to happen. Above the line AB the air is warm; the speed of sound is, $c_1 = 345$ m/s. Below AB the air is cool; the speed of sound is $c_2 = 340$ m/s. A sound of frequency 1 kHz passes from the warm air, across the boundary AB to the cool air. What is its wavelength in (a) the warm air, and (b) the cool air?

3 Calculate the refractive index for sound as it crosses the boundary AB (question 2).

4 If the sound waves in warm air reach the boundary with an angle of incidence $i = 25°$, what is the angle of refraction?

5 If a person walks from the warm air to the cool air, does he notice a change of pitch as he crosses the boundary?

6 Does he notice any change in the sound as he crosses the boundary? If so, what change does he notice?

The answers to the questions above tell us that *if* sound is refracted (we have not yet proved that it *can* be), the change of direction is likely to be very small for the range of temperature we usually get in air, under natural conditions. It is not surprising that we do not notice sound being refracted in everyday life. Also it is not easy to design experiments to detect and measure refraction of sound.

Yet there is an effect of refraction that we notice sometimes. It is often easier to hear sounds coming from a distance in the evening or at night. The effect is most often noticed when there is no wind to mix the layers of air. Then, at night, the air nearest the ground is cool. The air

Fig. 21.11   Refraction of sound on a still, cool evening.

above is warmer. Sound travels faster in the upper air. As sound travels into warmer layers (fig. 21.11), it is continuously refracted away from the normal, gradually changing its direction. The wavefronts are refracted toward the ground instead of spreading away higher and higher. Sounds coming from a distance are brought down towards the ground, so are heard clearly. By day, the air nearer the ground is usually warmer than the air above, the reverse effect happens, and sounds from a distance are less easily heard.

21.22 Interference

This effect can be easily demonstrated. Two loudspeakers are placed outdoors, well away from buildings. The loudspeakers are side by side, facing the same direction, and about 1 m apart. They act as two sources, as shown in fig. 21.7. The speakers are both connected to an audio-frequency signal generator set to give a sound of frequency about 300 Hz.

Questions

1 Why is the experiment done outdoors and away from buildings?
2 Describe what you expect to hear if you walk slowly across the area in front of the loudspeakers.

Evidence such as this gives further support to the idea (or hypothesis) that sound consists of waves having the same properties as the waves on the ripple-tank.

In the next investigation we produce another type of interference.

Instructions: investigating beats

1 Use two tuning forks of the same frequency, or two stretched wires (fig. 7.2) tuned so that they sound the same pitch, one should make the other resonate; test this.

2 Alter the frequency of one of the tuning forks slightly, by putting a piece of Plasticine on each of its prongs, near their tips. This reduces the frequency of the fork by a few hertz. If you are using wires, increase the length of one wire by a few millimetres.
3 Sound the two forks or wires together. What do you hear?
4 Use a microphone and oscilloscope to display the waves. What happens to the amplitude of the trace on the screen?
5 Count the number of beats per second.
6 Remove half of the Plasticine from both prongs of the fork; this will bring its frequency a little nearer the frequency of the other fork. With wires, decrease the length of the longer wire to make it nearer its original length. Sound both forks or wires together. Count the number of beats per second. Has the rate of beating changed?

When regular notes of steady pitch come from two similar sources, but they are of *slightly* different frequency, they interfere, causing beats. Figure 21.12 shows what happens. The diagram looks like several trains of waves, but we think that sound waves are longitudinal waves, which do not look like this. The diagram is really a graph showing the displacement of air molecules from their undisturbed positions. Their displacement is actually longitudinal but, by drawing it as a graph we make the waves look like transverse waves, which are easier to understand.

Graph A shows the displacement of a molecule when a note of 200 Hz is sounded. The length of time covered by the graph is 0.1 s. Graph B shows the displacement of the same molecule if the 200 Hz note is stopped and a note of 220 Hz is sounded. Again the length of time of the graph is 0.1 s.

Graph C shows the displacement of the same molecule when both sounds are made at the same time. Then the

219

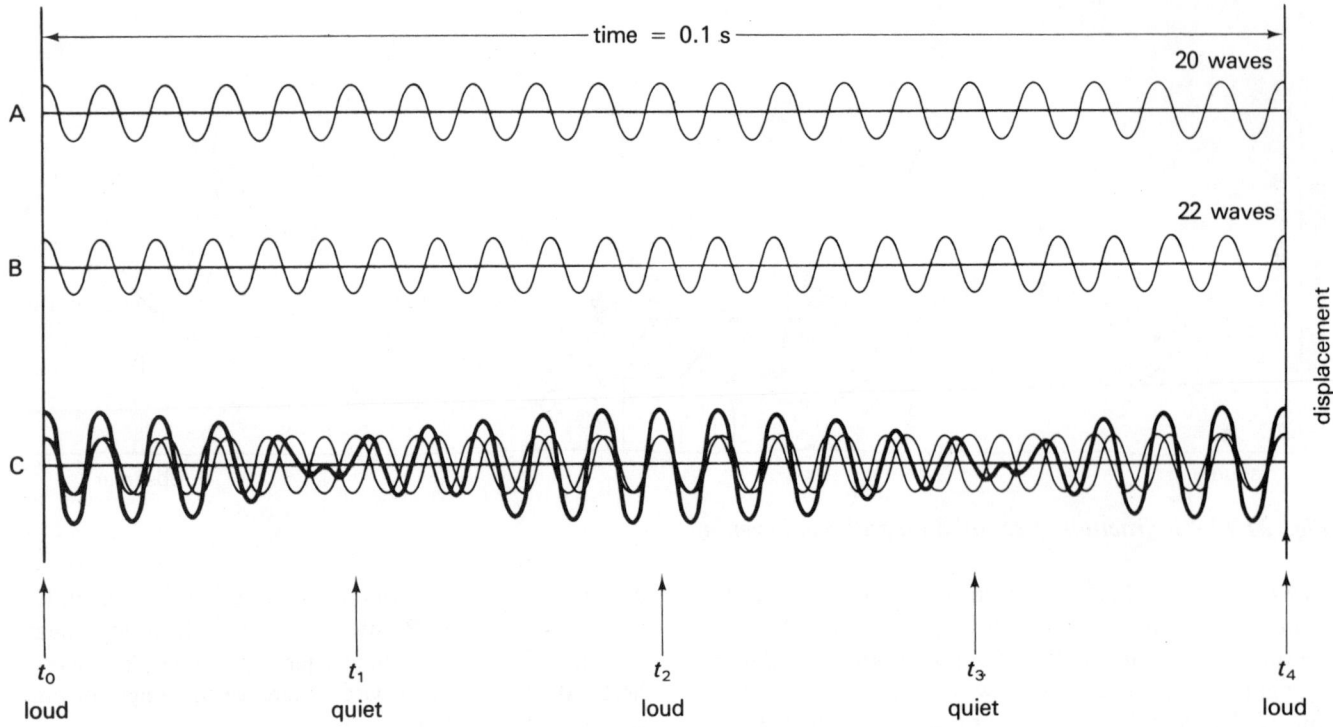

Fig. 21.12   Interference between two wave-trains of slightly different frequencies, producing beats.

molecule is acted on by forces from both sources. We can work out what happens to it by adding the displacements due to each force. The thin lines on graph C are the displacements from graphs A and B. The thick line is obtained by adding the displacements at each point in time. At t_0 the forces are interfering constructively; both act on the molecule in the same direction. The molecule is displaced a great distance; the amplitude of vibration is large; the sound is loud. At time t_1, sound A has made 5 vibrations but sound B has made $5\frac{1}{2}$ vibrations. Now the forces are in opposite directions; they interfere destructively; their effects on the air molecule cancel out; it does not move; no sound is heard. At time t_2, the sounds are in step again, interfering constructively, and producing a loud sound. At t_3, they are out of step again, and there is silence for a moment. At t_4 the sound is loud once again. So the sound becomes louder and softer at regular intervals; we hear beats. In the figure we have two complete beats in 0.1 s. We begin half-way through a loud stage and end half-way through a loud stage, with two quiet stages and one loud stage between. In 1 s there would be 20 beats. The rule is that the number of beats per second is the same as the *difference* between the frequencies of the two notes. With notes of 200 Hz and 220 Hz, the beat frequency is 20 Hz.

Questions
1 Explain why the beat frequency decreased at step 6 of the previous investigation.

2 If you were given a tuning fork which you knew had a frequency of exactly 440 Hz and a second fork which had a frequency close to but not exactly 440 Hz, how could you use beats to find the frequency of the second fork?

Beats are very useful when you need to tune two strings or two similar musical instruments to the same frequency. A difference of frequency of 2 Hz is too small to be detectable by ear, if the notes are sounded one after the other. But if you sound the notes together, and listen for beats, you can easily hear the beats at 2 Hz. Close tuning to a beat frequency of 1 Hz or less means that the two notes differ in frequency by less than 1 Hz.

21.23   Diffraction of sound

Experiments with the ripple-tank show that, for diffraction to occur, the gap must be small compared with the wavelength (fig. 21.10). If the gap is wider than the wavelength we obtain a beam, which spreads slightly into the shadow but does not radiate widely in all directions from the gap. If we could demonstrate diffraction of sound, we would have further evidence that sound is wave motion.

Instructions: demonstrating diffraction of sound
1 The apparatus is shown in fig. 21.13. The size of the gap can be adjusted by arranging the thick books. It should measure about 10 cm × 10 cm.

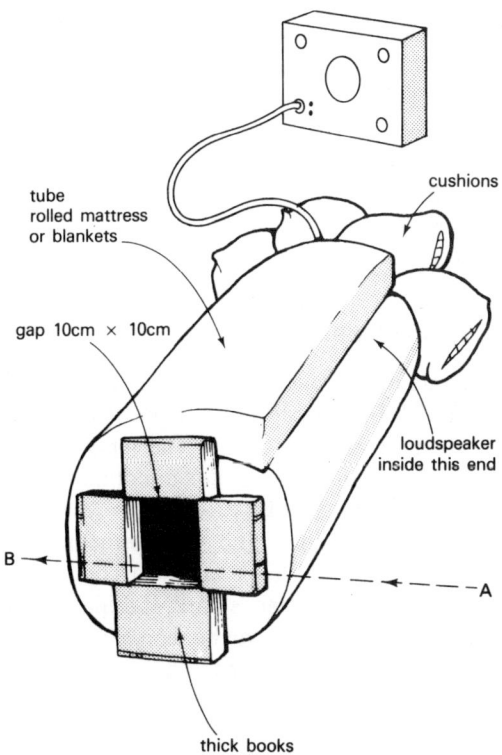

tube
rolled mattress
or blankets

cushions

gap 10cm × 10cm

loudspeaker
inside this end

B ←

← —A

thick books

Fig. 21.13   Diffraction of sound.

2 Set the audio-frequency generator to a frequency of 10kHz. Move your ear slowly from A to B. Describe what you hear coming from the gap. You may hear some sounds coming through the cushions and being reflected in the room; try to ignore this.

3 Repeat with frequency of 1kHz. What do you hear?

4 Repeat with frequency of 250Hz. What do you hear?

5 Put your ear directly in front of the gap, with sound of frequency 10kHz. Try placing small objects between the gap and your ear. Use a book, a ruler, a finger. Is it possible to reduce the volume of sound reaching your ear? Repeat using the lower frequencies.

6 Try reflecting the sound by placing a small object such as a book or a ruler in front of the gap. With your ear at point A, turn the reflector to obtain the loudest sound. Does this work equally well for all frequencies?

We have been able to explain the result obtained above by making use of what we know about the diffraction of waves in the ripple-tank. This is further confirmation of our hypothesis that sound is wave motion. When sound reaches a small gap('small' meaning'of width about equal to or less than the wavelength') it is diffracted. This explains why sounds are easily heard through windows and doorways. Most sounds in everyday life have frequency of 500Hz or more, which gives them a wavelength of 0.5m or more. The average wavelength of the sounds of the human voice is about 2m. Such sounds are easily diffracted. If a person is indoors speaking, the sound of his voice is diffracted in all directions after it has passed through the window. You do not need to be in a direct line of sight to be able to hear the person clearly. The same applies to sounds passing in through windows or doorways. Objects in the room, such as furniture and people are usually much less wide than the wavelengths of the sounds, and the waves are diffracted around these, so no shadows are formed (step 5). If somebody speaks to you and there is a pile of books between you and the person, the sound of the person's voice is diffracted around the books; you can hear the sound clearly. The same applies outdoors; tree trunks are no barrier to sound. You may not be able to *see* a person hiding behind a tree trunk but, if he makes a sound, you can easily hear him.

It follows from this that to obtain a sound shadow we *must* have objects larger than the wavelengths of sound. Objects must be several metres wide. They must also be firm enough to reflect or absorb the sound and not transmit it. A building makes a good sound barrier, provided there are no small gaps such as windows and doorways through which some sound can be diffracted.

Instructions: using a building as a sound barrier
1 Choose a corner of a building which has no windows or doorways. It should be at some distance from other buildings, to eliminate reflections.
2 Place a radio set close to one wall, and switch on to a music programme.
3 Walk around the corner so that the sound of the radio is very faint. You are in the sound shadow of the building.
4 Slowly move towards the corner. In what way does the quality of the music change as you slowly come out of the sound shadow? Which notes are lost when you are in the shadow? Which ones seem to be most easily diffracted around the corner?

Questions
1 If you are listening to music played by a high-fidelity record-player, with high quality speakers, where should you sit to hear the music perfectly?
2 If a band is playing on the other side of town, you can hear it, but the drums and low-pitch instruments sound much louder than the other instruments. Why is this?
3 You are creeping about in pitch darkness with a companion. You wish to speak to your companion, but do not want other people to be able to locate where you are standing, by locating the sound of your voice. Is it better to speak in your normal voice, but quietly, or is it better to whisper?

We have now studied sound in some detail and have found that in all our investigations it behaves as wave motion. Our hypothesis has been confirmed by every experiment and we have been able to explain many of the effects we notice in everyday life. We have also noted some points about diffraction that need to be summarised:

If waves reach a gap and the gap is small, the waves are diffracted from the gap.

If waves reach a reflector and the reflector is small, the waves are diffracted around the reflector; little reflection occurs.

Only a large object can produce a shadow, or cause reflection.

In the above statements, the word 'small' means 'about the same size as a wavelength, or smaller.' The word 'large' means 'larger than two or three wavelengths.' Whether a given object or gap is considered to be large or small depends on the length of the waves concerned.

21.30 The nature of light

21.31 Information from previous investigations

In chapter 5 we found that light is reflected and refracted; it has two of the properties of waves. Energy is required to produce light, so we may assume that there is transfer of energy when light is given off. These facts alone are not enough to *prove* that light consists of waves. An alternative hypothesis is that light consists of particles given off from the light source. The energy given to these particles by the source makes them travel in straight lines, according to Newton's first law, unless deflected. If they hit a smooth surface they bounce off (are reflected) like a bouncing ball and the angle of incidence equals the angle of reflection. If they enter a surface, for example passing from air to glass,

forces at the surface could make them change direction (refraction). Thus we can explain energy transfer, travelling in straight lines, reflection and refraction by either of two very different hypotheses: the wave hypothesis and the particle hypothesis. Neither hypothesis can be rejected on the evidence we have gathered so far. We must look for further evidence and hope that it will be conclusive.

21.32 Interference

To investigate interference of light we need the apparatus shown in fig. 21.14. The slits are very narrow; the distance between them is 0.5 mm or less. The lamp has a single vertical filament, but better results are obtained if a laser (page 273) is used instead.

The room is darkened and the lamp switched on. Several vertical bright lines are seen on the screen. Between the lines the screen is dark. If a filament lamp is used, about five bright lines can be seen; if a laser is used it is easy to see ten or more lines.

Questions
1 If light consists of particles that travel in straight lines away from their source, what would you expect to see on the screen?
2 How can you explain the fact that there are more than two bright lines on the screen?

It is hard to imagine two *particles* of light arriving at one point on the screen and cancelling each other to make

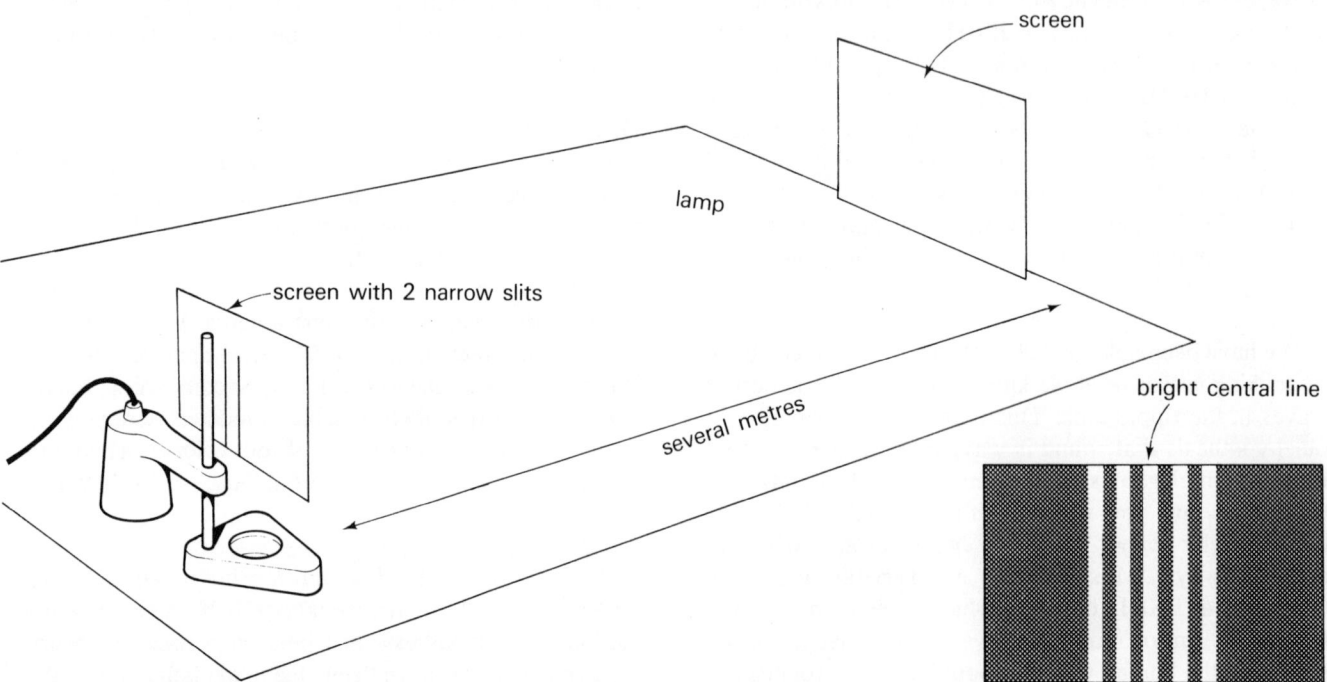

Fig. 21.14   Diagram showing the main parts of the apparatus used for demonstrating interference of light.

darkness. The particle hypothesis does not fit the facts of the experiment. Our knowledge of waves makes it easy to explain how two waves can arrive at the screen and produce darkness. The results of the experiment support the wave hypothesis. This experiment, first performed in 1801 by Thomas Young, was one of the first convincing demonstrations that light has the properties of waves.

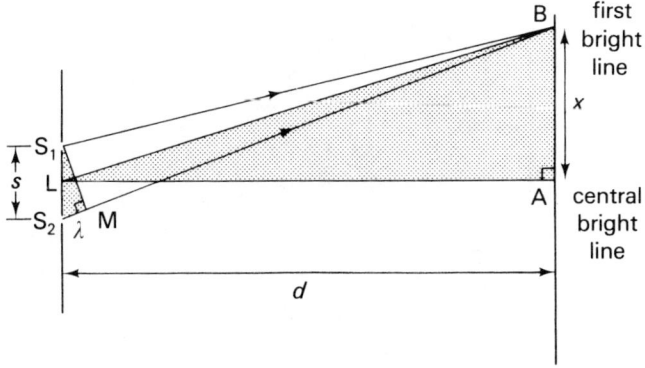

Fig. 21.15   Calculation of wavelength of light from results of interference experiment.

We can go further, and use the experiment to measure the wavelength of light. In fig. 21.15, light from two slits, S_1 and S_2 travels to the screen. Point A is equidistant from both slits; there we get constructive interference, making a bright line of light. The next bright line is at B; there the path from S_2 to B is exactly one wavelength longer than the path from S_1 to B. The two wave trains interfere constructively. In the figure a line S_1M is drawn perpendicular to S_2B. The distance S_2M is almost exactly equal to the difference between S_1B and S_2B. This is one wavelength, and S_2M has been marked 'λ' to indicate this. In the triangles S_1MS_2 and LAB,

$$\angle S_2S_1M = \angle BLA$$
$$\angle S_1MS_2 = \angle LAB \text{ (both right angles)}.$$

The triangles are similar and ratios between corresponding sides are equal.

In particular, $\dfrac{S_2M}{S_1S_2} = \dfrac{AB}{LB}$

or $\dfrac{\lambda}{s} = \dfrac{x}{LB}$

Since LB is very long and x is very short we can say that

$$LB \simeq LA = d$$

Within the errors of measurement of the experiment, this approximation makes no difference to the result. Substituting LB $= d$ in the equation, we get:

$$\frac{\lambda}{s} = \frac{x}{d}$$

or $\lambda = \dfrac{sx}{d}$

Distances s, x, and d can be measured; from these measurements we calculate λ. Noting that the bright lines are equally spaced, we normally measure the distance across five or ten bright lines, and divide this by five or ten. The result of the calculation usually gives a result in the region $\lambda = 5 \times 10^{-7}$m. This tells us that the wavelength of light is very short compared with that of sound. It explains why we need to use very narrow slits to diffract the light.

21.33   Diffraction

The experiment with Young's slits shows that light can be diffracted. Now we will go a stage further, to see what happens if we have a large number of slits, instead of only two. The slits are made by scratching fine parallel lines on a sheet of glass. Special machines are needed to do this, for the lines are very close together. Often there are 600 or more lines to the millimetre. A sheet of glass marked in this way is called a *diffraction grating*. Gratings are expensive; for use in school we make a *replica diffraction grating*. A glass grating is coated with a layer of collodion. This solution sets to a firm but flexible film that can be peeled off. It is transparent and has raised lines on it corresponding to the lines scratched in the glass. The collodion film is mounted between two thin sheets of plain glass and is used as a diffraction grating.

Instructions: investigating images made by a diffraction grating
1 Hold up a diffraction grating and look through it at a bright lamp, a point source lamp such as an automobile headlamp bulb, or a fluorescent tube. Can you see the lamp? What else can you see?
2 Still looking at the lamp, turn the grating clockwise. What happens to the images?
3 What can you say about the direction of the line of images and the direction of the lines of the grating?
4 Look at a point source or fluorescent tube reflected in the surface of a long-playing gramophone disc. Arrange the lamp and disc so that the angles of incidence and reflection are about 80°–85°. What can you see reflected in the recording area of the disc? Explain this effect.
   Figure 21.16a explains why we see so many images in a diffraction grating. Light passes through each slit and is diffracted: each slit acts as a source of light. Circular wavefronts spread out from each slit. The diagram looks most confusing, but if you pick up the book, hold it just below your eye level and look along the page from direction A, you can see more clearly what is happening. Rows of *parts* of *circular* wavefronts can be seen to merge

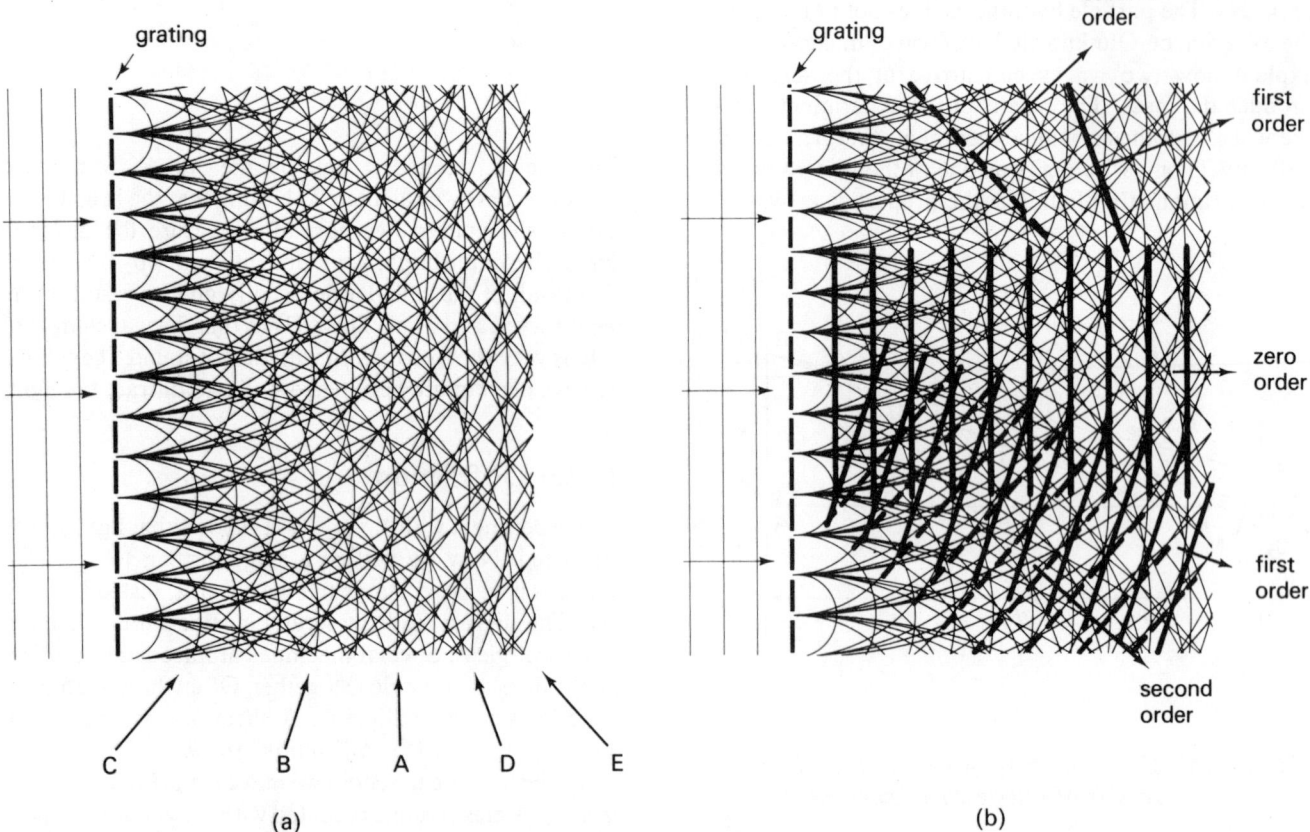

grating

grating

second order

first order

zero order

first order

second order

C   B   A   D   E

(a)

(b)

Fig. 21.16 The behaviour of light at a diffraction grating. In (b) parts of the combining wavefronts are made clearer by thicker lines.

together to make single *plane* wavefronts. In fig. 21.16b parts of these have been drawn with thicker lines so that you can pick them out more clearly. There are several sets of plane wavefronts formed in this way, one behind the other, one wavelength apart. In effect we have a train of plane wavefronts passing away from the grating in a direction perpendicular to the grating. These make the bright central image that we can see when we look through the grating.

Look again at fig. 21.16a, but from direction B. Can you see another train of plane wavefronts? Parts of these are drawn with thicker lines in fig. 21.16b. This train of waves leaves the grating and moves off in the direction shown. It produces an image to one side of the central image. Look again at fig. 21.16a from directions C, D, and E and pick out more trains of wavefronts. These all leave in different directions. Now you can see how a diffraction grating produces so many images.

A diffraction grating produces several distinct trains of wavefronts, which leave the grating as beams of light travelling in different directions. The central beam is called the zero order beam. Those on either side are called first order beams; the next are second order beams, and so on. In fig. 21.16a the first-order beams consist of the wavefronts we see from directions B and D. In fig. 21.17

an eye placed at A is in the zero order beam and sees the bright central image of the lamp. An eye placed at B is in one of the two first-order beams, and sees one of the first order images. The distance between wavefronts is λ (one wavelength) and the angle between the zero order beam and the first order beam is θ. The distances between slits in the grating is s. In the small shaded triangle,

$$\sin \theta = \frac{\lambda}{s}$$

or $\lambda = s \times \sin \theta$

We know s, for we know the number of lines per millimetre. We measure θ as shown in fig. 21.17. The source of light is a lamp with vertical filament. A colour filter in front of the light means that the light is monochromatic (only one wavelength present). One person looks through the grating and checks that the lamp is in line with the zero of the metre rule. Then a second person runs a pointer along the scale until the pointer reaches the spot at which the first-order image of the lamp appears to be. We are now able to calculate the angle θ, for, as the diagram shows,

$$\tan \theta = \frac{x}{D}$$

224

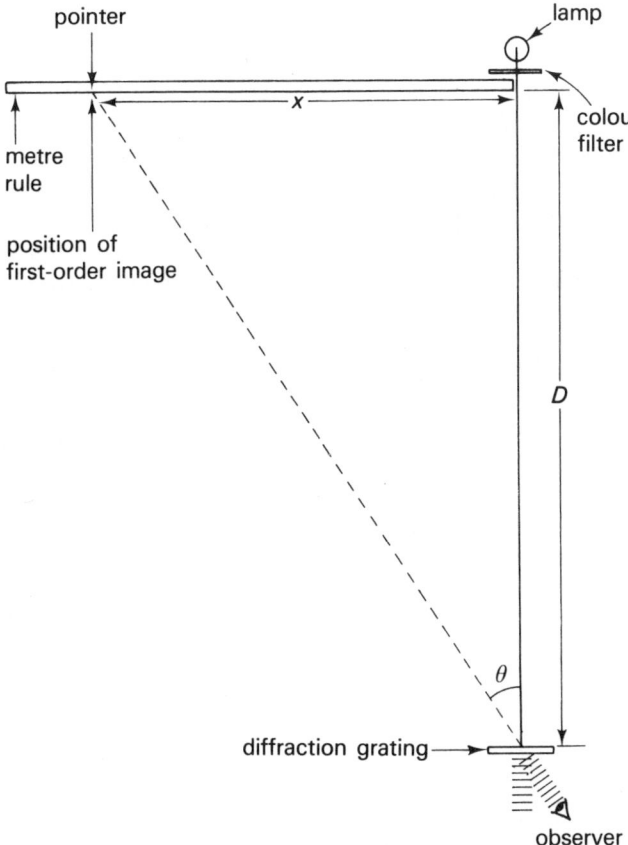

Fig. 21.17 Measuring the wavelength of light, using a diffraction grating.

Questions

1 An apparatus was set up as in fig. 21.17. The position of the first-order image of the lamp was found when a blue colour filter was in place. Distance D was 1 m, distance x was 0.257 m. What is the angle θ?

2 Given the angle θ calculated in 1, and that the grating has 500 lines per millimetre, what is the wavelength of blue light?

3 In the same apparatus a yellow filter was used. The position of the first order image was found and the distance x was 0.315 m. What is the wavelength of the yellow light?

4 A red filter was used in the same apparatus. The wavelength of the red light is 7×10^{-7} m. Calculate the value of x.

5 The colour filters are removed so that light of all colours passes through the grating. A series of overlapping first-order images is seen, in all spectral colours. Reading from left to right across the scale, what colours are these?

6 Which coloured light is dispersed (deflected) most by a diffraction grating?

7 Which coloured light is dispersed least?

8 Compare the dispersion of the different colours by a diffraction grating with dispersion by refraction in a triangular glass prism.

The diffraction grating has given us further evidence in support of the wave hypothesis of light. Its results can be explained by referring to what we know about the diffraction of waves. It also shows us that the difference between light of different colours is the result of their differing wavelengths. This reminds us of sound, where different wavelength gives different pitch.

The light coming from the Sun is white light, a mixture of different wavelengths. In the atmosphere are small particles. The red wavelengths, being longer, are diffracted around the particles. The red light passes straight on, through the atmosphere. Blue light has shorter wavelength. It is reflected and scattered by the particles. As a beam of sunlight passes through the atmosphere it loses more and more of its blue light by scattering sideways and backwards. It looks redder and redder. At sunset and sunrise, we see sunlight that has slanted through the atmosphere for a great distance. This is why sunlight looks so red at these times. In daytime, the blue light, scattered by particles in the air above us, reaches us from all directions. This is why the sky looks blue.

The information and experiments we have considered so far all support the wave hypothesis. We have used ideas about waves to explain the results of all the experiments, and they all make sense. We may therefore conclude that light has the properties of waves. Unfortunately, we may not make the simple statement that 'light is a kind of wave motion'. Other evidence, which we shall look at in 21.50, indicates that light also has the properties of particles.

21.40 The electromagnetic spectrum

Light requires no medium to travel in. It travels easily through a vacuum at a speed of approximately 3×10^8 m/s. Its speed through air is almost the same. It is difficult to imagine what kind of vibrations occur in a ray of light. It is easy to imagine ripples on the surface of water, and the vibrations of strings, springs and air. With light, there seems to be nothing to vibrate.

Light vibrations are electrical and magnetic. This is why we say that light is a type of electromagnetic wave. Think of light waves as a vibrating electrical force and a vibrating magnetic force. The two forces vibrate in planes at right angles to each other and to the direction in which the light is travelling. If this is so, we might expect that a source of light must be a vibrating electric charge or a vibrating particle with magnetic properties. This is true. Light is caused by vibrations inside atoms which generate the electrical and magnetic fields required to produce the forces. These vibrations become specially strong when the atoms are supplied with extra energy. In a fire, a heated lamp filament or the electrically energised gas of a neon lamp, the atomic vibrations are greatly increased. Energy is emitted as electromagnetic waves: light.

Light is just one type of electromagnetic radiation. There are several other types which differ from light by having a different wavelength. Wavelengths range from about 10^{-13}m up to 10km or more. We think of this as a vast spectrum, the electromagnetic spectrum. We cannot actually project the spectrum on a screen, for at different wavelengths the radiation has different properties. We need different kinds of diffraction grating or prism for different parts of the spectrum. Yet all types show the same general wave properties: they can all be reflected, refracted and diffracted by suitable equipment, and they all have the same speed. In fig. 21.18 the types of radiation are set out in order of wavelength, as if it were possible to produce a spectrum. From the diagram it looks as if each band of radiation is distinct, but really there are no sharp borderlines. Properties change *gradually* along the length of the spectrum, just as in the visible spectrum red merges gradually into orange, which merges into yellow and so on.

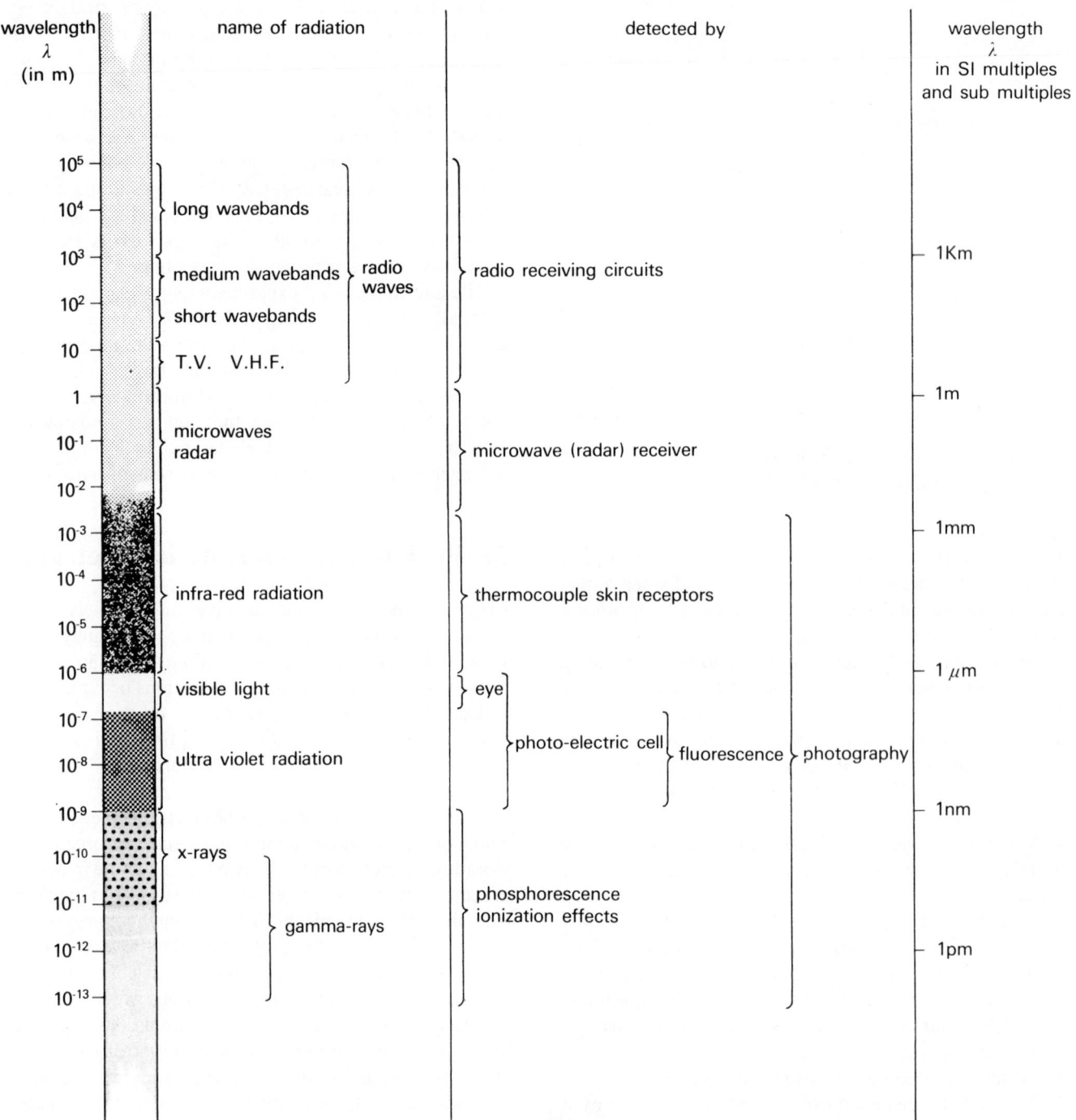

Fig. 21.18   The electromagnetic spectrum.

226

The longest electromagnetic waves, radio waves, are made when electrical vibrations are generated in the aerial of a radio transmitter. Currents flow to and fro along the aerial at high frequency. We usually describe a radio transmitter by its frequency, rather than by the length of the radio waves it produces. Given the frequency, we can calculate the wavelength. For example, if the frequency is 1 034 kHz, we use the equation $c = f\lambda$, and calculate

$$\lambda = \frac{c}{f} = \frac{3 \times 10^8}{1\ 034 \times 10^3} = 290\text{m}.$$

When radio waves reach a radio receiver, their electromagnetic force makes electrons vibrate in the receiver aerial. The radio set detects and amplifies these vibrations, (chapter 32).

Microwaves are also produced and detected electronically. Having shorter wavelength, they are more readily reflected by small objects. This gives them important uses in radar. Like radio waves, they can be directed by special reflectors that look very much like the reflectors used in headlamps and spotlights (fig. 21.20). Microwaves are absorbed by certain materials; their energy is transferred to the molecules as internal energy and the material is heated. Microwave ovens are used for cooking food and other purposes. The microwaves penetrate the material, which cooks evenly all the way through, not from the outside as in an ordinary oven. Microwave ovens are efficient in their use of electrical energy, and cooking times are shorter than those of ordinary electric cookers.

Infra-red radiation has a strong heating effect when absorbed. It is generated by fire and electric heating elements and by hot and warm objects generally. When we say that an object loses heat 'by radiation' (6.41) we mean that it is radiating infra-red radiation.

Visible light occupies only a very narrow band of the electromagnetic spectrum. This narrow band, from about 8×10^{-7}m to 4×10^{-7}m is very important to us. It gives us our view of the world around us.

Ultraviolet radiation is invisible to our eyes. Some insects are believed to be able to see ultraviolet wavelengths that are a little shorter than visible light. In the skin ultraviolet radiation causes the production of a brown pigment, melanin, and vitamin D. Several substances absorb ultraviolet radiation and then emit the absorbed energy as radiation in the visible band. These substances are used in making fluorescent paints and inks. They are used in posters and for novelty effects. In daylight they reflect the incident visible light and then emit an extra amount of visible light because of the ultraviolet radiation they are absorbing. They look very bright, or glow. Such substances can also be added to clothes-washing powders, to make clothes *look* 'whiter than white.' (This does not necessarily mean that the clothes are really clean!) Ultraviolet radiation does not easily penetrate glass; we

are not able to obtain a sun-tan indoors by sitting by a closed window. For cameras, microscopes and other optical instruments, special lenses made of quartz are used when photographs are to be made using ultraviolet radiation.

Ultraviolet radiation and all electromagnetic radiation of shorter wavelength have very destructive effects on living tissue. Excess exposure of the skin to ultraviolet radiation can lead to skin cancer. In most parts of the world the atmosphere absorbs most of the ultraviolet radiation in sunlight, so there is no harmful effect from being outdoors. In tropical areas, of altitude 2 000m or more, the constant exposure to strong ultraviolet radiation can be dangerous. Micro-organisms, such as bacteria, are readily killed by ultraviolet radiation, even at sea level. If washed clothes are hung outdoors in the sunlight to dry, this kills the bacteria on them. In the laboratory and hospital ultraviolet lamps are used to sterilise equipment. They are also used to keep the air free of bacteria in rooms in which any danger of infection by bacteria must be eliminated.

X-rays are produced when an electron beam bombards materials at high velocity (23.12). The rapid deceleration of the electrons generates electromagnetic forces, which produce the radiation. X-rays penetrate materials very easily. They are useful for examining the body for broken bones (fig. 21.19), for examining baggage for hidden weapons, for examining metals for flaws and fractures. They cause damage to living tissue and they kill small organisms. Damage to the cells of larger organisms includes damage to the chromosomes in the nucleus of the cell. This may have serious genetic effects. We should not be exposed to X-rays unless this is necessary for essential medical examination or treatment. People who work with X-rays take special precautions to avoid exposure (lead screens, for example) and have regular checks on their health. The killing effect of X-rays is used in medicine for treating tumours.

The wavelength of X-rays is so short that it is not possible to rule a diffraction grating fine enough to diffract them. We diffract them by using a crystal. The regular rows and columns of the atoms in the crystal lattice act as a fine diffraction grating to X-rays. When the rays are diffracted we can detect the patterns of lines and spots by a photographic film. By measuring and analysing the pattern we obtain, we learn a lot about the pattern and spacing of the molecules and atoms in the crystal. Thus X-rays are a very important tool for investigating the structure of crystals.

On the chart, X-rays and γ-rays overlap. There is no difference between these types in wavelength or in the effects they produce. The difference is that X-rays are made in an X-ray tube, and γ-rays are made during radioactive decay (20.10). The gamma rays with the shortest wavelengths are very penetrating.

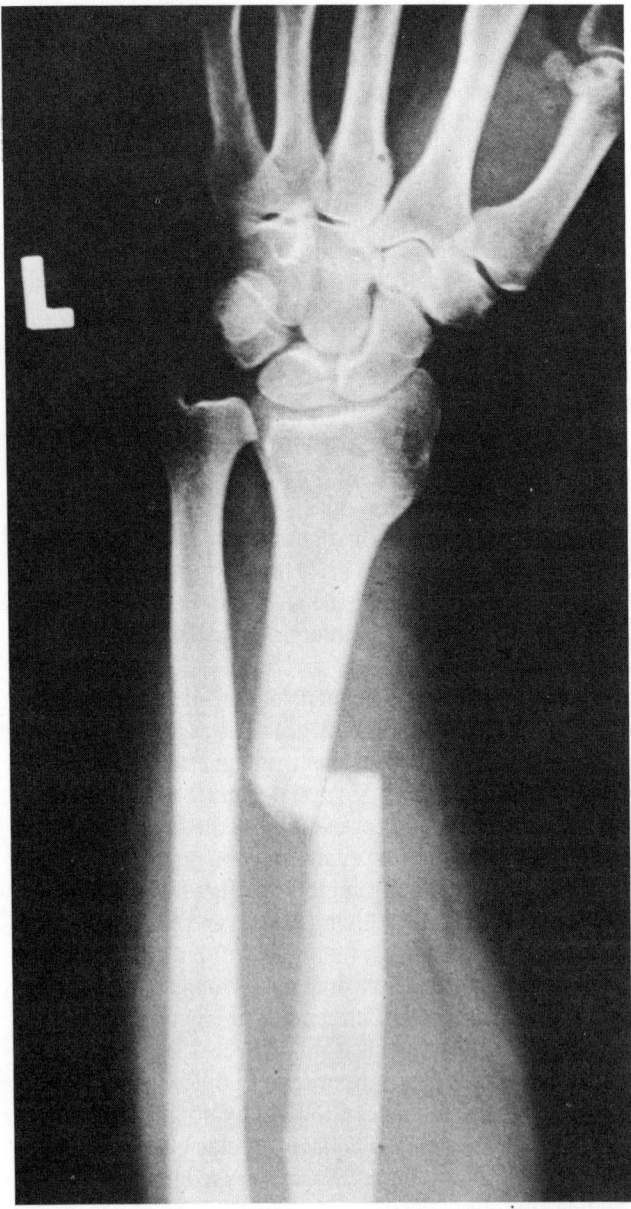

Fig. 21.19 X-ray radiograph showing a fractured fore-arm.

All the types of radiation discussed in this section have the essential property of wave motion: they are ways of transferring energy. They are generated by electromagnetic vibrations and they cause electromagnetic vibrations where they are received. From the Sun and stars we receive radiation over the whole electromagnetic spectrum. Radio-telescopes receive radio waves generated by electrical vibrations of atoms in distant galaxies. From the Sun we receive radio waves, infra-red, visible light and ultraviolet radiation. Special X-ray telescopes have been built to detect X-rays coming from distant bodies. By measuring and examining this radiation we are learning a lot about the nature of the Universe and the way in which energy is transferred from one part of it to another.

21.50 The photoelectric effect

Most things, but not everything, that we know about light leads us to accept the wave hypothesis. There is good evidence for the particle hypothesis too. We use an electroscope which has a small sheet of polished zinc attached to the plate. We give the electroscope a *negative* charge. The charge remains unaltered for many hours and the leaf remains diverging strongly. But as soon as we direct an ultraviolet lamp at the zinc sheet, the electroscope begins to discharge. If we charge the electroscope *positively*, the charge remains, even when we shine ultraviolet radiation on the zinc.

Questions

1 We know that some kinds of radiation discharge an electroscope by ionising the air around the plate (20.23). This does not explain the effect described above. Why not?

2 If a negatively-charged electroscope is losing its charge, what must be leaving the electroscope?

This effect is known as the *photoelectric effect*. When ultraviolet radiation falls on zinc, electrons are emitted by the zinc atoms. The same effect is observed with other metals. Some of them emit electrons when visible light is shone on them (29.20). A metal emits electrons provided that the wavelength of the radiation is *less than* a certain value, depending on the kind of metal. If the wavelength is too long, the effect does not occur. Using radiation of too long a wavelength, the intensity of radiation can be as strong as we can make it, and we may run the experiment for as long as we like, but no electrons are emitted. Yet weak radiation of a short wavelength causes electron emission immediately. This effect is hard to explain if we use the wave hypothesis. Let us try.

Imagine a beach with a rock on it. On the rock are some small animals. The rock is surrounded by the sea and waves hit it. If waves of long wavelength hit the rock, the animals are *not* knocked off the rock, even if the waves are high and strong, and even if they smash down on the rock for hours. Yet if the sea has only small ripples of *short wavelength*, animals are knocked off the rock continually. This sounds like nonsense. It *is* nonsense. The photoelectric effect can have nothing to do with waves.

Here is another story. Imagine a person on the beach, throwing very small stones at the animals. He might throw stones all day and hit the animals often, but the stones are too small to knock the animals off the rock. The stones do not have enough kinetic energy. Then another person begins to throw large stones at the animals. Any animal that is hit by a large stone is knocked off. The clue to success in knocking animals off the rock is the amount of energy transferred in one stone at one time. If an animal is hit on four separate occasions by small stones, it is not

228

knocked off; *one* hit with a stone four times the size is enough.

The photoelectric effect is better explained by the stone-throwing story than by the wave-action story. The stories are *models*. They have some features which help us explain and understand what really happens. A model does not need to be exactly like the real thing; for example, we are not suggesting that ultraviolet light consists of small pieces of stone.

Max Planck suggested that light is not emitted in a continuous stream but in small 'packets' of energy. This is rather like saying that water does not come out of a tap in a continuous stream, but as a number of molecules of water. There is a minimum quantity of water that you can get out of a tap: *one molecule*. If you need just a little more water than one molecule, you must have *two* molecules. You cannot open the tap and let one-and-a-half molecules come out. If you switch on a lamp, the minimum quantity of energy that can be emitted is *one quantum*. If you need just a little more energy than one quantum, you must have *two* quanta. You cannot switch on a lamp and let it emit one-and-a-half quanta.

A quantum of light energy is called a **photon**. A 100W room lamp emits about 10^{20} photons per second. With such large quantities of photons we are not worried by the fact that there is no such thing as a half-quantum, or any other fraction of a quantum.

The amount of energy in a quantum depends on the wavelength of the light. The shorter the wavelength, the greater the amount of energy in the quantum. A photon of ultraviolet radiation has more energy than a photon of red light. Photons of X-rays and γ-rays contain very high amounts of energy because of their very short wavelength. This explains the high penetrating power and the damaging effects of these radiations.

A definite minimum quantity of energy is needed to knock an electron out of a zinc atom. If a photon strikes the atom, an electron will be emitted, provided that the photon carries enough energy. A photon of ultraviolet light has enough energy. A photon of red light, with longer wave length, does not have enough. We may shine bright red light on the zinc for hours and hours. Millions of photons will arrive at the surface of the zinc plate every second. Yet no electrons will be knocked from the atoms. This corresponds to the person throwing the small stones at the animals. The energy required to knock out an electron *must* all arrive in one photon, the equivalent of the large stone.

Every photon of ultraviolet radiation has enough energy to knock an electron from a zinc atom. If we direct a low-intensity beam of ultraviolet radiation at the zinc, relatively few electrons are emitted. If we direct a high-intensity beam at the zinc, the number of photons arriving per second is increased and the number of electrons emitted per second is increased; the electroscope discharges more rapidly. Note that with a high-intensity beam (a bright light) we have a large number of photons passing per second; there is no increase in the amount of energy carried by any one photon. This amount of energy depends only on wavelength.

The wave hypothesis does not explain all the known facts, though it fits in very well with our knowledge of interference. The particle hypothesis, or quantum theory, fits the facts of photo-electricity, but we still have to think about *wavelength* when we are calculating the amount of energy carried by a *photon*. The best we can say is that neither hypothesis is enough to explain all that we know; light has some properties that are wave-like and other properties that are particle-like.

If light has particle-like properties and can interact with matter, does it produce forces? Everyday experience tells us that the force, if any, is extremely small. The radiation emitted by the Sun gives a pressure which is only about 10^{-6}Pa (or N/m²). Compare this with atmospheric pressure, 10^5Pa (or N/m²). This radiation pressure is thought to explain why the tails of comets point away from the Sun, both when the comet is approaching the Sun and when it is leaving. Radiation pressure 'blows' the low-density material of the tail; as the comet leaves the Sun it goes 'tail first'. This topic is referred to again in 26.15.

If waves can behave like particles of matter, can particles of matter behave like waves? This idea has been tested by experiment. A beam of electrons is passed through a pair of slits in an apparatus similar to that of fig. 21.14. A photographic film is placed where the screen should be, to detect the electrons. If the electrons pass through the two slits and travel in two thin beams towards the screen, we expect to get two lines on the film when developed. Instead, we find many bands, looking just like the bright and dark lines we obtain with visible light. The electron beam is being diffracted and it is showing interference. The electron beam is behaving just as if it has the properties of *waves*.

Results like these are hard to understand. Our wave models and particle models are just too simple to explain all our results. Physicists are trying to think of new models that will explain all the results satisfactorily. This is not easy to do; the models required are very complex. But in this way we are getting closer to an understanding of the nature of matter and energy and, through this, to an understanding of the nature of the Universe.

21.60 Spectra

When we pass an electric current through a tube containing neon gas, the gas glows red. We explain this as follows. When an atom is given energy, the energy is absorbed by its electrons. They become excited. They jump suddenly from one orbit in the electron cloud to a slightly different orbit, corresponding with their greater energy. After a while, they jump back to their original orbits, emitting the energy in the form of electromagnetic radiation. For each kind of atom there are several energy levels that the electrons can have. By jumping from one energy level to another, lower one they emit a quantum of energy. The amount of energy carried by the quantum depends on the difference in levels; the wavelength of the emitted light depends on the amount of energy carried by the quantum (21.50). Each kind of atom has its own set of energy levels, and emits photons of distinct wavelengths, depending on what changes of energy level are possible.

The spectrum from glowing neon does *not* consist of a continuous band of colour from red to violet. It consists of a number of thin lines, scattered through the region of visible light and beyond. Each line consists of light of a single colour and a single wavelength. The photons arriving all have exactly the same amount of energy. We can measure the wavelengths of these lines. We can learn to recognise the set of lines belonging to the element neon. We can do the same for other elements. Then we can identify an element by examining its spectrum.

If a quantity of a salt of sodium is heated strongly in a bunsen flame, the sodium is vaporised and emits a strong yellow light. In a spectrometer we find that the spectrum of sodium consists only of two lines very close together in the yellow region of the spectrum. If we are given a mixture of salts, this line can be looked for in the spectrum and if it appears we know that the mixture contains sodium.

Spectra tell us a lot about the composition of the Sun and stars. We cannot collect pieces of these bodies to bring back to Earth for testing. By looking at the spectra of light from the Sun and stars we can find out what elements they contain. The element helium was discovered in the Sun before it was discovered on Earth. The spectrum of the Sun was studied during an eclipse in 1881 by Pierre Janssen. The glowing gases in the Sun's outer region produced a complicated line spectrum. The sets of lines belonging to sodium, potassium and mercury were identified. Among the lines was a bright yellow one that did not correspond to any element known on Earth. The element was named 'helium', after '*helios*', the Greek name for 'Sun'. Later the element was discovered on Earth.

By giving energy to elements in the gas or vapour state, we can produce light of accurately known and unchanging wavelength. Today the SI metre is defined as a fixed number of wavelengths of light of one of the orange-red lines of the spectrum of krypton-86. By defining the metre in this way we have a standard that cannot rust or corrode, be lost or destroyed, or change with time.

Line spectra are produced by gases and vapours. Molecules produce spectra too but because a molecule is more complex than an atom the energy levels are not so sharply defined. Instead of a series of lines we get a series of narrow bands of colour: a band spectrum.

Fig. 21.20   The aerial of the satellite tracking station at Goonhilly.

If you look at the spectrum of the Sun (when not eclipsed) you get a continuous band of colours from red to violet. There is light of all wavelengths. This is because the light is being emitted by colliding molecules, not by single atoms. In the Sun, gases are very much compressed, so collisions of molecules are common. We get a continuous spectrum when we heat solids or liquids, the spectra from an electric lamp filament or an electric fire element are examples. If you look closely at the Sun's spectrum, you can see thin *dark* lines running across it. These dark lines correspond in wavelength with the sets of bright lines emitted by the elements. The central regions of the Sun are emitting a continuous spectrum, (white light). As the light passes out of the Sun it passes through gases which are cooler than the gases inside. Atoms of the cooler gases *absorb photons*. It was Albert Einstein who put forward the idea that just as atoms emit radiation in definite quanta or photons, they also absorb it in definite quanta. The photons they absorb are those which have just the right amount of energy to make their electrons change state from one energy level to a higher one. So a hot gas produces a line spectrum. A cooler gas through which white light is passing can give an *absorption spectrum*; the dark lines of this correspond exactly with those of the bright line spectrum, and we can identify the cool gases. This is how we have found out so much about the composition of the Sun and stars.

So far we have thought only of photons of the visible spectrum. There are quanta and spectrum lines at wavelengths outside the visible band of the electromagnetic spectrum. Extending our measurements into the infra-red and into the ultraviolet and X-ray regions gives much more information helpful in identifying elements. Even the radio region of the spectrum is used. Hydrogen gas emits radio waves very strongly at a few distinct frequencies, when excited. By searching the skies with radio-telescopes tuned to these frequencies we have detected huge clouds of hydrogen gas in Space. These clouds cannot be seen through ordinary telescopes; by making use of our knowledge of spectra, we have been able to add to our knowledge of the Universe.

Projects
1 Find out about plans to make use of the large amounts of energy contained in ocean waves, and to use this energy to generate electrical power.
2 Keep a notice-board in the laboratory where the latest information on astronomical discoveries can be put up. Search newspapers and magazines for photographs and items on radio-astronomy, X-ray stars, the 'big bang' theory of the origin of the Universe, orbiting telescopes and observations being made from satellites.

22 Electric charge

22.00 Charged conductors

In this section we study the behaviour of electrons as charge carriers. When we refer to positive charges we mean atoms from which an electron has been removed. All the effects described are due to the behaviour of electrons.

22.01 The distribution of charge on a conductor

1 You need hollow conductors on insulating stands (fig. 22.1). These can be made from sheet metal, or may consist of metal foil fixed to shaped wooden blocks.
2 The proof plane is a metal disc on an insulating handle. This is usually a flat circular disc, made of brass. Alternatively you can have a set of pieces of metal foil, all having the same area and shaped to fit the surface of the conductors in various places.
3 Charge a conductor by induction, or by connecting it briefly to a Van de Graaff generator (22.03).
4 Discharge the proof plane and an electroscope by earthing them.

5 Touch the proof plane against part of the surface of the conductor.
6 Touch the proof plane against the plate of the electroscope. Observe how much the leaf diverges.
7 Repeat steps 4 to 6 at different regions of the surface of the conductor. Draw a diagram to show where the surface has the greatest charge and where it has the least.
8 Repeat steps 3 to 7 for the other conductors.

Questions
1 Describe your results for the three conductors.
2 Make up a rule for the distribution of charge on the surface of a conductor. At which points is the charge highest?
3 If you have a charged cone-shaped conductor, what is the surface density of charge at the point of the cone?

Instructions: investigating the charge on the inside of a hollow conductor
1 Use a hollow metal sphere on an insulating stand. The sphere has a hole in it, so that a proof plane can be passed inside for testing. Instead you may use a metal can, open at one end, standing on an expanded polystyrene tile.

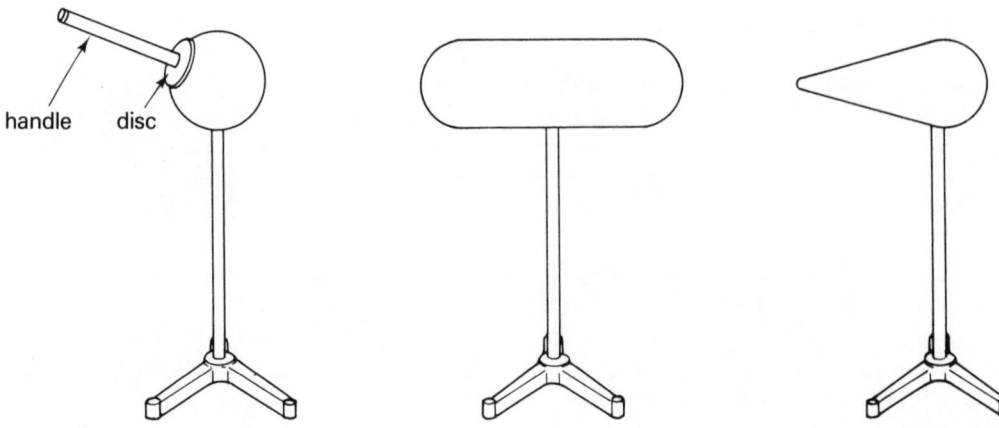

Fig. 22.1   Conductors of various shapes.

2 Charge the conductor by induction or from a generator.
3 Use a proof plane and electroscope to investigate surface density of charge inside and outside the conductor.

Questions
1 Describe your results.
2 What can be said about the *electrical potential* of the inner surface and the outer surface of the *conductor*?
3 Is there an electric field in the space inside the conductor?
4 A boy rides in an automobile on a hot, dry day. He gets out of the vehicle. When his foot touches the ground, he gets an electric shock. Explain this, using the information you obtained in the previous investigation.

22.02 Point discharge

Surface density is high at the edges and points of a charged conductor. The effect is demonstrated by the mill shown in fig. 22.2. When connected to a generator, the mill turns in the direction shown. The motion is due to *electric wind*. Around the points there is a strong electrical field, caused by the high surface density of charge. The field accelerates ions that are already present in the air in small numbers. The accelerated ions strike molecules in the air, making more ions. The positive ions are attracted to the points and are discharged there. The negative ions are strongly repelled, causing the electric wind. The reaction to this wind is an equal but oppositely directed force acting on the conductor (10.21). This makes the mill turn.

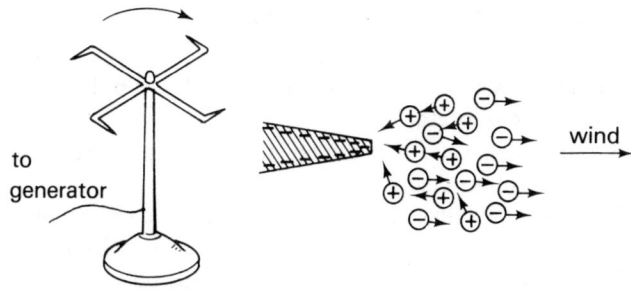

Fig. 22.2   A demonstration of the effects of point discharge.

The same principle explains how a lightning conductor works. The conductor is a thick strip of metal, able to carry a large current without melting. At its upper end there are spikes which point upward above the roof or tower. The lower end of the conductor is buried in the soil, attached to a large copper plate to make good contact. When a strongly charged cloud passes above the conductor, a strong electric field is created. The field is strongest around the points of the conductor. Ions are formed. If the cloud has negative charge, negative ions are attracted to the conductor. The negative charge flows to earth, as electrons. The air above the conductor now contains many positive ions; this creates what is called a *space charge* in the air above the conductor. This reduces the strength of the electric field, making it less likely that lightning will strike the building. If lightning does strike, the charge is attracted toward the spikes, and is carried safely away to earth through the conducting strip.

Instructions: another way of charging an electroscope
1 Discharge an electroscope.
2 Lay a sewing needle on the plate with its point projecting about 1 cm beyond the edge of the plate.
3 Charge a polythene or acetate strip.
4 Hold it near, but *not touching*, the point of the needle. Watch the leaf.
5 Remove the strip. Does the leaf fall?

Questions
1 If the strip is negatively charged, what charge is induced on an electroscope when the strip is brought near it?
2 Where is this charge most strongly concentrated?
3 What happens in the air around the point of the needle?
4 What happens to positive ions?
5 What happens to negative ions?
6 To begin with, the total charge on the electroscope (plate and leaf) is zero. When the strip is brought near, electrons flow down to the leaf; a positive charge is induced on the plate and a negative charge on the leaf. The charge on the plate is then discharged. When the strip is taken away, what charge is left on the electroscope?

22.03 The Van de Graaff generator

This is commonly used in schools for generating high electric potentials. Much larger versions are used in research laboratories for generating potentials of millions of volts required for accelerating atomic particles.

The principle of the school generator is illustrated by fig. 22.3. A rubber belt passes round two rollers. The lower roller is turned by a small electric motor. As the belt passes round the lower roller, it becomes charged by rubbing against it. The positive charge on the belt is carried up towards a sharp-pointed comb. The comb is connected to the inside of the large hollow metal sphere at the top of the generator. The positive charge on the belt induces a negative charge on the comb; the action is the same as in the method of charging an electroscope described above. Ions created in the gap between belt and comb have two effects. Negative ions are attracted toward the belt and discharge its positive charge. Positive ions are attracted toward the comb and are discharged by the electrons on it. The belt continues to move upwards, bringing more positive charge. More and more electrons flow from the sphere to the comb, where they discharge the positive

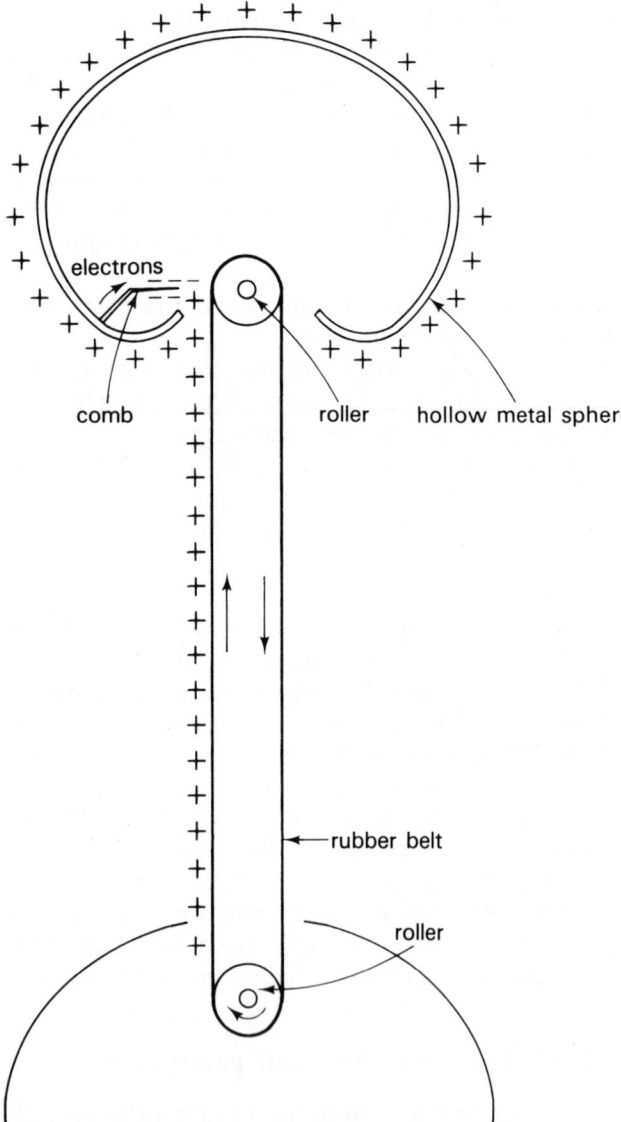

electrons

comb   roller   hollow metal sphere

rubber belt

roller

Fig. 22.3 The Van de Graaff generator.

ions. The flow of electrons from the sphere leaves its outer surface with positive charge. The longer the generator runs, the greater the charge on the sphere. In air a p.d. of about 30 kV between the sphere and an earthed sphere placed close by produces a spark about 1 cm long. A small generator can produce a spark about 7 cm long, which indicates that the p.d. between the generator sphere and earth is about 250 kV.

A large sphere has a gentle curve. It can hold a high surface charge, evenly spread. There is no region of extra high surface density where charge can leak away. The surface of the sphere is turned inwards around the hole in its lower surface. This reduces leakage of charge from its sharp edge.

Question

When a spark is made between the sphere and an earthed

sphere, a great deal of energy is set free. From where does this energy come?

22.10 Charge and potential

The electroscope is an instrument for measuring potential difference. The plate and leaf are both at the same potential (if they are not, a current flows from one to the other until they are). The case of the electroscope may be at a different potential. This is usually the potential of earth, which we call zero potential for convenience. If the leaf and case are at the same potential, the leaf does not diverge; there is no electric field between case and leaf. This can be shown by standing an electroscope on an expanded polystyrene tile, with its case connected to the plate by wire. The leaf does not diverge, whatever charge is given to the electroscope. In normal use, the plate is *not* connected to the case. When there is a p.d. between leaf and case the leaf diverges. The greater the p.d., the greater the angle of divergence.

In the instructions on page 232 we used the electroscope to compare *amounts of electric charge*, though this is not what an electroscope really measures. When a charge is transferred to the electroscope it alters the potential of the plate and leaf. The bigger the charge, the bigger the change in potential. The bigger the change in potential, the bigger the p.d. between leaf and case, so the bigger the angle of divergence. In the experiments we compared amounts of electric charge by observing their effects on the p.d. of the electroscope. In the next investigation we use the electroscope for its proper purpose: to measure *potential difference*.

Instructions: investigating the potential at points on a charged conductor

1 Connect a piece of insulated wire, bared at both ends, to the plate of an electroscope.
2 Wind a few turns of the wire around a rod of insulating material, to make a handle for the wire (fig. 22.4).
3 Discharge the electroscope. What is the potential of (a) its leaf; (b) its case?
4 Charge one of the conductors used on page 232.
5 Touch the wire at several places on the surface of the conductor. Note the divergence of the leaf.
6 Repeat, using the other conductors. Also test the *inside* of the hollow charged sphere. What rule can you state for the potential at different points on a charged conductor?

The investigations on page 232 showed that some parts of a conductor may have high surface density and other parts low surface density; charge is *unevenly distributed*. The last investigation showed that all parts of the conductor have the *same potential*. Clearly, *amount of charge* and *potential* are not the same thing. We must look more closely at the way they differ.

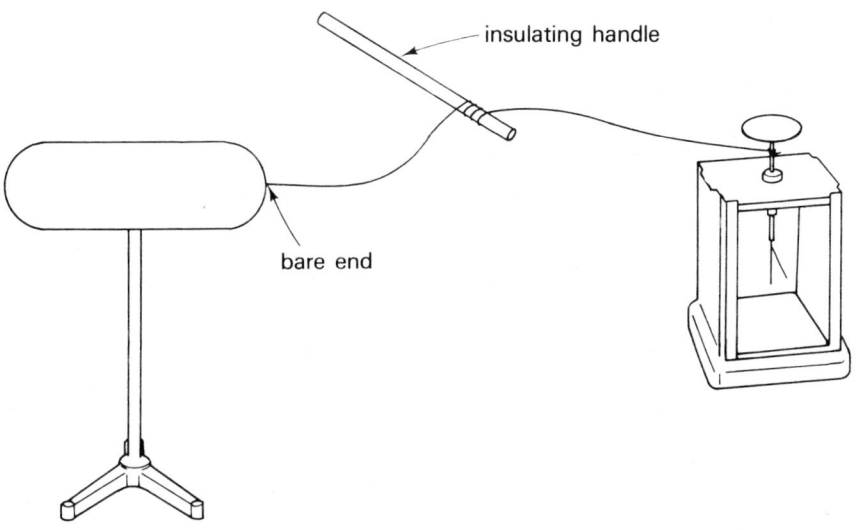

insulating handle

bare end

Fig. 22.4   Investigating the potential of a charged conductor.

In fig. 22.5 we have a positively charged body, A. It is a long way from other bodies; it is insulated, so that it cannot gain or lose charge. The body has *positive electric potential*. What we mean by this is that, if we want to bring another positively charged body B from a distance and move it towards A, we have to *do work*. We do work against the repulsive forces between the two bodies.

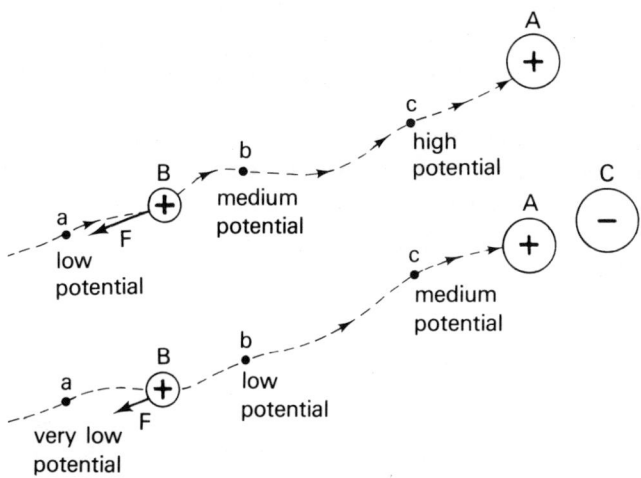

Fig. 22.5   Charge and potential (see text). The arrow marked F represents the force on B due to the electric field around A and around A with C.

As we bring B towards A, we give B electric potential energy. If we release B, it accelerates away from A, losing E_p and gaining E_k. The E_k which it gains is the work we did in bringing it towards A before releasing it.

As we bring B towards A we can mark certain points on its route where it has acquired definite amounts of potential energy. We need not put values to these, just call them low, medium and high potential. If any other charged

body is brought to these points the amount of work done depends on the amount of charge on the body (15.13). If we bring a body with unit charge (1 coulomb), the amount of work done is 1 J, if the potential at the point we bring it to is 1 volt.

Next, suppose that a third body, C, is placed close to A. This body has large negative charge. The *total charge* on A cannot be affected by putting C next to it, A is insulated and no charge can flow in or out of A. What happens to the potential of A? If we bring B towards A we now find that we need to do less work than before, because C *attracts* B. If less work has to be done, the potential at all points on the journey is less than it was before. By putting C next to A, we have reduced the potential of all these points (a, b, c, etc) and we have reduced the potential of A itself. By putting C next to A we have not altered the charge on A, but we have reduced its potential. Charge and potential are not the same thing; they are independent.

If C had the right amount of negative charge we might find that we need do *no work* in bringing a charged body towards A. This would mean that the potential of A is zero, even though it is charged. If C is very strongly charged its attractive force would bring B towards A; we would need to do work to take B *away from A*. Then we would say that A had *negative* potential, even when positively charged.

To sum up, a positively charged body has positive potential when it is away from other charged bodies. If other charged bodies are near, its potential may be positive, zero, or negative. Similar reasoning can be applied to uncharged and negatively charged bodies. This explains why it is possible for the charge on a body to be unevenly distributed, yet for all parts of it to be at the same potential. We can also see why the inside of a hollow conductor has the same potential as the outside, even though it carries no charge.

22.20 Storing charge

Instructions: finding out how much charge it is possible to store on a can

1 Use two metal cans, one whole and the other made by cutting a can in half. Or use two empty food cans (no lids), both about 7 cm diameter, one 10 cm tall and the other 5 cm or 6 cm tall.

2 Put the cans on the plates of two similar electroscopes. Discharge them.

3 Charge an electrophorus (19.11). Use the charged electrophorus disc to charge one of the cans. To do this, put the disc inside the can, so that it rests on the bottom. This transfers the charge to the can completely. The charge flows through the can to its outer surface and to the leaf of the electroscope.

4 Recharge the disc of the electrophorous: DO NOT RECHARGE THE SLAB. Use the disc to charge the other can. Why must the slab not be recharged?

5 Compare the divergence of the leaves in the two electroscopes. What can you say about the potential of each can?

6 Did you give both cans the same charge? Supply the missing word in this rule: 'For a given amount of charge stored, a large can has a _____ potential than a small can'.

Instructions: finding out how much charge it is possible to store on a flat plate

1 The two metal plates used in this investigation are sheets of aluminium foil fixed on two expanded polystyrene tiles (fig. 22.6).

2 Connect one plate A to the electroscope by a wire. Discharge the electroscope.

3 Charge A by induction so that the leaf diverges at a wide angle. You may need to charge your polythene strip strongly, or use a well-charged electrophorus, to produce a large induced charge on the plate.

4 Discharge the other plate B. Bring it close to A, but do not let them touch. What happens to the leaf? What change of potential does this show?

5 Connect B to earth by wire connected at its other end to a water pipe or buried in damp soil. What is the change in the potential of A?

6 Slowly move B away from A, still keeping B connected to earth. What is the change in potential of A?

7 Bring B close to A again. Now slowly slide B sideways, keeping the gap between A and B the same width. Note changes in potential of A.

8 Put B close to A again, with a gap about 2 cm wide between them. If A has lost charge during steps 3 to 7, recharge it until the leaf diverges widely. B must still be connected to earth.

9 Lower a sheet of glass between the plates. Take care not to touch either plate with the glass. Note the change in potential of A.

10 Repeat step 9 using other sheets of insulating material such as paper, card, polythene, expanded polystyrene, perspex. Note changes in potential of A.

We have seen that if a plate has a *fixed amount of charge*, the *potential* of the plate is decreased by:
(a) placing it close to an earthed plate,
(b) having as large an area of overlap as possible,
and (c) placing a sheet of non-conducting material between them.
From the previous investigations we can say that the potential of the plate can be reduced by:
(d) using as large a plate as is convenient.

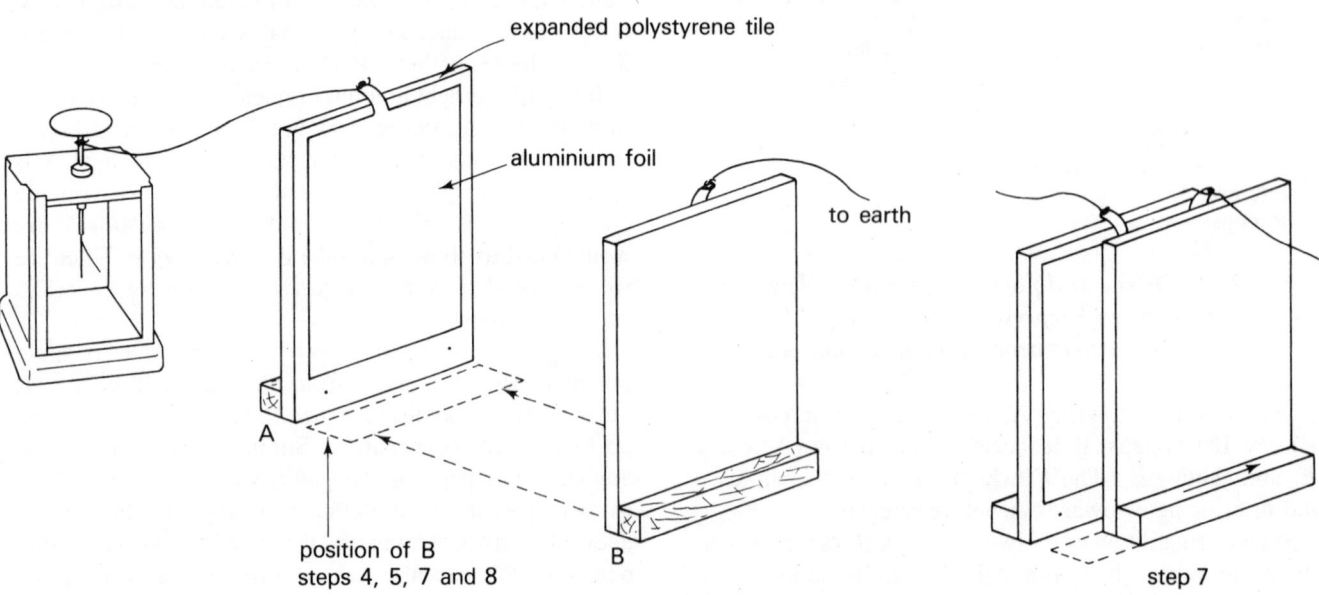

Fig. 22.6 Investigating the potential of a charged plate under various conditions.

This information helps us design devices for storing large quantities of electric charge, as will be explained below.

22.21 The capacitor

If we add charge to a metal plate, its potential rises. We can add more and more charge until the potential of the plate has risen to equal the potential of the device (electrophorus disc, Van de Graaff generator, high-tension power pack, etc.) that is supplying the charge. Then no more charge can be given to the plate. If we wish to store a large amount of charge on the plate, we must arrange that the potential of the plate rises a very small amount when a large amount of charge is given to the plate. The experiments of the previous section tell us how we can do this. We need two plates, one to store the charge on, the other plate connected to earth. The plates must have a large area, they must overlap fully, they must be as close together as possible (but not touching), and they must have a sheet of insulating material between them. Then when a large amount of charge is stored, the potential will remain low, allowing more and more charge to be stored. Eventually, after an extremely large amount of charge has been stored, the potential rises to that of the source of charge, and no more charge flows to the plate. A device which makes use of these ideas is the capacitor.

The most common kind of capacitor has plates made from metal foil. The insulation is provided by two sheets of waxed paper or films of plastic (polystyrene etc.). A connecting wire is joined to each plate and the sheets of metal and insulation are rolled tightly (fig. 22.7a). A large surface area of plate is contained in a conveniently small volume. The capacitor is enclosed in a metal can or plastic container, to keep out moisture (fig. 22.7b). To store large amounts of charge we use a special type, called an *electrolytic capacitor*. In this the plates are sheets of aluminium foil. Between them is a thin layer of fabric soaked in an electrolyte. After rolling and sealing, the capacitor has a current passed through it. This forms a very thin insulating layer of aluminium oxide on the plates and gives the capacitor the ability to store a large amount of charge at relatively low potential.

Electrolytic capacitors are marked to show which end should be connected to positive potential (fig. 22.7c). If the capacitor is connected the wrong way round, the electrolytic action is reversed and the capacitor destroyed.

A variable capacitor is shown in fig. 22.7d. This has special uses in tuning radio circuits (32.20). There are two sets of plates; members of one set are connected together and interleaved between the members of the other set, which are also connected together. Connecting the plates gives a large surface area. The gap between the plates is filled with air. One set of plates is fixed. The other set can be turned to vary the amount of overlap between the plates. This varies the effective area and thus the amount of charge stored at a given potential.

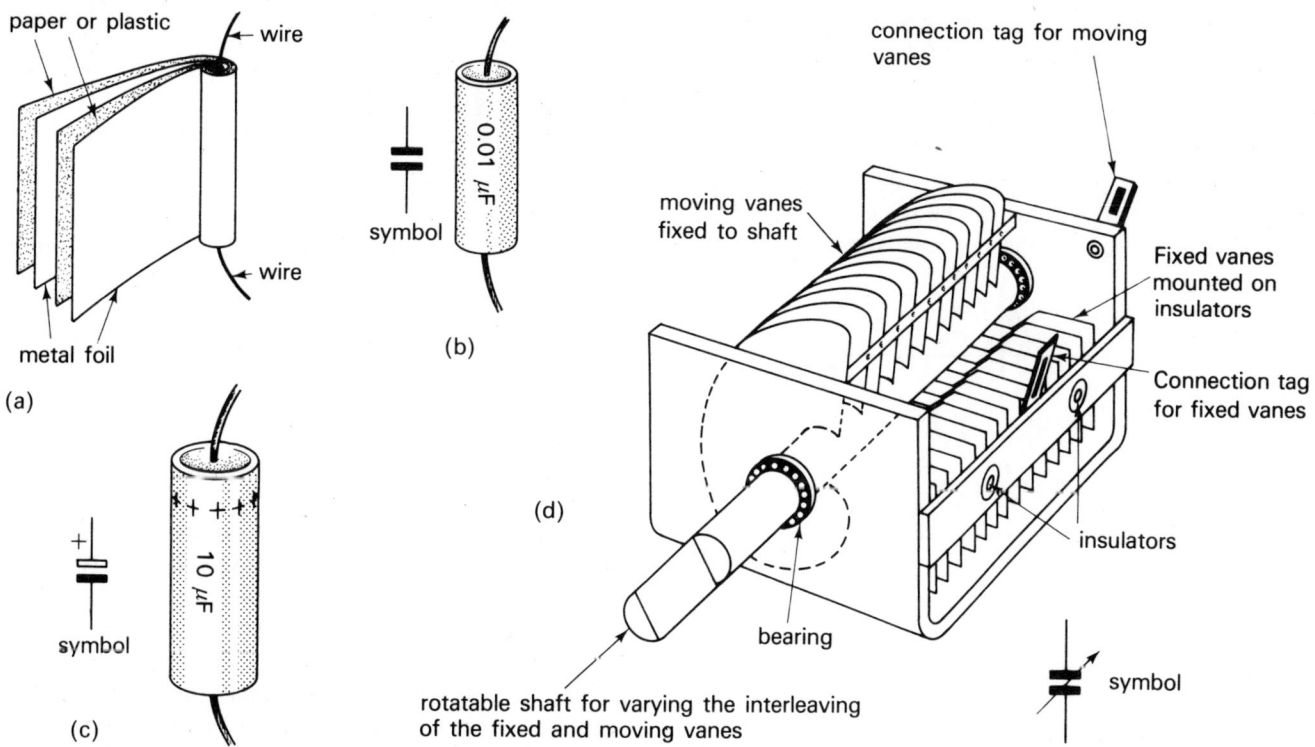

Fig. 22.7 *The capacitor: (a) how it is made; (b) typical capacitor; (c) an electrolytic capacitor, (d) variable capacitor.*

237

22.22 Capacitance

When a quantity of charge, Q, is transferred to the unearthed plate of a capacitor, the potential of the plate rises by an amount V. If V rises only slightly when Q is large, we say that the capacitor has high capacitance. We define the capacitance, C, as follows.

$$Capacitance = \frac{number\ of\ coulombs\ of\ charge\ transferred}{increase\ in\ number\ of\ volts\ of\ potential}$$

In the equation the unit of capacitance is *coulomb per volt*. This unit has been given a special name, the **farad** (symbol **F**), after one of the early workers on electricity, Michael Faraday. In symbols, we can write:

$$C = \frac{Q}{V}\ (C\ in\ F).$$

In practice we find that the farad is an inconveniently large unit. Generally we use the microfarad (symbol μF; $1\mu F = 10^{-6}F$) and the picofarad (symbol pF; $1pF = 10^{-12}F$). On some capacitors, especially those made a few years ago, the symbols MF or mF are used instead of μF. These symbols are now outdated and have been replaced by μF. This point is mentioned to help you identify the capacitors you use in the experiments. Use of MF, mF is not incorrect but merely outdated.

22.23 Experiments with capacitors

Instructions: charging and discharging a capacitor
1 Connect the circuit of fig. 22.8. The capacitor is the electrolytic type; make certain that it is connected the right way round. Wire A has a bare free end which may be touched against the positive terminal of the battery or the terminal of the lamp. A capacitor of $1\ 000\mu F$ capacitance is recommended. The value can be reduced to $250\mu F$, if one of high-capacitance is not available. If the only

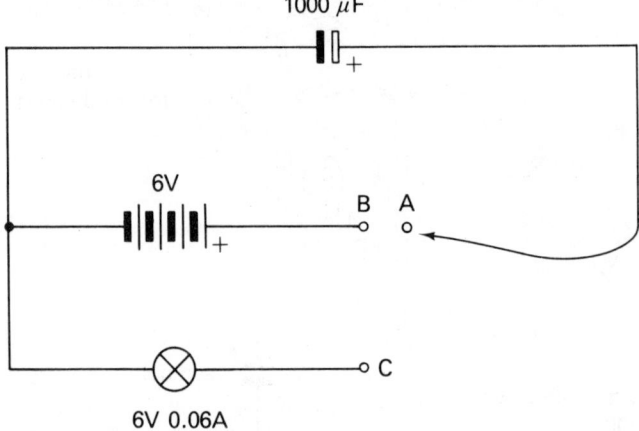

Fig. 22.8   Circuit for charging and discharging a capacitor.

238

capacitors available are of low value ($1\mu F$, etc), use a milliammeter instead of the lamp.
2 Touch A against the battery terminal B for about 5s. This charges the capacitor.
3 Immediately, touch A against the lamp terminal C. What happens?
4 Repeat steps 3 and 4, but wait for 10s between removing the wire from B and touching it against C. Repeat for longer times. How long can the capacitor hold enough charge to make the lamp flash?
5 Touch the wire against B for short times, to find out the length of the shortest time needed to charge the capacitor.

Instructions: further investigations with a capacitor
1 Connect the circuit of fig. 22.9. The capacitor is electrolytic; it can have any value from $100\mu F$ upward; connect it with correct polarity. The milliammeter is a centre-zero meter; it measures current flowing in either direction.

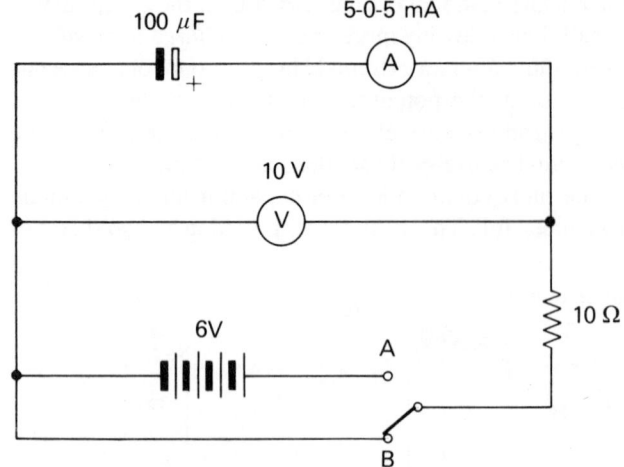

Fig. 22.9   Circuit for investigating slow charge and discharge of a capacitor.

2 Begin with the switch in position B. What does this do to the capacitor?
3 Switch to position A. What do you observe on the meters?
4 Switch to position B. What do you observe on the meters?

You may have noticed that in earlier sections we stated that one plate of the capacitor should be earthed. We have not earthed the capacitors in these experiments. The reason for earthing is to put one plate at zero potential. In the previous discussions, earth potential was to be taken as zero. In these investigations we take the potential of the negative battery terminal as zero. You could connect the negative battery terminal to earth, so that it and the plate are at earth potential; this would make no difference to the results.

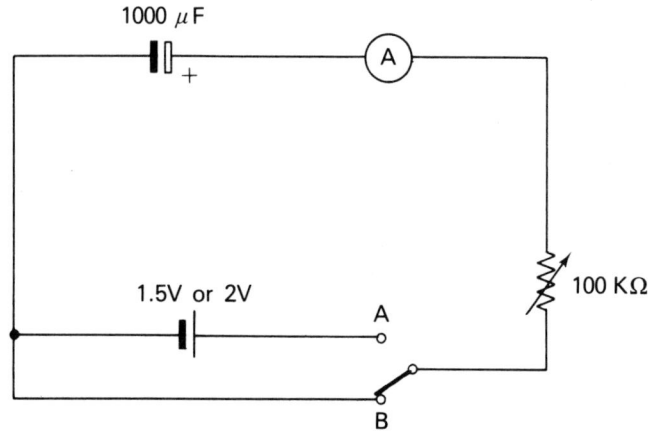

Fig. 22.10   Circuit for measuring capacitance.

Instructions: measuring capacitance

1 Connect the circuit of fig. 22.10.

2 Switch to B, to discharge the capacitor.

3 Switch to A. Adjust the variable resistor to make the current exactly 20μA. Switch to B, then back to A and make further adjustments to the resistor, so that the current is 20μA immediately it is switched on. As the capacitor charges, current will fall below 20μA, but that does not matter.

4 Now try a practice run. Switch from B to A; the current should be 20μA. Try to keep the current steady at 20μA by gradually reducing the resistance of the resistor. Do this for as long as you can; towards the end, when the capacitor is almost fully charged, current falls to less than 20μA, even when the resistor has been reduced to zero. Practise this several times. Your aim is to have an *average* current of 20μA. If you let current fall below 20μA and then adjust to exactly 20μA, this gives an average of *less than* 20μA. This is wrong. When adjusting, overshoot a little, so that the current is a little more than 20μA; during the experiment the needle should stay *around* (above and below) 20μA.

5 Switch from B to A and, holding current around 20μA, time how long it takes to charge the capacitor. Measure from the time of switching on until you can no longer control the current.

6 Repeat step 5 twice. Calculate the average time.

7 Use a voltmeter to measure the p.d. of the cell.

8 Repeat steps 4 to 6 for other charging currents eg. 15μA, 25μA and 30μA.

9 What quantity of electric charge flowed into the capacitor when the charging current was 20μA? You may ignore any current that flowed towards the end when the current was less than 20μA. (Hint: see chapter 14.)

10 Calculate the capacitance.

11 Repeat steps 9 and 10, for the results obtained with other charging currents, to see if these agree.

23 Electrons

23.00 Free electrons

Free electrons are those which have been detached from atoms, but are not being carried in conductors and are not deposited on the surface of a non-conductor. They are free in space or in a vacuum. If electrons are free, we can study their properties more easily. We can find out more about the nature of electricity.

23.01 Producing free electrons

A gas such as neon is put into a glass tube, under reduced pressure (fig. 23.1). At the ends of the tube are metal electrodes. When a high p.d. is applied to these an electrical discharge takes place in the tube. The gas glows. The colour of the glow depends on what gas is used. The gas glows because its atoms are given energy by the rapid flow of electrons through the tube. As pressure is further reduced, a dark space with no glowing gas appears near the cathode. Continued reduction of pressure makes the dark space spread along the tube, until the whole tube is dark. Then the glass begins to glow green, especially in the region furthest from the cathode. Early investigators wondered what caused the green glow. They also wondered what caused certain other materials to glow when placed in the tube. This led to the discovery of *cathode rays*. At the time they were given this name because they came from the cathode. Their real nature was not known. Later it was discovered that cathode rays are beams of free electrons.

23.02 Thermionic emission of electrons

The discharge tube described above has a *cold cathode*. It emits electrons because it is at high potential compared with the anode. Another way to produce free electrons uses a *hot cathode*. In the tube shown in fig. 23.2, the hot cathode is a tungsten wire, heated white-hot by passing current through it from a low-voltage supply. The energy given to the tungsten atoms by heating them allows electrons to escape for a short distance from the surface of the metal. This is called *thermionic emission*. The emitted electrons form a cloud around the cathode. When a p.d. is applied between the cathode and anode, the electrons from the cloud are repelled from the cathode and accelerate towards the anode. If the cathode heating current is switched off, there is no electron cloud and no stream of

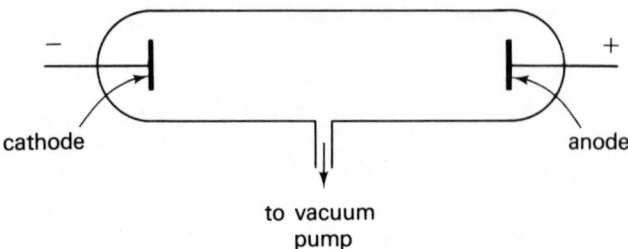

Fig. 23.1   Cold cathode discharge tube.

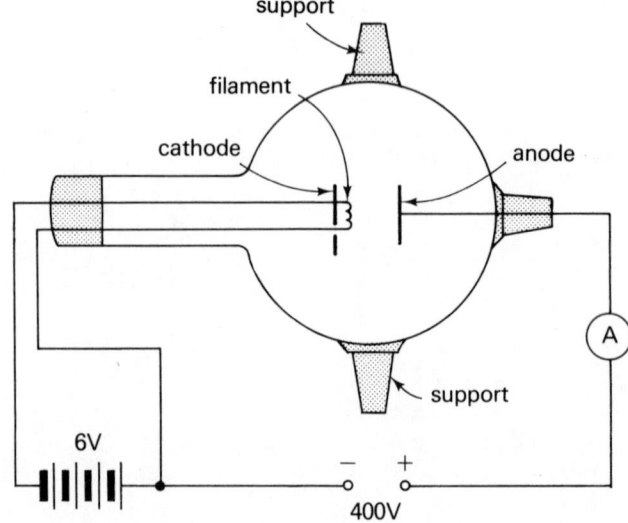

Fig. 23.2   Hot cathode discharge tube.

240

electrons through the tube. In many such tubes a flat metal plate is attached to the tungsten wire so that the electric field between cathode and anode is uniform.

Questions
1 Would a current flow through the tube if the reverse p.d. was applied, making the cathode positive?
2 Suggest a use for this type of tube (Hint: see fig. 19.20).
3 In which direction does conventional current flow in this tube?

The thermionic diode is a type of radio tube, often known as a radio 'valve'. The second name comes from its valve-like property of allowing current to flow in only one direction. Previously thermionic diodes were widely used, but are now largely replaced by semiconductor diodes. A typical thermionic diode is shown in fig. 23.3. The heater is a coil of thin tungsten wire. Close around this is the cathode, a cylinder of metal coated with a mixture of barium and strontium oxides. The oxides emit electrons readily when heated, giving a plentiful supply of electrons at relatively low temperature. The cathode is surrounded by another cylinder, the anode. The whole assembly is in a glass envelope containing gas at low pressure. This is sealed to a plastic base through which pass the pins for making connections to the circuit.

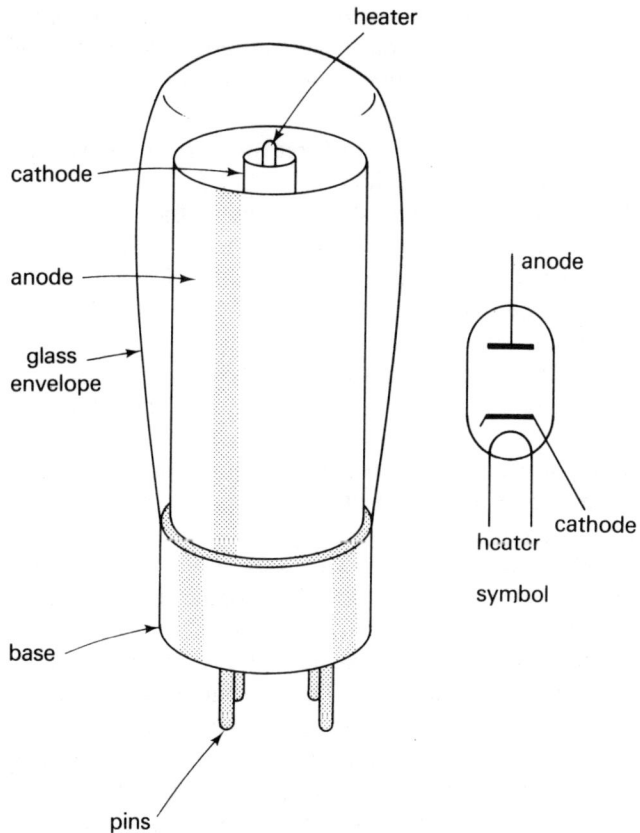

Fig. 23.3 Thermionic diode.

The method used is similar to that used in 19.41, in which we measured how much current flows when a known p.d. is applied.
1 Connect the circuit of fig. 23.4. Wait two or three minutes until the heater has warmed the tube.

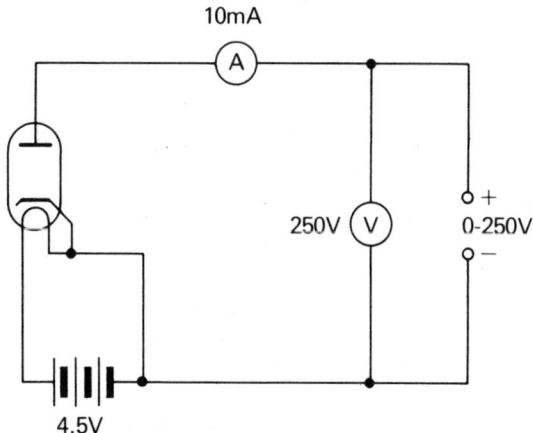

Fig. 23.4 Measuring the current through a thermionic diode.

2 Increase the anode voltage, V_A by steps of 10V, starting from 0V. For each value of V_A, read the anode current, I_A.
3 Continue to increase V_A by steps, until three increases bring no further increase in I_A.
4 Plot a graph to show the relation between V_A and I_A.

Questions
1 Does conduction through the diode obey Ohm's law?
2 Why do the final increases of V_A give no increase of I_A?
3 What would happen if the cathode was not heated as strongly?

23.10 Beams of free electrons

We use the principle of the thermionic diode to make a beam of electrons. The device is often called an *electron gun*. The tube shown in fig. 23.5 has an electron gun at one end. A low-voltage supply is used to heat the filament, the hot cathode. An electron cloud forms around this. The anode is a cylindrical metal can with a small hole in one end. A high p.d. is applied between cathode and anode. Electrons are attracted towards the positively charged anode and are accelerated to high speed by the strong electric field. Many of them hit the anode and then pass through the metal and out to the positive terminal of the power supply, as ordinary conduction electrons. A few electrons travel in the direction of the hole. When they

241

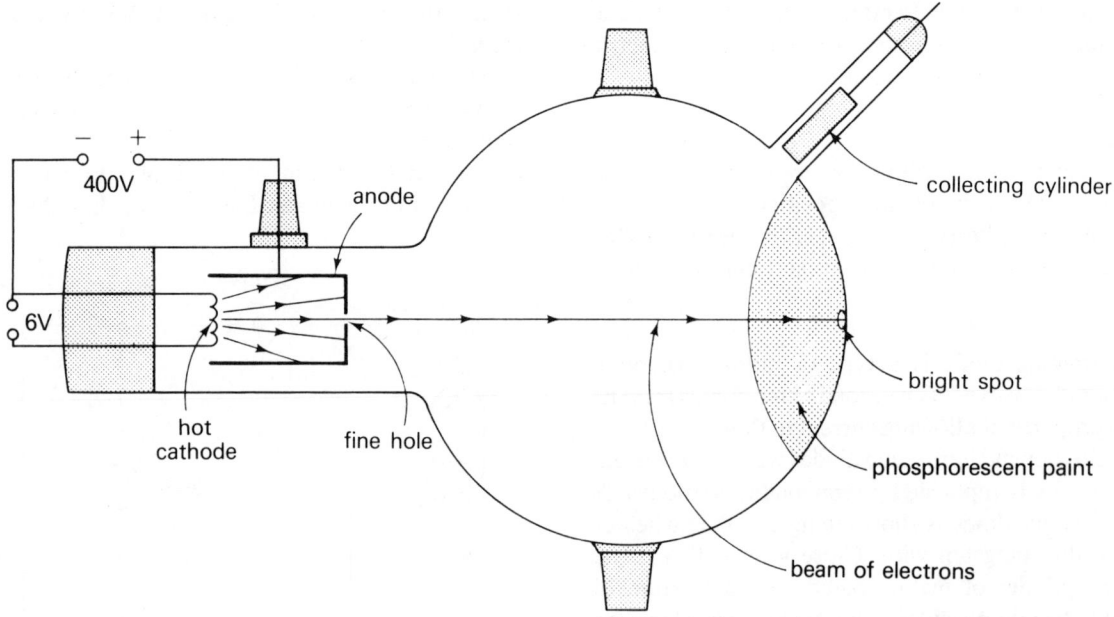

Fig. 23.5 Discharge tube for investigating the properties of a beam of electrons (Perrin's tube).

reach it, they pass straight through it at high speed and are soon far away from the influence of the electric field. They form a narrow beam of free electrons which travels across the tube and hits the opposite end. The glass there is usually coated with a special paint, called a *phosphor*. The energy of the electrons is absorbed by molecules of the phosphor, and is emitted as visible light, usually green or blue. The electron beam is invisible, but its effect can be seen where it hits the phosphor.

The beam travels straight across the tube. Gravity must be acting downwards on the electrons, accelerating them in a downward direction at 10m/s². They do not appear to lose height. This tells us that the *electrons are travelling at an exceedingly high speed.*

Some tubes have an electron gun with a larger hole; it makes a wider electron beam. In the centre of the tube is a sheet of metal, usually in the shape of a cross. When the beam passes along the tube the area of phosphor glows, except for a dark area: the shadow cast by the metal cross. This shadow is sharp and is the same size as the cross. This tells us that the *electrons travel in straight lines.* When a strong magnet is held to one side of the tube, the shadow becomes displaced and distorted. This tells us that *electrons are deflected by a magnetic field.*

A bar magnet placed to one side of the tube of fig. 23.5 demonstrates the deflection of an electron beam by a magnetic field. The magnet can be used to deflect the beam upwards, so that it strikes the metal collecting cylinder. If an uncharged electroscope has been connected to this cylinder, the leaf gradually diverges. The electroscope, when disconnected, is found to be negatively charged. This tells us that *electrons carry negative charge.*

23.11 The effects of magnetic and electric fields

In 9.20 we saw that an electric current exerts a force on a magnet. By Newton's *third* law we might reason that a magnet exerts a force on an electric current. Let us try out this idea, using a current flowing in a wire.

Instructions: studying the force on a current flowing in a wire in a magnetic field

1 Set up the apparatus shown in fig. 23.6a.
2 Connect the ends of the wire to the terminals of a 2V accumulator for a *fraction of a second*. Do not let the current flow for longer, for it is very large and will soon damage the cell.
3 Which way does the wire kick?
4 Reverse the magnet. Which was does the wire kick now?
5 Replace the magnet so that its poles are as marked in the figure. Apply a reversed current. Which way does the wire kick?
6 Which way will the wire kick when both magnet position and current are the reverse of those in the figure? Try this, to check your answer.

The results of the experiment are summarised in *Fleming's left-hand rule.* This is sometimes called the *motor rule* because it tells us about motion. The rule states that if we arrange the thumb and first two fingers of the left hand, so that they are at right-angles to each other, and if the **Fore-Finger** points in the direction of the magnetic **Field** (north to south), and the se**C**ond finger points in the direction of **C**onventional **C**urrent (positive to negative),

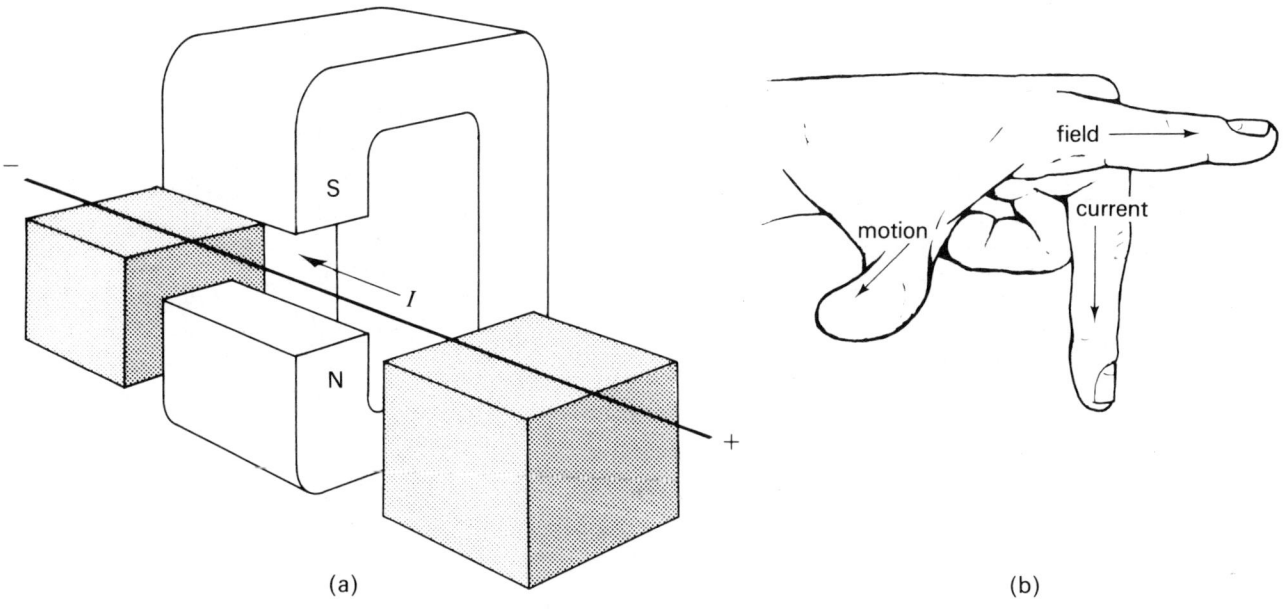

Fig. 23.6 (a) The effects of a magnetic field on a wire carrying an electric current (b) Fleming's Left Hand Rule.

the thu**M**b points in the direction of the **M**otion produced. Confirm this rule by revising the experiment (see fig. 23.6b).

The effect of a magnetic field on a beam of electrons can be investigated by using the discharge tube of fig. 23.5. If we place a bar magnet with its north pole against the nearest side of the bulb (as viewed in the figure) the field is directed down into the page. This deflects the beam downwards (towards the bottom of the page).

Question

Apply Fleming's left-hand rule to the above statement. What is the direction of the current? What does this tell us about the charge on the electrons?

The tube shown in fig. 23.7 has large coils to provide a uniform magnetic field by passing current through them. The electron gun has a narrow slit, making a ribbon-like electron beam. The screen is placed at a slight angle across this beam, so that the beam skims along the screen, making the phosphor glow along its track. In the figure we see how the beam might travel when a magnetic field is applied, directed down into the page (fig. 23.7b). Each electron in the beam has a downward force acting on it; it almost looks as if the beam is being deflected by gravity, but we know that gravity is not able to have a noticeable effect on a fast-moving electron in such a short distance. The magnetic field produces a force which causes constant acceleration of the electrons *at right angles to their direction of motion*. This is the effect noted in Fleming's left-hand rule. The force is always at right-angles to the direction of motion, so it acts in the same way as

centripetal force, as when an object is being whirled round on the end of a piece of string. The electrons follow a *circular path*. Measurements on the grid show that the beam is bent into part of a circle.

The screen is supported by two metal plates with connections to terminals. With the magnet coils switched off, we can apply a p.d. between these plates. If the upper plate is made negative to the lower plate, the electric field between them deflects the beam downwards. If we examine the shape of the curved track (fig. 23.7c), we find that it is *not* part of a circle. The force due to the electric field is a constant force acting perpendicular to the electrodes, from the upper electrode to the lower one. With the tube arranged as shown, this is a downward force. There is constant acceleration in a downwards direction as the electrons travel in the space between the electrodes. The electrons follow the same kind of path as a bullet shot from a gun. With increasing velocity downwards, their path becomes steeper and steeper. It is a *parabola*. Measurements on the grid prove this.

If we apply the magnetic field and electric field at the same time, and adjust their strength, we can make the beam travel its original straight path along the centre of the screen. The electrical and magnetic forces balance. Knowing the strengths of the fields we can calculate the velocity of the electrons and also the ratio e/m. The velocity is less than that of light, showing that cathode rays are not electromagnetic waves. The ratio e/m is the ratio of the electric charge, e, on the electron to the mass, m, of the electron. Results show this to be 1.76×10^{11} coulombs per kilogram. By experiment with electrolysis we can measure e/m for hydrogen ions (H^+). The value of e/m for

243

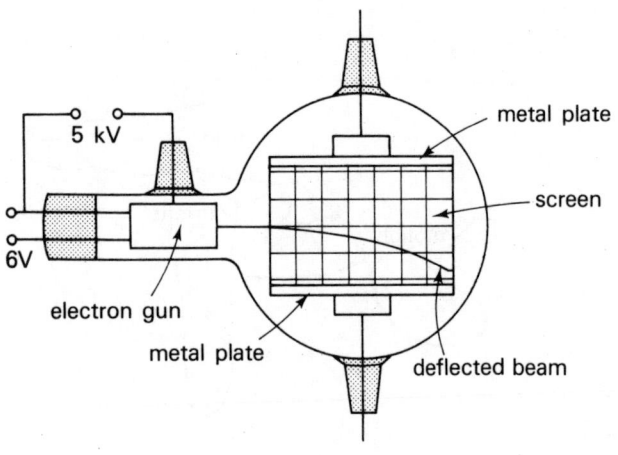

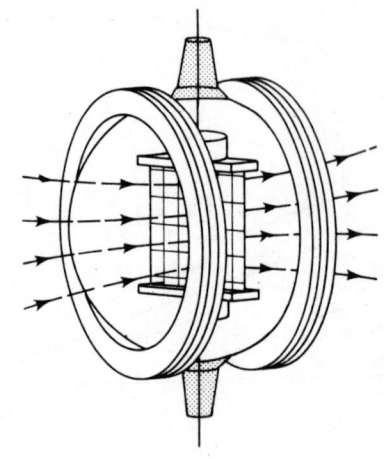

(a)

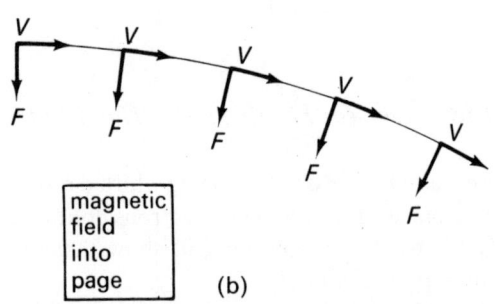

magnetic
field
into
page

(b)

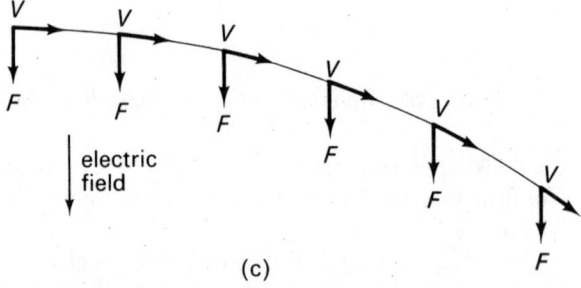

electric
field

(c)

Fig. 23.7 *Discharge tube for measuring the deflection of an electron beam in magnetic and electrical fields. (b) deflection produced by a magnetic field; (c) deflection produced by an electrical field.*

hydrogen ions is about 2 000 times smaller than the value for electrons. We know that their charges are equal and opposite. This suggests that the mass of the electron is 1/2 000 that of the hydrogen ion, which is a single proton.

The first experiments to measure e/m were performed by J.J. Thompson, using a method similar to that described above. He found that if he filled his discharge tube with different gases he always obtained the same value for e/m. He concluded that the cathode rays were alike and that the particles they are made of are found in every kind of matter.

The discharge tube is used to measure e/m. If we can measure e, we can then calculate m. An experiment by Millikan first gave the value of e. He arranged a pair of horizontal metal plates one above the other, with a gap between them. He could apply a p.d. between the plates, to create an electric field. He sprayed fine drops of oil into the space between the plates. With no electric field, the drops fell through the air at their terminal velocity. Many of the drops became charged by friction as they fell. Millikan placed an X-ray tube beside the space to ionise the air and help increase the number of charged drops. He then applied a p.d. between the plates and watched one of the drops, using a microscope. He could adjust the p.d.

until the drop was held motionless. The downward force of gravity was balanced exactly by the upward force of the electric field. All forces could be calculated. The electric force depended on the p.d. and on the amount of charge on the drop. He measured the charge on hundreds of different drops. He found that the charge was always a multiple of a certain quantity. He concluded that this smallest amount of charge must be the charge on a single electron. Some drops had one electron on them, some had two, some three, and so on; all had multiples of this single electron charge. Millikan's measurements gave the charge on the electron as 1.6×10^{-19}C.

Knowing e, we can calculate m:

$$\frac{e}{m} = 1.76 \times 10^{11}$$

$$m = \frac{e}{1.76 \times 10^{11}} = \frac{1.6 \times 10^{-19}}{1.76 \times 10^{11}} = 9 \times 10^{-31}\text{kg}$$

23.12 X-rays

In the X-ray tube (fig. 23.8), high p.d. is applied between the hot cathode and the anode. Electrons are accelerated

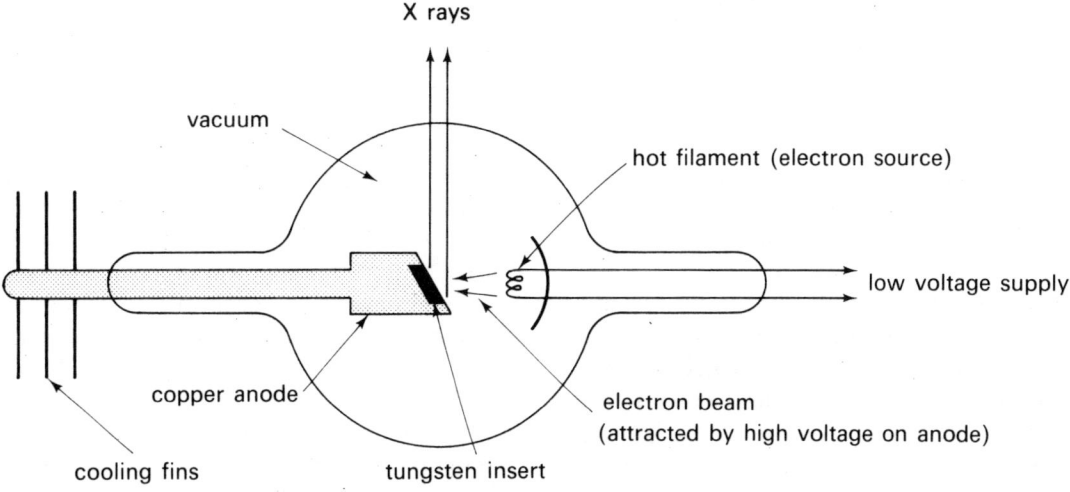

X rays

vacuum

hot filament (electron source)

low voltage supply

copper anode

electron beam
(attracted by high voltage on anode)

cooling fins

tungsten insert

Fig. 23.8   X-ray tube.

to extremely high speed. Their sudden deceleration as they strike the anode causes the emission of high-energy photons of short wavelength. These are in the X-ray band of the electromagnetic spectrum. In this process the anode becomes very hot. A special insert of tungsten receives the beam: tungsten has a high melting point. The remainder of the anode is made of thick copper which readily conducts heat to the cooling fins outside the tube. By increasing the current through the filament the number of electrons is increased; this increases the intensity of the X-ray beam. By increasing the p.d. between the cathode and the anode, the speed of electrons is increased, giving X-rays of shorter wavelength. These have greater penetrating power. In this way we can adjust the penetrating power of the rays to suit the material being examined. When examining a block of metal for fractures we need very penetrating rays. For examining an old oil painting to determine whether there is another painting hidden beneath the top layers of paint, only weakly penetrating rays are required.

23.13 The cathode ray oscilloscope

The oscilloscope has many points in common with the discharge tube of fig. 23.5. It has an electron gun and a screen coated with phosphor. By adjusting the p.d. applied to the gun, we adjust the brightness of the spot of light that appears on the screen. The cathode ray tube (fig. 23.9) has two pairs of electrode plates, the X-plates and Y-plates. If we alter the p.d. between the Y-plates the beam is deflected upwards or downwards on the screen. Varying the p.d. between the X-plates deflects the beam horizontally.

The input to the oscilloscope goes to an amplifier which controls the p.d. between the Y-plates. By adjusting the amplification we can arrange that an input potential of $+1\,\mathrm{V}$ (meaning that the potential of the input terminal is 1 V positive of the 'earth' terminal) results in an upward deflection of the beam, and the spot of light rises 1 cm above the centre of the screen. An input potential of $-1\,\mathrm{V}$

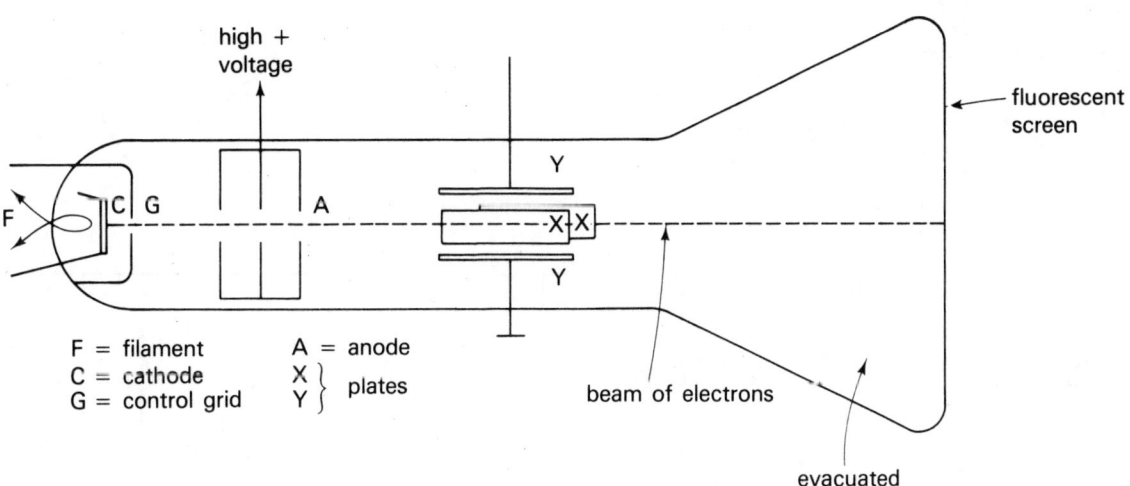

high +
voltage

fluorescent
screen

F

C G

A

Y

X X

Y

F = filament
C = cathode
G = control grid

A = anode
X }
Y } plates

beam of electrons

evacuated

Fig. 23.9   Cathode ray tube.

will then cause the spot of light to fall 1 cm below the centre of the screen. Other input potentials produce upward or downward deflections, in proportion. The setting of the Y-amplifier can usually be adjusted in steps. Input potentials of a range of values (100 mV, 200 mV, 500 mV and on up to say, 50 V) can be selected to give a vertical deflection of 1 cm on the screen. This allows us to use the oscilloscope to examine rapidly changing potentials from a wide variety of sources.

The oscilloscope has a time-base circuit which automatically applies a varying p.d. between the X-plates. The beam begins at the left of the screen and is deflected across the screen at a constant rate. On reaching the right side, it is brought back to the left side very quickly (flyback). During flyback the brightness is automatically reduced so that we cannot see the returning spot. The rate at which the spot sweeps across the screen is adjustable. The controls are usually marked according to the time taken for the spot to travel 1 cm horizontally. A typical range is from 1 μs/cm up to 1 s/cm.

If the time-base is switched on, but no p.d. is applied to the input terminal, the beam traces a straight line across the screen (fig. 23.10a). To make the beam fall exactly on the central line, there is usually a control to adjust the potential on the Y-plates. If a constant p.d. is applied to the input, the beam rises above the central line, the distance above the line depending on the setting of the Y-amplifier and on the value of the p.d. (fig. 23.10b). If, during the time the spot travels from left to right, the p.d. changes from zero to a positive p.d. and then to a negative p.d. and then back to zero, we obtain a trace similar to fig. 23.10c, d, or e. The exact shape depends on the rate at which p.d. changes at different instants. The shapes shown in the figure correspond to various forms of alternating potential. If applied to a circuit they would give alternating current. If the frequency of the time-base is known, we can measure the frequency of the alternating potential.

Figure 23.10f shows the trace obtained when the input is the potential from a half-wave rectifying circuit, using a

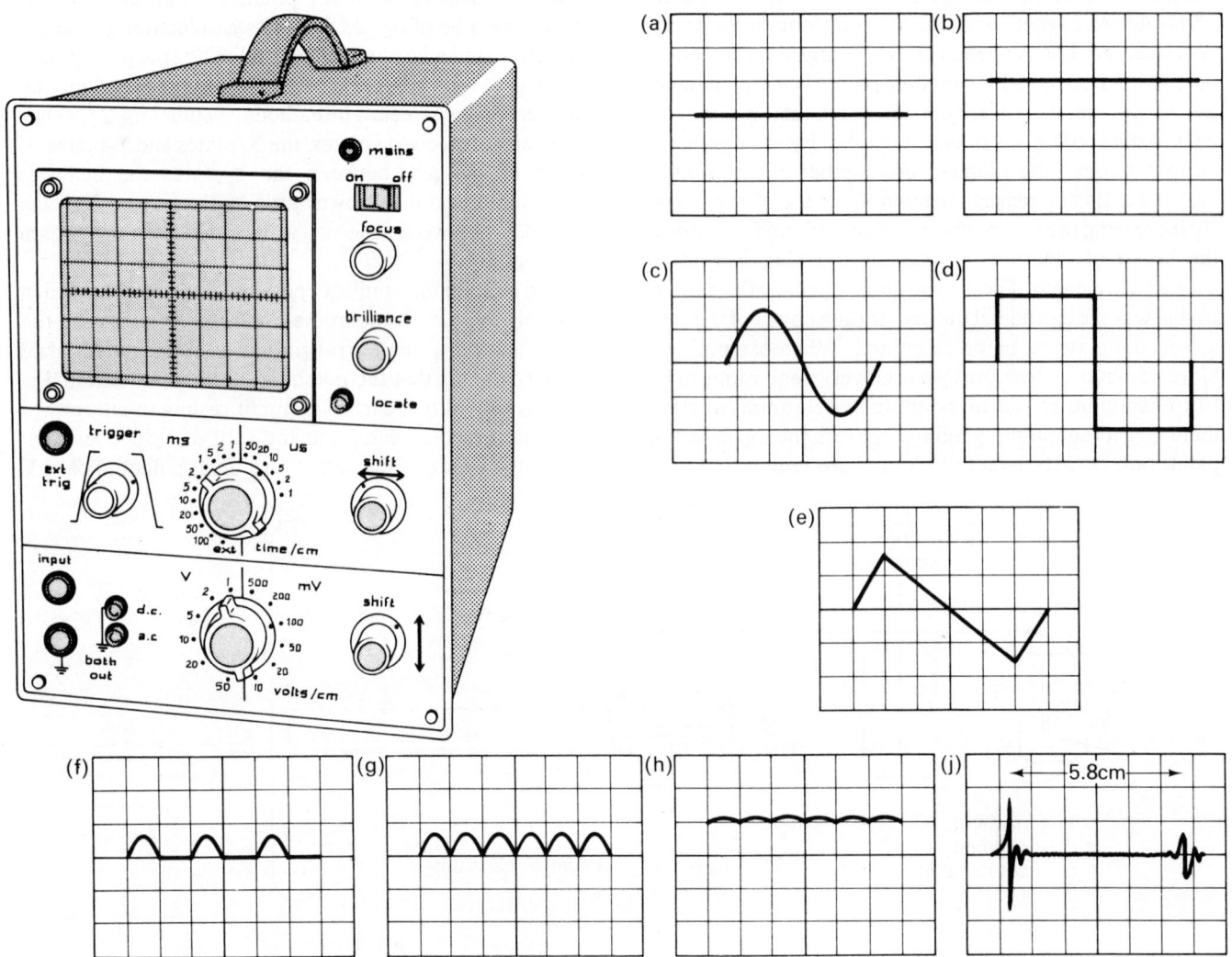

Fig. 23.10   *Cathode ray oscilloscope and various traces (see text).*

single diode. Only one-half of the a.c. cycle produces a potential. A full-wave rectifier produces a trace as in fig. 23.10g. If a capacitor of high value (500μF or more) is connected across the terminals of the rectifier, the trace appears as in fig. 23.10h. The capacitor smooths the output from the rectifier. Without a capacitor p.d. falls after it has reached its peak, as current flows in the circuit to which the rectifier is connected. With a capacitor in parallel with the circuit, charge can be stored when output p.d. is high and used to supply the circuit as p.d. begins to fall. This keeps the p.d. from falling low. The current from a smoothed rectifier output is used for supplying power to mains-operated radio sets and amplifiers. If the p.d. varied as much as in fig. 23.10g, their circuits would not operate properly. A loud hum, frequency 50Hz would be heard from the loudspeaker. When a loud hum of this kind is heard on such equipment it is frequently a sign that the smoothing capacitor has failed.

Since vertical deflection is proportional to the input p.d. we can use an oscilloscope as a voltmeter. It is especially useful for measuring voltages that are changing rapidly, or that exist for only a few milliseconds. An ordinary voltmeter would not have time to respond.

We can also use an oscilloscope for measuring short periods of time. For example, we clap hands beside a microphone connected to the input of the oscilloscope.

One metre from the microphone there is a smooth wall. The sound of the clap travels to the wall and is reflected back, reaching the microphone about 5.8ms later. If the spot is set to scan the screen at 1 ms/cm the spot travels 5.8cm across the screen in the time between the clap and the return of the echo. On the screen we see two sharp peaks (fig. 23.10j). We can measure the distance between these spikes accurately. This gives us an accurate measurement of the time and we can then calculate the speed of sound. This method is convenient because it can be used in the laboratory, away from the effects of wind. We can measure the speed of sound at different temperatures for there is no difficulty in heating or cooling the air in the laboratory.

23.14 The television tube

This is similar in structure to the tube of the oscilloscope. From the outside the most noticeable difference is that the screen is large and the tube is short. This gives a large picture yet fits into a cabinet that is not too deep. Control circuits in the set operate both the X-plates and Y-plates. The beam traces a regular pattern of horizontal parallel lines across the screen. The beam traces one set of alternate lines in a period of 0.02s, then traces the lines between these in a further period of 0.02s. Thus the whole

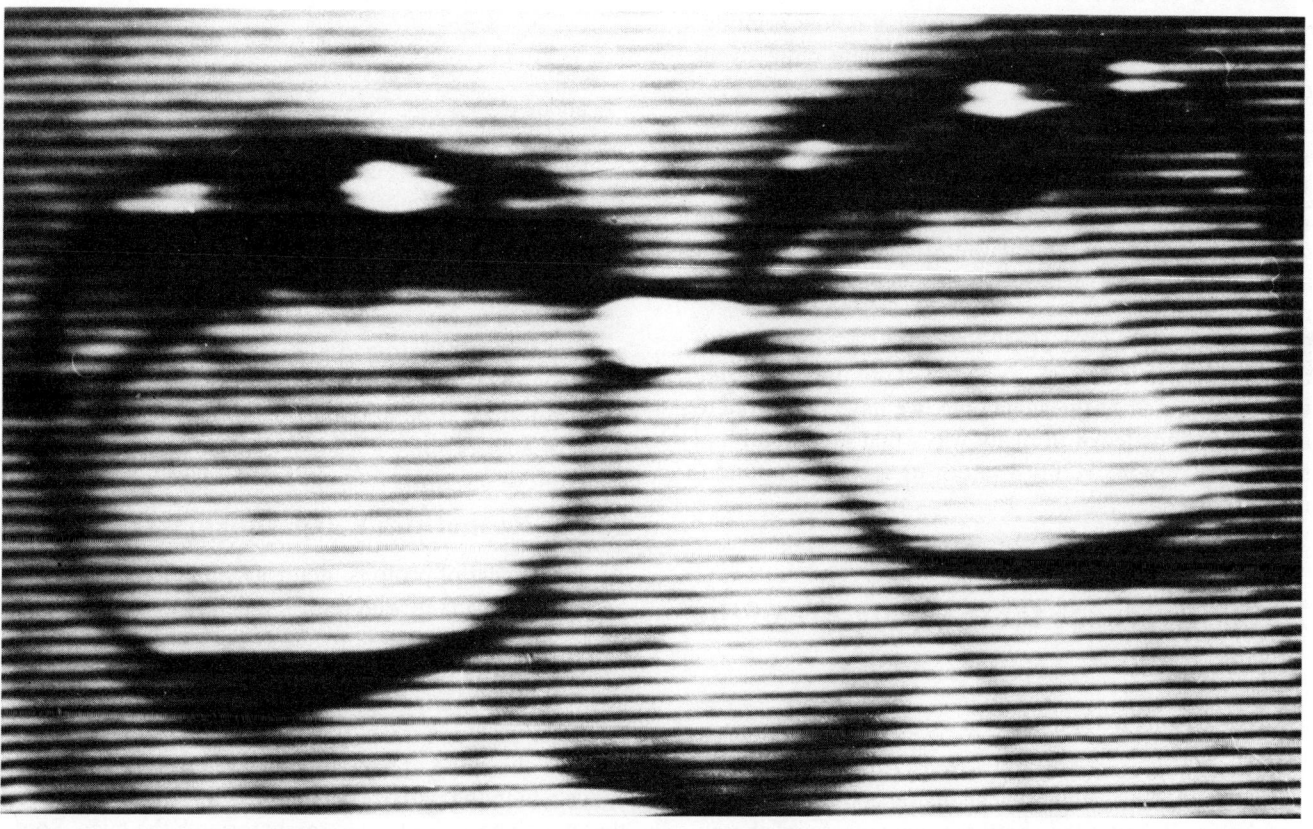

Fig. 23.11   Photograph of a television screen, showing the way the picture is made up from a large number of parallel lines, varying in brightness at points along their length.

pattern is traced in 0.04s, or 25 times per second. At the same time as the beam scans the tube, its brightness is varied. The brightness is controlled by signals from the television transmitting station, which also transmits signals to synchronise the scanning of the screen with the scanning of the television cameras. Variations in brightness produce a picture on the screen (fig. 23.11).

Colour television uses the three primary coloured lights, red, green and blue (5.31). The television camera forms three separate images corresponding to these three colours and information about the three images is transmitted simultaneously. In the receiver the colour television tube has three separate electron guns, one for each colour. These each form their own image on the screen, which is painted with phosphors to glow red, green or blue. Close behind the screen is a metal mask with a regular pattern of fine holes in it. The phosphors on the screen are in a matching pattern of tiny dots. These are so placed that the beam from the 'red' electron gun can hit only the red phosphor dots. It produces a red image. The mask prevents this beam from hitting the green and blue phosphor dots and making them glow. The beam from the 'green' gun can hit only the green phosphor dots; the beam from the 'blue' gun can hit only the blue phosphor dots. In this way each gun produces its own picture on the screen. The viewer sits too far away from the screen to be able to see the dots separately. The colours blend, just as they do in a colour photograph produced by three-colour printing. To the viewer the screen shows a complete range of all colours.

In both black and white and colour television, the screen is scanned 25 times a second. Each new picture is slightly different from the one before it and the one after it. This gives the same effect as a movie film. The brain combines its views of the separate still pictures; we do not see them as a series of still pictures, but as a continuously moving picture.

Electrons are a part of all matter and a knowledge of their properties is an important part of our understanding of the Universe. At the same time, we have been able to apply this knowledge in many ways that are useful and entertaining. Now we must return to some more fundamental topics.

23.20 Evidence for the structure of the atom

23.21 Electrons

The evidence comes mainly from experiments with cathode rays. This shows that all matter contains electrons and that in all matter they have the same mass and charge. This suggests that an electron is one of the fundamental particles from which matter is made.

23.22 Positive rays

These were discovered by Goldstein, one of the early workers with discharge tubes. In one of his tubes he placed the cathode half-way along the tube (fig. 23.12). The cathode had holes in it. The screen was painted with phosphor and was at the end of the tube opposite to the anode. When a discharge occurred in the tube, the screen glowed in spots, corresponding to the holes in the cathode. Rays were passing from the anode and through the holes in the cathode. He named these canal rays, but they were later re-named *positive rays*. Wein found that these rays are deflected by electrical and magnetic fields. They are deflected in the opposite direction to cathode rays, showing that they are positively charged. They are deflected much less strongly than cathode rays, showing that the particles are much more massive. The ratio e/m depends on which gas is in the tube. Calculations show that m has the same value as the mass of the atoms of gas. The positive rays are positive ions, formed in the region between the anode and cathode.

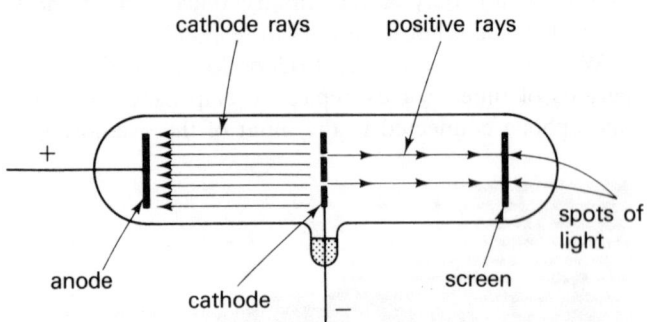

cathode rays   positive rays

+

anode   cathode   screen

spots of light

Fig. 23.12   Discharge tube for the production of positive rays.

Thomson calculated e/m for hydrogen ions. He found it about 2 000 times smaller than e/m for an electron. If we assume that the hydrogen ion is a single proton, and its charge is equal but opposite to that of an electron, the mass of a proton must be about 2 000 times greater than the mass of an electron.

23.23 Rutherford's experiments

Atoms consist of protons, neutrons and electrons. There are various ways in which these might be put together to make an atom. The hypothesis of the *nuclear atom* best fits the information we have available at present.

In an early experiment, Rutherford found that α-particles easily passed through a thin sheet of mica. Though the mica appeared to be solid, there was some space which the particles could pass through. He noticed that not all particles went straight through the mica. Some were slightly deflected. His hypothesis was that deflection occurred when an α-particle passed close to the nucleus of an atom. Both are positively charged and the repulsion would deflect the α-particle.

The experiment which first proved the existence of the nucleus was done by Geiger and Marsden at the suggestion of Rutherford. They used a radium source surrounded by screens to make a narrow beam of α-particles. In this beam they placed a very thin sheet of gold. The particles were detected by a screen of zinc sulphide, which they observed closely, through a microscope. The screen could be moved around the gold sheet to detect particles emerging in various directions. They found that:

(a) most of the particles passed straight through the gold,
(b) some particles were deflected through small angles,
(c) a few particles were deflected through large angles; they were turned almost in the opposite direction.

Figure 23.13 shows how these results were explained by Rutherford, on the hypothesis that the nucleus of the atom is small compared with the size of the whole atom.
(a) Particles which pass through the outer regions of the atom (the region of the electron cloud) are not deflected.

(b) Particles which pass nearer the nucleus are repelled as they pass by and are slightly deflected.
(c) Particles which are travelling directly towards the nucleus come very close to it; they are strongly repelled. These particles are the few that are deflected through a large angle.

The fact that only a few of the particles are strongly deflected shows that most of the atom consists of empty space. To make a large-scale model of an atom, we could take a tennis ball to represent the nucleus. Around this we have a cloud of the finest dust particles. To this scale the cloud is several *kilometres* in diameter. The cloud contains only a dozen or so particles (only one in the example of hydrogen!). This model shows what is meant when we say that an atom consists mainly of empty space. Yet matter feels firm when we touch it. This firm feeling comes not from the fact that there is solid material there, but from the fact that the small amount of matter is being held together by the exceedingly strong forces between the atom and the molecules of which it is composed.

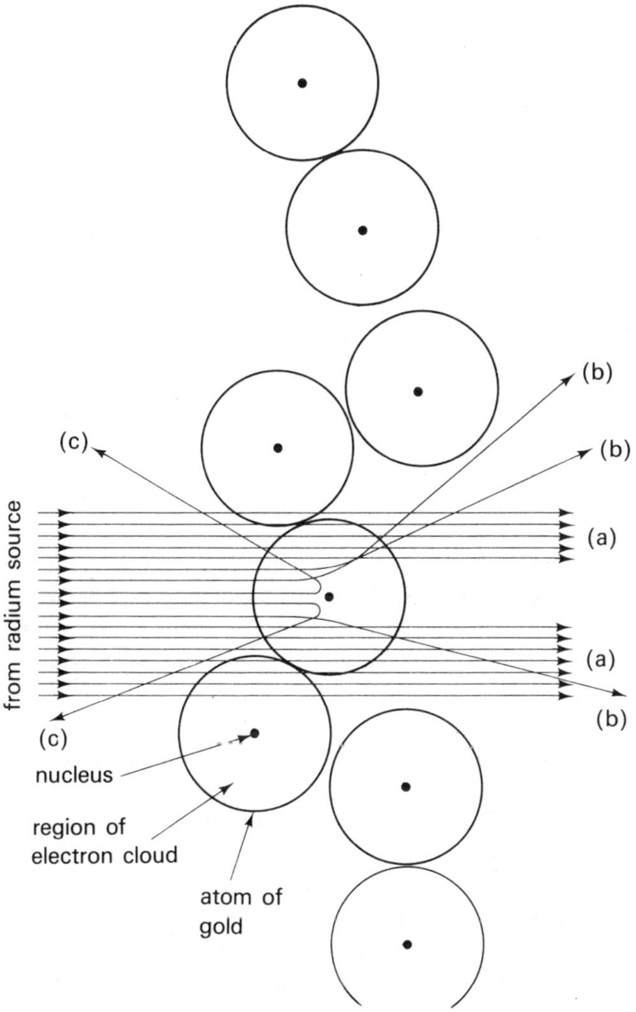

Fig. 23.13   Rutherford's explanation for the deflection of alpha-particles by atoms of gold.

Fig. 23.14   A modern electron-microscope in use.

PART G PHYSICS IN ACTION

24 Working with lenses

24.00 Cameras

24.01 Pinhole camera

Instructions: improving the pinhole camera

1 Make a pinhole camera as in fig. 5.6, but instead of the pinhole, make a larger hole (about 5 cm diameter). Close this hole with a black card or piece of aluminium foil glued or taped into place. Make a single pinhole (about 0.5 mm diameter) in the centre of the card or foil.

2 Aim the camera at an electric lamp, about 2 m away. Describe the image you see.

3 Make the pinhole larger (not more than 1 mm diameter). In what ways does this alter the image? Does this change improve the camera?

4 Make nine more pinholes (1 mm diameter) scattered all over the card or foil. What can you see on the screen now?

5 Get a lens from your teacher and slide it across in front of the camera, so that it covers all the pinholes. Watch the screen while you do this. What happens?

6 Make a large hole (about 5 cm diameter). Look at the screen, then slide the lens across in front of the hole. What do you see?

7 Move the lens slightly nearer to the screen and slightly further from the screen. What happens to the image?

Questions

1 What word is used to describe what happens to rays of light when they reach the lens?

2 What can you say about the direction of the rays of light after they have left the lens and have passed through the ten pinholes, at step 5?

24.02 Lens camera

For taking photographs, the lens camera needs some additions. Figure 24.1 is a section through a simple type of photographer's camera. Like your lens camera, it has a light-tight box and a converging lens. The screen is

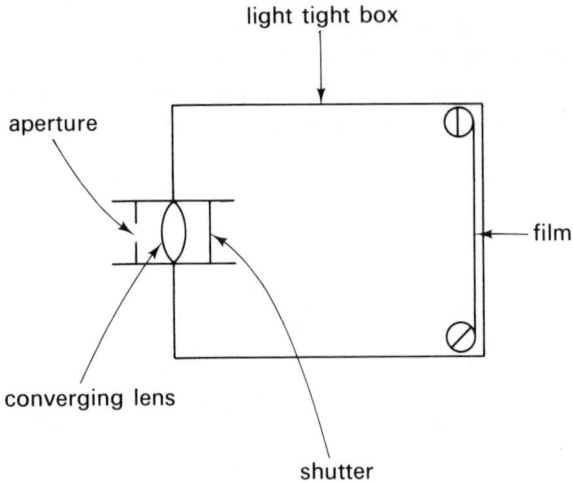

Fig. 24.1 *Section through a simple lens camera.*

replaced by film that is sensitive to light. It can be developed by using chemicals to make a photographic negative or perhaps a colour transparency. In all but the simplest cameras the lens can be moved to focus the image sharply. In reflex cameras there is a mirror which reflects the image into the viewfinder instead of letting it fall on the film. You can see the image as you focus it and know exactly what the final picture will look like. When you are ready to take the photograph the mirror is automatically moved to one side, so the image can fall on the film.

To 'take the photograph' the image is allowed to fall on the film for a short period of time, usually 0.02 s or less. During this time, light makes changes in certain molecules in the film; the effects of this are seen when the film is developed. To allow the film to be exposed to light for short periods there is a shutter which opens and closes rapidly when the shutter release button is pressed. In many cameras it is possible to control the length of time for which the shutter is open; the time required depends on the amount of light shining on the scene and the sensitivity of the film. The aperture is a metal plate with a hole in it; the diameter of the hole can be altered, allowing more or less light to pass into the camera. Later we will note another effect of varying the diameter of the hole or aperture. The aperture is sometimes called the 'stop'. Its diameter is marked on the camera as an aperture scale consisting of a series of numbers (5.6, 8, 11, 16 and so on). The larger the number, the smaller the aperture and the smaller the amount of light reaching the film.

The cine camera (or movie camera) is a special form of lens camera used for taking motion pictures. When the camera is running the shutter opens for a short period 18 times per second. The film is moved through the camera by a clockwork or electric motor, but is held still every time the shutter opens. In this way, 18 photographs are taken every second, showing stages in the motion of the subject. The camera records a sequence of still pictures. When the film has been processed it is projected as a sequence of still pictures as the film moves through the cine projector. The shutter of the projector opens as each picture is centred on the projector lens. Our eyes receive a sequence of still pictures; because the pictures are seen one after another in rapid succession we get the *illusion* of motion. In most amateur cameras the picture frequency is 18/s; some run at 16/s. Professional motion-picture cameras and projectors operate at 25 pictures per second, and so does the television set (23.14); this gives smoother action with less flickering of the screen image, but it is relatively more expensive because of the extra length of film needed.

While on the subject of cameras and projectors we should look at the slide projector (fig. 24.2). In many ways this is like a lens camera working in reverse. The slide is placed at AB; it is illuminated by a lamp, L. A mirror, C, and lenses, D and E, collect as much light as possible from the lamp; they direct it through the slide to light it evenly and brightly. In most projectors the lamp has high wattage and gives out a lot of heat. To avoid overheating the slide,

there is an electric motor which drives a fan, F, to blow cool air through the projector case. The light passes through the projector lens, P, to the screen. The lens focusses the light to make a sharp image on the screen, several metres away.

Questions
1 Explain why the slide is placed upside down in the projector.
2 Explain why a powerful lamp and a system of mirrors and lenses, D and E, are needed.

24.03 Light rays and lenses

At the beginning of this chapter you worked with rays of light in three dimensions. Your teacher may use a smoke-filled box to show you rays passing through the air after they have been refracted by a lens. It is easier to work in only two dimensions as you did in chapter 5, using ray streaks across the surface of a sheet of paper. For this you need special *cylindrical* lenses. Their surfaces are curved only one way (fig. 24.3). When you use them with a ray box or with a lamp and a screen with slits, you see the rays passing along one plane. This gives you the same view as in most of the diagrams which show how lenses work.

Instructions: working with converging lenses
As you follow these instructions draw diagrams to illustrate your results.
1 Arrange a ray-box or lamp and comb screen (many slits) as in fig. 24.3. You should see a fan of diverging ray-streaks. From what point do the streaks diverge?
2 Put a cylindrical converging lens in the path of the fan of rays. In which direction are the rays refracted? Are there rays that pass through the lens but are not refracted? If you cannot find any, shift the lens slightly sideways and note what happens.

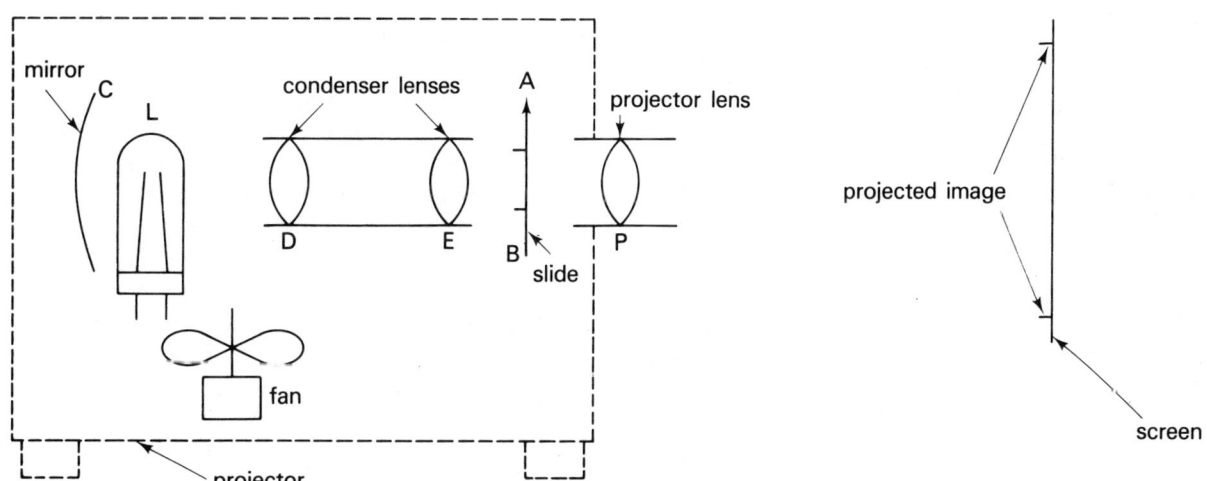

Fig. 24.2   Section through a slide projector.

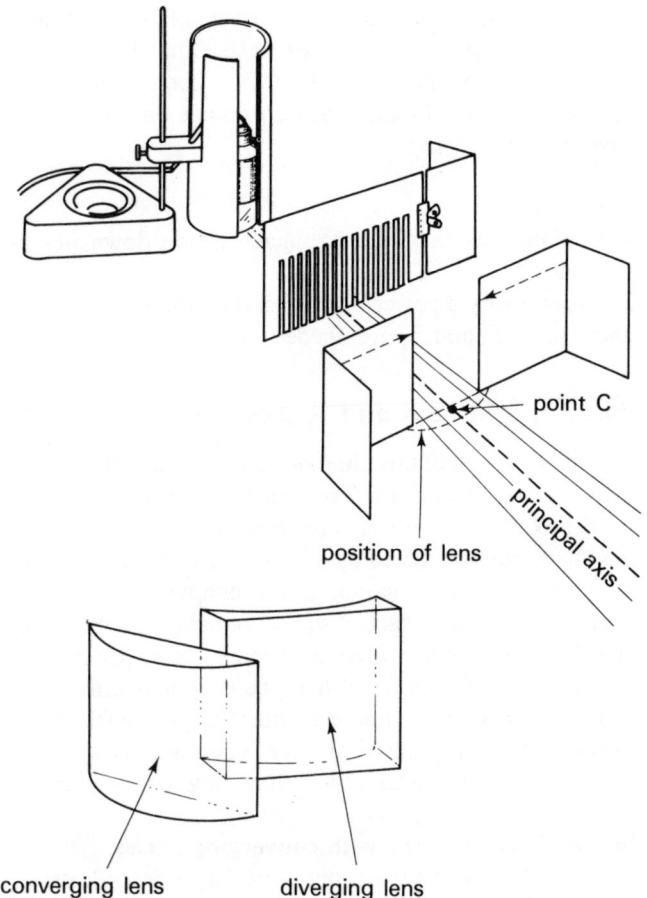

Fig. 24.3 *Investigating the action of a converging lens.*

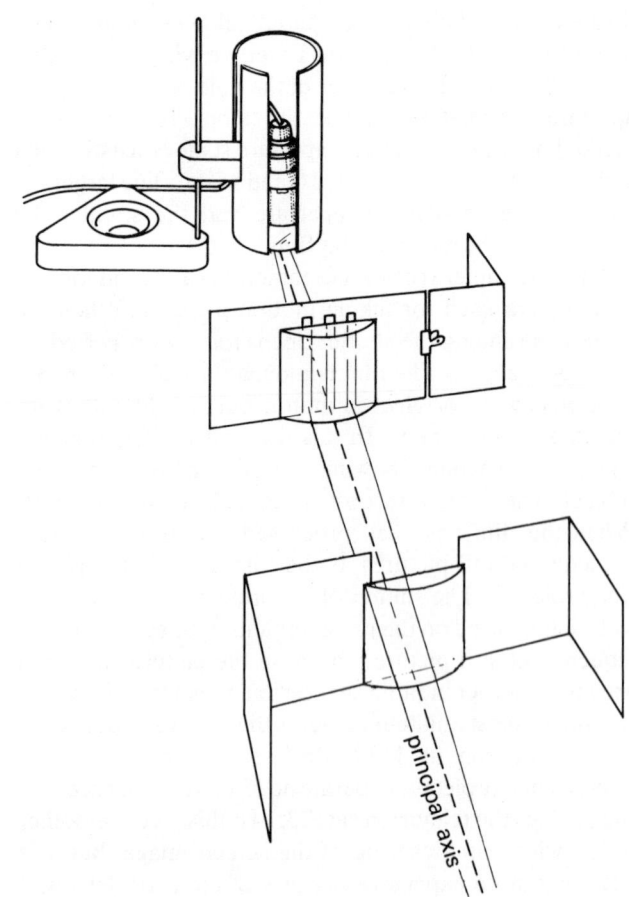

Fig. 24.4 *Investigating the effect of a converging lens on parallel rays.*

3 The rays converge towards the principal axis. Do they all converge towards the same point?

4 In fig. 24.3 the screens at the side of the lens allow rays to reach the whole lens. Reduce the gap (or aperture) between the screens, by sliding them slowly together in the direction shown by arrows. Watch what happens as the outer rays are cut off by the screens. Explain the action of the variable aperture of a lens camera (fig. 24.1).

5 Replace the comb screen with a screen that has three slits. Arrange this so that the central ray streak passes exactly along the principal axis of the lens, with the other two streaks symmetrical on either side of the axis. These should now converge exactly at the same point on the axis. Measure the distance along the principal axis between this point and the lens. The distance is measured to the optical centre of the lens (point C in fig. 24.3).

6 Replace the lens with another which is curved more strongly or less strongly. Find the point at which the rays converge and measure its distance from the optical centre. Is this distance greater than or less than the distance measured at step 5?

7 Move the lens a little closer to the screen. In what way does this affect the distance from the optical centre of the point at which the rays converge?

Instructions: investigating the effect of a converging lens on parallel rays

Rays coming from a distant object are almost parallel. In this investigation we do not use a distant lamp, as the rays would be too faint to see. Instead, we place a converging lens close in front of the screen (fig. 24.4). It refracts the rays and makes them parallel.

1 Place the lens just in front of the screen. Move it slightly backwards or forwards until the emerging rays are parallel. Check that they are parallel by measuring the distance between them at several points along their length.

2 Put a converging lens in the beam of parallel rays, so that its principal axis is exactly in line with the central ray. The rays converge at a point on the principal axis, as in the previous instructions. When the rays incident on a lens are *parallel* we call this point the principal focus. Measure the distance between the optical centre and the principal focus. This distance is called the *focal length*.

3 Measure the focal length of other converging lenses in the same way.

The results obtained above lead us to a useful rule:

Any incident ray that is parallel to the principal axis of a lens is refracted through the principal focus.

Light can pass in either direction along the paths of the ray streaks. We can state another rule:

Any incident ray that passes through the principal focus of a lens is refracted in a direction parallel to the principal axis.

We shall use these rules often, beginning in the next experiment.

Instructions: working with light from two distant sources

1 Set up two lamps with three-slit screens and converging lenses to make two beams of parallel rays (fig. 24.5a).
2 Mark a point C on the paper to locate the optical centre. Direct the beams so that their central rays converge exactly on C. These represent beams of light coming from two distant sources, both incident on a lens, which has its optical centre at C.
3 Put a converging lens at C, so that its principal axis is exactly in line with the central ray from Lamp 1. If F is the principal focus, where is I_1 the image of lamp 1?
4 Where is I_2, the image of lamp 2?
5 Describe the path of the central ray from lamp 2.

6 Move lamp 2 a little further from lamp 1, keeping the central ray directed at C. Is the answer to the question above always the same answer?

This demonstration can be compared with a lens camera pointing at a distant object (fig. 24.5b). Rays from lamp 1 represent rays coming from the bottom of the tree. Rays from lamp 2 represent rays coming from the top of the tree. The image of the top of the tree is formed *below* the image of the bottom of the tree. The result is an inverted image of the tree. This confirms the result of the work with the pinhole camera. It also gives us another useful rule:

Any incident ray that passes through the optical centre continues in a straight line.

Instructions: working with diverging lenses

1 Set up a lamp, screen and converging lens to give a beam of three parallel ray streaks, as in the previous instructions.
2 Place a diverging cylindrical lens in the beam with its principal axis in line with the central ray. From where do the diverging rays appear to come?

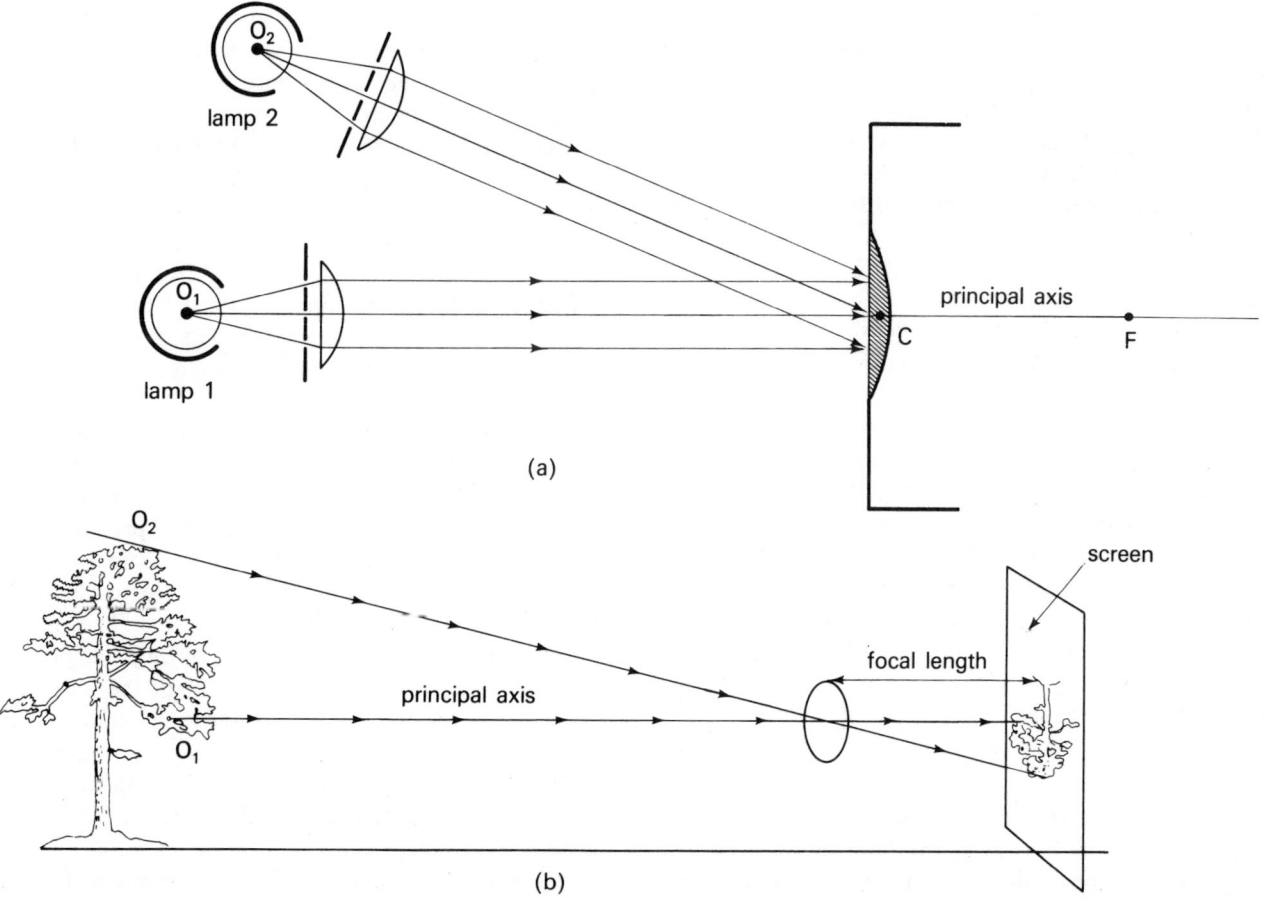

(a)

(b)

Fig. 24.5 (a) Investigating the action of a converging lens on light from two distant sources; (b) Application of the same idea to explain the action of a lens camera.

3 To locate the point exactly, lay a ruler along one of the diverging rays; draw along the line with a pencil to mark the path of the ray. Repeat for the other diverging ray and the central ray.

4 Remove the lens. Continue the drawn lines until you find the point from which they diverge. Where is this point?

5 Measure the focal length of the lens.

6 Experiment with the rays until you are able to complete these statements of rules:

(a) *Any incident ray parallel to the principal axis is refracted as if it had come through* _____ .

(b) *Any incident ray which is directed at the principal focus is refracted and emerges* _____ .

(c) *Any incident ray through* _____ *continues in a straight line.*

24.04 Measuring focal length

Now that we have some rules for the refraction of rays of light in cylindrical lenses we apply these rules to ordinary (spherical) lenses.

Instructions: a rough way of measuring the focal length of a converging lens

1 Hold a converging lens in one hand and a sheet of white card in the other.

2 Focus an image of a distant window on the card. Better still, if the day is bright, focus the image of an even more distant object such as a tree, or a cloud, or the Sun.

3 Ask your partner to measure the distance between the lens and the card.

Questions

1 What is the focal length of the lens?

2 Why must you focus on a *distant* object, not a near one?

In the previous instructions you made an image on the card; rays were reflected from the card to your eye; you saw the image on the card. This is a *real* image. So far, we have made real images by using a converging lens to converge the rays toward the principal focus. You will have noticed that, after converging to the principal focus, the rays diverge again as they travel on towards the edge of the paper. In effect, the real image at the principal focus *acts as a source of light.* If a real image is a source of light, we should be able to see it, without the need to focus it on card. Try the following experiment.

Instructions: locating the real image

1 Stand several metres away from a window or electric lamp. Hold a converging lens in front of you at arm's length.

2 With the other hand, hold a piece of thin paper (typist's copy paper or tracing paper) between your eyes and the lens. Move the paper backwards and forwards until you see a sharply focussed image of the window or lamp. You are looking from behind the paper screen; the image is visible because some of its light passes through the thin paper.

3 Focus your eyes on the image on the paper.

4 Slowly move the paper sideways. Hold the lens still and keep the paper at the same distance from the lens, so that the image remains sharp as you withdraw the paper. Keep your eyes focussed on the edge of the paper; as the paper is moved away and the lens becomes visible, do not adjust focus to look at the lens. With practice, you are able to see the image of the window or lamp located at a definite position in space between your eyes and the lens. This is what we mean by a real image.

5 Hold a diverging lens between your eyes and a window or lamp. Where does the image appear to be? Is it possible to focus this image on a screen? In what ways does the image differ from the image formed by the converging lens at step 3?

Instructions: an accurate method for measuring the focal length of a converging lens

1 Place a lamp (preferably with 'pearl' glass) behind a screen in which there is a hole with cross-wires (fig. 24.6a). The wires are to help you focus the image sharply.

2 A short distance away, place a vertical plane mirror parallel to the screen.

3 Place the lens close in front of the mirror. The lens must be vertical and parallel to the screen.

4 Alter the distance, d, between lens and screen until a sharp image of the cross-wires is focussed on the screen, as seen in the lower picture, fig. 24.6a. If you have difficulty in focussing, find the focal length roughly as before, then place the lens so that its distance from the screen, d, is equal to the approximate focal length. Adjust the distance for sharp focus.

5 Measure d. Assuming that the rim of the lens is below the optical centre, distance d is the focal length of the lens.

The explanation is in fig. 24.6b. Rays diverge from the cross-wires. If the cross-wires are at the principal focus, the rays are refracted parallel to the principal axis. The parallel rays strike the mirror and are reflected back along the direction from which they have come. They are still parallel and therefore are refracted through the principal focus by their second passage through the lens. They converge on the cross-wires. To see the position of sharp focus we displace the lens slightly so that the image is focussed on the screen *beside* the cross-wires. This produces only a small error. The distance from lens to cross-wires is the focal length of the lens.

A lens of focal length 1 m converges a beam of parallel rays to a point exactly 1 m from the optical centre of the lens. A second lens of focal length 0.1 m converges the rays to a point 0.1 m from the optical centre. We say that

254

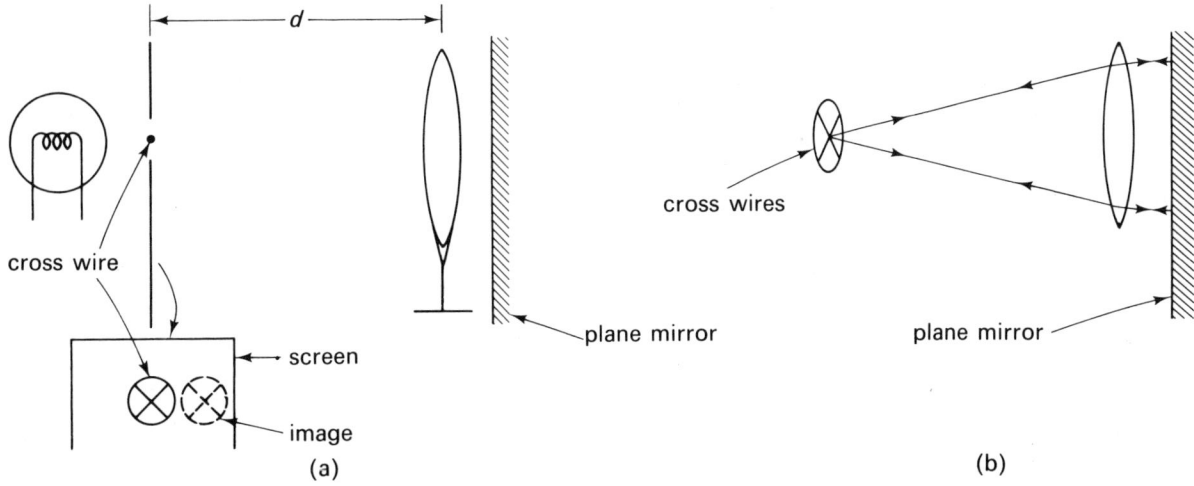

Fig. 24.6 Measuring the focal length of a converging lens.

the second lens is more powerful than the first lens, because it has a greater converging action on the rays. This leads us to the idea of the *power* of a lens. This has nothing to do with power as the rate of converting energy. It is not measured in watts. The power of a lens is defined by:

$$\text{power} = \frac{1}{\text{focal length}} \; (\text{power in } \textbf{dioptres}).$$

Focal length must be measured in metres. The unit of power, symbol **D**, is not part of SI. It is useful to know of it, for lenses are often marked to indicate their power. If a lens is marked 2D it means that its power is 2 dioptres. This tells us

$$\frac{1}{\text{focal length}} = 2$$

We calculate that focal length, $f = \frac{1}{2} = 0.5$ (f in m)

Similarly, a lens of power 5D has focal length 0.2 m. For converging lenses, power is written as a positive quantity. For diverging lenses, power is negative. A lens marked −10D is a diverging lens of focal length 0.1 m.

24.05 Ray diagrams for lens calculations

In fig. 24.7 the lens is represented by a vertical line marked with triangles at its ends. This gives us an exact indication of where refraction occurs. The triangles point outwards to indicate a converging lens, and inwards to indicate a diverging lens. Having drawn the principal axis and the position of the lens, point X, where these cross, is the optical centre. At a distance from C which depends on the scale of our drawing, we mark the principal focus, F. Since light can pass through the lens in either direction we mark F on both sides.

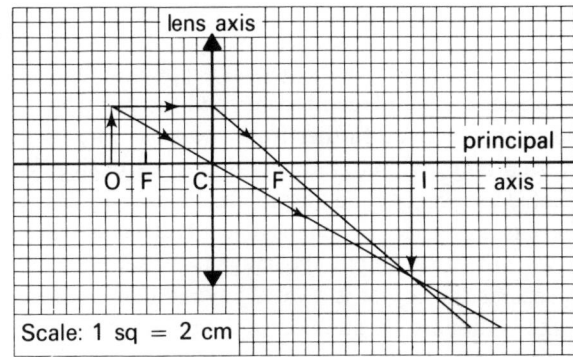

Fig. 24.7 Ray diagram showing the action of a converging lens when the object is just beyond the principal focus.

In fig. 24.7, focal length is 10 cm. The arrow, O, represents an object 8 cm high, placed 15 cm from the lens. The problem is to locate the position of the image and to find out what its features are. We draw two rays coming from the point of O and use the rules we have discovered:

(a) *A ray parallel to the axis is refracted through the principal focus*

Draw a ray from the point of O, parallel to the principal axis as far as the lens axis. From the point where it meets the lens axis, draw a ray, passing through F on the other side of the lens, towards the right-hand edge of the paper.

(b) *A ray through the optical centre continues in a straight line*

Draw another ray from the point of O, going straight through the optical centre.

We now have two rays from the same point of the object. These meet where the image of the point of O is located. A similar construction for pairs of rays from other parts of O tells us the location of other points in the image

255

I. We need not draw these other points. We know that a ray from the base of O passes straight along the principal axis; the image of the base of O must also be on the principal axis, at the same distance from the lens as the other points of the image. We know the position of both ends of I, and can draw it as an arrow, marked I in the figure. If we measure I we find it is 16cm high; it is *magnified*. It is 30cm from the lens. It is *inverted*. The two rays we have drawn, cross at the point of I and diverge again beyond it; this tells us that the image is *real*.

Questions
1 Construct a diagram like fig. 24.7 in which O is 20cm from the lens which has focal length 10cm. Locate I. Measure its size and distance from the lens. What type of image is formed?
2 Construct a diagram like fig. 24.7 in which O is 50cm from the lens which has focal length 10cm. Report on the features of I.

The examples above show what happens to the image if the object is first placed just beyond the principal focus and is then moved further from the lens. The image is inverted and real but gradually decreases in size. When the object is just beyond the principal focus (fig. 24.7), the image is magnified. When the object is at a distance equal to twice the focal length, the image is the same size as the object. As the object is moved beyond that distance the image is diminished and becomes smaller the further the object is taken from the lens. What happens to the image if the object is taken a long way from the lens? Figure 24.8 supplies the answer. The rays reaching the lens from the point of O are parallel (or near enough to parallel). They converge on a point in the plane of the principal focus. The image is inverted, real and diminished. Note that as the object is taken further and further from the lens the image gets closer to the lens, finishing at the principal focus when

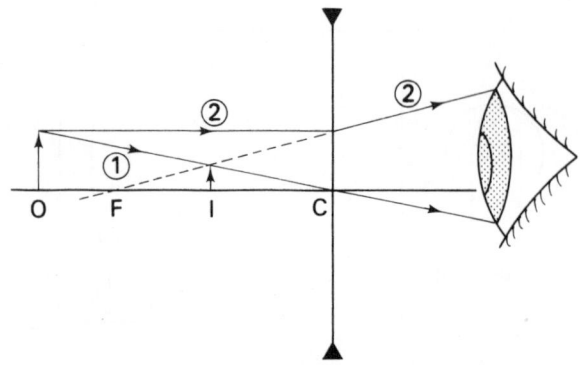

Fig. 24.9   Diagram showing the action of a diverging lens.

the object is at great distance. In the investigation on page 254 we used this fact, for we focussed on a distant object in order to locate the principal focus and measure focal length.

We can draw diagrams to solve problems involving diverging lenses. The diverging lens of fig. 24.9 has focal length 4cm. The object is 1cm high and is 5.3cm from the lens. First we draw ray 1, which passes through the optical centre of the lens. Then we draw ray 2, parallel to the principal axis. When it reaches the lens, it is refracted and *appears* to have come from F, as shown by the dashed line in the diagram. Both rays *appear* to have come from the point of the image I but, in fact, only one of the rays has passed through that point. We now locate the position of the point of the image, I, where the rays appear to cross, or more accurately, from where they appear to diverge. By measurement, we find that I is 2.2cm from the lens and is 0.42cm high. The image is diminished and erect (the correct way up). There is no point on the diagram where the two rays actually meet, except at the object itself. There is no point at which these rays converge to form a real image. Therefore the image must be virtual.

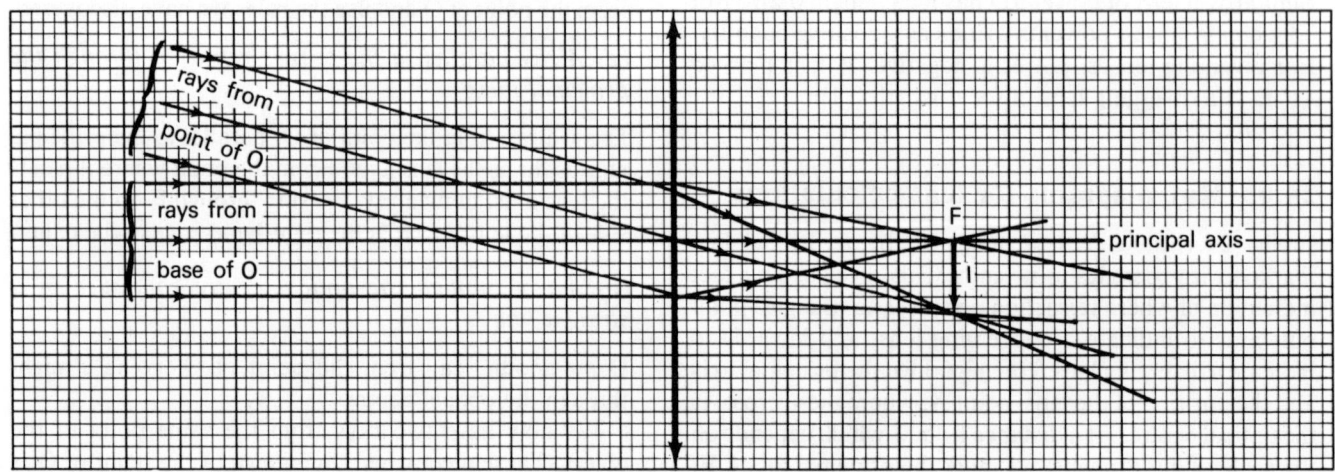

Fig. 24.8   Ray diagram showing the action of a converging lens when the object is at great distance.

Questions

1 Construct two or three diagrams like fig. 24.9 to investigate what happens to I as the distance between O and the lens is altered. What are the features of *all* images formed by diverging lenses?

2 A boy focusses an image of distant clouds and trees on a screen. Then he moves the lens to focus a sharp image of a window frame. The window is 4m from the lens, the power of the lens is 1D. In what direction and how far does the boy move the lens as he changes focus from the distant objects to the window frame? If a caterpillar 3cm long is crawling on the glass of the window, what is the length of the image of the caterpillar on the screen?

3 A girl has a camera which has a lens of focal length 24mm. She takes a photograph of a flower, placed 90mm from the lens. When the film is developed the image of the flower is measured; it is 7.5mm in diameter. What was the diameter of the flower?

4 A boy looks at a poster through a diverging lens. The poster is 0.6m high and is 1.1m from the lens. The image appears to be 0.1m high. What is the focal length of the lens? What is its power, in dioptres?

24.10 The eye

24.11 Structure

To study the structure of the eye, examine a dissected ox eye and look at large-scale models. The section shown in fig. 24.10 shows that the eye is like a camera in some ways; in other ways it is different. It is like a camera in having a strong light-proof container. It is like a camera in having a light-sensitive layer: in the camera there is

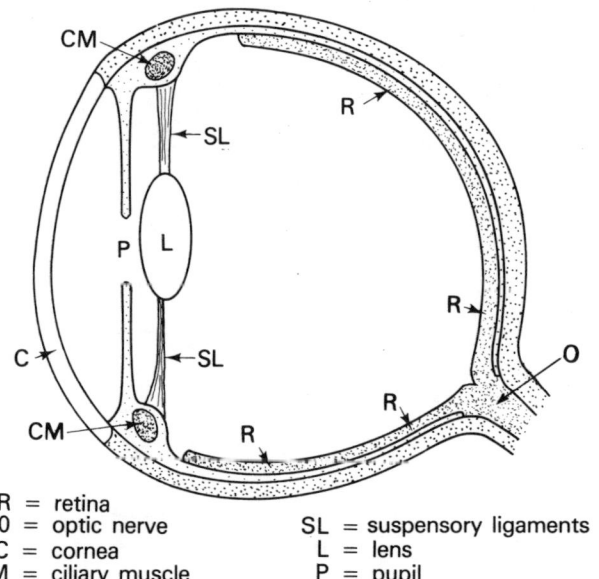

R = retina
0 = optic nerve
C = cornea
CM = ciliary muscle
SL = suspensory ligaments
L = lens
P = pupil

Fig. 24.10 Section through the eye (see text).

photographic film; in the eye is the retina, R. This consists of millions of nerve cells sensitive to light. Film is exposed once only and then developed; the retina operates continuously. Through the optic nerve, O, it sends nerve messages to the brain, giving us the sensation of sight. The images of objects in the world around us are focussed on to the retina by the transparent materials which make up the main bulk of the eye. At the front is the cornea, C, a tough but transparent layer. Its front surface is strongly curved, so it acts as a converging lens of short focal length. The space between the cornea and the lens, L, is filled with watery fluid. The space between the lens and the retina is filled with a jelly-like fluid. Both the fluid-filled spaces act as lenses. In this way the eye is very unlike a camera which contains only air, and uses only the lens for focussing. In front of the lens is a thin muscular sheet, often coloured brown. In some people it is blue, or greenish in colour. This is the iris. In the centre of the iris is a hole, the pupil, P. From the outside, the pupil looks like a black spot, because the interior of the eye is darker than the world outside. When we look at a house on a bright day, the windows appear black for the same reason. The aperture of the pupil is controlled by the brain, being increased in diameter when light intensity is low and being decreased in size when light intensity is high, as in bright sunlight. In this way it acts like the aperture of a camera. The retina is sensitive to the amount of light falling on it so it can detect images and thus tell us the shapes of objects, even when light is dim, as in the evening. In brighter light the retina is sensitive to colours too, particularly near its centre. In this region the nerve endings are closely packed together, giving us the ability to see things in fine detail.

24.12 Accommodation

The lens has liquid on both sides of it, so the refractive index at its surfaces is low. Refraction at its surface is not as great as it would be if the lens had air on both sides. The main function of the lens is not to focus images of the outside world but to *adjust* the focussing action of the cornea and fluid-filled spaces. The lens allows us to alter focus so that distant objects or close objects can be brought into sharp focus, as required. We call this action *accommodation*. The lens is slightly flexible; its shape can be altered. It is held in place by suspensory ligaments, SL. These are under slight tension, due to the slight excess pressure of the liquids in the eyeball. Tension holds the lens tense, making its surfaces only slightly curved. This makes its focal length relatively long; distant objects are sharply focussed by the combined action of cornea, fluids and lens. Around the suspensory ligament is a ring of muscle, the ciliary muscle, CM. When this muscle contracts the diameter of the ring decreases; this releases the tension in the suspensory ligament; the tension in the lens is released and it becomes rounder. Its focal length is

257

thus decreased. Objects close to the eye come into sharp focus.

Note that in the normal relaxed state the eye is focussed on distant objects. Muscular effort is needed to focus on a near object. This is why long periods of reading or other close work can tire the eye muscles. A period spent outdoors, looking at distant objects gives the ciliary muscles time to relax. When the ciliary muscles are fully contracted and the lens has its most rounded shape, the eye is focussed at its closest possible distance. This is the *least distance of distinct vision*. For the average person, this distance is about 25 cm. Objects closer to the eye than 25 cm cannot be sharply focussed.

24.13 Defects of the eye

People with *long sight* can focus sharply on distant objects, but not on objects as close as 25 cm. The eyeball is too short; the image of a near object is formed behind the retina (fig. 24.11a). Spectacles consisting of converging lenses are used to correct this defect (fig. 24.11b). Rays from close objects then enter the eye as if they had come from a greater distance and can be focussed on the retina. Long sight also occurs in some older people when the lens becomes unable to accommodate enough to focus on near objects.

The opposite defect is *short sight*, in which the eyeball is too long. Images of distant objects are formed in front of the retina (fig. 24.11c). Only near objects can be sharply focussed. Spectacles with diverging lenses correct this defect (fig. 24.11d). Rays from distant objects then enter the eye as if they had come from a short distance and can be focussed on the retina.

24.20 Magnifiers

To see something as clearly as possible we usually go as close to it as possible. If it is small, we pick it up and hold it close to our eyes. The smallest distance that we can hold it *and still see it clearly* is about 25 cm. This gives us the largest possible sharp image on the retina. The *visual angle* (fig. 24.12) is as large as possible. How can we increase the visual angle yet still keep the image sharply focussed? We need a magnifier.

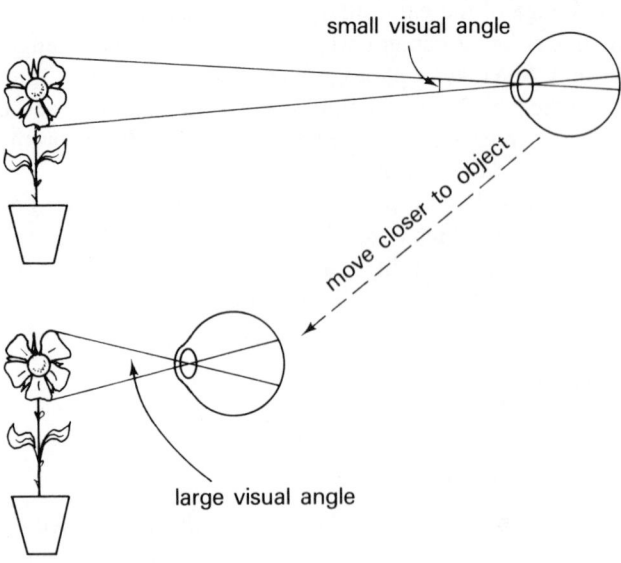

Fig. 24.12   Showing how visual angle varies with the distance of the object.

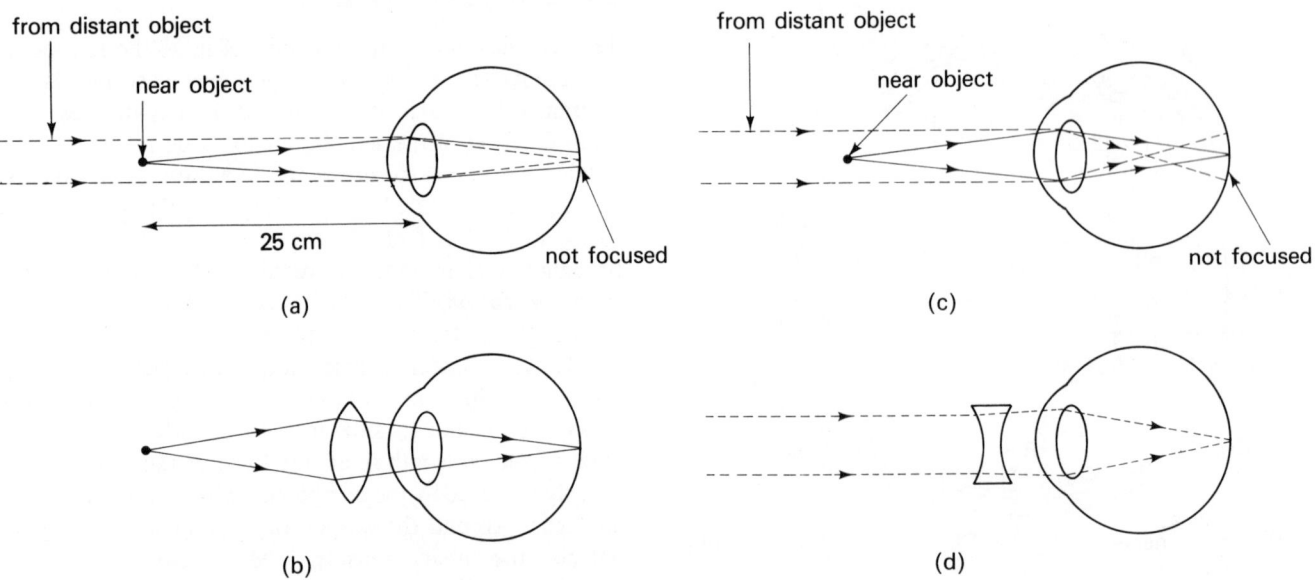

Fig. 24.11   Defects of the eye and their correction: (a) long sight; (b) long sight corrected by a converging lens; (c) short sight; (d) short sight corrected by a diverging lens.

258

24.21 Simple magnifiers

Instructions: using a converging lens as a magnifier

1 Use a converging lens of focal length about 10 cm. Hold this *close* in front of one eye.

2 Bring a finger slowly toward the lens, watching its image as you do so. Find out how close you can bring it, while still keeping it in sharp focus.

3 Ask your partner to measure the distance between your finger and the lens. Is this distance greater or less than the focal length of the lens?

4 Do not alter the focus of your eye. Take the lens quickly away. Is the finger sharply focussed now? Why not?

5 Replace the lens close to your eye. Hold a sheet of graph paper as close as possible yet still sharply focussed. Its image appears to be further from your eye than the paper. Roughly how far away does it appear to be?

6 Put a piece of graph paper on the bench. Look down on it with one eye (no lens) and move your head closer until the paper is at the least distance of distinct vision. Place the lens over the other eye; bring another piece of graph paper as close as possible to the lens to give the largest possible sharp image. With practice, you can see both sheets of graph paper at once. Compare the direct view (no lens) with the magnified view; by matching squares work out how much the graph paper is magnified by the lens.

Figure 24.13 shows how the lens was used to help increase the visual angle. The object is placed closer to the lens than the principal focus. Rays from the point of O diverge after being refracted. They appear to come from a point behind the lens. The lens gives an upright *virtual* image. Fig. 24.14 shows the position of the image, which is *magnified*. To compare visual angles we can use their *tangents*. For small angles like these the value of the tangent is close to the size of the angle expressed in radians. A **radian** is the SI unit of angle. (One radian is equal to approximately 57° on your protractor). Looking at O without a lens we can get no closer than 25 cm, a point 8.5 cm to the right of F in fig. 24.14. From this point the

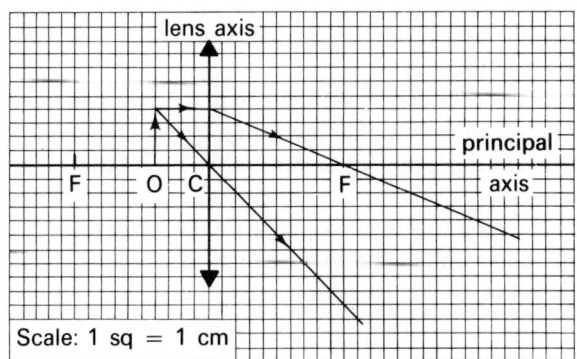

Fig. 24.13 Ray diagram showing the action of a converging lens when the object is between the lens and the principal focus.

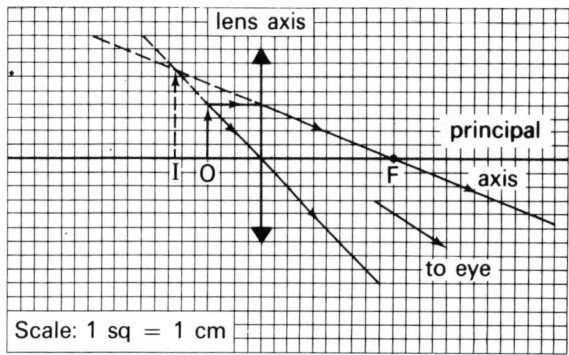

Scale: 1 sq = 1 cm

Fig. 24.14 Ray diagram showing the virtual image formed by a converging lens when the object is between the lens and principal focus.

visual angle is $4/25 = 0.16$ rad. When we view O through the lens and O is 4 cm from the lens, I is 6.5 cm from the lens and its height is 6.5 cm. To view this most closely we bring the eye forward, so that it is 25 cm *from the image*. At this distance the visual angle of the image is $6.5/25 = 0.26$ rad. By using the lens we have increased the visual angle from 0.16 rad to 0.26 rad.

Questions

1 In fig. 24.14 the object is 4 cm from the lens. Is it possible to increase the visual angle by placing O at some other position? Draw a diagram like fig. 24.14 but with I 25 cm away from the lens. If we assume that the eye is *close* behind the lens, this puts I at the least distance of distinct vision, where it can be seen sharply and its visual angle is as large as possible. Where must O be placed? What size is I? What is its visual angle?

2 Given the answers to question 1, what is the best procedure for using a converging lens as a magnifier?

3 An object 8 cm high is placed at the principal focus of a converging lens, focal length 20 cm. Draw a diagram to show where the image is formed. Describe the image.

4 If you were given a number of good quality converging lenses, all of different focal length, which one would you choose to obtain the greatest magnification? Draw a sketch to explain your choice.

24.22 Compound microscope

There is a practical limit to the use of a single converging lens as a magnifier. In theory we use a lens of very short focal length, but in practice it is not easy to make such a lens that is free of defects. The image obtained has poor quality. Another way of obtaining high magnification is to use two converging lenses: a compound microscope.

Instructions: making a model of a compound microscope

1 For the object to be magnified, use a piece of graph

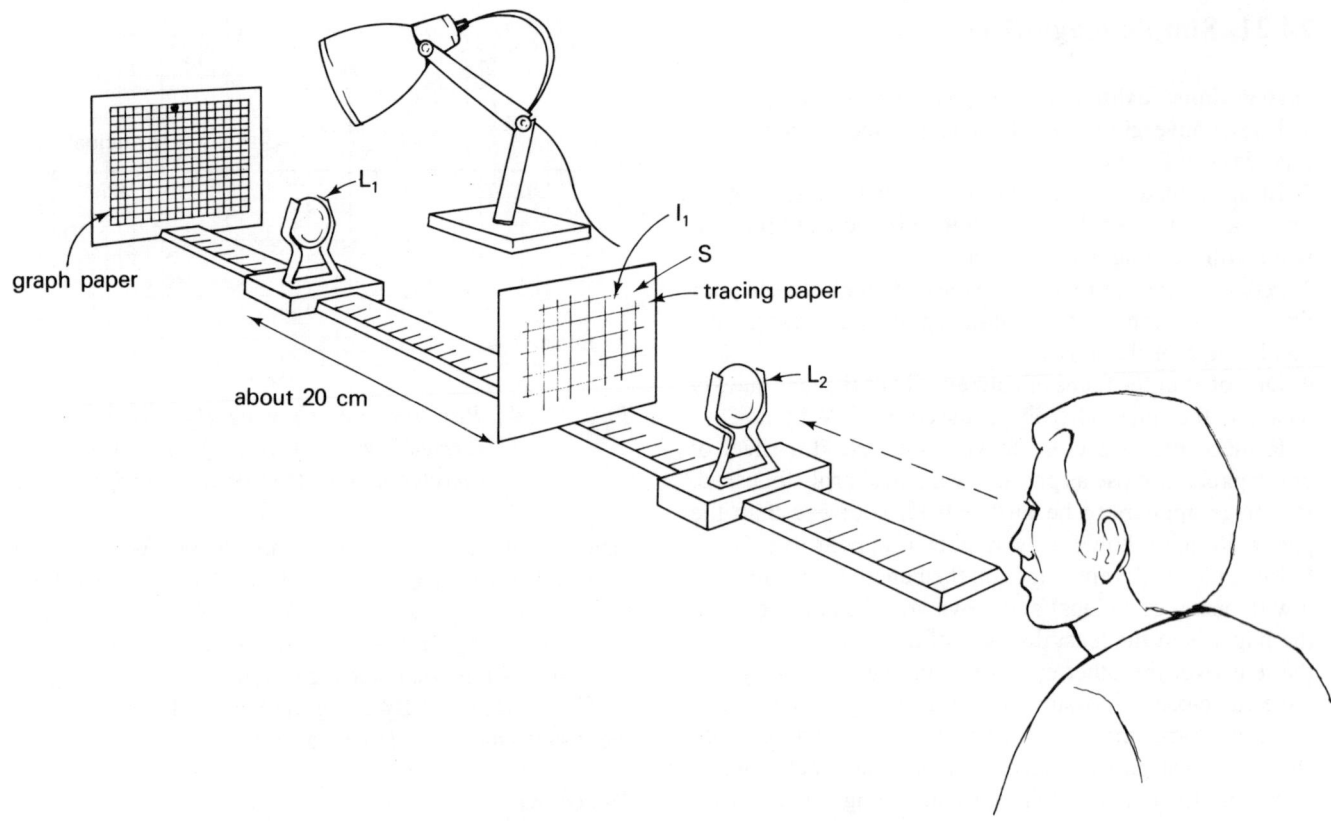

Fig. 24.15   Making a model compound microscope.

paper mounted on a vertical card; a light shines on the card (fig. 24.15).

2 Converging lens L_1 has focal length about 5 cm. Put this in front of O and focus an image I_1 on screen S. The screen is made from typing paper or tracing paper. Describe I_1.

3 Lens L_2 has focal length about 10 cm. Move this towards S until, with your eye *close to* L_2, you can see a magnified image of S, with I_1 showing through it. You are using L_2 to obtain a magnified view of I_1. This magnified image is I_2. Describe I_2.

4 Remove S. Adjust the position of your eye, possibly moving it back a little to get a better view. Look for the highly magnified image. Estimate its magnification by comparing it with the view of the graph paper seen directly with your other eye. Is the magnification greater than you obtain with a single lens?

If you examine a real microscope, you will see that L_1 is represented by a complex lens system of short focal length. This is the *objective*. The focal length is usually between 4 mm and 10 mm. With such short focal length it is close to the object when in use and does not need to have a wide aperture. At the other end of the microscope tube is the eyepiece, corresponding to L_2 of the model. This has two or more lenses combined to make a high-quality magnifier of the real image formed by the objective lens system. The virtual image formed by the eyepiece is about 25 cm from the eye.

Instructions: making a ray-streak model of a microscope

1 Use a lamp, a three-slit screen and two cylindrical converging lenses of short focal length to make a model to show how the compound microscope works.

2 Put one lens (the objective) close to the screen to focus a real image of the filament. The rays converge at some point on the principal axis, forming the image, I_1, of the filament.

3 Collect the rays diverging from I_1 and refract them through the second lens (the eyepiece). As they emerge they must be diverging, but diverging less than before refraction. If your lenses are flat on one side, you will obtain the best result by placing them so that their curved sides face together.

4 Move the lamp so that its filament is 2 cm to one side of the principal axis of the 'microscope'. The filament now represents the 'point' of an arrow-like object that is being magnified.

5 Use a ruler to draw in the ray-streaks. Then remove the lenses and continue the diverging rays backwards to locate the position of the virtual image formed by L_2. How far is this from the principal axis? Calculate the magnification of the microscope.

6 Now label the parts of your drawing so that you will have a ray diagram showing how a compound microscope works.

24.30 Telescopes

These are for increasing the visual angle of objects at a great distance, when we are not able to go closer to the object to see them in better detail.

Instructions: making a model telescope

We use two converging lenses as in the model of a compound microscope, but it is more convenient if the lenses are mounted to slide along a beam or strip of wood about 1 m long. The beam is best held horizontally in a stand, at eye level.

1 The object is a lamp bulb, shining at the far end of the laboratory.

2 For the objective, use lens L_1, focal length about 20 cm.

3 Focus the image of the lamp, I_1, on to screen S. Describe I_1. In what way does it differ from I_1 in the microscope?

4 Lens L_2 has focal length about 7 cm. Bring this to the screen. Adjust it until you see a magnified image of the screen, just as you did with the model microscope. You can see I_2, the magnified image of I_1. Keep both eyes open; look at the object with one eye while you focus the other eye on I_2. Your eyes are both accommodated for distant vision. As you focus L_2 you will make I_2 appear to be *as far away as the lamp*. It is *not* at the least distance of distinct vision. What is the distance between S and L_2?

5 Remove the screen. Look for the virtual image I_2. You may need to move your eye a slight distance forwards or backwards to get the best view. View the image with one eye and the lamp with the other; estimate the magnification. For a more accurate estimate, replace the lamp by a scale marked in centimetres; shine a lamp on this to make it easier to see.

Instructions: making a ray-streak model of a telescope

1 Use a lamp, three-slit screen and converging lens to produce a beam of three parallel rays. These represent rays coming from a distant object.

2 Converging lens L_1 has focal length 20 cm. Use this to produce a real image, I_1. How far behind the lens is the real image formed?

3 Collect the rays diverging from this image, by refracting them through L_2. This represents the eye-lens and has focal length about 7 cm. Adjust the position of L_2 so that the emerging rays are parallel. An image seen through the lens is apparently at a great distance: the same distance as the object. What is the distance between I_1 and L_2?

4 The ray streaks show the path of rays which come from some part of the object that is exactly on the axis of the telescope. To see what happens to rays from other parts of the object (not on the axis), move the lamp, slits and converging lens about 2 cm to one side, but turn them so that the central ray still goes through the optical centre of L_1.

5 Sketch in the rays with pencil so that you have a diagram to show how a telescope works. Measure θ_1, the angle between the central ray from the object and the axis of the telescope. This is the visual angle of the object. Measure θ_2, the angle between the central emerging ray and the axis. This is the visual angle of the image. Calculate the magnification of the telescope, $M = \theta_2/\theta_1$.

6 If the focal length of L_1 is f_1, and the focal length of L_2 is f_2, what is the distance between L_1 and L_2?

7 If the perpendicular distance between I_1 and the principal axis is x, we can say that $\theta_1 = x/f_1$ and $\theta_2 = x/f_2$. This approximation is allowed because, when angles are small, $\theta \simeq \tan\theta$. We can express the equation for magnification in another way:

$$M = \frac{\theta_2}{\theta_1} = \frac{x}{f_2} \times \frac{f_1}{x} = \frac{f_1}{f_2}$$

If you could use different lenses in the telescope, what changes would you make to get higher magnification?

8 For high magnification the two lenses must be a long distance apart. If the telescope must be portable, this is not convenient. In what two ways can this problem be overcome?

Most telescopes have an objective lens of large diameter, often 10 cm or more. This collects much more light than the pupil of the eye, which is only a few millimetres in diameter. The image seen through a telescope may appear much brighter than the direct view of the scene. To an astronomer the increase in brightness is often much more important than high magnification. By using a telescope he can see millions of stars that are invisible to the unaided eye. He can also see clouds of faintly luminous gas, that cover quite large areas of the sky, yet are invisible because they are so faint.

The image of the telescope is inverted. This is no disadvantage to astronomers but for general use an extra lens is added. This inverts the inverted image produced by the objective. The erect image is then magnified by the lens of the eye.

Projects

1 In this chapter we have studied cylindrical and spherical lenses. Find out about other kinds of lens; make charts, try your own experiments and get all the information you can about them. Here are some clue words to help you find the information: Fresnel lens, coastal light-house, spotlight, overhead projector, quartz lenses, ultraviolet microscopes, lenses for microwaves, magnetic and electrical lenses for electron beams, electron microscope.

2 Make and test a 'sound' lens, using a balloon filled with carbon dioxide.

25   Working with reflectors

In chapter 5 we studied reflection of light from plane mirrors. Now we look at reflection from other kinds of surface.

25.00   Converging mirrors

These are sometimes called concave mirrors.

Instructions: investigating reflection from a wide, converging mirror

1 Use a lamp and comb of slits to make a fan of rays, as in fig. 24.3, but without the lens and screens.
2 Place a large converging mirror in the fan. This should be a cylindrical mirror specially made for use with this kind of apparatus; it should be almost a complete semi-circle. Alternatively, cut a narrow strip of sheet aluminium and bend it into a semicircle.
3 Adjust the position of the mirror. Try to make the reflected rays converge to one point. Can you find any position in which this happens?
4 Take another screen with a single wide slit in it; this slit is wide enough to cover all except three of the slits in the comb. Place this wide slit in front of the comb. As you slide it slowly across the front of the comb, light from three slits passes through a different set of three slits each time the comb is moved. For each set of three slits, mark the point where the three rays converge (or focus). As you slide the wide slit and mark each point, the line of dots traces a *caustic curve*. Join the row of dots to show the shape of the caustic curve. This mirror does not give a sharply focussed image. What could be done to make it give a sharper image? (Hint: see instructions page 251, step 4).

Instructions: investigating reflection from a converging mirror with narrow aperture

1 Use a lamp and a screen with three slits. In front place a converging lens to produce a beam of three parallel rays, as in fig. 24.4. You also need a cylindrical converging mirror of narrow aperture.
2 Recall the topic of converging lenses (24.03 to 24.05). Decide which is the principal axis of the mirror. Place the mirror so that its principal axis lies along the central incident ray streak.
3 The centre of the mirror is called the pole (it corresponds with the optical centre of the lens). Observe the reflected rays and decide where the principal focus is located. Measure the distance between the principal focus and the pole. What do you think this distance is called? Does this type of mirror produce a real image or a virtual image when reflecting parallel rays?
4 Use the apparatus to discover the answer to these questions. What happens to an incident ray that:

(a) is parallel to the principal axis?
(b) strikes the pole at an angle θ to the principal axis?
(c) passes through the principal focus before striking the mirror (not necessarily at the pole)?
(d) passes through the centre of curvature of the mirror before striking the mirror (not necessarily at its pole)? The centre of curvature is on the principal axis and its distance from the pole is *twice* the focal length.

If you look at a reflection of the room in a small spherical converging mirror, you can see that the images of objects are real inverted and located in front of the mirror. This applies if the objects are further from the mirror than its principal focus. If the object is closer than the principal focus, the image is virtual, erect and magnified. This is made use of in a shaving mirror or make-up mirror which has a focal length of about 0.5 m. To use the mirror, you put your face close to it; the image is virtual and erect as in an ordinary plane mirror, but is magnified.

Questions
1 Draw a ray diagram to show how a converging mirror forms an inverted image of a distant object. Is the image magnified or diminished? (Hint: use the rules given above.)

2 A converging mirror has focal length 20 cm. An object 5 cm high is placed 40 cm from the pole of the mirror. Draw a ray diagram. Describe the image produced.

3 An object is placed 15 cm in front of a converging mirror of focal length 10 cm. Describe the image. If the image is 2 cm high, what is the height of the object?

4 A girl's face is 0.2 m from the pole of a make-up mirror of focal length 0.5 m. By what amount is it magnified?

Large converging mirrors have been built to collect sunlight and concentrate its energy on a small area. Solar furnaces are used to produce extremely high temperatures for experimental purposes and in the generation of electricity. On the smaller scale, solar ovens are used for cooking as a possible way of avoiding the use of fuels.

Instructions: investigating reflection from a parabolic converging mirror

1 Set up a lamp with comb and converging lens to produce a large number of parallel ray streaks.

2 Place a parabolic mirror of wide aperture in the beam. Does this converge the rays to one point?

3 Mark the position of the principal focus. Measure the focal length of the mirror.

4 Remove the lamp, comb and lens. Place a point source of light at the principal focus of the mirror (fig. 25.1). Use a small flashlamp bulb. Place a small card beside the lamp so that no light can go directly from the lamp to the comb. Light passing through the slits is all reflected from the mirror. What do you notice about the rays as they come through the slits?

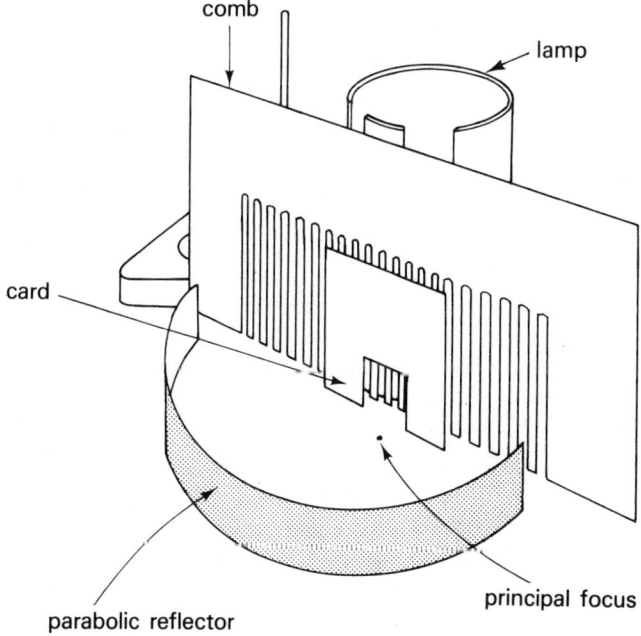

Fig. 25.1 *Investigating the properties of a parabolic reflector.*

When we need a wide beam of parallel rays, we use a parabolic reflector. Reflectors of this kind are found in automobile headlamps, in flashlamps and in searchlights. We also use parabolic mirrors in reflecting telescopes, to receive parallel rays from distant stars and focus them to a sharp image. Such a mirror does the same job as the objective lens of a telescope (24.30). The mirror is at the end of the telescope tube furthest from the star and the image is formed at some position along the tube, in front of the mirror. To be able to observe the image we usually place a small plane mirror in the tube to reflect the converging rays out to the side of the tube. An eyepiece mounted on the side of the tube is used to magnify the image. The advantage of using a mirror for an astronomical telescope is that it is easier to make a large mirror than to make a large lens. Mirrors of extremely large diameter have been made; that at the observatory on Mount Palomar, USA, is more than 5 m in diameter. Such a mirror collects a lot of light, making it possible to view faint objects, such as distant galaxies. An ordinary mirror has its silvering on the back surface. The main image is formed by reflection at this surface, but a second fainter image is formed by reflection at the front surface of the glass. Such a double reflection is not acceptable in astronomical work. To avoid it, the mirrors of telescopes are coated on the *front* surface with a thin layer of a highly reflective metal, usually aluminium.

Parabolic reflectors are also used for radio waves (fig. 21.20) and sound waves. For sound waves the mirror may be made of plastic. A microphone is placed at the principal focus. Sounds coming from points along the principal axis are reflected and directed towards the microphone. Sounds coming from other directions are focussed on other points, but not on the microphone. When recording bird-song or making other wild-life recordings the reflector is aimed at the animal to be recorded. This is then clearly heard, but the noises of any other animals close by are heard only faintly.

25.10 Diverging mirrors

These are sometimes called convex mirrors.

Instructions: investigating reflection from a diverging mirror

1 Set up a lamp with comb and converging lens to make a beam of many parallel rays.

2 Place a diverging mirror in the beam. What happens to the reflected rays?

3 Draw lines along some of the ray-streaks. Remove the mirror and continue the lines back. Do they meet at one point or many points? What kind of image does this mirror form?

4 Measure the focal length of the mirror.

If you use a small spherical diverging mirror and look at objects in the room, you will see that the image is always a short distance behind the mirror, it is always erect and diminished. If you compare the view in a diverging mirror and a plane mirror of the same size, it is clear that the diverging mirror gives a much wider field of view. For this reason, diverging mirrors are frequently used as 'rear-view mirrors' in automobiles.

25.20 Total internal reflection

In chapter 5 we investigated the refraction of a ray of light as it entered or left a block of glass or a tank full of water. Now we will study more closely what may happen to the ray as it leaves the denser medium (the glass or water) and passes into the less dense medium (the air).

25.21 Light leaves a dense medium

Instructions: finding the direction of the emergent ray
1 Use a raybox to produce a single, narrow ray-streak. Alternatively use a lamp with a screen which has one slit; place a converging lens in front of the slit to focus the ray-streak to a narrow parallel-sided beam.
2 You need a semicircular block of glass or perspex (fig. 25.2), with its lower surface painted white, so the ray-streak can be seen as it passes through the block. Put the block on the paper and draw round it, to mark its position.

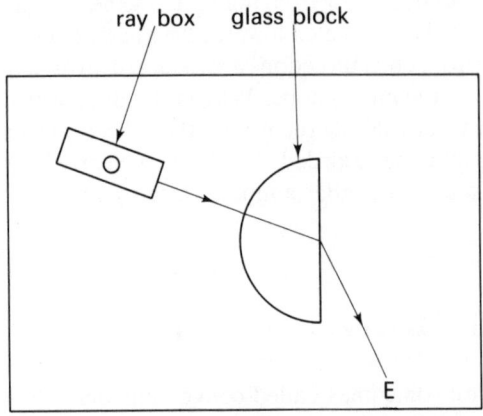

ray box   glass block

E

Fig. 25.2 Plotting the direction of the emergent ray.

3 Remove the block from the paper and mark the position of the central point of its straight side. All rays directed at this point through the curved side enter the block along a normal; as shown in the figure, the ray is not refracted where it enters the block. This makes it easy to aim a ray-streak directly at the centre of the flat side. Draw a normal at this central point and lines at angles to the normal to show the paths of rays *incident* on the flat side (see the table at step 4).

4 Aim the ray-streak so that it passes along the lines you have drawn and is incident on the flat side at various angles (in the figure, for example, the angle of incidence, $i = 20°$). Lay a ruler along the emergent ray, E; measure the angle between this ray and the normal, the angle of refraction (in the figure, $r = 31°$). Enter your results in a table:

$i°$	$r°$	sin i	sin r	$_G n_A = \dfrac{\sin i}{\sin r}$
0 (the normal)				
5				
10				
15				
20				
25				
30				
35				
40				
45				

Questions
1 Why is it not possible to calculate $_G n_A$ when $i = 0°$?
2 What is your average value of $_G n_A$?
3 Were you able to measure r when i was $45°$? If not, what was the difficulty?

Instructions: finding the path of light when i is more than $45°$
1 Use the same apparatus as before. Draw incident rays for which $i = 45°$, $50°$, $60°$ and $80°$.
2 Aim the ray-streak along the paths indicated by the lines. What happens to the ray when it strikes the flat surface of the block?
3 For each angle of incidence, measure the angle of reflection. What rule tells you the relation between i and r?

25.22 Critical angle

In fig. 25.3a we see a ray incident on a glass-air surface being refracted on leaving the glass. In fig. 25.3b, i has been increased and r has increased by a greater amount. In fig. 23.5c, i has been further increased, making r increase to $90°$. The emergent ray skims the surface without actually leaving it. The value of i at which this happens is called the *critical angle*. We usually give this the special symbol, c. In fig. 23.5d the angle of incidence is greater than c, and the ray is reflected. In your investigations you may have noticed that a small part of the light is reflected when i is less than c. Most of the light is refracted and makes a bright emergent ray, but a faint ray is seen reflected from the surface; for this ray, which is reflected internally, the usual rule applies, $i = r$. There is always

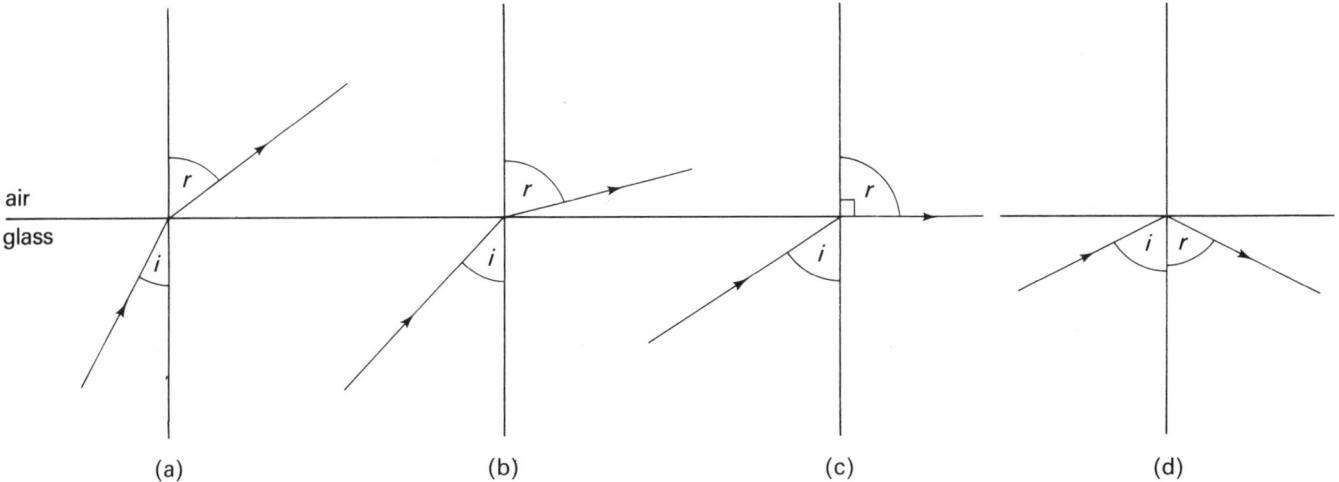

air

glass

(a)   (b)   (c)   (d)

Fig. 25.3   The refraction of a ray of light leaving a denser medium.

some internal reflection, whatever the value of i but, when i is greater than the critical angle, there is no emergent ray; *all* the light is reflected. We say there is *total internal reflection*.

For the critical ray, $i = c$

and $_G n_A = \dfrac{\sin i}{\sin r} = \dfrac{\sin c}{\sin 90°} = \dfrac{\sin c}{1}$

so $_G n_A = \sin c$ or $_A n_G = \dfrac{1}{\sin c}$

The value of c depends only on the refractive index at the surface.

Questions

1 If the refractive index at the glass-air surface in fig. 25.3 is $_G n_A = 0.66$, calculate c.
2 A boy measured the refractive index of water, using a fishtank containing water (5.21). He found that $_A n_W = 1.33$. What is the critical angle at the water-air surface?
3 For refraction at a diamond-air surface, $c = 24.4°$. Calculate $_D n_A$.

Diamond has an extremely high refractive index, giving it a low critical angle for refraction into air. Rays of light inside a diamond are internally reflected many times before emerging. This fact explains the exceptionally brilliant appearance of a properly-cut diamond. Measuring critical angle gives us another way to measure refractive index, as explained in the following instructions.

Instructions: measuring critical angle

1 Put a glass or perspex prism on a sheet of paper. Draw round it to mark its position. Push an object pin O into the paper, exactly on the outline of the prism, in the position shown in fig. 25.4a. Push the prism close against the pin.
2 The figure shows three rays of light from O. Ray OA is

totally reflected. Looking from A, we see a strong reflection of O. Ray OB is mostly refracted (OB′); only a small part of the light is reflected to B. Looking from B, we see a weak image of O. Between A and B is the *critical ray*, which you have to locate. Move your head from side to side. Look for the position in which the image of O *suddenly* becomes faint, as you move your head *from right to left*. Push two pins, P_1 and P_2, into the paper in line with the direction in which the sudden change in the brightness of the image occurs. Then, if you look from the right of the pins (fig. 25.4b) you see three pins in a row (P_2, P_1 and a strong reflection of O). As your head moves across towards the left the three stay in line and the reflection of O remains strong. Just as your eye crosses the line $P_1 P_2$, pin O goes out of sight behind the other two pins. As your eye continues to the left, O reappears, but now it is seen faintly reflected (fig. 25.4c).
3 Adjust the positions of P_1 and P_2 until the effect described above is very clear and occurs exactly on the line $P_1 P_2$.
4 Remove the prism and pins from the paper. Using a ruler join $P_1 P_2$; continue the line to meet the outline of the prism at L (fig. 25.4d).
5 Because O is reflected, the laws of reflection can be used to locate the position of its reflected image I. Draw a normal ON from O to the reflecting surface. Continue this line on the other side of the surface. Measure ON and mark I so that ON = IN.
6 The ray appears to have come straight from I to L, but really it has been reflected at M during its journey. Draw LI, cutting the surface at M.
7 Draw OM. We have now traced the path of the ray completely. Because $i = r$, the angle between OM and ML is $2c$. Measure this angle and calculate c, the critical angle.
8 Calculate the refractive index from glass to air, $_G n_A$ and the refractive index from air to glass, $_A n_G$.

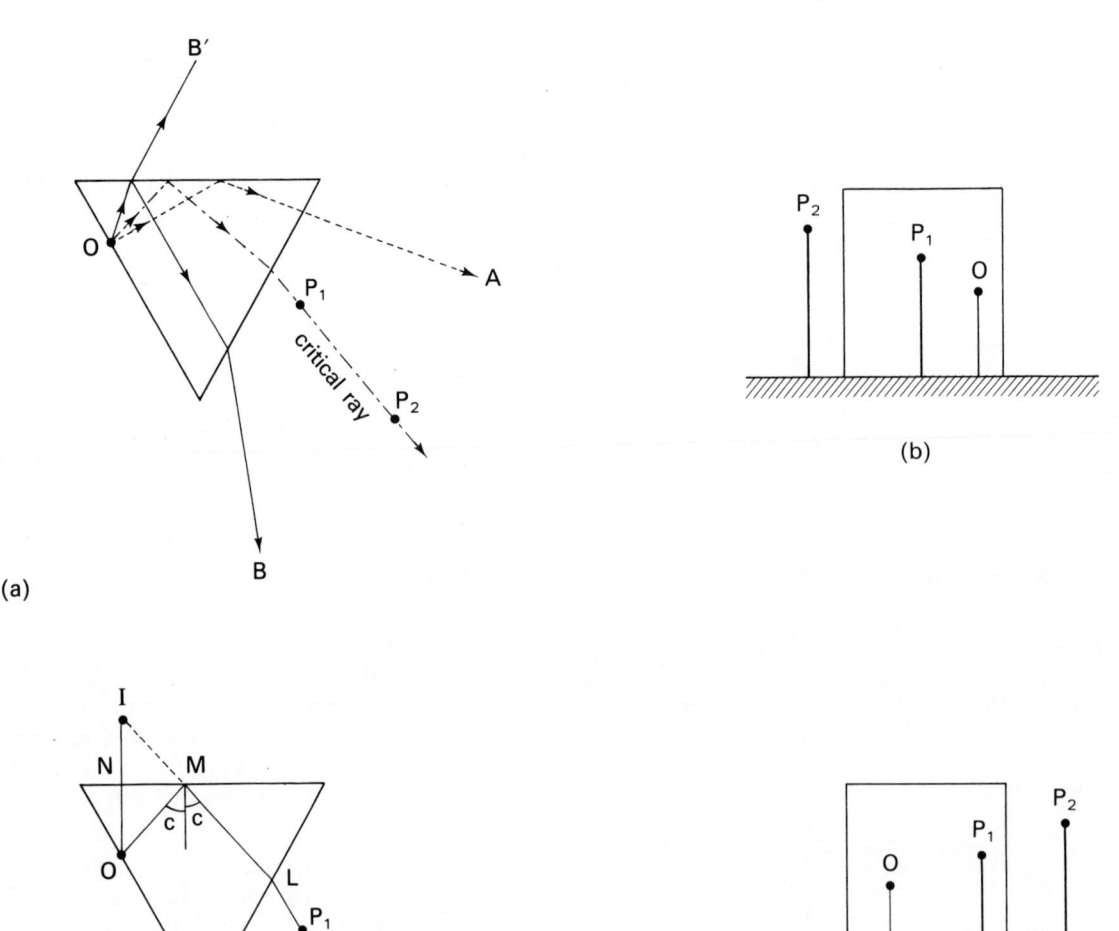

(a)

(b)

(d)

(c)

Fig. 25.4 Measuring critical angle.

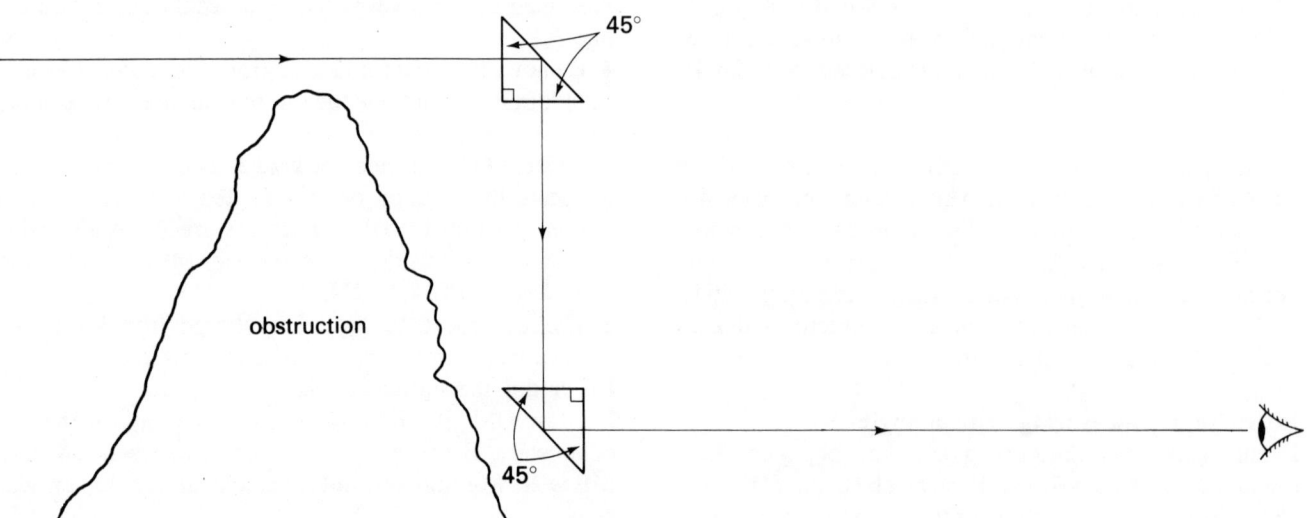

Fig. 25.5 Using totally reflecting prisms in a periscope.

25.23 Using total internal reflection

If we have a right-angled prism, with the other two angles 45° each, we can use the prism in the same way that we use a plane mirror. Compare fig. 5.15 with fig. 25.5. Light entering and leaving the prism goes along the normal so is not refracted. This means that there is no dispersion of different wavelengths, no coloured edges on the images. It strikes the reflecting surface at 45°, an angle of incidence a little greater than the critical angle. Therefore it is totally reflected. One advantage of using a prism instead of a silvered mirror is that the reflecting surface does not become tarnished or spotted with age. A silvered mirror also has the disadvantage referred to previously; a second weak image is reflected in the front glass surface, spoiling the quality of the main reflected image. This does not happen with a reflecting prism. A small amount of light is reflected as the light enters and leaves the prism, but this is reflected *back* along the path of the beam; it does not continue with the beam and so cannot spoil the image. For this reason reflecting prisms are widely used in optical instruments such as periscopes and camera-viewfinders.

Binoculars are really two telescopes fixed side by side so that both eyes may be used. To overcome the inconvenience of the long tube required for high magnification (24.30), the path of light inside the binoculars is folded. Each prism has two reflecting surfaces, turning the beam through 180° (fig. 25.6). This arrangement in effect makes the binocular tube three times longer than its actual length. Unless a third lens is used, the image produced by a telescope is seen inverted. In binoculars, the prisms are arranged to give an erect image without the need for a third lens.

A modern use for total internal reflection is the *light-pipe*. If we shine light in at one end of a narrow rod of glass or transparent plastic, the light travels along inside the rod and passes out at the other end. The rays travel in a direction that is almost parallel with the axis of the rod. Even if the rod is curved, rays which strike the surface of the rod are totally reflected. No light leaves the rod except at the far end. We can bend the rod to carry light into places where it is not convenient to put lamps or mirrors. This idea is used in devices used for examining the human body. Light from a lamp is carried along a light-pipe (often a flexible material is used) to internal cavities, such as the stomach. A second light pipe is used to carry the light reflected from the interior of the stomach to the eye of the surgeon or to a camera. Light-pipes can also be used in decorative dials on radio sets, on advertising displays in shops, and in ornamental lamps for the home.

A light-pipe need not be wide; it can be as thin as a hair. The use of glass fibre for transmitting light is a relatively new development. The glass required for transmitting light for long distances must be exceptionally clear. It is drawn into thin fibres a kilometre or more long. The fibre is coated with an outer layer of glass of a different type, which has lower refractive index. Total internal reflection occurs at the surface between the main fibre and its coating. The cover of this book carries a photograph of a coil of this optical fibre. At the lower part of the picture blue light from a krypton laser is directed into one end of the fibre. The red appearance of the coil is caused by fluorescence of the material of the fibre as it absorbs some of the krypton laser light. The absorption of blue combined with red fluorescence means that light emerging from the fibre has a distinctly red appearance, which can be seen as the spot of red light at the end of the fibre, a little above and to the right of the picture centre. This light has travelled over a kilometre along the fibre yet is still bright enough to be clearly seen.

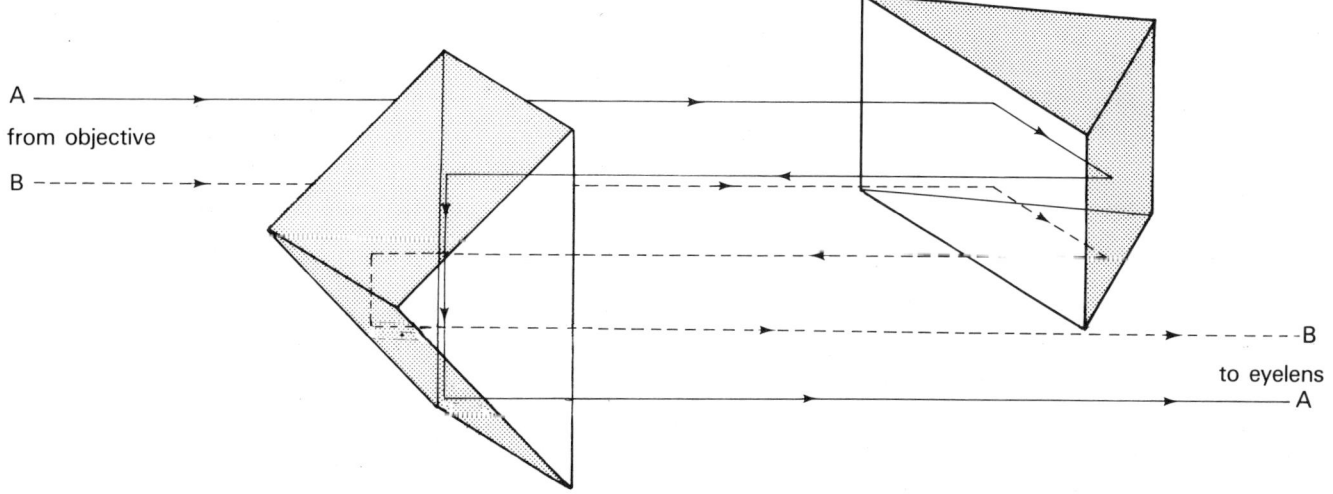

Fig. 25.6   *Using totally reflecting prisms in binoculars. Two reflections occur in each prism. An identical set of prisms is placed between the other objective and eye lens.*

An important application of fibre optics is in tele-communications. The light beam is used for carrying messages along a single fibre. The beam of light is modulated in a similar way to a beam of radio waves (32.10). Messages are passed from a transmitter at one end of the fibre to a receiver at the other end. Many such fibres can be included in a single cable; the coating on the fibre prevents light passing from one fibre to its neighbours. The cables also have the advantage that they are not affected by electrical interference as is an ordinary telephone cable carrying messages as electrical currents. In addition the fineness of the fibres makes it possible to carry thousands of telephone conversations along one cable at the same time. Fibre optics will play an increasing part in telecommunications of the future.

Look for more examples of the use of total internal reflection, yourself.

26 Using electricity

26.00 Heat from electricity

When a current flows along a wire, free electrons move between the atoms in the lattice (fig. 19.7). They collide with them frequently. This increases the energy of the atoms. They vibrate more vigorously. There is an increase in the internal energy of the metal. Its temperature rises. The additional energy is then transferred from the heated metal to its surroundings, there is a flow of *heat*. In the process which has been described, electrical energy has been converted to heat.

26.01 The heating effect of an electric current

If current flows in a wire and the temperature is relatively low, the wire does not glow. All the electrical energy is converted to heat. We made use of this fact when we measured specific heat capacity (16.31) and specific latent heat (18.05). Electricity is a convenient way of supplying measurable quantities of energy to the water. Now we will study the heating effect in more detail.

Instructions: investigating the heating effect of a current passing for differing lengths of time

1 Use a heater coil made by winding nichrome wire around a pencil. When the coil is made, slip the pencil out of it. Use a length of wire to make the total resistance about $2.5\,\Omega$. Connect the coil in series with an ammeter (reading to 5A), a switch and a variable resistor, resistance about 10Ω and able to carry a current of at least 4A.
2 The coil is to be used to heat water in a polystyrene beaker or cup, capacity about $250\,cm^3$. This has a lid with two holes: one hole is for the heater coil, the other for a thermometer. Stand the cup on a tile of expanded polystyrene and, if possible, surround it with felt or expanded polystyrene.
3 Connect the circuit to a power supply able to deliver 4A at 12V. Six accumulator cells connected in series provide a suitable supply.

4 Weigh 0.2kg water and put this in the cup. Make sure that the heater coil is completely below the water surface. After a few minutes, measure and record the temperature of the water.
5 Switch on the current and begin timing. Immediately adjust the variable resistor so that the current is 2A. Make further adjustments, when needed, to keep the current at 2A for the whole trial.
6 Read the water temperature after 2min, 4min, 6min, 8min and 10min. As you take each reading, stir the water with the thermometer.
7 Plot a graph showing the relation between the *time* during which the current flowed and the *rise in temperature*. What is the relation between the time and the *rise* in temperature?

Instructions: investigating the relation between heating effect and size of current

1 Use the same apparatus as above. Run four trials, renewing the water at the beginning of each trial. For each trial use 0.2kg water and switch on the current for 5min. The current should be held at a different value in each trial: 1A, 2A, 3A, and 4A. Record the *rise* in temperature in each trial.
2 Plot a graph to show the relation between current and rise in temperature. What is the relation between current and rise in temperature? (Hint: compare your curve with those on page 398.)

Instructions: investigating the relation between heating effect and resistance

1 Use the same apparatus as before. Run five trials, renewing the water at the beginning of each trial. For each trial use 0.2kg water and switch on a current of 2A for 5min. The resistance of the heater coil should be different at each trial; cut off wire to reduce its length after each trial, so that its resistance for the five trials is $2.5\Omega, 2\Omega$, $1.5\Omega, 1\Omega,$ and 0.5Ω. Record the rise in temperature in each trial.

2 Plot a graph to show the relation between resistance and rise in temperature. What is the relation between resistance and rise in temperature?

If we ignore loss of heat to the surroundings, rise in temperature depends on the amount of heat supplied to the water. This equals the amount of electrical energy converted to internal energy. If this is represented by the symbol W, we can write the results of the previous investigations like this:

$$W \propto t \qquad W \propto I^2 \qquad W \propto R$$

These can be combined in one relation:

$$W \propto I^2Rt$$
$$\text{or } W = kI^2Rt \ (k \text{ is a constant})$$

If we choose the right units for our measurements, we can make $k = 1$, and then

$$W = I^2Rt$$

In this equation W is measured in joules, I in amperes, R in ohms and t in seconds. These are the 'right units'; let us see why. The 'right units' were chosen when we first defined the volt (15.13). Using this definition, we found that when an electric charge Q flowed through a p.d. V in time t, it lost potential energy:

$$E_p = VQ = VIt \ (E_p \text{ in J, by definition of } V)$$

This equation is derived in 15.21. Later (19.20), the volt was used in defining the ohm:

$$R = \frac{V}{I}$$

from which we get $V = IR$.

When the charge flows through a heater coil immersed in water, the loss of p.e. of the charge, E_p, equals the amount of work, W, done on the water in raising its internal energy. When W joules of work is done on the water, it gains U joules of energy. Since there is no loss of energy, $W = U$. The change of symbol from W to U indicates that the energy has changed *in form*. It has changed from *work* to *internal energy*. Thus the energy conversions in this investigation can be expressed as:

$$E_p = W = U$$

We have also shown that $V = IR$, so we can now use the earlier equations for E_p to write a new set of equations:

$$U = E_p = VIt = IR \times It = I^2Rt$$
(U in J, by definition of V and R).

This gives us the same relation as we found by investigation of the heating effect. The reasoning above explains why we can make $k = 1$ when we work in SI units.

The relation for U can be written in several forms:

$$U = VIt = I^2Rt = \frac{V^2}{R} t \ (U \text{ in J})$$

These are known as *Joule's Laws*. He discovered them in 1841 by investigations similar to those you have just done, using a coil immersed in water. From them we can obtain a set of equations for power, P:

$$P = VI = I^2R = \frac{V^2}{R} \ (P \text{ in W})$$

Note *For the investigations* it was not necessary to know how many joules are required to raise the temperature of the water by a given amount. In other words, we did not need to know the specific heat capacity of water. This has been measured by methods that do not involve electricity, but depend on converting mechanical energy into internal energy. The experiment with lead shot (16.30) is one such experiment. For precise measurements, the apparatus required is complex. When we have accurately measured c for water, we can pass a measured current through an immersed heater coil and use our known value c to calculate V. We can see if the value indicated by the voltmeter actually agrees with the calculated value of V. If not, and if we have eliminated all experimental error, the voltmeter is wrong. It is not really reading joules per coulomb (= volt). Methods such as this can be used to *calibrate* a voltmeter. Then we can use the calibrated voltmeter in other investigations. We connect it across the terminals of an immersed heater coil, measure V, and use this value with I and t to calculate how much electrical energy has been converted to internal energy. This is the way you used the voltmeter in your measurements of specific heat capacity and specific latent heat.

Questions

1 A heater coil is immersed in water, mass 0.5 kg, temperature 20°C. A current 1.5 A flows for 5 min, during which the temperature rises to 30°C. Calculate the p.d. across the heater coil. You may ignore heat losses to the coil itself, the container and the surroundings. Take the specific heat capacity of water to be 4 200 J/kg K.

2 A current 3 A flows for 1 min through a coil resistance 35 Ω. What quantity of electrical energy is converted into heat?

3 If the current had been 6 A in the previous question, what quantity of electrical energy would be converted?

4 A lamp has resistance 576 Ω when operating at full brightness. It is connected to a 240 V mains. What is the power of the lamp?

26.02 Using the heating effect

At home the heating effect is used in cookers, toasters, hot-plates, kettles, clothes-irons, room heaters and in many other devices.

Questions

1 A diagram of an electric clothes-iron is shown in fig. 26.1. It is called an iron, but is it really made from iron?

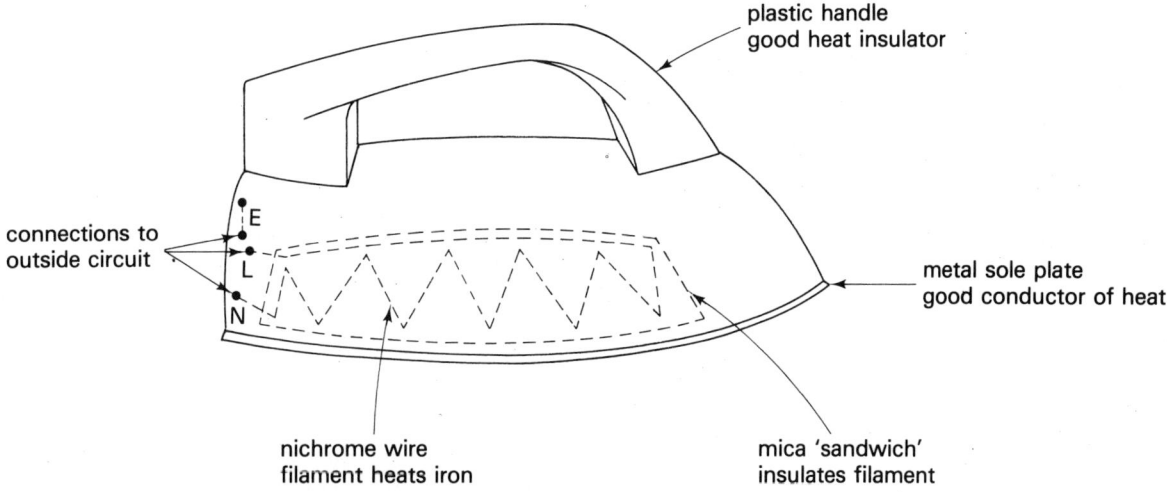

Fig. 26.1   Diagram of an electric clothes-iron.

What metal is the sole plate usually made from? Why is this better than iron? What sort of material is mica? What properties of mica are important for use in an iron? What do the letters L, N, E stand for? If the iron is rated at 1 000 W and operates on a 240 V supply, what current passes through it when operating? What is the resistance of the filament? If it takes 1½ min to iron a shirt, what quantity of heat is produced in that time?

2 Figure 26.2 shows a simple room-heater. In what forms is energy given off when a current flows through the filament? What kind of material is nichrome? What properties of nichrome make is suitable for use as a filament? What properties of fire-clay make it a good support for the filament? What shape is best for the polished reflector? Why do we need a protective guard on this heater? The heater is switched on for 1 h; it is rated at 750 W and the mains supply is 240 V. What quantity of electrical energy is converted in that time?

3 Find out as much as you can about other types of electric room heater. Describe how they work and what are their advantages and disadvantages.

Another application of the heating effect is illustrated in fig. 26.3. This is the hot wire ammeter. The current that is to be measured flows through a thin wire and heats it. As

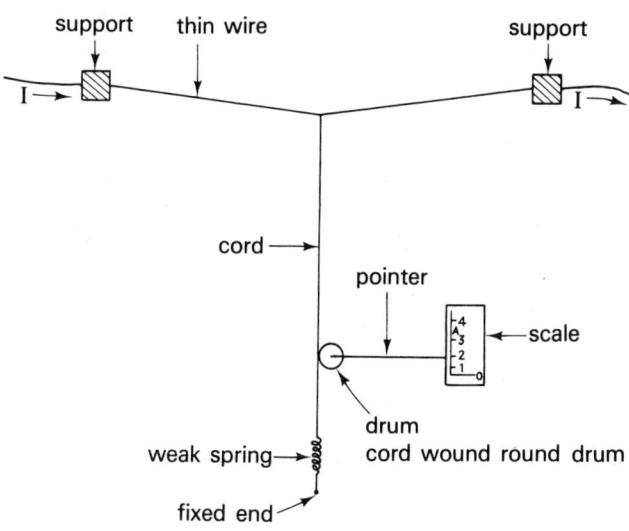

Fig. 26.3   Demonstrating the principle of the hot-wire ammeter.

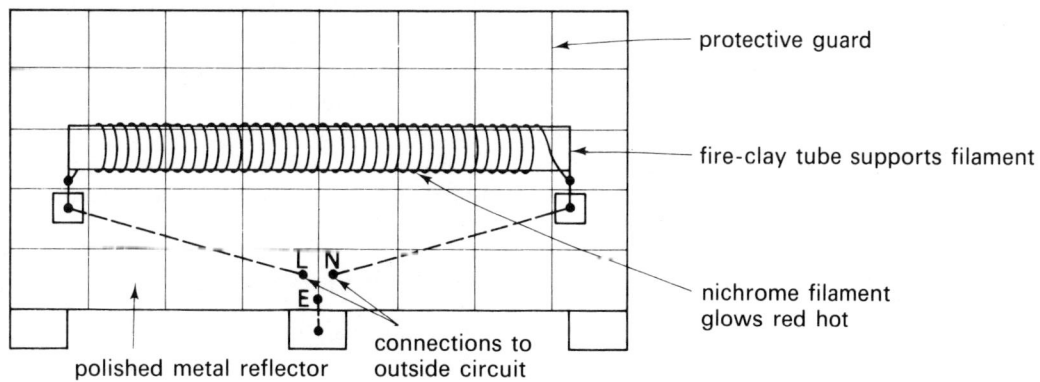

Fig. 26.2   Diagram of an electric room-heater.

the wire gets hot, it expands. The spring and cord pull the centre of the wire downwards. The cord is wound round a drum, which turns and moves a pointer across a scale. The scale is calibrated, indicating current in amperes. The chief advantage of this kind of ammeter is that it works with alternating current or direct current. The direction in which the current flows through the wire has no effect on the amount of heat produced. Other types of meter (26.21) do not have this advantage, which is important for certain applications. The main disadvantage is that because the heating effect varies with the *square* of the current, the scale of the meter is not uniform (see figure). The scale is not easy to read. The meter is less sensitive to small changes of current when the current is low.

26.10 Light from electricity

26.11 Filament lamps

A typical filament lamp is shown in fig. 26.4. The filament glows 'white hot'; at this high temperature much more of the electrical energy is set free in the form of visible light, less is set free in the form of heat. The filament is tungsten, which has high melting point (3.30). In earlier filament lamps the filament was carbon which also has high melting point. Filaments were made by burning strips of bamboo or cotton thread to convert them to carbon. The original tungsten lamp had a vacuum inside the envelope to prevent oxidation of the tungsten at high temperature. This allowed the tungsten to evaporate gradually and the glass became blackened. Modern lamps are filled with argon which reduces evaporation of tungsten.

The quartz-iodine lamp produces light of high intensity. The envelope is made of quartz which can stand high temperature without melting. The lamp is filled with iodine vapour; this combines with evaporated tungsten, forming tungsten iodide which remains as a vapour and

does not condense and blacken on the envelope. Lamps of this type are used in projectors, in powerful microscopes and in some automobile headlamps.

Instructions: measuring the power of a lamp
1 Use a lamp rated to operate on a 6V or 12V supply. Connect it to the supply in series with an ammeter. Connect a voltmeter in parallel with the lamp.
2 Switch on the current. Measure current, I. How many coulombs of electric charge are passing through the lamp per second?
3 The voltmeter tells you how many joules of electrical energy are being converted to heat and light for each coulomb passing through the lamp. Read the voltmeter and calculate the rate of energy conversion.
4 What is the power of the lamp? Does this agree with the marking (if any) on the lamp?

The method above measures *power input* to the lamp. Power output consists of heat and light. For a modern filament lamp the power output in the form of light is only 7% of the electrical power input. This is a low efficiency, yet the early carbon filament lamps were less efficient; only 2% or 3% of the electrical energy appeared as light. Lamp efficiency is increased by using tungsten which operates at high temperatures, for at high temperatures a greater fraction of the electricity is converted to light. Further increase in efficiency is obtained by making the filament compact; the wire is coiled and the coiled wire is coiled again. The coiled-coil reduces heat loss through convection in the argon. Quartz-iodine lamps operate at even higher temperatures and are more efficient than ordinary filament lamps.

26.12 Discharge lamps

Neon discharge tubes (fig. 23.1) can be made in all kinds of shapes for use in advertising displays; small neon lamps are used as indicator lamps. The lamp used to indicate that a mains socket is switched on is generally a small neon lamp. As the electrical discharge passes through the gas the excited neon atoms emit light in the red region of the spectrum. If other gases are used, lamps emitting other colours can be made. Discharge lamps filled with sodium vapour or with mercury vapour are widely used for street lighting. The sodium vapour lamp gives a yellow light. The mercury vapour lamp gives a bluish-green light. These lamps are about five times as efficient as the coiled-coil filament lamp, so they have low running costs.

26.13 Fluorescent lamps

Their efficiency is about three times that of the filament lamp; for this reason they are widely used in offices, shops, factories and homes. Heat production is relatively low, as you can tell by placing your hand on the tube after it has

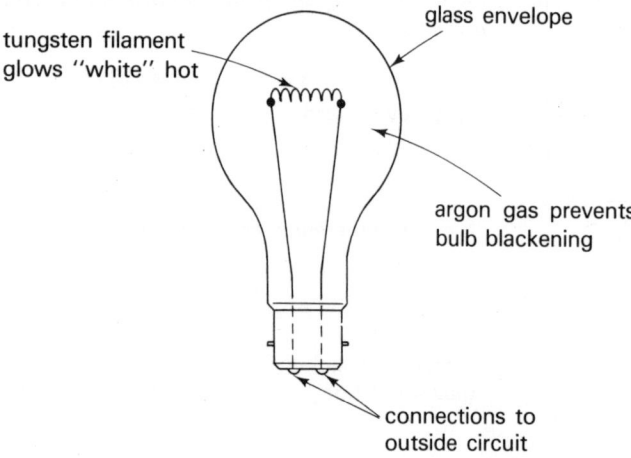

tungsten filament glows "white" hot

glass envelope

argon gas prevents bulb blackening

connections to outside circuit

Fig. 26.4   An electric lamp.

been operating for a few minutes. The tube is a discharge tube; it contains mercury vapour and argon. During discharge mercury vapour emits not only wavelengths in the blue and green regions of the spectrum, as mentioned in the previous section, but also a considerable amount of ultraviolet radiation. The inner surface of the tube is coated with a fluorescent material that absorbs ultraviolet radiation and emits the absorbed energy in the form of light of visible wavelengths. Most tubes give a bluish-white light that is suitable for offices and factories. For the home, many people prefer a 'warmer' light, that contains a higher proportion of wavelengths from the red end of the spectrum. A special mixture of fluorescent materials is used to obtain this. Fluorescent tubes may also be coated with materials that give light of distinct colours, such as red, yellow, green or blue; these are used for ornamental purposes and advertising.

The tubes are usually about 1 m long and about 4 cm in diameter. Light is radiated evenly in all directions from this large surface; the tube casts few dark shadows, so it is ideal for illuminating working areas. Fluorescent lamps are more efficient than filament lamps in the use of electric current, and they last longer. They have higher installation costs but they are cheaper to operate, which more than compensates for this.

26.14 Light-emitting diodes

In certain types of semi-conductor diode (19.41) the semiconductor emits light when its atoms are excited by an electric current passing through it. Light-emitting diodes emit red, yellow or green light, depending on the material from which they are made. They are small in size (small ones are less than 1 mm in diameter) and are often used as indicator lamps. They are not easily broken and have long life. The illuminated figures in pocket calculators, digital watches and digital clocks are usually made from light-emitting diodes (LED, for short). Each digit consists of seven bar-shaped diodes arranged in a pattern similar to an '8'. We have already mentioned the displays in calculators and digital clocks. No doubt you can think of many other examples.

Groups of these diodes are lit at the same time to make up any numeral or many of the letters of the alphabet. In another form, several dozen diodes, the size of a pin's head, are arranged in rows and columns and lit to display any numeral or letter. LEDs have low efficiency so they are unlikely to be used for lighting homes or work places. One of their great advantages is that they light up in a very short time (about 50 ns) when switched on; they go dark equally quickly when switched off. By contrast a filament lamp takes several tenths of a second to respond; a discharge lamp or fluorescent tube takes several seconds or even minutes. The rapid response of the LED has many applications.

26.15 Lasers

One form of this is the ruby laser, which consists of a specially grown ruby crystal in rod shape. The ends of the crystal are flat and parallel. One end is silvered, the other end is partly silvered. Light (photons) inside the crystal is reflected backwards and forwards between the two end surfaces. A little escapes at the sides and a small proportion escapes through the half-silvered end. The crystal is surrounded by the spiral tube of a high intensity discharge lamp. Photons from this lamp enter the crystal and excite the atoms there. Normally an atom absorbs a photon, becomes excited and a few microseconds later emits a photon of its own special wavelength. The exact period between absorption and emission is not fixed; in a group of excited atoms, photons will be emitted at random intervals, like the emission of radiation from a radioactive material. In the laser crystal the emitted photons are reflected backwards and forwards inside the crystal. More and more photons are absorbed from the discharge lamp and the number of emitted photons increases. As the number of photons increases inside the crystal, the atoms are stimulated to emit more photons. All atoms are stimulated to emit at the same time, instead of at random. The result is an extremely high rate of emission, causing a short burst of light of high intensity to pass out through the half-silvered end of the crystal. The photons in this beam are lost from the crystal, but are soon replaced as more photons from the discharge tube are absorbed by the crystal atoms. Lasers are highly efficient energy converters; a gas laser converts 40% of the supplied energy into light energy.

The laser beam contains photons that are all of one wavelength. The beam is very intense and it can be focussed to make an exceedingly narrow beam of even higher intensity. A large amount of energy is contained in the beam and this can be concentrated on a very small area, giving the beam important applications as a heating and cutting tool. In this use a pulse of 100 MW can be delivered for a period of a few microseconds. In beams of this intensity, light pressure is sufficient visibly to move small particles of plastic. By directing laser beams on particles from opposite sides, pressure up to 10^{17} Pa (or N/m²) have been produced. This is 10^{12} times greater than atmospheric pressure. This has applications in the study of extremely high pressures on the structure and behaviour of materials.

The light from the beam has photons of only one wavelength. It is monochromatic. This means that its speed can be very accurately measured. It can be focussed to a beam which travels for long distances yet remains a narrow beam of high intensity. Laser beams may be directed across to reflectors placed several kilometres away and the time taken for the return journey measured. The distance of the reflector is calculated with an accuracy of a few millimetres. Small changes in the distance, caused by movement between parts of the Earth's surface can be detected. Such information is valuable in the study of geological changes and the investigation of the causes of earthquakes. The *Apollo XI* expedition placed a reflector on the Moon's surface. Pulses of laser light are directed from Earth towards this reflector; the time for the pulse to travel from Earth to Moon and back is measured, and the distance between Earth and Moon is calculated with high accuracy. This technique is being used to discover if the distance changes over a period of time.

The high intensity of the beam leads to danger. The human eye is designed to focus light so, when the beam enters the eye it is focussed to a fine point on the retina. The high light intensity at that point damages the nerve cells permanently, resulting in blindness at that point. A laser beam should never be allowed to enter the eye. Care must be taken that reflections from small shiny surfaces nearby do not accidentally reach the eye. The effect has been used in eye surgery to treat patients whose retina has become loose from the inside of the eyeball, which causes great lack of clearness of vision. If a laser beam is directed into the eye, it can be used to fuse the retina to the eyeball, to fix it in position.

With the development of fibre optics for telecommunication (25.23) miniature lasers are being used to send light for distances of several kilometres along the fibre. Light from the laser can be modulated to carry messages along the fibre. The cover of this book shows light from a xenon laser passing along a fibre 1 km long. The xenon laser is one of the gas lasers, which operate on the same principle as the ruby laser. The photons are emitted by atoms of xenon, contained in a tube with parallel silvered ends. The atoms are excited by passing an electrical discharge through the gas.

26.20 Force from electricity

When a moving electron passes through a magnetic field which acts at right-angles to the direction of its motion, a force acts on the electron, deflecting it sideways (23.11). Fleming's left-hand rule tells you the direction of deflection (fig. 23.6). If the electron is travelling in a wire and is deflected by a magnetic field the force is transmitted to the wire. The wire moves.

26.21 Meters

The basic type of electrical meter is the *galvanometer*. It measures current.

Instructions: making a model galvanometer
1 Take a piece of thin insulated wire about 1 m long. Wind it loosely around a small box (matchbox) to make a rectangular coil (fig. 26.5). Remove the box from the coil; bind the sides of the coil with sticky tape, so that the ends of the wire are at the top and bottom of the coil. Secure them by winding them once around the coil, at the centre of the top and bottom; form the remainder of the ends of

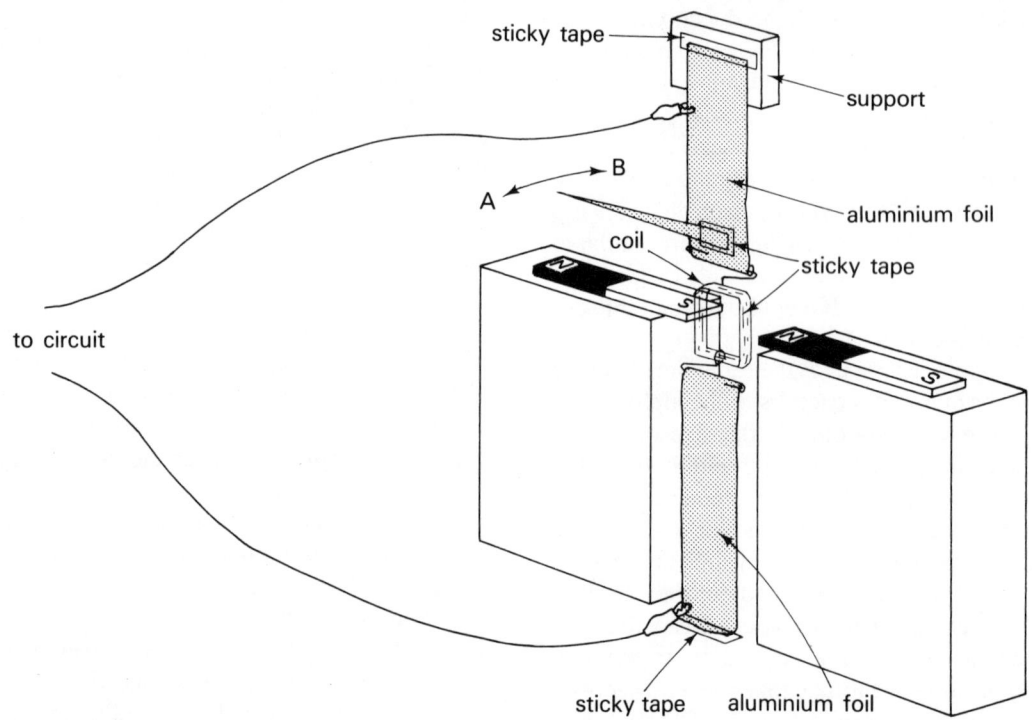

Fig. 26.5 A model galvanometer.

the wire into flat hooks as shown. Remove the insulation from the hooked ends of the wire.

2 Cut two pieces of aluminium foil about 1 cm wide and 10 cm long. Roll one end of each around a hook and bend the end of the hook over to stop the foil unwinding and to make good electrical contact.

3 Attach the upper end of the upper strip to a stand or other support (use sticky-tape). Fix the lower end to the bench, (use sticky-tape).

4 Cut a narrow pointer from foil; tape this to the upper foil strip, close to the hook.

5 Attach wires to the foil strips, using crocodile clips. Connect the other ends of the wires in series with a 12 V battery, a 12 V lamp and a switch.

6 The magnetic field is provided by two bar magnets, placed as shown in the figure. Alternatively you may use a large horseshoe magnet or two home-made electro-magnets (fig. 9.8). The poles of the magnets must be as close as possible to the coil, yet allow it to turn freely.

7 Switch on the current. If the lamp does not light, you will know that you have not made good electrical contact between the coil and strips. If the lamp lights, watch the pointer when you switch on. Which way does it move? Does this agree with the left-hand rule?

8 Reverse the current. Observe the direction of motion of the pointer.

9 Connect a second lamp in the circuit, in *parallel* with the other lamp. Switch on and observe the movement of the pointer.

10 Connect a third lamp in parallel with the other two lamps. Switch on and observe the motion of the pointer. What is the effect of having extra lamps in the circuit? Explain this.

Questions

1 When the current is switched on, electromagnetic force is produced, making the coil turn. It turns through a small angle, then stops. Why does it stop turning?

2 Explain why the angle of turn varies with the current in the coil.

3 If the coil was completely free to turn, how far would it turn?

4 How could this galvanometer be made more sensitive, so that it would work with smaller currents?

Examine the mechanism of a laboratory galvanometer. The main parts are shown in fig. 26.6a. The coil is supported on two bearings, which may be jewelled for hard wear. The opposing couple is provided by two control springs. Current flows through these springs from the external circuit to the coil and back. The magnet is usually a powerful horseshoe magnet. Its poles are usually shaped as in fig. 26.6b so that they are close to the coil as it turns. The space between the poles is filled with a cylinder of soft iron (the core); the gap between the poles and core is just wide enough to allow the coil to turn. The iron core helps make the magnetic field strong; more important, it makes the field *uniform*. Magnetic lines of force pass radially across the gap between poles and core. Even when the coil has turned through a small angle, the field is still perpendicular to the plane of the coil. The electro-magnetic couple does not decrease as the coil turns; the marks on the scale of the meter are uniformly spaced.

A galvanometer measures small currents of a few microamperes. To make a highly sensitive galvanometer we suspend the coil so that it turns easily; we fix a tiny mirror to the coil and shine a beam of light on this. The

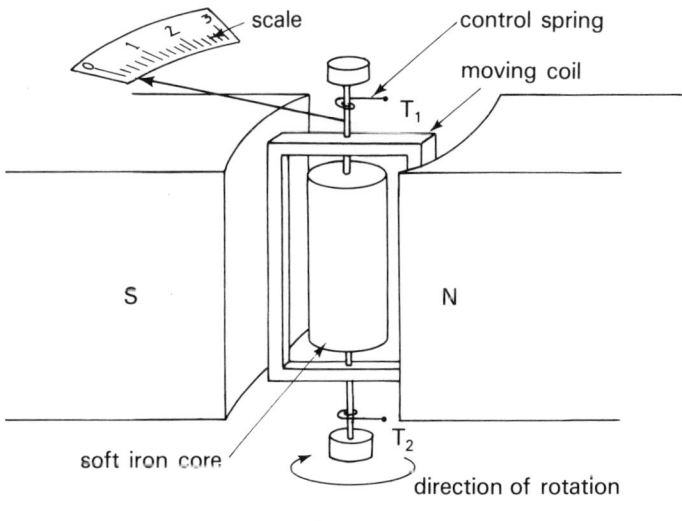

(a)

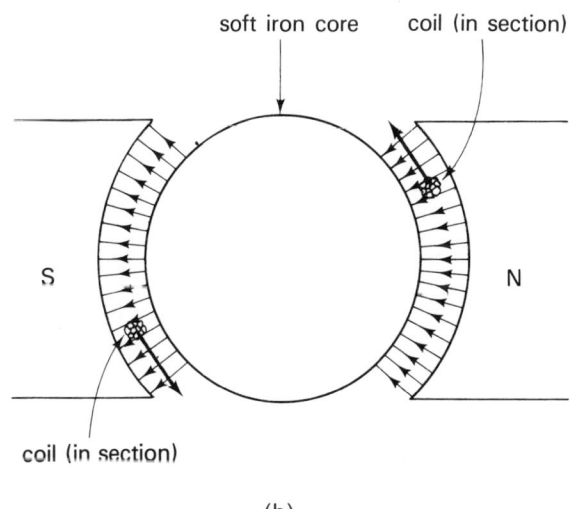

(b)

Fig. 26.6 Moving-coil galvanometer: (a) structure; (b) action, the fine lines with arrows represent the magnetic field in the gap between the poles and core, the thick arrows represent the forces producing the couple which turns the coil.

beam of light is reflected on to a scale placed at a distance. A slight change of angle of the coil makes the reflected spot of light move a measurable distance along the scale. A mirror galvanometer can detect currents as small as $2\,\mu A$. Galvanometers such as these are highly *sensitive*, they can detect small currents. Galvanometers *may* also be *accurate*, the reading indicated is close to the true value of the current flowing through the instrument. Note the difference between the words 'sensitive' and 'accurate' and try to use the correct word. An accurate galvanometer is one that has been carefully calibrated by comparing its readings with those of a standard meter known to be accurate, perhaps a standard current balance. Mass-produced galvanometers are not usually calibrated individually. They are made to a standard design and then checked to see that they agree closely enough with a standard meter. If their reading is within 2% of the reading of the standard meter, they are passed to be sold. If you are measuring a small current with such an instrument and the reading is $3\,\mu A$, its true value may be between $2.94\,\mu A$ and $3.06\,\mu A$. A meter may lose accuracy after a period of use, especially if it is badly handled. The magnetic field becomes weaker; the control springs change in springiness; the electrical resistance of the coil may be altered if it has been overloaded with currents that are too high. Delicate instruments such as galvanometers must be stored safely, handled carefully and used sensibly.

An *ammeter* is a galvanometer that is converted for measuring currents larger than those normally measured by a galvanometer. A resistor, called a *shunt resistor* is wired in parallel with the coil of the galvanometer (fig. 26.7a). The resistor is normally out of sight, inside the case of the ammeter. Some of the ammeters used in schools have the shunt resistor outside the case, connected across the terminals of the meter. Shunt resistors of different values are used to alter the sensitivity of the meter, as explained below.

A shunt resistor has low resistance, often less than 1 ohm. When in circuit, most of the current flows through the shunt resistor. Only a small fraction flows through the coil of the galvanometer. The galvanometer measures this small fraction, but its *scale is marked to show the total current* flowing through the coil and shunt resistor. In a typical ammeter, the galvanometer has 'f.s.d. 1 mA' (when 1 mA flows, its pointer is fully deflected across the scale). This current, I_G, is the maximum that can be measured by the galvanometer. The resistance, R_G, of the coil of the galvanometer, is 75Ω. To convert the galvanometer to an ammeter with f.s.d. 10 mA we must choose the shunt resistor so that when 10 mA (I_A) flows into the ammeter terminal, only 1 mA goes through the galvanometer coil, and the remaining 9 mA goes through the shunt resistor.

To calculate the value of the shunt resistor, R_S, use this equation (19.31):

$$\frac{I_G}{I_S} = \frac{R_S}{R_G}$$

In this example, $R_S = \dfrac{I_G}{I_S} \times R_G = \dfrac{1}{9} \times 75$
$$= 8.33$$

By adding the shunt resistor we have reduced sensitivity; the scale of the galvanometer must now be changed, so that when the pointer is fully deflected, it indicates '10 mA'.

Questions

1 In the ammeter referred to in the example above, $I_A = 4\,mA$. Calculate I_G and I_S. When the current is 4 mA, what number does the pointer indicate?

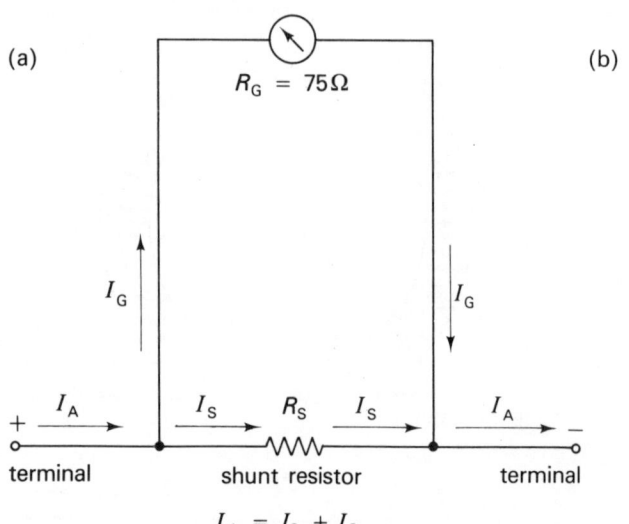

(a)

$R_G = 75\Omega$

I_G

I_G

$+$ terminal

I_A

I_S R_S I_S

shunt resistor

I_A

$-$ terminal

$I_A = I_G + I_S$

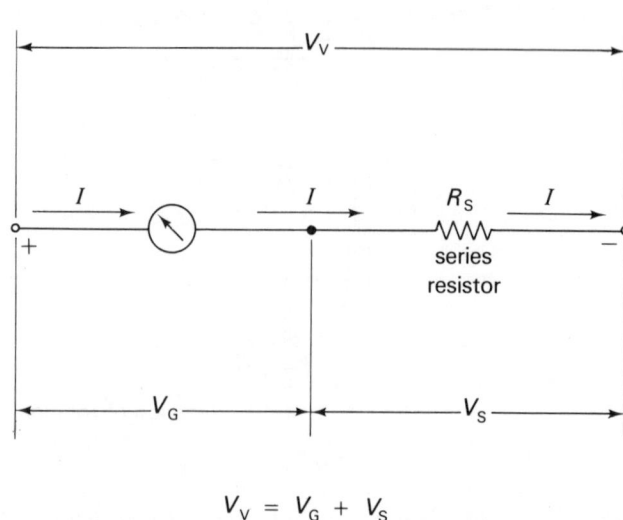

(b)

V_V

I

I

R_S

I

series resistor

$+$

$-$

V_G

V_S

$V_V = V_G + V_S$

Fig. 26.7   *Converting a galvanometer to (a) an ammeter and (b) a voltmeter.*

2 What value has the shunt resistor if the meter has f.s.d. 100mA? (Assuming we use the same galvanometer as above). What change must be made to the galvanometer scale?

3 What value shunt resistor is needed to give f.s.d. 10A?

4 What is the total resistance of the ammeter (galvanometer coil + shunt resistor) in each of the above questions?

5 As a general rule, what can be said about the resistance of an ammeter?

6 In use an ammeter is always connected in *series* with the other components of the circuit. Why do we do this?

7 Why must an ammeter have low resistance?

8 The shunt resistor reduces the sensitivity of the galvanometer. Does the shunt resistor have any effect on the accuracy of the ammeter?

A *voltmeter* is a galvanometer that has been converted to measure potential difference. Figure 26.7b shows that the galvanometer has a resistor in series with it. This *series resistor* is normally inside the instrument case. In some types of voltmeter used in schools, the series resistor is attached to one terminal of the galvanometer; different resistors can be used to vary the sensitivity of the instrument.

A galvanometer really measures *current*; here we use it to measure *p.d.* For making a voltmeter we generally use a galvanometer with a coil of relatively high resistance (for example, $1\,300k\Omega$) and sensitive to small current (for example, f.s.d. $100\mu A$). With a coil such as that described, carrying a current $100\mu A$, the p.d. between its ends is

$$V_G, = I_G R_G, = 100 \times 10^{-6} \times 1.3 \times 10^3 = 130 \times 10^{-3} = 130mV.$$

We could use the galvanometer just as it is, marking its scale to read 130mV at f.s.d. It makes a sensitive voltmeter, better called a *millivoltmeter*. To measure larger p.d. we must reduce its sensitivity.

When we connect a series resistor ($R_S = 98.7k\Omega$) to the galvanometer, their total resistance is $100k\Omega$. To deflect the meter fully, we need a current $100\mu A$ to flow through both meter and resistor. The p.d. across the two is

$$V_v = I_v R_v = 100 \times 10^{-6} \times 100 \times 10^3 = 10V.$$

If we apply p.d. 10V across the terminals of the voltmeter, the p.d. across the resistor is 9.87V. The remaining p.d. is 0.13V (or 130mV) across the galvanometer. This is just enough to make a current $100\mu A$ flow through its coil and make the pointer swing to full scale deflection. The galvanometer scale is renumbered and renamed, so that at f.s.d. it reads '10 *volts*'.

Questions

1 What value series resistor is needed to convert the galvanometer ($R_G = 1\,300k\Omega$; f.s.d. $100\mu A$) to a voltmeter to measure p.d. up to 100V?

2 What series resistor is required to convert the galvanometer (same resistance and f.s.d.) to a voltmeter to measure p.d. up to 250V?

3 In general, what can be said about the resistance of voltmeters?

4 When we use a voltmeter, we always connect it in *parallel* with one or more of the components of the circuit. The voltmeter is not really part of the circuit. Why do we do this?

5 Why is it important for the resistance of a voltmeter to be high?

A good example of the importance of the high resistance on a voltmeter is shown by the circuit used for measuring resistance (fig. 19.13). The voltmeter measures p.d. across R. The ammeter measures the current through the circuit and we assume that the current going through the ammeter is exactly the same as the current going through R. Suppose the galvanometer of the voltmeter is like the one we have taken as typical in the examples above. Its needle indicates a p.d. of 6V (the e.m.f. of the battery); this means that the current flowing through it is $60\mu A$. If R is, say, $1k\Omega$, the current through it is 6mA (= $6\,000\mu A$). The ammeter reads the *total* current, $I + I_v = 6\,060\mu A$. This is an error of 1%, which we may ignore.

If R is less than $1k\Omega$ the current through it is more than $6\,000\mu A$, and the error is less than 1%. It can be ignored.

If R is more than $1k\Omega$, the error is more than 1%. For example, if $R = 100k\Omega$, $I = 600\mu A$. R has the same resistance as the voltmeter; equal currents flow through each. The ammeter shows the total current, $1\,200\mu A$, but the true current through R is only $600\mu A$. The error is 100%. This is why this circuit is *not* suitable for measuring resistance when R is more than about $1k\Omega$.

To overcome this problem, we connect the voltmeter to measure p.d. across *R and the ammeter*. The ammeter now measures only the current flowing through R. The p.d. across R and the ammeter is a little higher than the p.d. across R alone, but the resistance of the ammeter is low compared with that of R. The p.d. across the ammeter is so small that we may ignore it.

A *multimeter* consists of a galvanometer with a number of shunt resistors and series resistors. The resistors can be connected in turn to the galvanometer, usually by turning a switch that has many positions. We can use the meter as an ammeter or as a voltmeter, with several different values for f.s.d. The multimeter may also contain circuits for measuring resistance and a rectifying circuit which allows us to measure a.c. voltages and current. Such an instrument is very useful for radio engineers and for people who have electronics as a hobby.

The meters described so far have all been based on the *moving coil galvanometer*. When a current passes, the *coil* moves. Another type of meter is the *moving iron meter*, shown in fig. 26.8. When a current passes through

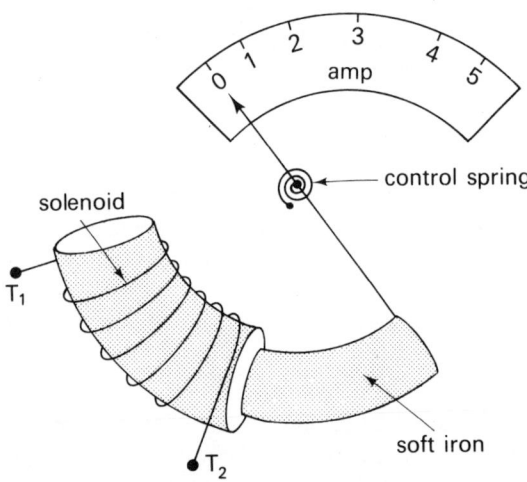

Fig. 26.8   Diagram of a moving-iron meter.

the solenoid it generates a magnetic field. This induces magnetism in the soft iron. Attraction between the field of the solenoid and the induced soft iron magnet pulls the soft iron into the coil. When the attractive force is balanced by the opposing force due to the control spring the needle comes to rest. The greater the current, the more the pointer turns. The scale may be marked to indicate either current or p.d., for this device can be used as part of an ammeter or a voltmeter. The moving iron meter is simple and cheap to make, but is less accurate than a moving-coil meter. The magnetic effect is complex, so the scale is often not marked uniformly. The soft iron must be specially shaped if a uniform scale is required. The main advantage of this meter is that it can measure both d.c. and a.c. Whatever direction the current flows, and whatever pole is produced at the end of the solenoid nearer the soft iron, an opposite pole is induced in the soft iron. Always, the soft iron is *attracted*, even if the direction of current changes rapidly, as in mains a.c. By contrast, the coil of a moving-coil meter is unable to change position rapidly; the pointer remains at zero, even though the alternating current through the meter is high.

26.22   Direct current motor

If a coil is free to turn in a magnetic field, and if we supply the coil with current flowing in the right direction, we can make the coil turn as long as current is supplied. In other words, we have an electric motor.

Instructions: making a model electric motor

1   Cut grooves in a large cork (fig. 26.9a). Make a hole in the cork and push a piece of glass tube through it. Alternatively, use plastic tube cut from a used ball-point pen. This part of the motor is called the *armature*.

2   Wind about 4 m of fine insulated wire around the cork, making an *armature coil* (fig. 26.9b).

3   Bring the ends of the wire to the same end of the coil. Remove the insulation for 2 cm at the end of each wire. Bend the wires back about 1 cm from the end. Place the ends on opposite sides of the glass tube. Hold them in place with two rubber rings. The rings are cut from a piece of narrow rubber tubing. This completes the *commutator*.

4   Thread a piece of stiff wire through the tube. It should be a fairly loose fit; the cork should spin easily and smoothly on the wire. Bend both ends of the wire, taking care not to crack the glass tube, and push the ends into holes already made in the wooden base-board (fig. 26.9c).

5   Make the two contacts or *brushes* by removing the insulation from the ends of two pieces of wire. Wind them around drawing-pins (thumb-tacks) pushed into the base-board. Wind them around two more pins, or around the supporting-wire. Adjust their ends to make contact with the commutator wires, one contact on each side.

6   Place two strong bar magnets as shown, to provide the magnetic field. Alternatively use a large horse-shoe magnet or a pair of electromagnets made as in fig. 9.8. The magnetic poles must be as close as possible to the armature coil, yet not prevent it from turning freely.

7   Connect the brushes in series with a 6 V battery, an ammeter and a variable resistor. The resistor (about 10 Ω) is required to limit the current to 5 A.

8   Spin the armature to make it start turning.

9   Try the effect of reversing the direction of the current.

10   Try the effect of reversing the direction of the magnetic field.

11   This motor is needed in chapter 29; keep it assembled for use later.

Questions

1   Think of ways of making the motor more powerful.

2   Will this motor operate on an a.c. supply? Give reasons.

Figure 26.10 shows the principle of the d.c. motor. The left-hand rule tells us that a downward force acts on the side of the coil that is nearer the north pole. The side of the coil nearer the south pole is acted on by an upward force. As in the galvanometer, the two forces make a couple, turning the coil anti-clockwise. Just as the coil reaches the vertical position, the brushes lose contact with the commutator. The flow of current stops. The armature continues to turn, because of its inertia. Contact between brushes and commutator is made again and current flows in the coil, but in the opposite direction. The direction of current *in the coil* has been reversed, *but* the coil has made a half turn so the current still flows away from us in the side nearer the north pole and towards us in the side nearer the south pole. The coil continues to turn in an anti-clockwise direction.

The figure shows another way of thinking about this, which you may find easier. The current flowing through

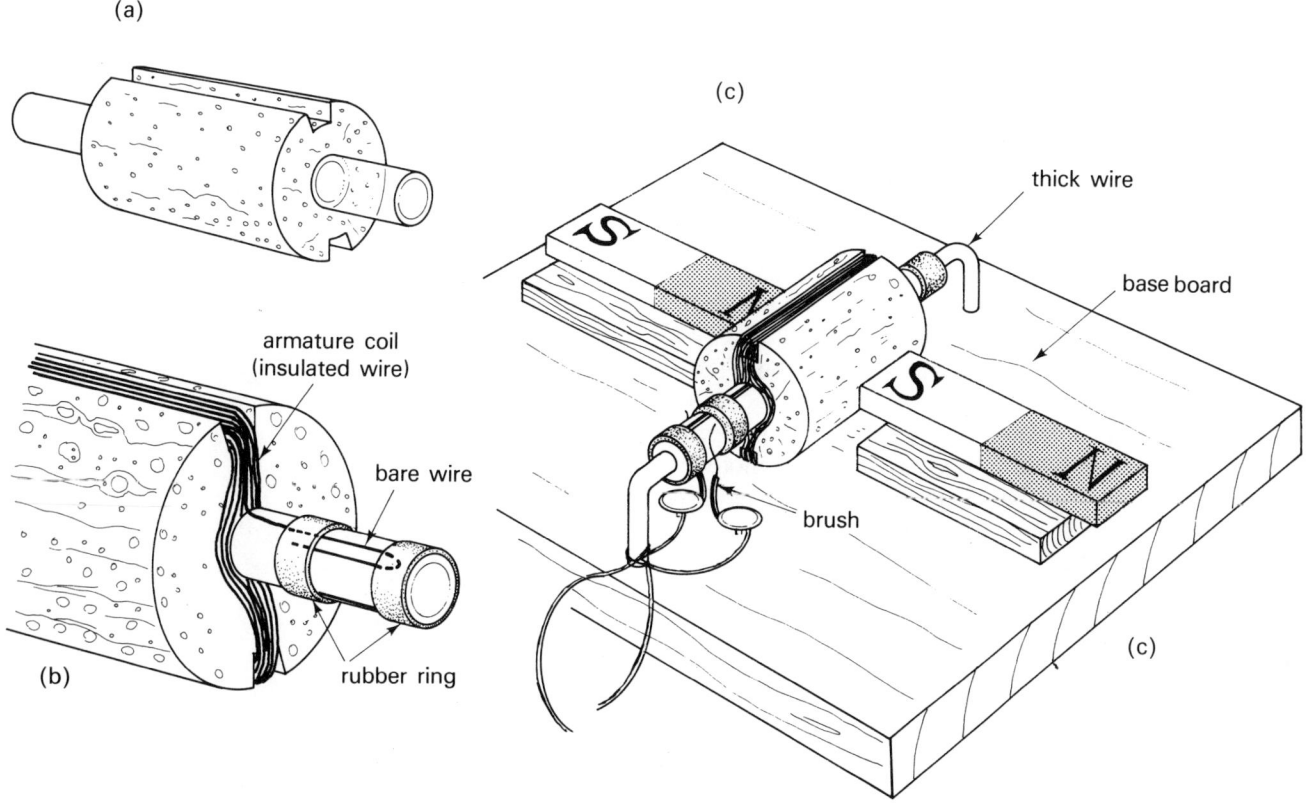

(a)

(b)

(c)

armature coil
(insulated wire)

bare wire

rubber ring

thick wire

base board

brush

Fig. 26.9   A model d.c. motor.

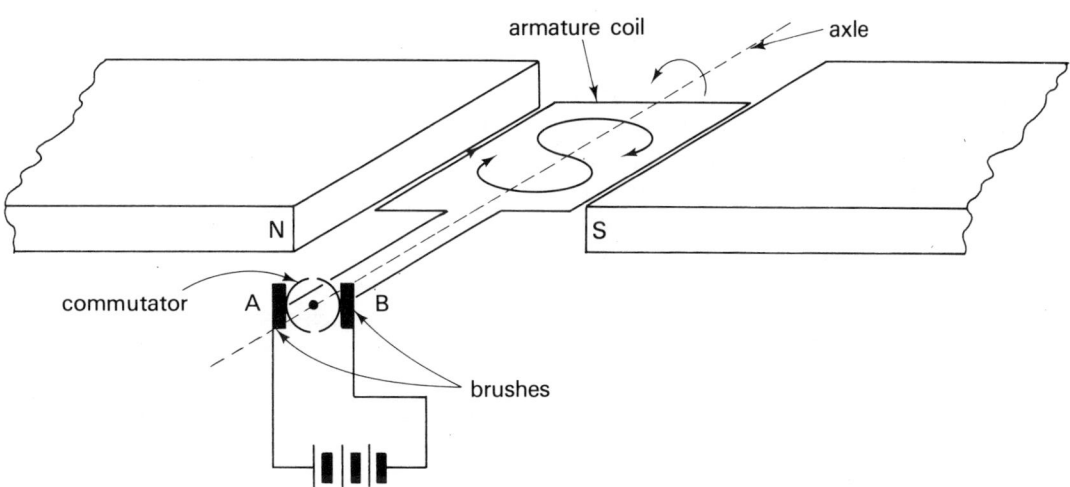

armature coil

axle

N

S

commutator

A

B

brushes

Fig. 26.10   Diagram of a d.c. motor.

the coil creates a magnetic field, with south pole upper-most. This is repelled by the south pole of the magnet. The coil is forced to turn anti-clockwise. The north pole on the underside of the coil is repelled by the north pole of the magnet, with the same result. As the coil turns over the commutator reverses the direction of current in the coil. The south pole is again created uppermost; it is repelled and the coil continues to turn.

The motor you have constructed has very low efficiency. It takes a large current; it has high power input. Yet it is not able to produce great turning force; its power output is low. A factory-made motor has many of the features mentioned in the answer to question 1 on page 278. In addition it may have more than one coil, each with its own connection to a commutator with several segments. This gives much smoother running. Look at a simple d.c. motor to see how it is made and why it is more efficient than the motor you made. Then measure its power in the way described below.

Instructions: measuring the power of a d.c. motor

1 Connect a low-voltage motor to a circuit connected as fig. 19.13; the motor replaces the resistor.

2 Switch on; read current, I, while the motor is running. How many coulombs pass through the motor per second?

3 Read the voltmeter to find the number of joules of electrical energy being converted into rotational energy and internal energy for each coulomb passing through the motor.

4 Calculate the rate of conversion of energy into motion and internal energy. This is the power of the motor. Express this in watts. (Hint: if in difficulties, revise the instructions on page 272).

5 Switch on again. While the motor is running, press your finger or a piece of cork against the axle, to make it run slowly. Measure the power of the motor. Explain the change in its power.

6 Attach a thread to the axle so that the thread is wound on to the axle as the motor rotates. Hang a load from the thread. Measure the time taken for the load to be lifted a measured distance. Calculate the work done. Calculate the useful output power of the motor.

7 Calculate the efficiency of the motor.

26.23 Loudspeakers

In fig. 23.6 a current is passed along a wire in a magnetic field. The wire moves. The amount and direction of its movement depends on the amount and direction of the current in the wire. Instead of wire, you may use a thin strip of aluminium foil. Instead of d.c. from a battery, you may use the varying current from an audio generator or a radio set. It is essential to connect a resistor in series with the strip, so that the current passing is not greater than that which the generator can safely supply. When the generator or radio is switched on, a current varying in amount and direction flows through the strip. The strip moves according to the current. It vibrates creating sound waves in the air. Electrical energy is converted to motion and then to the wave motion we call sound. This demonstrates the working of a loudspeaker.

The structure of a loudspeaker is shown in fig. 26.11. The current passes through a coil of many turns. The coil is attached to a cone made from thick paper. The edge of the cone is shaped to make it very flexible and is attached to a supporting ring. The flexible edge allows the cone to move. The coil is in a strong magnetic field produced by a magnet with soft iron poles attached. The gap between the poles is small, making the field strong. When the coil has a varying current passed through it, electromagnetic forces make the coil vibrate. As seen in the diagram, the

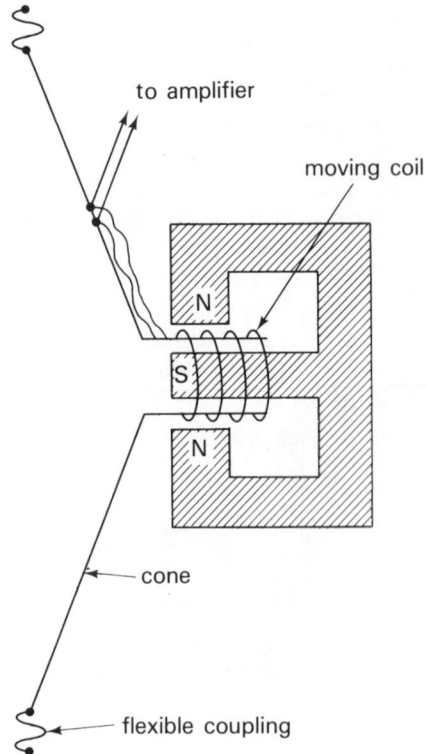

Fig. 26.11 *Diagram of a moving-coil loudspeaker.*

vibrations are in a left-right direction. This vibration is transmitted to the cone. The cone vibrates and this causes the air on both sides of it to vibrate, producing sound.

26.30 The chemical effects of electricity

26.31 Electrolysis and the flow of current

In 19.12 we studied electrolysis and its applications. Let us set out what we *know* about electrolysis.

(a) It occurs when a current passes through an electrolyte.

(b) Action occurs at the electrodes; we see bubbles of gas, or a deposit of metal.

We *explain* what we know by saying that:

(a) The substances in solution in the electrolyte become ionised.

(b) Ions are electrically charged and are attracted towards electrodes when a p.d. is applied.

(c) Ions *may* be discharged at electrodes, forming bubbles or deposits.

(d) In doing this, ions act as charge carriers; a current flows through the electrolyte.

We are not able to *see* ions moving through the solution. The explanation fits the facts, *but* it would be more convincing if we could test our ion hypothesis in some way.

We say that ions are formed when an electron is lost

from an atom or part of a molecule, creating a positively charged ion. The electron is transferred to another atom or part of a molecule, creating a negatively charged ion. We know that all electrons have the same amount of negative charge (23.10). If we assume that only one electron or some fixed number of electrons is transferred in ionisation, all ions should have the same amount of charge, positive or negative. When a current flows, the number of ions passing to the electrodes in a given time should be related to the amount of charge passing through the electrolyte in that time. In short *number of ions varies with charge*. If m is the mass of gas set free, or metal deposited, and Q is the amount of charge passing in the same time, and if our ion hypothesis is true, we expect that:

$$m \propto Q$$

By definition of current (chapter 14) $Q = It$
If our hypothesis is true, we expect to find that:

$m \propto I$ (if we make t constant)
and $m \propto t$ (if we make I constant)

Let us perform two *experiments*, to test the hypothesis.

Instructions: an experiment to test the prediction that m varies with t

1 You need a copper *voltameter*, consisting of a glass container or jar and two copper electrodes. It is best if there is one copper plate for the cathode with a plate on either side of this for the anode. The two anode plates are connected by wire. With this arrangement, copper is deposited on both sides of the cathode. We can use a large current and deposit a large amount of copper in a given time. What advantage is this?
2 Clean the cathode by rubbing it with emery paper, then rinse it in water.
3 Connect the voltameter in series with a 4V power supply (battery of accumulators or power pack), an ammeter (f.s.d. 1A), a variable resistor (5Ω or more) and a switch.
4 Fill the voltameter with copper sulphate solution to which a little sulphuric acid has been added. Switch on and adjust the resistor until 1A flows through the voltameter. Switch off; do not alter the setting of the resistor.
5 Remove the cathode. Wash it in water, rinse it in distilled water and then dry it. Find its mass to the nearest 0.01g. Record its mass, m_1.
6 Replace the cathode in the voltameter. Switch on the current for exactly 10min. During this time read the current occasionally and adjust the resistor to keep the current constant at 1A.
7 Remove, wash and dry the cathode. Find its new mass, m_2. What change of mass is noticed? How do you account for this?
8 Replace the cathode and repeat steps 6 and 7 twice more. Do the results agree with the prediction that m varies with t?

Instructions: an experiment to test the prediction that m varies with I

1 Use the same apparatus as in the previous experiment.
2 Run the voltameter three times, for exactly 15min each time, but with different current. Suitable current values are 1A, 0.75A and 0.5A. Do the results agree with the prediction that m varies with I?

The two experiments may also be performed using a Hofmann voltameter. This is a glass apparatus with two platinum electrodes. It is filled with acidified water. Two tubes above the electrodes collect the gases (oxygen and hydrogen) set free when a current passes through the electrolyte. The tubes are graduated to measure the gas volumes. If we assume that the mass of gas is proportional to its volume, we can show that mass varies with time and current.

The experiments support the hypothesis that conduction in electrolytes is by charged ions. They give us a definite relation between charge carried and the mass of matter set free from solution. The hypothesis has been well established and is known as *Faraday's first law of electrolysis*.

26.32 Accumulators

When we speak of an accumulator, we usually mean a lead-acid cell. If you examine an old lead-acid cell, you will see a number of plates made of lead. In modern cells these are made from an alloy of lead and antimony. The plates are alternately cathodes and anodes, sandwiched together, but not touching. The cathodes are all electrically connected and so are the anodes. The plates are in the form of grids, with a special paste in the spaces. In a new cell, the paste contains oxides of lead. The electrolyte is a mixture of sulphuric acid and water of the right density. A current is passed through the cell. By electrolysis, the lead oxides in the paste on the positive plates become converted to lead (IV) oxide. On the negative plate they become converted to lead. The cell is now said to be *charged*.

When we charge a capacitor we put extra electrons on one plate and there is a shortage of electrons on the other. There is electrical potential energy due to the field between the plates. In the cell there has been a chemical change; there is potential chemical energy due to the different chemical substances formed on the two sets of plates. When the terminals of the cell are connected to an external circuit, the chemical potential energy is converted to electrical energy.

A lead-acid accumulator is able to deliver a large amount of charge at a high rate (a large current). The cell may have a *capacity* of 30 ampere-hours, for example. Note that this capacity is not the same as capacitance. A capacity of 30 ampere-hours means that it can supply a current of 1A for 30h before becoming discharged. Or it can supply a current of 2A for 15h, or 4A for 7½h, and so on. If the current is very high, the length of time becomes rather less than expected from the capacity rated in ampere-hours.

As the cell supplies current, chemical changes occur. The lead (IV) oxide and lead both become converted to lead (II) sulphate. The acid becomes dilute. We can use a hydrometer (4.31) to measure the density of the acid and so tell how much charge remains in the cell. Small battery-testing hydrometers are graduated to indicate directly whether the cell is fully charged, half-charged or discharged.

The cell is re-charged by connecting it to a battery-charger. This consists of a transformer (29.30) which has an output voltage lower than mains voltage.

The output is rectified (fig. 19.20). If the output voltage is 20V, ten cells are connected in series and to the battery charger. The positive output terminal of the charger is connected to the positive terminal of the first cell in series. Current flows through the cells in the opposite direction to the current the cell supplies. A resistor in the charger limits the current to a safe level. During re-charging, the lead (II) sulphate is converted to lead (IV) oxide and lead again and the density of the acid increases. Towards the end of re-charging, bubbles of oxygen and hydrogen are produced at the plates, indicating that the process is nearly complete.

The e.m.f. of a fully charged cell is 2V, and remains at that value for most of the period of use. The internal resistance of the cell is extremely low. Thus almost all the e.m.f. is available for driving current through the external circuit. High currents can be obtained, up to 200A in theory (but see below). The main reason for the low internal resistance is the large surface area of the plates and the small distance between them. To prevent the plates from actually touching and causing a short-circuit, thin spacers made from plastic are placed between the plates.

High current damages the plates; they become bent and the paste may fall from the grid. For this reason, the terminals of a lead-acid cell should never be joined together with a piece of wire. Before switching on a circuit, it must be checked to see that there are no short-circuits, and to make sure that when switched on the current will not exceed a few amperes.

To keep cells in good condition they need attention. The water evaporates from the dilute acid and must occasionally be replaced. From time to time *distilled* water is added through stoppered openings at the top of the cell. Cells must be kept charged if they are to be taken out of use for a while. They must be recharged once monthly; if this is not done, lead (II) sulphate is converted to a white crystalline form which cannot be converted back to lead or lead (IV) oxide by re-charging.

Another type of re-chargeable cell is the nickel-alkaline cell (or 'Nife' cell). This is less likely to be damaged by bad handling. It has a longer life than the lead-acid cell. It can supply large currents without damage to its plates. It can be left in the discharged state for long periods. These cells are expensive but are useful for emergency lighting in hospitals and ships, because they do not need attention when they are not being used. Their main disadvantage is that their e.m.f. is only 1.2V.

26.33 Fuel cells

In *dry cells*, the Leclanché and similar cells, we supply chemicals. Chemical energy is converted to electrical energy and the chemicals are used up. One of the electrodes becomes thinner and must eventually be renewed. In *storage cells* or accumulators, we do not need to supply new chemicals except when the cell needs renewing after a number of years, but we must have a supply of electricity for recharging the cell regularly.

In *fuel cells* a fuel is oxidised and the chemical energy released is converted directly to electrical energy. During this process we use up fuel and oxygen, but the electrodes remain unchanged. Fuel and oxygen are fed into the cell while it is operating and electrical energy is released immediately.

In electrolysis a current is passed through acidified water; oxygen and hydrogen gases are set free. If we can *supply* oxygen and hydrogen to a cell, we ought to be able to run the process backwards and *generate* electricity. In practical fuel cells, oxygen can be supplied by bubbling air past an electrode. The fuel may be hydrogen or some other oxidizable substance such as methanol. Many types of fuel cell are being developed. They are efficient at converting chemical energy into electrical energy; twice as efficient as the most efficient power plants. Instead of burning fuel to drive an engine to turn a generator to generate electricity, we can convert the energy of the fuel to electrical energy in *one step*.

One of the problems of electrically-powered vehicles (including automobiles) is to provide them with enough stored electricity to travel for useful distances. If we store the electricity in lead-acid cells, we need many cells. This means great expense and a heavy load of cells to be carried in the vehicle. Also, the continual vibration on the journey shakes the paste from the plates and the cells soon need replacing. Alkaline cells are not so easily damaged by vibration but these too are large. If we can use fuel cells, we overcome all these problems. We store the energy in *compact* chemical form, as fuel. This is just what we do in an ordinary petrol-driven vehicle when we fill the tank with petroleum. One kilogram of petroleum 'stores' about 50MJ; this is about 50 times as much energy as can be stored in an automobile battery. Remember too that the battery weighs considerably more than 1 kg. The fuel cell is not an energy store but an energy *converter*. The fuel cell and electric motor have the same function as the petroleum driven engine of an ordinary automobile. To store electrical energy we need many storage cells. To convert the energy of fuel into electricity requires only a few fuel cells. This saves much weight and space in the vehicle and it can make long journeys without the need for re-fuelling.

If reliable and efficient fuel cells can be developed they will have many applications in electrically-propelled vehicles. They produce no polluting substances. It has also been suggested that surplus electrical power that is generated at times of low demand (night, warm season) could be used to produce hydrogen from water, by electrolysis. This hydrogen, stored in cylinders under pressure represents a store of energy. This could be used as fuel for modified internal combustion engines (chapter 28) or for generating electrical power in fuel cells. In this way we could supply energy for driving machines and vehicles when the oil supplies of the world begin to become exhausted at the end of the century.

27 Flight

27.00 Pressure in moving fluids

Instructions: investigating the effect of motion on air pressure

1 Take two sheets of thin paper about 20cm × 30cm. Hold them by their top edge between the thumb and forefinger of each hand, so that they hang vertically. Bring your hands together so that the sheets face together, are parallel and there is a gap about 4cm wide between them.

2 From above, blow gently downwards into the gap. What happens to the sheets of paper?

3 Blow harder. Blow as hard as you can. What happens to the paper?

4 Use your knowledge of air pressure to explain what is happening. You have demonstrated the main principle on which flight depends. It is called *Bernoulli's principle*. Stated simply, it is:

The pressure in a moving fluid varies inversely as the speed of the fluid.

The principle applies to all kinds of fluid, to liquids as well as to gases. For example, imagine two boats, side by side in the water with a narrow gap between them. A current of water flows past the boats; it is faster in the gap. The reduction in pressure forces the boats together. We can demonstrate the reduction of pressure by passing water along a tube, which has side-tubes attached to act as manometers. In fig. 27.1a the tube is uniform in diameter. Pressure drops slightly along the tube because of forces of viscosity. In fig. 27.1b the volume V_t of water passing through the tube in a given time must be the same for all parts of the tube. This means that the speed of the water must be greater in the narrower part of the tube. The manometers show that the pressure is lower in the narrower part of the tube. It is difficult to explain this without using long mathematical proofs. Yet the idea behind it is simple. If the water in the narrow tube has increased speed, it has increased kinetic energy. The law of conservation of energy tells us that gain in energy of one

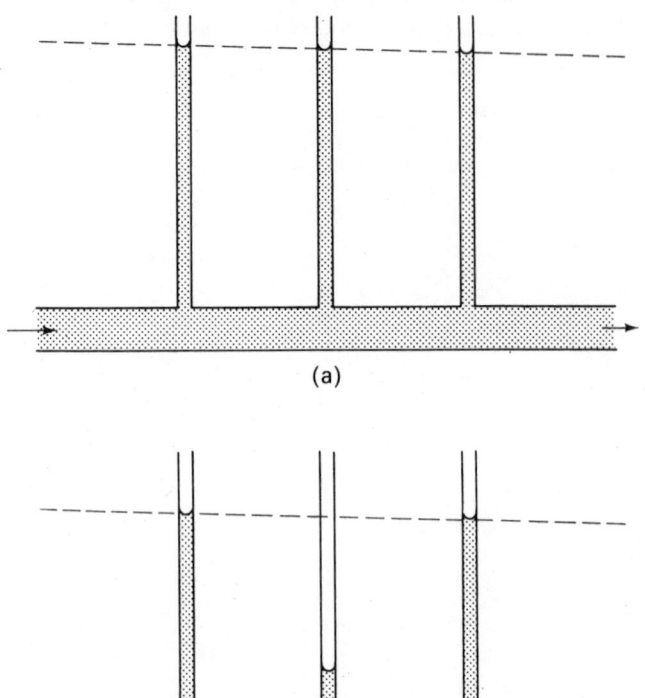

Fig. 27.1 *Demonstration of Bernouilli's principle.*

form must be balanced by loss of energy of another form. In the narrow tube, the gain of kinetic energy is balanced by the fall in pressure. Pressure of the molecules of water striking the walls of the tube represents potential energy. The amount of potential energy lost through reduced pressure equals the amount gained by increased speed of flow. A number of commonly used devices such as the insecticide spray, the bunsen burner and the filter pump make use of Bernoulli's principle (see project 1).

284

27.10 Aerofoils

Instructions: applying Bernoulli's principle to the aerofoil

1 Take a piece of paper about 20cm × 30cm. Make a sharp straight fold across it, about 5cm from one end.
2 Curve the larger part of the paper as in fig. 27.2. The easiest way to do this is to hold the paper firmly and draw it over a sharp edge, such as the edge of a table.

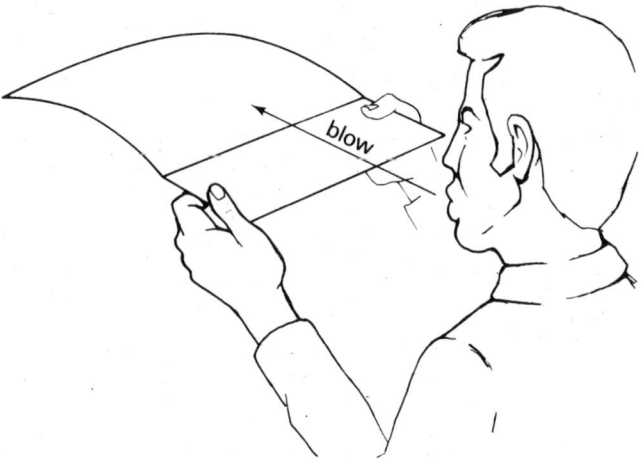

Fig. 27.2   A simple aerofoil.

3 Hold the paper horizontally. The curved portion will hang down more than is shown in the figure.
4 Blow across the top surface of the paper. What happens to the curved part?
5 Explain the effect by using your knowledge of pressure.
6 If the effect is the result of Bernoulli's principle, what can you say about the speed of air over the top surface?

The effect that you have just demonstrated explains the action of the *aerofoil*. Figure 27.3 shows the behaviour of air as it flows past an aerofoil, seen in section in the figure. The thin lines show the path taken by particles of the moving air. We call these *streamlines*. The same effect occurs if the air is still and the aerofoil is moved forward through the air. At the front edge of the aerofoil, the *leading edge*, the air divides. It passes above or below the aerofoil and meets again at the *trailing edge*. The large curvature of the upper surface gives the air above the aerofoil a greater distance to travel before it meets the air that has passed below the aerofoil. The air passing above the aerofoil must have greater speed than the air passing below. Greater speed means lower pressure; the pressure on the upper surface must be less than the pressure on the lower surface. The resultant pressure is *upward*; we call it *lift*.

The effect can be demonstrated by making model aerofoils and placing them in a moving air-stream (see project 2). By fixing small pieces of wool to the aerofoil and by holding pieces of wool in the air-stream the

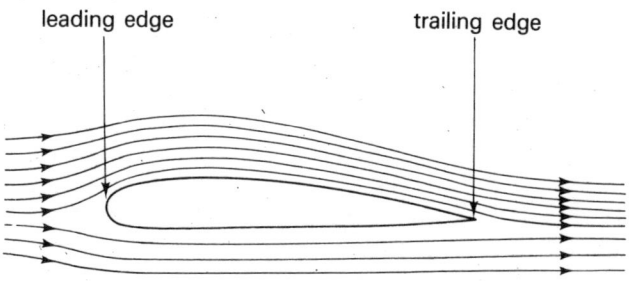

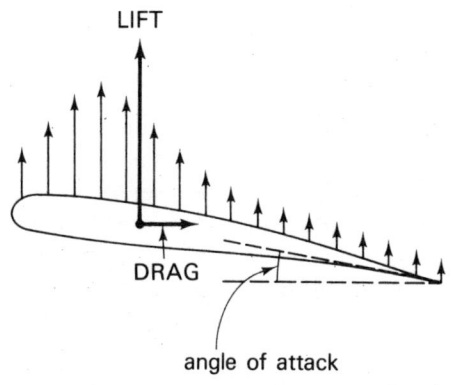

Fig. 27.3   Airflow past an aerofoil.

streamlines can be shown to be like those of fig. 27.3. If the model aerofoil is attached to a suitable balance the amount of lift can be measured. The main use of the aerofoil is to give lift to aeroplanes. The wings of aeroplanes are shaped like aerofoils in section. Other applications in aircraft are the blades of propellors and the rotor blades of helicopters.

Pressure is not uniform over the aerofoil surface. It is higher in some areas (where speed is least) and lower in other areas (where speed is greatest). In fig. 27.3b we see arrows representing the lift at various points between the leading and trailing edge. These arrows represent the *difference* between the pressures on the lower and upper surfaces. These forces may be replaced by a single resultant force, the lift. This acts through a point in the aerofoil called the *centre of lift*. Because lift is not uniform, the centre of lift is usually situated nearer to the leading edge than to the trailing edge.

Another force acting on an aerofoil is *drag*. This is a backwardly directed force caused partly by viscosity (fig. 17.9) and partly by air disturbance. Drag is also caused by the fuselage and other parts of the aeroplane. Drag is an unwanted force. It makes it harder for us to propel an aerofoil or an aeroplane forward. We must use energy in a way that is not useful. Aeroplane designers usually aim to obtain as much lift and as little drag as possible.

The amounts of lift and drag of a given aerofoil moving at given speed depend on the angle of attack (see project

3). As the angle is increased lift increases, but when a certain angle is reached, further increase may bring decreased lift. The air is no longer able to flow smoothly over the upper surface: it breaks away and spins in small whirlwinds (vortices); Bernoulli's principle does not apply to *turbulent* flow. There is loss of lift and considerable increase in drag. For normal flight the angle of attack is set so that the greatest possible lift is obtained without high drag.

When an aeroplane is about to land, its speed is reduced. This reduces the lift. To avoid losing lift, the pilot tilts the aeroplane slightly backwards, to increase the angle of attack. With a steep angle of attack and slow air-speed a point is soon reached at which the air flow becomes turbulent. It breaks away from the upper wing surface. Lift decreases rapidly. We say that the aeroplane is *stalled*. If this happens when the aeroplane is only a few metres above the ground, there is no way for the pilot to recover control. The aeroplane loses lift and crashes to the ground. An aeroplane may also stall when it is climbing steeply, high above the ground. As it falls it gains speed giving the pilot a chance to recover control before it comes dangerously near to the ground. In modern aeroplanes, flaps and other types of air brake are used for reducing speed when landing and there is less risk of stalling.

The aerofoil is also used in many types of sailing boat. In many of the older types of boat, which are 'square-rigged', the sail is set at right-angles to the length of the boat (fig. 27.4a). The pressure of wind coming from behind drives the boat forward. This type of vessel can go only in the direction that the wind blows. Modern sailing boats (and *some* of the traditional ones) use the aerofoil principle. The sail is set so that there is a small angle between it and the length of the boat (fig. 27.4b). The sail is curved, like an aerofoil. The wind travels further and faster as it passes in front of the sail. This produces a force (equivalent to lift) acting in a forward-sideways direction. The sideways action of the force is balanced by the action of the water on the keel of the boat. This prevents the boat from moving sideways. There is nothing except a small amount of drag to prevent the boat from moving forward. The boat can sail across the direction of the wind and in a direction almost opposite to that of the wind. By sailing a zig-zag course into the wind (tacking) the boat can travel in a direction opposite to that of the wind. For sailing with the wind the sail is let out at right-angles to the length of the boat and acts like the sail of fig. 27.4a. With this type of sail the boat can go anywhere, no matter which way the wind is blowing. The figure shows a small additional sail, the jib, which is used on many sailing boats. Its main job is to direct the flow of air on to the front surface of the main sail. It helps make the air flow smoothly over that surface, so that there is no turbulence. This makes the aerofoil effect greater, producing a greater forward force on the boat, even when wind speed is low.

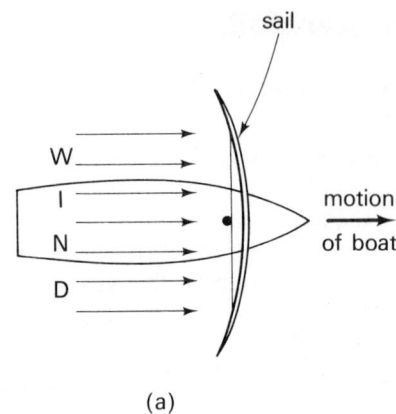

(a)

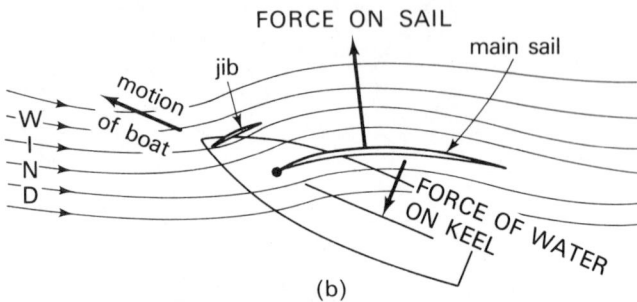

(b)

Fig. 27.4   Sailing boats: (a) square-rigged; (b) fore-and-aft rigged.

Questions
1 What is a hydrofoil? How is it used?
2 A ball such as a tennis ball or golf ball, is thrown through the air. When he threw the ball, the thrower made it spin. What effect does this have on the motion of the ball through the air?

27.20 Forces acting on an aeroplane in flight

The four forces are shown in fig. 27.5. Lift and weight are much greater than thrust and drag, commonly ten times as great.

For *constant speed*, thrust from the propellor or jet engine must equal drag. It is opposite in direction and acts along the same line, so there is no couple. During acceleration, thrust is made greater than drag by increasing engine power. During deceleration thrust may be decreased by reducing engine power or by altering the angle of the propellor blades; drag may be increased by increasing the angle of attack, or by lowering flaps mounted on the wings. Some naval aeroplanes have a small parachute that can be released from the rear to increase drag when landing on the flight deck of a carrier ship. This gives the rapid deceleration needed for landing in a short distance.

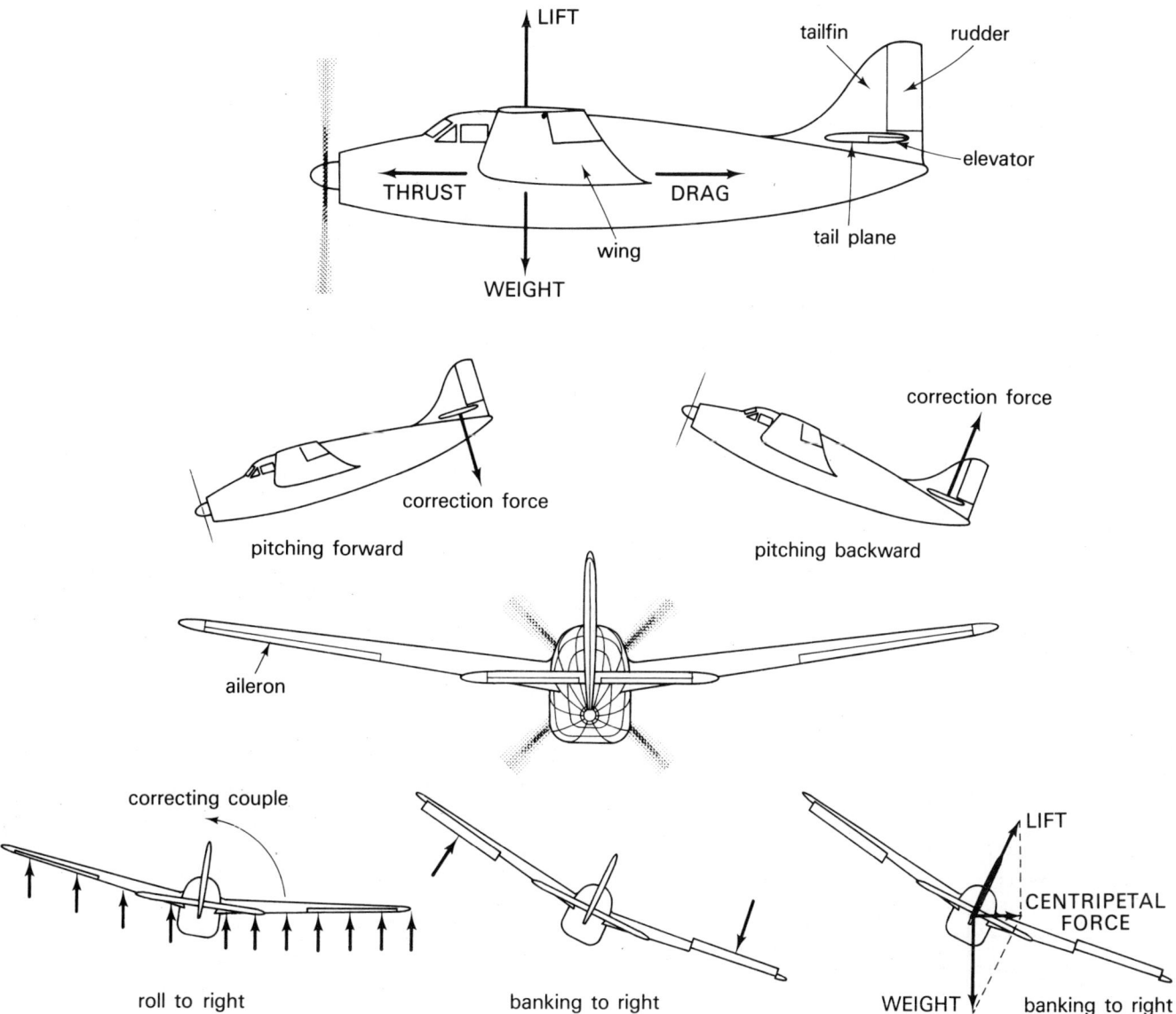

Fig. 27.5 Forces acting on an aeroplane in flight.

For *constant altitude*, lift and weight must be equal and opposite. They must also act along the same line; if they do not, the couple will make the aeroplane pitch forwards or backwards. For stability the centre of lift should be just above the centre of gravity.

Questions
1 Explain why the aeroplane is stabilised by having its centre of lift directly above the centre of gravity.
2 When there are only a few passengers in a light aeroplane they are usually asked to spread themselves evenly over the seating area, and not to occupy only the front seats or only the rear seats. Can you explain this?
3 What precautions must be taken to give high stability when loading heavy cargo in a light aircraft?

The tailplane is to give stability. If the aeroplane pitches forwards, there is a downward force on the tailplane which turns the aeroplane back to its level position. The reverse occurs when the aeroplane pitches backwards. Any slight change of position is corrected as soon as it begins to occur. The rear portion of the tailplane is hinged to form *elevators*, which may be moved up and down under the control of the pilot.

Questions
1 What use does the pilot make of elevators?
2 Why are the tailplane and elevators at the rear end of the aeroplane, at a great distance from the centre of gravity?
3 What is the use of the tail fin?
4 The rear portion of the tail fin is hinged to form a rudder which may be moved from side to side, under the control of the pilot. What use does the pilot make of the rudder?

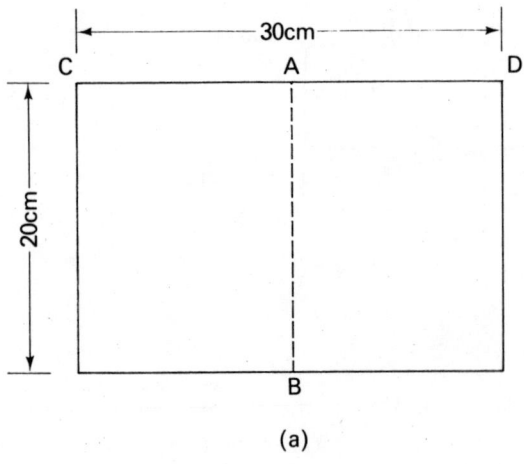

(a)

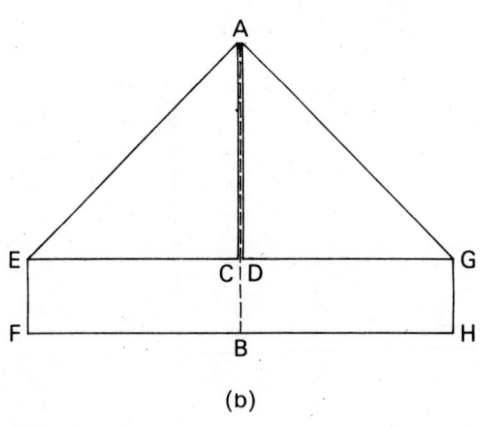

(b)

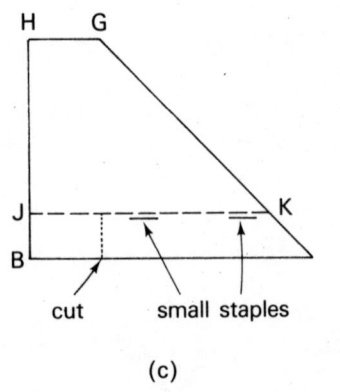

cut   small staples

(c)

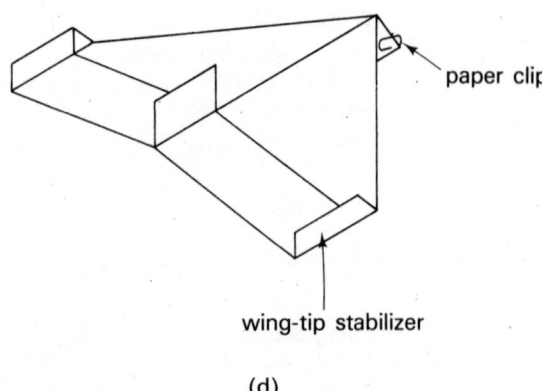

paper clip

wing-tip stabilizer

(d)

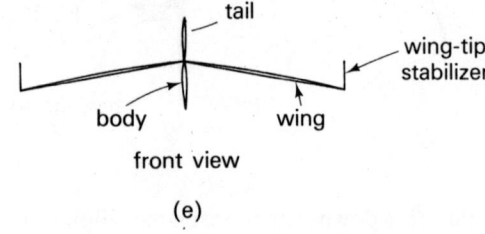

tail

body   wing

wing-tip stabilizer

front view

(e)

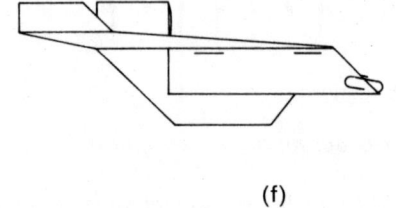

(f)

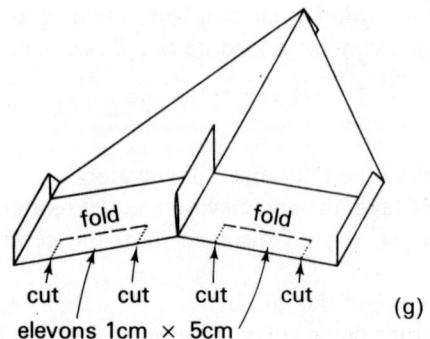

fold   fold

cut   cut   cut   cut   (g)

elevons 1cm × 5cm

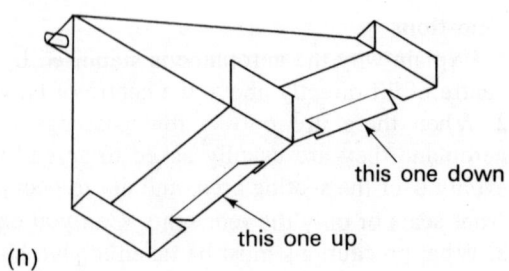

this one down

this one up

(h)

Fig. 27.6 Making a controllable model glider.

If you view an aeroplane from the front or rear, you usually find that the wings are not in a straight line. They are angled slightly upwards. This is to increase stability. If the aeroplane rolls to the right, the lift on the right wing is increased; the lift on the left wing is decreased; the effect is to restore the aeroplane to its proper position. The reverse happens if the aeroplane rolls to the left. At the rear edge of each wing, toward the wing tip, there are small flaps, the *ailerons*. These can be turned up or down, under the control of the pilot. As shown in the figure, these are used to make the aeroplane roll to either side. This motion is used when turning or *banking*. The reason for doing this is the same as that for banking the corners of a road used by fast automobiles, such as a race-track or motorway.

If the road is not steeply banked at the corners, lack of sufficient centripetal force causes a vehicle to skid off the road. Banking creates an inward force on the vehicle (a force acting down the slope) which is sufficient to allow the vehicle to negotiate the corner. In the same way an aeroplane skids sideways when turning left or right, and loses stability and control, unless it banks. Banking directs some of the lift inwards, providing centripetal force necessary for keeping the aeroplane on its curved path.

Instructions: controlling a model glider

1 Take a sheet of stiff paper, A4 size, or about 20cm × 30cm. Fold it exactly in half, along line AB (fig. 27.6a). Then open it out flat again.

2 Fold along AD and AG, so that corners C and D meet on the mid-line AB (fig. 27.6b).

3 Fold along AB again, so that EF and GH come together.

4 Fold along JK, which is parallel to and about 3cm from AB (fig. 27.6c). Fold the other wing similarly.

5 Make a cut where shown. Push the tail up between the wings (fig. 27.4d). Staple where shown in fig. 27.4c, using the smallest staples you can get, or use glue.

6 Fold the wing-tips up. What is the action of these upward-folded surfaces?

7 Place a paper-clip on the nose, to bring the centre of gravity forward.

8 Try launching the glider. Do not throw it quickly. Just throw it gently and smoothly on a slightly downward path. Watch how it flies after you have released it.

9 If on launching the glider climbs, loses speed rapidly and then crashes to the ground, what has happened? How may this be corrected?

10 If on launching the glider pitches forwards and dives steeply, what is wrong? How may this be corrected?

11 If the glider climbs a little, loses speed, then dives gaining speed, then climbs losing speed and so on, what is wrong? How may this be corrected?

12 Adjust the glider until it glides steadily along a gently descending path. If the glider turns to right or left, how can this be corrected?

13 Now add the control surfaces, or *elevons*. These are in the trailing edge of the wings (fig. 27.6g). Make *cuts* where the dotted lines are drawn. Fold up and down where the dashed lines are drawn, to crease the paper and make the 'hinge' flexible.

14 Try the effect of various settings of the elevons. At first do not bend them too far up or down. The angle between the wings and elevons should be between 5° and 10°. Fly the glider with both elevons up, both down, left one up and right one down, and with the right one up and the left one down. Each time launch the glider and watch where it goes. Explain the action of the elevons.

The glider demonstrates another way of controlling the direction of flight of an aeroplane. When both elevons are moved in the same direction (both up or both down), they do the job of elevators. The aeroplane climbs or dives. When they are moved in opposite directions (one up and one down) they do the job of ailerons *and rudder*. The aeroplane banks as it turns. Instead of five control surfaces, we need only two. This method of control is used in many of the modern flying-wing aeroplanes, such as the Aérospatiale/BAC *Concorde*. Your glider has the general shape and downwardly-directed wings of *Concorde* and also Tupolev *Tu-144*. In these aeroplanes the elevons are used for climbing, diving and turning (with banking) to

Fig. 27.7   Concorde in British Airways livery.

right or left. They have a rudder, but this is not used for steering. Its main use is for making slight adjustments to keep on course when flying in a strong crosswind.

Projects

1 Collect and examine as many devices as you can find that make use of Bernoulli's principle. Find out how they work. Draw large diagrams to show how they work. Make a display for the laboratory. You should include: an insecticide spray, a bunsen burner, a laboratory filter-pump, an automobile carburettor, and any other devices you can think of.

2 Make an aerofoil from paper or card. Place this in an airstream produced by an electric fan or blower. You will need to place a honeycomb grid in front of the fan so that the airstream does not whirl around; it must flow straight past the aerofoil. You can make the honeycomb grid by slotting pieces of thin card together. Design and build a balance (make it from wood, card, or use a constructional set) to measure the lift and drag on the aerofoil. Find out how these vary when the angle of attack is changed. Find out which angle gives the largest lift/drag ratio.

3 Study various designs of kite, including traditional designs made locally and the modern manœuvrable kites. Which designs make use of Bernoulli's principle? Where and in what direction are forces acting on the kite? Consider forces due to wind, weight and the tension of strings. If the kite has a tail, what is its purpose? Work out ways of improving the designs, perhaps by using curved surfaces instead of flat ones. You could build examples of your best designs and test them.

4 Investigate flying models made from paper. There are many designs that you and your friends can share among you. Try modifying these to improve them, using your knowledge of the principles of flight. Try adding control surfaces; cut flaps in the wings or tail, or stick flaps of paper or aluminium foil on the model. Models can also be made from cardboard and balsa wood.

28 Engines

Engines are used to propel vehicles such as automobiles, boats, aeroplanes and spacecraft. Many of the engines used in vehicles can also be used for driving machinery such as lathes, presses, cranes and pumps. In this chapter we are mainly concerned with engines in vehicles.

To accelerate a vehicle we must supply it with energy; when the vehicle is travelling at uniform speed we still need to supply energy to replace that lost through friction and the resistance of air or water. The energy may be supplied from outside the vehicle. For example, catapults are sometimes used to launch aeroplanes; cable-cars and lifts are pulled along by ropes moved by engines and electric trains pick up energy from the electrified track or overhead conductor.

For most vehicles it is more convenient and gives greater mobility if the energy supply is taken on board before the journey begins and is stored to be used during the journey. An automobile carries a supply of petroleum in its tank. Although electricity is ideal in many ways for powering factory tools, lifts and other machinery driven by fixed motors, it is less useful for driving vehicles. The most commonly used source of energy for the engines of vehicles is chemical fuel.

28.00 Energy from chemical fuels

Instructions: measuring the amount of energy in a liquid chemical fuel

1 Support a metal can a few centimetres above the bench. Surround it with screens to stop draughts of air from blowing on it.

2 Pour exactly 1 kg water into the can.

3 Measure the temperature θ_1, of the water.

4 Pour exactly 1 cm^3 ethanol or other liquid fuel into a shallow dish. It is best to use a small syringe to measure out the quantity exactly. Place the dish underneath the can and set light to the fuel. Some fuels burn better if you put a small wick of cotton wool in the dish.

5 While the fuel burns, stir the water in the can. Record the highest temperature reached, θ_2. Take care not to blow out the flame accidentally before all the fuel has burned away.

6 Calculate the amount of heat given to the water. Calculate how much energy is in 1 kg of fuel.

7 There are several serious errors in this method. Say what they are.

8 What quantity of heat was used in raising the temperature of the can? Calculate this. Recalculate the amount of energy in the fuel.

Solid fuels are used for providing energy for driving steam engines and certain types of gas turbine. Wood and peat produce relatively little energy. Lignite or brown coal is a low grade fuel, producing only 23 MJ per kilogram. Of the coals used as fuel, bituminous coal contains a high proportion of hydrocarbon compounds which easily evaporate and burn with a yellow flame. They contain about 31.5 MJ per kilogram. Anthracite contains more fixed carbon which burns with hardly any flame, but is hard to set alight. It gives relatively more energy than bituminous coal: 35 MJ/kg. Coke, which is coal from which the hydrocarbon materials have been burned away contains burnable carbon; its energy content is relatively low: 28 MJ/kg.

Liquid fuels are generally richer in energy and more convenient to store and handle, particularly in a vehicle. The fuels obtainable from crude petroleum are petrol (46.5 MJ/kg), paraffin or kerosine (44 MJ/kg) and the fuel oils used in diesel engines (44 MJ/kg). A similar type of oil is burnt to heat water in oil-fired central-heating systems. Among the by-products of coal-gas production, benzene and toluene can be used in internal combustion engines (28.10); their energy yield is 40 MJ/kg. Shortage of mineral oil supplies means that other forms of energy supply are being investigated. By fermenting vegetable matter, using processes similar to those of brewing or wine-making, we can obtain ethanol, which contains 30 MJ/kg. This supply of ethanol could be almost unlimited.

Gaseous fuels are not suitable for vehicles owing to the difficulty of storing large volumes of gas. Town gas, made from coal, and natural gas have very low energy content (less than 20MJ/kg). Vehicles have been driven by producer gas made from anthracite in a special generator. This can be mounted on the vehicle or towed behind it. The energy content is low (4.8MJ/kg) so this method of fuelling vehicles has been little used.

28.10 Internal combustion engines

As long as the world supply of oil remains available, petrol is a very convenient fuel for the engines of vehicles. By small changes of design it is possible to operate many kinds of petrol engine on other fuels, such as ethanol, hydrogen gas, methane or coal gas. In all the engines to be described in this section the fuel is burned inside the cylinder of the engine. The energy released is used to drive a piston which is forced to move along inside the cylinder. Contrast this with an external combustion engine, such as a steam engine. In this, fuel is burnt to heat water to generate steam. The steam is then taken into the cylinder where it does work in moving the piston.

28.11 Four-stroke petrol engine

This is the type of engine most commonly used in automobiles. The cycle of its operation is shown in fig. 28.1. It has four stages.

(a) *Intake* of petrol vapour from the carburettor, mixed with air. The inlet valve is open, the piston moves down, sucking in the mixture.

(b) *Compression* of the mixture as the piston moves upwards. The valves are both closed for the whole of this stage.

(c) *Expansion* of hot gases as the mixture is ignited by the spark from the sparking-plug. This is the stage at which chemical energy is converted into kinetic energy. The piston is pushed downwards with great force. The valves remain closed.

(d) *Exhaust*, as the burnt gases are pushed out of the cylinder through the open exhaust valve.

The cycle is repeated. The energy released at stage c is transmitted to the crankshaft, on which is a flywheel. This continues to rotate during the other three stages. Mechanisms are geared to the crankshaft to open and close the valves and to operate the sparking-plug at the correct times. An engine consisting of a single cylinder produces energy during only one stage of its cycle. Even with a flywheel, the action of such an engine is not smooth. In most automobile engines there are four such cylinders, all connected to the same crankshaft. The cylinders fire in turn. At any given time, one of the cylinders is producing energy. Part of this energy is transmitted along the crankshaft to the pistons of the other three cylinders and is used to drive them through stages a, b and d. The remainder of the energy is used to drive the vehicle.

28.12 Diesel engine

Like the four-stroke engine, this has a cylinder, a piston, a crankshaft, and two valves at the top of the cylinder. Instead of a sparking-plug, there is a fine jet through which the fuel (diesel oil) is sprayed into the cylinder. Like the four-stroke engine, the diesel engine has four stages in its

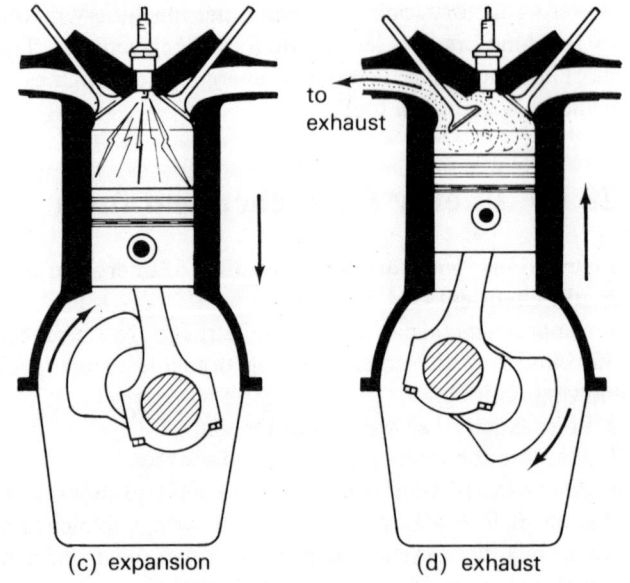

exhaust valve   inlet valve

from carburettor

← cylinder

piston

connecting rod

crank shaft

(a) intake   (b) compression   (c) expansion   (d) exhaust

to exhaust

Fig. 28.1   Operation of the four-stroke petrol engine.

cycle of operation. There are some important differences at each stage.

(a) *Intake* of air as the piston moves downwards. The inlet valve is open.

(b) *Compression* of *air* as the piston moves upwards, the valves being closed for the whole of this stage. The amount of compression is much greater than that in a petrol engine. Compression makes the air extremely hot. The piston does work on the air as it pushes against it, compressing it. This energy appears as greatly increased internal energy of the molecules in the air. Just as the piston gets to the top of its motion, a small amount of fuel oil is sprayed into the cylinder.

(c) *Expansion* of hot gases. The high temperature makes the mixture of sprayed fuel and air ignite. The hot gases expand, driving the cylinder down. This is the stage at which chemical energy is converted to kinetic energy. The valves remain closed during this stage.

(d) *Exhaust*, as the burnt gases are pushed out of the cylinder through the open exhaust valve.

The cycle is repeated. As in the four-stroke petrol engine, a diesel engine has several cylinders driving one crankshaft. Mechanisms geared to the crankshaft open and close the valves and spray oil into the cylinder at the correct times. The lack of sparking-plugs and the electrical circuits needed to produce the sparking currents means that a diesel engine is less complicated than the four-stroke engine. In addition the fuel is cheaper. The diesel engine is particularly efficient when driving heavy vehicles carrying heavy loads. It is commonly used in heavy trucks, railway locomotives and buses.

28.13 Rotary engines

In the four-stroke engine and the diesel engine the piston moves along inside a cylinder and *changes direction* four times during one cycle of operation. We call this *reciprocating motion*. The connecting rod and crankshaft convert this to rotary motion, for driving the wheels of the vehicle. Reciprocating motion is inefficient because work has to be done to accelerate the piston at the beginning of each stroke and decelerate it at the end of each stroke. Energy is wasted, except at stage c where it is used to compress the gases. The conversion of one type of motion into another type of motion requires extra mechanical parts in the engine, and means extra friction, and additional engineering problems. It is much simpler if we can make an engine in which rotary motion is produced directly. The *Wankel* engine is one such design. It is used in a number of automobiles. The engine is illustrated in fig. 28.2. A four-stroke engine has many moving parts: piston; connecting rod; crankshaft and the complex mechanism operating the valve. This engine has only two moving parts: the triangular rotor and the output shaft. There is much less chance of parts wearing away and the engine

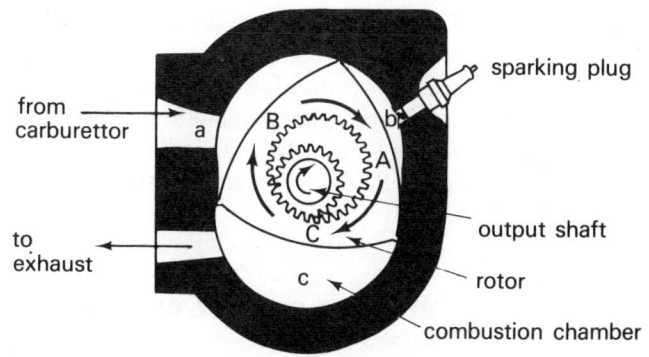

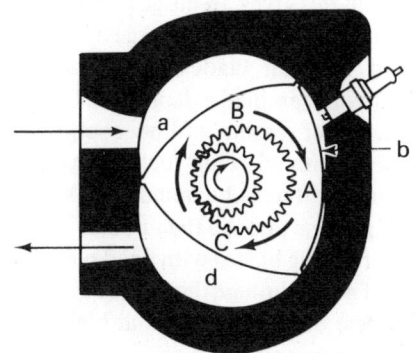

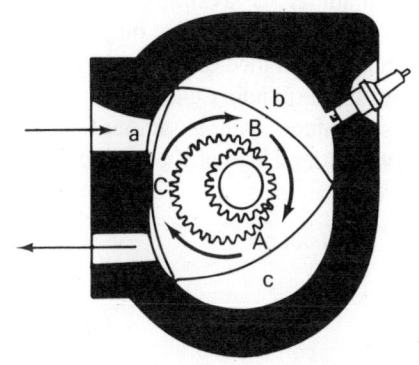

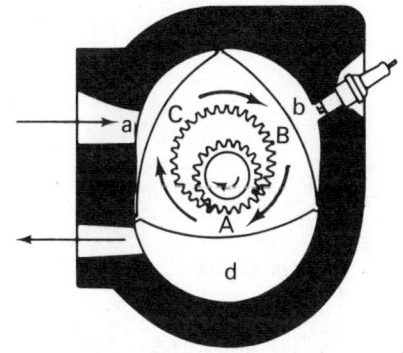

Fig. 28.2 Operation of the Wankel rotary engine. The numbers inside the combustion chamber indicate the four stages of the combustion cycle, (a) intake; (b) compression; (c) expansion; (d) exhaust.

293

breaking down. The tips of the rotor stay in contact with the walls of the combustion chamber as the rotor rotates. In effect it divides the combustion chamber into three compartments. In each compartment a different stage of the four-stage cycle is taking place, as indicated by the numbers in the figure. Study the diagram to see how this engine operates.

28.20 Gas turbines and jet engines

These, too, are engines that operate only by rotation, with the advantages that this gives. In the gas turbine (fig. 28.3) air is drawn in and compressed by a large number of metal blades mounted like fan blades on a rotating shaft. Between each row of fan blades is a row of fixed blades attached to the outer casing of the compressor. The compressed air is mixed with fuel sprayed into the combustion chamber. The hot expanding gases produced by the burning fuel pass through the turbine. This consists of another set of rotating blades with fixed blades between them. The flow of gases through the turbine gives energy to the rotating blades; the turbine turns at high speed. The turbine is on the same shaft as the compressor and a small portion of the energy of rotation is used to compress the air supply. The remainder is available for driving any machinery or vehicle to which the output coupling is joined. The turbine is designed to extract *as much energy as possible* from the expanding gases.

Gas turbines produce a large amount of power in relation to the weight and size of the engine. This makes them very suitable as engines for aeroplanes. The turbo-prop engines, in which a gas turbine drives conventional aeroplane propellors, has been used in a number of aeroplane designs. The high reliability of this type of engine is another factor of great importance in their use in aeroplanes. Gas turbines are also used widely in high speed boats (including hydrofoils), for driving electric generators, for driving pumps on oil-pipelines, for automobiles and trucks, for railway locomotives, for hovercraft and for racing cars. In 1964 the world land speed record was broken by Donald Campbell driving *Bluebird* at 403 miles per hour. *Bluebird* was powered by a Bristol-Siddley gas turbine engine. In some of the uses mentioned above another feature of value is that the engine can be easily operated by remote control.

Gas turbines can be adapted to run on almost any gaseous fuel, including coal gas, natural gas, blast furnace gas and the waste gas produced in oil-fields. It can also run on liquid fuels such as kerosine.

The jet engine is similar to the gas turbine in many ways. Air is taken in and compressed by a rotary compressor. Fuel (kerosine) is sprayed into the combustion chamber, which is generally directly behind the compressor, not to one side as in the gas turbine shown in the figure. The hot expanding gases pass through a short turbine which extracts *just enough* energy to drive the compressor. When the gases leave the turbine they still have high velocity. They are directed out at the rear of the engine through a specially shaped jet. The high momentum of this jet gives high forward momentum to the jet engine and to the aeroplane or other vehicle to which it is attached

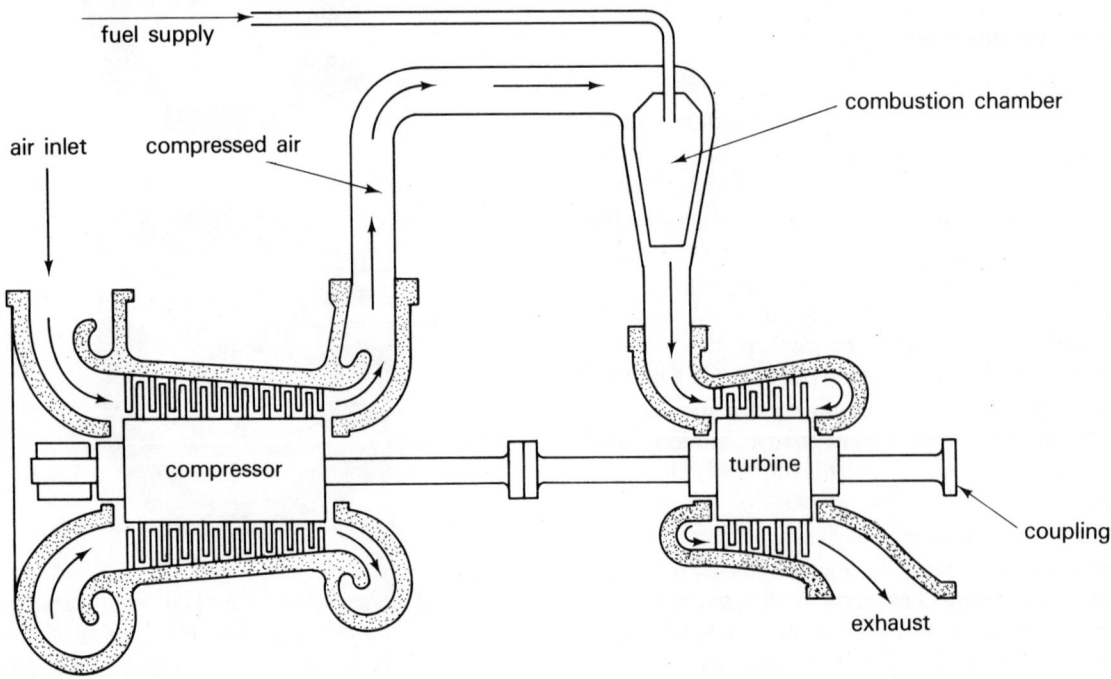

Fig. 28.3 A simple type of gas turbine.

(10.40). The high ratio of power output to engine weight and size is one of the main reasons why this type of engine is so widely used in aeroplanes.

28.30 Rocket engines

Rocket action depends on rapid production of hot expanding gases which pass at high velocity from an opening at the rear of the engine. In the rockets used in firework displays and for distress signals the 'engine' is nothing more than a stout tube of card or similar material, packed with solid chemicals which burn when ignited, producing a jet of hot gas that escapes from the rear end of the tube at high velocity (10.40). In rockets designed to carry satellites into Space there must be a specially designed combustion chamber where the fuel (kerosine, or hydrogen, for example) is mixed with pure oxygen and burned at high temperature. Leading from the rear of the combustion chamber is a nozzle through which the expanding gases pass at high velocity. The metals and alloys from which these engines are made must be able to withstand high temperatures and pressures. To reach high altitude and to accelerate to a velocity at which the rocket is able to orbit the Earth or move away into Space, a large amount of energy is required. The large amounts of fuel and oxygen required to obtain this energy are stored in the rocket in compact form, as *liquid* oxygen and hydrogen.

29 Producing electricity

In the previous chapter we have seen how chemical fuels, natural or man-made, are used as sources of energy for powering engines. Another widely used source of energy is electricity. For many purposes it has advantages over chemical fuels, as will have been realized from the study of its uses in chapter 26. Electricity does not occur naturally in a usable form. It must be generated by converting other forms of energy (including the energy from chemical fuels) into electrical energy. In this chapter we see the many ways in which this can be done. Some are already widely used. Some may become more widely used in the future. Others are more likely to remain just a matter of scientific interest.

29.00 Electricity from mechanical energy

Mechanical energy is readily available; it can be obtained from chemical fuels by using some of the engines described in chapter 28. How may this be converted into electrical energy? Most of the methods used depend on the relation between motion, a magnetic field and an electric current (26.20).

29.01 Using an electric motor in reverse

Instructions: generating electricity by using a d.c. motor

1 Use the d.c. motor you made by the instructions on page 278, or a small d.c. motor like those used in electrically-powered models.
2 Connect a galvanometer or microammeter across the terminals of the motor. If possible, use a meter with centre zero.
3 Spin the armature. Does the current depend on the direction in which you spin the armature? Does the current depend on the speed with which the armature is turned?

By spinning the armature we are making a conductor move through a magnetic field. Current is generated. The motor acts as a generator, or dynamo. To find out how it works we need a simpler arrangement of conductor and field.

29.02 Electromagnetic induction

Instructions: generating electricity by using a coil and magnet

1 Use a coil about 5-10 cm long and 2-3 cm in diameter, with a few hundred turns. You can make one by winding insulated wire round a wooden rod or piece of bamboo.
2 Connect the coil to a galvanometer or microammeter. If possible, use a meter with centre zero.
3 Lay the coil on the bench or fix it horizontally in a wooden clamp.
4 Bring the north pole of a bar magnet towards one end of the coil. Then hold it close to the end of the coil, without moving coil or magnet. Watch the meter and note the current.
5 Take the magnet away from the coil. Note the current.
6 Experiment with the coil and magnet. Find out what you must do to generate the biggest possible current.
7 Try the effect of using different coils. Or change the number of turns in the coil, if you have wound it yourself. What must you do to a coil to increase the current?
8 Try using different bar magnets, some weak, some very strong, and others of various strengths between. What is the relation between magnetic field strength and current?
9 In the steps above, the coil is fixed and you move the magnet. What is the effect of fixing the magnet and moving the coil instead?
10 Put a soft iron core inside the coil. Does this affect the amount of current generated?

Michael Faraday worked with magnets and coils and got results similar to those you have just found. Current is generated only when there is movement. The amount of current depends on the speed of movement of magnet or coil, the strength of the magnet and the number of turns in the coil. An iron core gives larger currents. All these different effects may be explained by using one idea. We must think about the lines of magnetic force coming from the bar magnet and passing through the coil and among the turns of wire in the coil. We say that these lines are *linked* with the circuit which contains the coil. *Faraday's law of electromagnetic induction* states:

When there is a change in the magnetic field linked with a circuit, an electromotive force is induced in the circuit. The strength of the e.m.f. varies with the rate of change of the magnetic field linked with the circuit.

As we found in an electric cell, an e.m.f. causes a flow of electric charge, a current.

Questions
1 According to Faraday's law, if there is no *change* of magnetic field, there is no current. What evidence have you to support this idea?
2 What happens to the amount of magnetic field linked with the coil when the magnet is brought closer to the coil?
3 What effect does this have, if Faraday's law is true?
4 How do you explain the fact that there is a current when you bring the magnet towards the coil, and there is a current when you take it away?
5 Why does a stronger magnet produce a greater current?
6 Why does a coil with more turns generate greater e.m.f.?
7 Why does a soft iron core make the e.m.f. greater?

Instructions: investigating the direction of induced current
1 Use the apparatus you used for the instructions on page 296.
2 Figure 29.1 shows the direction of the current when the magnet is moved in the direction of the arrow. Confirm that this happens with your magnet and coil.
3 Find out the direction of current when the magnet is moved away from the coil.

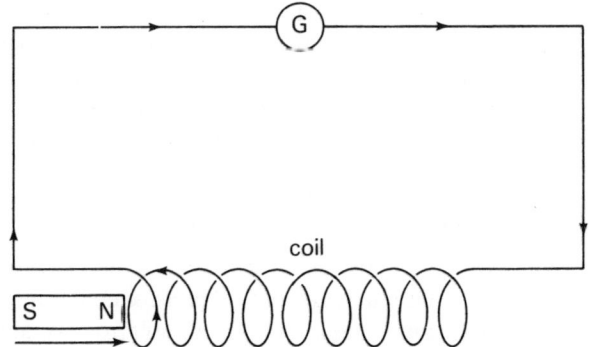

Fig. 29.1   Electromagnetic induction.

4 Find out the direction of current if the magnet is moved towards the coil with its south pole nearer the coil. What happens when the magnet is moved away?

Questions
1 Figure 29.1 shows a current flowing in the coil. If this coil was an electromagnet coil, which pole would be at the end of the coil nearer the magnet?
2 What would be the action of the pole of the coil on the pole of the magnet?
3 When we take the magnet away, a reverse current is induced. What pole does the current produce at the end of the coil nearer the bar magnet?
4 What is the action of this pole on the magnet?
5 Answer questions 1 to 4 for the currents produced when the south pole of the magnet is moved towards and away from the coil. It might help if you draw diagrams to show currents, the poles produced in the coil by the current and the action between coil and magnet.
6 Make up a rule about the direction of the current in the coil.

The investigations lead us to a statement which is known as *Lenz's law*:

The direction of the induced current is that which opposes the motion or change of linked magnetic field which is producing it.

Note that the law includes not only *motion* (which we have just investigated) but any other method of changing the field linked with the coil (which we shall investigate next). To move the magnet towards or away from a coil we have to do work against the magnetic field of the coil. We have to do extra work that we would not need to do if the coil was not there. The extra energy supplied by us is an input of energy to the magnet/coil system. It reappears as the extra electrical energy in the circuit, the induced e.m.f. This causes a current to flow and the energy is then mainly converted to internal energy. At all stages, the law of conservation of energy has been obeyed. To generate electricity, *work must be done.*

Instructions: investigating the choke, an application of Lenz's law
1 Set up the circuit shown in fig. 29.2a. L1 and L2 are identical lamps.
2 The choke, C, is a coil of several hundred turns of insulated wire. A soft-iron core can be put in the coil when needed. Measure the resistance of the choke.
3 Choose a resistor, R, that has the same resistance as the choke.
4 Connect the circuit to a low voltage d.c. supply. Use the coil *without* a core in it.
5 Switch on. Observe the brightness of the lamps. Does this confirm that the resistances of C and R are equal?

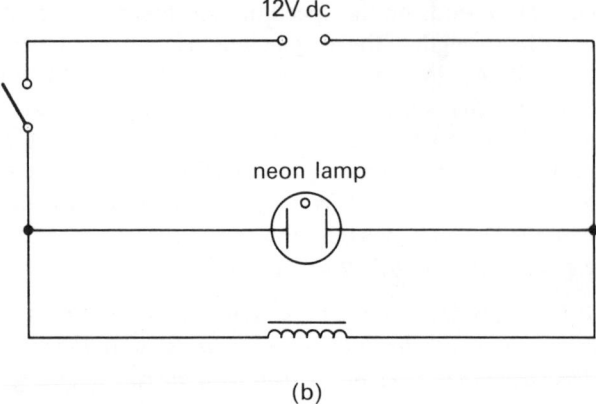

battery or a.c. power pack

12V dc

C   L1

R   L2

(a)

neon lamp

(b)

Fig. 29.2   Effects of inductance in a coil: (a) the action of a choke (b) induction of high e.m.f.

6 With the current still switched on, push the core into the coil. How does this effect the brightness of the lamps?
7 Switch off. Connect the circuit to an a.c. supply of the same voltage as the d.c. supply. Repeat steps 4 to 6.

Questions
1 Alternating current changes direction regularly; it changes 100 times a second, if you use a power-pack that operates on the 50Hz mains supply. At each change of direction the current is zero; then it increases *rapidly* (for 0.005s or less). If the current in the coil increases rapidly, what must happen to the magnetic field in the coil?
2 The field produced by the coil is linked with the coil. If the field increases rapidly, this has the *same effect* as a bar magnet brought quickly towards the coil. According to Lenz's law, what will happen?

Every time the e.m.f. of the external power supply tries to make a current flow through the coil, the induced e.m.f. produces a current in the opposite direction. The coil opposes any attempt to make a current flow through it. In effect, it has *high resistance*. This is why the lamp did not light. With many turns and a soft-iron core, the induced e.m.f. is large enough to almost completely oppose and cancel the external e.m.f. This effect occurs only with a.c. because e.m.f. is induced only when there is a *change* of magnetic field (Faraday's law). With a.c. the field changes many times each second. With d.c. there is constant external e.m.f., constant current flowing through the coil and no change of magnetic field. No e.m.f. is induced. The coil behaves simply as a resistor of very low value. This effect is called *self inductance*.

Questions
1 A coil and switch are connected to a battery. Is there any way in which we can change the amount of current flowing through the coil?
2 What will happen when the amount of current is changed?

The effect mentioned in the answer to question 2 above is demonstrated by the circuit of fig. 29.2b. The supply is 12V *d.c.* The neon lamp needs a p.d. of 70V or more to make it discharge so it does *not* glow when the switch is turned on. When the switch is turned off, the sudden disappearance of the magnetic field induces an e.m.f. in the coil. The field disappears almost instantly just as if a powerful magnet had been taken away from the coil at high speed. The rate of change of linked field is high. This induces a high e.m.f. which produces a p.d. greater than 70V across the neon lamp. The lamp flashes brightly.

The effect is soon over, but it demonstrates that the induced e.m.f. is much higher than that of the battery. This principle is made use of in the induction coil sometimes used in the laboratory for producing high p.d. for operating discharge tubes. It is also used in the *ignition coil* of the four-stroke engine. In the latter, two coils are wound around one core. The 12V d.c. supply from the engine battery sends a current through one coil. When the current is switched off by a contact-breaker, a high e.m.f. is induced in the other coil. The contact-breaker is geared to the engine crankshaft; contact is broken at the exact moment when a spark is to be produced in the cylinder (fig. 28.1). The induced e.m.f. gives a p.d. of about 5 000V at the sparking-plug, igniting the fuel-air mixture. The ignition coil produces high p.d. using only a low p.d. from the battery to do so. In the induction coil the principle is the same except that a mechanism similar to that of an electric bell (fig. 9.13) is used to make and break contact many times per second. A capacitor connected across the coil becomes charged to high p.d. by the induced e.m.f.

29.03 Alternating current generator

The a.c. generator (fig. 29.3) is similar to the d.c. motor, except that it has slip rings instead of a commutator. The coil has many turns and is wound on a soft iron armature. In the figure a single turn is shown for simplicity. Lenz's law tells us which way the current flows. The upper face of

298

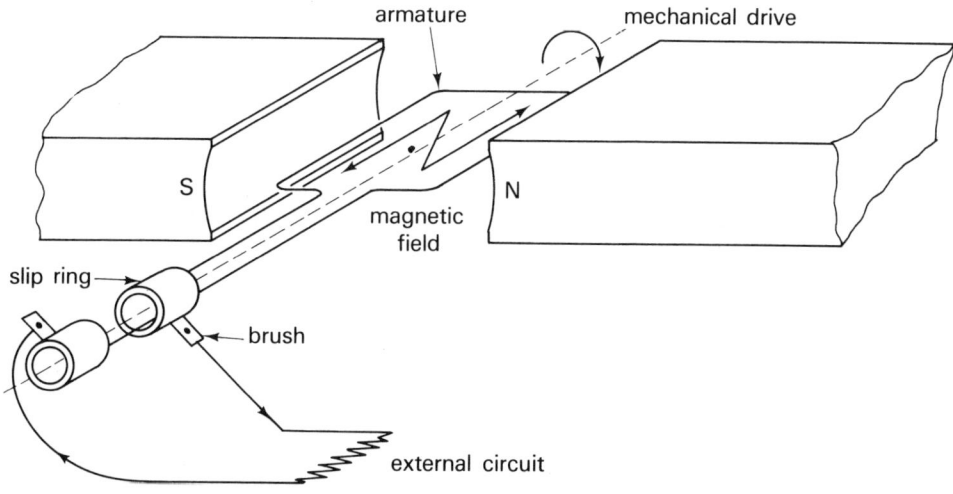

Fig. 29.3 *Alternating current generator.*

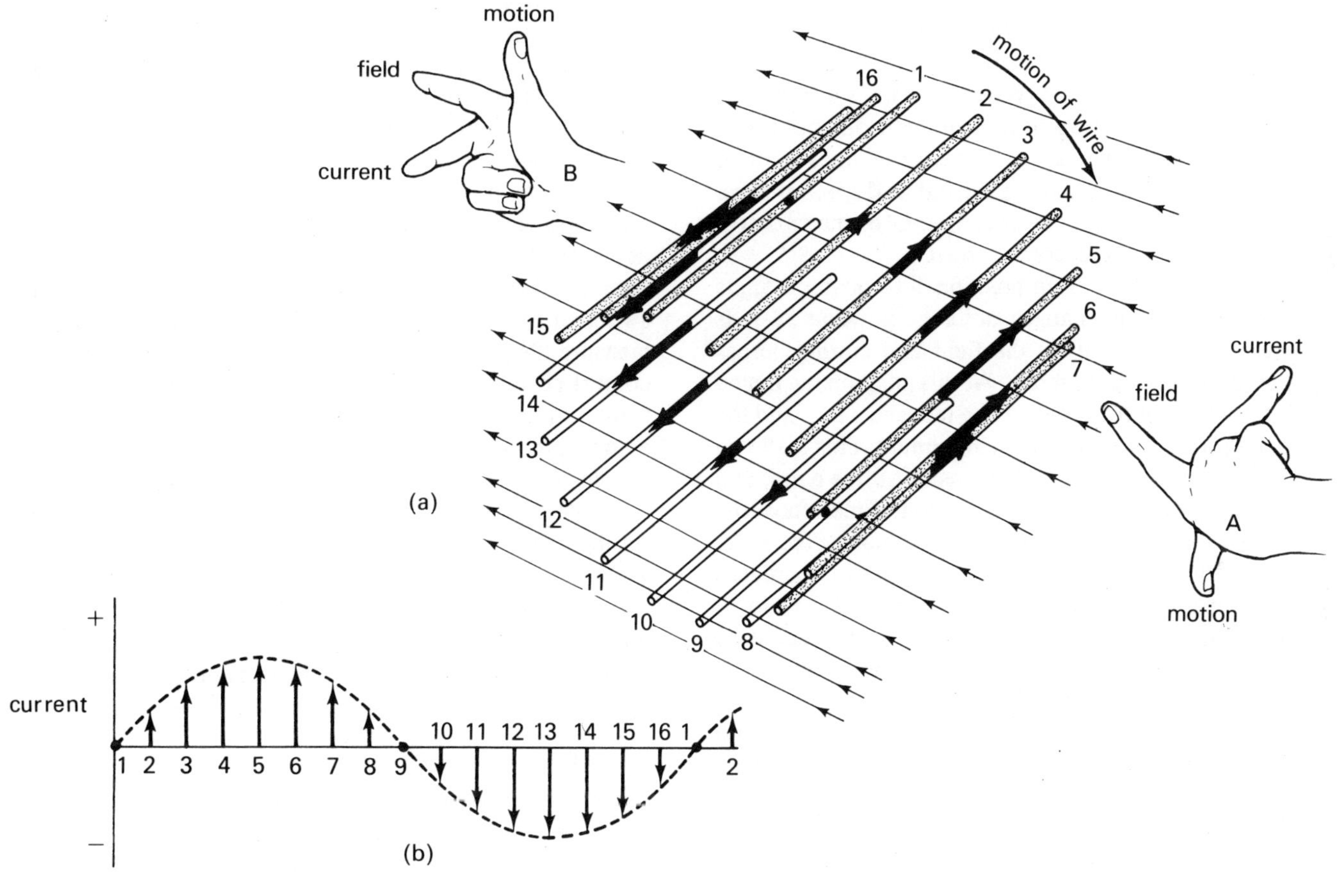

motion

field

current

B

16 1

2

3

4

5

6

7

motion of wire

current

field

A

motion

15

14

13

12

11

10

9 8

(a)

+

current

1 2 3 4 5 6 7 8 9

10 11 12 13 14 15 16 1

2

−

(b)

Fig. 29.4 *(a) The induction of e.m.f. and current in one wire of an a.c. generator as it is moved through the magnetic field; (b) showing the form of the a.c.*

the coil is turning towards the north pole of the magnet. The current induced in the coil must produce a north pole on the upper face of the coil, to *oppose* the mechanical drive. Extra work has to be done to turn the armature; this extra work appears as induced e.m.f.

To follow the action of the generator in more detail, it is useful to know *Fleming's Right Hand Rule* (or Dynamo rule). This is shown in fig. 29.4a. Compare it with the left hand rule (fig. 23.6). If you hold both hands in front of you, with thu**M**bs (**M**otion) and First Fingers (**F**ield) pointing in the same two directions, your seCond fingers (Current) point in opposite directions. This is the result of the effect noted in Lenz's law. The induced e.m.f. opposes change.

In fig. 29.4a we follow what happens in a single wire of *one* side of the coil while the armature makes one complete revolution (360°). Starting at position 1, with the wire at the top, the wire travels down past the north pole (positions 1 to 9). The current induced flows in the direction indicated by the right hand rule as at A. As the wire travels up past the south pole (positions 9 to 16 and then back to 1), the current flows in the opposite direction, as indicated by the right hand rule at B.

Questions

1 At what positions does the direction of the current reverse?

2 How many times does the current reverse during one revolution of the armature?

Faraday's law of electromagnetic induction tells us that the size of the induced e.m.f. varies with the rate of change of the linked magnetic field. When the wire is in positions 1 and 9, it is moving more-or-less *along* the lines of force. This means that the wire is cutting across no lines of force; the linked field is not changing; the induced e.m.f. is zero. The graph of current against time (fig. 29.4b) shows that at positions 1 and 9 the current is zero.

As the wire moves from position 1 to position 5, its direction of motion changes until, at position 5, it is moving perpendicularly to the field and cutting across lines of force at maximum rate. At this position the linked field is changing most rapidly. The induced e.m.f. and the current it produces are both at a maximum. This is shown on the graph. From position 5 to position 9 the e.m.f. and current decrease, because the direction of motion of the wire moves away from being perpendicular to the field. From position 9 to position 13, the current increases again, but in the reverse direction. From positions 14 to 16 and returning to position 1, the e.m.f. and current decrease to zero. In fig. 29.4b the arrows represent the current in the wire when it is in each position. If we join the tips of these arrows, the graph obtained shows us how the e.m.f. and current change during one revolution of the armature. This is alternating current.

Questions

1 If you were designing an a.c. generator, what features would you make use of to get as large a current as possible?

2 What instrument would you use to measure the output of an a.c. generator so that you can study how its output changes as the armature rotates?

Instructions: investigating the heating effect of a.c.

1 Connect the circuit of fig. 29.5. The oscilloscope displays the p.d. across the lamp.

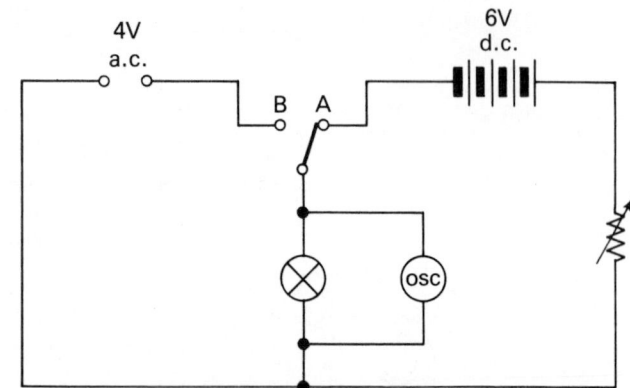

Fig. 29.5   Comparing the heating effects of a d.c. supply and an a.c. supply; the a.c. supply may be either a mains power-pack or an audio-frequency oscillator.

2 Turn the switch to position B, then to A. Adjust the variable resistor until the lamp appears to have the same brightness with the switch in either position. Change the switch rapidly several times, to check that there is no change in brightness. The d.c. and a.c. are having the *same heating effect* in the lamp.

3 Switch to position A. Switch the Y amplifier to its 2V/cm range. The oscilloscope trace is a horizontal line above the centre of the screen, indicating a steady d.c. potential. Measure this. If your oscilloscope does not have switched ranges on the Y amplifier, use a voltmeter to measure the p.d. across the lamp. Then adjust the Y-gain control so that the vertical displacement of the trace is some convenient amount. For example, if the p.d. is 3.6 V, adjust to make the trace 1.8 cm above the centre of the screen. Then you know that *for this setting of the control* the scale for measuring voltage is 2 V/cm. Do not alter the setting of the control during the measurements of a.c.

4 Switch to position B. Adjust the X-control to produce a trace like fig. 29.6. Your trace need not have the same number of waves as this. Measure the *peak values* of the alternating p.d. These are the highest and the lowest p.d.

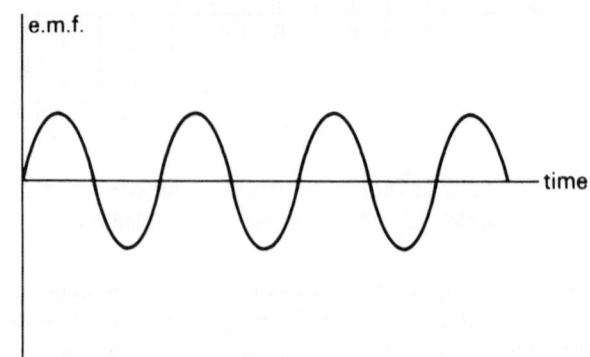

Fig. 29.6   Variation of e.m.f. from an a.c. generator.

shown on the trace. To do this, measure distances on the screen and convert to voltages. Are the peak values the same size as the steady p.d. measured at step 3?

5 Adjust the X-control to a suitable value. Calculate the frequency of the a.c. When we say that there is an alternating e.m.f. of 10V, what do we mean? Does the e.m.f. alternate between +10V and –10V, or is it an alternating e.m.f. that has the same heating effect as a steady e.m.f. of 10V?

If we mean the latter, its peak values are greater than +10V and less than –10V. An a.c. generator gives an e.m.f./time curve like fig. 29.6. The trace on an oscilloscope has the same shape. This is the shape we get when we plot a graph of $\sin \theta$ against θ for angles running from zero to 360° and more. We call this a *sine curve*. If an alternating e.m.f. varies as sine curve, the steady d.c. with equal heating effect is about 0.7 times the peak values (ignoring sign).

So if we operate a heater by a 10V a.c. (where the peak values are +10V and –10V), we could get the same heating effect with a 7V d.c. supply. When we specify the mains voltage as 240V a.c., this is actually the value of d.c. giving the same heating effect. The peak voltages of that mains supply are +339V and –339V. The constant e.m.f. which has the same heating effect as an alternating e.m.f. is called the *root mean square* e.m.f. (written r.m.s., for short). The same term is used to describe currents and p.ds. For the 4V a.c. supply used in the demonstration, you could write $V_{peak} = \pm 5.7V$, and $V_{r.m.s.} = 4V$.

The relation between V_{peak} and $V_{r.m.s.}$ (or between I_{peak} and $I_{r.m.s.}$) depends on the exact shape of the e.m.f./time curve. For a sine curve,

$$V_{r.m.s.} \approx 0.7 V_{peak}$$

For curves of other shapes (square wave, triangular wave, etc) the factor is different, though always less than 1.

29.04 Generating direct current

Filament lamps and heaters work just as well on a.c. as on d.c. They rely on the heating effect of the current and this is not affected by the frequent changes in direction of current. It is also easy to build motors that work on a.c. The motor of a mains-powered electric clock runs at constant speed because it is designed to respond to the rapid alterations of current. At the power station the alternation is maintained at *exactly* 50Hz, so that clocks and other timing devices can use a.c. to control their speed accurately.

Radio sets, tape-recorders, oscilloscopes and many other devices cannot operate on a.c. They need steady potentials. When such equipment is powered from the mains it uses a transformer (29.30) and rectifier to produce d.c. of suitable voltage. If it is to be portable, it may make use of batteries. If large amounts of current and relatively high voltages are needed, it is often better to use a d.c. generator (fig. 29.7). Compare this with the d.c. motor (fig. 26.10) and you see that these are exactly the same. This is why you could use your model motor to generate electricity (instructions page 296). The output from a d.c. generator appears as fig. 29.8 when viewed on an oscilloscope. It has the same appearance as a.c. that has been rectified by a full-wave rectifier (fig. 19.20d). In the generator the rectifying action is carried out by the commutator. As the current reverses in the coil, the connections between the coil and the brushes are reversed.

Questions

1 In what way does the d.c. from a d.c. generator differ from that obtained from a battery?

2 In what way can you obtain a reasonably steady e.m.f. from a d.c. generator?

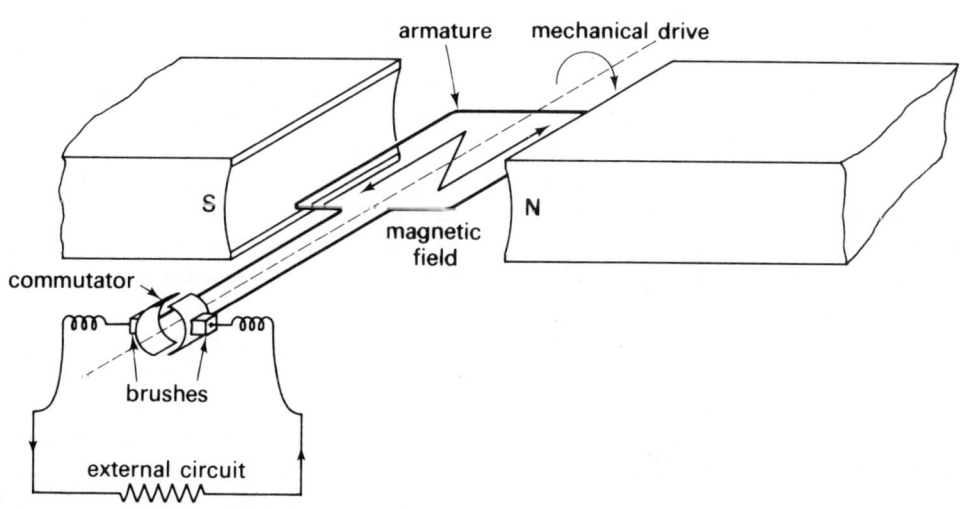

Fig. 29.7   Direct current generator.

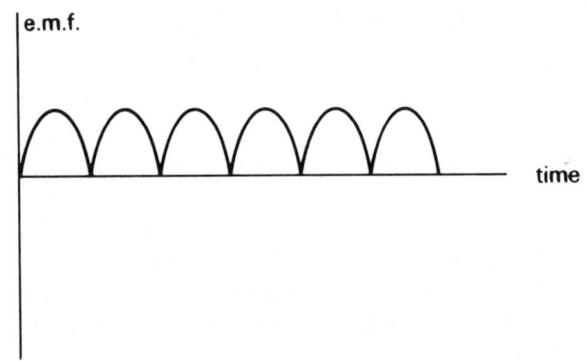

Fig. 29.8   Variation of e.m.f. from a d.c. generator.

3 If $V_{peak} = \pm\ 20V$ in fig. 29.8, what steady e.m.f. from a battery produces the same heating effect?

When we turn the armature of a d.c. motor we generate an e.m.f. (instructions page 296). It does not matter if we turn the armature by spinning it with our fingers, or if we make it turn by passing a current through its coil. If the armature turns, an e.m.f. is produced and this causes a current to flow. If we apply an e.m.f. to a motor, using a battery, the armature turns and an e.m.f. is produced in it. This e.m.f. *opposes* the e.m.f. of the battery, as required by Lenz's law. We call this the *back e.m.f.* The faster the motor is turned, the greater the back e.m.f. When we first switch on the current from the battery the motor is still; there is no back e.m.f. The motor begins to turn slowly; the back e.m.f. is low. As the speed increases the back e.m.f. increases. When the motor reaches its full speed, back e.m.f. reaches its maximum value.

When the motor has reached full speed, it should not need any further supply of energy to keep it running. According to Newton's first law, it has been accelerated to a constant speed of rotation and should continue rotating at the same speed for ever. If this were really so, the back e.m.f. would be exactly equal to the e.m.f. of the battery. They would be opposite in direction. The p.d. across the terminals of the motor would be zero. No current would flow, no more work would be done. This would happen if we could make a perfectly frictionless motor, which had wires with no electrical resistance and which was not being used to drive any other machinery.

In real motors there is friction in the bearings. The wires have resistance to the flow of electricity. Both these effects mean that energy is converted to internal energy. The bearings and wires get hot. If the motor is to be kept running, energy must be supplied to replace the energy lost as heat. We must also supply energy to replace that converted to heat, or other forms, in any machinery to which the motor is connected. When the motor is running the terminal p.d. (p.d. between the terminals of the motor) is just enough to supply the energy needed to replace losses through friction and heat production, as well as losses in attached machinery.

$$\text{terminal p.d.} = \text{applied e.m.f.} - \text{back e.m.f.}$$
(from battery etc)

If there were no losses (a perfect motor), terminal p.d. would be zero, as explained in the paragraph above. Then applied e.m.f. and back e.m.f. are equal. Motors are designed so that the coils are able to carry the currents which pass through them when the motor is running at full speed. If friction is low, and only a small load is being driven by the motor, back e.m.f. is normally almost equal to applied e.m.f. and terminal p.d. is low. With low p.d., only a small current flows through the coil; the wires can be thin. We can wind many turns of thin wire, to make a motor of high power. If the motor jams or is prevented from turning, back e.m.f. is zero. Then terminal p.d. is the full applied e.m.f. The current through the coils is then much larger than the coils are designed to carry. They quickly overheat and may burn out. If a motor becomes jammed while power is applied, the power must be turned off as soon as possible. This can easily happen with model railways or automobiles. If these become jammed on the track or against some obstruction, do not try to release them by increasing power, switch off immediately. Motors on lathes and power-tools are often provided with a *starting resistor*. It is connected in series with the motor when it is first switched on. It limits the terminal p.d. to a safe value. As the motor increases speed and back e.m.f. is increasing, the resistance of the resistor is reduced (either by the operator or automatically) and the motor gradually reaches maximum speed. Then the resistor is switched out of the circuit.

29.05 Moving-coil microphone and carbon microphone

The moving-coil microphone is similar in structure to a moving-coil loudspeaker (fig. 26.11). A small moving-coil loudspeaker is often used as a microphone in 'intercom' systems used in homes and offices. Sound waves strike the diaphragm, which vibrates. The coil vibrates in the magnetic field. As it does so an e.m.f. of varying strength and direction is induced in the coil. If you connect a moving-coil microphone to an oscilloscope, you can see how the e.m.f. varies when sounds of different kinds are made.

The p.d. from the terminals of a moving-coil microphone are small, generally only a few millivolts. To operate a loudspeaker they are amplified by a suitable electronic circuit (31.20). The magnetic cartridge of a record-player works in a similar way to the moving coil microphone. In this the stylus vibrates as it passes along the irregularly-shaped groove on the disc. The vibrations are transmitted to a moving coil in which varying e.m.f. is induced.

Magnetic microphones are *generators* of electricity. A cheap type of microphone, the carbon microphone, works in an entirely different way. It consists of a small container filled loosely with grains of carbon. There are two plates of carbon at opposite ends of the container; the terminals of the microphone are connected to these. The microphone acts as a variable resistance. To use it, it must be connected *in series with a battery* (or some other power supply). When sounds are made the vibrations shake the grains of carbon. This varies the resistance between the two plates of carbon. As the resistance varies, the current flowing through the microphone varies. The p.d. across its terminals varies. This varying p.d. can be amplified and used to drive a loudspeaker. The quality of sound reproduction of this microphone is poor compared with that obtained from a moving-coil microphone; in addition, it is much less sensitive. However, it is cheap and strong and has several important uses, such as in telephones.

29.06 Crystal microphones

When we press on certain kinds of crystal, a p.d. is created between opposite surfaces of the crystal. This is another way of converting mechanical energy into electrical energy. Unlike the other methods described so far in this chapter, it does *not* depend on electromagnetic induction. The p.d. generated may be large enough to make a spark. One form of gas-lighter has a finger-operated mechanism to apply pressure to a crystal. The faces of the crystal are connected to a spark-gap. When the trigger is pressed, the spark lights the gas.

In a crystal microphone, sound vibrations produce rapidly changing pressure on the crystal. In crystal pick-up cartridges, the vibrations of the stylus are transmitted to the crystal and cause variation of pressure on it. The varying p.d. (a few millivolts) corresponds to sound vibrations in the microphone, or to the irregularities in the groove on a recorded disc. The p.d. is amplified for reproduction of the sounds through a loudspeaker.

29.10 Electricity from chemical energy

Under this heading we include cells of all kinds (9.00, 26.32, 26.33) and the burning of fuel to drive an engine to turn an electrical generator (chapter 28, 29.03, 29.04). In this chapter we have studied the e.m.f. produced by electromagnetic induction. In this section we will study the e.m.f. produced by the conversion of chemical energy to electrical energy in a cell.

In chapter 9 we measured the e.m.f. of a cell by connecting a voltmeter across its terminals. We later learned (15.13) that the value of the e.m.f. in volts, tells us the number of joules set free when 1 coulomb passes around a circuit to which the cell is connected. This means a *complete* journey round the circuit, including passing through the cell. The law of conservation of energy applies, so this amount of energy must be the same as the amount of chemical energy converted into electrical energy in the cell.

Instructions: investigating the terminal p.d. of a cell or battery

1 Connect a high-resistance voltmeter across the terminals of a battery of four dry cells, connected in series. What is the p.d. between the terminals of the battery? Is this the same as the e.m.f.?

2 Keep the voltmeter connected. Also connect a lamp across the terminals. Measure the terminal p.d. What energy conversions are occurring in this circuit? How much chemical energy is being converted to electrical energy for each coulomb passing round the complete circuit?

3 How much electrical energy is being converted to light and heat in the lamp, at step 2. How do you explain the difference between the amount of energy converted in the battery and the amount converted in the lamp?

4 Connect a $100\,\Omega$ resistor in place of the lamp. Measure terminal p.d. How much energy is converted to internal energy in the resistor for each coulomb flowing through it? What other conversions of energy are occurring in the circuit and what is their amount, in J/C?

5 Connect a $10\,\Omega$ resistor in place of the $100\,\Omega$ resistor. Measure terminal p.d. Answer the same questions as in step 4. Take care that you do not keep the resistor connected to the battery for longer than is necessary to read the voltmeter; current is high and the battery will quickly discharge.

The investigation shows that when the resistance in the circuit is low, current is high and the passage of this higher current through the cell means a greater conversion of electrical energy to internal energy *in the cell*. This is necessary to overcome the *internal resistance* of the cell. As a result there is a reduction of the terminal p.d. of the cell. We sometimes say that we have 'lost volts'. If the internal resistance of the cell is similar in amount to the resistance of the circuit to which it is connected, its terminal p.d. may drop to a much lesser value than its e.m.f. If it has low internal resistance (as, for example, in the lead-acid cell), we find that there is relatively little fall in terminal p.d., even though the circuit may have low resistance.

Although we have said that we can ignore the current flowing through the voltmeter, it does draw a current from the battery, and there is a small amount of energy conversion in the voltmeter and in the battery, even when only the voltmeter is connected to the battery. Therefore we cannot measure the e.m.f. exactly. It is slightly greater than the terminal p.d.

29.20 Electricity from other sources of energy

Instructions: making a thermocouple

1 Twist together two wires, A and B, of different metals or alloys (fig. 29.9). Use various pairs of these metals or alloys: iron; copper; constantan; nichrome; platinum; manganin.

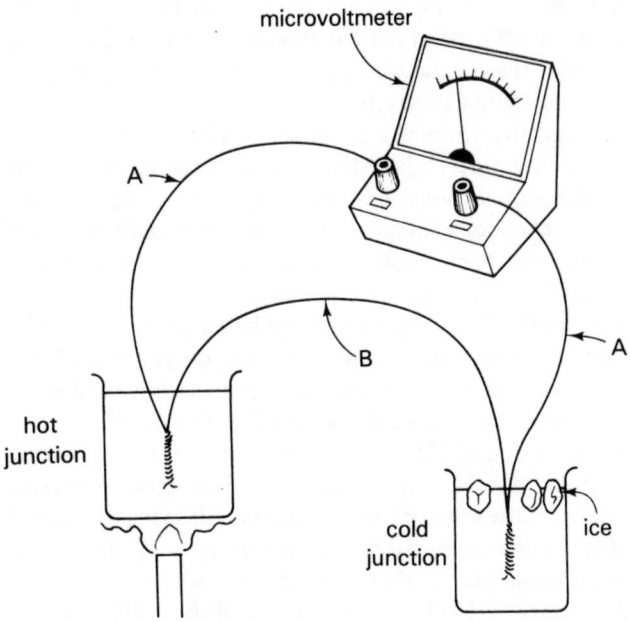

Fig. 29.9 Investigating the action of a thermocouple.

2 A thermocouple has two junctions: the hot junction and the cold junction. Keep the cold junction in iced water. Heat the hot junction in various ways. Measure the p.d. of the thermocouple, using the microvoltmeter.

3 Find out how you can obtain the greatest p.d.

4 Put both junctions in boiling water. Does heating create a p.d.?

If we keep the cold junction at a fixed temperature the p.d. varies with the temperature of the hot junction. We can use a thermocouple to measure temperature. If we heat the hot junction to known temperatures (for example, put it in steam at 100°C) the scale of the microvoltmeter can be marked directly in °C or K. Thermocouples are used for measuring high temperatures for they can be made from metals or alloys which have high melting points. A thermocouple made from platinum and a platinum-rhodium alloy is used for measuring temperatures up to 1 750°C. Several thermocouples may be connected in series, with their hot junctions close together and their cold junctions close together. The total e.m.f. is the sum of their separate e.m.fs. Such a device is called a *thermopile*. It is very sensitive to small changes of temperature and is often used for detecting and measuring infra-red radiation.

In the photoelectric effect (21.50) the energy of photons is converted into electrical energy. A *photoelectric cell* has a glass bulb containing a large metal plate (cathode) to receive light. There is also a small metal plate or rod (anode). A battery maintains a p.d. of about 100V between cathode and anode. When light falls on the cathode, electrons are emitted; they are attracted towards the anode. The current through the cell increases. In this way, variations in the amount of light falling on the cathode control the amount of current through the cell. The current is small; by amplifying it we can operate intruder alarms and fire alarms. We can use the photoelectric cell to detect objects passing by on a factory conveyor-belt; pulses of current through the cell are counted electronically, so the objects can be counted. The photoelectric cell is also used to detect variations in the intensity of light passing through the sound-track of a cinema film; the amplified current corresponds to the original recorded sound.

The photoelectric cell generates a small e.m.f. and needs an external power supply. The *photovoltaic cell* needs no power supply. When you connect a photodiode to a microvoltmeter and expose it to light from a lamp, it has an e.m.f. of a few tens of microvolts. A photodiode is simply a semiconductor diode (19.41) that has a *glass* capsule, so that light can reach the surface where the *n*-type and *p*-type materials join. Some ordinary diodes are in glass capsules that have been painted black to keep out the light; if you scrape the paint off one of these, it can be used as a photodiode. When light falls on the p-n junction it creates electron-hole pairs, by the photoelectric effect. This creates an e.m.f. between the *p*-type and n-type materials. Photodiodes can be used for the same purposes as the photoelectric cell (see above). They are cheaper and simpler to use, as they do not need an external power supply.

Another type of semiconductor cell is the *silicon cell*. In bright sunlight it generates an e.m.f. of 0.6V. Connected to a millivoltmeter, it is used as an exposure meter by photographers. Large numbers of such cells wired in series and parallel are used to make a *solar battery*. The total e.m.f. of such a battery may be 30V or more and its power output as much as 100W. Solar batteries are used to provide electrical power for spacecraft and satellites. They may also be used for converting sunlight to electricity for use in the home.

29.30 Transformers

A transformer consists of two coils, called primary and secondary, wound round a core of soft iron (fig. 29.10). The current in the primary coil is changing direction at high frequency (at 50Hz if connected to the mains). The magnetic field reverses direction at the same frequency.

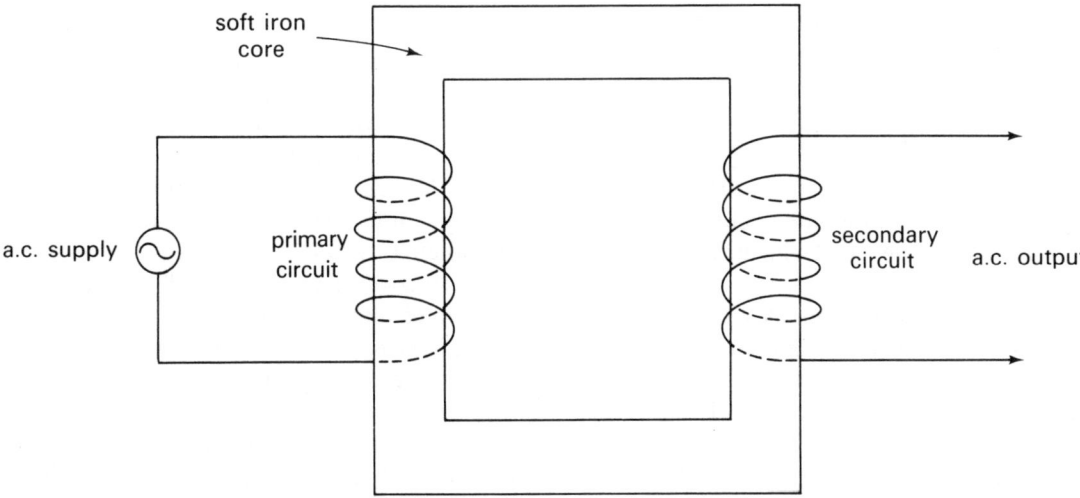

Fig. 29.10   The principle of a transformer.

This changing field induces an e.m.f. in the secondary coil. The size of the secondary e.m.f. depends on the size of the e.m.f. of the a.c. supply to the primary coil. It also depends on the number of turns in the coils:

$$\frac{E_S}{E_P} = \frac{n_S}{n_P}$$

E refers to the e.m.fs, n is the number of turns in the coil; the suffixes P and S mean primary and secondary.

For example, the primary coil of a transformer has 1 200 turns; it is connected to the mains (240V a.c.). If the secondary coil has 300 turns, the e.m.f. induced in it is:

$$E_S = \frac{n_S}{n_P} \times E_P = \frac{300}{1200} \times 240 = 60V$$

In this example we are using the transformer as a *step-down transformer.* to produce an e.m.f. that is lower than the e.m.f. of the supply. Transformers are often used to transform the mains voltage to a lower value for operating low-voltage devices such as projector lamps and electric door-bells. In mains-powered radio sets and amplifiers the voltage is stepped down to 9V a.c. and then rectified to give 9V d.c. as required to operate the circuits. Small transformers (with rectifiers) are used in power packs which provide 9V d.c. for supplying transistor radio sets and electronic calculators instead of using a battery.

A transformer may have more than one connection to its secondary coil. At various points along the coil, wires may be brought to terminals on the outside of the transformer. We say the coil is *tapped.* If the coil with 300 turns (see example above) is tapped at its fiftieth turn, we obtain an e.m.f. of 10V in the portion between this tapping and the nearer end of the coil. We obtain an e.m.f. of 50V between the tapping and the other end of the coil. We still can obtain an e.m.f. of 60V from the whole coil. If there are several tappings on the coil it is possible to select a range of e.m.fs. from a single transformer by connecting to the correct pairs of tappings. Transformers of this kind are often used in power-packs to supply outputs of a range of voltages. It is also possible for a transformer to have two or more separate secondary coils, each giving a different e.m.f.

In another transformer we may have a primary coil with 1 200 turns (as before) and a secondary coil with 3 600 turns. If the primary is connected to the mains (240V a.c.), the output from the secondary is:

$$E_S = \frac{3\ 600}{1\ 200} \times 240 = 720V$$

The transformer is being used as a *step-up transformer.* Transformers are used in this way to increase the e.m.f. from power stations, before the current is fed into the power transmission lines (29.40).

Questions
1 A transformer has a primary coil with 500 turns. Its secondary coil has 2 500 turns. The input to the primary coil has e.m.f. 120V a.c. What is the output e.m.f. from the secondary? Where would you tap the secondary to obtain an e.m.f. of 360V?

2 The input e.m.f. of a transformer is 240V a.c.; its output e.m.f. is 960V a.c. The secondary coil has 744 turns. How many turns has the primary coil?

3 Why does a transformer give no output when its primary coil is connected to a d.c. supply?

A step-up transformer produces a larger e.m.f.; does this give us something for nothing? By now you will have realised that energy is always conserved. It is not possible to obtain a higher e.m.f. without doing additional work. Suppose a transformer has an e.m.f. 720V at its secondary terminals. These are connected to a circuit in which a current 0.5A flows. The power of the secondary circuit is:

$$P_S = I_S V_S = 0.5 \times 720 = 360W$$

If we measure the current in the primary coil, which is connected to the mains we find that the current is 1.5 A. Power in the primary circuit is

$$P_P = I_P V_P = 1.5 \times 240 = 360 \, \text{W}$$

Measurements on transformers show that $P_S = P_P$. The rate of doing work is the same for both circuits connected to the transformer. Just as we found with the lever and the pulley, we have not gained energy.

Reference to levers and pulleys reminds us that we lose a part of the work we put into a machine. Friction and the weight of parts of the machine mean that the amount of *useful* energy obtained from the machine is less than the energy we put into it. The efficiency of any machine is less than 100%. The same applies to transformers. In both coils, some of the electrical energy is converted to internal energy and escapes as heat. More energy is wasted because electric currents are induced in the core itself. These currents heat the core and more energy is lost as heat. The currents flow across the width of the core (in and out of the paper in fig. 29.10). They are called *eddy currents*. To reduce these currents the iron core is made of layers of sheet iron (or laminations); before assembling the transformer, each layer is coated in paint or varnish. When stacked together to form the core, the layers are not in electrical contact. Currents cannot flow across the core, but the magnetic lines of force can run around the core, as usual. Eddy currents are much reduced and waste of energy is prevented. The direction of magnetisation of the core is reversed 100 times every second (on a 50 Hz supply); energy is required to do this, giving another reason for loss of power in a transformer. In spite of all these ways that energy may be lost, a well-designed transformer has high efficiency.

29.40 The distribution of electricity

Electricity is carried in wires from power stations to homes and other places where it is used. Even if wires are thick and made from materials with low resistance (copper or aluminium), the wires are many kilometres long and the resistance of the power lines cannot be ignored. As electricity passes along the lines, energy is converted to internal energy and wasted as heat. The heating effect of a current varies with the square of the current (26.01). If we keep current as small as possible, loss of energy will be as small as possible. The power is given by

$$P = IV$$

For a given value of P, if we want to make I as low as possible, we must make V as high as possible. This is why high voltages are used on power transmission lines.

In the power station, electricity is generated at high voltage; often this is 10 kV or more. It is then stepped up by

transformer to an even higher voltage (66 kV, 132 kV, or even as much as 400 kV) for transmission by power line. In towns and villages there are step-down transformers. Usually the voltage is stepped down for local distribution and finally stepped down by a second set of transformers to 240 V; each transformer supplies a small group of houses. Note the importance of transformers in the distribution of electric power. Note also that only alternating current can be distributed in this way. Direct current cannot be transformed and would have to be distributed at 240 V, with big heat losses in transmission. This is one of the main advantages of using alternating current for mains power supplies.

Let us work an example to see how high-voltage transmission saves energy. Suppose that the resistance of the wire between the power station and your home is 1 Ω. The power needed in your home is 6 kW. If the transmission voltage is 240 V, the current in the wire is:

$$I = \frac{6\,000}{240} = 25 \, \text{A}$$

The rate of conversion of electrical energy to internal energy in the wire is:

$$P = I^2 R = 25^2 \times 1 = 625 \, \text{W}$$
(over 10% of the power actually used in your home)

Now suppose that the supply is transformed to 132 kV for transmission. We ignore the short part of its journey after it has been stepped down again, close to your home.

$$I = \frac{6\,000}{132\,000} = 0.045 \, \text{A}$$

$$P = 0.045^2 \times 1 = 0.002 \, \text{W}$$
(0.00003% of the power used)

By transforming the supply, energy loss has been reduced from over 10% to 0.00003%. This is a big saving in power.

The electricity supply enters your house through a thick cable. It must then be distributed to the electrically powered devices in the house. A typical home wiring system is shown in fig. 29.11. The main service cable consists of two wires. They are called *live* (L) and *neutral* (N). The potential of the neutral wire is very close to earth potential, since it is earthed at the power station. The potential of the live wire alternates between +339 V and −339 V relative to the potential of the neutral wire. This is equivalent to 240 V r.m.s. (29.03). When a lamp is connected to the wires, current flows from the live wire to the neutral wire or from the neutral wire to the live wire, reversing direction 100 times per second.

The main fuse is able to pass a high current. If this is exceeded, the fuse wire becomes hot and melts. This cuts off the electricity supply to the house. This is an essential safety precaution. If there is a short-circuit in the wiring in

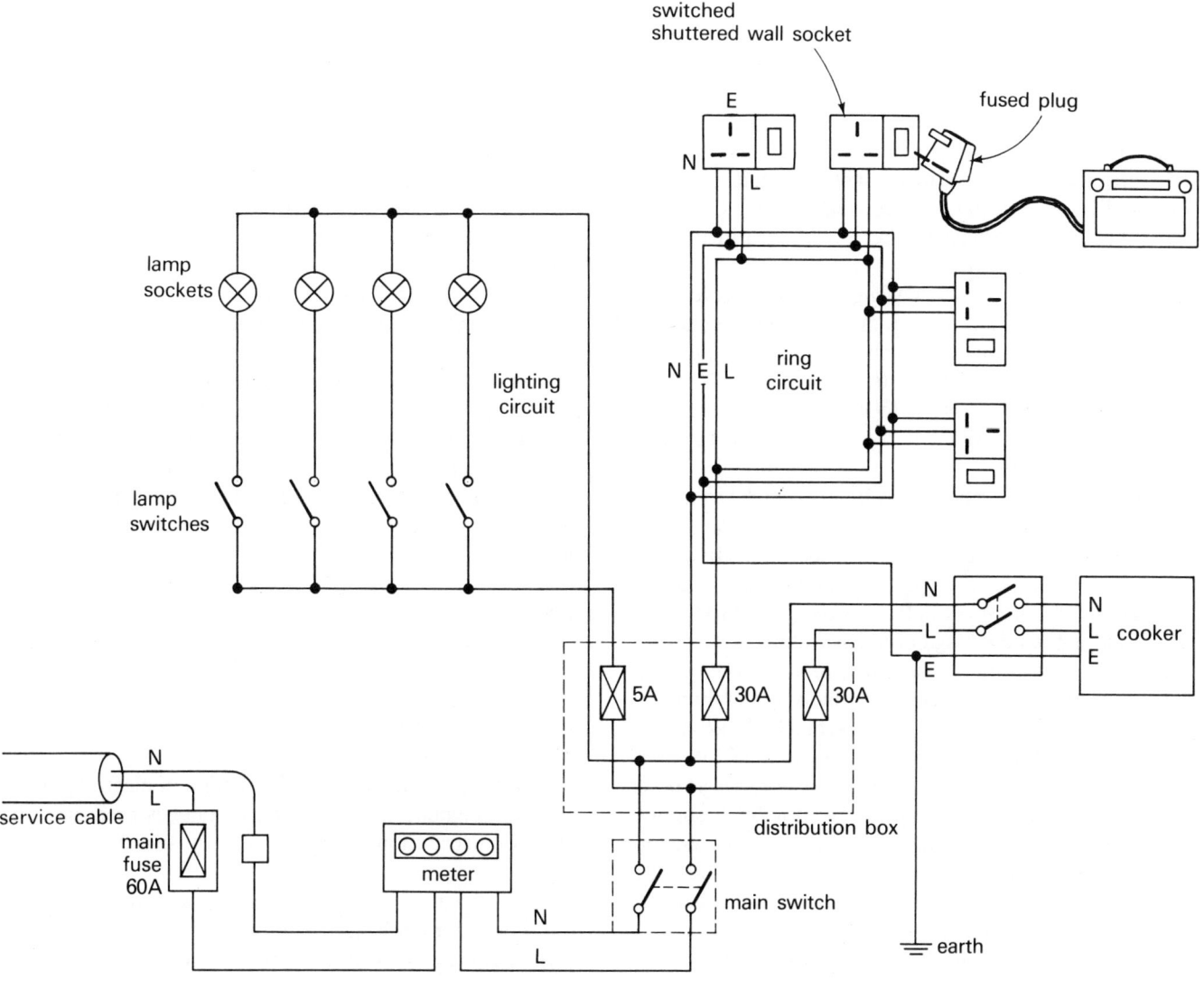

Fig. 29.11   A typical home wiring system L = live, N = neutral, E = earth.

the house, or in one of the items of electrical equipment in the house, a large current would flow. This might overheat the wiring or the equipment and start a fire.

The current flows from the main fuse to the electric meter. This is a special kind of electric motor. It is connected to a mechanism with dials to record the number of times the motor rotates. The speed of rotation of the motor varies with the power being used in the house.

From the meter, wires lead to a distribution box which contains a main switch. When open this disconnects both wires (L and N) of the main supply from the wiring in the house. The distribution box also contains fuses or cut-outs in the live connections to the chief sub-circuits. Sub-circuits usually include lighting, cooker, and a ring-circuit to all wall sockets. The fuses are rated according to the maximum current that is likely to be required in the sub-circuits. Instead of a wire fuse there may be a cut-out; this becomes hot when the current flowing through it is too

high; it automatically switches off the current when this occurs. When the cause of the trouble has been dealt with, and the cut-out has cooled, the current can be switched on again.

In modern wiring systems the supply to wall sockets is made through a ring circuit. A heavy-duty three-wire cable (live, neutral and earth) runs from the distribution box to all rooms in the house and back to the box again. Sockets are connected at points along this ring circuit. If an item of equipment which needs large current is plugged into one of the sockets, the current can reach the socket by passing round both sides of the ring. There is less danger of overheating the wires.

Questions

Study Fig. 29.11 and answer these questions

1 Why is the fuse in the lighting circuit of lower value than those in the ring circuit and the cooker circuit?

2 How much is the power rating of a typical electric room heater? How much current does it require? What would happen if the heater is connected to one of the lighting sockets?

3 Why should a 30A fuse not be used on the lighting circuit?

4 Why are the switches in the lighting circuit connected on the 'live' side of the lamp socket?

5 What is the purpose of the 'earth' wire in the ring circuit and cooker circuit?

6 Wall sockets vary in design, but most have three terminal pins: live, neutral and earth. Usually the plug and socket are made so that when you push the plug in, the first pin to make contact is the earth pin. Why is this?

7 The figure specifies *switched sockets*, but not all houses have these. What is the advantage in having a switch on the socket?

8 If the socket has a built-in switch, to which terminal of the socket is the switch connected?

9 The figure also specified *shuttered* sockets. In these a plastic shutter closes the openings in the socket when the plug is pulled out. What is the advantage of this?

10 What is the advantage of a *fused* plug?

11 The fuse in a fused plug must be able to carry the current required, but not a current much greater than this. Common values of fuses are: 1A, 3A and 13A. What fuse would you choose for a plug connected to (a) a 2kW room heater; (b) a table-lamp; (c) a slide projector with a 500W lamp. Assume that all are operating on a 240V supply.

12 Cookers and other single items that require a large current usually have their own circuit with separate fuse. Why is the cooker not wired to the ring circuit, with a 60A fuse in the ring circuit?

13 The cooker is usually connected to a switch panel on the wall. This has a double switch, disconnecting both live and neutral wires. Why is there a *double* switch?

14 What should the householder do before washing the cooker?

The electric meter is used to measure how much electrical energy has been used in the house during a given period. The unit of electrical energy is the *kilowatt hour* (symbol, kWh). It is often referred to as a 'unit'. The unit or kWh is used for calculating how much we have to pay to the generating authority.

If a lamp rated at 100W is switched on for 2h, it is converting energy at the rate of 100W for 2h. We say that the amount of electrical energy used is 200 watt hours. The same lamp runs for 10h and converts 1 000 watt hours or 1kWh. A spotlight lamp rated at 500W uses this amount of energy when run for only 2h.

To say that a lamp operates at 100W is another way of saying that it converts electrical energy at the rate of 100J/s. In 10h there are 36 000s. In 10h the lamp converts 36 000 $\times$ 100 = 3 600 000J, or 3.6MJ. Thus 1 kWh = 3.6MJ. The electric meter is a *joulemeter*. It measures the amount of electrical energy that has been used in the circuit to which it is connected. During this process it uses a very small fraction of this energy for driving its mechanism. In the home the dial of the meter is marked in kWh, or 'units' for every 3.6MJ. The kWh is not really a new way of measuring energy. It is a convenient unit for recording and charging for the quantities of electrical energy used at home, in the factory and in the office.

Projects

1 Try to arrange a visit to your local power-generating station. Find out what equipment is used and how much power it generates. Draw a plan of the station to show how it operates. Try to get information from which you can calculate the efficiency of the station.

2 Study maps showing the distribution of electric power in your neighbourhood. Mark the voltages used at each stage of distribution and note where the transformers are located.

3 Study the distribution of power in your home or school. You could draw a diagram similar to fig. 29.11. **Under no circumstances must you attempt to open fuse boxes or to interfere in any way with the wiring.** You can get all the information you need simply by looking.

4 As a class project with your teacher, test household or school equipment to see that the plug is properly wired, that the switch (if any) is in the live wire and that metal parts are properly earthed. To do this you must **first disconnect the item completely from the mains.** If your checking shows faults, repairs must be made by experts. **Under no circumstances must you try to repair the fault yourself.**

30 Using materials

In this chapter we look at some of the many materials we use in everyday life for making the things we use. Many of the properties of materials can be understood by referring to what was learned in chapter 17 about the structure of matter.

30.00 Investigating the properties of materials

30.01 Appearance

Knowing the appearance of materials is important if we want to be able to recognise materials easily. It is especially important for geologists. Artists and craftsmen who use materials in a decorative way are also very interested in appearance.

Instructions: classifying materials by appearance
1 Use a lens to help you examine the appearance of an assortment of materials. Look at rocks, metals, timber, glass, plastics, fibres (natural and synthetic), leather and building materials of many kinds.
2 Set out the information you collect, in a table:

| Material | Colour | Surface | | Notes |
		Texture	Lustre	

3 Enter the colour of each material in the second column. If there is more than one colour, list all colours. In the **'Notes'** column, explain how the colours are arranged; for example, 'mainly brown with specks of white', 'grey with greenish streaks'.

4 Under **'Texture'** describe the surface as rough, smooth, shiny, crumbly, or use any other simple descriptions that you can think of.
5 Under **'Lustre'** describe the surface as metallic, glassy, dull, crystalline.
6 Under **'Notes'** add any other details which help give the material its distinctive appearance. If it is crystalline, mention the shape, colour and lustre of the different crystals that are seen in it. If the material comes from plants or animals, make notes on graininess, pores, or other interesting features.

30.02 Density

Methods of measuring density are described in 1.12. Density depends on the mass of the molecules of the substance and how close together they are placed in their crystal lattice. Materials of low density, such as wood and aluminium have many advantages for construction. Wooden beams, posts and boards are used in building houses, boats and bridges, and for making carts and other vehicles. Wood is sometimes used for aircraft frames and in the bodywork of automobiles. As we shall see later, wood has great strength. Its combination of strength and low density (less than $1\,000\,kg/m^3$ for most timbers) has made it an ideal building material, used by man for thousands of years. Aluminium is a metal of low density ($2\,700\,kg/m^3$); when alloyed with other metals it gains strength. It is used for frameworks for which lightness is the most important factor, for example, the frames of aircraft. It has good resistance to corrosion (30.10), making it useful for constructing parts of buildings that are exposed to the weather (roofs, window frames). It conducts electricity well. It is less dense than copper (and cheaper too) so is becoming widely used for electricity transmission lines. Many plastics combine low density with high strength, good electrical insulation, resistance to corrosion and various other useful properties.

High density is not often thought of as a useful property.

Lead is the densest common material (11 300kg/m³). It has been widely used in the past for roofing and water-pipes because of its excellent resistance to damp and corrosion and the fact that it is easy to melt to form watertight joints. It is used *in spite of* its high density. Recently, lead has been put to a new use; its high density makes it an excellent material for absorbing the radiation from radioactive materials (20.26).

30.03 Tensile strength

Tensile strength is the ability of a material to resist a force that is pulling on it. The tension produced is resisted by intermolecular forces. The amount of tension that can be resisted depends on the kinds of molecule in the material and the way they are arranged in the lattice.

Instructions: measuring tensile strength

The method is similar to those of the instructions on page 158, except that you do not need to measure extension. Wires, rods or thin sheets of the material are attached to a firm support above; weights are hung from the lower end. There is no need for long pieces of material, a piece a few centimetres long is sufficient. Having chosen some materials for testing (see list below):

1 Measure the area of cross-section. If this is not uniform, measure where the material is narrowest, for it is there that it is likely to break.

2 Attach the material to the upper support. Add weights gradually until it breaks.

3 Calculate the *tensile breaking stress*

$$= \frac{\text{force due to weights}}{\text{area of cross-section}} \text{ (in Pa or N/m}^2\text{)}.$$

4 Group the materials you test, according to their tensile breaking stress. Which types of material have high tensile strength? Which have low tensile strength? Think of examples of materials that are used because of their high tensile strength; what are they used for?

Materials that can be tested include:

Metals: including copper, steel, magnesium, aluminium, lead, solder (an alloy). These are usually available as wire or thin sheet (eg. aluminium kitchen foil, steel spring etc). As mentioned later, tensile strength partly depends on the way in which the material has been prepared; for example, when a metal is drawn to make it into wire, its tensile strength becomes increased.

Wood: tensile strength depends on whether tension is along the grain or across it; try both, if possible. You may need to use very narrow splinters of wood.

Fibres (natural or synthetic): cotton or wool thread, jute string, silk, nylon, terylene.

Fabrics: cotton, wool, nylon, terylene. It is interesting to compare the strength of a fabric when it is wet with its strength when it is dry.

Plastics: strips of polythene, PVC, polystyrene and others.

Other materials: various kinds of paper and card (wet and dry), glass rods and fibres, leather, rubber, catgut.

30.04 Rigidity

If we clamp a strip of material vertically in a vice, and apply a force acting sideways at the top of the strip, the strip is being acted on by two forces. There is the force we apply, at the top; there is the force exerted by the jaws of the vice, preventing the strip from moving sideways when we push on it. This force is equal to the force we apply at the top; it acts in an opposite direction; but it does not act along the same line. These two forces are described as *shearing forces*. The action of shearing forces is to *bend* the material out of shape. The forces act to distort the crystal lattice, to alter the spacing and pattern of the molecules. Intermolecular forces resist the shearing forces to an extent that depends upon the nature of the molecules and the lattice. Some materials are not easily bent out of shape, we say they are *rigid* or *stiff*; other materials are easily bent, we say they are flexible.

Instructions: comparing the rigidity of different materials

1 Obtain some strips of metal, wood and plastic. If possible they should be nearly the same size: about 20cm long, 2cm wide and 0.3cm thick.

2 Measure the area of cross-section of each strip.

3 Fix the strip vertically in a vice, with 7cm extending above the jaws of the vice.

4 Wind a short piece of wire around the strip, about 2cm from the top. The essential point is that the length of strip between the top of the vice and the wire should be 15cm, or some other constant amount, for all trials.

5 Attach a dynamometer to the wire.

6 Fix a horizontal scale beside the strip, level with the wire. Note the position of the strip with reference to this scale.

7 Pull horizontally on the dynamometer until the strip has been pulled sideways a distance of 1cm (or some other constant distance for all trials). Note the force exerted to produce this amount of bending. Record results in a table:

Material	Area of cross section A in m²	Force F in N	Force/Area N/m² or Pa

8 Compare the values of force/area for different materials. Which are the most rigid? Which are the least rigid?

9 Give examples of the use of materials that are rigid.

10 Give examples of structure in which great rigidity could be a disadvantage.

11 Give examples where low rigidity in a material is useful.

30.05 Ductility and brittleness

When a material is under tension and has passed its elastic limit, one of two things may happen.

(a) The molecules begin to slide past one another. The substance enters a stage of *plastic deformation*. It behaves almost as if it is liquid, though it is cold and not near its melting point. This stage was noted in the instructions on page 159. Materials which behave in this way are called *ductile*.

(b) Increased tension has little effect on the length of the piece of material. There is no plastic deformation. The applied force overcomes the intermolecular force at some weak point in the lattice, which suddenly gives way. This increases the amount of tension per unit area on nearby regions; the break spreads rapidly. The material snaps suddenly. Such materials are called *brittle*.

Brittle materials can be snapped by tension or by shearing forces. You may have found materials (see instructions page 310) that break almost as soon as you try to bend them. They may be able to remain stiff against a large force, but with a slight increase in the force they snap.

Some plastics are flexible and rubbery, but become hard and brittle below a certain temperature. This is the effect of changes in the structure of the lattice at that temperature. Perspex and polystyrene, for example, are normally below the temperature at which they are brittle. They are suitable for making containers for food, and other items from cassette tapes to pills, for kitchenware, for the transparent covers for record players, for making 'glasses' for watches and clocks, for fish-tanks and thousands more purposes. Plastics such as polyethylene (polythene) and polypropylene are normally above the brittle temperature; they are normally flexible. Containers made from these plastics are not rigid, but are almost unbreakable; they do not shatter or crack like the containers made from polystyrene. They can be used for making sealable food containers in which a tightly-fitting cap or lid is slightly stretched when in position, making an air-tight, water-tight seal. Propropylene containers become brittle below 0°C, so they must not be used for storing foods in a refrigerator.

Ductile materials can be drawn into wires and threads quite easily. The process of drawing them to make a wire increases their tensile strength to double or more.

Questions

1 Name some brittle substances.

2 What useful property have brittle substances? Name some uses of brittle substances.

3 Name some ductile materials. Where are they used?

Instructions: investigating the effect of repeated bending of a strip of metal

1 Place a strip of metal in a vice, as in the previous instructions.

2 Bend the metal as far as it will go.

3 Bend it as far as possible the other way.

4 Repeat the bending one way, and then the other, counting the number of times you bend it.

5 Eventually the metal breaks, usually close to the jaws of the vice.

6 Repeat for different metals. Use strips of the same thickness, if possible.

A strip of metal can be bent one way and another at the same place a few times and can be bent back to its original shape. With repeated bending it eventually breaks. This is called metal *fatigue*. Different metals and alloys show different rates of fatigue. An aeroplane frame is subject to continual vibration and changes of forces acting on it. Metal fatigue can cause failure of parts of the frame after many hours of flying. Failures due to fatigue have caused serious accidents to aeroplanes. Designers take special care to discover in what parts fatigue is most likely to occur, and to strengthen those parts. Fatigue can occur in a bridge, with the continual changes of forces as traffic passes across.

30.06 Hardness

Instructions: investigating the hardness of materials

1 Collect small pieces of materials for testing: lead, mild steel, stainless steel, aluminium, brass, copper, iron, zinc, glass, wood, plastics. If possible the pieces should be squares about 2 cm × 2 cm cut from a sheet of the material, and with sharply-cut corners.

2 Try to scratch each piece with a corner of each of the other pieces. After you have tried to make a scratch, rub the scratched surface before you decide if it has become scratched or not. What looks like a scratch may really be a row of particles from the scratching material, left on the surface. If a distinct groove is seen (use a lens to help you make sure) the scratch is confirmed.

3 By repeated testing, it is possible to arrange the materials in order of hardness. List your materials in order, beginning with the hardest: this is not scratched by any of the other materials. At the bottom of the list is the softest material; this is scratched by all the others.

Hardness is an important property of materials that must resist wear. Materials used for making the bearings of machines and for making cutting tools must have a high degree of hardness. There are no exact physical units of hardness, but it is easy to arrange materials in order of hardness, as you have done above. The hardest substance known is diamond. Its hardness depends on the way its carbon atoms are arranged in the lattice; when arranged differently, as in graphite, a relatively soft material is produced. Diamond is often used as a cutting material in industry. Tungsten is a hard metal. This is made into a tungsten-steel alloy often used for making high-speed cutting tools such as the bits of drills. These keep a sharp edge when cutting hard substances, even when the tool is red-hot. Most drill bits used in the workshop are made from steel alloys of other types, which are harder than normal steel. For drilling brick and stone we use masonry drills, tipped with tungsten steel. The ultra-high-speed drill used by the dentist is tipped with tungsten. Tungsten carbide is another very hard substance; the material is used in the outer layers of drills to give them good cutting edges. The same material is used in the tiny balls used at the tip of ball-point-pens. Another carbide, silicon carbide, often known as carborundum, is a hard gritty substance used for polishing metal and glass.

Questions

1 Can you think of any other ways in which we make use of the hardness of diamond?

2 What hard *stone* is used for making long-lasting bearings?

3 What other hard gritty substances are used as abrasives, for polishing and grinding?

Hard substances are usually rigid substances. Rigid substances are usually brittle. All three features are really different effects of the firm structure of the crystal lattice of this kind of material. Though a diamond stylus may resist the wear caused by playing many discs, it can be chipped easily if the pick-up head is accidentally dropped on to the disc or turntable. From his earliest days, man has made cutting tools from hard stone by making use of their brittleness. When two stones are struck together, they fracture. The edges formed are sharp and hard wearing. They make good axe-heads, spear-points and arrow-heads.

If hard materials are brittle materials, we can use measurements of hardness to tell us how brittle a substance is, without needing to break it.

Instructions: testing materials for hardness

1 Use a steel ball about 1 cm diameter. Place the sample to be tested on the bench or floor. Drop the ball on it from a fixed height, say, 1 m.

2 Measure the diameter of the hollow made where the ball hits the sample.

3 Use this method for testing the materials listed in the previous instructions (except glass). Test several kinds of

wood to see which is the hardest. For what purposes are these hard woods most often used? For what purposes are the softest woods most often used?

30.07 Other properties

Under this heading we list several properties of materials that are referred to at other points in the book:

Thermal expansivity see 6.20, 18.30;
Thermal conductivity see 6.42, 18.03;
Magnetic properties see 8.30;
Electrical conductivity see 9.10, 19.10;
Water-absorbing ability see project 1 at the end of this chapter.

Two other important properties are the ability to resist corrosion and fire. The basis of these is chiefly chemical, so we will not consider them here. One type of corrosion has an electrical cause and is described in the following section.

30.10 Electrolytic corrosion

In fig. 30.1a a sheet of copper is fixed to a wooden framework by iron nails. The nails and copper are covered by water. This may be only a thin layer of dampness. We can compare this with an electric cell. (fig. 30.1b). The copper sheet and iron nails are electrodes. The film of water is the electrolyte. Although pure water would not act in this way, water exposed to the atmosphere soon gains particles of dust and soluble materials which form a dilute solution of salts and give it the ionisable materials needed for a true electrolyte.

The combination of copper sheet, iron nails and film of water is an electric cell. A p.d. is created between the copper and the iron. The copper and iron are in contact; this means that the cell is like one in which the electrodes are joined externally by a low-resistance wire. Current readily flows in the direction indicated by the arrows. In 19.12 it was explained that in the Leclanché cell the zinc electrode gradually dissolves; it is corroded. In the copper/iron cell of fig. 30.1, the flow of electric current leads to corrosion of the iron nails. A way of avoiding this is to fix the copper sheet to the framework with *copper* nails. Then there will be no cell, no current and no corrosion.

Cells of this kind occur wherever two or more different metals are part of a structure and where there is water or dampness to act as an electrolyte. The effect also occurs with boats, if the hull and propellor are of different metals; the salt sea water acts as the electrolyte. Painting the hull is one way of reducing this type of corrosion, since paint is a non conductor. A better method is described later in this chapter.

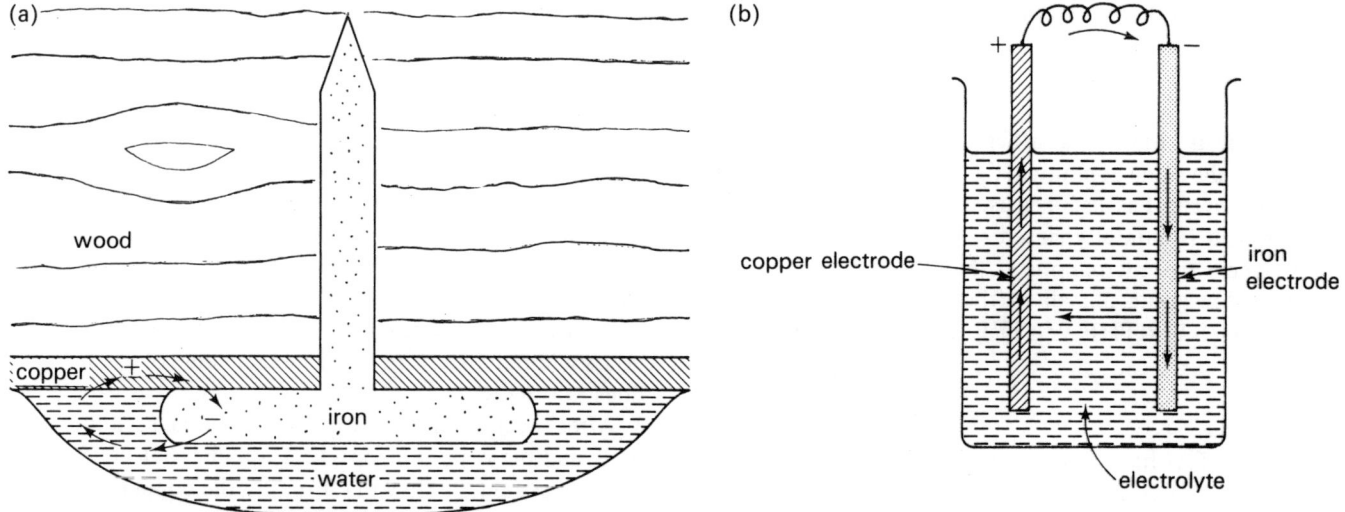

Fig. 30.1 *Electrolytic corrosion: (a) practical example (b) the equivalent cell*

Instructions: making an unusual kind of electric cell

1 Set up the apparatus shown in fig. 30.2. The electrodes can be two ordinary iron nails. The exact concentration of the electrolyte does not matter; make up enough solution for both beakers in one large beaker, then fill both beakers of the cell from this supply. It is important that the concentration of electrolyte in the beakers is equal. Connect the electrodes to a microammeter.

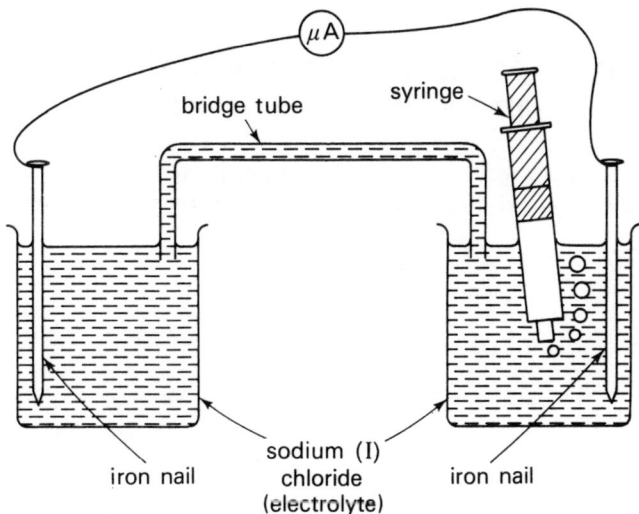

Fig. 30.2 *An unusual kind of cell.*

2 Note the reading of the microammeter, if it is not zero.

3 Bubble air into *one* beaker only, taking care not to let the bubbles pass up the bridge tube. For bubbling, use a plastic syringe or a pipette with squeeze-bulb. Gentle bubbling is all that is needed. Watch the ammeter as you blow air into the electrolyte. If necessary, reverse the connections to the microammeter so that the current flows through it in the correct direction.

4 Continue to blow air for several minutes. Watch the ammeter.

5 Blow air into the *other* beaker for a few *seconds*. Watch the meter.

6 If you have an aquarium aerator, use this to bubble air continuously into the first beaker for several hours. If you have no aerator, bubble air into the beaker as before and look at the meter from time to time. If the current is low, blow more air into the beaker. The meter must be permanently connected, so that current flows through the cell the whole time. After three or four hours, remove the nails, place them side by side and compare their appearance.

Questions

1 When you bubble air into one beaker, which electrode becomes positive?

2 Is the effect reversed when you bubble air into the other beaker?

3 What difference is there in the appearance of the two nails at the end of the test?

4 An unpainted iron gate-post is half-buried vertically in the soil. During rainstorms the upper half becomes wet; rain runs down the post and wets the soil around the lower part. What happens to the post?

5 An iron pipe runs below ground. Some regions of the soil are very water-logged; other regions are damp but there is some air in the soil. What happens to the pipe?

6 A steel plough has lumps of mud on it. It is kept in a damp place. What will happen to the steel?

Electrolytic corrosion is a serious problem whenever metals are used, because most places are damp for short periods at least. Corrosion occurs when two different metals are used and in a single metal if oxygen supply is uneven. It can also occur on the surface of steel and some

313

other metals if there is a difference of composition of the metal at different points on the surface. Steel must always be kept as dry as possible.

Instructions: making another unusual cell

1 Set up the same apparatus (fig. 30.2) but use tap water. If the apparatus has just been used with salt solution, make sure that it is carefully rinsed.
2 Connect the microammeter. Note its reading.
3 Tip a few grams of sodium (ɪ) chloride into *one* beaker. DO NOT STIR THE WATER. Note the reading of the meter.

Questions

1 What effect have you observed?
2 In what way could this effect be of practical importance?
3 What is the reason for having the beakers linked by a bridge tube, rather than having both electrodes in one beaker?
4 Why were you warned not to stir the water?

30.11 Preventing corrosion

A coat of paint is *not* a satisfactory way to prevent corrosion of underground pipes. The film of paint is easily scratched when the pipes are laid; even a small hole in the film becomes a site for corrosion. One method of preventing corrosion is to cover the iron surface with a thin layer of zinc. Sheets of galvanized iron (zinc-covered iron) are used for roofing and we use pipes, wire and fencing materials made from galvanized iron. The coating of zinc is not perfect. Where there are holes, the iron, zinc and water form a cell. In this zinc-iron cell, the iron electrode is *positive*. Corrosion occurs, as usual, at the negative electrode. The zinc corrodes. The coating of zinc gradually corrodes away. If it is thick enough to begin with, it lasts many years. We are sacrificing the zinc to preserve the iron. As long as some of the zinc remains, the iron does not rust.

Instructions: preventing iron from rusting

1 Set up the apparatus shown in fig. 30.3. Note the direction of the current. This is a copper-iron cell; if this was not connected to the power supply, which way would current flow?
2 Leave the apparatus with current flowing for several hours without disturbing it. The exact value of the current does not matter.
3 After several hours look in the dish in the region of the nails and the copper electrode. Explain what you see.

By opposing the flow of current in the copper-iron cell we are making the copper corrode instead of the iron. The copper is sacrificed to protect the iron. This method is used to protect underground pipes and storage tanks from corrosion. Sacrificial electrodes are placed in the soil near the surface. These are usually made from magnesium or graphite. They are connected to the positive terminal of a power supply. The pipe or tank is connected to the negative terminal.

The pipe or tank is protected; the sacrificial electrodes gradually corrode. When they have corroded they can be replaced cheaply. The same principle is used to protect the hull of a boat from corrosion.

30.20 Making the best use of the properties of materials

The more we learn about the properties of materials and the physical and chemical reasons for these properties,

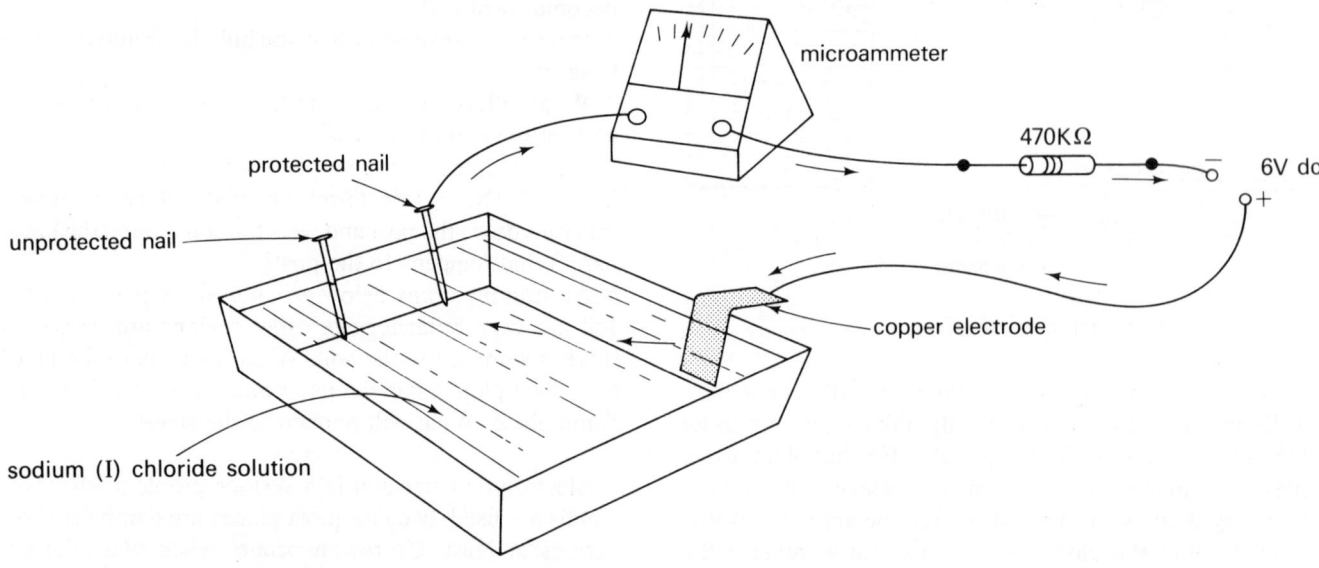

Fig. 30.3   Protecting iron from corrosion.

the more we are able to use our knowledge to produce materials suited to any need. Some of this knowledge has been acquired by man by trial and error; it was acquired hundreds of years ago and has been handed down since then. Recently, our detailed study of the molecular structure of matter has given us the ability to make materials that have properties which would have been thought impossible years ago. In this century the invention of plastic materials has been a major contribution of chemistry.

Iron is one of our most widely used materials; it is available in many forms such as cast iron, wrought iron, pure iron, various grades of steel (iron + carbon) and many alloys of steel. These vary in brittleness, hardness, ductility and tensile strength. It is possible to choose from these to obtain just the right material for any particular purpose. We can use cheap cast iron for fencing posts. For making cutting tools, cast iron is too brittle; we use a high-carbon steel. For many other purposes (nuts, bolts, girders, tubes) we use mild steel and medium grades of steel. These have strength but are soft enough to be hammered into shape or machined to make the immense variety of parts that we need for building houses, boats, bridges, automobiles, aircraft, typewriters, household tools and gadgets. For special purposes there are many alloys of steel: vanadium and tungsten steels that are shock-resistant and used for armour-plating and engine crankshafts; tantalum steel that is highly resistant to corrosion and is used in making wires and plates to be surgically inserted in the human body to repair broken bones; chromium steels that are resistant to corrosion and generally known as 'stainless steel' used for cutlery, household goods, automobile parts and many other purposes: manganese steel that is very resistant to abrasion, so is used for the jaws of mechanical shovels and railway crossing-points; nickel steel (invar) that expands very little when heated and is used for making surveyor's tapes and the pendulums of clocks; aluminium-nickel-cobalt steel (alnico) that is capable of being permanently highly magnetised and is used for the powerful magnets of loudspeakers and similar devices.

Aluminium has low density and is easily worked. It has many uses, particularly for making pots and pans used in the home. Its lightness makes it suitable for the frame-works of aircraft, automobiles and parts of buildings, but for some of these purposes it does not have sufficient rigidity. Alloys of aluminium have been devised to improve its properties. Duralumin (aluminium-copper) combines low density with increased strength.

In all these examples the mechanical properties depend on intermolecular forces. These depend partly on the pattern of the crystal lattice. A piece of metal consists of a large number of crystals or grains joined together. Though the intermolecular forces within a single crystal may be strong, the boundary between one crystal or grain and the next may prevent the full intermolecular force from acting. The strength of the metal is less than is expected from our knowledge of intermolecular force alone. The effect of the grain structure of metal depends on the size of grains and the nature of materials that lie between one grain and another. The situation is very complex. The size of grains and the nature of their boundaries is greatly affected by the conditions under which the metal or alloy is processed. The processing of the material also affects the regular structure of the lattices of the crystals. If the lattice is irregular (a molecule missing from its place, or an extra molecule for which there is not room, or a molecule of some impurity replacing one of the molecules of the lattice) strains and tensions are set up in the lattice. These may make the lattice structure weaker or stronger. Know-ledge of these effects and how to use them to the best advantage is an essential part of making the best use of materials. To obtain the greatest strength the material should be in the form of a single crystal, with no irregularities in its lattice. Then we should obtain the maximum strength that is possible theoretically. To produce materials of this form, we must grow the crystals under carefully controlled conditions, rather as you grow crystals in chemistry lessons. It is difficult to grow crystals that are both large and perfect, but success has been achieved in producing fibres of carbon having high tensile strength. The way these are used is described in 30.21.

30.21 Integrated materials

Instructions: investigating the effects of defects in a material

1 Prepare two strips of polythene sheet, both the same size, about 20 cm long and 5 cm wide. Cut the long edges of the strips very carefully, using a sharp knife, so that there are no irregularities in them.

2 Cut a small V-shaped notch half-way along one long edge of one of the strips. The notch should be about 5 mm wide and 5 mm deep.

3 Hang the straight-sided strip vertically from a clamp. Use a bull-dog clamp or other clamp that grips the whole of the strip evenly. Place a similar clip on the lower end of the strip.

4 Hang weights from the lower clip. Add more weights gradually. Find the greatest weight that the strip will support before it tears and breaks.

5 Repeat steps 3 and 4, using the notched strip. Does notching affect the ability of the strip to support weight?

6 Does the notch effect depend on the angle at the point of the notch? Try notches of various shapes on newly cut pieces of polythene; test them to find how great a weight they will support without tearing. What shape of notch is least liable to result in tearing? What shape is most likely to result in easy tearing? If the strip has a single tear or scissor-cut made in one side, about 5 mm long and perpendicular to the side, what can you do to this to stop it from spreading when a load is applied?

7 What could you do to the polythene sheet so that any tears that formed in it would not spread across the sheet?

The notch effect demonstrated above shows how a small defect in the surface can lead to complete failure of a strip or bar of material. The notch may be large to begin with, as in the tests, but even a small surface defect, such as a scratch can be the beginning of a crack that leads to failure. On the smaller scale, a defect in the crystal lattice, far too small to be seen with a microscope, can grow when the material is under strain, and lead to failure. This is why it is important in the preparation of many kinds of material to avoid irregularities and defects in the surface. Any one of these is a possible site of breakdown of the material when under strain. You may have noticed that sometimes the V-shaped notch becomes rounder and more U-shaped as it grows. This often happens with plastic materials and with many metals. When this happens, the notch becomes *less* likely to spread, the material becomes stronger. This is what happens when a wire is stretched until it reaches the point of plastic flow. At that stage notches become more rounded and less likely to spread. This is why copper that has been drawn out to make wire has greater tensile strength than a bar of copper that has not been drawn.

The last question above leads to another approach to the strengthening of materials. By incorporating fibres or wires or other materials with high tensile strength, we are able to increase the tear-resistance of the polythene sheet. Sheet plastic materials are liable to tear once a small defect, split or notch has appeared. Owners of cheap plastic raincoats find that any small tear soon develops into a large one. If the plastic sheet is reinforced with a mesh of thin threads, its resistance to tearing is much improved and it still has its essential properties of flexibility and water-resistance.

The example above shows how we make the best of materials by combining two different materials, so that we may obtain in one *integrated material* the useful properties of both. Fibre reinforcement has been used for many years in those parts of the world where people mix straw with clay when making bricks. The straws are fibres and their tensile strength improves the mechanical properties of the brick. Straw is a natural material and is itself a complicated structure containing fibres of plant origin which provide its main strength. Other natural products such as timber, bone and skin have fibres embedded in them, providing strength and rigidity (timber, bone) or flexibility and tear-resistance (skin, leather). These are natural examples of integrated materials.

Modern fibre-reinforced materials often make use of fibres of very great tensile strength. Carbon fibres can be made with a tensile strength up to $23 \times 10^3 \mathrm{MN/m^2}$ or MPa. Compare this with steel ($3 \times 10^3 \mathrm{MN/m^2}$ or MPa) and nylon ($70 \mathrm{MN/m^2}$ or MPa) which are normally thought of as being strong materials. The great strength of the carbon fibres (or 'whiskers') is due to their having an almost perfect crystal structure, with no surface defects. Carbon 'whiskers' are difficult and expensive to make, so are used only for special purposes. For cheapness we use glass fibre. When the softened glass is drawn out to thin fibres, its surface becomes almost free of defects and its tensile strength rises to around $3 \times 10^3 \mathrm{MN/m^2}$ or MPa. Drawn metal wires have similar strength and are used in similar ways.

Much of the recent development of fibre-reinforced materials came from the need to produce light strong materials for constructing spacecraft and aircraft. Plastics reinforced by glass fibre are widely used. Tyres of automobiles may be reinforced with nylon fibres or steel wire. The handles and frames of squash rackets and the handles of golf clubs are fibre-reinforced. Plastic or glass fibres are used to reinforce wall-plaster. These are just a few of the many ways in which we use integrated materials.

Questions
Below are some more examples of the use of integrated materials. What are the useful properties of each member of the pair? What can the integrated material do, that the members could not do separately? Think of as many uses as you can for each type of integrated material.
(a) thin sheets of wood + glue (plywood)
(b) steel sheet + layer of tin on both surfaces (tinplate)
(c) concrete + steel rods embedded in concrete (reinforced concrete)
(d) particles of diamond + hard rubber (grinding wheel)
(e) glass + wire mesh embedded in glass (wired glass)
(f) wood + layer of paint on the surface (painted wood)
(g) thin steel wire + thick layer of aluminium around steel (electric power line)
Think of some more examples of your own to add to this list.

30.22 Beams and girders

Nearly always, we wish to make a structure (house, bridge, aircraft, bicycle, chair) that is light yet strong. We can use materials of low density and high strength, of which many examples were given in the previous section. We can also produce a light, strong structure if we shape the materials in the right way.

Instructions: investigating the strength of girders and other structures
1 Cut about ten sheets of thin card or thick paper, about 10 cm wide and 30 cm long. They must all be cut from the same kind of card or paper.
2 Make several different kinds of girder or other structure, as in fig. 30.4. Corrugated sheets are made simply by

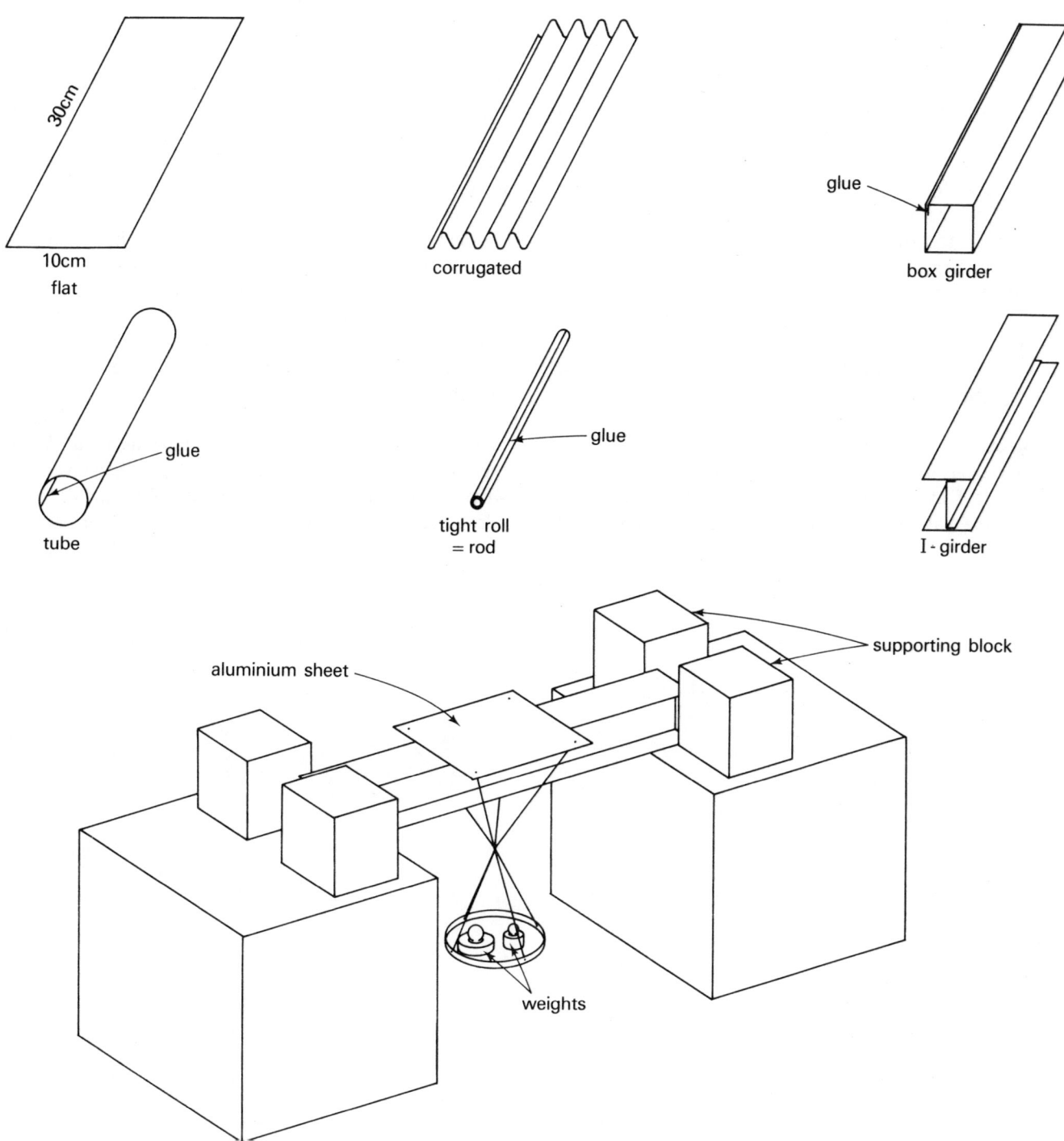

Fig. 30.4 *Investigating the strength of girders and other structures.*

folding the sheet. You can try the effect of different numbers of corrugations (2, 3, 4 or 5). The box girder may be square in section or it may be rectangular; you could try two or three different shapes of box. The tight roll is to represent a solid rod; roll it as tightly as you can before glueing the outer edge to secure it. The I-girder is made from three strips, cut from a single sheet of card or paper. Two of the strips are of equal width, for the top and bottom rail of the girder; the third strip is for the 'web', the middle piece which joins the top and bottom rails. You can try making girders with wide rails and narrow web, or with narrow rails and wide web, but remember that you may use only *one* piece of card 10cm × 30cm to make the whole girder. If you follow these instructions you will have a number of different structures, all made from the same amount of card and all having the same weight. Some have a little more glue than others, but this is a difference we must ignore.

3 When the glue is dry, support one of the structures between two blocks. It may be necessary to hold its ends upright between two blocks or in a clamp. This is to stop it from falling over sideways. To test the structure, a weight is to be hung from it; the weight must be evenly distributed. A sheet of aluminium with threads attached can be used for many of the structures. With the cylinder and rod, it is better if the aluminium is bent to a curve to fit the top surface. In this way we can hang the weights from the structure without altering its shape.

4 Add weights to the pan, a little at a time, until the structure collapses.

5 Which of the structures is the strongest?

6 Look around you for examples of the use of girders, corrugated sheets and tubular frames.

Projects

1 Investigate the water-holding ability of common building materials. Devise your own methods. Why is it generally bad for the walls and roofs of buildings to be damp? What is done to keep out rain from above and water soaking upwards from the damp soil below?

2 Carry out some long-term tests on corrosion. Fix copper sheet to wooden frames, using iron nails and copper nails. Also fix iron sheet with nails of both kinds. Leave these exposed to the weather and observe the stages in corrosion.

3 Carry out long-term corrosion tests to find out how seriously a steel sheet is corroded when it is muddy and not cleaned. Another cause of corrosion is damage to the surface caused by impact. Carry out tests to see if hammering can cause rapid corrosion of metal sheets.

4 Carry out a survey of corrosion in automobiles. Find out in what parts it occurs, and why. What can be done to prevent it?

5 Find out about methods used locally in the working of metals. What methods are used to shape the metal? What types of heat treatment are given, why are they given and what are their effects on the properties of the metal? What alloys are used, and what are their special properties?

31 Transistors

31.00 The transistor as a switch

In this section you use a transistor to see what it does. An explanation is given in 31.10.

The transistor is made from semi-conductors. It has three terminals or wire connections, called *base, collector* and *emitter*. We will call them b, c, and e, for short.

Instructions: switching with a transistor

1 Connect the circuit shown in fig. 31.1. Instead of the transistor type named, you can use any other type. If you use another type, your teacher will tell you which wires are b, c, and e. Different types of transistor have different arrangements of connection wire. Before switching on any circuit, check that your connections to the transistor are correct, as a wrong connection can destroy a transistor instantly.

2 Begin with the switch turned off. Does current flow through lamp L1? Does current flow through lamp L2?

3 Turn the switch on. Does current flow through L1? Does current flow through L2?

4 With the switch *on*, remove L2 from its socket. What happens? Was L2 carrying a current at step 3? If so, what can you say about this current?

When the switch is on, the base is connected to the positive terminal of the battery, through the resistor and lamp. The base has positive potential. This can be measured.

Instructions: measuring the potential of the base

1 Use the same circuit as fig. 31.1 except that, instead of the $1\text{k}\Omega$ fixed resistor, use a variable resistor. This can have a maximum value between $10\text{k}\Omega$ and $100\text{k}\Omega$. Connect a voltmeter between points A and B. A high resistance voltmeter with f.s.d. 1 V or 2 V is preferred..

2 Switch on, with the lamp L2 in its socket. Vary the setting of the resistor. What happens to the potential of the base?

3 When the potential is low, what happens to L1?

4 When the potential is high, what happens to L1?

5 Adjust the resistor until you find the potential at which L1 is *just* switched on. Measure this potential.

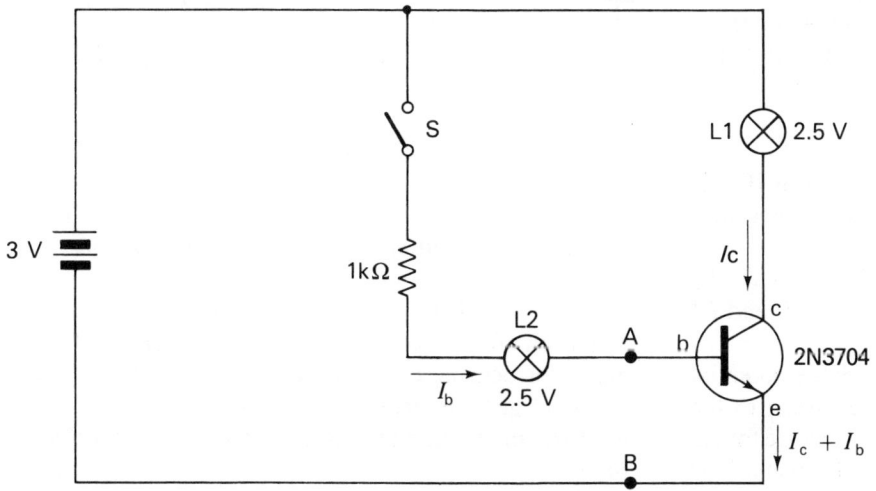
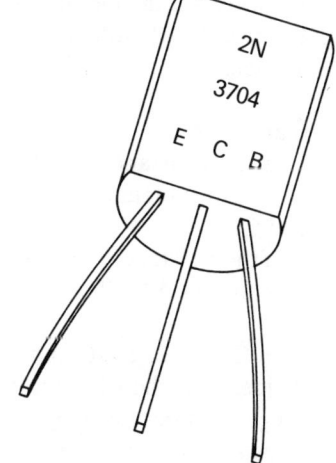

Fig. 31.1   *Circuit for investigating the action of an n-p-n transistor.*

When a transistor is being used as a switch we can say:
(a) The *base current* (I_b) is small, and
(b) the *collector current* (I_c) is large.
By using the switch, S, we switch a small current on or off. This makes the transistor switch a large current on or off.

There is not much practical use for a circuit like the one we have just used. If we wish to switch L1 on or off, we can simply wire an ordinary switch in series with it. We do not need to use a transistor. But a mechanical switch has to be operated by hand. A transistor can be operated in many other ways. The instructions below demonstrate one example of the use of a transistor switch.

Instructions: using a light-controlled switch

1 Connect the circuit of fig. 31.2. There is no lamp in the base circuit; in the previous investigations it was there to demonstrate that I_b is small. LDR is a light-dependent resistor, made of semi-conductor material. Its resistance depends on the amount of light falling on it. LDR and the fixed resistor R1 make up a potential divider (19.31).

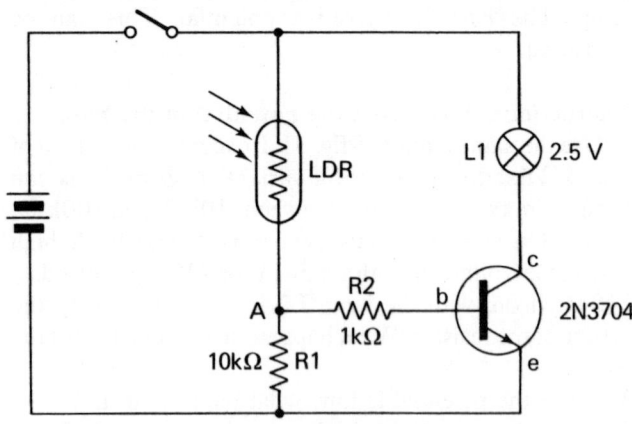

Fig. 31.2 Circuit of a light-operated switch.

2 Place LDR so that it receives bright light.
3 Switch on the circuit. What happens to the lamp?
4 Cover LDR with your hand, or a sheet of card. What happens?

Questions

1 If the resistance of LDR in bright daylight is 100Ω, what is the potential at A? What effect does this have on I_b and I_c?
2 If the resistance of LDR in darkness is $20k\Omega$ what is the potential at A? What effect does this have on I_b and I_c?

In fig. 31.2, LDR and R1 act as a variable potential divider. By choosing the value of R1 according to the range of values taken by the LDR, we can arrange for L1 to be switched on or off at any given light intensity. You can use a voltmeter connected at A and B to measure the potential at A as the light intensity on LDR is varied. If R1

is replaced by a variable resistor, it is easy to set this circuit to operate at any required light intensity.

This circuit uses a transistor to switch on a lamp. It could switch on a warning lamp when it 'sees another light'. For example it could detect the light from flames if the building catches fire. Or it can detect the light from the headlamps of an automobile approaching the house at night. Instead of L1 we could have another electrically powered device, such as a bell or buzzer, a small electric motor, a counter (to count people or objects passing by) or a relay (fig. 9.11). If a relay is used, it can switch on devices that require large currents: a room lamp, street-lamps, syrens, pumping motors and so on. The small transistor used in the demonstration circuit will burn out if I_c is larger than about 0.8A, but larger power transistors can carry *direct* currents of 30A or more (depending on the type of transistor); then a relay may not be required.

Questions

1 If we put a thermistor (19.50) in place of the LDR of fig. 31.2, what happens as temperature increases?
2 Redesign the circuit so that the lamp is switched off as temperature increases.
3 In the redesigned circuit (question 2), the lamp and thermistor are placed close together in a small box. What do you expect to happen when the circuit is switched on? How could this circuit be put to practical use?

Instructions: making a delay switch

1 Connect the circuit of fig. 31.3.

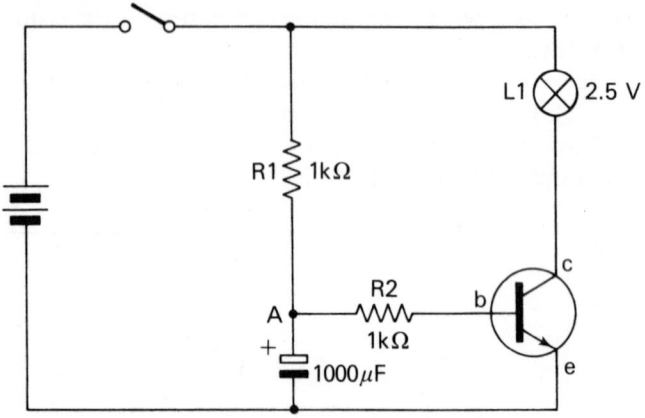

Fig. 31.3 Circuit of switch with delayed action.

2 Switch on. What happens? Explain.
3 Switch off. A second or so later, switch on again. The capacitor became charged during step 2, why does the lamp not glow immediately you switch on?
4 Make an alteration to the circuit to decrease the delay time.
5 Make an alteration to the circuit to increase the delay time.

320

31.10 How a transistor works

In this book we use only one type of transistor, a junction transistor. The description of the structure and action of a transistor applies only to this type. Other types of transistor, which are less commonly used than junction transistors, work in different ways.

Inside the metal or plastic case of a transistor is a tiny chip of silicon, or some other semi-conductor material. This is so small that there is room for a dozen or more chips inside the case. Some devices do have more than one transistor inside. The chip forms the base. It has two electrodes attached to it, one on either side. One is the emitter; the other is the collector. In the practical investigations in this book we use n-p-n transistors. In these the base is made from p-type material (19.13) and the electrodes are made from n-type material. In effect, the base is a thin layer of p-type material sandwiched between two layers of n-type material. This is why we call it an n-p-n transistor. Transistors can also be made that have p-n-p construction.

In the n-p-n transistor, the distance between the two n-type layers is small, just a few micrometres. This is a very thin sandwich.

A p-n junction acts as a diode (fig. 19.20). Current can flow from b to e; this is base current, I_b. A current *cannot* pass from e to b, or from c to b. So no current can pass from c to e. How can I_c flow through the transistor?

To see how this is possible, we must remember that the charge carriers in n-type material are electrons. They move in the opposite direction to conventional current. I_b is actually a flow of electrons which *enters* the transistor through the emitter and *leaves* through the base. While the current is flowing through the p-type material of the base, charge is carried as holes, but this is not of importance in this description. The main point to consider is that as the electrons enter the emitter they are accelerated by the p.d. between base and emitter (b is positive with respect to e). At the same time there is a *much bigger p.d.* between c and

e (c is positive with respect to e). As the electrons flow from e to b, they pass close to c (because b is thin). They are attracted and accelerated by the high p.d. between c and e. They overcome the effect of the diode between b and c, they pass through to c. The number of electrons flowing from e to c is much larger than the number flowing from e to b: I_c is much larger than I_b. But the number flowing to c depends on how many were originally being accelerated to b: I_c varies with I_b. In this way the changes in the size of the small current I_b control the changes in the size of the large current, I_c. We can use a small current to control a much larger one.

31.20 A transistor as an amplifier

Measurements (project 3) confirm the theory that has been outlined above. There is a definite relation between I_c and I_b; over a large range of values, I_c varies with I_b. For many types of transistor I_c is between 20 times and 50 times as great as I_b. For special high gain transistors, I_c may be several hundred times I_b. This means that a transistor can be used as *current amplifier*. As an example of this we will use a transistor to amplify the current from a microphone. This can be a moving-coil microphone (29.05) or a crystal microphone (29.06). As an alternative we can use the moving-coil or crystal pick-up of a record player. The currents produced from these devices are far too small to operate a loudspeaker directly. If we use their varying output current to change the base current of a transistor, we can operate a loudspeaker from the varying collector current. Then we can obtain all the power we need, as the current that operates the loudspeaker comes *from the battery*. We use the transistor to regulate this according to the amount of current from the microphone.

Instructions: using a transistor as an amplifier
1 Connect the circuit of fig. 31.4. The exact value of the capacitor does not matter, it may be between $100\mu F$ and

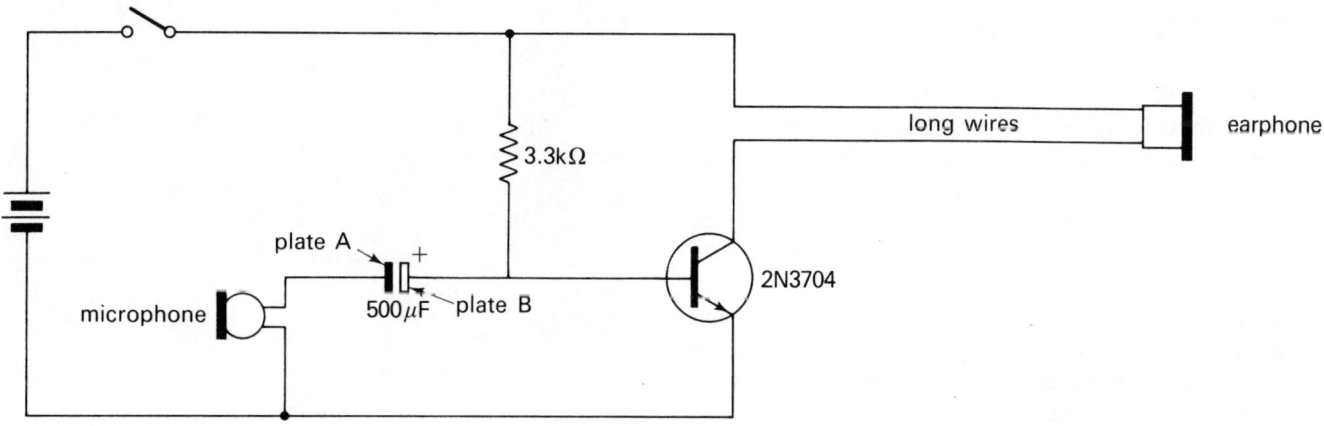

Fig. 31.4   An amplifier circuit using a single transistor.

1 000µF. You may obtain better results if you alter the value of the resistor after a few trials. It is best to use a moving-coil microphone, but you can use a small moving-coil loudspeaker, a crystal microphone or a record-player cartridge. The earphones are the type normally used with transistor radio sets (either a double earphone, or a single ear-plug). Instead you may use a small loudspeaker, provided that it has resistance of 50-100 Ω. A low-resistance speaker takes too high a current and overloads the transistor. The wires between the earphones and the rest of the circuit should be long enough to lead to the next room, or outdoors.

2 Switch on. Is a base current flowing? Is a collector current flowing? What can be said about these currents? What evidence have you that a collector current is flowing?

3 Ask someone to speak into the microphone while you listen at the earphones.

Questions

1 If the microphone is generating a varying p.d. across its terminals, what effect does this have on the capacitor?

2 If plate B suddenly has a higher potential, what happens to I_b?

3 If plate B suddenly has a lower potential, what happens to I_b?

4 What is the effect of these variations in I_b?

If you connect the microphone *directly* to the earphone, you *might* hear a faint sound from the speaker, but you would probably hear nothing. By using a battery, with a transistor to control the current, we have amplified the microphone current and obtained an audible sound from the earphone. The amplifier is a simple one. To obtain greater amplification we could use a second transistor in place of the earphone. This would control a larger collector current which could be used to operate a large loudspeaker.

31.30 Using transistors

In the early days of electronics, radio sets and amplifiers used thermionic 'tubes' or 'valves' similar to the diode shown in fig. 23.3. These were large, heavy and fragile. They needed high currents (about 0.5 A for *each* tube) to heat the filament. They produced much heat when operating. They needed high potentials to make currents flow to the anode.

By contrast, semi-conductors are exceedingly small and light, and they are not easily broken. They do not require a heater current. They produce little heat (except when they are used for controlling large currents) and they can be used with very small potentials. In your investigations you used only 3 V and most portable radio sets operate from a

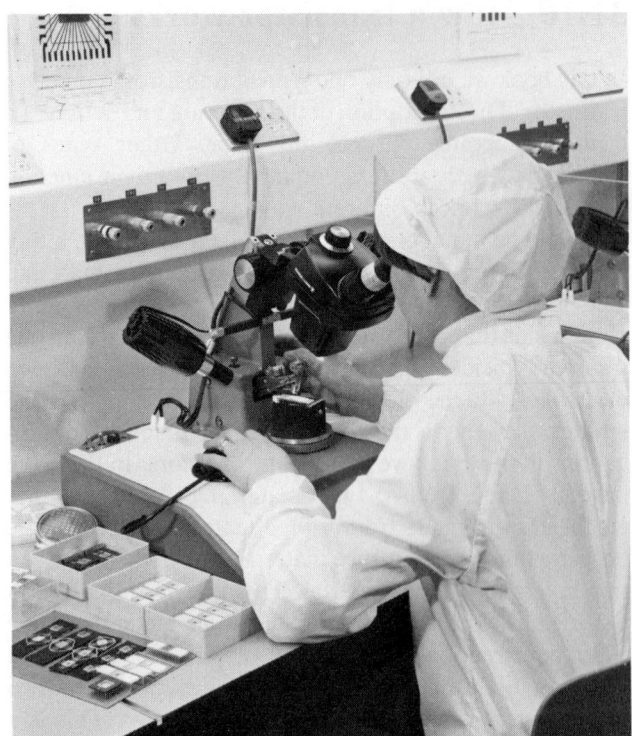

Fig. 31.5 *Technicians using microscopes to assemble integrated circuits. (top)*
An integrated circuit as seen under the microscope. (bottom)

322

9V battery. Semi-conductors can be mass-produced and are cheap. All these facts mean that we can use semi-conductors to build elaborate circuits with high standards of performance, yet they require little power and can be packed into a small space. We have portable radio and television sets, pocket calculators and digital watches. All these contain dozens if not thousands of transistors, yet they operate from ordinary dry batteries.

The active part of a transistor is almost microscopically small; dozens or hundreds of transistors can be made on one chip, complete with the connections between them. A complete circuit, consisting of diodes, transistors and resistors can be made on a chip of silicon only 1mm square. We call this an integrated circuit. In the more complex integrated circuits the chip measures a few millimetres across and on these we can make almost all the components needed for a radio receiver, a HiFi amplifier, an electronic clock or a small computer. The invention and development of semi-conductor devices has led to immense advances in space exploration, tele-communications and complex computing. Without these developments much that we know about the structure and nature of the Universe and the atom would still be unknown.

Projects

1 Design and build a circuit, based on fig. 31.2, that switches the light *on* when the LDR is in darkness or dim light and *off* when the LDR is brightly lit. The circuit could have some useful applications in the home.

2 Design and build a circuit based on fig. 31.2 to operate a small motor or relay. You might be able to think of a way to use light to control the operation of an electrically driven model. If the transistor is to be used to control a device that has electromagnets in it (a motor, relay, bell) you must take a special precaution. When the current through an electromagnet or coil is switched *off* a high e.m.f. is induced (fig. 29.2b). This could make a large current flow through the transistor and destroy it. To avoid this, connect a diode in series with the device. Arrange the diode so that the normal collector current does *not* pass through it, connect the cathode of the diode to the wire that goes to the positive battery terminal: connect the anode to the wire that goes to the emitter of the transistor. The e.m.f. induced when current is switched off is in the *reverse* direction (Lenz's law) so that the current flows through the diode and not to other parts of the circuit.

3 Investigate the relation between I_b and I_c in more detail. Use a circuit like fig. 31.1. In place of L1 use a milliammeter; in place of L2 use a microammeter; in series with the $1k\Omega$ resistor connect a $100k\Omega$ variable resistor. Adjust the resistor to obtain various values of I_b; read I_c. Plot a graph to show the relation between them. The gradient of this line is the *gain* of the transistor.

4 Popular magazines on electronics contain many simple projects that you could build.

32 Radio

Radio is a form of electromagnetic radiation, with frequency in the approximate range 150 kHz to 100 MHz. To understand how a radio transmitter works, we must first study how high-frequency electrical vibrations (or oscillations) are produced.

32.00 Oscillators

A simple mechanical oscillator or vibrator is made by hanging a block from a spring (fig. 32.1). To make it oscillate, we must give the system some energy. We do

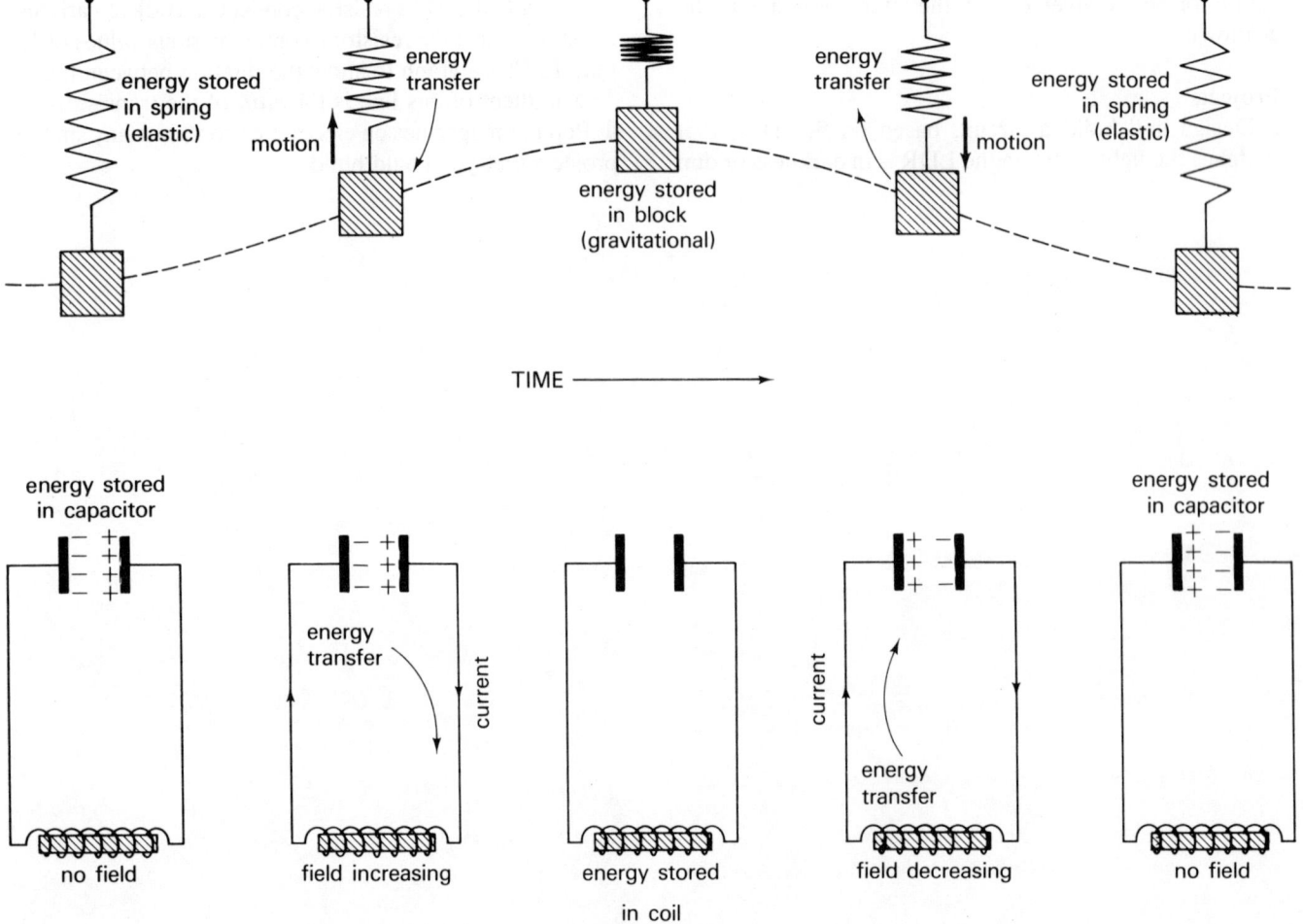

Fig. 32.1   Oscillating systems: block and spring; capacitor and coil.

this by pulling the block down by hand and then releasing it. It oscillates up and down. Energy is stored alternately in the spring (elastic potential energy) and in the block (gravitational potential energy). As the system oscillates, energy is transferred from spring to block, and from block to spring, and back again, indefinitely. Energy transfer involves motion of the block and spring. If no energy was lost during transfer, the system would oscillate for ever. Motion of the block and spring causes some of the potential energy to be converted to internal energy of the spring and the molecules of air around the system. Energy is lost, the amplitude of the oscillations gradually becomes smaller. Eventually the system comes to rest.

An electrical oscillator is made by joining a capacitor to a coil. To make it oscillate, we must give the system some energy. We do this by supplying a charge to the capacitor. The system then oscillates as shown in fig. 32.1. Energy is stored alternately in the capacitor (p.d. between its plates) and in the coil (magnetic field). As the system oscillates, energy is transferred from capacitor to coil, and from coil to capacitor, and back again, indefinitely. Energy transfer involves the flow of electric current. If no energy was lost during transfer, the system would oscillate for ever. Flow of current causes some of the energy of the system to be converted to internal energy in the capacitor, the coil and the connecting wires.

The figure shows how current flows from the capacitor, producing a field in the coil. When the capacitor has discharged (middle picture), the field is large, but it disappears instantly because the p.d. of the capacitor is zero. An e.m.f. is induced in the coil to oppose the disappearance of the field (Lenz's law); this makes a current flow, so recharging the capacitor. The capacitor has reversed charge at each oscillation.

Questions

1 What factors affect the frequency of oscillation of the block on the spring?

2 Is the frequency of this system affected by the amplitude?

3 What factors might affect the frequency of oscillation of the capacitor-coil circuit?

4 Do you think the frequency of this system is affected by amplitude?

5 How can you alter the frequency of the block-spring system?

6 How can you alter the frequency of the capacitor-coil system?

7 If we wish to supply energy to the block-spring system, to make it oscillate continuously, what is the best way of doing this? (Hint: see 7.40.)

8 If we wish to supply energy to the capacitor-coil circuit, to make it oscillate continuously, what is the best way of doing this?

9 In the description of the capacitor coil circuit it was said that energy could be given to the circuit by supplying charge to the capacitor. Think of another way of supplying energy to the circuit.

32.10 Radio transmitters

Any sudden discharge of electricity can produce electromagnetic radiation of radio frequency. A discharge of lightning produces a loud crackle on a radio set. Small sparks around the home may also produce noise (called interference) on a radio set. Sparks are made by the turning on and off of switches, especially those carrying heavy current. Sparks in thermostat switches in cookers and refrigerators, sparking at the contacts of electric bells and the sparks between the brushes and commutator of an electric motor are all liable to produce radiation that is detected by a radio set. Some of the early radio transmitters were based on the principle of spark production. Modern transmitters work differently. They consist of a number of secitons, which we shall consider separately:

(a) *Oscillator* This consists essentially of a capacitor and coil circuit, similar to the one we have just studied. The capacitor and coil are chosen so that the circuit oscillates at frequencies of several kilohertz or megahertz. These are the radio frequencies. If this circuit is connected to an aerial wire or antenna, an oscillating electric field is created in the wire. Electrons in the wire oscillate at the same high frequency. The rapidly changing velocity of the electrons results in the emission of electromagnetic waves: radio waves. These have the same frequency as the oscillator.

(b) *Feedback amplifier* The oscillator cannot supply energy to the aerial unless it receives a supply of energy to make up for its losses. As was explained in the answers to questions 7 and 8, the energy supply to the oscillator must be in resonance with the oscillator. To do this we take a *small fraction* of the oscillating current from the oscillator; the remainder goes to the aerial. The small fraction of current that we take is fed into an amplifier (31.20). The output of the amplifier is a *large oscillating current*; it has the same frequency as the oscillator, and it is always at the same stage of oscillation. This large current is then fed into the oscillator to supply the energy it needs to make it oscillate continuously. We have taken a small oscillating current, amplified it and *fed it back* into the oscillator. This process is called *feedback*, so the amplifier is a *feedback amplifier*. The power required to keep the oscillator in action is provided by the power supplied to the amplifier. Together, the oscillator and feedback amplifier produce an oscillating current of *constant frequency and amplitude*. This is called the carrier wave.

(c) *Modulator* If a radio transmitter consisted only of an oscillator and feedback amplifier, it would produce only a continuous train of radio waves of constant frequency and amplitude. Like a continuously-lit electric

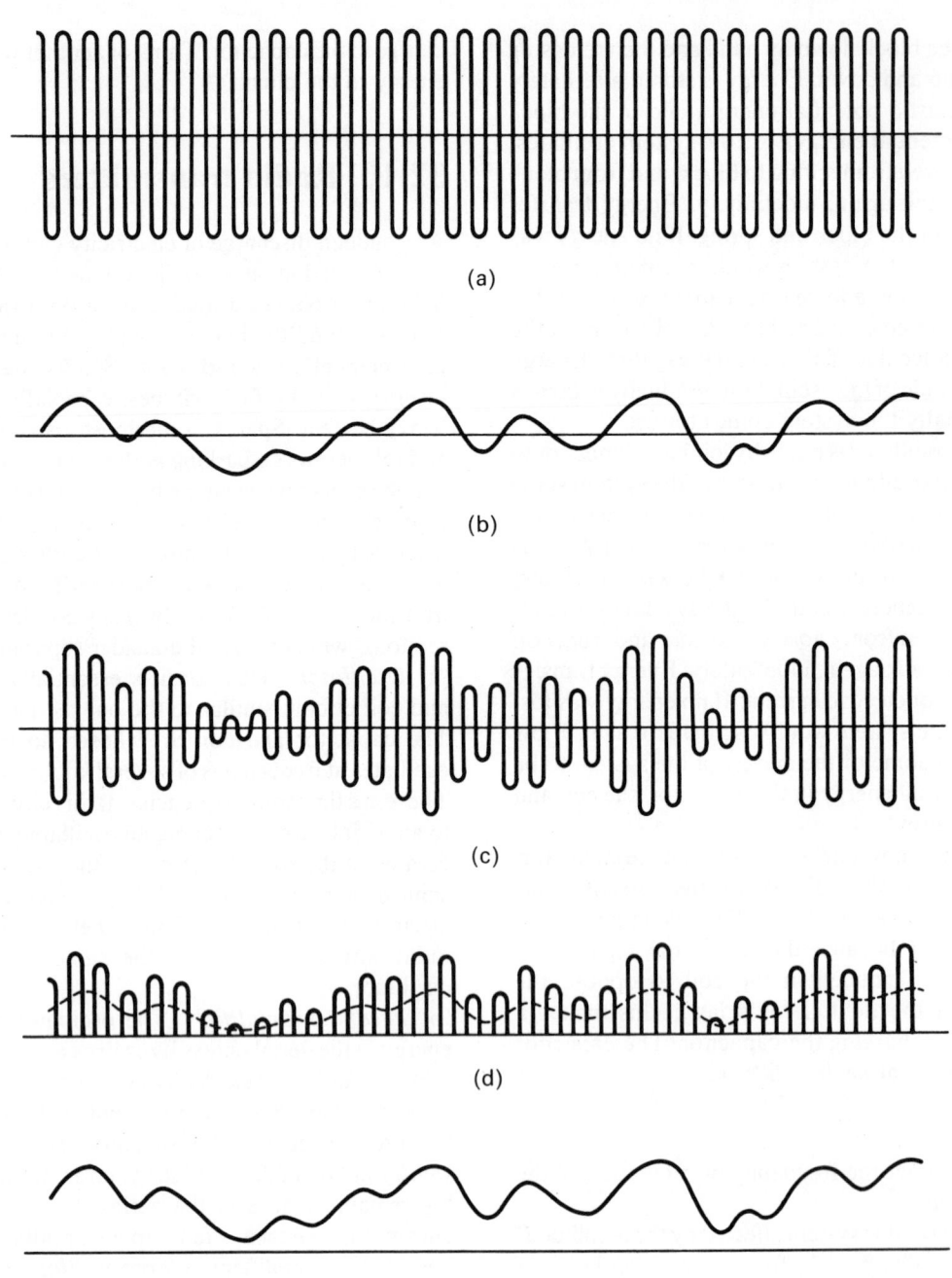

Fig. 32.2 *Wave forms in radio transmission and reception: (a) carrier wave at radio-frequency (b) sound wave at audio frequency, and varying e.m.f. of amplifier output (c) carrier wave modulated at audio-frequency, as transmitted and received (d) rectified modulated carrier wave represented by current passing through earphone of Fig. 32.3 dashed line represents the response of the earphone and the sound heard (e) wave form of I_b and I_c in Fig. 32.4, representing the original sound.*

lamp (light of constant frequency and amplitude) such a transmitter could not send out much information. We can use a lamp for sending information in many ways: for example a red lamp instructs traffic to stop; a light in a window means that someone is at home; a flashing lamp may be used to send a message by morse code. We could use a radio transmitter in a similar way; we could switch it on and off to send code messages. We can use it much more

effectively for sending information, though, if we add another section to the transmitter. This is the modulator. In fig. 32.2a we see the carrier wave, the radio-frequency oscillations from the oscillator. In fig. 32.2b we see a sound wave. This has lower frequencies: 30 Hz to about 12 kHz; we call these audio-frequencies. It is not electromagnetic. The graph could represent the motion of molecules of air through which the sound is passing. It could also represent the varying e.m.f. of the output of an amplifier, after the sound has been detected by a microphone and the small microphone e.m.f. has been amplified. In a radio transmitter, this audio frequency e.m.f. is fed into a modulator. The carrier wave too is fed into the modulator. Then the modulator modulates the *amplitude* of the carrier wave to follow the wave form of the audio-frequency e.m.f. In fig. 32.2c we see the modulated carrier wave.

This amplitude-modulated carrier wave is then fed to the aerial. The radio waves from the aerial have the same form. Returning to our idea of a light, we can imagine the light shining with constant frequency (colour) but varying amplitude (brightness). Using suitable equipment, a beam of light *can* be modulated to carry information (speech, music, TV pictures). Modulated laser light is used for transmitting information along optical fibres over long distances (25.23, 26.15).

Before leaving the subject of modulation it should be noted that other types of modulation of the carrier wave are possible. In one form of radio transmission amplitude is constant and frequency is modulated instead. Frequency modulation (FM) has several advantages over amplitude modulation (AM) but we cannot go into the method here.

32.20 Radio wave detection

Electromagnetic waves of radio wavelength spread in all directions from the aerial of the transmitter. For some purposes they may be reflected by a parabolic reflector to make a beam of radio waves, aimed at a receiving station many kilometres away. Reflectors are used for transmissions to and from telecommunications satellites and for microwave radio links used in telephone systems.

When the electromagnetic radiation reaches a metal object it produces exceedingly small electric currents in it. We use a long wire (the aerial) to catch the radio waves. The aerial carries currents produced by all the radio transmissions that reach it. If we tried to listen to this, we would hear dozens of stations at once. To pick out the currents caused by one radio transmission and to convert this to sound, we use a radio receiver. This consists of a number of sections:

(a) *Tuner* This has a capacitor-coil circuit. We can vary the frequency of oscillation of this circuit (usually by varying the capacitance) so that it has the same frequency as the radio station we wish to receive. Then it resonates with the oscillating currents in the aerial produced by waves of that frequency. The small amount of energy of the oscillating currents is readily transferred to the capacitor-coil circuit. Currents oscillating at other frequencies are not able to make the oscillator resonate; they can have no effect on the radio receiver. In this way the receiver is *tuned* to receive radiation of only the one frequency we wish to receive. The oscillating current in the capacitor-coil circuit has the form of fig. 32.2c.

(b) *Rectifier* If we connect an earphone across the capacitor we hear nothing. Though the current is oscillating, it is doing so at radio frequency (150 kHz or more) which is above the frequency that the ear can detect. If the current is rectified we obtain the wave-form of fig. 32.2d. The earphone connected to the rectifier responds to the *average* current; its vibration is represented by the dashed line. This vibration is transferred to the air around and so we hear a reproduction of the original sound.

(c) *Amplifier* The varying current from the rectifier is enough only to operate a small earphone. To operate a loudspeaker, the current must be amplified.

Instructions: making a simple radio receiver

1 Connect the circuit shown in fig. 32.3. The aerial is an insulated wire, at least 25 m long. Support it horizontally outdoors, as far above the ground as possible, and as far from buildings as is convenient.

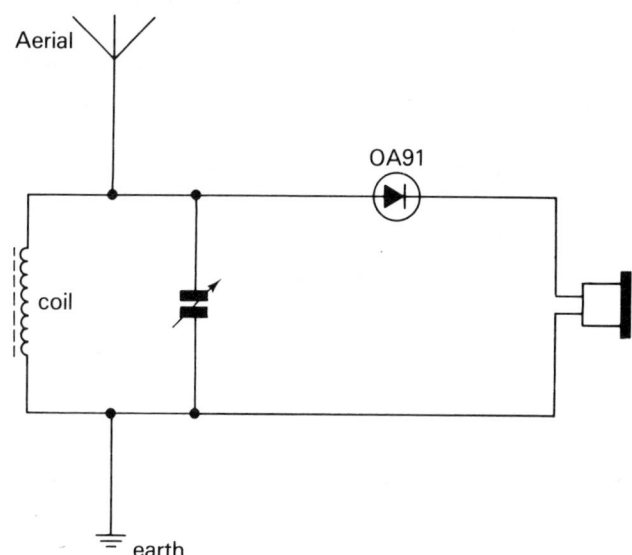

Fig. 32.3 Circuit of a simple radio receiver.

2 The earth connection may be made to a water-pipe, or bury a strip of metal in the soil and connect the earth wire to this. In dry weather make the soil around the earth plate wet by pouring water on it.

3 Make the coil by winding about 80 turns of fine insulated wire on to a ferrite rod. Ferrite is a magnetic material which improves the properties of the coil.

4 The capacitor is a variable one, capacitance a few hundred picofarads. If you cannot obtain a variable capacitor, try a number of fixed-value capacitors in the range 10 pF to 470 pF; one of these will probably be just right for tuning the receiver to your local radio station.

5 The diode, type OA91, is a *germanium* diode; other types of germanium diode are suitable.

6 When the circuit is assembled, listen carefully with the earphone while you alter the setting of the capacitor.

7 If you get no result, try using a coil made from only 40 turns.

8 If you still get no result, it may be because your nearest station is not powerful enough for such a simple circuit to receive. Do not be disappointed, but read the next paragraph. Keep the circuit assembled, ready for the addition of an amplifying stage.

This simple receiver has no battery to supply energy. The only source of energy is the action of electromagnetic waves when they are caught by the aerial. If you are a long way from the transmitter, the amount of energy reaching the aerial is small. The receiver *is* working, but the currents in the earphone do not produce a sound loud enough to hear. The currents need to be amplified.

Instructions: making a radio receiver with amplifier

1 The circuit is shown in fig. 32.4. The part to the left of the dashed line was constructed by following the previous instructions. The amplifier section is similar to the amplifier of fig. 31.4. It uses a single transistor; you may use the 2N3704 used in chapter 31, or one specially designed for audio-frequency amplification, such as 2N2926, BC108, or BC109.

2 When the circuit has been built, switch on the battery. Listen with the earphone while altering the setting of the variable capacitor.

3 Base current I_b comes from two sources; what are they, and in what way do they differ?

4 What can you say about the size of I_c?

The fixed capacitor helps improve the quality of the sound by removing the radio-frequency part of the oscillating current. Its effect is to make the current like fig. 32.2e. Only the audio-frequency modulation passes to the transistor for amplification..

This simple receiver still has several faults. If you live near to two or three strong transmitters which have frequencies close together, you may hear all stations at once. The tuning circuit is resonating to all their frequencies. More complex circuits can be made which will tune more exactly to only one frequency. You are not likely to hear foreign stations with this receiver; the amount of energy reaching your aerial is too small, even after amplification. Most radio receivers have further amplifying stages. Then distant stations can be received and enough current is available to drive a loudspeaker. Although a commercially-made radio set is likely to be much more complicated than the one that you have made, the essential stages of tuning, rectification and amplification are just the same as in your set.

Project

List the uses of radio, including the ways in which it helps save life and its uses in education. Are there any bad uses to which radio has been put?

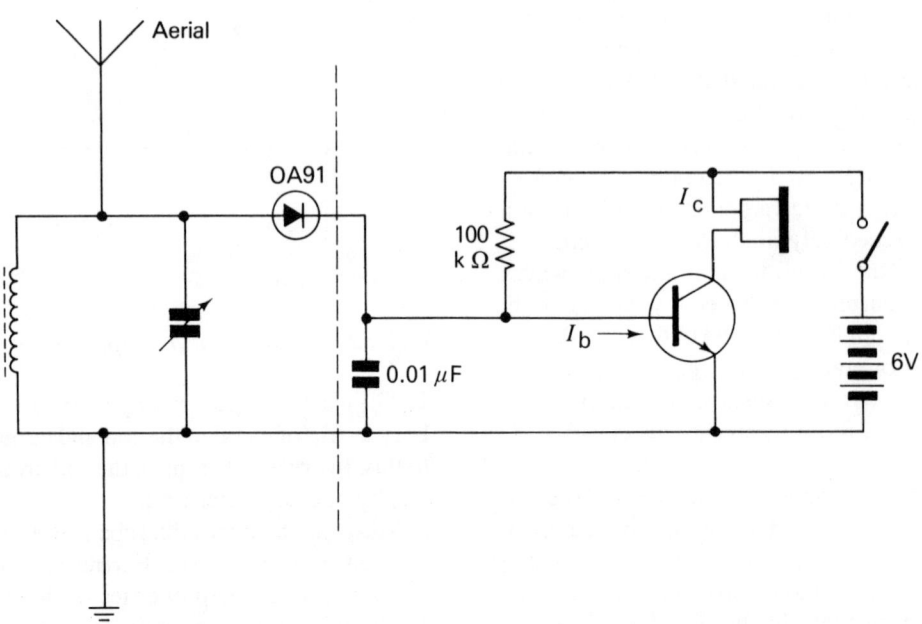

Fig. 32.4 Circuit of a radio receiver with one stage of amplification.

33 Nuclear power

The world's supply of fuel oils is expected to run out at the beginning of the twenty-first century. There will still be coal, but this is not a convenient source of energy for some purposes and many countries have no coal. People are looking for new sources of energy. It is hoped that it will be possible to make greater use of sunlight (solar energy), of the wind and of the ocean waves. Nuclear energy is becoming well developed as an energy source. Many nuclear power stations have been built for the generation of electricity. Submarines and warships using nuclear energy to drive them have been in service for several years. Most spacecraft use solar energy to power them but some that are being sent to the outer planets (Jupiter, Saturn, Uranus, Neptune and Pluto) will travel so far from the Sun that too little light energy can reach them. These are powered by small nuclear power plants. In spite of the possible dangers of the use of radioactive materials, and the problems of disposing of the radioactive waste materials from nuclear power plants, it seems that nuclear power will continue to be an important source of energy.

33.00 Matter into energy

When radium decays radioactively, it becomes radon:

$$^{226}_{88}\text{Ra} \rightarrow \,^{222}_{86}\text{Rn} + \,^{4}_{2}\alpha$$

The figures in the top row of this equation add up correctly. These are nucleon numbers which are integers (obtained by 'counting' the number of protons and neutrons). If they did *not* add up correctly, it would mean that whole neutrons or protons had been created or destroyed in the reaction. This does not happen. If we measure mass accurately, using a mass spectrometer (fig. 20.1), we find that masses cannot be accounted for *exactly*. During the reaction we have lost 0.006 u for each atom of radium that decays. The equation shows that the law of conservation of matter is not true. Matter *can* be destroyed.

When the radium atom decays, energy is set free. Some of this appears as the kinetic energy of the α-particle. Some appears as quanta of γ-radiation given out from the radium nucleus as it decays. This energy appears to have come from nowhere. It seems that the law of conservation of energy is not true, either. We have found that energy *can* be created.

Scientists try to discover the laws by which the Universe operates. Laws such as the conservation of matter, the conservation of energy, the conservation of momentum and many others have been discovered. They have been useful to us, helping us to understand and explain how the Universe works. For almost everything that happens in the Universe the laws hold true. Or at least we can use the laws *as if* they were true, even when we know that they are not. Careful measurement has shown that the laws of conservation of matter and energy are not exactly true when we are concerned with nuclear changes. In everyday life and in much of physics we can still use these laws. The fact that a small amount of matter is destroyed in a *nuclear* reaction is not important when we are thinking about a *chemical* reaction, or about a physical process that has nothing to do with nuclear changes. However, for nuclear reactions we must remember that the laws do not apply exactly. We must find different laws.

Mass of radium nucleus	226.025 u*	Mass of radon nucleus	222.016 u
		Mass of alpha-particle	4.003 u
Total mass on left of equation	226.025 u	Total mass on right of equation	226.019 u

* u is the *atomic mass unit*; it is $\frac{1}{12}$ of the mass of the carbon nuclide, $^{12}_{6}$C.

In the decay of radium, we have lost matter and gained energy. This happens with many other nuclear reactions. It seems that matter is being converted into energy. This is hard to understand. We never see it happen in everyday life, or in the school physics laboratory. Yet it was predicted by Einstein and has since been proved by measurements of the kind described above.

Einstein suggested that the conversion of matter into energy would be represented by a simple equation:

$$\triangle E = \triangle mc^2 \ (\triangle E \text{ in } J)$$

In this equation, $\triangle E$ is the *change* in energy (the amount of energy created), $\triangle m$ is the *change* in matter (the amount of matter destroyed, measured in kg) and c is the speed of electromagnetic radiation (3×10^8, in m/s).

Questions

1 How much energy is created when 1g of radium decays?

2 If we burn 1 kg of petroleum, we can obtain about 5×10^7 J. This is obtained by chemical action; there is no loss in mass. If we could convert the whole of the mass of 1 kg of petroleum into energy, according to Einstein's equation, how much energy would we obtain?

The amounts of energy that can be obtained by nuclear reactions are far greater than can be obtained by burning fuels in the ordinary chemical way. This is how nuclear power promises to provide vast quantities of energy from relatively small quantities of nuclear fuel.

33.10 Nuclear fission

One nuclear reaction of special interest is that which occurs when uranium-235 is bombarded with neutrons. At first, the neutron is taken into the nucleus, forming another isotope of uranium:

$$^{235}_{92}\text{U} + ^{1}_{0}\text{n} \rightarrow \ ^{236}_{92}\text{U}$$

The neutron is represented by $^{1}_{0}$n, indicating that its nucleon number is 1, with no electric charge. The absence of charge means that it is able to approach and enter the positively-charged nucleus without being repelled. The isotope produced by the reaction above is unstable. Soon the nucleus breaks apart. This happens in various ways, giving various products. One typical result is:

$$^{236}_{92}\text{U} \rightarrow \ ^{141}_{56}\text{Ba} + ^{92}_{36}\text{Kr} + ^{1}_{0}\text{n} + ^{1}_{0}\text{n} + ^{1}_{0}\text{n}$$

The result is the production of nuclei of barium and krypton and *three neutrons*. Mass is lost; energy is created and set free. In this reaction, the nucleus splits into pieces. Another name for 'splitting' is 'fission', so this reaction is an example of *nuclear fission*.

The nucleus of uranium-236 can split in various ways, sometimes giving three neutrons, sometimes giving two neutrons. The essential point is that we began with *one* neutron bombarding the nucleus of uranium-235 and as a result we have set free energy and produced *more than one* neutron.

If we have a block of uranium-235 and a single neutron enters the nucleus of one atom, we later have three neutrons. These could enter the nuclei of three nearby atoms. Each of these disintegrates, giving us a total of nine neutrons. These could enter nine nuclei and produce 27 neutrons, and so on. At each stage the number of neutrons increases threefold (slightly less, if some atoms produce only two neutrons). At each stage the number of nuclei disintegrating increases by threefold and so does the amount of matter being converted to energy. If the process is allowed to continue all the atoms in the block could be affected and disintegrate. An enormous amount of energy would be released. There would be a nuclear explosion.

If we have a *large* block of uranium, it needs only a single neutron to set off a series of reactions as described above. We call it a *chain reaction*. In a fraction of a second the number of neutrons would have increased to millions and the explosion would occur. Luckily, we do not normally have *large* blocks of uranium. Most of the neutrons that are formed escape immediately from the sides of the block. They do not enter the nuclei of other uranium atoms. The chain reaction does not occur. There is no explosion.

With a large block we get a chain reaction, followed by explosion. With a small block, there is no explosion. There is a *critical mass* of uranium; if the block contains more uranium than this critical mass, explosion is likely; if the block contains less, there is no explosion. The first atomic bomb worked on this principle. In the bomb were two blocks of uranium-235. Each block was *less than* the critical mass. They were held apart in the bomb; no chain reaction could occur. When the bomb was dropped, a small explosive charge was fired inside the bomb to force the two blocks together very quickly. As soon as they touched, their combined mass was *greater than* critical mass. There was a chain reaction, causing an explosion that destroyed a whole city.

The energy released by nuclear fission can be used for destruction. It can also be used to generate electric power; it can help us in many peaceful ways. Used in this way 1 kg of uranium can give us as much energy as a million kilograms of coal. Power generation is described in 33.30.

33.20 Nuclear fusion

In nuclear fission, heavy nuclides such as uranium or radium break into lighter nuclides, with loss of mass and creation of energy. Another type of nuclear reaction occurs when light nuclides combine to give heavier nuclides.

There is loss of mass and creation of energy. This type of reaction is called *nuclear fusion*.

An example of nuclear fusion is the fusion of two isotopes of hydrogen:

$$^2_1\text{H} + {}^3_1\text{H} \rightarrow {}^4_2\text{He} + {}^1_0\text{n}$$

These two isotopes of hydrogen are sometimes known by special names, deuterium (hydrogen-2) and tritium (hydrogen-3).

For this reaction to occur the nuclei must come so close together that fusion can occur. Under normal conditions, this is extremely unlikely to happen. Both are positively charged. The closer they come together the stronger the electrical repulsion between them. They can meet and fuse only if they are approaching each other at extremely high velocity. Then they make contact and fuse before the repulsive force is able to decelerate them and send them flying apart again. High velocities mean high temperatures. Reactions such as these occur only when temperature is high, as high as 10^8K. Such temperatures do not naturally occur on Earth. In the interior of the Sun we find temperatures as great as 10^8K, and so these *thermonuclear reactions* can occur there. The Sun, like other stars, consists mainly of hydrogen. In the interior of the Sun temperature and pressure are high. Thermonuclear reactions occur easily, producing the enormous quantities of energy that are given out by the Sun and stars. The reaction converts hydrogen to helium (first discovered in the Sun, 21.60) and part of the mass of the Sun is converted to energy.

Hydrogen fusion is the basis of the hydrogen bomb. To obtain the high temperature required, we explode a uranium fission bomb. This starts a hydrogen fusion reaction. The amounts of energy generated by the hydrogen fusion reaction are greater than those obtained from fission reactions. From one kilogram of hydrogen we can obtain as much energy as from three million kilograms of coal.

If methods can be found for *controlling* the thermonuclear reaction we may be able to use it as a source of energy for generating electricity. To produce the required high temperature we could send an intense electrical discharge through the vessel containing the hydrogen. One of the problems is to produce high temperatures, yet not melt and vaporise the equipment in which the hydrogen is contained. If a thermonuclear power station can be developed, it would have one great advantage over a power station using nuclear fission. Nuclear fission requires heavy nuclides as nuclear fuels, for example uranium. These are rare and they are found in only a few countries of the world. Deuterium occurs naturally in all water; it can be extracted from the sea or other water supply in any part of the world. Tritium can be made by nuclear reaction. A thermonuclear power station works on fuels that are relatively common and available.

33.30 Generating electricity from nuclear energy

Many types of power station have been designed and put into operation. In this description we deal with only the main features of design. There are many variations. The early types of nuclear power station made use of the fission of uranium-235 (33.10). The core of the nuclear reactor (fig. 33.1) consists of a block of graphite (carbon). The purpose of this is to reduce the velocity of the neutrons. Only neutrons with a relatively low velocity are able to enter the nucleus of the uranium-235 atom. The block of graphite measures several metres in each direction. There are tubular channels running through the block. In these are placed rods of uranium, each contained in an aluminium tube. The nuclear reaction occurs in the uranium and the surrounding graphite makes sure that neutrons entering a uranium rod are slow enough to make the reaction occur. Among the uranium rods are other rods, made of boron steel. This material absorbs neutrons. By varying the number of boron steel rods placed in the reactor the rate of reaction can be controlled. The aim is that for every *one* neutron that enters a uranium nucleus, *one* neutron should be allowed to go on to enter the next uranium nucleus. The reaction produces two or three neutrons, so of these *all but one* must be absorbed by the boron steel. If the number of rods is adjusted so that on average only one neutron escapes to take part in the next reaction, then the reaction proceeds steadily and energy is generated steadily. If the number of rods is too great, too many neutrons are absorbed; the reaction dies out. If the number of rods is too few, a chain reaction occurs and the reactor explodes. For fine control of the rate of reaction, the rods may be partly lowered into the channels in the reactor core. As a safety precaution, boron steel rods that are not fully lowered are held by electromagnetic catches. If it is found that the reaction is going too fast, the rods can be dropped quickly into their channels in the core, so shutting down the reaction.

The energy emitted by the nuclear reaction causes an increase in the internal energy and other materials in the reactor. In some designs of reactor carbon dioxide gas is pumped through channels in the core and is heated by this internal energy. The hot gas is pumped out of the core to a heat exchanger, where it is used to boil water. This produces steam that can be used to drive a turbine. The turbine drives an electric generator. By these stages the energy from nuclear fission is converted to electrical energy.

The reactor described above requires uranium-235. This isotope is rare; natural uranium consists mainly of uranium-238. Less than 1% of natural uranium is uranium-235. For use in the reactor the uranium has to be treated to increase the percentage of uranium-235. This

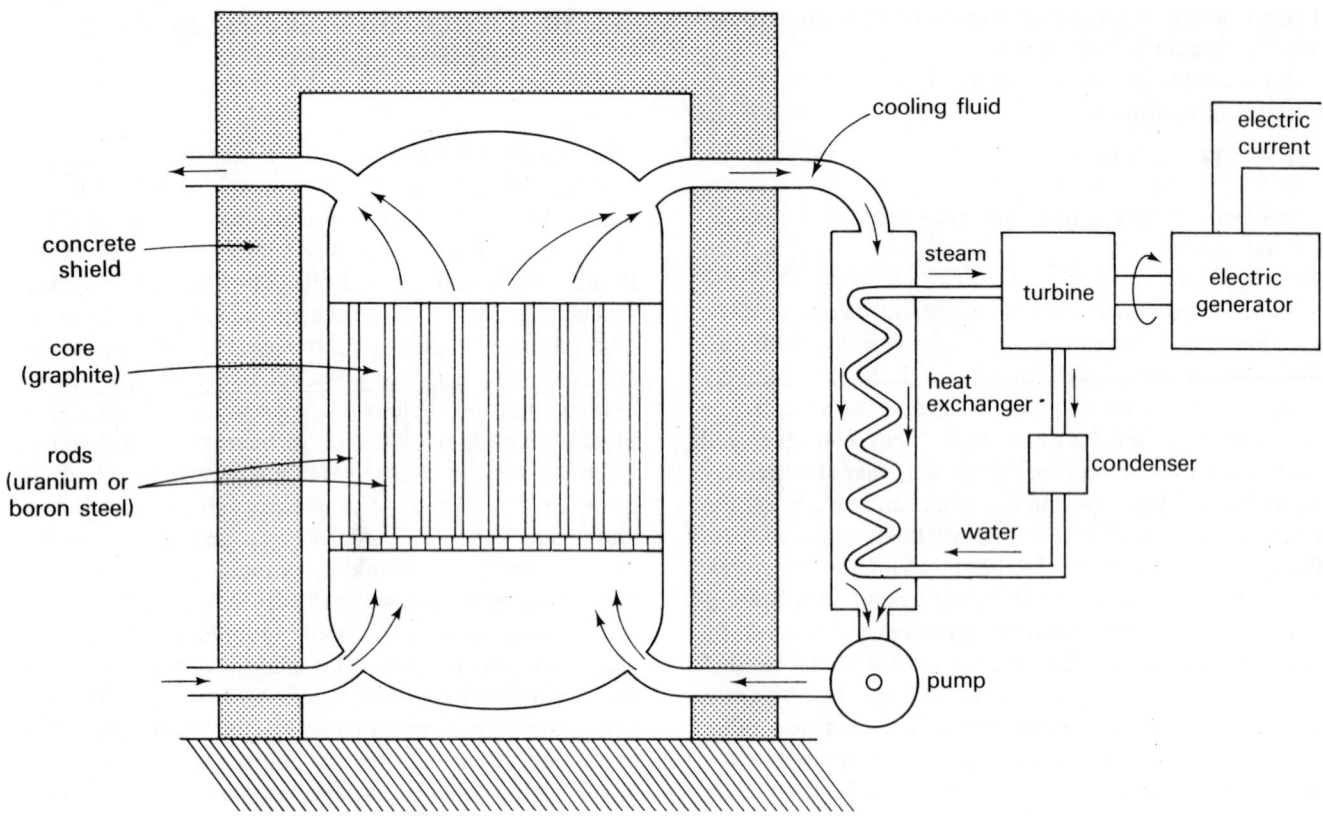

Fig. 33.1 Diagram of a nuclear reactor equipped for generating electricity. The mechanism for inserting and removing rods from the core is not shown. Several heat-exchangers are connected to one reactor, though only one is shown here.

makes the nuclear fuel expensive to prepare. A more recently developed reactor, the *breeder reactor*, uses the common uranium isotope, uranium-238. The reaction requires faster-moving highly energetic neutrons to break the uranium-238 nuclei. During the reaction uranium-238 is converted to plutonium-239. This is a nuclear fuel. It can be purified and used in the reactor as a source of nuclear energy. The breeder reactor breeds its own fuel. It is therefore much more economical to run than the uranium-235 reactor. In some designs of reactor the cooling fluid is liquid sodium (melting point, 97.8°C)

which is pumped through the reactor and then to a heat exchanger where water is boiled to produce steam for the turbo-generator.

The story of the development of nuclear power plants shows how our knowledge of the structure of atoms and the forces connected with them can be used to produce a plentiful supply of electric power for the benefit of us all. Our knowledge has also been used in terrible ways. The way in which scientific knowledge is used depends totally on the will and the wisdom of those people who control it.

Part A   Measuring

1   Length, mass, time

1.01 Measuring length

Instructions: measuring lengths
Vernier readings (fig. 1.3):   (a) 15.4mm, (b) 76.8mm, (c) 54.6mm (nearer to '6' than to '7'), (d) 30.5mm (did you forget to record the zero?).
Micrometer screw gauge readings (fig. 1.5):   (a) scale reading is 2.50mm, plus 29 divisions (= 0.29mm), giving total result = 2.50 + 0.29 = 2.79mm. (b) scale reading is 3.50mm, plus 1 division (= 0.01mm), giving total result = 3.50 + 0.01 = 3.51mm.
3 Measure the distance travelled on level ground in a large number of steps. For example, if 50 steps are taken, measure the total distance, and divide this by 50, to find the average length of 1 step. Steps which are longer than average are cancelled out by steps which are shorter than average.
4 Make a pile of 10 (or more) pieces of paper, *all cut from the same sheet*. Measure the thickness of the pile, and divide by 10 (or more).

1.02 Measuring mass

Instructions: finding the mass
3 To weigh loose powders and similar substances, weigh a suitable empty container. Then put some of the powder in the container and weigh again. The weight of the powder can be calculated as follows:

$$\frac{\text{Mass of}}{\text{powder}} = \frac{\text{Mass of powder}}{\text{and container}} - \frac{\text{Mass of empty}}{\text{container}}$$

4 If you have many thumb-tacks *all alike*, count out 10 or more (as many as 100, if you have that many); put them together on a balance; weigh them. Divide the total mass by the number of tacks. This tells you the mass of a single tack.

1.12 Density

Instructions: measuring a property of iron
6 Whatever the size or shape of the piece of iron, the value of m/V is nearly the same for every piece. For cast iron the value is close to 7 000. Other types of iron have been differently prepared and have a higher value, between 7 800 and 7 900. To keep the calculations simple, we will assume a value of 7 000 in the discussions which follow.

Instructions: measuring the same property for other materials
3 Different materials have different values of m/V, but if there are several pieces made from the *same* material they all have the *same* value. The value of m/V does not depend on the size or shape of the piece but on the *material* of which it is made. m/V is a property *of the material*. Here are some values for common materials:

Material	m/V	Relative Density
aluminium	2 700	2.7
glass	2 400	2.4
lead	11 300	11.3
paraffin wax	900	0.9
stone	2 700	2.7 (more or less depending on type)
wood	700	0.7 (average: from 0.2 for balsa, up to 1.2 for ebony)

Questions page 7
1 0.07kg (70g).
2 1.806kg.
3 0.63kg.
4 $V = 1\,886\,\text{cm}^3$; $m = 1\,886 \times 0.4 = 754.4$g   or 0.7544kg.
5 $V = 33.5\,\text{cm}^3$; $m = 33.5 \times 1.14 = 38.19$g   or 0.03819kg.

Instructions: measuring the density of a liquid
Density of liquids: water 1 000kg/m³; glycerine 1 300kg/m³, oils 800-950kg/m³, depending on which oil used; methylated spirit 800kg/m³; turpentine 850kg/m³. You can see that most liquids have a density close to that of water. The liquid metal, mercury, has a very high density, 13 585kg/m³.

Instructions: measuring the density of air
2 No, it is full of air.
6 The answer depends on the air temperature and on your height above sea level. Your result is probably close to 1.2kg/m³, or possibly slightly less.

1.13 Units of pressure

Instructions: discovering the effect of pressure
6 They are all roughly equal to one another. It is not easy to judge when the card is exactly level with the foam, so you may have some figures that are not close. Taking an average over the whole class, you should find that W/A is the same for all.

Questions page 10
1 20kg.
2 1 030kg/m³.
3 2 000 000kg.
4 Mass of ethanol =152g; volume of castor oil =160cm³.
5 This depends on the size of the room. For a classroom or laboratory measuring 10m × 10m × 3m, and air density of 1.2kg/m³, the mass of air is 360kg.
6 0.48g (assuming density = 1.2kg/m³).
7 18 000kg.
8 1.95m³.
9 The exact answer depends on you. If your weight is 500N and your sole area is 0.025m², the pressure is $P = 500/0.025 = 20 000$ N/m²(Pa).
10 21 000 N/m²(Pa).
11 3 150 000 N/m²(Pa); a very high pressure, that can easily damage a floor made of wood or plastic.
12 Weight of block is 0.78N (a) $P = 195$ N/m²(Pa), (b) $P = 780$ N/m²(Pa) (c) $P = 1 950$ N/m²(Pa).

2 Using measurements

2.00 Levers

Instructions: balancing load and effort
4 For each trial, $xL = yE$, except for small errors of measurement. This rule is a very useful one. If you know *any three* of the quantities E, y, x, and L, you can always calculate the fourth one.

Instructions: a lever with a pivot at one end
2 For each trial xL is approximately equal to yE, but generally they are not *exactly* equal.
3 yE is usually slightly greater than xL. This is because the weight of the lever itself is adding to every reading of E. If the lever is light and the weights used are heavy, this error is not serious.
4 Force L.

Instructions: a third kind of lever
2 The rule still applies: $xL = yE$.
3 There is a small error due to the weight of the lever.
4 Force E.

2.10 Pendulums

Instructions: factors affecting the period of swing (T) of a pendulum
3 The longer the pendulum, the longer the period.
4 Period is not affected by amplitude, (provided amplitude is not large).
5 Period is not affected by the mass of the bob.

2.20 Pulleys

Instructions: using systems of pulleys
2 E is always less than L, in theory, at least. If the pulleys are heavy compared with the load, and if they have not been oiled and do not turn easily, extra effort is needed. For a given system, the greater the load the greater the effort. With light pulleys and cord we can forget about the effort needed to lift the pulleys and to turn the wheels. Then we can say that if we double the load we must double the effort; if we treble the load we must treble the effort, and so on. In short, we say E varies with L. Another way of saying this is: L/E is constant (see page 398).
When we pull on the cord, a force, tension, acts along the whole length of the cord. Fig. 34.1 shows this for system 4. The tension, T, equals the effort, E, with which we pull on the cord. The load, L is supported by *two* sections of the cord, with a force of T acting *in each* section. The total upward force equals the total downward force, so $L = 2T$. But $T = E$, so $L = 2E$. This means $L/E = 2$. You should have obtained a value close to 2 in the investigation. This is a *theoretical* value, since it does not allow for the extra effort required to lift the weight of the lower pulley and to turn both pulley wheels. In practice, the effort required is *more* than E, so L/E is a little *less* than 2. Systems 2 and 3 (fig. 2.6) have the lower pulley supported by two sections of the cord; $L/E = 2$ for these systems.

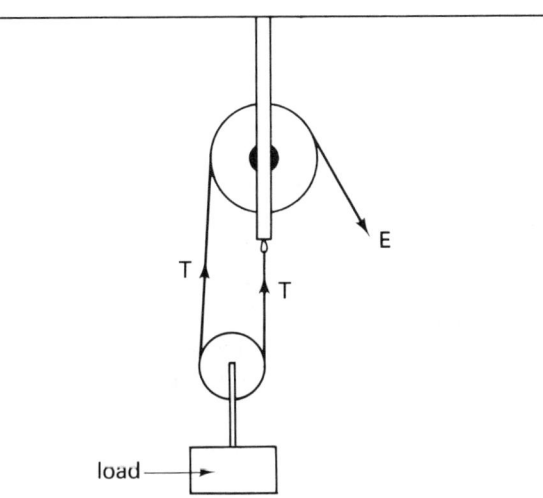

Fig. 34.1   Forces in a pulley system.

System 5 has three sections of cord supporting the lower pulley: $L = 3T = 3E$; $L/E = 3$. The same applies to system 6. In system 7 there are four sections of cord supporting the lower pulley and load, so $L/E = 4$. The rule is: count the number of sections of the cord supporting the lower pulley and load: L/E equals this number of sections (in theory, but it is slightly less in practice).

3 One possible design is in fig. 34.2. This is the best design with two double-pulley blocks. If you have a treble-pulley block, you can make a system in which the effort is downward. This is generally easier to use.

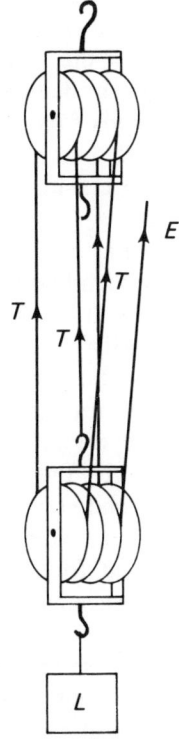

Fig. 34.2   A pulley system with $L/E = 5$.

2.30 Air pressure

Instructions: finding the direction of air pressure
4 The card and water remain in position. The atmospheric pressure pushing upwards on the card is more than enough to support the weight of the water and the card. The important point of this experiment is that it shows that atmospheric pressure acts upwards. The experiment with the can shows that it acts sideways too. Pressure in a fluid (liquid or gas) acts *equally* in *all* directions.

2.31 Measuring atmospheric pressure

Questions page 15
1 Multiply the total area of the can (not forgetting the bottom) in m² by 100 000. This gives the total force on the can, in N.
2 The can was open and atmospheric pressure acted inside it, pushing outward on the walls of the can with a force exactly equal to the inward force.
3 200 000N: the weight of a teak log 1m diameter and 30m long! Pressure acts downwards, sideways, and upwards, i.e. in all directions, so there is no overpowering downward force, to push you towards the floor. Your lungs contain air at atmospheric pressure (plus or minus a very small amount as you breathe in and out). This pressure balances the great load of atmospheric pressure on the outside of your chest. The body also contains blood and other fluids at a pressure slightly greater than atmospheric pressure. This produces an outward pressure, balancing atmospheric pressure on the skin.

2.32 Measuring gas pressure

Instructions: using a manometer
2 They come to the same level. The pressure on the surface of the water in the open tube is exactly the same as the pressure on the water in the tube connected to the flask. In other words, the pressure in the flask is atmospheric pressure.
3 The level rises in the open tube and sinks in the other. The pressure in the flask is greater than atmospheric pressure. The extra pressure forces the water down until it is balanced by atmospheric pressure and the extra weight of the taller column of water.
4 Measure to the lowest point on the water surface. (fig. 1.7).
5 The level sinks in the open tube and rises in the other. The pressure in the flask is less than atmospheric. Atmospheric pressure forces the water down until it is balanced by the pressure in the flask and the extra weight of the taller column of water.
6 Because the pressure due to the weight of this column of

335

water must be subtracted from atmospheric pressure to give us the pressure of the gas in the flask.

9 Changes of level in the manometer would have been very much smaller. The reason for this is explained in 13.00. We use mercury in a manometer when we wish to measure large pressure differences; for measuring small pressure differences water is used.

3 Measuring temperature

3.00 Simple thermometers

Questions page 19

1 The mark nearest the top of the card indicates the highest temperature. The mark nearest the bottom of the card indicates the lowest temperature.

2 The hottest place is the one in which the water expanded most. Its level is the one marked nearest the top of the card. The coolest place is the one marked nearest the bottom of the card.

Questions page 20

1 It is sensitive to much smaller changes in temperature. It responds much more quickly to temperature change.

2 If both are heated the same amount, air expands much more than water.

3.11 The fixed points

Questions page 21

1 22.5°C.

2 (a) 132 mm above the zero mark; (b) 22 mm below the zero mark.

3 $T = \dfrac{56 \times 100}{102} = 54.9 \ °C.$

3.30 Melting points

Questions page 22

1 Solution A, approximately –2°C; solution B, approximately –4°C.

2 When salt is dissolved in water the melting point of the salt-water-ice is decreased. The greater the amount of dissolved salt, the greater the decrease. Other investigations show that this effect is found with other substances dissolved in water.

3 If the ice contains dissolved impurities, its melting point is below 0°C; the zero mark on the thermometer would be wrongly placed.

3.40 Boiling points

Questions page 23

1 Solution A, approximately 102.3°C; solution B, approximately 104.6°C.

2 When sugar is dissolved in water the boiling point is raised. The greater the amount of dissolved sugar the greater the rise. Other experiments show the same effect for other substances dissolved in water.

3 The water is gradually boiled away and the concentration of the solution becomes higher. Since boiling point depends on the concentration of the solution, there is a gradual increase of boiling point.

4 If the bulb is *in* the water, and there are impurities dissolved in the water, the water boils at a temperature higher than 100°C. The 100°C mark on the thermometer would be wrongly placed.

4 From measurements to laws

4.10 Friction

Questions page 26

1 There is no friction. If there *was* a friction force, the book would move without being pushed!

2 Yes, there is friction to oppose that component of the force of gravity (or the weight of the book), which is acting towards the lower edge of the table top. Friction acts towards the upper edge and by being equal and opposite to this force, prevents the book from slipping.

3 Yes there is friction, but now it is not enough completely to oppose the gravitational force (weight of book).

4 Usually one with a shiny cover. Some plastic covers are shiny but may also be slightly sticky. This is a different kind of effect which we shall not deal with here.

5 Both slide equally easily.

6 You push backwards on the ground with your foot. This would make your foot slip backwards along the ground, (it *will* do if the ground is slippery). Friction helps your foot, or the sole of your shoe, grip the ground. You can push against the gripping foot and your body moves forward.

7 This is the reverse of what happens in question 6. You push forward on the ground, friction helps your foot grip the ground, and by pushing against this you can bring your body to a halt.

8 Friction between the case and the ground makes it harder to shift the case; you must supply the force needed to overcome friction. As you push, the friction between your feet and the ground stops your feet from sliding away backwards. If you have smooth-soled shoes you may not be able to push the case, for the friction between case and ground may be greater than you can have between your shoes and ground. Changing to rough-soled shoes might help make the job easier.

9 The wheels *roll* over the ground; they do not slide. The only friction now is in the bearings of the wheels. These are specially made to have very low friction.

10 You could use rollers (poles or logs) instead of wheels. You could reduce friction between the ground and case by greasing or wetting the ground, or by scattering loose dry sand on it.

11 They are more likely to burst suddenly, though this has nothing to do with friction. Tyres are made of rubber or some similar substance that does not slide easily on other surfaces. Tyres have a tread which is specially designed to have two effects:

(a) to grip the road surface firmly; high friction between tyre and road,

(b) to squeeze away any water that is on the road surface. A wet road is very slippery, low friction. A modern tyre has channels in the tread so that the water is thrown out sideways. The tread comes into full contact with the road surface with no water to reduce the friction.

12 Probably the bearings need oiling, or greasing. Oil and grease are often used in machines to hold the moving surfaces apart and so reduce friction.

13 Friction between the brake-blocks and the rim of the wheel is a force which makes the wheel turn more slowly and stop. Friction between the tyres and the ground is a force which stops the wheels from skidding (they will skid if you brake sharply when you are on a smooth dry dusty road). Since the wheels are turning more and more slowly and they do not skid, the bicycle must gradually slow down and stop.

14 It can be a nuisance; it can be a help. More about this in 4.12.

4.11 Measuring friction

Instructions: experiments on friction

2 As you add weight to the loop, friction increases from zero up to the maximum value, which you then record. The maximum value reached is not affected by the area of contact. Hypothesis (c) is true. Hypotheses (a) and (b) are false. The maximum value reached is greater if the force of contact is greater. Hypothesis (d) is true. *While you are adding weights*, friction equals the total weight added, so you could say that at that stage hypothesis (f) is true also. Hypothesis (e) is never true. Similarly (h) is true for the maximum force of friction; (i) is true for forces less than this, when we are loading the weights; (g) is never true.

Instructions: experiments on sliding friction

4 If only a few grams weight is removed, the block does not

stop sliding. *Once sliding has begun*, the block continues to slide, even when the weight in the loop is *less than* limiting friction. Dynamic friction is less than static friction. You may notice this effect if you are trying to climb a muddy slope. If you move steadily and push slowly back with your feet, you are not pushing more strongly than limiting static friction. Your feet do not slide. You are able to climb the slope. If you try to move quickly, the greater force exceeds limiting friction. Your feet begin to slide. Then you have only dynamic friction to help you and this is less than static friction. You are not able to grip the slope well enough to climb it. With an automobile on a slippery slope, low gear and slight acceleration are much more effective for making the vehicle begin to move. If you run the wheels at full speed, they spin uselessly, having only dynamic friction to help them grip the mud.

4.12 The laws of friction

Questions page 27

1 Examples are too many to be fully listed here but include soles of shoes, tyres, brakes, the clutch of an automobile, rubbing wood to produce fire, lighting matches, driving-belts, corks in bottles, nails in wood, rubber feet on furniture, non-slip floors. (There are many other examples from sports equipment.)

2 Special asbestos compound makes a high-friction, hard-wearing surface in automobile brakes; special rubber for brake-blocks of bicycles; hydraulic system (13.10), to exert great contact force in automobile brakes; special design of tyre surfaces; roughened surfaces on control knobs and levers to assist good grip (all kinds of machinery), rubber handle-grips on bicycles, etc.

3 Friction in all types of machinery causes waste of energy, production of unwanted heat, bearings and other parts to become worn so that they do not work properly. Much noise is produced, which is a nuisance.

4 If moving parts of machines are made as light as possible, contact forces are small and friction is less. Nylon gears are an example of this. Oil or other types of lubrication reduce friction between surfaces. When you are pushing a glass tube into a hole in a cork or rubber stopper, wet both the tube and the inside of the hole; water lubricates the surfaces and the tube goes into the hole much more easily.

4.21 Measuring upthrust

Instructions: investigating the weight of an object in water

4 Equation required is $\rho = m/V$ (1.12). Multiply V by density of water to get mass of water displaced. Since density of water is $1\,g/cm^3$, we multiply by 1. The *number* of cm³ of submerged can = the *number* of cm³ of water displaced = the *number* of grams of water displaced = the *number* of grams-force of upthrust.

6 Apparent loss in weight (or upthrust) is always approximately equal to the weight of the water displaced.

4.22 Archimedes' principle

Instructions: finding the relative density of an object

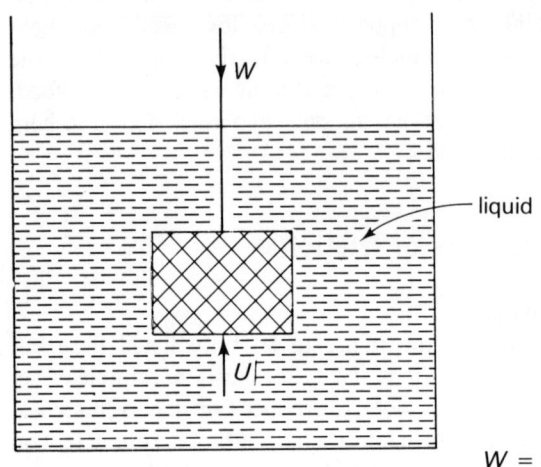

Fig. 34.3 Demonstrating Archimedes' principle.

W = weight
U = upthrust

Weight of solid
$W = m_1 g$ where m_1 = mass of the solid

$W = \rho_1 Vg$ where ρ_1 = density of the solid

Upthrust on solid
U = volume × density × g

$U = \rho_2 Vg$ where ρ_2 = density of the liquid

$$\frac{W}{U} = \frac{\rho_1 Vg}{\rho_2 Vg} = \frac{\rho_1}{\rho_2}$$

and if ρ_2 = density of water then

$$\frac{W}{U} = \frac{\rho_1}{\rho_2} = \text{relative density of the solid}$$

5 Check the result you obtain against a table of relative densities.

Instructions: finding the relative density of a liquid

Let U_1 = upthrust in liquid = $\rho_3 Vg$

U_2 = upthrust in water = $\rho_2 Vg$

$$\frac{U_1}{U_2} = \frac{\rho_3 Vg}{\rho_2 Vg} = \frac{\rho_3}{\rho_2} = \text{relative density of the liquid}$$

when ρ_2 = density of water

4.30 Floating

Questions page 29

1 Yes, this is true for all levels. Archimedes' principle still applies. The figures obtained should be very close to those obtained in the instructions on page 28.

2 Upthrust = W_1. This is a special case of Archimedes' principle.

Questions page 30

1 Volume of log = $100/700$ = $0.143\,m^3$. When log plus load equals 1 430N, the log is completely submerged. One boy sitting on the log will almost completely submerge it.

2 If 800N of wood is removed, this can be replaced by 800N of useful load, e.g., two boys. It would then float in the water at the same level as a solid log, unloaded.

Instructions: measuring the relative density of expanded polystyrene

5 $W = W_2 - W_1$.

6 Apparent loss = $W_2 - W_3$.

7 Same weight as apparent loss, $W_2 - W_3$.

8 The relative density is about 0.027.

4.31 Hydrometer

Instructions: making a simple hydrometer

3 A quantity of water having the same volume as this, weighs exactly the same as the whole rod (plus nail). The floating rod must displace a volume of water or oil which weighs as much as the rod (plus nail). The density of oil is less than the density of water. The volume of oil displaced must be greater than the volume of water displaced. The rod must sink further into the oil to displace a larger volume of it. The mark made when floating in oil is above the mark made when floating in water.

5 It is used for measuring the density of liquids.

4.32 Floating in air

Questions page 33

1 Weight of air displaced is $10 \times 0.000001 \times 1.2 \times 10$ = 0.00012 N. Upthrust = 0.00012N.

2 0.26988N.

3 Weigh it in a vacuum.

4 The upthrust is so small compared with the weight of the object that we can ignore it.

5 Weight of hydrogen = $1\,000 \times 0.09 \times 10$ = 900N. Total weight = 10 900N.

6 Upthrust = $1\,000 \times 1.2 \times 10$ = 12 000N.

7 The downward force of the balloon's weight is 10 900N; the upthrust is 12 000N. The upward force is greater than the downward force, so the balloon rises.

8 To balance upward and downward forces we need a load of $(12\,000 - 10\,900) = 1\,100$N. The balloon could carry a man and some equipment.

9 Usually bags of sand are carried in the basket of the balloon. To make the balloon rise, some of the sand is thrown out. This reduces total weight, making it *slightly* less than upthrust. The balloon rises slowly.

10 If some of the hydrogen gas is allowed to escape through a valve in the balloon the total volume of the balloon is reduced. Upthrust is reduced, making it *slightly* less than weight. The balloon sinks slowly.

11 Hydrogen is inflammable. If it is mixed in certain proportions with air, it is explosive. This has been known to cause several serious accidents in hydrogen-filled aircraft.

12 Helium can be used; it is not inflammable; it is more expensive than hydrogen; its density is twice that of hydrogen, so it cannot lift such great loads. Hot air is very cheap and available anywhere; at 120°C its density is $0.9\,\text{kg/m}^3$ so it cannot carry a heavy load. Using very thin balloon fabric, large passenger-carrying hot air balloons can travel for several hundred kilometres.

PART B Discovering Physics

5 Discovering about light

5.03 A pinhole camera

Questions page 37

1 The picture is upside-down (inverted).

2 Rays of light come from the scene, pass through the pinhole and hit the screen, making the picture there. Each ray of light travels in a straight line. Therefore rays from the top of the scene (sky, clouds, Sun, tree-tops) come down towards the pinhole and after passing through it continue downwards and hit the bottom of the screen (fig. 5.7). Rays from the lower part of the scene (grass, soil, people's feet) come in an upward direction, through the pinhole and travel on to hit the top of the screen. Similarly rays from the left of the scene hit the right of the screen, and rays from the right of the scene hit the left of the screen. The effect is to give an inverted picture.

3 From each part of the scene, light travels through the hole and makes a tiny spot of light on the screen. If the hole is larger the spots of light are larger. More light reaches the screen *but* the larger spots overlap with spots of light from nearby parts of the scene. The picture is no longer as sharp as it was with a small pinhole.

5.11 The laws of reflection

Instructions: investigating reflection in a plane mirror (1)

5 When seen from above, the incident ray and reflected ray (or the sections of the cord which represent these rays) appear to lie in a straight line. To put it more precisely, they are *in the same plane* (fig. 5.9).

Instructions: investigating reflection in a plane mirror (2)

7 Point C is on the mirror (line AB), since this is where reflection occurs. If it does not come on or very near to the

line AB, this is probably due to inaccurate sighting or drawing, or because the glass of your mirror is too thick. If necessary, repeat the instructions to see if both lines really meet on AB.

10 They are equal.

12 Angle of incidence equals angle of reflection. In short, $i = r$. This is another law of reflection.

Instructions: finding the position of the image

3 The rays are not parallel; they *diverge* (spread out) from the source, the filament.

5 They appear to diverge from a point that is behind the mirror.

9 Line OI is perpendicular to AB. In other words, O and I lie on the same normal. O and I lie at equal distances from the mirror. These are not new laws of reflection; they follow from the two laws already stated.

Instructions: finding the nature of the reflected image

5 The image is the same distance from the mirror as the object letter. All parts of it are the same distance from the mirror as corresponding parts of the object. The result is a left-to-right reversal of the letter, looking like this: Я. This is called *lateral inversion*. Mathematicians say that the letter has been transformed by a reflection about AB. Object and image are alike in being equal in size.

6 A, H, I, M, O, T, U, V, W, X, and Y. These are bilaterally symmetrical, so object and image look alike.

5.12 Using the laws of reflection

Instructions: locating reflections in two mirrors

3 The toy is usually called a *kaleidoscope*. Small pieces of material, brightly coloured and differently shaped, are scattered on a white surface. They are viewed from between two mirrors set at 60°. The pieces and their five reflections form a symmetrical pattern. Most kaleidoscopes rotate, so that the pieces continually change their arrangements and a never-ending series of new patterns is produced.

5.20 Refraction

5.21 The laws of refraction

Instructions: investigating refraction at an air-glass surface
4 They are parallel.
8 The greater i, the greater r. i is always bigger than r. There is no easy way of saying how much bigger i is. It is not always twice as big, or three times as big, or bigger by any constant amount such as '10° bigger than r'.
10 It is more-or-less constant, approximately 1.5; sin i varies with sin r.

Instructions: investigating refraction in other materials
2 Values for some common liquids are: water, 1.33; glycerine, 1.47; kerosine, 1.48; ethanol, 1.36.

5.22 The effects of refraction

Instructions: investigating refraction through a triangular glass prism
1 White.
2 The white streak of light is spread out sideways, and has a coloured edge. One edge is reddish orange and the other edge bluish violet. If you hold the prism and needle in exactly the right position you will see a complete set of colours, in order: red, orange, yellow, green, blue, indigo (a sort of bluish violet) and violet.
4 When the ray of white light strikes the nearest surface of the prism it is refracted. It passes through the prism and is refracted again as it leaves. The second refraction is in the *same direction* as the first (compare with the rectangular block, where the second refraction is in the opposite direction and returns the ray to its original direction). When light is refracted, lights of different colours are refracted by different amounts. Lights of different colours become separated out or scattered (*dispersed*) by a prism. 'White' light is a mixture of lights of different colours, they are dispersed by the prism, and leave it travelling in slightly different directions.
5 At the first surface, sin i is the same for all colours as they are all in the same ('white') beam. Red light shows the least refraction; r is greatest for this colour so sin r is greatest and ANG is least. Conversely ANG is greatest for violet light. When we quote the refractive index of a material we should state the colour of the light being refracted. Usually we do not state this, and it is assumed that we refer to refraction of yellow light (as from a sodium lamp, 26.12).

5.30 Colour

5.31 Coloured light

Instructions: investigating reflection of coloured light
3 (a) The paper appears to have the same colour as the light: (b) the paper appears to have the same colour as the light; it looks the same as when it is viewed in white light: (c) the paper appears black: (d) the paper appears to have its own colour, but darker, perhaps very dark, even black.
4 (a) White paper reflects light of any colour: (b) the coloured paper reflects all the light that shines on it, since this is of the same colour: (c) black paper does not reflect light, it absorbs all colours: (d) if the colours are not pure spectral colours (as is probable) the coloured light is a mixture that *may* contain a small amount of light that is reflected by the paper. If the light and paper are of fairly pure spectral colours, the light may contain no colours that the paper reflects; the paper absorbs all light falling on it, so appears black. This may happen, for example, when green paper has red light falling on it.

Instructions: mixing coloured lights
4 Red and green give yellow, red and blue give magenta, blue and green give cyan, red, blue and green give white.

6 Discovering about heat

6.10 Thermal expansion of liquids

Questions page 49
1 and **2** Acetone (most expansion), ethanol and methylated spirit, xylene, glycerine, water (least expansion).
3 For a given change of temperature, ethanol expands much more than mercury. Thermometers containing ethanol are more sensitive to small changes of temperature than thermometers which contain mercury.
4 It expands a lot when heated, but with a boiling point of only 56.5°C it is not of much use for putting in a thermometer.

6.11 The thermal expansion of water

Instructions: investigating the thermal expansion of water (1)
3 At first the water contracts as it cools; this is as expected. As the temperature falls to near freezing point, the water expands; the end of the column moves *up* the tube. The water continues to expand until it freezes.
4 As water freezes, it expands a lot, increasing its volume

by about 9%. This is a very unusual property. If you did not remove the thermometer from the freezing mixture, the water in the thermometer would expand as it became ice and would crack the glass.

Instructions: investigating the thermal expansion of water (2)

2 As soon as heating begins the water level goes down. After a few seconds it begins to rise and continues to rise as long as heating is continued. It appears as if the water contracted when it was first heated, but this is not true. If you repeat the procedure several times, always beginning with water at various temperatures above 4°C, this effect is always noted. It is not likely to be explained by saying that water contracts if it is at room temperature and is then heated. A better hypothesis is needed. You may have decided on this one: when the thermometer is heated, the glass bulb is the first part to become hotter. It expands; it becomes bigger; it can hold more water; water passes into the bulb to fill the extra space; the level in the tube falls. Soon afterwards the water begins to get hot too; it expands normally. The expansion of water is greater than the expansion of glass; the water level in the tube rises. This illustrates the fact that the expansion of solids is less than that of liquids when both are heated the same amount.

6.20 Thermal expansion of solids

Instructions: investigating the thermal expansion of metals

1 It does not pass through the ring after heating; it has expanded. Later it falls through the ring; the ring becomes heated by the hot ball; gradually the ring expands while the cooling ball contracts; when both reach the same temperature the ball can pass through the ring again. The change is reversible; it is a physical change.

2 When heated, the bar cannot be put into the slot or into the hole; it has expanded. The failure to fit the slot shows that it has expanded in *length*; the failure to fit the hole shows that it has expanded in *diameter*. Expansion takes place in *all directions*. When cool it fits again, showing that expansion is reversible; it is a physical change.

3 When the rod is heated, the pointer moves to the right, showing that the rod is expanding. When the rod cools, the pointer moves to the left and returns to its original position. This shows that the rod has contracted to its original size. The change is reversible: a physical change.

Instructions: heating the bimetallic strip

2 The strip becomes curved. When heated equal amounts brass expands more than iron. To allow for this, the strip must bend so that the brass is on the *outside* of the curve.

3 The metals return to their original size and the strip becomes straight again.

Questions page 52

1 If the glass is thin, heat is more easily transferred from the outside to the inside. There is less temperature difference, less difference in expansion and forces tending to break the glass are not as strong.

2 Even when the outside is very hot and the inside is very cold, the outside has not expanded much relative to the inside. Forces tending to break the glass are therefore not strong. Items made from borosilicate glass can be made very thick, yet will not crack. This means that they are more resistant to rough handling and less likely to be cracked by mechanical shock. In the home, oven-ware and table-ware can be made of this glass. We can have glass frying-pans which do not crack even when heated over a flame.

6.41 Radiation

Questions page 53

1 The black one.

2 The pad of newspaper or the cork mat prevents heat from passing from the surface to the can.

3 A dull black surface absorbs better than a white surface.

Questions page 53

1 The black one.

2 Newspaper pad or cork mat reduces conduction to the table.

3 Air heated by contact with the can would be blown away, taking some heat with it. The effect of radiation would be less easy to detect.

4 The lid on the can reduces loss by convection from the top surface. Some convection losses occur through the air coming in contact with the sides of the can. This cannot be avoided, but should be equal for both cans.

5 A dull black surface radiates better than a white one.

6 Good absorbing surfaces are also good radiating surfaces. This is a useful rule about heat radiation and absorption. The opposite applies: bad absorbing surfaces are also bad radiating surfaces. White and shiny metallic surfaces are bad at absorbing radiation; they *reflect* most of the radiation that reaches them. We can say they are bad absorbers and *good reflectors*. It follows that good reflectors are also bad radiators.

7 A shiny metal surface is a bad radiator. A teapot made from shiny metal loses little heat by radiation; the tea stays warm for longer.

8 A white surface is a bad absorber. When the Sun shines, most of the heat is reflected away from the building; little is absorbed into the walls. The building stays cool inside.

9 For the same reasons given above, white clothes are cooler to wear than black or darkly coloured clothes, outdoors during daytime, at least. At night, black clothes radiate heat to the surroundings, so are cooler, though possibly *too* cool for night wear.

6.42 Conduction

Questions page 54

1 All the metals are good conductors. Of these copper, aluminium and bronze are the best, in that order. Glass is much the worst.

2 Glass is a very bad conductor. When water is poured into the jar, the inside gets hot immediately and expands. Little heat is conducted to the outside of the jar; it stays cool and does not expand. Differences of expansion cause cracking.

3 Heat is conducted along the metal handle when the pan is used. This makes it too hot to hold comfortably.

Instructions: testing some good insulators

5 and 6 It is not possible to give a list which tells you the 'correct' order. A lot depends on the exact type of material used. Your own results tell you the correct order. Your own investigations will show how good insulators are used locally. Your investigations may indicate materials which are not widely used locally, but which might be better (and possibly cheaper) than the ones people normally use.

Instructions: finding out whether the thickness of the material affects conduction

2 A double layer gives better insulation. The thicker the layer, the better the insulation. With double layers of some of the materials, the wax may never melt.

3 When arranged as in fig. 6.8c, the glass shows the effects of thickness, giving better insulation with a double layer than with a single layer. With the arrangement as in fig. 6.8d, the drops on the double layer melt as quickly as they do with the first arrangement. The drops on the single layer take a very long time to melt; they may never melt. The explanation is that in the second arrangement, we have a gap. This greatly reduces the conduction of heat. The gap is not *empty*; it contains air. We conclude that air is a very good insulator.

Instructions: testing water as a conductor

6 Water is not a good conductor. Its ability to conduct heat is about the same as that of glass. This means that it is a worse conductor than metals, but a better conductor than most of the materials tested in the discussion on page 54. It appears that water conducts heat much better upwards than downwards. This is not possible. If you put tea-leaves or other small pieces of material in the water, you can see that when water is heated it rises. If you heat the bottom of the tube, water rises to the top, taking the heat with it. This is heat transfer by convection. If you heat

water that is at the top of the tube, it cannot rise further; there is no convection; only conduction can carry heat to the bottom of the tube and water is not a good conductor.

6.43 Convection

Questions page 57

1 Water in the pipe is cooler than water in the tank. It loses heat to the rooms through which the pipes pass, especially if there is a radiator. The cooled water sinks to the lower pipe and returns to the tank: hot water flows into the upper pipe and radiator to take its place. Just as convection carries heat from a hot place (the boiler) to a place that is not as hot as the boiler (the tank), it also carries heat from the tank (hot) to the radiator (not as hot as the tank).

2 This is not a good name. It does radiate heat, if you put your hand to one side of a radiator you can detect radiant heat coming from it, but it loses most of its heat by convection. A 'radiator' is made to have a large surface area so that it makes good contact with the air. It is shaped so that air currents can freely pass up past the surface of the 'radiator', convecting the heat away. A better name would be a 'convector'.

3 As temperature changes in different parts of the system, the pipes and the water expand and contract. If the system is completely sealed, high pressure might burst the pipes. The expansion pipe is open and the water level in it can rise or fall to allow for changes in the volume of the system.

4 Usually in the storage tank. The heating element is inside the tank, surrounded by water: an immersion heater. All the heat produced by the electric current is conducted from the heater into the water, and little is wasted.

5 The hot air and gases produced by the candle flame rise; they leave the box through opening A. This draws a current of cool air down through B to replace the rising air. If smouldering string is held over B, the convection current carries the smoke down through B, through the box, and out through A. If you place your hand over B, no fresh air can enter the box, what do you think will happen then?

6.44 Science and technology

Questions page 58

1 The vacuum (conduction impossible, for there is nothing there to conduct heat); the cork; the thin glass at the top of the neck is the place where the inner and outer glass containers are joined, as this connection is made of glass (bad conductor) little heat can be conducted across it.

2 The cork. The outer glass container is at room temperature so there is no convection from this surface. Convection is not possible in the vacuum, as there is no fluid to circulate.

3 The silvering on the glass. If hot food is in the flask, the silvering on the inner flask reflects radiant heat back into the food. Since this wall is in contact with the food, it gains heat by conduction. It does not radiate much, as silver is a bad radiator. Some heat is radiated across the vacuum (this is the only way that heat can travel across a vacuum) but is mostly reflected back again by the silvering on the outer glass container. The same kind of reasoning explains why heat cannot pass into the flask from outside when there is cold food in the flask.

4 Often there is a plastic stopper (bad conductor), usually hollow and filled with air (bad conductor).

7 Discovering about sound

7.00 Making a noise

Questions page 60

1 Something is moving, it shakes or vibrates very quickly. If you put your finger in the right place, you can easily feel the vibrations.

2 The vibrating part makes the air around it vibrate. The vibrations spread through the air to your ears.

3 Large vibrations make loud sounds. Small vibrations make soft sounds.

4 Rapid vibrations make high-pitch sounds. Slow vibrations make low-pitch sounds.

7.20 The transmission of sound

Questions page 62

1 You can *see* the hammer hitting the gong.

2 Sound is transmitted along the wires between the bell and the top of the jar.

3 Coiled wires (instead of straight wires); or put a battery inside the jar and hang the bell and battery from a rubber band.

4 The sound gradually increases in loudness and reaches its normal loudness when the air in the jar reaches normal pressure.

5 No. There is no atmosphere to transmit sounds from one astronaut to another. To talk to each other they use radio transmitters and receivers built into their space suits.

Instructions: investigating the reflection of sound

6 The angle of incidence equals the angle of reflection. A beam of sound spreads out more than a ray of light. The angles cannot be measured so accurately but they can be measured closely enough to show that the same law of reflection applies both to light and to sound. This experiment also shows that the other law of reflection is also true for sound; that the incident ray, the normal and the reflected ray all lie in the same plane (fig. 5.10).

7.30 The speed of sound

Questions page 64

1 The exact result depends on local conditions (see later). An average figure is $c = 346 \text{m/s}$ for dry air at 25°C.

2 From you to the wall the sound is carried along with the wind; it travels faster. On the return journey its speed is slower. The effects cancel out almost exactly, so can be ignored.

3 $c = 346 \text{m/s} = 1\,246 \text{km/h}$. Aeroplanes travelling at speeds greater than this are called 'supersonic'.

4 No. Loudness does not affect the speed of sound. If you listen to people singing in the distance, the loud notes and the sound of loud instruments such as drums do not reach you ahead of the soft ones. If they did, the notes would reach you out of order and the tune would sound wrong.

5 No. Pitch (or frequency) does not affect the speed of sound. The reason is similar to that given for question 4.

7.40 Resonance

Instructions: finding out the nature of resonance

3 The secret is to hit the bob in time with its natural period of swing. Each little hit has a very small effect but, if each hit is delivered at just the right time, the effects of one hit after another add up. Each hit helps the pendulum to swing a little further in the direction in which it is already swinging. The pendulum gradually swings with greater amplitude. The effect is just the same as if you are pushing someone on a swing. If you push in time with the period of the swing, the person swings higher and higher. If you push away when the swing is coming towards you, you make it swing less. It may stop swinging altogether.

9 The best rate is that which is exactly equal to f. When two systems have exactly the same period we say they are in *resonance*. Here, one system is the pendulum and the other system is you, pushing the pendulum with a paper strip. When two systems are in resonance, energy can be transferred from one system to the other, a little at a time, with very little waste. In this experiment, energy can be transferred if you hit the pendulum at rate $f/2$, but then you hit the pendulum only on alternate swings, so the rate of transfer of energy is halved; it takes twice as long to make the pendulum swing high. When the rate of hitting is not f or is irregular, there is no resonance and no increase of amplitude of the pendulum.

Instructions: investigating resonance

5 Their periods are nearly equal, so the swinging of A is able to transfer energy to all three lighter pendulums. They all *begin* to swing in time with A; but B is slightly faster and D is slightly slower and after about 10 or 20 swings they are out of step with A. Energy from A is acting to make them go in one direction but the pendulums are trying to swing in the opposite direction. We say they are *out of phase* with A. The result is that the amplitude of B and D decreases, and after a few more swings they both come to rest. Then they start to swing again, reach a maximum amplitude, get out of phase and come to rest again. Only C, which has exactly the same period as A, can resonate with A continuously. You may have found that after a long time even C comes to rest; this is because you have not made A and C exactly the same length. This is difficult to do, but a slight adjustment of the length of C could improve its resonance with A.

Instructions: investigating resonance with sound
4 They jump when the frequency of the sound you are making is exactly the same as the frequency of vibration of the air in the jar. The air resonates with your voice, the pieces gain energy from your voice and jump. When the frequencies are not the same, the air in the jar does not resonate with your voice and the pieces lie still.
5 A note of the same frequency: blowing makes the air in the jar vibrate at its natural frequency. This is the frequency at which it resonated in step 4.
6 Large jars contain greater volumes of air. They vibrate more slowly (2.10). They resonate at lower pitch.
7 They are equal.

Instructions: investigating resonance in strings
2 The string resonates with your voice. Energy is transferred to the string. After you have stopped singing, the string continues to vibrate for a second or so, giving out the note which you hear. This has the same frequency (pitch) as the note you sang.

8 Discovering about Magnets

8.00 Magnetic poles

Questions page 67
1 Yes, it always comes to rest in *exactly* the same direction.
2 This direction is roughly north-south (the reason why it is not *exactly* north-south is explained later).
3 Yes, one of its ends always points north (approximately) and the other always points south.
4 A metal stand attracts the magnet and might affect the direction in which it comes to rest.

Questions page 68
1 Usually a magnet is strongest at the ends of the bar, or at least within half a centimetre of the ends. Used or faulty magnets may have poles in other places. If you find a magnet like this, put it aside, as it is unsuitable for these experiments.
2 Mark N and S at or near the ends of the bar, just above the longest chains of nails.

Instructions: investigating the behaviour of magnetic poles
7 North *repels* north; north *attracts* south; south *attracts* north; south *repels* south. In short: like poles repel, unlike poles attract.

Questions page 69
1 The bar does not ever *repel* the magnet. Either end of the bar *attracts* either end of the magnet. In general, magnets are attracted by magnetic materials. Only the *like* poles of *two magnets* repel each other.
2 Bring one end of the bar up to the north pole of a magnet and then bring it up to the south pole. If it attracts both poles, it is unmagnetised. If it attracts one pole and repels the other, it is magnetised. Only repulsion is a true test of magnetisation.

8.10 Making magnets

Questions page 70
1 It becomes magnetised; small iron nails are attracted by it now.
2 The poles are at the ends of the bar.
3 The north pole is at the end further from the bar magnet (on the right of the figure). The south pole is at the end nearer the bar magnet (on the left in the figure).
4 When a magnet is close to a bar of magnetic material, such as iron, it *induces* magnetic poles in it. The pole of the magnet which is nearer to the bar induces an *unlike* pole in the region of the bar nearest to it. A like pole is induced at the opposite end.
5 If the bar is made of soft iron (such as a nail) the magnetism is not permanent. It disappears immediately the bar magnet is taken away. If the bar is made of steel, much of the magnetism disappears when the magnet is removed, but it remains weakly magnetised.
6 The bar magnet induces an unlike pole in the part of the bar nearest to it. Unlike poles are attracted. So the magnet and bar are attracted together. The closer the bar magnet goes to the bar, the stronger the induced pole and the greater the force of attraction between them.
7 If we hold one end of an unmagnetised bar close to the *north* pole of a hanging magnet, a *south* pole is induced in the bar and the magnet is *attracted* to it. If we hold the bar close to the *south* pole, a *north* pole is induced and again

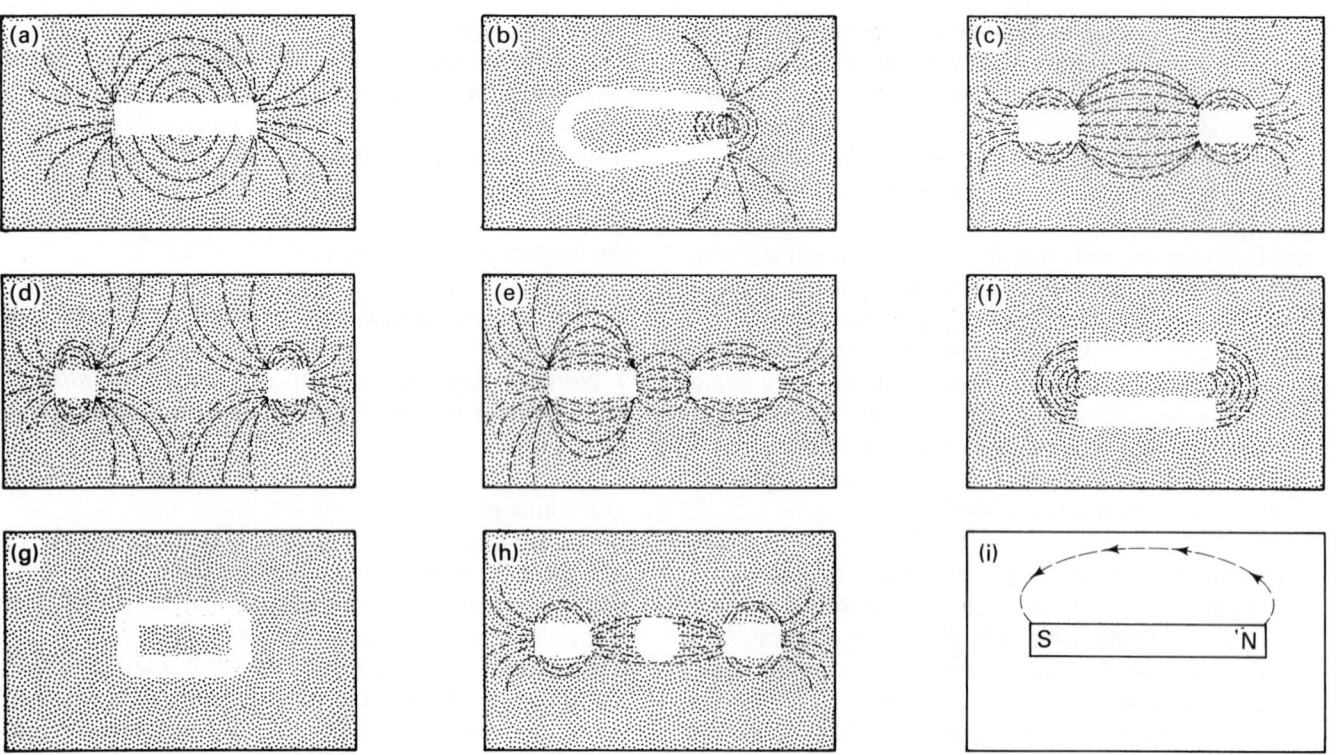

Fig. 34.4 *The magnetic fields obtained from the arrangements shown in Fig. 8.7.*

these are unlike and therefore *attracted*. Either way, we get *attraction*. Repulsion can only occur with *like* poles close together; this cannot happen with induced magnetism, but only when the bar is already magnetised by some other means.

8.20 Magnetic Fields

Instructions: investigating the magnetic fields of bar magnets and horseshoe magnets
Magnetic fields: your results probably look something like fig. 34.4.
These results are discussed in the main text.

8.30 Magnetic materials

Questions page 73
1 No, you cannot separate the poles of a magnet by cutting the magnet in two. The cut halves each have a north and south pole.
2 Every time you cut the magnet the pieces all have north and south poles and are perfect magnets. The south poles of each piece lie at the end which was nearer the south pole of the original magnet. As you cut smaller and smaller pieces you may find that their field is too weak for you to detect, but this does not mean that it does not exist.

Sensitive instruments can detect fields in very small pieces of metal. In the experiments with iron filings we made use of the fact that each tiny speck of iron becomes a magnet. So where does the cutting stop, how small can we go? This discussion is continued in the main text.

8.40 The Earth's magnetic field

Questions page 78
The fields are shown in fig. 34.5.
1 The field is strongest near the poles of the magnet, especially in a region around the north and the south of the poles.
2 The field is weakest to the east and west of the magnet at a short distance from the magnet.
3 The neutral points of the field are east and west of the magnet.
4 The field is strongest to the east and west of the magnet. It is strong very close to the poles, but rapidly becomes weaker as we go north or south. The neutral points are north and south of the magnet.
5 It would not point in any particular direction as there is no magnetic force acting on it.
6 The two fields would look exactly the same as the two you have already plotted but on a larger scale. The neutral points would be further from the magnet.
7 No. The situation shown in fig. 8.15 cannot happen, this diagram was drawn to remind you what the fields are

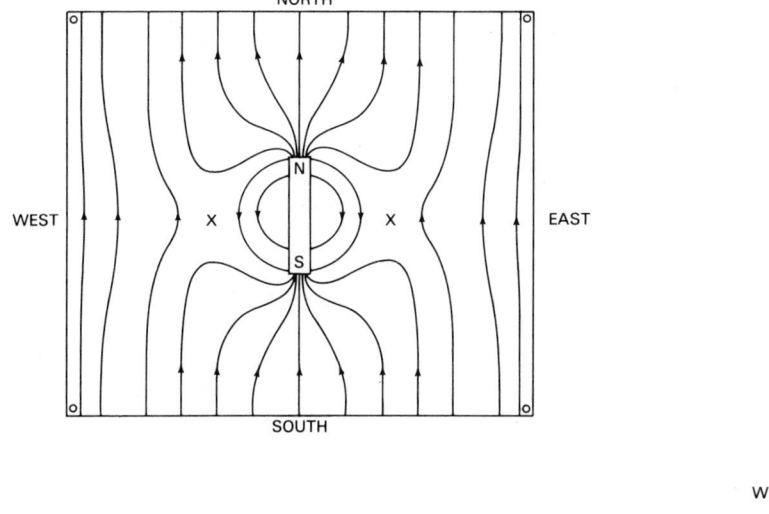

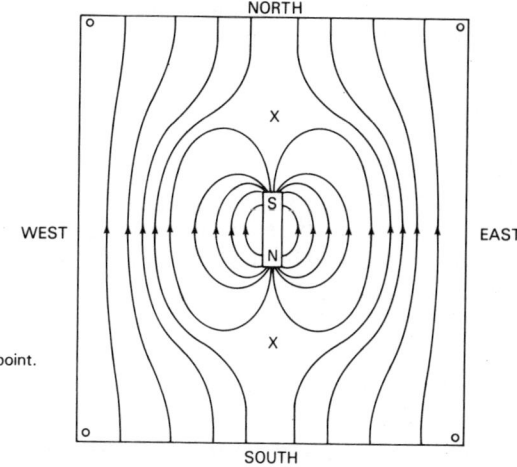

X   denotes a neutral point.

Fig. 34.5   The Earth's field in the region of a bar magnet. Neutral points are indicated by X.

like *separately*. If two fields act at one place there are two forces acting on the needle. These two forces are combined to give one force (chapter 11). This means one line of force which has a direction intermediate between those of the two original lines. If the forces are exactly equal and are in exactly opposite directions, they cancel exactly and the result is *no* force: a neutral point.

9   Discovering about Electricity

9.00   Electric cells

Questions page 80

1 The exact e.m.f. you obtain depends partly on how pure the electrode materials are and how clean they are. You may find that you need to scrape their surfaces with a file to make them bright and shiny. The cells with biggest e.m.f. are those with carbon as the positive electrode and with zinc or aluminium as the negative electrode.

2 Carbon.

3 If you ensure that the electrodes are clean and you use a high e.m.f. cell (carbon–zinc, for example), a lamp can be made to glow for a while, though it may not glow very brightly.

9.10   Conduction and insulators

Questions page 81

1 All of the *metals* used as electrodes and *carbon*. *Copper* is also used as the connecting wire and for most of the wiring inside the voltmeter. The terminals of the voltmeter may be made of *brass* (an alloy of copper and zinc), possibly coated with *nickel*. Copper wire is sometimes coated with *tin*. All these metals are electrical conductors and so is the *electrolyte*.

2 There are many insulators or non-conductors. Those which were part of your equipment for the experiment on cells include: glass, the plastic covering (insulation) on the wire and the plastic or wood casing of the voltmeter.

Questions page 82

1 The best conductors are the metals. Of substances which are not metals the only common good conductor is carbon. The 'lead' of a pencil is not actually lead but composed mainly of graphite, which is a form of carbon. This is therefore a good conductor. Many electrolytes are good conductors, but note that pure water is a very bad conductor; it is the dissolved substances that conduct the electricity, as we shall see in more detail later.

2 Plastics, glass, mica, paraffin wax and cotton thread are widely used as insulators (see discussion in main text).

3 Metals are good thermal conductors and good electrical conductors. The reason for this is discussed in 18.00 and 19.10.

9.21 The effect of a current in a straight wire

Questions page 83

1 Yes, the direction is reversed when the current is reversed.

2 Yes, the direction is reversed when the wire is changed from above to below.

3 The entries in the third column are: west, east, east, west. You may be able to make up a short easily-remembered sentence to help you remember this; you might even put it in verse! If not, return to the main text and study the rule there.

Question page 84

The results of Oersted's experiment give the clue to the answer. If the current is reversed the magnetic field is reversed. There will still be circular lines of force, but the direction of the force will be clockwise.

9.30 Using electromagnetism

Instructions: making and using an electromagnet

5 Its magnetic properties are demonstrated by the way it affects a compass needle. Its poles are determined by the direction of current flow, according to the rule given in fig. 9.8c.

6 The magnetic field is strong enough to move a freely suspended object such as a compass needle, but is not normally strong enough to lift even a light object such as a pin.

8 The iron core greatly increases the strength of the coil. The *coil* with its *core* makes up an *electromagnet*. It can now pick up several pins and other small iron or steel objects.

9 The objects fall from the electromagnet. Some may remain attached for they may already be magnetised or have become magnetised by the electromagnet. Usually these can be easily shaken off when the current is off, though they cannot be as easily shaken off when the current is flowing.

10 You could try using more batteries so that the electric current was greater. You could try winding more turns on the coil. Both of these methods give a stronger electromagnet.

Instructions: using the electromagnet

4 By using your finger to push the spring away from the contact.

7 The electromagnet attracts the thumb-tack on the spring. This pulls the spring away from the contact. The lamp goes out. When the button is released the spring returns to its original position and the lamp comes on again.

8 Put the contact on the other side of the spring so that when the spring is attracted towards the coil it touches the contact.

Instructions: making a relay operate itself

3 Before the current flows, the spring and contact are touching. As soon as the button is pressed, current flows through the coil; this attracts the spring, which is pulled away from the contact. Now the current does not flow, so the electromagnet loses its magnetism. The spring is not attracted and returns to its original position. It touches the contact, current flows again and the whole sequence is repeated. The switching on and off occurs many times a second, making the spring vibrate at a frequency of 100 Hz or more. This causes the buzzing sound.

Questions page 90

1 The effect is to make the two electromagnets into one, like a horseshoe magnet. Magnetic lines of force run straight from one core to the other through this piece of soft iron. They have only a very small air gap to cross to get to the armature. Since most of the path of the lines of force is in soft iron the strength of the field is much greater (your experiment of page 71 showed that the field in air is much weaker than a field in iron), so the bell works more strongly.

2 Soft iron becomes magnetised quickly when in a magnetic field and loses its magnetism quickly when the field is removed. In the relay, buzzer and electric bell, the spring must be able to return to its original position quickly when the current is switched off. If we used steel instead of soft iron, this would become permanently magnetised and remain attracted to the coils. Then the devices would not work.

3 Silver does not oxidize easily, so the contacts remain bright in spite of the sparking. The bell runs for many hours and the contacts do not become oxidized.

Part C   Investigating Forces

10 Motion and momentum

10.01 Speed

Question page 93

Until the end of the fourth second it moves with uniform speed, 5m/s. During the next three seconds it moves with uniform speed, 10m/s. During the next (8th) second it is stationary. During the final two seconds its speed is a uniform 2m/s. Its average speed for the first five seconds is 6m/s. For the whole journey the average speed is 5.4m/s.

Question page 94

Speed-time data are given in Table A1:

We are assuming that the object is able to change its speed instantaneously at the times where two speeds are given above. This makes the mathematics easy, but it does not happen in real life. The speed-time graph of these numbers has three rectangles, of areas 20, 30 and 4, giving a total area 54. The total distance is 54m.

10.04 Acceleration

Question page 97

Look at table A2 and table A3 which you will find at the bottom of the page.

This gives the graph plotted in fig. 10.8.

Table A1

Time taken t in s	0	1	2	3	4	5	6	7	8	9	10
Speed u in m/s	0 and 5	5	5	5	5 and 10	10	10	10 and 0	0 and 2	2	2 and 0

Table A2

Period (seconds)	0 to 1	1 to 2	2 to 3	3 to 4	4 to 5	5 to 6	6 to 7	7 to 8	8 to 9	9 to 10
Displacement s in m	2.5	7.5	12.5	17.5	20	20	18.75	16.25	13.75	11.25

If we add these cumulatively, we get:

Table A3

Time t in s	0	1	2	3	4	5	6	7	8	9	10
Displacement s in m	0	2.5	10	22.5	40	60	80	98.75	115	128.75	140

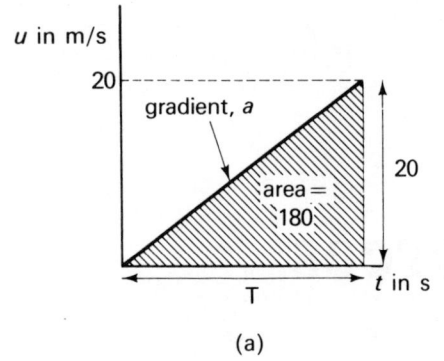

(a)

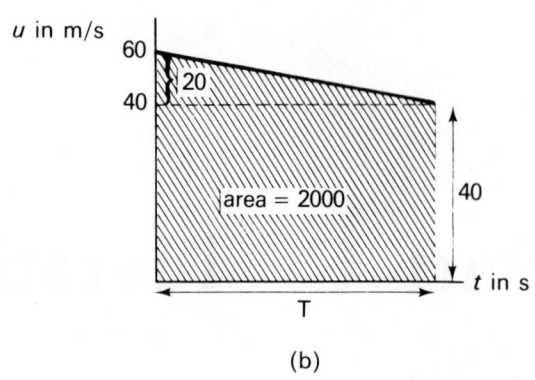

(b)

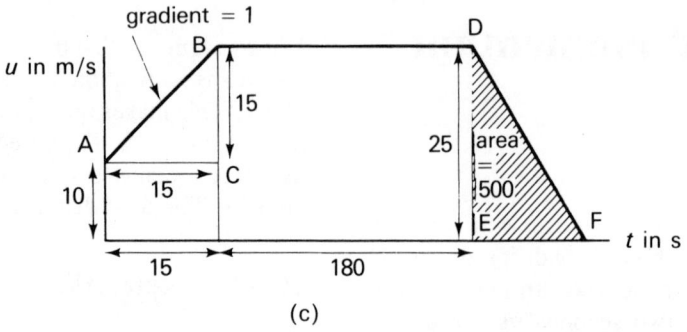

(c)

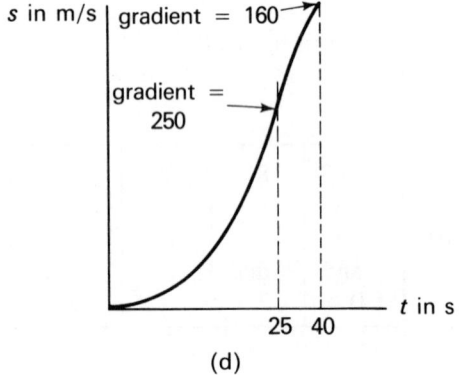

(d)

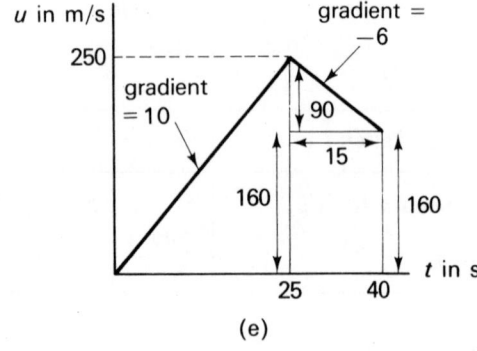

(e)

Fig. 34.6 Sketch-graphs for working the answers to problems on motion.

Questions page 98

1 Draw a *simple* sketch, like fig. 34.6a. It need not be drawn to scale; it is just to help you think things out. The sketch is a velocity-time graph and we make use of the rule that the area under the graph tells us the displacement. We know the height and the area of the triangular area under the line. For a triangle, area $= \frac{1}{2} \times$ base $\times$ height.

So, $180 = \frac{1}{2} \times$ base $\times$ 20.
This gives base $= 180 \times 2 \times \frac{1}{20} = 18$.
This tells us that the time is 18 s. Next we use the rule that gradient tells us the acceleration.

The gradient of the line is $20/18 = 1.11$.
Acceleration is 1.11m/s^2.

2 Draw a simple sketch, not to scale (fig. 34.5 b).

Total area $=$ area of rectangle $+$ area of triangle
$2\,000 = (T \times 40) + \frac{1}{2}(T \times 20)$
$\qquad = 40T \times 10T = 50T$
$\qquad T = 2\,000/50 = 40$

Time for deceleration, T, is 40 s.
Gradient is $-20/40 = -0.5$.
Deceleration is 0.5m/s^2.

350

3 A sketch of the displacement-time graph is given in fig. 34.5c. Acceleration from A to B is 1 m/s^2, so the gradient is 1, and $BC = AC = 15$. Velocity increases by a total 15m/s, reaching a maximum velocity of $10+15 = 25\text{m/s}$. The area of triangle DEF is 500, its height is $DE = 25$, so its base $= \dfrac{2 \times 500}{25} = 40$. Time for deceleration is 40s.

Total area under curve $= (10 \times 15)+(\frac{1}{2} \times 15 \times 15)$ $+(180 \times 25) \times 500 = 5\,262.5$
Displacement from sign to gate is 5 262.5m, or 5.2625km. Total time of journey $= 15 + 180 + 40 = 235$s. Average velocity $= 5\,262.5/235 = 22.39$m/s.

4 The displacement-time graph is an irregular line. The gradient varies, so velocity is not uniform. The gradient is greatest between the first and second posts when the butterfly flies 1m in 2.5s, a velocity of 0.4m/s. The line zig-zags, showing that velocity increases and decreases several times; therefore acceleration is not uniform during the journey.

5 The displacement-time graph (fig. 34.5d) had gradient increasing from zero up to about 25s. At that point the curve is steepest and straightest. Cut-out was at 25s. A tangent drawn at 25s has gradient 250, so velocity is 250m/s. The gradient at 40s is 160, so final velocity is 160m/s. The velocity-time curve (fig. 34.5e) has gradient 250/25 for the period of motor drive, acceleration 10m/s. After cut-out, the gradient is –90/15 indicating deceleration of 6m/s^2.

10.05 Equations of motion

Instructions: investigating the motion of a trolley
5 Divide its length (in m) by the time taken (0.2s).
6 Plot a velocity-time graph; measure its gradient.
7 The acceleration is greater on a steeper slope.
9 Friction in the bearings; friction in the ticker-timer; air resistance.

Instructions: investigating uniform motion of a trolley
2 The force of gravity acts to make the trolley accelerate down the slope; the frictional forces act to decelerate the trolley as it runs down the slope. With the correct slope these forces are equal and opposite; they cancel; there is neither acceleration nor deceleration; the trolley moves with uniform velocity, the velocity depending on how hard it is pushed to start it moving.

10.11 Mass

Instructions: finding the velocity of a trolley when pushed
6 Exactly twice.
12 They are more or less equal.

Questions page 102
1 About 2/3 of the velocity of the single trolley.
2 Place an empty container for the rice on the trolley. Place a 1kg mass inside this container; it would be best to use the international prototype kilogram but it would be very unlikely that you would be allowed to have this, so you would use an accurate copy of this instead. Perform steps 3 to 5 of the instructions and calculate the velocity of the trolley. Next remove the 1kg mass and put rice in the container. Repeat, adding or subtracting rice each time until the velocity of the trolley is the same as when it had the 1kg mass on it. Then you know that the quantity of rice in the container has a mass of exactly 1kg.
3 The apparatus is used in exactly the same way. The behaviour of the trolley depends only on mass, not on weight. Whether you are on Earth, on the Moon or anywhere else in the Universe, the mass of the trolley, the 1kg mass and the rice are all unaltered, and the same quantity of rice will be measured by this method.

10.12 Momentum

Questions page 102
1 0.28kg m/s south. Did you remember the direction?
2 Momentum $= 600$kg m/s east. Momentum of truck is 500kg m/s east. The momentum of the bicycle and rider is greater than that of the truck.
3 Its velocity increases by a total of 40m/s west. Its momentum increases by $100\,000 \times 40 = 4\,000\,000$kg m/s west.
4 Use a sketch of velocity-time graph, or the alternative displacement-time equation to find that the velocity of the ball is 8m/s south, to begin with. It then has momentum 0.8kg m/s and loses all of this while rolling.
5 It has zero velocity.

Instructions: finding the effects of different amounts of pushing force
5 The stronger the push, the greater the velocity, and therefore the greater the momentum given to the trolley.

Instructions: investigating impulse and acceleration
4 It should be, if you are able to keep the end of the cord exactly level with the front of the trolley. Then the force on the trolley is constant, producing uniform acceleration.
5 This is approximately double the acceleration obtained before.
6 Acceleration varies with the number of cords used; or, acceleration varies with the force.

Instructions: accelerating different masses with constant force
4 Doubling the mass halves the acceleration; trebling the mass gives one-third the acceleration. This confirms the equation.

10.20 Force

Questions page 105

1 Gain in momentum = 4 000 000kgm/s west. Impulse required = 4 000 000kgm/s west, or changing units, 4 000 000Ns west. Force acted for 20s, so its size is 4 000 000/20 = 200 000N west. This impulse was delivered to the train by the friction between the track and the wheels of the train, acting west, to stop the wheels from slipping along the track. The force could also be written as 200kN (kilonewton) west. You could also have calculated this by using the equation $F = ma = 100 000 \times 2 = 200 000$N west.

2 The ball lost momentum equal to 0.8kg m/s south. Impulse required = 0.8kg m/s *north*. Force = 0.08N north.

3 Momentum gained = 3 000kg m/s upwards. Impulse = 3 000Ns upwards. Force of motor = 600N.

4 Vector directions are all along the track in the direction of motion of the model. Mass of model = 0.05kg. Momentum gained = 0.05 × 5 = 0.25kg m/s. Impulse = 0.25Ns. Time = impulse/force = 0.25/2 = 0.125s.

5 Momentum lost = 0.25kg m/s, assuming no loss through friction or air resistance. The whole of this is lost when the model is brought to rest in the sand. Impulse of sand on model is the impulse required to oppose the whole momentum, 0.25Ns in the opposite direction to the velocity of the model. The impulse on the sand is 0.25Ns in the direction of the velocity of the model. We do not know the time of the impulse as it is too short to measure easily; therefore we cannot calculate the force.

10.21 Newton's laws of motion

Instructions: investigating collisions between two trolleys

4 Trolley B has zero momentum, for its starting velocity is zero. The total momentum before collision equals the total momentum after collision.

Instructions: investigating elastic collisions

3 Trolley B has zero momentum to begin with. The total momentum before collision equals the total momentum after collision.

Questions page 106

1 Multiply the increase of velocity of B by its mass (in trolley units); this gives the impulse action on B in the same direction as the velocity of A and B.

2 Multiply the decrease in velocity of A by its mass (in trolley units); this gives the impulse acting on A in the opposite direction to the velocity of A and B.

3 For each trial, the impulse of B on A is exactly equal in size to the impulse of A on B, but acts in an exactly opposite direction.

Questions page 107

1 To bring the ball to rest your hands must apply an impulse equal to the ball's momentum, but oppositely directed. If the time of impulse is short, the force must be great; if the time is long, the force is less. If you swing your hands away as the ball lands in them, you make the time of impulse longer, so the force your hands must exert on the ball, and the force the ball exerts on your hands, is much reduced. It does not hurt as much.

2 If you let your knees bend, you lengthen the time of impulse. This makes the force of impact less.

3 A heavy head has greater momentum; when it hits the nail there is more momentum to be lost, producing a greater impulse on the nail. With a firm support, the head of the hammer is brought to rest in a fraction of a second, so the force is very great, and is more effective in driving the nail into the wood.

4 When the truck slowly towed the automobile back on to the road, the rate of increase of momentum was relatively small. The force required (= tension in the tow-rope) was below that which the rope could stand. When the truck was accelerated rapidly a greater force was required to increase the momentum of the automobile rapidly. This was more than the rope could withstand, so it broke.

5 To begin with, the total momentum of the girl and rock is zero. If momentum is conserved, the total must remain zero. When the rock is thrown its momentum is 6kg m/s forwards. To conserve momentum the girl's momentum must become −6kg m/s forwards, that is to say, 6kg m/s *backwards*. This means that she moves backwards with velocity 0.1 m/s.

6 The initial momentum of the ball is 1.4kg m/s in its original direction. Its final momentum is 1.68kg m/s in the opposite direction, equivalent to −1.68kg m/s in the original direction. Total change of momentum is (−1.68 − 1.4)kg m/s which equals −3.08kg m/s in the original direction. Therefore the impulse of the racket on the ball is 3.08Ns in the opposite direction. The impulse of the ball on the racket is 3.08Ns in the original direction.

7 Total momentum before firing is zero. Momentum of bullet after firing is 0.6kg m/s forwards. Pistol has momentum −0.6kg m/s forwards, equivalent to 0.6kg m/s backwards. Initial velocity of recoil is 0.4m/s backwards. Note how small this is compared with the velocity of the bullet. Use a graph or the velocity-distance equation to find that, given $v = 0.4$m/s backwards, $u = 0$m/s, $s = 0.01$m, gives $a = 20$m/s² backwards. $F = ma = 1.5 \times -20 = -30$N backwards. The force required to check recoil is 30N forwards.

Instructions: exploding trolleys

6 Momentum is not created by the explosion. The total momentum is zero throughout the investigation.

Instructions: investigating recoil

4 The trolley is accelerated in the opposite direction to that in which the object is thrown.

7 Use a less massive object and a more massive trolley. Use a more massive object and a less massive trolley. In both cases, use a stronger elastic in the catapult.

11 Forces in action

11.01 The addition of vectors

Questions page 110

1 7.65 m 44° north of east.

2 You cannot find the answer by simply halving the size of the direction of the answer above. Draw a completely new diagram. Answer: 12.1 m 14.5° north of east.

3 64 m/s 9.5° north of west.

Instructions: finding resultant force

13 Part of the tension in the strings is due to the weight of the dynamometers. From step 11 onwards you have an extra dynamometer and its extra weight will affect the tensions slightly. This is why it is suggested that you should work with tensions of at least 10N.

Questions page 112

1 195N 5° to the boy's side of the 'forward' direction.

2 The donkey can resist a force of 195N, so the children cannot move it. The donkey does not move.

3 The resultant force of the children is 195N forwards; the frictional force of the ground on the donkey's hooves is 190N backwards. The resultant of these is 5N forward. The donkey slides slowly forwards.

4 They could reduce the angles between their ropes and the forward direction. If they would both pull almost straight ahead, their resultant would be almost 220N forwards, which is more than enough to move the donkey.

Questions page 112

1 First draw a *scale diagram* of the string between its supports, with the object hanging from the string, as in fig. 11.4a. This gives you the angles (fig. 34.7a). Measure these. Draw the triangle of forces (fig. 34.7b) to scale. Side AB could be 10cm long, representing 50N, the weight of the object. We know all the angles and have to

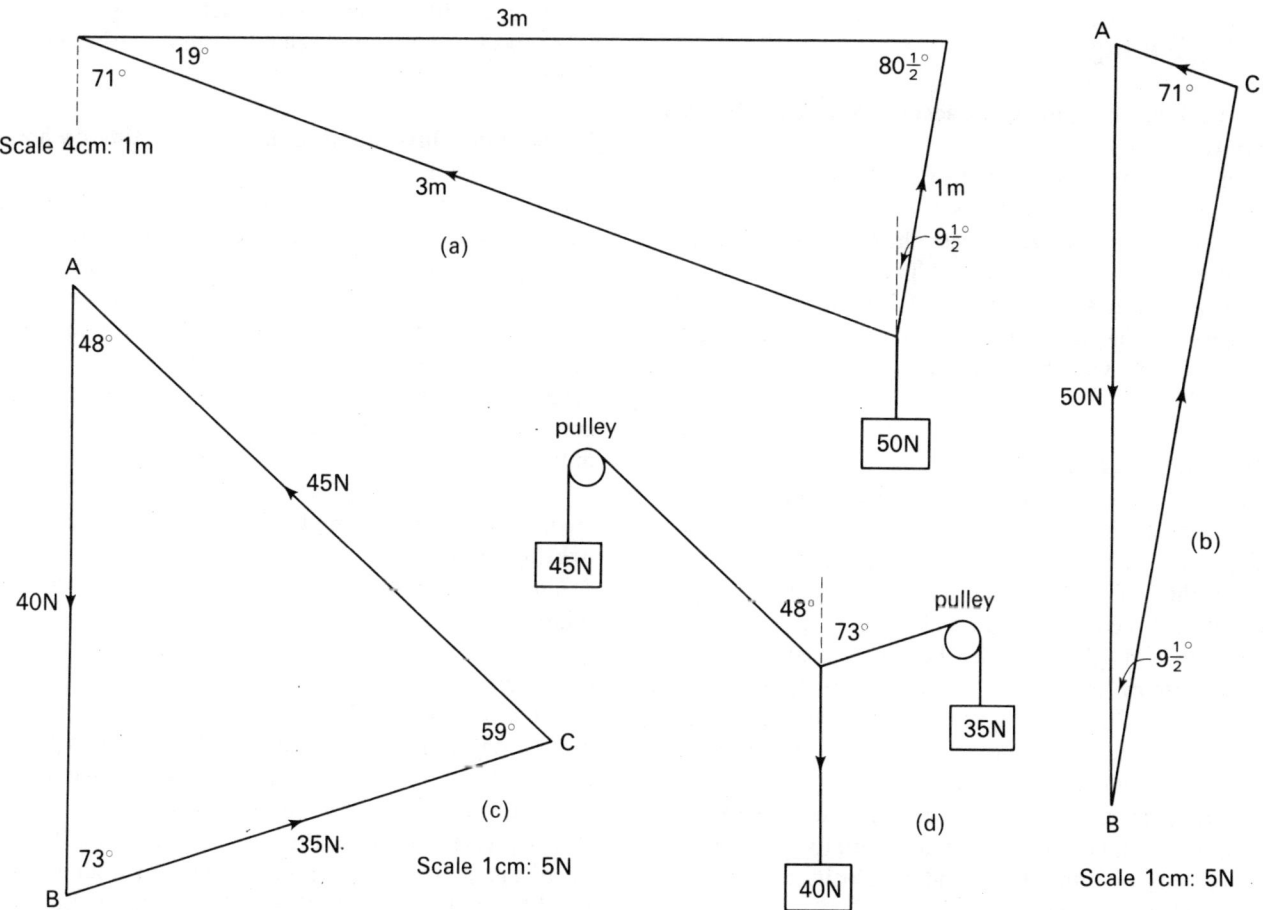

Fig. 34.7 Triangles of forces: solutions to problems.

353

find the lengths of the other two sides. Draw a line from B at an angle of 9.5° to AB. Draw a line from A at an angle of 71°. The lines meet at C. Mark in the arrows and check that they 'go round' the triangle in the same direction. Measure BC and CA. BC is 9.58 cm long, representing a force of 47.9 N in the short string. CA is 1.68 cm long, representing a force of 8.4 N in the long string.

2 The forces are in equilibrium, so we may construct a triangle of forces. Draw line AB vertically (on the page), to represent the downward force of the object weighing 40 N (fig. 34.7 c). We know the lengths of the three sides and have to find the three angles. With compasses set to radii corresponding to 35 N and 45 N, find the position of point C. Draw the lines AC and BC. Draw the arrows and check that they 'run round' the triangle. Measure the angles. Construct a diagram using these angles, to show how the strings are arranged. The angles depend only on the sizes of the forces, they do not depend on the positions of the pulleys. If the pulleys are closer together, or change their relative levels, it makes no difference to the angles where the strings meet. If you cannot see why this should be, set up some weights and pulleys and check that it really does happen.

11.12 Falling

Instructions: measuring the acceleration due to Earth's gravity

5 The acceleration is 9.8m/s^2. Measurement error, especially the effect of friction in the ticker-timer will probably make your result less than this. In what way might you improve your result? Try it.

6 Acceleration is the same for all falling bodies, no matter what weight they have or of what material they are made. If you use a heavier object the error due to friction is less in proportion.

Questions page 115

1 In answering this question and others in this group you can use sketches of velocity-time graphs, or the equations of motion (10.05): $u = 20 \text{m/s}$ upwards; $v = 0 \text{m/s}$ (when the ball has reached maximum height and is just reversing direction to begin its fall); $a = g = -10 \text{m/s}^2$ *upwards*. This gives $t = 2 \text{s}$; height reached, $s = 20 \text{m}$. On the return journey, $u = 0 \text{m/s}$, $a = 10 \text{m/s}$ downwards, $s = 20 \text{m}$; this gives $t = 2 \text{s}$. The velocity on returning to the girl's hand is $v = -20 \text{m/s}$. Note that this kind of result is general for objects thrown vertically upwards (ignoring air resistance): (a) time taken to reach maximum height = time taken to return to ground again; (b) velocity on return to ground = velocity at launching, but in opposite direction.

2 $u = 0 \text{m/s}$; $a = g = 10 \text{m/s}^2$ downwards; $t = 5 \text{s}$. Therefore $s = 125 \text{m}$.

3 Velocity on reaching ground is $v = 50 \text{m/s}$. Momentum is 5 kg m/s. Impulse required to bring it to rest is 5 Ns upwards. Its impulse on the ground is 5 Ns downwards.

4 For the coconut, $u = 0 \text{m/s}$; $a = 10 \text{m/s}^2$ downwards, $s = 24.2 \text{m}$. Time to reach ground is $t = 2.2 \text{s}$. For the boy, $u = 0 \text{m/s}$, $a = 4 \text{m/s}^2$ away from the tree, $t = 2.2 \text{s}$. So $s = 9.68 \text{m}$ away from the tree.

5 The horizontal velocity of the ball has no effect on the downward acceleration of the ball due to gravity. The ball takes exactly the same time to reach the ground as if it were simply dropped. In the downward direction, $u = 0 \text{m/s}$, $a = g = 10 \text{m/s}^2$; $s = 1.8 \text{m}$. Time to reach ground $= 0.6 \text{s}$.

6 Horizontal velocity of ball is 20 m/s east and is uniform; $t = 0.6 \text{s}$, so $s = 12 \text{m}$ east of the boy.

11.20 Orbits

Questions page 116

1 All discs hit the floor at the same time.
2 The disc from the notch furthest from the pivot.
3 The one with the greatest horizontal velocity.
4 The disc from the notch nearest the pivot.
5 The disc with the least horizontal velocity.
6 Increase the horizontal velocity with which it leaves the barrel of the gun.

Instructions: investigating the forces acting on an orbiting object

2 An outward force, pulling on your finger. Your finger pulling inwards to prevent the cork and elastic flying off. An outward force on the elastic stretches it, so that the cork moves in a larger circle as you increase its speed.

3 When you let go the cork flies off in a straight line (almost, at first, but it becomes curved as the cork accelerates downwards under gravity). As you let go, the cork is no longer held in its circular path by your finger pulling on the elastic.

Instructions: measuring the forces acting on an orbiting object

4 Circumference of orbit $= 2 \pi r$. If there are n orbits per minute, distance covered by stopper is $2 \pi r n$. Speed is $2 \pi r n / 60 = \pi r n / 30 \text{m/s}$.

6 The greater the speed, the greater the centripetal force. It needs a greater force to alter the direction of a fast-moving object. In fact the centripetal force varies with the square of the speed (see later) but this equipment is not accurate enough to show the relation exactly.

7 The greater the radius, the lesser the centripetal force. It needs a greater force to make an object turn around a sharp bend (a circle of small radius). In fact the force varies inversely as radius (see table at back of the book).

8 $F = ma = mv^2/r$. Therefore $a = v^2/r$. Note that the *acceleration* depends only on speed and radius, NOT on the mass of the object.

Questions page 119

1 The satellite exerts an equal but opposite gravitational pull on the Earth. In fact the satellite does not simply circle round the Earth. They *both* circle around a point which lies somewhere between them. But the Earth is so much more massive than any artificial satellite that this effect is not measurable. With the Moon and Earth, the effect can be detected. As the Moon goes round the Earth it produces regular 'wobbles' in the Earth's orbit around the Sun. Many stars actually consist of two stars of roughly equal mass, we call these binary stars. The members of a binary pair orbit around each other, like partners dancing a waltz.

2 The water stays in the bucket because the centripetal force exerted by the base of the bucket on the water is great enough to keep the water in circular motion. This centripetal force is the reaction to the thrust of the water on the base of the bucket.

3 It would travel in a curved path, but less curved than when it is fired at 8 055 m/s. Therefore it would leave the Earth and travel off into Space. The velocity above which an object is able to escape from the gravitational attraction of the Earth (or Moon or other planet) is known as the *escape velocity*. For Earth, the escape velocity is approximately 8 055 m/s.

4 No. The second equation in instructions page 118, step 8 does not contain *m*. (See page 354)

11.30 Satellites

Questions page 120

1 At 400 000 000 m from Earth, $g = 9.8/60^2 = 0.0027$ m/s².

$$T = 2\pi\sqrt{\frac{400\,000\,000}{0.0027}} = 2\pi \times \sqrt{148\,148\,000\,000}$$
$$= 2\pi \times 384\,900 = 2\,418\,000\text{s}.$$

This comes to almost exactly 28 days.

2 $T = 2\pi\sqrt{\dfrac{23\,460\,000}{0.078}} = 2\pi\sqrt{300\,000\,000}$
$$= 2\pi \times 17\,320 = 108\,810\text{s}.$$

This comes to 38.2 h, compared with the observed period, 30.3 h.

3 For a period of 24 h, $T = 86\,400$ s.

$$86\,400 = 2\pi\sqrt{\frac{r}{0.225}} \quad \text{or} \quad 86\,400^2 = \frac{4\pi^2 r}{0.225}$$

$$7\,464\,960\,000 = 175.5\,r$$

$$r = \frac{7\,464\,960\,000}{175.5} = 42\,500\,000 \text{ (approx.)}$$

This is the radius in metres. The radius is 42 500 km, giving a height above Earth's surface of about 36 000 km.

4 $r = 7\,270$ km. Gravitational attraction is not as strong at this distance and must be calculated. Radius of orbit is 1.14 times Earth radius; $g = 7.5$ m/s². $T = 2\pi\sqrt{7\,270\,000/7.5} = 103$ min. Collision with small particles in space and air resistance (still appreciable at this altitude) gradually reduces velocity, though the effect is extremely small at first. As v is reduced, r must be reduced, to keep F equal to the weight of the satellite. The radius of orbit decreases. This process continues, and, as the satellite orbits closer to Earth, it meets increasing air resistance. The deceleration increases, and r decreases more and more rapidly. Eventually friction between the satellite and atmosphere generates so much heat that, unless special precautions are taken, the satellite melts and 'burns up' as it enters the denser regions of the atmosphere on its last few orbits.

12 Turning forces

12.00 Moments

Questions page 121

1 0.8 Nm anti-clockwise (this can be called a *positive* moment, thus 0.8 Nm).

2 0.8 Nm clockwise (this can be called a *negative* moment, thus – 0.8 Nm).

3 The sum is zero. In adding moments we total all the anti-clockwise moments by simple addition; then we total all the clockwise moments. The difference between these two totals is the sum of all the moments.

4 1.8 Nm anti-clockwise, or + 1.8 Nm.

5 1.2 Nm anti-clockwise, or + 1.2 Nm.

6 3 Nm anti-clockwise, or + 3 Nm.

7 It would turn in an anti-clockwise direction, because of the moments of the two forces.

8 + 0.6 Nm and –1.2 Nm. Sum = –0.6 Nm.

9 It would turn in a clockwise direction because the sum of moments is negative.

10 When a lever is in equilibrium, the sum of moments about the pivot is zero.

11 The upward force at the pivot is not shown in the diagram. Ignoring the weight of the lever, the force at P is 6 N upwards. The total upward force equals the total downward force and cancels it. There is thus no resultant force, either upwards or downwards, so the lever does not accelerate upwards or downwards as a whole.

12 Moments about A: *anti-clockwise*: 1.8 and 1.2, total 3 Nm; *clockwise* 3 Nm (the moment of the force on the pivot). Total moment about A is zero.

Moments about B: *anti-clockwise* 0.6 and 0.6 (force of pivot), total 1.2Nm *clockwise* 1.2Nm. Total moment about B is zero. If you take moments about *any* point on the lever the total moment is zero, providing you take all forces into account, including the force of the pivot. Since total moment is zero, the lever does not turn.

Questions page 122

1 Calculate moments about P. From left to right these are −1.8, + 0.4, + 1.4. Total moment about P is zero. The lever is in equilibrium.

2 Size of resultant upward force must be 16N. It must act so that its moment about P is the same as the total moment of the forces it replaces. Thus its total moment must be −1.8 + 1.4 = − 0.4. Therefore its distance from P must be 0.4/16 = 0.025. Since it must be clockwise, it must act at a point 0.025m to the left of P.

3 This must be sufficient to balance the total of upward and downward forces, including the weight. We have 2N + 14N upwards, and 1N downwards. Total of these is 15N upwards. Force of pivot on beam must be 15N downwards.

4 Total downward force must be 16N. Its moment about P must be equal to the moment of the two forces it replaces. Their moment about P is 0.4Nm (the force *through* P has no moment about P). The distance of the resultant from P must be 0.4/16 = 0.025m. Since it must be anti-clockwise it must act through a point to the left of P.

5 They are equal in size but opposite in direction. Their moments about P are equal in size but opposite in direction. They cancel each other and the lever is in equilibrium.

6 Total upward force is 10N; total downward force is 10N; these are equal; the lever as a whole neither rises nor falls.

7 For convenience, take moments about X, the left-hand end of the lever. The moment of A is zero, the moment of B is 3Nm. The resultant force must be 10N with moment 3Nm about X. The distance from X must be 3/10 = 0.3m.

8 For convenience, take moments about Y. Moment of C is zero, moment of W is 0.5Nm. Resultant force must be 10N with moment 0.5Nm about Y. Distance from Y must be 0.5/10 = 0.05m.

9 They are equal in size and opposite in direction, but they do not act along the same line. The left-hand end of the lever will rise, and the right-hand end will fall; the lever will rotate, in a clockwise direction.

Instructions: investigating forces on a lever in equilibrium

5 They are equal in size, opposite in direction.
7 They are equal in size, opposite in direction.
9 They are equal in size, opposite in direction.

12.10 Centre of gravity

Instructions: investigations with hanging shapes

7 They all meet at one point, or at least in a very small area of the card. If the card is a regular shape they meet on one or more of the axes of symmetry. If the card is L-shaped, for example, the meeting-point is generally not on the card.

8 The card balances when the finger-tip is just below the meeting-point of the lines. If the meeting-point is not on the card, it is not possible to balance the card horizontally on a finger-tip. If you make a wire support which ends at the position in space where the lines would meet, and if you tape this wire to the card, you can balance this structure by placing your finger under the end of the wire.

Instructions: making a microbalance

9 Weigh several sheets of graph paper. Find their total area; calculate the weight of graph paper in grams per square millimetre. Calculate the weight of a 'square' of area 4mm².

12.20 Equilibrium

Questions page 125

1 Stable: resting vertically on its square-cut end; this is not really very stable unless the pencil is a short fat one. Neutral: lying on its side. Unstable: balanced on its sharpened point.

2 Neutral. If the sphere is rolled to another position, it will remain there.

3 Unstable. If the marble is displaced a little towards the rim of the saucer, it will roll down the side and off the saucer.

4 Stable. If the bob is displaced to one side, it swings back again and eventually comes to rest vertically below the point of support.

13 Force and pressure

13.00 Fluid pressure

Instructions: finding the direction of pressure in a liquid

2 As far as you can tell, the pressure is equal in all directions.

3 As they emerge from the holes, the jets are perpendicular to the surface. This suggests that the pressure acts perpendicular to the surface.

13.10 Pascal's principle

Instructions: transmitting pressure by water
4 As one piston is pushed in, the other moves out.
6 The result depends very much on the area of the two pistons and also on frictional forces between pistons and barrels. The weight required is much less than 20N.
7 A little extra weight will start the 20N weight accelerating upwards.
8 The answer depends on the relative diameters of the syringes. If 1cm^3 of water is forced out of B, the same quantity must enter A. The distance apart of the graduation marks tells you what answer to expect.

Questions page 130
1 Pressure $= 100/0.01 = 10\,000$ Pa (or N/m^2).
2 Force $P = 4 \times 10\,000 = 40\,000$N. This is 400 times the applied force, illustrating the high increase of force that is possible with hydraulic systems.

13.20 Units for measuring atmospheric pressure

Questions page 132
1 Difference of water levels $= 0.05$m.
Difference of pressure $= \rho gh = 1\,000 \times 10 \times 0.05 = 500$Pa (or N/m^2)
Pressure inside $= 96\,000 + 500 = 96\,500$Pa (or N/m^2)
2 Difference in pressure $= 50\,000$Pa. For water,

$$h = \frac{50\,000}{1\,000 \times 10} = 5\text{m}.$$

For mercury, $h = \dfrac{50\,000}{13\,600 \times 10} = 0.37$m,

or 370mm. With pressures of this size it is much more convenient to use mercury as the manometer liquid.
3 $h = \dfrac{100\,000}{1\,000 \times 10} = 10$m.
This is too long a tube to be practicable.

13.30 Archimedes' principle

Instructions: making a Cartesian diver
5 It sinks, wait 10-20s to give it time to accelerate downwards.

6 It rises.
7 Use your own words, but include terms such as pressure; density; Pascal's principle; Archimedes' principle; upthrust; weight of water displaced.

14 Investigating electro-magnetic force

Instructions: demonstrating electromagnetic force
6 Current in same direction in each strip: they move towards each other. Current in opposite directions: they move apart.

Instructions: finding the relation between force and current
10 Make a heavier rider and begin again, placing the rider nearer the needle. Alternatively, make a coil with fewer turns.
11 If the rider has mass m and is (at step 5) distance x from the needle, its original moment is mxg. If it is shifted by y from this position, its moment becomes $m(x+y)g$. The increase in its moment is myg. If you measure the mass of the rider and the distance y, you can calculate the actual value of the moment in Nm. The direction of this moment is clockwise (as seen in the figure). Since the beam is again balanced, the extra moment due to electromagnetic force is myg anti-clockwise. Since mg is constant, the electromagnetic force varies with y.
12 The current is less (approximately half); the electromagnetic force is less.
13 The current is less than when one or two lamps were in circuit. The electromagnetic force is less too.
15 The greater the current, the greater the electromagnetic force. The electromagnetic force varies with the current.

Questions page 137
1 $t = 10\,800$s;
$Q = It = 0.4 \times 10\,800 = 4\,320$C.
2 $t = 97\,200$s;
$Q = It = 0.06 \times 97\,200 = 5\,832$C.
3 $t = 300$s;
$I = 0.01$A;
$Q = It = 3$C.

Part D   Investigating Energy

15 Work, energy and power

15.00 Work

Questions page 138

1 2.5J.

2 2 100J or 2.1kJ. Note that since the man's weight acts vertically downwards, he does work against this force in climbing the ladder and we need only take the vertical distance into account in the calculation.

3 The ball is accelerated to 2m/s in 0.5s, so the acceleration is 4m/s^2. $F = ma = 0.1 \times 4 = 0.4$N. Distance travelled by the ball before it leaves his hand is 0.5m. So work done is 0.2J.

4 Acceleration is 10m/s^2, so force, $F = 1$N. Since distance travelled by the ball before it leaves his hand is now only 0.2m, work done is 0.2J. This is exactly the same result as before. The time taken to throw the ball makes no difference to the amount of work done. You did not need to be told the time in order to calculate the work done, though it is easier to handle the equations if you use a definite time instead of just using '*t*'.

15.01 Work done by machines

Questions page 139

1 Velocity ratio is a *ratio* between two distances moved in the same time. Therefore it is a *number* only, with no units. The same applies to mechanical advantage.

2 V.R. = 4.

3 The curve slopes up steeply from zero-load, zero M.A. For small loads, M.A. is much less than 4, owing to the effort required to lift the lower pulley block and to rotate the pulley wheels; none of this counts as *useful* work done. As load increases, this non-useful work becomes less in proportion to useful work, so M.A. gradually increases. For heavy loads, M.A. almost reaches 4.

4 No, the slope of the M.A.-load graph becomes less and less as load increases. M.A. rises towards, but never quite reaches, its maximum theoretical value 4.

5 Efficiency is another ratio, so has no units. The ratio is multiplied by the number 100, so converting it to a percentage, but the result is still a number.

6 The efficiency-load curve has the same shape as the M.A.-load curve, since with constant V.R., efficiency is proportional to M.A. At low loads, efficiency is low. At high loads, efficiency approaches 100%.

7 No. The useful work done by a machine can never exceed the work put into the machine.

Questions page 140

1 Engineer's vice, automobile jack and some types of crowbar.

2 Using the machine makes it just possible to lift or move a load that it would be impossible to move without the machine. Such machines are usually very cheap. For example, the inclined plane may just be an ordinary plank of wood borrowed for the purpose.

3 The answer depends on how well the machines are constructed and maintained and how well they are lubricated. Machines with no moving parts, such as a screwdriver are almost 100% efficient, since there is no friction to overcome and the effort needed to rotate the screwdriver is very small compared with the great effort needed to rotate the screw. Of the more complicated machines, a well-kept bicycle has high efficiency. High efficiency in complex machines requires light-weight construction and special low-friction bearings; these add to expense.

4 You can change the velocity ratio of the machine. With heavy load, such as when climbing a hill, or riding into a head-wind, a low velocity ratio and correspondingly high mechanical advantage means that the rider can produce the effort needed to propel the bicycle, though it moves forward at low speed. On a level road, in top gear, the machine has a high velocity ratio. A relatively small effort is required to propel the machine at high speed.

15.11 Kinetic energy

Questions page 141
1 $v = 90 \text{km/h} = 25 \text{m/s}$. $m = 1\,000\,000 \text{kg}$.
$E_k = \frac{1}{2} mv^2 = 312\,500\,000 \text{J} = 312.5 \text{MJ}$.
2 Final velocity $= 10 \text{m/s}$. $E_k = 100 \text{J}$.

15.12 Potential energy

Questions page 142
1 The moving bob and string give motion to the air around them. This is what we usually call 'air resistance', but we can think of it as the extra kinetic energy of the air particles. Friction at the bearings of the pendulum generate heat, so some of the pendulum's mechanical energy is converted to thermal energy.
2 Its velocity is greatest when its E_k is greatest, which is when its E_p is least; at the bottom of its swing, position B in fig. 15.3.
3 If the weight of the bob is 10N, its mass is 1kg. Use the equation $E_k = \frac{1}{2} mv^2$.
$v^2 = 2E_k/m$. E_k at the bottom of the swing $= E_p$ at the top of swing $= 0.1 \times 10 \text{J} = 1 \text{J}$. So $v^2 = 2/1 = 2$. $v = \sqrt{2} = 1.414 \text{m/s}$, horizontally to right or left. (Velocity is a vector; did you remember to state its direction?)

15.20 Power

Questions page 145
1 From velocity-time graph or the appropriate equation (10.05) we find $a = 0.4 \text{m/s}$.
2 Force $= ma = 800 \text{N}$.
3 At 36km/h ($= 10 \text{m/s}$) the point of application of the force of 200N moves 10m in 1 second. Work done $=$ force $\times$ distance $= 200 \times 10 = 2\,000 \text{J}$. This takes 1 second, so rate of working is 2 000 joules per second. In other words, power, $P = 2\,000 \text{W} = 2 \text{kW}$.
4 Weight of box $= 100\,000 \text{N}$. Work done $= 100\,000 \times 5 = 500\,000 \text{J}$ or 500kJ. Thus $E_p = 500 \text{kJ}$. 500kJ in 200s $= 2\,500 \text{J/s}$. Power of men $= 2\,500 \text{W}$ or 2.5kW. 500kJ in 600s $= 833 \text{J/s}$. Power of men $= 833 \text{W}$. If the box falls, all its E_p is converted to E_k, by the time it reaches the ground, so $E_k = 500 \text{kJ}$.
5 Work $=$ force $\times$ distance $= 450 \times 20\,000 = 900\,000 \text{J}$, or 9MJ (megajoules, millions of joules). Note that the *work* done depends only on force and distance; the velocity of the locomotive does not matter. For calculating *power* we must always take time into account: in 1 second the train travels 5m, so the work done *in 1 second* is $450 \times 5 = 2\,250 \text{J}$. Rate of working $=$ power $= 2\,250 \text{J/s}$ or 2 250W or 2.250kW.

15.21 Electrical power

Questions page 146
1 0.06A means 0.06C per second.
2 6V means 6J per coulomb, so for 0.06C the amount of energy converted is $0.06 \times 6 = 0.36 \text{J}$.
3 0.36J/s means 0.36W.

Questions page 146
1 $P = 0.7 \times 6 = 4.2 \text{W}$.
2 $I = P/V = 3\,000/240 = 12.5 \text{A}$.

16 Using energy

16.10 Conservation of energy

Questions page 147
1 Momentum before and after impact are very nearly equal. This conforms to the law of conservation of momentum. Since there is no external horizontal force which can increase or decrease momentum (except for friction and the effects of the tape-timer which are small), momentum cannot be changed.
2 The kinetic energy after collision is less than that before collision. Some kinetic energy has been lost. It appears that the law of conservation of energy has been disobeyed. The only explanation is that the kinetic energy has been *converted* to one or more of the other types of energy. A little was converted into the sound vibrations which you heard when the trolleys collided. Most will have been converted to internal energy. It requires very precise experiments to trace where the 'lost' energy goes; but it *can* all be accounted for, and the law *is* obeyed.

16.20 Conversion of energy

Questions page 148
1 Chemical energy in muscle of hand, converted to kinetic energy of muscle, which is transferred to bones of fingers and to the match. Kinetic energy of match rubbing on rough edge of matchbox is mainly converted by friction to internal energy of match-head and matchbox surface. Increase in internal energy causes chemical action to begin, converting chemical energy to light and heat. Heat is transferred to wood, increasing its internal energy. This starts another chemical action and chemical energy of the wood, which burns, converting its chemical energy to light and heat. The chemical energy of the wood originally came from the Sun, where it was created as nuclear energy. It crossed Space as radiant energy (light), was absorbed by

the leaves of a tree and there converted into chemical energy as part of the material of which the tree was made. Similarly the chemical energy of the muscle came from food, which came from plants or other animals, and so originated in the Sun.

2 The Sun's radiant energy causes water to evaporate from seas, lakes, vegetation and soil. It collects high in the sky, acquiring potential energy. Then the rain falls, converting potential energy into kinetic energy. The rain collects into rivers and would continue to flow down, losing all its kinetic energy, but we make it pass through turbines, converting a lot of its kinetic energy into mechanical kinetic energy of the turbines. These are connected to generators which convert it to electrical energy. The electrical energy can be distributed by wires to houses, offices and other work places where it can be used; it can be converted to light, to heat or to mechanical energy. All this energy comes from the Sun.

Instructions: some experiments on energy conversion
Discussion of energy conversions.

1 In a trial experiment, with a bob, mass 100g, the vertical height was 0.02m to begin with and 0.003m after 6min. These correspond to E_p starting at 0.02J and reduced to 0.003J after 6min. A loss of 0.000047J/s; energy is lost at the rate of 0.000047W, or 0.047mW. Losses are due to energy transferred to air particles around the bob and to friction in the cord at the point of support, this can all be considered as internal energy.

2 Figures you obtain depend so much on the type of motor used. The power output will generally be quite a small fraction of the power input. Much of the energy is lost as friction in the bearings which convert it to internal energy in the moving parts of the motor. The remainder is lost as internal energy in the wires due to the resistance they offer to electricity passing through them.

3 When the weight is used to generate electricity, we find once again that output is considerably less than input; perhaps only 10% or so of input. Again there are frictional losses and losses due to electrical resistance. Again, most of the lost energy is converted to internal energy.

4 Toys have roughly-made bearings and are usually not lubricated. If gears are used in the mechanisms, they are inaccurately shaped. For these reasons friction is high and much energy is lost as internal energy. In a model plane the rotating propellor creates a large disturbance in the air around. Much of this energy makes no contribution to driving the plane forward. Air resistance is another factor which diverts energy from its useful purpose. Also, a fraction of the force produced by the propellor is redirected by the wings of the aeroplane to produce lift, a downward force which keeps the plane in the air. For these reasons and as a result of friction in the bearings of the propellor, much of the energy stored in the wound elastic is not available to drive the plane forward. Its forward distance

is much less than you would expect from your calculation of the stored energy.

5 For an average person, the total for 24 hours comes to between 8 500kJ and 10 500kJ which means that power is between 0.1 and 0.12kW (around 100W, the power rating of an electric lamp burning day and night!) For vigorous activities, such as running upstairs, the rating is 2 000kJ/h, or just over 0.5kW. This high rate of energy production cannot be maintained for long periods.

6 A single thermocouple produces very little power, though we can join several together and get a useful output (29.20). A great amount of the energy from the candle is wasted because it cannot all be concentrated on the thermopile. Efficiency is very low indeed. In hammering, the energy conversions are: chemical energy in muscle *to* kinetic energy of muscle, arm and hammer *to* internal energy of thermopile, *to* electrical energy in circuit, *to* kinetic energy in pointers of meters.

7 The more the blades are turned at an angle to their direction of motion, the more air must be moved and the greater the resistance to the rotation of the mill. The weight falls more slowly.

8 There are no moving parts here, so the energy conversion can be reasonably efficient : 10% or more depending on the conditions of operation, and the exact type of cell used.

16.30 Conversion of other forms of energy into internal energy

Instructions: measuring the conversion of mechanical energy into internal energy

4 Distance fallen is l; calculate $v^2 = 2gl$ (10.05). Calculate $E_k = \frac{1}{2} mv^2$. The kinetic energy is converted into internal energy in the pellets and in the stopper and lower end of the tube.

5 Quick working cuts down loss of heat from the pellets. Total energy is 100 times that calculated at step 4. Call this U.

6 Temperature rise, $\theta = \theta_2 - \theta_1$.

7 Total internal energy gained is U. We assume that *all* the kinetic energy becomes internal energy of the pellets; thus this experiment has large errors.

8 U/θ = heat capacity (the amount of internal energy required to raise the temperature of *this particular mass* of lead pellets by 1 degree).

9 For 1kg of lead, the amount of energy required is 130J.

16.31 Specific heat capacity

Instructions: measuring specific heat capacity by the conversion of electrical energy into internal energy
7 $P = IV$.

8 Total energy converted, $U = Pt$ (t measured in seconds).

9 Specific heat capacity, $c = U/m\theta$, but since $m = 1\,kg$ in this experiment, $c = U/\theta$. For aluminium, $c = 900$; for brass, $c = 380$; for copper, $c = 400$; for iron, $c = 460$; all values of c in J/kg K.

10 The heater itself, the insulating material, the wires leading to the heater. In spite of all precautions some heat is lost to the air during heating. There may also be inaccuracies owing to the block not being at uniform temperature in all parts.

Instructions: measuring the specific heat capacity of a liquid

2 Amount of heat required to raise its temperature 1 K is its heat capacity, $q = m_1 c$.

3 $m = m_2 - m_1$.

9/10 See the corresponding steps in the previous instructions.

11 The amount going to the liquid is $U_1 = U - q$.

12 $c = U_1/\theta m$. For water, $c = 4\,200$; for ethanol, $c = 2\,300$; for glycerol, $c = 2\,400$; for turpentine, $c = 1\,800$; all values of c in J/kg K.

Part E Investigating matter

17 The structure of matter

17.01 Evidence for molecules

Questions page 156
1 The particles move slightly in a very irregular zig-zagging way. To see this effect best you must not try to watch several particles, concentrate on one particle and watch it for several minutes. If all the particles seem to be streaming in the *same* direction, this is caused by a water-current across the slide and is *not* Brownian movement. Probably you have tilted the microscope or are pressing on the slide. With Brownian movement, the particles are all moving in different directions and their motion is very jerky.
2 The water consists of molecules of water, all moving around at high speed. When these molecules hit a carbon particle, they give it momentum. If the particles are much larger than the carbon particles you have on the slide, they move very little, for their mass is so much larger than that of water molecules. Also, with larger particles it is most likely that they will be hit by several water molecules at once, on all sides, with the result that there will be no gain of momentum in any particular direction. The carbon particles are small enough to be accelerated when hit by water molecules and are likely to travel a visible distance before being hit by another molecule and made to change direction. Though we cannot see the water molecules we can see their effect on the carbon particles. The particles are pushed this way and that by the continuous bombardment which they receive from the water molecules. This experiment is evidence that the water consists of molecules and that these molecules are moving rapidly.
3 This is similar to that shown by the carbon particles in water, but is much more vigorous. The reason is similar to that given in the answer to question 2: the smoke particles are being bombarded by molecules of the fluid in which they are suspended i.e. air.

4 Both experiments show the same effect but the effect in smoke is clearer. Possibly this is because the smoke particles are lighter than the carbon particles from ink and because the molecules of air move faster.

17.02 The size of molecules

Questions page 157
1 The result comes to about 1 nm (if you do not remember what this means, refer to the beginning of Part E).
2 The spread out drops are not often circular, so there are errors in measuring the diameter. We are assuming that all drops are almost equal in volume. We are assuming that the molecules spread out to form a single layer. We are ignoring the fact that part of the volume of the film is accounted for by the space *between* the molecules. In spite of these possible errors, the result shows that molecules are very small. You have measured the length of the molecule. Its thickness is much smaller. These molecules are larger in size than most molecules.

17.11 Solids

Questions page 159
1 The wire recovers its previous lengths at each stage of reduction of load.
2 Extension varies with the load. The extension-load graph is a straight line at this stage. If the elastic limit is not exceeded, the wire has *elasticity*, the ability to recover its original shape when the load is removed.
3 The load pulls the molecules further apart than their normal distance; this causes the attractive force between molecules to increase, tending to pull the molecules back to their original distance; the effect is an increase in tension in the wire, just balancing the load.
5 The force due to load is now great enough to overcome the attractive forces and there is slipping between the layers in the crystal lattices and also between crystals.

Now, a relatively small increase in load can cause a large extension. If the load is removed at this stage, the wire does not return to its original length. As load is reduced the curve follows a line parallel to the original line (dashed line in fig. 17.7) and indicates a permanent stretching when load is reduced to zero.

6 The very slightest addition of load causes a big extension; the extension continues for several seconds, even if the load is slightly reduced. At this stage the wire is becoming very narrow at certain points; this means that the force per unit area of cross-section becomes much higher, causing more narrowing and an *even higher* force per unit area. Once this stage has been reached, the narrow parts become even narrower. They rapidly become so narrow that the wire breaks. If you examine the wire at the break, you can see this narrowing effect.

17.12 Liquids

Instructions: investigating diffusion in a liquid
6 At first the dark red colour is only at the bottom of the water. Gradually it spreads upwards, advancing at a rate of only a few centimetres per day. After about a week the whole of the liquid is coloured red, paler because it is now diluted. The molecules of potassium manganate(vii) mix with the water (they dissolve) and are pushed around among the water molecules. Gradually they spread (diffuse) through the water until they are evenly scattered through-the whole of the water.

7 Diffusion continues. Molecules of potassium manganate(vii) and molecules of water are always moving, but since there is no longer a concentration of potassium manganate(vii) at the bottom and a concentration of water at the top, we can no longer *see the effects* of the diffusion. This investigation must not be done in sunlight because this might cause convection currents and carry the coloured solution quickly to the top of the beaker. Convection *currents* are a *mass movement* of molecules, like a crowd of people all running away from a fire. In a 'still' solution, in which diffusion is occurring, the movements of the molecules are in all directions at once, like people in a market-place.

Instructions: supporting a steel needle on water
5 It *appears* as if there is a thin transparent film at the water surface. The film sags slightly under the weight of the needle; it looks as if the needle is being supported by an elastic film.
6 When the drop of detergent is added, the needle immediately falls to the bottom.

Instructions: investigating further examples of surface tension
3 The film which is not burst pulls the thread tightly

toward its own side. The thread takes the shape of part of a circle.
4 When E is burst, the loop is pulled into a perfect circle.
5 In general, the rule is that the unburst film contracts until its area is as *small as possible* and the threads are *arcs of circles*. A circle is the figure which has the largest possible area within a given perimeter. This means that the area on the side of the thread *away* from the film is as large as possible, so the area of the film is as small as possible.

Instructions: comparing viscosities of different liquids
3 Of the liquids listed, golden syrup is the most viscous, 100% glycerol is the next most viscous, then castor oil, then 95% glycerol. Of the liquids listed, water is by far the least viscous. You must have found it impossible to measure the terminal velocity of the ball in water. This method of measuring viscosity is one that is used in industry for testing very viscous liquids. To measure the viscosity of less viscous liquids, other methods can be used. For example, by placing the liquid in a container with a very fine hole at the bottom and measuring how long it takes for a given quantity of liquid to run out through the hole. You could try this method, using a can with a very fine hole in the bottom. The weight, W, depends on the radius of the ball and the density of the material it is made from. In your experiments you used the *same ball* for all tests, so W is constant. If other people do the test using a ball of different size, or made from some other material, they will obtain a different set of results, though the *relative* values for different liquids should still be in the same order. Upthrust, U, depends on the weight of liquid displaced by the ball. This depends on the density of the liquid. The liquids used have *different* densities, so U is not the same for all tests, and this is a source of error when comparing different liquids. However, the error is not large in this comparative test and can be ignored.

17.13 Gases

Instructions: investigating diffusion in a gas
4 Diffusion is complete in a few minutes. When the gas and air are at the same temperature, the gas is denser than air. It would remain in the lower jar, with the air floating above it, *but* diffusion occurs. Molecules of nitrogen dioxide spread among the molecules of the gases present in air. Soon both jars contain the same mixture of molecules. Diffusion is much quicker in gases than in liquids because of the higher velocity of gas molecules.

Instructions: investigating the effect of increasing the pressure in a gas
5 pV has about the same value for all pairs of readings. This suggests the rule: pV is constant. This rule is true, provided that temperature is constant.

6 If $pV = k$, and k is a constant number, we can rearrange the equation to get

$$V = k \times \frac{1}{p}$$

If this relation is true a graph of V against $1/p$ will give a straight line. The line passes through the origin and has gradient k. If you plot the graph it is easier to judge if the law is true or not, for you can usually tell if the points lie roughly along a straight line. Any slight experimental errors may make some points lie a little above or below a truly straight line, but in spite of this, the points will be obviously in a straight line, not in any sort of curved line. If you find this when you plot your graph, you can then believe that $pV = k$.

Instructions: further investigations on the effect of pressure on the volume of a gas
5 The same rule applies to pressures that are less than atmospheric.

18 Matter and heat

18.04 Change of state

Questions page 169
1 It may have been a few degrees below $0°C$ to begin with, but quickly rises to $0°C$ and then the ice begins to melt.
2 It remained at $0°C$ for the whole of the time that the ice was melting.
3 As soon as the ice had completely melted its temperature began to rise above $0°C$; with the flame, the temperature rise was quicker than before. It rose until it reached $100°C$, then remained steady at $100°C$ while the water boiled, showing no sign of increasing further.
4 The water would have *boiled* completely away.
5 The water would have *evaporated*. The rate of evaporation depends on the room temperature, but after a long enough period all the water would have gone.

18.05 Latent heat

Instructions: measuring the specific latent heat of fusion (melting) of ice
7 Total electrical energy, in joules $= VIt$.
8 $l = VIt/m$. The value is $336 \times 10^3 J/kg$.

Instructions: measuring the specific latent heat of vaporization of water
2 $l = VIt/m = 2.26 \times 10^6 J/kg$.

Instructions: measuring the melting point of paraffin wax
2 Though the wax is losing heat, the heat it is losing is *latent heat*, for it is solidifying. Its temperature does not fall until it has become solid; then it falls again until it reaches room temperature. While it is solidifying, you can read its temperature accurately and so find its melting point, which is between $38°C$ and $56°C$, depending on the exact composition of the wax. This method is a useful one for finding the melting-point of pure substances. Wax is often not pure, so you may find that the temperature is not exactly steady as the wax solidifies, but drops slowly for a while as the various waxes in the mixture become solid at slightly different temperatures.

18.20 Evaporation and cooling

Instructions: investigating cooling by evaporation
3 A mistiness of water droplets appears on the outside of the beaker. This has condensed from water vapour in the atmosphere. This shows that the surface of the beaker has become cool.
4 The water freezes; when you lift the beaker, the block remains frozen to it.

18.32 Expansion of a gas at constant pressure

Instructions: measuring the expansion of a gas at constant pressure
6 The value of the coefficient, α is approximately $0.004/K$ (or $0.004/°C$).
7 $-273°C$, though with the simple method used, your result may not be close to this.

18.33 The effect of temperature on pressure

Instructions: measuring the pressure coefficient of air at constant volume
5 The value of β is similar to the value of α, found in the previous instructions. The equation defining β has a form very similar to that which defines α.
7 Their kinetic energy at $-273°C$ is zero. They are completely still.

Questions page 176
1 Volume is constant; use the equation for the pressure law.

$p_2 = p_1 T_2/T_1 = 99\,000 \times 276/300 = 91\,080$ Pa (or N/m^2).

2 $V_2 = p_1 V_1 T_2/p_2 T_1 = \dfrac{10^5 \times 10 \times 253}{0.5 \times 10^5 \times 298}$

$= 16.98 m^3$.

19 Matter and electricity

19.11 The behaviour of electric charge

Instructions: removing electrons from atoms

3 The strips spread apart (diverge), usually alternate strips stand out at opposite sides, so that each strip is as far away from its neighbour as it possibly can be.

4 The strips are attracted towards the wall.

5 After they have contacted the wall they diverge less than they did before. This suggests that the strips were electrically charged but that some of the charge is transferred to the wall.

6 When you put a finger near a strip, the strip is attracted towards the finger.

7 When two sheets of strips are brought together, sheets and strips all repel each other very strongly.

Questions page 178

1 Some additional force must be acting on the strips to make them diverge. This is an electric force. It is a repulsive force between strips.

2 The force is attractive.

3 Both sheets of strips were charged by the same method, so both should be charged with the same kind of electric charge. They repel strongly. This suggests a rule: *like charges repel.*

4 No, there is no means of telling if the charge is negative or positive. Evidence from other investigations shows that the charge is *negative*. By rubbing the polythene we have transferred electrons from the cloth to the polythene.

5 In chapter 8 we found that with magnets, *like poles repel.*

Instructions: investigating positive and negative charge

2 The two polythene strips repel one another; they are similarly charged (both negatively).

3 The two acetate strips repel one another; they are similarly charged.

4 The polythene strip and acetate strip strongly attract one another; they must be oppositely charged. The acetate has a positive charge. By rubbing acetate we have transferred electrons from the acetate to the cloth. *Unlike charges attract.* This reminds us of the rule for magnets: *unlike poles attract.*

Instructions: using the gold leaf electroscope

If the charge brought near is like the charge on the electroscope, the divergence is increased; if it is unlike *or if the object is uncharged*, it is decreased.

Questions page 180

1 Because an *un*charged object can also make the divergence decrease.

2 Charge the electroscope positively (polythene) and look for divergence increasing.

3 Charge the electroscope negatively (acetate) and look for divergence increasing.

4 The greater the divergence the greater the amount of charge.

Instructions: charging by induction

2 The electrons are repelled by the charged rod. They flow through the cans, which conduct them easily, being made of metal. They flow from the can nearer the rod and into the can further from the rod.

4 The nearer can has become positively charged; the further can has become negatively charged.

5 You cannot have a north magnetic pole on its own; any magnetized object has a north pole and a south pole. By contrast an object may be negatively charged or positively charged and need not have both kinds of electric charge on it.

Instructions: another way of charging by induction

4 When you touched the can, the electrons, being repelled by the negatively charged rod, were able to flow away through your finger and your body to earth. This left the can positively charged, even when the rod was removed from it.

Instructions: charging an electroscope by induction

4 When the plate is touched, electrons are repelled to earth, leaving the plate positively charged. Since the plate and case of the electroscope are both earthed, there is no field between them and the leaves do not diverge. When the finger is taken away the positive charge is redistributed evenly over the plate and leaves; the leaves then diverge. In practice this method of charging an electroscope is found to be much more reliable than the method of rubbing the plate.

Question page 182

Rubbing charges the slab negatively (if ebonite or polythene). When the disc is placed on the slab, the electrons in it are repelled towards the upper surface; this leaves the lower surface positively charged. When the disc is earthed, the electrons flow to earth, leaving an overall positive charge on the disc. This remains when the disc is lifted off the slab. When the disc is on the slab, its positive charge is neutralised by the negative charge on the slab; no energy is available for doing work. As you lift the disc away from the slab *you are doing work* against the attractive force between the disc and the slab; you are increasing the potential energy of the charges on the disc. Thus your mechanical energy is converted into electrical potential energy. So therefore the electrophorus does not provide electrical energy for nothing, the energy has been provided by *you.*

19.12 Conduction in electrolytes

Instructions: investigating the effect of passing a current through sodium chloride solution
5 The greater volume collects over the cathode; it is approximately twice the volume collected over the anode.
6 There is small explosion, or 'pop'. This suggests that the gas is hydrogen.
7 The string glows brightly or even bursts into flames. This suggests that the gas is oxygen.

Questions page 184

1 The hydrogen ion, being positively charged, is attracted towards the negative electrode, the cathode. There it gains an electron from the cathode and becomes discharged. The hydrogen atoms at the cathode combine in pairs, forming H_2, or hydrogen gas. The story of the hydroxyl ion is a little more complicated. Being negatively charged, it is attracted towards the anode; there it gives up its electron to the anode and becomes discharged. Then it combines with other discharged hydroxyl ions: $4OH \rightarrow O_2 + 2H_2O$. The effect of these actions is that water is gradually decomposed (or electrolysed) to give hydrogen and oxygen gases.
2 Na^+ ions move towards the cathode; Cl^- ions move toward the anode. In weak solutions they are not discharged, but remain in solution. Thus, as electrolysis proceeds, the solution becomes stronger.

Instructions: finding out the effect of passing a current through copper sulphate solution
6 No bubbles appear, except perhaps for bubbles of air coming out of solution, as they often do; you will see these on the glass too.
7 The cathode becomes covered with a bright deposit of new copper.
8 m_A decreases; m_C increases; their total remains unchanged.
9 It remains unchanged.
10 Copper has been transferred from the anode to the cathode, through the electrolyte. Cu^{2+} ions have been attracted to and deposited on the cathode. Since the density of solution remains unchanged, and since the anode has lost mass, the simplest explanation is that Cu^{2+} ions have been formed from the anode and have moved across to the cathode, under the action of the electric field. This is what actually happens; the copper atoms of the anode dissolve in the electrolyte solution and, as they do so, they each leave two electrons behind in the anode, so becoming positively charged. The electrons flow away through the anode and connecting wire to the cell. Electrons flow from the cell, through the connecting wire to the cathode and become attached to Cu^{2+} ions arriving there, and discharge them as they are deposited. The other ions (SO_4^{2-}, H^+ and OH^-) move towards the electrodes under the action of the field, but are not discharged. They remain in solution.

Instructions: an experimental electrolysis of copper sulphate solution
Copper is deposited at the cathode; oxygen gas is given off at the anode; the electrolyte gradually becomes converted to sulphuric acid (it loses its blue colour and can be tested with blue litmus).

19.20 Ohm's law

Instructions: investigating the relation between potential difference and current
4 The ratio V/I is constant, within the accuracy of the measurements. If you use exactly 1 m of 34 S.W.G. wire, the ratio is about 11. A simple rule is: *current varies with p.d.*

Instructions: investigating currents and potentials in circuits
2 The total equals V_{AE}. In words, the total of p.d.s across parts of the circuit equals the p.d. across the whole circuit.
3 The current has the same value at all points. There is a single path and current can neither enter nor leave this path.
7 At step 5 you probably obtained a value around $50\,\Omega$. At step 6 the calculated value is $100\,\Omega$. The lamp, when connected in the circuit, is not operating under the conditions for which it is rated. The p.d. across its terminals is much less than 6 V. You will have noticed that it does not glow brightly, its filament is not as hot as it is under normal operating conditions; this makes its resistance less than under normal operating conditions.
8 The totals should be the same, by either method of calculation.

19.30 Resistance

Instructions: measuring resistance by substitution
4 Most resistance boxes do not have steps smaller than $1\,\Omega$. If we have a resistance that is $15.6\,\Omega$, we can tell that it is more than $15\,\Omega$ and less than $16\,\Omega$, but this is the limit of the accuracy of our answer. If we say it is $16\,\Omega$, we have a 3% error. For accuracy better than 1%, R_x must be more than $100\,\Omega$.

Instructions: measuring the resistance of a length of wire
5 Resistance varies with length.
6 Wind a coil of bare constantan wire around a non-conducting rod or strip, such as a plastic tube or strip of

cardboard. Connect one end to the circuit. Make a slider (e.g. a wire paper-clip) that can be moved along the coil and fixed to any part of it. Connect this to the circuit. As the slider is moved along the coil the length of wire between slider and the end of the coil is varied and so is the resistance.

Instructions: making more measurements of the resistance of wire

4 R varies inversely with A. The greater the cross-sectional area, the less the resistance.

5 Their total area of cross-section is double that of a single wire. The resistance of the two parallel wires is half that of a single wire of the same length. This agrees with the rule previously stated.

6 The resistance of three wires in parallel is one-third of the resistance of a single wire.

19.31 Series and parallel connections

Instructions: investigating cells in series and parallel

2 Total p.d. = p.d. of one cell $\times$ number of cells. Cells must be connected negative to positive, as shown.

3 I_r equals the sum of currents I_1, I_2 etc.

5 Total p.d. = p.d. of one cell. The number of cells makes no difference to the p.d. produced.

Instructions: finding out more about cells in series

3 Total p.d. = total of p.d.s of individual cells. In symbols, $V = V_1 + V_2 + V_3 +$ etc. If cells of different types are connected in parallel, the total p.d. is the p.d. of the cell with greatest p.d. Cells with lower p.d. have an external p.d. applied to them, causing a current to flow through them in the reverse direction. The effect is to electrolyse the contents of the cell and they are damaged perhaps by excess gas-production or by corrosion of the zinc casing. Even with cells of the same kind (as in the previous instructions) cells should not be left connected in parallel for long. Identical cells may be in varying states of freshness and their p.d.s. not *exactly* equal. The cells with lower p.d. can become damaged after a while. The only exception to this statement are the cells designed to be re-charged, such as lead-acid accumulators and nickel-iron (NiFe) cells. If reverse current is applied to these, they become re-charged. This is what we normally do when we connect such cells to a battery charger. The operation of a battery charger needs care. The amount of current passed into each cell must not be higher than a certain value. In general, cells should not be connected in parallel, and there is no reason for wanting to do so.

Instructions: investigating resistances in series

3 The total resistance = the sum of the resistances of the individual resistors.

Instructions: investigating resistances in parallel

5 The total resistance is *less* than that of the least resistor. With more than one resistor in parallel the current has more than one conductor to convey it from the positive terminal of the battery to the negative terminal; there is less resistance to its flow. The rule is not obvious at this stage; try the next investigation.

Instructions: investigating currents through resistors in parallel

6 $\dfrac{I}{V} = \dfrac{1}{R}$; $\dfrac{I_1}{V} = \dfrac{1}{R_1}$; $\dfrac{I_2}{V} = \dfrac{1}{R_2}$

7 If $\dfrac{I}{V} = \dfrac{I_1}{V} + \dfrac{I_2}{V}$, as stated at step 6, we can substitute the values given in step 6 by their equivalent values, $\dfrac{1}{R}$, $\dfrac{1}{R_1}$ and $\dfrac{1}{R_2}$. This gives the equation, $\dfrac{1}{R} = \dfrac{1}{R_1} + \dfrac{1}{R_2}$

This is the rule for combining two resistors in parallel.

8 They are equal. This point is referred to again later.

Questions page 191

1 (a) $356\,\Omega$;

(b) $\dfrac{1}{R} = \dfrac{1}{56} + \dfrac{1}{100} + \dfrac{1}{200} = 0.018 + 0.01 +$ $0.005 = 0.033.$ $R = 1/0.033 = 30.3\,\Omega$.

2 $\dfrac{1}{R} = \dfrac{1}{10} + \dfrac{1}{1000} = 0.1 + 0.001 = 0.101;$ $R = 1/0.101 = 9.9\,\Omega$. Note that a high value resistor connected in parallel with a low value resistor has very little effect; most of the current flows through the low value resistor.

3 Connect the $50\,\Omega$ and $100\,\Omega$ resistors in series, with total resistance $150\,\Omega$. In parallel with this pair, connect the $1\,500\,\Omega$ resistor.

4 $R_2 = I_1 R_1/I_2 = (0.005 \times 100)/0.002 = 250\,\Omega$.

5 $1/R = 0.1 + 0.05 + 0.04 + 0.083 + 0.167 = 0.44;$ $R = 1/0.44 = 2.27\,\Omega$.
Currents are: 0.2, 0.1, 0.08, 0.167 and 0.33 A.

19.40 Other relations between p.d. and current

Questions page 192

1-3 At low p.d. there is no conduction. As the p.d. exceeds about 0.6 V, conduction begins; as p.d. increases, I increases at an ever increasing rate; the curve slopes more steeply at higher p.d. The effect on resistance is that resistance decreases as p.d. increases. With a silicon diode, at p.d. 1 V, the current is around 30 mA, giving a resistance $33\,\Omega$. This figure varies according to the type of diode used and is higher for germanium diodes. Conduction in the diode does *not* conform to Ohm's law.

4 Conduction in the reverse direction is very slight and hardly measurable, even though we are using much higher p.ds. and a much more sensitive current-measuring instrument. The resistance to current passing from cathode to anode is several millions of ohms.

Questions page 193

1 With positive of battery to A, current flows readily through the diode, lighting the lamp brightly. With positive to B the resistance of the diode is very high and virtually no current flows, so the lamp does not light.

2 The bulb lights brightly in either position. The direction in which the current flows through the ammeter and through the bulb is the same, whichever way round the battery is connected.

3 From C to E; F; ammeter; lamp; H; G; D. From D to G; F; ammeter; lamp; H; E; C. In *both* sequences it flows from ammeter to lamp.

19.50 Measuring temperature electrically

Questions page 194

1 The resistance increases with increasing temperature.

2 When cold, the resistance of the filament is low. When current begins to flow, it is much larger at first owing to the very low resistance. If the lamp is old, this large current may melt the filament. If the filament can stand this initial surge of current, its resistance increases as it gets hotter and the current is reduced to a lesser amount, which will not melt the filament.

Part F Physics of the Universe

20 Atoms

20.23 The electroscope

Instructions: discharging an electroscope with ions
4 Radiation ionises molecules in the air, forming ions charged positively and negatively. Depending on the charge carried by the plate, oppositely charged ions are attracted to the plate and discharge it. To use this effect, charge the plate to a known potential, then measure the time taken for it to be discharged.

20.24 The cloud chamber

Questions page 201
1 Trails appear very rapidly, as straight greyish lines across the chamber. They disappear gradually after about 10s; this is caused by air movements, diffusion, and re-evaporation of alcohol from the cloud droplets. A bent track is sometimes seen; it is only slightly bent; the bend is caused by an α-particle coming close to the nucleus of an atom of one of the molecules of the air and being repelled by it. Trails extend most of the way across the chamber, but not all reach the far side. These α-particles vary in the amount of energy they have; they are released by the decay of several different nuclides in the source, the various products of the decay of radium.
2 The tracks are fewer and fainter; you need good lighting to see them clearly. The α-particles are absorbed by the foil, but the β-particles can pass through foil. They are less massive, so are more easily deflected than α-particles; their trails show sharper bends. They have high energy and can pass out through the walls of the chamber. Trails of β-particles were there at step 4, but not easily seen among the thicker trails of α-particles.
3 The thicker foil absorbs both α-particles and β-particles; no trails are seen. Although gamma-rays from the source penetrate the foils easily, we see no trails from γ-rays. They are not charged particles, so produce no trail. On their way they interact with atoms, knocking out electrons; *these electrons* may then produce trails. In a beam of γ-rays a fine cloud of short thin electron trails appears; the trails run in all directions. They cannot be seen in this type of cloud chamber.

Questions page 201
1 The trails are deflected to one side. If the north pole is uppermost, the trails are deflected towards the right (as seen with the source at the side of the chamber which is nearer to you. Using the left-hand rule, which we shall study experimentally later (23.41), we deduce that α-particles are positively charged.
2 The trails are curved in the opposite direction. This shows that β-particles have opposite (negative) charge. Their trails are more strongly curved, indicating that they are less massive than α-particles.

20.25 The spark-counter

Questions page 202
1 Sparking is irregular. See page 202 for the explanation.
2 As the source is raised the rate of sparking decreases only slightly at first. Then at a distance of a few centimetres there is a sudden decrease as the source is raised only a small amount. At that distance the α-particles are all absorbed by the air. If the particles all have about the same energy they all have about the same maximum range in air. When the source is beyond this range, no particles reach the spark counter.
3 (a) The particles are easily absorbed by solid materials, such as paper; the rate of sparking decreases noticeably.
(b) Aluminium foil absorbs all the α-particles; sparking ceases.

Questions page 202
1 Decay is irregular, like the sparking of the spark counter.
2 No; what will happen is a matter of chance.

3 The average rate is about one decay for every four throws.

4 With such a small number of atoms you may easily obtain quite a different average.

5 Because you would begin with only five atoms instead of ten, the chance of a decay at the first throw is now only half what it was at the beginning of the experiment. You would probably need to run the experiment for very many throws before the final atom decays.

6 The average rate could be about one decay for eight throws. Again, with such a small number of atoms, it is quite possible that you will achieve something totally different.

20.26 The Geiger-Müller tube

Questions page 203

1 Knock electrons from atoms, producing positive and negative ions.

2 They are accelerated, gaining kinetic energy.

3 Produce an even greater number of ions, like an avalanche.

4 An electric current passes briefly: a discharge.

20.30 Half-life

Questions page 205

1 Throw number represents the passing of time.

2 About one-sixth of the number of dice being thrown. With 100 dice an average of 17 should 'decay', leaving about 83 'undecayed'.

3 About one-sixth of the number of dice being thrown. If 83 are left from the first throw, one sixth of these (14 dice) should decay, leaving 69 dice to be thrown at the third throw.

4 At some throws slightly more than one-sixth may decay, but at other throws slightly fewer may decay, so there is a tendency for variations to become smoothed out. On average, if we let one-sixth decay at each throw, the number should fall like this: 100, 83, 69, 58, 48. In four throws the number of atoms becomes half its original number (approximately).

5 If we continue to subtract one-sixth at each throw, the numbers fall like this 48, 40, 33, 28, 23. In four throws (a total of eight throws from the beginning) the number of atoms becomes a quarter of the original number. So, though the first four throws reduce the number to half of the original, the next four throws *do not* cause all the remaining atoms (the remaining half) to decay. Only *half of those that are left* decay during the second four throws. In the next four throws *half of those that are left* will decay, and so on. The rate of decay becomes less and less as the number of undecayed atoms becomes less and less.

21 Waves

21.00 Features of waves

Instructions: making waves in a thread

4 Waves.

5 Downwards from the bob to the free end of the thread.

6 The same as that of the pendulum; if the pendulum is a little less than 15 cm long, the frequency is about 0.75 Hz.

7 At the end attached to the pendulum the amplitude is the same as that of the pendulum. Further along the thread, amplitude decreases owing to air resistance and the weight of the thread.

8 It moves from side to side, horizontally.

9 The cork has kinetic energy, transferred to it from the bob, along the string. Wave motion always brings about a *transfer of energy*.

21.02 Transmission of waves

Questions page 210

1 They are all horizontal. All particles of the thread move horizontally, from side to side, as was demonstrated at step 8, in the previous instructions.

2 Vertically. The direction of motion of the waves is *perpendicular to* the direction of motion of the particles of thread.

3 Point A has zero velocity; it has maximum displacement to the left. It accelerates towards the right, increasing in velocity towards the right and its displacement decreases. At zero displacement (from the 'rest' position) its velocity towards the right is a maximum. Then it decelerates, coming to rest when it reaches its maximum displacement to the right. It then accelerates towards the left, reaches maximum leftward velocity at zero displacement then decelerates and comes to rest again at its original displacement. The cycle is then repeated. In short it is following the motion of the bob, a little later in time and with slightly less amplitude.

4 Yes, but each point is slightly later in time and moving with slightly less amplitude than the point immediately above it.

5 They are at exactly the same stage in their cycles.

6 A and G; D and I.

7 These distances are all equal. We call this distance the *wave-length*.

8 Two.

Instructions: investigating another type of wave motion

4 The bottom of the can bulges outwards and inwards. The centre of the bottom moves to and fro in a direction *along* the direction of the string. The can must be pulling and pushing against the end of the string. Voices and noises are heard at the other end of the string, so energy must be transferred.

Instructions: investigating waves in a slinky spring
1 The directions are perpendicular.
5 The molecules move backwards and forwards in a direction *along* the direction of the spring. The waves move in the same direction.

Instructions: measuring the speed of waves
4 No.
5 No.
6 While your hand is making two waves, the beginning of the first wave must have time to travel to the far end of the spring. To travel a distance s, the waves needs time t. If its speed is c, then $t = s/c$. The hand makes two waves in this time; the hand makes one wave in time $T = t/2 = s/2c$. The number of waves per second, or frequency must be $2c/s$. Fit your own measurements into these equations.
8 Time for one wave is $t/3 = s/3c$. Frequency, $f = 3c/s$ (f in Hz).
9 $f = 4c/s$ (f in Hz).
10 In all the calculations we have used the equation $f = nc/s$, where n is the number of wavelengths to be fitted into the spring. To fit in one wavelength, $n = 1$, and $f = c/s$. When one wavelength is to be fitted in we can say that $s = \lambda$. We use the symbol λ for wavelength. This gives us the equation $f = c/\lambda$. This can be rearranged to give another equation, $c = f\lambda$.

21.11 Reflection

Questions page 214
1 Their direction is changed. Note that waves always move in a direction perpendicular to the wavefront. The arrows in the figure show the directions of incident (i) and reflected (r) wavefronts. If you make measurements on the screen of the ripple-tank, you can discover if the angle of incidence and angle of reflection are equal.
2 There is no change. This can be checked by measurement on the screen.
3 There is no change. Check this by looking steadily at one point on the screen where incident waves are passing. If the stroboscope is run to make the waves appear to move slowly, you can clap your hands in time with the dark bands as they pass this point. Your partner can do the same for a point in the reflected beam. Your clapping and that of your partner have the same frequency.
4 They are circular.
5 From a point behind the reflector, on the same normal as the source; the source and this point are equal distances from the reflector.
6 No. This statement can be checked, as above.

7 Energy must be conserved. As the wavefront becomes shorter, the amount of energy per centimetre of front must increase. The result is increased amplitude of the waves. Increased amplitude gives higher E_p at troughs and crests, and higher E_k between, like a pendulum swinging with large amplitude. On the screen the contrast between bright and dark bands increases in the region around F.
8 As the wavefront becomes wider, energy is conserved and the amount of energy per centimetre of front becomes less. Amplitude decreases.
9 They will be plane waves. Try this on a ripple-tank.

21.12 Interference

Questions page 216
1 The insect is in calm undisturbed water; it does not rise or fall. Yet as it looks at the water surface around it, it sees violent waves on all sides.
2 No. Strong forces are acting, but there are two sets of forces acting equally strongly but in opposite directions at all times. The water surface remains stationary under the action of these balanced forces. Wave energy is conserved and is transferred through these regions to the regions beyond.

21.13 Refraction

Questions page 216
1 The angle of incidence.
2 Yes; the wavelength is reduced.
3 No. Check this by the clapping method mentioned on this page (21.11).
4 The equation relating speed, frequency and wavelength is $c = f\lambda$ (page 214). We know that f is unchanged and that λ is reduced. Therefore c is reduced in shallow water.

21.14 Diffraction

Question page 218
The two gaps act as two point sources. They make circular wave trains the same frequency in step with each other. In effect they act just like the two-pronged vibrator of fig. 21.7. They produce the same pattern of regions of constructive interference and destructive interference. A demonstration on a ripple-tank shows this effect, though the waves are much weaker than when a two-pronged vibrator is used.

21.21 The nature of sound – information from previous investigations

Questions page 218
1 (a) 0.034m, (b) 1.33m, (c) 6.8m.

2 (a) 0.345m, (b) 0.340m.

3 $n = c_1/c_2 = 345/340 = 1.015$.

4 $\sin r = n \times \sin i = 1.015 \times 0.4226 = 0.4289$. $r = 25.4°$.

5 No. Pitch is an effect of frequency, which is not affected by refraction.

6 He might notice a slight change in the direction from which the sound appears to come, but this is not likely as the change is only 0.4°.

21.22 Interference

Questions page 219

1 To eliminate reflected sounds, which cause complicated effects through interference with the sounds coming directly from the loudspeakers.

2 At places where interference is constructive a loud sound is heard, frequency 300 Hz. At places where there is destructive interference, no sound is heard. You could try this and walk around the area, marking the ground where you stand when you hear the sounds loudest. Then mark the ground where you hear nothing. The marking should give a pattern similar to the dashed lines of fig. 21.7.

Instructions: investigating beats

3 The sound increases and decreases in loudness several times a second. You might describe it as a 'throbbing' sound. These regular pulses of loud and faint sound are called *beats*.

4 The amplitude of the trace increases and decreases in time with the beats.

6 There are fewer beats per second. The rate of beating may become so slow that it is hard to detect by ear; the oscilloscope trace shows it more clearly, for you can measure the amplitude as it changes slowly.

Questions page 220

1 The notes had been made closer in frequency.

2 Sound the two forks together and listen for beats. Count the number of beats per second. This tells you the *difference* between forks. For example, if you hear beats at 4 Hz, the second fork has frequency 444 Hz *or* 436 Hz. Reduce the frequency of the second fork by putting Plasticine on its prongs. Test for beats again. If the beat frequency is now more than 4 Hz, you know that the frequency of the second fork (without Plasticine) is 436 Hz; if the beat frequency is less than 4 Hz, you know its frequency is 444 Hz.

21.23 Diffraction of sound

Instructions: demonstrating diffraction of sound

2 When the ear is directly in front of the gap, the sound can be heard loudly. With the ear to one side it is faint; probably most of what is heard then is coming indirectly through reflection from objects in the room. This indicates that the sound waves are in a beam coming straight through the gap and not spreading.

3 The sound is heard almost equally loudly at all points along the line AB.

4 As at step 3, the sound is heard equally loudly at all points. In steps 3 and 4 the sound spreads out after it has passed through the gap. This suggests that it is being *diffracted*. At steps 3 and 4 the wavelength of the sound is about 1.4 m (240 Hz) and 0.34 m (1 kHz). It is much longer than the width of the gap. When the gap is about as small as or smaller than the wavelength, diffraction occurs. When the note is 10 kHz (step 2) it has wavelength 0.034 m, or 3.4 cm. This is less than the width of the gap (10 cm) so there should be no diffraction. This is exactly what is found. The high-frequency note gives a narrow beam of sound.

5 With high frequency (10 kHz) it is possible to obtain a 'sound shadow' using a relatively small object. With lower frequencies, the waves are diffracted around the object. There is no sound shadow, so little effect is noticed if such objects are placed between the gap and the ear.

6 With high frequency, the effects of reflection are very clear. Waves of small wavelength can be reflected by small objects; a clear-cut reflected beam can be detected. With lower frequencies and longer wavelength, there is no clear effect. The waves pass by the reflector and re-form beyond it. For reflection to occur, the reflector must be about as big as a wavelength, or larger.

Instructions: using a building as a sound barrier

4 When you are in the sound shadow you lose most of the high-pitch notes (high frequency and short wavelength). These are not diffracted far around the corner. In the shadow you may still hear the low-pitch notes, for these have longer wavelength and are diffracted more. Therefore as you move towards the corner, the music at first consists mainly of bass notes (thumping drums, etc); only when you come into direct line with the radio can you hear the high-pitched whistles, flutes, violins etc.

Questions page 221

1 It is best to sit exactly in front of the speaker. If you sit to one side of the speaker, you will hear the low-pitch notes just as easily, for they are diffracted and spread more or less equally in all directions. You will not hear the high-pitch notes with full loudness though, for they do not spread as much.

2 You hear mainly the low-pitch notes, such as drums because the houses block the waves of high-frequency notes. The gaps between the houses are too large to diffract high-frequency notes.

3 Speak normally but softly; it is much easier to locate the source of a whisper. The frequencies of the ordinary speaking voice are so easily diffracted by gaps that it is difficult to locate the speaker. This effect is made use of by ventriloquists; when you see the mouth of the doll moving and the mouth of the ventriloquist is *not* moving, you have the illusion that the doll is speaking; your ears cannot locate the source of sound accurately enough to prove that the illusion is false. In the same way, the loudspeaker of a TV set is beside the screen and the loudspeaker at a cinema may be beside the screen, yet the sound *appears* to come from the actors on the screen. By contrast, a high-pitched sound such as a whisper is very slightly diffracted and is much easier to locate.

21.32 Interference

Questions page 222
1 Two lines of equal brightness (assuming the slits are of equal width). The lines would be to either side of the centre of the screen. The centre of the screen would be dark; the shadow cast by the material between the slits.
2 The light is diffracted from the two slits, as in fig. 21.7. It spreads out and the central area of the screen is lit by light diffracted by both slits. The two beams interfere. We get bright lines on the screen where interference is constructive between dark lines where interference is destructive. The centre of the screen has a *bright* line (an equal number of wavelengths from both slits).

21.33 Diffraction

Instructions: investigating images made by a diffraction grating
1 You can see the lamp. In a row on both sides of the lamp you can see several other images of the lamp. The edges of the images are coloured.
2 The rows of images turn clockwise around the lamp.
3 They are perpendicular.
4 You see the main reflection and several other images on either side of it. The fine grooves and ridges in the recording area are acting as a diffraction grating.

Questions page 225
1 $\tan \theta = 0.257$. $\theta = 14° 25'$.
2 $s = 1/500 \text{mm} = 0.002 \text{mm} = 2 \times 10^{-6} \text{m}$.
   $\sin \theta = 0.2500$.
   $\lambda = 2 \times 10^{-6} \times 0.25 = 0.5 \times 10^{-6} = 5 \times 10^{-7}$.
3 $\tan \theta = 0.315$. $\theta = 17° 29'$. $\sin \theta = 0.3004$ (call this 0.3).
   $\lambda = 2 \times 10^{-6} \times 0.3 = 0.6 \times 10^{-6} = 6 \times 10^{-7} \text{m}$.
4 $\sin \theta = \dfrac{7 \times 10^{-7}}{2 \times 10^{-6}} = 3.5 \times 10^{-1} = 0.35$.
   $\theta = 20° 29'$. $\tan \theta = 0.3736$. $x = 0.374 \text{m}$.

5 The distances, x, of the images for colour filters used in the questions above are red (0.374m), yellow (0.315m), blue (0.257m). These images are arranged in this order from left to right (as seen against the metre rule of fig. 21.16). The three colours have been sorted out in the same order as they appear in the spectrum. Other colours are sorted out in the same way. If the lamp is a vertical thin filament the first-order image of this lamp is a spectrum running from red (on the left) through orange, yellow, green, blue, indigo to violet (on the right).
6 Red.
7 Violet.
8 In a spectrum produced by refraction, red is dispersed least and violet is dispersed most, the exact opposite of a diffraction spectrum.

21.50 The photoelectric effect

Questions page 228
1 Only a negatively charged electroscope is discharged in this way. Ionisation makes ion pairs, positive and negative. If discharged by ionisation, it does not matter what charge the electroscope is given.
2 Electrons.

22 Electric charge

22.01 The distribution of charge on a conductor

Questions page 232
1 *Sphere*: charge evenly distributed. *Cylinder*: most of the charge around the ends. *Pear-shaped*: charge evenly distributed over the half-spherical part; very high charge near the pointed end.
2 The more sharply the surface curves, the greater the amount of charge per unit area. The amount of charge per unit area of surface (per m²) is the *surface density*.
3 A sharp point is a region of especially high charge; the surface density at the point is far higher than on the rest of the cone.

Questions page 233
1 There is charge on the outer surface; there is no charge on the inner surface.
2 The potential of the inner surface *equals* the potential of the outer surface. This is a conductor and if there was p.d. between the inner and outer surfaces, charge (current) would flow until p.d. was reduced to zero. There is a small p.d. while the conductor is being charged, but this disappears almost instantly when charging is complete.

3 The potential at all points on the inner surface is equal; there can be no p.d. across the space inside the conductor; there can be no electric field.

4 Friction between vehicle and air charges the vehicle; the rubber tyres prevent the charge from leaking away. The potential of the vehicle is increased and the potential of all objects inside the vehicle (a hollow conductor), is increased to the same value. The boy has high potential compared with earth. If he stays in the vehicle, he does not notice this. When his foot touches the ground, the p.d. makes current flow and he receives an electric shock. The same thing could happen when people get out of an aeroplane, but special conducting tyres are used on the wheels of aeroplanes, so that the aeroplane is discharged before the passengers get out.

22.02 Point discharge

Questions page 233
1 Positive.
2 At the point of the needle.
3 Ions in the air are accelerated, creating more ions.
4 They are attracted towards the charged strip and are discharged by the charge on it.
5 They are attracted towards the needle and are discharged by the charge on it.
6 Negative. Check the correctness of this reasoning by testing for the sign of the charge on your electroscope.

22.03 The Van de Graaff generator

Question page 234
From the electric motor, or from the battery or power supply to the electric motor. A certain amount of energy is required to overcome the mechanical friction in the moving parts of the generator. Additional energy is required (a) to separate the positive charge on the belt from the negative charge on the lower roller and (b) to move the positively charged belt towards the positively charged sphere. Work must be done and appears as the electrical potential energy of the charged sphere. When the sphere is discharged to earth, this potential energy is set free.

22.10 Charge and potential

Instructions: investigating the potential at points on a charged conductor
3 (a) zero (with reference to earth); (b) zero (earthed). The p.d. is zero; the leaf does not diverge.
6 The potential is the same at all points on the surface of a charged conductor. This applies to the inner and outer surfaces of hollow conductors. We have already reasoned

that this must be true, because if a p.d. exists current flows until p.d. is zero. Now we have shown by measurement that there is no p.d. between different points.

22.20 Storing charge

Instructions: finding out how much charge it is possible to store on a can
4 The amount of charge induced on the disc depends on the amount on the slab. If the disc is recharged the amount of charge on it is not the same as at step 3. You would then be giving different amounts of charge to each can, so the experiment would be meaningless.
5 The electroscope with the small can shows more divergence than the electroscope with the large can. This shows that the small can has higher potential than the large can.
6 Both were given the same amount of charge. For a given amount of charge stored, a large can has a lower (or smaller) potential than a small can.

Instructions: finding out how much charge it is possible to store on a flat plate
4 The leaf falls slightly; the potential of A has become slightly less.
5 The leaf falls much more; the potential of A is now much lower than before.
6 The leaf slowly rises; the potential of A is rising.
7 As the overlapping area is reduced, the potential of A rises.
9 The leaf falls, showing a fall in potential of A.
10 Similar falls in potential are produced.

22.23 Experiments with capacitors

Instructions: charging and discharging a capacitor
3 The lamp flashes, showing the discharge of the charge stored in the capacitor.
4 The capacitor stores charge for several minutes. In good conditions the charge is stored for hours. Eventually the charge leaks away: point discharge from the ends of the wires, especially if the air is damp; leakage through the insulating layer, especially if this is not perfect, which is likely in an old electrolytic capacitor.
5 No matter how quickly you touch A to B, the capacitor becomes fully charged almost instantly (compare with the next investigation).

Instructions: further investigations with a capacitor
2 This discharges the capacitor.
3 P.d. across the capacitor increases rapidly at first, then less and less rapidly as it reaches its maximum value. The maximum value is p.d. between the terminals of the

battery. In this experiment the capacitor does not charge in an instant. The resistor allows only a small current to flow (a few coulombs per second) so charging takes several seconds. At first, when the p.d. across the capacitor is low, current is at its maximum; later, when p.d. is high and the capacitor is almost fully charged, current falls. When the capacitor is fully charged, current is zero.

4 P.d. falls rapidly at first, then more slowly. It is zero when the capacitor is fully discharged. Current is at a maximum to begin with, then falls and becomes zero when the capacitor is fully discharged.

Instructions: measuring capacitance
9 $Q = It$. Calculate Q from measurements of I and t.
10 $C = Q/V$. Calculate C from your measurements of V and your figure for Q.

23 Electrons

23.02 Thermionic emission of electrons

Questions page 241
1 No. Electrons are emitted only by the heated cathode. Electrons produced there are attracted back into the cathode, so no current can flow.
2 This tube allows current to flow in one direction only. It is a diode. Like the semiconductor diode, it can be used for rectifying alternating currents.

3 The conventional current flows from anode to cathode, the opposite direction to that in which the current carriers (electrons) flow.

Questions page 241
1 No. The slope of the graph varies, indicating a varying value of the ratio V_A/I_A.
2 The rate at which the electrons are removed from the cloud around the cathode is equal to the maximum rate at which electrons are set free from the cathode by thermionic emission. The diode is *saturated*.
3 The answers to the previous questions suggests the answer to this one. Find the answer by measurement: use three cells or fewer for the supply to the heater.

23.11 The effects of magnetic and electric fields

Instructions: studying the force on a current flowing in a wire in a magnetic field
3 To the right (as viewed in the figure).
4 To the left.
5 To the left.

Question page 243
The current flows from right to left. The rule applies to conventional current. The electron beam is known to consist of particles flowing in the opposite direction (that is, left to right). Therefore the particles must have negative charge.

Part G   Physics in Action

24 Working with lenses

24.01 Pinhole camera

Instructions: improving the pinhole camera
2 It is smaller than the lamp; it is upside down (inverted); it is faint.
3 The image is brighter, but not as sharp as before. Brightness cannot be improved without losing sharpness.
4 Ten images of the lamp, arranged in the same pattern as the pinholes.
5 The ten images are replaced by a single image in the centre of the screen. It is inverted, much brighter and sharper than the single image of step 3 and about the same size.
6 At first there is a large patch of light that does not look like an image of a lamp. With the lens in place there is a bright inverted image of the lamp. Now that the pinhole has been replaced, we have a *lens camera*.
7 The sharpness changes slightly. In one position, the image has maximum sharpness; it is *focussed*.

Questions page 250
1 Refraction.
2 The ray passing through the central pinhole is not refracted by the lens. All other rays are refracted towards the central ray. When the lens is in the focussed position all rays are refracted towards the same region of the screen, making a single bright image. They *converge*, so this is called a *converging lens*.

24.02 Lens camera

Questions page 251
1 The lens makes an inverted image, as in the camera. To obtain a picture the right way up, the slide must be put in the projector upside down. In fig. 24.2 the image has A' at the top; the slide has the corresponding part A at the bottom.

2 The light passing through the slide must be spread over a much larger area of screen. For example, if the slide is projected as an image 30 times as wide and 30 times as high, the area of the image is 900 times the area of the slide. The intensity of light on the screen is only 1/900 of the intensity of light at the slide. Actually it is less, for there is loss of light between slide and screen. To make the light intensity on the screen bright enough to see, the intensity on the slide must be considerably higher.

24.03 Light rays and lenses

Instructions: working with converging lenses
1 From the point on the paper immediately below the filament of the lamp.
2 Rays are refracted towards the principal axis. Any ray that runs along the principal axis before it reaches the lens is not refracted; it passes straight through and continues along the principal axis.
3 No. Those which go through the outer regions of the lens are refracted more than those which go through central regions. This is a fault in the lens. It can be *corrected* if the lens is made with a slightly different shape, but this is expensive to do. The effect can be *reduced* if we cut off the outer rays by using screens.
4 With small aperture, the outer rays are cut off; the remaining rays converge more exactly on a single point, giving a sharper image. To obtain a sharper picture with a camera the aperture must be small; we then have the problem that the amount of light getting to the film is not enough to affect the film. When light is dim the aperture must be opened wide to expose the film sufficiently; we must then accept the fact that the image is not perfectly sharp.
6 If the lens surfaces are more strongly curved than before, the distance is less; if the surfaces are more gently curved, the distance is more.
7 They converge at a point further from the optical centre.

The converging rays are forming an image of the filament. There is a relation between the distance of the filament from the lens and the distance of the image from the lens.

Instructions: working with light from two distant sources

3 At F.

4 At a point just below F (as seen in the figure).

5 It passes straight through the optical centre of the lens; it is not refracted.

6 The answer to question 5 is the same, whatever the position of lamp 2.

Instructions: working with diverging lenses

2 From a point between the source of light and the diverging lens.

4 The point is on the principal axis. This is the focal point and its distance from the optical centre is the focal length of the lens.

6 (a) The principal focus; (b) parallel to the principal axis; (c) the optical centre.

24.04 Measuring focal length

Questions page 254

1 The distance measured by your partner is the focal length of the lens. For a rough method, we can forget about the fact that the distance should really be measured to the optical centre, which is inside the lens about half-way between its two surfaces.

2 The definitions of principal focus and focal length depend on the incident rays being parallel rays. Rays are parallel only if they come from an infinitely distant object. Rays from the Sun and clouds are so nearly parallel that we can think of them as parallel. Rays from nearer objects, such as the window frame, are not parallel; to focus these the lens must be moved a little further from the card.

Instructions: locating the real image

5 The image appears to be behind the lens, between the lens and the window or lamp. It is not possible to focus the image on a screen. It is not real; we call it a *virtual image*. It is upright. At step 3 the image formed by the converging lens is inverted. Later we shall see how to form a virtual upright image, using a converging lens.

24.05 Ray diagrams for lens calculations

Questions page 256

1 Height is 8cm (image and object have same size); it is 20cm from the lens (image and object are equal distances from lens); it is real and inverted. These are the features of the image when the distance of the object from the lens is exactly twice the focal length.

2 I is 12.5cm from the lens. It is 2cm high (it is smaller than O, a *diminished* image); it is inverted and real.

Questions page 257

1 A ray parallel to the principal axis is always refracted along the same path no matter from what distance it comes. It always appears to come from somewhere along the dashed line of fig. 24.9. Therefore I must always be somewhere between the lens and F; it must always be diminished, virtual and erect.

2 Power of 1D, means focal length = 1m. When the lens is focussed on distant objects it is 1m from the screen. A ray diagram similar to fig. 24.7 gives the position of I, 1.33m from the lens. To change focus the lens must be moved away from the screen a distance 0.33m. The ray diagram shows that I has one-third the size of O; the image of the caterpillar is 1cm long.

3 A ray diagram similar to fig. 24.7 shows that O is 2.7 times larger than I. The diameter of the flower is 20.25mm.

4 Draw a diagram similar to fig. 24.9. Draw ray 1. Find the position where the height of I is 0.1m (to scale). You can then draw the dashed line and find where it cuts the principal axis (F). Measure focal length, 0.22m. Power = 1/0.22 = 4.5D.

24.21 Simple magnifiers

Instructions: using a converging lens as a magnifier

3 Less.

4 No; it is nearer to your eye than the least distance of distinct vision.

5 About 25cm, if your eyesight is normal.

Questions page 259

1 O is 7.1cm from the lens. The size of I is 14cm, giving magnification of 14/4 = 3.5 times. The visual angle is 14/25 = 0.56 rad. Examination of the diagram shows that this is the largest obtainable visual angle.

2 Hold the lens as close as possible to the eye; bring the object slowly towards the lens until the image is at the least distance of distinct vision.

3 The image is at infinity, for the two rays are parallel after they have been refracted. The image is magnified, virtual and erect.

4 The lens with the shortest focal length. If you draw fig. 24.15 again, but make the focal length 5cm, you will see why.

24.22 Compound microscope

Instructions: making a model of a compound microscope

2 Real, inverted, magnified, located on S.

3 Virtual, inverted, more magnified than I_1, located at approximately the least distance of distinct vision from your eye.

4 The magnification should be much greater, but it is difficult to estimate owing to defects in the lenses. Real microscopes use complex lens systems that are designed to reduce defects.

24.30 Telescopes

Instructions: making a model telescope

3 Real, inverted, diminished. In the microscope I_1 was magnified.

4 Roughly the focal length of L_2.

Instructions: making a ray-streak model of a telescope

2 The rays are parallel, so they converge on the principal focus; 20cm behind L_1.

3 Rays leaving the lens are parallel; therefore they must diverge from the principal focus of L_2. The distance from I_1 to L_2 is 7cm.

6 The answers to questions in steps 2 and 3 show that the total distance is $f_1 + f_2$.

7 Make f_1 as large as possible and f_2 as small as possible. You could replace L_1 with a lens, focal length 1m, and replace L_2 with a lens, focal length 2cm. Magnification would then be $100/2 = 50$.

8 Use a collapsible tube, with sections that slide inside each other. Use prisms to 'fold' the path of the rays between L_1 and L_2; this is done in binoculars.

25 Working with reflectors

25.00 Converging mirrors

Instructions: investigating reflection from a wide, converging mirror

3 There is no position like that described. The rays converge along a pair of *curved lines* which show up as bright curves in front of the mirror. We call these *caustic curves*. You often get the same effect if sunlight or a bright lamp is reflected from the inside surface of a coffee cup, on to the surface of the coffee.

4 Reduce the aperture of the mirror, or use a mirror that is not as wide. The alternative is to alter the shape of the mirror slightly; it is part of a circle, a better shape will be described later.

Instructions: investigating reflection from a converging mirror with narrow aperture

2 The principal axis is the normal to the centre of the mirror.

3 The principal focus is the point on the principal axis where parallel incident rays converge. The distance between this point and the pole is the focal length of the mirror. The image produced is real, since rays converge to the point and diverge again.

4 (a) reflected through the principal focus; (b) reflected from the pole at an angle θ on the other side of the principal axis; (c) reflected parallel to the principal axis; (d) reflected back along its path, through the centre of curvature.

Questions page 262

1 Diminished.

2 The image is inverted, real, the same size as the object and located at the same distance from the pole. Note that the object is at exactly twice the focal length from the mirror, where the centre of curvature is located.

3 Inverted, real, magnified, located 30cm from the pole. Magnification is two times, so the image is 4cm high.

4 The image is 0.33m behind the lens; its magnification is 1.67 times.

Instructions: investigating reflection from a parabolic converging mirror

4 The rays are parallel.

25.10 Diverging mirrors

Instructions: investigating reflection from a diverging mirror

2 They are reflected back and diverge after reflection.

3 Yes. The rays appear to diverge from a point; the principal focus. A virtual image is formed.

4 This is the distance between the pole and the principal focus.

25.21 Light leaves a dense medium

Questions page 264

1 When $i = 0°$, $r = 0°$, and $\sin i = \sin r = 0$. The ratio cannot be calculated; there is no refraction and Snell's law does not apply.

2 For glass: $_Gn_A = 0.66$; for perspex, $_Pn_A = 0.67$.

3 As i is increased by equal steps, r increases by larger and larger steps. When $i = 40°$, $r \approx 77°$. If i is slightly greater than this, r increases to about $90°$; the emergent ray just skims along the surface. The problem is that when $i = 45°$ or more, there is *no emergent ray*.

Instructions: finding the path of light when i is more than 45°

2 It is reflected; then it passes back through the block emerging through the curved surface without being refracted. It is not refracted at the curved surface because it is directed along the normal to the curved surface.

3 $i = r$. The flat surface acts like a plane mirror.

Questions page 265

1 41.3°.

2 48.7°.

3 2.42.

26 Using electricity

26.01 The heating effect of an electric current

Instructions: investigating the heating effects of a current passing for differing lengths of time

7 The graph is a straight line passing through the origin (provided that you plotted *rise* in temperature, not actual temperature). From this we can state a rule: *rise in temperature varies with time.*

Instructions: investigating the relation between heating effect and size of current

2 The shape of the curve is like that of the graph for 'A varies with B squared.' This suggests that rise in temperature varies with current squared. You can check this roughly by noting if the rise in temperature for a current of 2A is *four times* that produced by a current of 1A. The rise for a current of 4A should be sixteen times that for a current of 1A. This does not allow for the fact that with higher current we get higher temperatures and greater loss of heat to surroundings; the rise of temperature is not as high as it should be theoretically. Another way of confirming the relation is to plot a graph of rise against I^2. This should be a straight line passing through the origin.

Instructions: investigating the relation between heating effect and resistance

2 The graph is a straight line passing through the origin. The rule is: *rise in temperature varies with resistance.*

Questions page 270

1 $W = mc(\theta_2 - \theta_1) = 0.5 \times 4\,200 \times 10 = 21\,000\text{J}.$

$Q = VIt = V \times 1.5 \times 300 = 450V\text{J}.$

$V = 21\,000/450 = 46.7\text{V}.$ A voltmeter connected across the heater should indicate this value.

2 $Q = I^2Rt = 3^2 \times 35 \times 60 = 18\,900\text{J}.$

3 $Q = 6^2 \times 35 \times 60 = 75\,600\text{J}$ (*four* times the amount in the previous question, but the current is only *twice* as great).

4 $P = V^2/R = 240^2/576 = 100\text{W}.$

Questions page 270

1 In modern clothes-irons, the metal is aluminium or an alloy of aluminium; it is less dense, so makes the clothes-iron less tiring to use; it conducts heat well from the element to the clothes and the sole is at even temperature with no 'hot spots' that might scorch the clothes; it does not corrode with dampness. Mica is a mineral (rock); it splits naturally into very thin sheets, so has a shape suitable for putting in a clothes-iron; it is not affected by high temperature; it is a non-conductor of electricity. L, N, and E are the connections to the mains supply wires live, neutral and earth (see 29.40). $I = 4.167\text{A}.$ $R = 57.6\Omega.$ $W = Pt = 90\,000\text{J}.$

2 Heat by conduction and convection; infra-red radiation; visible light mainly in the red wavelengths. An alloy of nickel and chromium; it is resistant to corrosion, even when hot. If we compare wires of equal thickness, we find that nichrome wire has higher resistance than most other commonly used metals or alloys; we need a relatively short length of nichrome wire to make a filament with the resistance needed for a heater coil. Fireclay can stand high temperatures; it is electrically non-conducting. To direct a beam of radiant heat the reflector is usually parabolic in section, with the heater filament at the principal focus. The filament operates at red heat; at this temperature it is able to set fire to clothes, paper and many other inflammable objects that contact it; it also injures the skin; the guard prevents objects and people from accidentally touching the heater. $Q = It = 750 \times 3\,600 = 2\,700\,000\text{J}$ or 2.7MJ

Instructions: measuring the power of a lamp

2 An ampere is the name for 'coulomb per second'.

3 Rate of energy conversion $= IV$.

4 This is the rate of energy conversion, calculated at step 3. Agreement between your result and the marking may not be exact because of variations in manufacture.

26.21 Meters

Instructions: making a model galvanometer

7 If the (conventional) current flows in a clockwise direction (viewed as in the figure) and if the field is from N to S, electromagnetic force acts horizontally on the two sides of the coil. On the left-hand side it acts in a forward direction; on the right-hand side it acts in a backward direction. The two forces create a couple, which makes the coil turn to bring the pointer nearer to A.

8 The couple is in the opposite direction; the pointer is turned towards B.

10 Adding lamps in parallel has the effect of adding resistances in parallel; the total resistance of the circuit is decreased; the current is increased; the pointer turns through a greater angle.

Questions page 275

1 As the coil turns, the foil is twisted. The twisted foil produces a couple that *opposes* the electromagnetic couple. The greater the angle of twist, the greater the opposing couple. The coil turns until the two couples balance.

2 The stronger the current, the greater the electromagnetic couple, the more the coil must turn before the electromagnetic couple is balanced by the opposing couple.

3 It would turn through 90°. The plane of the coil would then be perpendicular to the magnetic field. The two forces of the electromagnetic couple would then act *along the same line*; being oppositely directed they would cancel and exert no turning action.

4 Magnets of greater strength; thinner, longer or narrower foil (weaker opposing couple); more turns of wire in the coil (current passes a greater distance in the magnetic field).

Questions page 276

1 $I_G/I_S = 1/9$, as before. $I_G = I_S/9$. $I_A = I_S + I_G = I_S + I_S/9 = \frac{10}{9}I_S = 4$. $I_S = 3.6$mA. $I_G = 0.4$A. As before, one-tenth of the current goes through the galvanometer; nine-tenths goes through the shunt resistor. The pointer points to the 0.4mA mark, but this is re-numbered '4mA' to indicate the *total* current flowing through galvanometer *and shunt resistor*.

2 $I_G = 1$mA. $I_S = 99$mA. $R_S = 1/99 \times 75 = 0.758\,\Omega$. The scale is re-numbered to read '100mA' at full deflection; numbers on the rest of the scale are altered in proportion.

3 $I_G = 1$mA. $I_S = 9\,999$mA. $R_S = 1/9\,999 \times 75 = 0.0075\,\Omega$, or 7.5m$\Omega$.

4 For 10mA f.s.d., total $= 7.5\,\Omega$. For 100mA f.s.d., total $= 0.75\,\Omega$. For 10A f.s.d., total $= 7.5$mΩ.

5 The total resistance of an ammeter is low.

6 An ammeter is used for measuring how much current flows round the circuit. To do this, all the current must go through the ammeter; the ammeter must be in series with the rest of the circuit.

7 If its resistance was high, it would reduce the current flowing through the circuit. It would reduce the current that it was measuring. With low resistance, an ammeter has little effect on the current; measurement is accurate.

8 Yes, the resistance of the shunt must be accurate; if it is not, the fraction of current going through the galvanometer will be too large or too small.

Questions page 277

1 V_G must be 130mV, as before; I must be 100μA, as before. If $V_V = 100$V, $V_S = 100 - 0.13 = 99.87$V.

$$R_S = \frac{V_S}{I_S} = \frac{99.87}{100 \times 10^{-6}} = 0.9987 \times 10^6 \approx 1\text{M}\Omega$$

2 $V_S = 249.87$; $I = 100\,\mu$A

$$R_S = \frac{249.87}{100 \times 10^{-6}} = 2.4897 \times 10^6 \approx 2.5\text{M}\Omega.$$

3 It is high, usually several hundred kilohm.

4 We use the voltmeter to measure p.d. between two points in the circuit.

5 A low resistance would carry a large current between the points to which the voltmeter is connected. This would reduce the p.d. that we are trying to measure.

Questions page 278

1 More turns on the armature coil; stronger magnets; a soft iron armature increases the strength of magnetic field through the coil; as in the galvanometer; improved bearings to reduce friction; shape the poles as in fig. 26.6, to give a stronger magnetic field with more uniform turning force.

2 No. The couple on the coil would be changing in direction as the current changed direction. The opposite couples would cancel, so the armature would not turn.

26.31 Electrolysis and the flow of current

Instructions: an experiment to test the prediction that m varies with t

1 The larger the amount of copper deposited, the greater the percentage accuracy of the weighing.

7 The mass has increased by about 0.2g. Copper has been deposited on the cathode.

8 Results confirm that $m \propto t$. The hypothesis is supported.

Instructions: an experiment to test the prediction that m varies with I

2 Results confirm that $m \propto I$. The hypothesis is supported.

27 Flight

27.00 Pressure in moving fluids

Instructions: investigating the effect of motion on air pressure

2 Their free lower ends move towards each other, reducing the width of the gap.

3 The harder you blow, the closer the sheets come together. They almost touch.

4 The air pressure in the gap must be lower than the air pressure outside. The reduction of air pressure between the papers must be caused by the motion of air.

27.10 Aerofoils

Instructions: applying Bernoulli's principle to the aerofoil

4 It rises. The harder you blow (up to a certain amount), the more it rises.

5 The pressure on the upper surface is lower than the pressure on the lower surface.

6 The air speed over the top surface must be greater than that over the lower surface.

Questions page 286

1 A hydrofoil has a section similar to that of an aerofoil, but it is designed to operate in water. Hydrofoils are fixed on supports beneath a high-speed boat. When the boat moves quickly, the hydrofoil lifts the boat. The hull of the boat is lifted above the water level; the hydrofoil remains below water level. The advantage is that there is little drag, since the hull is not in the water; this means that the boat can travel at very high speed. The hull is above the level of the smaller waves, so its motion is much smoother than that of a conventional boat.

2 One side of the ball is moving forwards faster than the rest of the ball (call this the 'faster side'); the opposite side is moving slower than the rest of the ball (call this the 'slower side'). On the faster side, the air is being given speed *in the direction of motion of the ball*, by skin friction (fig. 17.9); the speed of the air past the ball is reduced. The pressure is increased. On the opposite side the reverse happens; pressure is decreased. The ball is deflected towards the slower side. If the ball is spinning on a vertical axis, it is deflected right or left. If it is spinning on a horizontal axis perpendicular to its direction of motion, it falls faster or less fast than it would fall under the action of gravity alone. Players of tennis often make use of this effect to confuse opponents. Golf players give spin to the ball to give it lift, so that it travels further.

27.20 Forces acting on an aeroplane in flight

Questions page 287

1 If there is a sudden gust of wind to make the aeroplane pitch forwards or backwards or roll sideways, lift and weight form a couple which tends to *restore* the aeroplane to its level position. The aeroplane is like the toy of fig. 12.7; the aeroplane is balanced on its centre of lift, just as the toy is balanced on the point of the needle. The centre of gravity is below the point of support (centre of lift) giving stable equilibrium.

2 This is to bring the centre of gravity just below the centre of lift. If the passengers occupy only the front seats, the centre of gravity is shifted forwards and the weight of the aeroplane (plus passengers) does not act along the same line as lift; this creates a couple and makes it difficult to control the angle at which the aeroplane is flown.

3 Heavy cargo must be positioned so that the centre of gravity is just below the centre of lift.

Questions page 287

1 If they are moved downwards, the extra upward pressure makes the aeroplane pitch forward, reducing angle of attack. Lift decreases, the aeroplane decends. If moved upwards, the aeroplane pitches backwards, the angle of attack increases, lift increases; the aeroplane climbs. The elevators have another use: the pilot can set the elevator at some small angle up or down to produce a *constant* force upwards or downwards. This force creates a couple which is used to balance any couple caused by the lift and weight forces not being exactly in line.

2 If the distance is great, the moment of the forces acting on the tailplane and elevators is large. Only a small force is required to alter the angle at which the aeroplane flies.

3 This works like the tailplane, giving stability of direction. Any slight change of direction to left or right is immediately corrected.

4 By turning the rudder to left or right, the aeroplane is turned to left or right by forces acting in a sideways direction on the rudder.

Instructions: controlling a model glider

6 They act in the same way as a tail fin, to give stability of direction.

9 It has stalled. The centre of gravity is probably too far behind the centre of lift; the glider pitches backwards, climbs, loses speed and stalls. Use a heavier paper clip or more than one clip.

10 The centre of gravity is too far forward. Move the paper clip back away from the nose, or use a smaller clip or a few staples near the nose.

11 The centre of gravity is still too far back, but not far enough back to make the glider stall completely. Take action as in step 9.

12 Check the folding of the wing-tips to make sure it is symmetrical. If not, it is probably better to start again and make a new glider. Check also that the nose or tail fins have not become bent or curled during the earlier trials. Most gliders of this type fly exactly straight ahead without adjustments, because of the large tail fin, the wing-tip fins and the keel-like body.

14 *Both elevons up*: a downward force on the elevons makes the glider pitch backwards; it climbs; usually it loses speed and then stalls. In a powered aeroplane, extra power from the engine would maintain speed while climbing.

Both elevons down: an upward force on the elevons makes the glider pitch forwards; it dives. If you stand on a chair and hold the glider high above your head to launch it, you may be able to make it do a half-loop, turn upside down and fly close to the ground upside down.

One elevon up and one down: the glider turns towards the side on which the elevon is turned up; it also rolls over towards the same side; it banks correctly. In fig. 26.7h the glider banks and turns to the left. There is a downward force on the left elevon, and an upward force on the right elevon; these forces make the glider bank. The raising of the left elevon caused increased pressure in the air above the left wing; there is decreased pressure in the air above the right wing. The increased pressure over the left wing pushes the tail fin towards the right; the reduced pressure over the right wing helps this action. The tail fin is forced towards the right, making the glider turn to the left.

28 Engines

28.00 Energy from chemical fuels

Instructions: measuring the amount of energy in a liquid chemical fuel

6 For calculations, see 16.31. The amount of energy in 1 kg of liquid fuels is in the range 30 MJ to 47 MJ. Your answer was probably far smaller than this.

7 Much of the heat does not go to the water; instead it goes to the can, the dish in which the fuel is burning, the thermometer and (by convection) to the air. Your estimate of the amount of energy in the fuel is probably low.

8 Heat to the can = mass of can $\times$ specific heat capacity of metal of can $\times$ rise in temperature.

29 Producing electricity

29.01 Using an electric motor in reverse

Instructions: generating electricity using a d.c. motor
3 If the direction of spin is reversed, the current direction is reversed. The greater the speed, the greater the current.

29.02 Electromagnetic induction

Instructions: generating electricity by using a coil and magnet
4 A current is generated as you bring the magnet nearer the coil. When both are held still, there is no current.

5 As you move the magnet away from the coil, a current is generated. It flows in the reverse direction to the current generated at step 4.

6 The largest current is generated when you move the magnet as fast as you can towards or away from the coil. Current is generated only if there is motion.

7 Increased number of turns gives greater current.

8 A stronger magnetic field gives greater current.

9 This makes no difference to the results.

10 When a soft iron core is used, currents are greater.

Questions page 297
1 When the magnet and coil are not moved, there is no current.

2 More lines of magnetic force become linked with the coil. The linked magnetic field increases; there is *change*.

3 An e.m.f. should be induced; a current should flow. This is confirmed by observations.

4 Both movements cause a *change* in the linked field; both movements induce e.m.f.

5 It has more lines of force; as the magnet moves, the linked field changes at a greater rate.

6 The more turns, the more places in the coil linked with the field; linkage is stronger. Any motion produces a greater rate of change of linked field.

7 Lines of force from the magnet tend to pass into the core, because the core becomes magnetised by induction (8.10). More lines are diverted into the coil and are linked with it. There is greater rate of change when the magnet is moved. This means that greater e.m.f. is generated.

Instructions: investigating the direction of induced current
3 The current is reversed.

4 With poles reversed and magnet moving towards the coil, the current is the reverse of that shown in the figure. When the magnet is taken away, the current is the same as in the figure.

Questions page 297
1 A north pole (see fig. 9.7c).

2 The coil and magnets have like poles close together. They repel each other. The magnetic field of the coil opposes the motion of the magnet. The repulsion means that *extra work must be done* to move the magnet towards the coil.

3 A south pole.

4 The coil and magnet have unlike poles together. They attract each other. The magnetic field of the coil opposes the motion of the magnet. The attraction means that *extra work must be done* to move the magnet away from the coil.

5 The answer to question 1 is now 'a south pole'; the answer to question 3 is now 'a north pole'. The answer to questions 2 and 4 are the same as before.

6 When the magnet moves, the current induced in the coil produces a magnetic field in the coil which opposes the motion of the magnet.

Instructions: investigating the choke, an application of Lenz's law

5 The lamps should have similar brightness (if not, check that they really have the same rating).

6 When the core is pushed in, there is no effect on the brightness of the lamps.

7 Without the core, there is very little difference in brightness; when the core is pushed in, the lamp in series with the coil becomes dim; it may give out no light at all.

Questions page 298

1 The magnetic field is zero as the current changes direction; then it increases rapidly in strength.

2 An e.m.f. is induced in the coil to *oppose* the increasing magnetic field. This produces a current *opposite in direction* to the current that is already flowing in the coil.

Questions page 298

1 By switching the current on or off.

2 When the current is switched on, there is sudden increase in current flowing through the coil. The induced e.m.f. opposes this increase. The effect of this is that the current does not increase instantly, but gradually. Even so, the rate of increase is too quick for you to notice by looking at a lamp connected in series with the coil. As the current is switched off, the sudden disappearance of the magnetic field induces an e.m.f. to oppose its disappearance. This makes a current flow in the coil for a short time after the supply current has been switched off.

29.03 Alternating current generator

Questions page 300

1 At positions 1 and 9. This is when the plane of the coil is perpendicular to the direction of the magnetic field

2 Twice.

Questions page 300

1 Large numbers of turns in the coil; high speed of rotation of the armature; strong magnetic field; soft iron armature.

2 Cathode ray oscilloscope. Connect the generator to a resistor of suitable value. The alternating e.m.f. produces an alternating p.d. across this resistor. The internal resistance of the coil is low, so p.d. is very nearly equal to e.m.f. The CRO is connected across the resistor; its display shows how the p.d. varies with time. You can measure the p.d. ($\approx$ e.m.f.) by measuring the trace on the screen. Then, knowing the value of the resistor, you can convert all p.d. measurements to current values, if required.

Instructions: investigating the heating effect of a.c.

4 No. Its peak values are greater (ignoring sign) than the steady value of the d.c. Alternating current is near its peak values for only a short length of time. At other times it is much lower. The average alternating current flowing (ignoring its direction, which does not matter) has the *same heating effect* as the steady d.c.

5 If this is from a mains-powered-pack, the frequency is 50 Hz (or 60 Hz in some countries). If the a.c. is from an audio-frequency oscillator, your result can be checked against the setting on the scale of the oscillator.

Questions page 301

1 The e.m.f. from a battery is constant, that from the generator varies regularly between zero and V_{peak}.

2 Connect a capacitor across the output terminals. If this has sufficiently high capacitance, it can store charge when the e.m.f. is close to V_{peak} and can supply charge to the external circuit when e.m.f. is low. It *smooths* the output of the generator.

3 $V_{\text{r.m.s.}} \approx 14\,\text{V}$. A battery with e.m.f. 14 V is needed.

29.10 Electricity from chemical energy

Instructions: investigating the terminal p.d. of a cell or battery

The exact values obtained in this investigation depend on what type of battery you use, how old it is and on the exact values of resistor used. Typical results are given here, to help you see how to work out your own results.

1 Terminal p.d. = 6.2 V. If we ignore the small current that flows through the voltmeter (26.21), which is less than a milliampere, we can say that with no current flowing our measurement of p.d. *represents* the e.m.f., the force available to drive current around any circuit to which the battery may later be connected. We will take the e.m.f. to be 6.2 V in the discussions of the answers below.

2 Terminal p.d. = 5.8 V. Energy conversions are: chemical energy to electrical energy in the cell; electrical energy to internal energy in any part of the circuit that has electrical resistance, which means all parts of the circuit. Internal energy is produced in the lamp (appears as heat and light), in the connecting wires (very small amount of heat) and in the cell (but not enough heat to be detected when you put your fingers on the cell). If the e.m.f. is 6.2 V, the amount of chemical energy converted is 6.2 J/C (by definition, because this is what we mean by 'volt').

3 To find the amount of energy converted in the lamp (ignore the connecting wires), we need to know the terminal p.d. *across the lamp*. In this circuit this is the same as the p.d. across the battery, 5.8 V. Therefore the amount of electrical energy converted to internal energy (then to heat and light) is 5.8 J/C. We still have to account for the difference, 0.4 J/C. This amount of energy cannot be lost for the law of conservation of energy applies. This energy can be accounted for as electrical energy converted into internal energy in the battery; the battery becomes hotter.

4 Terminal p.d. $= 5.9$V. Energy converted in resistor is 5.9J/C. Chemical energy converted to electrical energy in battery is 6.2J/C; electrical energy converted into internal energy in the battery accounts for the difference, 0.3J/C.

5 Terminal p.d. $= 4.9$V. Energy converted in resistor is 4.9J/C. Chemical energy converted in the battery is 6.2J/C; electrical energy converted to internal energy in the battery accounts for the difference, 1.3J/C. If the resistor is connected for several minutes the battery becomes warm to the touch.

29.20 Electricity from other sources of energy

Instructions: making a thermocouple

3 The bigger the temperature difference between the junctions, the bigger the p.d. (except at very high temperatures, higher than you are likely to have used). The effect is greater with certain pairs of metals; good results are obtained with an iron-constantan thermocouple.

4 There is no *difference* of temperature, so no p.d. Heating alone is not enough, there must be a difference of temperature.

29.30 Transformers

Questions page 305

1 600V a.c. At its 1 500th turn.

2 186 turns.

3 With d.c., the magnetic field of the primary coil does not change. This constant field is linked to the secondary coil. Because it is not changing, no e.m.f. is induced in the secondary coil. A pulse of current is induced in the secondary coil when the d.c. supply to the primary coil is switched on or off. Why?

29.40 The distribution of electricity

Questions page 307

1 A 5A fuse on a 240V supply allows a power consumption of 1 200W. This is enough for 12 lamps of 100W each, or 15 fluorescent tubes of 80W each.

2 Many heaters are rated at 2kW. On a 240V supply the current is just over 8A. If connected to a lighting socket, the high current would burn out the fuse of the lighting circuit. This is a safety factor, since the wiring to the lighting sockets is relatively thin and is unsuitable for carrying currents as great as 8A.

3 If 30A was allowed to flow through the wires of the lighting circuit, they would over-heat; there would be risk of fire.

4 When the switch is off, the live terminal of the socket is completely disconnected from the wiring of the house. The other terminal is neutral, at earth potential. There is no danger of electric shock when removing or replacing a lamp in the socket. If the switch was in the neutral wire, the live terminal would be connected to the live of the mains, even when the switch was off. There would be danger of shock by accidentally touching this terminal. But *people make mistakes*; the switches in your house may have been put in the wrong wire. **Never touch any terminals of electrical equipment** while they are connected to the mains, **even if the switch is off.** Before you examine any electrically-powered device, make sure that it is *unplugged* from its socket. Always leave the examining of wiring and sockets to experts.

5 The earth wire runs to the ground just outside the house, or to a cold-water pipe that comes directly from the ground. It is at earth potential and provides an easy path by which electric current can flow to ground. Any item of electrical equipment with metal casing or other exposed metal parts has these connected to the earth wire of its three-wire cable. If there is an accident in which the metal parts come into contact with the live wires of the equipment, current flows to earth. A person handling the equipment is saved from serious electrical shock. If the current is large, the fuse blows and the operator knows that the equipment is faulty. Metal parts may accidentally touch the live wires in old or faulty equipment, especially if the insulation has broken off the live wires, owing to age. If equipment (including new equipment) is accidentally dropped or knocked over when the power is switched on, the damage might easily bring metal parts into contact with live wires.

6 If there is any fault or short-circuit in the equipment, the earth connection is made first, preventing the equipment being unsafe while the plug is being pushed into the socket.

7 Safety: if a plug is pushed in or pulled out while the socket is live, there is danger of touching the live pin with the fingers. Convenience: some devices do not have built-in switches and it is convenient (as well as safer) to have a switch in the socket.

8 To the live terminal, for safety (see question 4).

9 Safety: it prevents people (especially young children) from putting thin metal objects (or their fingers) into the live socket.

10 It reduces the risk of damage to the equipment. If a short-circuit develops, the fuse burns out immediately, cutting off the supply. If the supply were not cut off damage might become worse; the equipment might catch fire and be totally destroyed.

11 (a) this carries 8.3A, use a 13A fuse; (b) with a 150W lamp this carries 0.625A, use a 1A fuse; (c) this carries 2A, use a 3A fuse.

12 We would need to wire the whole ring circuit with cable able to carry *more* than 60A, to allow the cooker to be used at the same time as room heaters, electric kettles and other items requiring large currents.

13 The cooker can be completely disconnected from the mains. The cooker is permanently wired to the wall panel; it must be possible to disconnect it completely when it is to be serviced.

14 Turn off the cooker switch, so that it is completely disconnected from the mains.

30 Using materials

30.03 Tensile strength

Instructions: measuring tensile strength

4 Most metals have high tensile strength, with exceptions such as lead. Steel has very high tensile strength; steel wires are used for holding up bridges and masts and in musical instruments such as guitars and pianos. Wires made of other metals cannot be made tight enough to give high-pitched notes without breaking under the high tension required. Catgut has high tensile strength; it is used in musical instruments, for bow-strings and in surgery (sutures). Other fibres of high strength are silk, nylon and terylene, which are used for making fabrics. An automobile safety-belt must have high strength. Ropes made from various fibres (especially nylon and terylene) are used for mooring boats, tethering cattle and horses, hoisting sails, towing vehicles and securing parcels. It is interesting to note that the thread of a spider's web has tensile strength as great as that of many metals.

Instructions: comparing the rigidity of different materials

9 Frameworks for buildings (steel, concrete), sea walls, harbours, springs in clocks and balances (when you pull on a spiral spring you actually *twist* the wire, so this is an effect of rigidity, not of tensile strength).

10 Where there is much vibration, which is absorbed if the structure is not rigid; or where there may be sudden large forces, which must be absorbed. Examples are the frameworks of aeroplanes and frameworks of buildings in earthquake areas. Such structures need to be made from *elastic* materials which recover their shape when the external forces are removed. In an automobile a rigid body is dangerous; if the bumpers (fenders), bodywork and steering-wheel are made from materials that collapse and bend easily, much of the energy of impact in a crash is used for bending these. This leaves less energy available to harm the passengers. The vehicle may become damaged beyond repair, but the passengers may be saved.

11 Soft rubber for absorbing vibration (shock absorbers; flexible materials for clothing, curtains, tents, hoses, furnishings; many-stranded flexible cables and wires, having high tensile strength with low rigidity).

30.05 Ductility and brittleness

Questions page 311

1 Most rocks, baked clay, china, glass, brick, concrete, some plastics (perspex, polystyrene), some woods, cast iron.

2 Most brittle materials are very rigid. They *keep their shape*, even when forces acting on them are large. This is important for houses, bridges, cups, plates, laboratory glassware, machine parts and the plastic cases of radio sets, clocks and other equipment. But if the maximum force is exceeded even for an instant, the material breaks and is often damaged beyond repair; for example, damage to buildings in earthquakes, damage to crockery and glassware when dropped.

3 Most commonly used metals are ductile. They can be formed into wires; copper and aluminium are good examples.

30.06 Hardness

Questions page 312

1 Diamond is used for the stylus of the pick-up of a phonograph, or record-player; it does not wear out of shape as it is moved along the kilometres of groove on the discs. Sapphire is a slightly cheaper but less hard-wearing stone used for the same purpose. Diamond is used in glass-cutters and in 'pencils' for writing or engraving on glass.

2 Sapphire is used for bearings in watches and some clocks. Sapphire bearings are sometimes called 'jewels'. The knife-edge of a chemical balance is usually made from agate, another hard stone.

3 An oxide of aluminium, called corundum, or emery, is used as a powder or paste, in solid blocks (grindstones) or coated on to paper or cloth (emery paper). Silicon dioxide in the form of sand is used for sand-blasting or is coated on paper (sand-paper).

30.10 Electrolytic corrosion

Questions page 313

1 The electrode in the beaker into which air is bubbled becomes positive, compared with the electrode in the other beaker. A current flows from the electrode in the aerated beaker, through the meter, to the other electrode, then through the electrolyte (by way of the bridge tube) and back to the positive electrode again. The e.m.f. is a few millivolts. This effect is due to the *oxygen* in the air.

The electrolyte around one electrode has plentiful oxygen; the electrolyte around the other electrode has not; this creates a p.d. between electrodes. In the instructions, strong salt solution is suggested to give an easily measurable current. The same effect occurs with ordinary tap water, though p.d. and current are smaller.

2 Yes. Blowing air into the other beaker reduces p.d. If blowing is continued, p.d. is reversed.

3 You may notice that the nail in the non-aerated beaker (negative electrode) is duller in appearance than the other one, but this may not be easy to see until several hours have passed. Much earlier than this, you can see a powdery deposit of orange-brown iron (II) oxide on the bottom of the beaker, just below the nail. This is evidence of corrosion.

4 A few centimetres below the soil surface the supply of oxygen is less than that above ground. Dampness contains dissolved material, so acts as an electrolyte. The situation is shown in fig. 34.8. The iron above ground becomes a positive potential compared with the iron below ground. Current flows as shown. The negative electrode becomes corroded, the post corrodes *just below* the soil surface. Check this by examining old iron posts in damp soil or leaving an iron nail in salt solution for a few days.

5 Oxygen is absent in the water-logged soil. Sections of the pipe in this soil become negative electrodes; sections in the air-containing soil become positive electrodes. Current flows along the pipe and back through the damp and wet soil. The pipe corrodes in the water-logged soil.

6 The parts of the surface covered by mud have little oxygen around them. As in the previous examples, corrosion occurs where oxygen supply is low. It is important to keep metal surfaces clean, so that this effect cannot occur.

Questions page 314

1 At first, current is zero. When salt is tipped into one beaker, the electrode in that beaker becomes positive; a current flows.

2 When the concentrations of electrolyte are unequal, potentials are unequal; there is p.d. and current flows. The positive electrode is the nail in the beaker in which salt is tipped. The negative electrode corrodes. You can check this by letting the cell operate for a few hours. Iron pipes underground pass through regions of soil in which the concentration of dissolved salts varies. This often happens, and leads to corrosion of pipes in the regions of soil where salt concentration is low.

3 This type of cell depends on having the electrodes surrounded by electrolytes of different composition (in the instructions on page 313, much oxygen and little oxygen; in the instructions on page 314 much salt and little salt). Yet there must be an electrical connection between the two; the electrolyte in the bridge does this while keeping the two electrolytes from mixing too easily.

4 Stirring increases the amount of dissolved oxygen; in this cell we wish to keep the amounts of oxygen the same in both beakers.

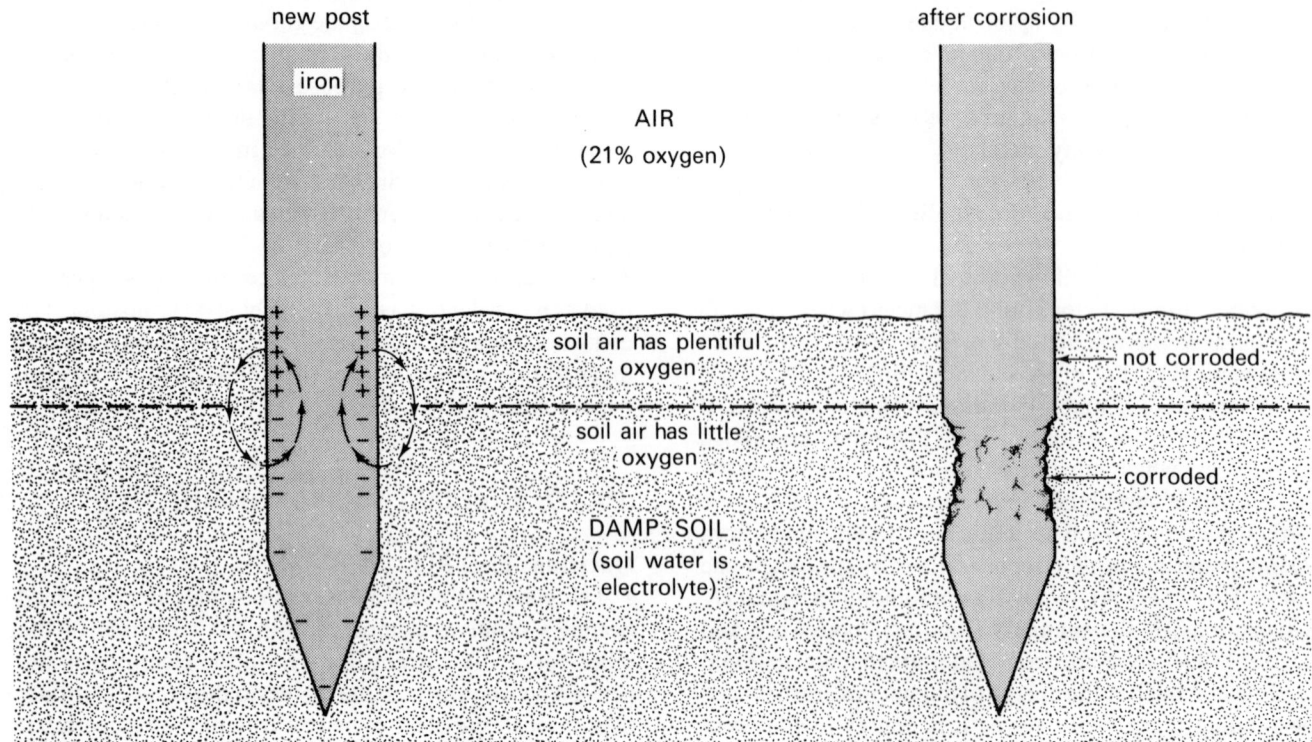

Fig. 34.8 Corrosion of an iron gate-post, in damp soil.

30.11 Preventing corrosion

Instructions: preventing iron from rusting
1 It would flow in the opposite direction to that shown in the figure.
3 The protected nail remains shiny. The unprotected nail is corroding; orange-brown iron (II) oxide can be seen in the electrolyte around the nail and on the bottom of the dish below the nail. The copper electrode is corroding; blue copper (II) chloride is seen around it.

30.21 Integrated materials

Instructions: investigating the effects of defects in a material
5 The notched strip tears with a lower load than the un-notched strip.
6 Notching has made the strip narrower and the force per unit area is greater than for an un-notched strip. The additional strain acts at the point of the notch and the material splits at this point, making the notch larger. With a larger notch the strain is even greater, and so the tear spreads rapidly across the material. The angle of notch does affect the ease with which it spreads. If the angle at the point of the V is small, a tear spreads quickly from the point, even with a small load; the additional strain is concentrated in the small area around the point of the notch. A fine slit or cut has the smallest angle, so spreads with a low load. A wider angled notch can withstand a higher load without spreading. A U-shaped notch, with a well *rounded* end is least likely to spread. To stop a tear or cut from spreading, its end must be made round: cut away the film around it to make it into a U-shaped notch; or use a punch to punch out a circular area around the point of the slit. If you have not tried these methods, try them now, to see if they work.
7 The material can be strengthened by fixing something to it to take the additional strain without giving way. You could glue a thin wire or thread parallel with the long edges. A narrow strip of self-adhesive tape stuck along each long side would help stop a notch from beginning to spread. If the polythene has a wide-mesh fabric fixed to it, it is prevented from tearing in any direction, unless strong forces are applied to it. You may have made suggestions of your own that are just as good as any of these.

30.22 Beams and girders

Instructions: investigating the strength of girders and other structures
5 Girders and tubes are much stronger than the other structures tested. The strength of a girder depends on the relative width of rails and web; the relation is complex. I-girders are resistant to bending in one direction, but can be easily bent in a perpendicular direction (you could find out more about this by a practical test). Tubes are equally resistant to bending in any direction. The strength of a single sheet of material is much increased by making it corrugated.
6 There are so many examples. Here are a few: *Girders*: building frameworks, bridges, cranes, automobile chassis; *Tubular frameworks*: steel furniture, bicycles, iron lamp-posts and telegraph posts; *Corrugated sheets*: corrugated iron for roofing, corrugated transparent plastic sheet for roofing, corrugations in parts of the bodywork of automobiles and aeroplanes (look for corrugations in the floor of an automobile, or in the floor of the luggage compartment), plastic packing boxes, corrugated cardboard.

31 Transistors

31.00 The transistor as a switch

Instructions: switching with a transistor
2 Both lamps are off. No current can flow through L2 as the switch is open. It is possible that a small current is flowing through L1, but it is too small to make the lamp glow.
3 L1 glows brightly; a large current is passing through it. L2 does not glow; maybe no current is flowing, or maybe there is a current that is too small to make the lamp glow.
4 L1 goes out. There has been a change in the flow of currents in the circuit caused by removing L2 from its socket. We get the same change by turning the switch off. This indicates that a current was flowing through L2 at step 3. It was a small current, not enough to make the lamp glow.

Instructions: measuring the potential of the base
2 When the resistor is set to a high value, the potential of the base is low. When the resistor is set to a low value, the potential of the base is high.
3 L1 is off; it does not glow.
4 L1 is on; it glows brightly.
5 If the transistor is a silicon transistor, as specified in Fig. 31.1, the potential of the base is about 0.6 V when the current begins to flow through L1. To give enough current to make the lamp glow visibly the potential may need to be a little higher than 0.6 V.

Instructions: using a light-controlled switch
3 The lamp lights. If it does not, check the connections of the circuit. If it still does not light, replace R1 with a resistor of higher value. LDRs vary in resistance according to their type and yours may need R1 to be higher than $10 k\Omega$.

4 It goes out. If the lamp does not go out, even when the LDR is closely covered with a black cloth, replace R1 with a resistor of lower value.

Questions page 320

1 Potential at A is $3 \times \dfrac{10\,000}{10\,100} = 2.97\,V$.

This makes I_b sufficient to turn the transistor on. I_c is large enough to light the lamp.

2 Potential at A is $3 \times \dfrac{10\,000}{30\,000} = 1\,V$.

I_b is not large enough to switch the transistor on. I_c is zero or low; the lamp does not light.

Questions page 320

1 As temperature increases, the resistance of the thermistor decreases; potential at A increases; I_b increases; the transistor is switched on; thus the lamp, L1 is switched on.

2 Connect the thermistor in place of R1 and connect R1 in place of the LDR (fig. 31.2). As temperature increases the potential at A decreases.

3 At first the lamp lights. Then the heat from the lamp warms the thermistor. This turns the transistor off. The lamp goes out and, as soon as the interior of the box cools down, the thermistor increases in resistance again; the transistor is switched on again and the lamp lights again. In this way the lamp is turned on and off indefinitely. The temperature inside the box rises slightly above and falls slightly below a certain fixed value; this is a simple *thermostat* (see fig. 6.4) but this one does not use a mechanical switch.

Instructions: making a delay switch

2 The lamp does not light when the switch is turned on. After a few seconds it lights, brightening gradually. When the switch is turned on, current flows through R1 and gradually charges the capacitor. Potential at A increases slowly. Then it reaches the correct value, the transistor is switched on.

3 During the brief period the switch was off, charge leaked away from the capacitor, through the base and emitter of the transistor. The capacitor became discharged as the potential at A was low. The switch was ready to repeat its delaying action.

4 Use a capacitor of lower capacitance, or replace R1 with a resistor of lower value, so that the charging current is increased.

5 The best way is to use a capacitor with higher capacitance; or connect a second capacitor in parallel. If you try to increase delay time by increasing the value of R1, the current may be so small that it leaks away through the transistor; then the capacitor will never become charged to a potential that is high enough to switch the transistor on.

31.20 A transistor as an amplifier

Instructions: using a transistor as an amplifier

2 I_b can flow through the resistor to b and then to e and back to the negative terminal of the battery (conventional current). If I_b flows, I_c flows. Both currents are *steady*; they do not change in size. The evidence for the collector current is the 'click' you hear when you switch on.

Questions page 322

1 As p.d. varies, charge is added to or taken away from plate A. The potential of A varies rapidly in a way corresponding to the vibrations of the sound entering the microphone. This induces corresponding variations in the potential of B.

2 Current flows *from* B. This current is added to the steady current flowing through the resistor. I_b is thus increased.

3 Current flows *to* B. Some of the current flowing through the resistor goes to B, the remainder is I_b, and goes to the transistor. I_b is reduced.

4 As I_b varies, there are corresponding variations in I_c. Variation of I_c causes sound to be emitted from the loudspeaker (26.13).

32 Radio

32.00 Oscillators

Questions page 325

1 The springiness of the spring; the mass of the block.

2 No, unless the amplitude is exceptionally large.

3 The capacitance of the capacitor; the structure of the coil, especially the number of turns and the presence of a core of soft-iron or other ferromagnetic material. The resistance of the wires is also a factor, but of less importance in the type of circuit shown in the figure.

4 If we can compare the two systems and find them to be similar in all ways, it is reasonable to think that frequency is not affected by amplitude, if amplitude is not large. This assumption is correct.

5 Use a different spring; alter the mass of the block; surround the system by water or oil.

6 Alter the capacitance of the capacitor; alter the number of turns in the coil; remove the core or replace it with a different one.

7 One way is to hit the block gently, *in time with its oscillations*. We supply energy in small amounts to replace the energy that is lost (also in small amounts) at each oscillation. If we tap the block gently, each time it moves down, our finger is in *resonance* with the system; all the energy we supply is transferred to the system. The situation is similar to that of pushing somebody who is on a swing.

8 The same principle applies to the capacitor-coil system. We must supply small amounts of charge to the capacitor *every time it is charging*. The supply of charge must be in resonance with the circuit.

9 We could have another coil wound on the same core as the coil of the oscillator circuit. A pulse of current through the coil produces a magnetic field linked to the circuit; an e.m.f. is induced; the circuit is given energy.

32.20 Radio wave detection

Instructions: making a radio receiver with amplifier
3 (a) A small *steady* current flows through the $100k\Omega$ resistor; this keeps the potential of the base above $0.6\,V$, so that a small collector current flows. (b) A *varying* current flows from the diode, varying as the oscillating current in the tuning circuit. This is added to the steady current to make I_b. I_b varies according to the audio-frequency modulation of the received radio waves.

4 It is much larger than I_b. Its size varies with I_b. Thus I_c varies as the audio-frequency modulation. This drives the earphone.

33 Nuclear power

33.00 Matter into energy

Questions page 330
1 When 1g decays the amount of matter destroyed is $0.006/226.025 = 2.65 \times 10^{-5}g = 2.65 \times 10^{-8}kg$.
$E = 2.65 \times 10^{-8} \times (3 \times 10^8)^2 = 23.85 \times 10^8 J$.
The loss of matter is less than 30 micrograms, yet this amount of energy, as electricity, would keep a 1 000W spotlight bulb or room-heater running day and night for 4 weeks.
2 $E = 1 \times (3 \times 10^8)^2 = 9 \times 10^{16} J$. This is about 1.8×10^9 times more energy than can be got from the same mass of petroleum by burning it.

Revision questions

(Answers to numerical problems are on p.396)

Part A   Measuring

1 Measurement of length, mass, time

1.1 Write out in full the units represented by these symbols: m, mg, d, s, mm², Pa, min, km³, dm, kPa.

1.2 Which are the most suitable units to use for stating (a) the mass of a door-key, (b) the volume of air in a room, (c) the mass of a man, (d) the density of glass, (e) the volume of a drop of water, (f) the area of a farm?

1.3 The density of castor oil is 950kg/m³. Calculate the mass of 1.6 litres of castor oil.

1.4 A relative density bottle together with its stopper has mass 16g. Some pieces of lead shot are put into the bottle; the total mass of bottle, stopper and shot is now 129g. Next the bottle, still containing the shot, is filled with water, and the total mass is 169g. Finally the bottle is emptied, then refilled with water only; the total mass is now 66g. Calculate the relative density of lead.

1.5 A petrol can holds exactly 20 litres of fluid. Its mass is 1kg. What is the total mass when the can is completely filled with petrol, with relative density 0.8?

1.6 The can in question 1.5 has a flat, rectangular, bottom surface, measuring 25cm × 17cm. When the can is full of petrol, what pressure is exerted by the can on the flat floor on which it is standing?

2 Using measurements

2.1 What is meant by the term *lever*? A lever 1.2m long is balanced half-way along its length and is horizontal. A 6kg mass is hung at a point 0.3m from one end of the lever. Where must a 9kg mass be hung from the lever to keep it balanced horizontally? Draw a diagram to explain your answer.

2.2 A pendulum consists of a bob, mass 250g, hanging on a thin cord 1m long. Its period is 2s, and its amplitude of swing is 0.1m. How could you change the pendulum to make it swing with a period of 1s?

2.3 Draw a pulley system for which load/effort = 4.

2.4 Explain the working of (a) a simple mercury barometer and (b) an aneroid barometer. What are the advantages and disadvantages of each instrument?

2.5 Explain the working of (a) a liquid manometer and (b) a Bourdon gauge. What are the advantages and disadvantages of each instrument?

3 Measuring temperature

3.1 You are given a mercury-in-glass thermometer, but it has no scale marked on it. You have no other thermometer. Describe what you would do to mark your thermometer to show temperatures in the Celsius scale.

3.2 List the reasons why mercury is a very suitable liquid for use in thermometers. What are the advantages and disadvantages of ethanol for use in thermometers?

3.3 In northern countries, salt is put on the roads during winter, to keep them clear of ice. Explain the reasons for this.

4 From measurements to laws

4.1 Explain, by giving one example of each, what is meant by the terms *hypothesis*, *experiment* and *law* (or *principle*) in science.

4.2 A hollow ball, mass 2kg, and made of plastic, floats completely immersed in methylated spirit. The density of the plastic material is 1 250kg/m³ and the density of methylated spirit is 800kg/m³. Calculate the internal volume of the ball.

4.3 State *Archimedes' principle*. A cube of wood is floated in water in a measuring cylinder. When the wood is put into the water, the water level rises from 50cm³ to 67cm³. Next the same piece of wood is placed in a measuring cylinder containing methylated spirit. It sinks and the level of the spirit rises from 60cm³ to 80cm³. Calculate the relative density of the wood.

Part B Discovering Physics

5 Discovering about light

5.1 State the *laws of reflection* and describe an experiment to demonstrate one of these laws.

5.2 A ray of light enters the plane, polished surface of a block of perspex with an angle of incidence of 30°. The air-to-perspex refractive index is 1.5. Calculate the angle of refraction.

5.3 Explain why a swimming pool appears to be less deep than it really is. A boy looked into a river and saw a fish apparently 0.36m below the surface. He tried to spear it, but missed. What was the real depth of the fish? (Air-to-water refractive index = 1.333)

5.4 Draw a diagram to explain what is meant by an *eclipse of the Sun*. Why is the Sun not eclipsed every month?

5.5 Draw a ray diagram to show how a spectrum from a ray of white light is formed by a triangular glass prism. A small object that has a yellow colour when seen in ordinary daylight is moved across the spectrum from one end to the other. Describe and explain the changes in its appearance that occur, assuming that the demonstration is done in a darkened room.

6 Discovering about heat

6.1 Explain, giving examples, the difference between a *chemical change* and a *physical change*. Give three examples of the way in which we make use of physical change.

6.2 Give an example of convection currents. Explain how heat is transferred by convection.

6.3 Describe, with the help of a diagram, a demonstration that a rod of metal expands when heated.

6.4 When water freezes its density changes from 1 000kg/m³ to 920kg/m³. What is the increase in volume when 100cm³ of water freezes? Explain the practical importance of this expansion.

7 Discovering about sound

7.1 A person drops a piece of rock on the floor. You are standing a few metres away and hear the noise made by the rock as it hits the floor. Explain how the energy of the falling rock reaches your ears as sound. What happens to your ear-drum when the sound reaches it?

7.2 Explain why we hear thunder several seconds after we have seen the flash of lightning from a distant storm. If the thunder is heard 9s after the lightning is seen, calculate the distance of the storm, given that the speed of sound in air is 330m/s.

7.3 A person stands 100m from a large wall and claps regularly, the claps being in time with the sound of their echoes reflected from the wall. The speed of sound is 340m/s. Calculate the frequency of clapping.

7.4 Explain what is meant by *resonance*. Give two practical examples of how we make use of resonance.

8 Discovering about magnets

8.1 You are given a bar magnet and a bar of steel of similar size and shape. Explain what you would do to discover if the bar of steel is a magnet.

8.2 What are the differences between the magnetic properties of steel and soft iron? Describe a practical application of the magnetic properties of these materials.

8.3 With the aid of diagrams explain what is meant by the terms *geomagnetic pole*, *declination*, *dip*, and *magnetic meridian*.

8.4 Describe two ways of making an unmagnetised bar of steel into a bar magnet. Explain what changes occur in the steel as it becomes magnetised.

8.5 Draw a diagram of the magnetic field in the region of two bar magnets that have been placed so that there is a neutral point between their ends.

8.6 Draw a diagram of the magnetic field in the region of a bar magnet arranged so that it lies along the magnetic meridian with its north pole pointing towards the North Geomagnetic Pole. What direction is taken up by small compass needles placed at the neutral points? What would be the effect on the field if the bar magnet were replaced by a stronger one?

9 Discovering about electricity

9.1 With the help of a diagram describe the structure of a dry Leclanché cell. In what way is polarization of the cell avoided?

9.2 State *Ampere's rule*. Describe how it can be demonstrated in the laboratory.

9.3 With the help of a simple diagram, explain the action of an electric bell.

Part C Investigating Forces

10 Investigation motion and momentum

10.1 Explain, by giving examples, the differences between a *scalar quantity* and a *vector quantity*.

10.2 A body of mass 25 kg is travelling with a uniform velocity of 25 m/s. For a period of 10 s a force of 50 N acts on it, in the direction in which it is travelling. Calculate the velocity of the body at the end of the period of 10 s and the distance it travels during that time. What is the impulse of the force on the body?

10.3 State the *principle of conservation of momentum*. How may this principle be used to explain the action of a rocket?

10.4 A bullet with mass 8 g is fired at a block of wood, mass 92 g, that is resting on a smooth horizontal table. As it strikes the block the bullet has horizontal velocity 250 m/s. Calculate the velocity of the block with the bullet embedded in it. The block then strikes a soft cushion and is decelerated to rest while travelling a distance of 10 cm. What is the rate of deceleration? Assuming that deceleration is uniform, what impulse does the cushion exert on the block?

10.5 A girl standing on ice-skates has mass 80 kg (including clothes and skates). She is holding a metal ball, mass 10 kg. She throws the ball forward with velocity 4 m/s. What velocity does the girl attain from this action?

10.6 A rocket, mass 50 kg, is travelling at uniform velocity of 100 m/s. Its rocket motor is then fired to accelerate it by 15 m/s² while it travels a distance of 1 km. By how much does its momentum increase during this firing?

11 Investigating forces in action

11.1 A small object is acted on by force 10 N north and force 5 N east. Find the direction of the resultant force. The object has mass 5 kg and is free to move. Calculate its acceleration.

11.2 List the forces acting on an object that is falling freely in still air. Explain what is meant by *terminal velocity*.

11.3 A piece of rock has mass 5 kg. What is its weight (a) on the surface of Earth and (b) on the surface of the Moon? It is thrown horizontally with a velocity of 5 m/s. What is its momentum (c) on the surface of Earth (d) on the surface of the Moon? (For Earth take $g = 10$ m/s²; for Moon take $g = 1.6$ m/s²).

11.4 A train is travelling north along a straight track. A bird flies overhead eastward *across* the track. To the engine-driver the bird appears to be travelling in a direction 60° south of east with velocity 5 m/s. What is the velocity of the train relative to the bird? What is the velocity of the bird relative to the Earth? What is the velocity of the train relative to the Earth?

12 Investigating turning forces

12.1 Explain what is meant by the term *moment*. What is the SI unit of moment? If several forces are acting on a body, not all along the same line, yet the body is not moving or turning, what can you say about the moments of these forces?

12.2 A lever AB, length 10 m, mass 5 kg has its centre of gravity half-way along its length. It is pivoted at B. A mass 20 kg is attached to the lever 2 m from end B. Draw a diagram to show the forces acting on the lever. What force must be applied at end A to hold the lever stationary? ($g = 10$ m/s²).

12.3 You are given a triangular piece of thin card; its sides measure 15 cm, 20 cm and 35 cm. Describe how you would find the position of the centre of gravity of the card by practical investigation. Draw a diagram to scale to show the position of the centre of gravity of this piece of card.

12.4 You are shown a horizontal flat table on which there is a small metal ball and a large watch-glass or saucer. Explain how you would use these items to demonstrate the three types of equilibrium.

13 Investigating force and pressure

13.1 State *Archimedes' principle*. A piece of metal weighs 20 N in air, but when hung on a thread and completely immersed in water it weighs only 17.5 N. Calculate the relative density of the metal.

13.2 A U-tube manometer contains glycerine (propane-1, 2, 3 triol), which has density 1 260 kg/m³. One end is open to the atmosphere, the other is connected to a sealed cylinder. If the liquid level in the open side of the manometer is 5 cm higher than the level in the other side, what can be said about the pressure in the container? ($g = 10$ m/s²). If another manometer connected to the same container has its U-tube twice the diameter of the first one, and contains the same liquid, what is the difference in level of the liquid in its two sides?

14 Investigating electromagnetic force

14.1 A small button-cell, used in the light-meter of a camera can supply a current of 3.5 mA for 100 h. Assuming that the cell supplies a constant current for the whole of its life, what quantity of electric charge does it supply before it becomes exhausted?

14.2 Describe a simple demonstration of the force between two parallel conductors when they are carrying an electric current. What is the effect of reversing the direction of the current in one of the conductors?

14.3 A current of 2.5 A is flowing in a circuit. For how long should the circuit be switched on so that exactly 60 C of charge flows?

Part D   Investigating energy

15 Work, energy and power

15.1 What is meant by the terms *work*, *energy* and *power*? In what units are these quantities expressed in SI?

15.2 What is meant by the *velocity ratio* of a machine? Show how the velocity ratio of an inclined plane can be calculated.

15.3 A pulley system has an efficiency of 80%. To lift a load of weight 1 600 N requires an effort of 500 N. Calculate the velocity ratio of this system. For what reasons may the efficiency of the pulley system be less than 100%?

15.4 If the pulley system of question 15.3 is being used to lift the load of 1 600 N at a velocity of 0.2 m/s, what is the power developed by the engine providing the effort?

15.5 A truck is being driven along a straight, level road at uniform velocity of 20 m/s. The mass of the truck is 2 000 kg. The engine is developing 5 kW. Calculate the total force resisting the motion of the truck. If the engine is turned off and the truck decelerates to rest, how much energy does it lose?

15.6 Explain what is meant by *kinetic energy* and *potential energy*. Give three examples of each.

16 Using energy

16.1 State the principle of *Conservation of Energy*. Name five different forms of energy and give three practical examples of the way in which one form of energy is converted into another.

16.2 A quantity of water, mass 2 kg, in a well-insulated container is at temperature 15°C. A cube of copper, mass 0.25 kg, is first heated to 180°C, then dropped into the water. The water is stirred rapidly and the temperature rises to a maximum value, exactly 17°C. In this operation water gains heat and the copper loses heat. Are the two amounts equal? If not, what quantity of heat is unaccounted for? Say what you think might have happened to this heat. (Assume that the specific heat capacity of copper = 0.4 kJ/kg K and that the specific heat capacity of water = 4.0 kJ/kg K.)

16.3 A copper electric kettle contains 1.5 kg of water, the mass of the kettle is 1 kg. The heating element is rated at 2.5 kW. How long will it take to heat the water from 20°C to 90°C? (Assume specific heats as in question 16.2).

Part E   Investigating matter

17 The structure of matter

17.1 Write out in full the units represented by these symbols: mW, kPa, μF, K, pV, MN, mHz, kC, pA, ms.

17.2 Express the following quantities in standard form: 356 kg, 145 000 000 mm, 0.25 s, 25.472 μF, 0.00025 Pa, 1 258 000 000 000 N, 25 462.56 kg.

17.3 Describe two ways in which you could demonstrate surface tension.

17.4 What is meant by *Brownian Movement*? What does this indicate about the structure of matter?

18 Matter and heat

18.1 Describe two ways in which energy may be given to molecules. Explain what is meant by the terms *internal energy* and *transfer of energy*.

18.2 Explain what is meant by the term *latent heat*.
An electric heater is used to melt a block of ice, mass 1.5 kg. If the heater is powered by a 12 V battery and a current of 20 A flows through the coil, calculate the time taken to melt the block of ice. (Take the specific latent heat of fusion of ice to be 336 000 J/kg).

18.3 State *Charles' law* for a gas at constant pressure. A balloon is filled with 175 m³ of gas in the early morning, when the temperature is 18°C. Calculate the volume of the gas, later in the day, when the temperature reaches 25°C. (Assume that atmospheric pressure remains constant).

19 Matter and electricity

19.1 Draw circuit diagrams to show how a 12 V battery may be used to operate (a) 4 bulbs, marked 3 V 0.06 A and (b) 3 bulbs, each marked 12 V 100 mA. In each example, calculate the total current supplied by the battery.

19.2 Describe how you would measure the resistance of a coil of wire believed to have a resistance of about 20 Ω . Mention the ranges of any measuring instruments that you would use.

19.3 State *Ohm's law*. Three resistors, with resistances 250 Ω , 500 Ω and 1 kΩ are connected in series. A 6 V battery is connected to either end of the chain of resistors. Calculate the potential difference between the ends of each of the resistors. Calculate the current flowing through each resistor.

19.4 The three resistors of the previous question are now connected in parallel and the same battery is connected across their ends. Calculate the potential difference across each resistor, and the current flowing through each. If the three resistors are to be replaced by a single resistor having the same effective resistance, what should its value be?

19.5 Describe two ways in which temperature may be measured electrically.

19.6 Describe three examples in which conduction of electricity does not obey Ohm's law. Explain why the law is not obeyed in each case.

Part F Physics of the Universe

20 Atoms

20.1 Define the terms: *nucleon number, proton number, element, nuclide* and *isotope*.

20.2 A radioactive substance has a half-life of 3 hours. Draw a graph to show how the activity would change over a period of 24 hours.

20.3 A radioactive material is said to emit alpha and beta radiation, but not gamma radiation. Describe the way in which you would try to confirm this.

20.4 A radioactive isotope of phosphorous, $^{32}_{15}P$, decays to produce a nuclide of sulphur with nucleon number 28. Write out the equation for this change. What instrument would you use to detect the radiation produced? The half-life of $^{32}_{15}P$ is about 14d. If a sample contained 100g of the phosphorus nuclide, what mass of phosphorous would remain after 70d?

21 Waves

21.1 What is meant by the *electromagnetic spectrum*. Name the main types of radiation in this spectrum and list briefly the features of each.

21.2 Describe how you would use a ripple-tank to demonstrate (a) the reflection and (b) the refraction of plane wave-fronts.

21.3 A loudspeaker is emitting a note of exactly 256Hz. You have a tuning fork that is marked with this frequency but may differ from 256Hz by a few hertz. Describe how you would measure the exact frequency of the fork. If the speed of sound in air is 330m/s what is the wavelength of the sound emitted at 256Hz by the loudspeaker?

21.4 What is the *photoelectric effect*? What does this tell us about the nature of electromagnetic radiation?

22 Electric charge

22.1 Define the *capacitance* of a capacitor. What factors affect the capacitance of a parallel-plate capacitor?

22.2 Describe, with the aid of diagrams, the action of the Van de Graaff generator. What is the advantage of using a sphere to store the charge?

22.3 A capacitor, capacitance $500\,\mu F$, is being charged with a constant current of $10\,\mu A$. How long does it take for the potential difference across its terminals to rise from 2V to 10V?

23 Electrons

23.1 What are *cathode rays*? Describe one practical use to which our knowledge of cathode rays has been put.

23.2 Describe *Millikan's experiments*. How can the measurements he made be combined with measurements made with a discharge tube so that we can calculate the mass of an electron?

23.3 Describe the action of an X-ray tube. In what ways do we make use of X-rays?

Part G Physics in action

24 Working with lenses

24.1 Explain the difference between (a) a converging lens and a diverging lens; (b) a real image and a virtual image.

24.2 Draw a diagram to show the structure of the human eye. Which of its parts have the same action as corresponding parts in a photographic camera? What is meant by short-sightedness? By means of a diagram show how spectacles can be used to overcome short-sightedness.

24.3 Draw a ray-diagram to show how (a) a compound microscope or (b) an astronomical refracting telescope works. Add brief notes to explain the essential features of the optical system you have drawn.

24.4 A screen is placed 5m from the converging lens of a slide projector. The image of the slide is 20 times as wide as the slide. Find by a scale drawing the distance between the lens and the slide, and the focal length of the projector lens.

24.5 An object 50mm long is 150mm from a converging lens, focal length 80mm. It is perpendicular to the principal axis of the lens and has one end on the axis. Draw a diagram to show the lens and object, and the image that is produced. Describe the nature of the image, and state its distance from the lens.

25 Working with reflectors

25.1 What is meant by *total internal reflection*? Describe a demonstration of this, using a glass block, and show how the critical angle can be measured. If the critical angle at an ice-air surface is 55.3°, what is the refractive index going from air to ice?

25.2 A converging mirror has focal length 20cm. An object is placed 15cm in front of the mirror, on the principal axis. Draw a ray diagram and find the position and magnification of the image.

26 Using electricity

26.1 A heating coil carries a current of 5A for 10 min. In this time 240kJ of electrical energy are converted to thermal energy. What is the power of the coil? What is the potential difference across the terminals of the coil? What quantity of electric charge passes through the coil in this time?

26.2 Given a microammeter with full scale deflection 50μA, how would you convert it into a voltmeter with full scale deflection 10V? Assume that the resistance of the microammeter is 5Ω. Illustrate your answer by a diagram.

26.3 Given a microammeter full scale deflection 100mA, coil resistance 0.5Ω, how would you convert it into an ammeter with full scale deflection 10A? Illustrate your answer with a diagram. What are the advantages and disadvantages of a moving-iron ammeter, compared with a moving-coil ammeter?

26.4 Describe briefly four different ways of producing light from electricity. Comment on the advantages and disadvantages of each.

26.5 State *Faraday's first law of electrolysis*. Describe how you would demonstrate this law practically.

27 Flight

27.1 Name three surfaces of an aeroplane (excluding the wings) that act upon the air around them and describe the action of each.

27.2 State *Bernoulli's principle*. Explain how the principle is applied to (a) an insecticide spray, and (b) an aerofoil.

28 Engines

28.1 Draw a simple diagram of a 4-stroke petrol engine. Briefly describe what happens at each of the 4 stages of its operation.

28.2 Draw a simple diagram of a diesel engine. Describe briefly what happens at each stage of its operation.

28.3 Explain the action of the gas turbine and the jet engine, using diagrams to make your explanation clearer. Carefully contrast the differences between these two types of engine and mention the advantages of each.

29 Producing electricity

29.1 State *Lenz's law of electromagnetic induction*. A coil has its terminals connected to a galvanometer, with zero in the centre of scale. Describe what happens when a bar magnet is brought slowly towards the coil, with one pole pointing toward the coil. What happens when the magnet is held close to the coil, but without moving it? What happens when the magnet is *quickly* taken away?

29.2 Draw a simple diagram of an a.c. generator. On your diagram show the direction of movement of the armature and the direction of induced currents. By means of a graph show how the current produced varies during one complete turn of the armature, marking the positions of the armature on your graph.

29.3 How would you charge a 6V lead accumulator? Draw a diagram of the circuit you would use and explain how it works.

29.4 Explain the differences in action and the different applications of a photoelectric cell and a photovoltaic cell. What practical applications have these two types of cell?

29.5 Explain how a moving-coil microphone works. State briefly how the crystal microphone and carbon microphone differ in their action.

29.6 A battery has e.m.f. 6V. When connected to a circuit which has total resistance 27Ω, the potential difference between the terminals of the cell is 5.5V. What is the internal resistance of the cell? If the current is allowed to flow for 10s, how much energy is converted in the external circuit?

29.7 Explain the reason for transmitting electrical power at high voltage, and as *alternating* current instead of *direct* current. An electricity transmission line has resistance 2Ω, and the transmission voltage is 132kV. What is the rate of conversion of electrical energy into internal energy in the cable when the cable is carrying a current of 0.1A?

30 Using materials

30.1 What properties are needed in a material that is to be used for (a) making the handle of an axe, (b) making the blade of an axe, (c) the roof of a house, (d) a container to hold vegetables pickled in vinegar, (e) a container to hold a radioactive isotope that emits gamma-rays.

30.2 What is meant by *electrolytic corrosion*? Describe three ways in which this can be caused, and describe how each may be avoided or reduced.

30.3 What is the *notch effect*? How can this effect be prevented from weakening the strength of materials? List 4 examples of integrated materials and state briefly the properties and main uses of the integrated material.

31 Transistors

31.1 Draw a circuit diagram to show how an *n-p-n* transistor may be connected to switch a large current by means of a relatively small controlling current. Show the paths taken by the controlled and controlling currents, and give their usual names.

31.2 Draw a diagram to show an *n-p-n* transistor connected for use as a current amplifier. What are the advantages of using a transistor for amplifying, compared with using a thermionic valve (or tube)?

32 Radio

32.1 Describe the action of an oscillating circuit made from a capacitor connected to a coil. In what way can we supply energy to the circuit so that it continues to oscillate at constant amplitude?

32.2 Draw a diagram to show the main sections of a radio transmitter. What is meant by *modulation* in connection with radio transmission?

32.3 Draw a diagram to show the main sections of a radio receiver. Describe how the receiver is made to detect one particular radio station out of the many that are being received in the aerial of the receiving station.

33 Nuclear power

33.1 Describe the operation of a nuclear power generator that obtains its energy from nuclear fission. What are the problems of safety concerned with nuclear power generation?

33.2 Explain, giving examples, what is meant by the term *nuclear fusion*. What would be the advantage of a power generator based on nuclear fusion?

Answers to numerical questions

1.3 1.52kg

1.4 11 300kg/m^3

1.5 17kg

1.6 4kPa (or 4kN/m^2)

2.1 0.4m from the other end

2.2 Reduce its length to 0.25m

4.2 0.0009m^3

4.3 0.85

5.2 19.5°

5.3 0.48m

6.4 8.7cm^3

7.2 2.97km

7.3 1.7Hz

10.2 45m/s, 350m, 500Ns

10.4 20m/s horizontally in the same direction as the bullet was travelling before impact, 2km/s^2 horizontally backward, 2Ns

10.5 0.5m/s backward

10.6 5kNs in the direction of its travel

11.1 11.2N 63° 26′ north of east, 2.24m/s 63° 26′ north of east

11.3 (a) 50N, (b) 8N, (c) 25Ns, (d) 25Ns

11.4 5m/s 60° north of west, 2.5m/s east, 4.3m/s north

12.2 65N upward

13.1 8

13.2 It is 630Pa (or N/m^2) more than atmospheric pressure, 5cm

14.1 1 260C

14.3 24s

15.3 4

15.4 100W

15.5 250N 400kJ

16.2 300J

16.3 179.2s

17.2 3.56×10^2kg. 1.45×10^5m, 2.5472×10^{-7}F, 2.5×10^{-1}s, 2.5×10^{-4}Pa, 1.258×10^{12}N, 2.546256×10^4kg.

18.2 34.375 min

18.3 243.05m^3

19.1 (a) 0.06A, (b) 300mA

19.3 0.86V, 1.71V, and 3.43V; 3.43mA through each

19.4 6V across each; 24mA, 12mA and 6mA, 143Ω

20.4 3.125g

21.3 1.289m

22.3 400s

24.4 0.25m, 0.238m

24.5 Inverted, real, magnified, 57mm long, 171mm from lens

25.1 1.216

25.2 –60cm, times 4 virtual

26.1 400W, 80V, 3kC

26.2 200kΩ

26.3 5.05mΩ

29.6 2.45Ω, 11.2J

29.7 0.02W

Symbols for electrical diagrams

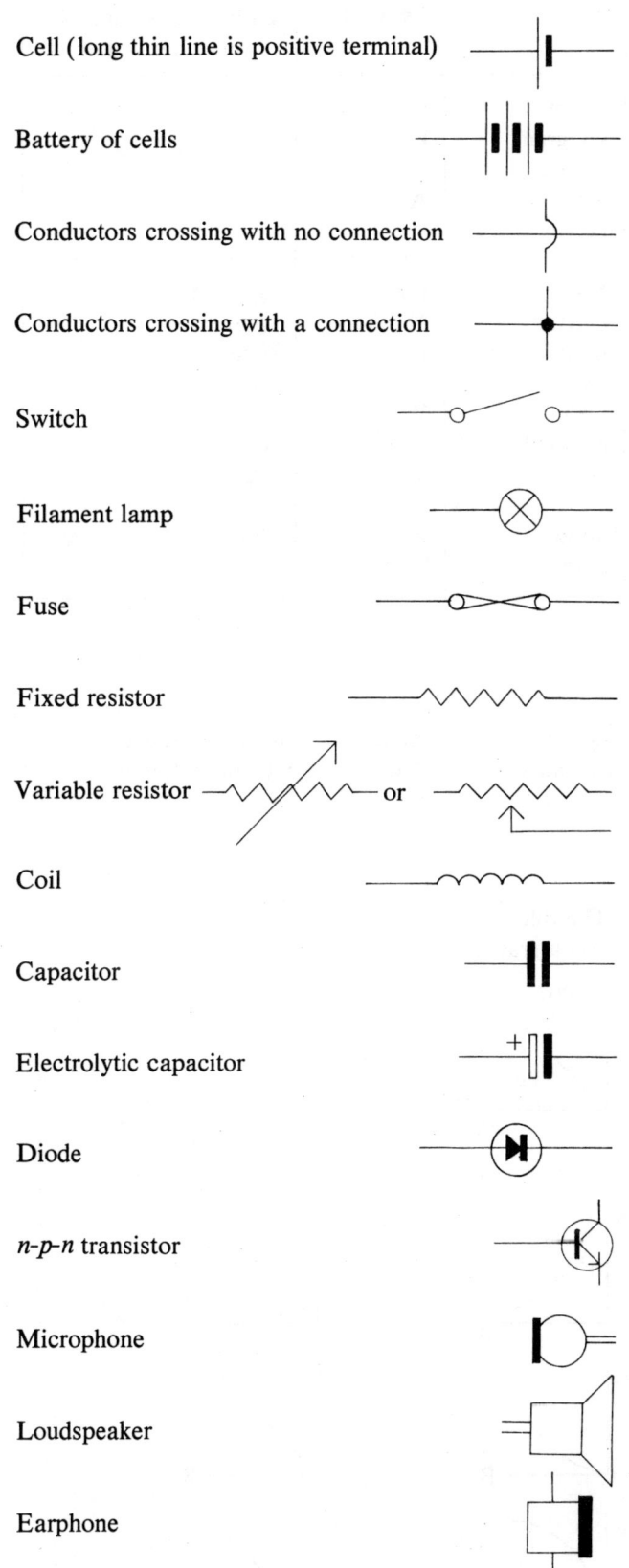

Cell (long thin line is positive terminal)

Battery of cells

Conductors crossing with no connection

Conductors crossing with a connection

Switch

Filament lamp

Fuse

Fixed resistor

Variable resistor — or

Coil

Capacitor

Electrolytic capacitor

Diode

n-p-n transistor

Microphone

Loudspeaker

Earphone

Resistor colour code

Colour	Value	Multiplying factor
Black	0	× 1
Brown	1	× 10
Red	2	× 100
Orange	3	× 1 000
Yellow	4	× 10 000
Green	5	× 100 000
Blue	6	× 1 000 000
Violet	7	
Grey	8	not used
White	9	

The resistor has four coloured bands. The first and second tell you the first two figures of the resistance (use 'value' column of table); the third band tells you how much to multiply the figures you get from the first two bands. For example:

'green, blue, orange' *means* $56 \times 1\ 000$ *making* $56\ 000\ \Omega$
'red, red, red' *means* 22×100 *making* $2\ 200\ \Omega$
'brown, green, black' *means* 15×1 *making* $15\ \Omega$

The fourth band indicates how near the true resistance is to the value marked by colours:

silver   – within 10%
gold   – within 5%
red   – within 2%
brown   – within 1%
no fourth band – within 20%

This table is to help you identify resistors; use the table, but do not learn it.

SI units

	Quantity	Name	Symbol	Unit (in base units or other derived units)
Base units	length	metre	m	
	mass	kilogram	kg	
	time	second	s	
	electric current	ampere	A	
	temperature	kelvin	K	
Derived units (special names)	frequency	hertz	Hz	/s
	force	newton	N	kg m/s²
	pressure	pascal	Pa	N/m²
	work, energy, heat	joule	J	Nm
	power	watt	W	J/s
	electric charge	coulomb	C	A s
	electric potential, pd., e.m.f. }	volt	V	J/C
	electric capacitance	farad	F	C/V
	electric resistance	ohm	Ω	V/A

Abbreviations

a.c.	alternating current	r.m.s.	root mean square	f.s.d.	full scale deflection
d.c.	direct current	s.w.g.	standard wire gauge	p.d.	potential difference
e.m.f.	electromotive force				

Relations between two quantities

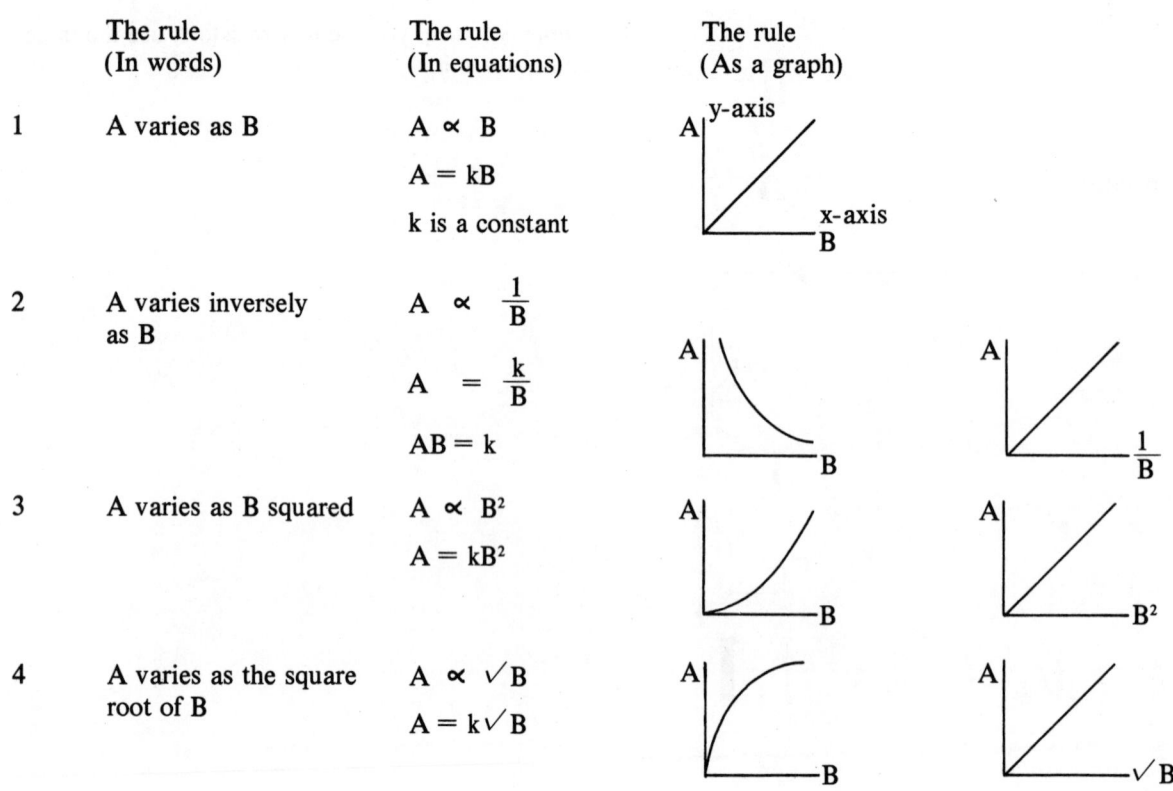

	The rule (In words)	The rule (In equations)
1	A varies as B	$A \propto B$ $A = kB$ k is a constant
2	A varies inversely as B	$A \propto \dfrac{1}{B}$ $A = \dfrac{k}{B}$ $AB = k$
3	A varies as B squared	$A \propto B^2$ $A = kB^2$
4	A varies as the square root of B	$A \propto \sqrt{B}$ $A = k\sqrt{B}$

LOGARITHMS

	0	1	2	3	4	5	6	7	8	9	1	2	3	4	5	6	7	8	9
50	6990	6998	7007	7016	7024	7033	7042	7050	7059	7067	1	2	3	3	4	5	6	7	8
51	7076	7084	7093	7101	7110	7118	7126	7135	7143	7152	1	2	3	3	4	5	6	7	8
52	7160	7168	7177	7185	7193	7202	7210	7218	7226	7235	1	2	3	3	4	5	6	7	7
53	7243	7251	7259	7267	7275	7284	7292	7300	7308	7316	1	2	2	3	4	5	6	7	7
54	7324	7332	7340	7348	7356	7364	7372	7380	7388	7396	1	2	2	3	4	5	6	6	7
55	7404	7412	7419	7427	7435	7443	7451	7459	7466	7474	1	2	2	3	4	5	5	6	7
56	7482	7490	7497	7505	7513	7520	7528	7536	7543	7551	1	2	2	3	4	5	5	6	7
57	7559	7566	7574	7582	7589	7597	7604	7612	7619	7627	1	2	2	3	4	5	5	6	7
58	7634	7642	7649	7657	7664	7672	7679	7686	7694	7701	1	1	2	3	4	4	5	6	7
59	7709	7716	7723	7731	7738	7745	7752	7760	7767	7774	1	1	2	3	4	4	5	6	7
60	7782	7789	7796	7803	7810	7818	7825	7832	7839	7846	1	1	2	3	4	4	5	6	6
61	7853	7860	7868	7875	7882	7889	7896	7903	7910	7917	1	1	2	3	4	4	5	6	6
62	7924	7931	7938	7945	7952	7959	7966	7973	7980	7987	1	1	2	3	4	4	5	5	6
63	7993	8000	8007	8014	8021	8028	8035	8041	8048	8055	1	1	2	3	3	4	5	5	6
64	8062	8069	8075	8082	8089	8096	8102	8109	8116	8122	1	1	2	3	3	4	5	5	6
65	8129	8136	8142	8149	8156	8162	8169	8176	8182	8189	1	1	2	3	3	4	5	5	6
66	8195	8202	8209	8215	8222	8228	8235	8241	8248	8254	1	1	2	3	3	4	5	5	6
67	8261	8267	8274	8280	8287	8293	8299	8306	8312	8319	1	1	2	3	3	4	5	5	6
68	8325	8331	8338	8344	8351	8357	8363	8370	8376	8382	1	1	2	3	3	4	4	5	6
69	8388	8395	8401	8407	8414	8420	8426	8432	8439	8445	1	1	2	3	3	4	4	5	6
70	8451	8457	8463	8470	8476	8482	8488	8494	8500	8506	1	1	2	2	3	4	4	5	6
71	8513	8519	8525	8531	8537	8543	8549	8555	8561	8567	1	1	2	2	3	4	4	5	5
72	8573	8579	8585	8591	8597	8603	8609	8615	8621	8627	1	1	2	2	3	4	4	5	5
73	8633	8639	8645	8651	8657	8663	8669	8675	8681	8686	1	1	2	2	3	4	4	5	5
74	8692	8698	8704	8710	8716	8722	8727	8733	8739	8745	1	1	2	2	3	4	4	5	5
75	8751	8756	8762	8768	8774	8779	8785	8791	8797	8802	1	1	2	2	3	3	4	5	5
76	8808	8814	8820	8825	8831	8837	8842	8848	8854	8859	1	1	2	2	3	3	4	5	5
77	8865	8871	8876	8882	8887	8893	8899	8904	8910	8915	1	1	2	2	3	3	4	4	5
78	8921	8927	8932	8938	8943	8949	8954	8960	8965	8971	1	1	2	2	3	3	4	4	5
79	8976	8982	8987	8993	8998	9004	9009	9015	9020	9025	1	1	2	2	3	3	4	4	5
80	9031	9036	9042	9047	9053	9058	9063	9069	9074	9079	1	1	2	2	3	3	4	4	5
81	9085	9090	9096	9101	9106	9112	9117	9122	9128	9133	1	1	2	2	3	3	4	4	5
82	9138	9143	9149	9154	9159	9165	9170	9175	9180	9186	1	1	2	2	3	3	4	4	5
83	9191	9196	9201	9206	9212	9217	9222	9227	9232	9238	1	1	2	2	3	3	4	4	5
84	9243	9248	9253	9258	9263	9269	9274	9279	9284	9289	1	1	2	2	3	3	4	4	5
85	9294	9299	9304	9309	9315	9320	9325	9330	9335	9340	1	1	2	2	3	3	4	4	5
86	9345	9350	9355	9360	9365	9370	9375	9380	9385	9390	1	1	2	2	3	3	4	4	5
87	9395	9400	9405	9410	9415	9420	9425	9430	9435	9440	0	1	1	2	3	3	4	4	5
88	9445	9450	9455	9460	9465	9469	9474	9479	9484	9489	0	1	1	2	2	3	3	4	4
89	9494	9499	9504	9509	9513	9518	9523	9528	9533	9538	0	1	1	2	2	3	3	4	4
90	9542	9547	9552	9557	9562	9566	9571	9576	9581	9586	0	1	1	2	2	3	3	4	4
91	9590	9595	9600	9605	9609	9614	9619	9624	9628	9633	0	1	1	2	2	3	3	4	4
92	9638	9643	9647	9652	9657	9661	9666	9671	9675	9680	0	1	1	2	2	3	3	4	4
93	9685	9689	9694	9699	9703	9708	9713	9717	9722	9727	0	1	1	2	2	3	3	4	4
94	9731	9736	9741	9745	9750	9754	9759	9763	9768	9773	0	1	1	2	2	3	3	4	4
95	9777	9782	9786	9791	9795	9800	9805	9809	9814	9818	0	1	1	2	2	3	3	4	4
96	9823	9827	9832	9836	9841	9845	9850	9854	9859	9863	0	1	1	2	2	3	3	4	4
97	9868	9872	9877	9881	9886	9890	9894	9899	9903	9908	0	1	1	2	2	3	3	3	4
98	9912	9917	9921	9926	9930	9934	9939	9943	9948	9952	0	1	1	2	2	3	3	3	4
99	9956	9961	9965	9969	9974	9978	9983	9987	9991	9996	0	1	1	2	2	3	3	3	4

LOGARITHMS

	0	1	2	3	4	5	6	7	8	9	1	2	3	4	5	6	7	8	9
10	0000	0043	0086	0128	0170	0212	0253	0294	0334	0374	4	9	13	17	21	26	30	34	38
11	0414	0453	0492	0531	0569	0607	0645	0682	0719	0755	4	8	12	16	20	24	28	32	36
12	0792	0828	0864	0899	0934	0969	1004	1038	1072	1106	4	7	11	15	18	22	26	29	33
13	1139	1173	1206	1239	1271	1303	1335	1367	1399	1430	3	7	10	13	16	19	23	26	29
14	1461	1492	1523	1553	1584	1614	1644	1673	1703	1732	3	6	9	12	15	18	21	24	27
15	1761	1790	1818	1847	1875	1903	1931	1959	1987	2014	3	6	8	11	14	17	20	22	25
16	2041	2068	2095	2122	2148	2175	2201	2227	2253	2279	3	5	8	11	13	16	18	21	24
17	2304	2330	2355	2380	2405	2430	2455	2480	2504	2529	2	5	7	10	12	15	17	20	22
18	2553	2577	2601	2625	2648	2672	2695	2718	2742	2765	2	5	7	9	12	14	16	19	21
19	2788	2810	2833	2856	2878	2900	2923	2945	2967	2989	2	4	7	9	11	13	16	18	20
20	3010	3032	3054	3075	3096	3118	3139	3160	3181	3201	2	4	6	8	11	13	15	17	19
21	3222	3243	3263	3284	3304	3324	3345	3365	3385	3404	2	4	6	8	10	12	14	16	18
22	3424	3444	3464	3483	3502	3522	3541	3560	3579	3598	2	4	6	8	10	12	14	15	17
23	3617	3636	3655	3674	3692	3711	3729	3747	3766	3784	2	4	6	7	9	11	13	15	17
24	3802	3820	3838	3856	3874	3892	3909	3927	3945	3962	2	4	5	7	9	11	12	14	16
25	3979	3997	4014	4031	4048	4065	4082	4099	4116	4133	2	3	5	7	9	10	12	14	15
26	4150	4166	4183	4200	4216	4232	4249	4265	4281	4298	2	3	5	7	8	10	11	13	15
27	4314	4330	4346	4362	4378	4393	4409	4425	4440	4456	2	3	5	6	8	9	11	13	14
28	4472	4487	4502	4518	4533	4548	4564	4579	4594	4609	2	3	5	6	8	9	11	12	14
29	4624	4639	4654	4669	4683	4698	4713	4728	4742	4757	1	3	4	6	7	9	10	12	13
30	4771	4786	4800	4814	4829	4843	4857	4871	4886	4900	1	3	4	6	7	9	10	11	13
31	4914	4928	4942	4955	4969	4983	4997	5011	5024	5038	1	3	4	6	7	8	10	11	13
32	5051	5065	5079	5092	5105	5119	5132	5145	5159	5172	1	3	4	5	7	8	9	11	12
33	5185	5198	5211	5224	5237	5250	5263	5276	5289	5302	1	3	4	5	6	8	9	10	12
34	5315	5328	5340	5353	5366	5378	5391	5403	5416	5428	1	3	4	5	6	8	9	10	11
35	5441	5453	5465	5478	5490	5502	5514	5527	5539	5551	1	2	4	5	6	7	9	10	11
36	5563	5575	5587	5599	5611	5623	5635	5647	5658	5670	1	2	4	5	6	7	8	10	11
37	5682	5694	5705	5717	5729	5740	5752	5763	5775	5786	1	2	3	5	6	7	8	9	11
38	5798	5809	5821	5832	5843	5855	5866	5877	5888	5899	1	2	3	5	6	7	8	9	10
39	5911	5922	5933	5944	5955	5966	5977	5988	5999	6010	1	2	3	4	6	7	8	9	10
40	6021	6031	6042	6053	6064	6075	6085	6096	6107	6117	1	2	3	4	5	6	8	9	10
41	6128	6138	6149	6160	6170	6180	6191	6201	6212	6222	1	2	3	4	5	6	7	8	9
42	6232	6243	6253	6263	6274	6284	6294	6304	6314	6325	1	2	3	4	5	6	7	8	9
43	6335	6345	6355	6365	6375	6385	6395	6405	6415	6425	1	2	3	4	5	6	7	8	9
44	6435	6444	6454	6464	6474	6484	6493	6503	6513	6522	1	2	3	4	5	6	7	8	9
45	6532	6542	6551	6561	6571	6580	6590	6599	6609	6618	1	2	3	4	5	6	7	8	9
46	6628	6637	6646	6656	6665	6675	6684	6693	6702	6712	1	2	3	4	5	5	6	7	8
47	6721	6730	6739	6749	6758	6767	6776	6785	6794	6803	1	2	3	4	5	5	6	7	8
48	6812	6821	6830	6839	6848	6857	6866	6875	6884	6893	1	2	3	4	4	5	6	7	8
49	6902	6911	6920	6928	6937	6946	6955	6964	6972	6981	1	2	3	4	4	5	6	7	8

NATURAL SINES

Degrees	0' 0°.0	6' 0°.1	12' 0°.2	18' 0°.3	24' 0°.4	30' 0°.5	36' 0°.6	42' 0°.7	48' 0°.8	54' 0°.9	1	2	3	4	5
45	.7071	7083	7096	7108	7120	7133	7145	7157	7169	7181	2	4	6	8	10
46	.7193	7206	7218	7230	7242	7254	7266	7278	7290	7302	2	4	6	8	10
47	.7314	7325	7337	7349	7361	7373	7385	7396	7408	7420	2	4	6	8	10
48	.7431	7443	7455	7466	7478	7490	7501	7513	7524	7536	2	4	6	8	10
49	.7547	7558	7570	7581	7593	7604	7615	7627	7638	7649	2	4	6	8	9
50	.7660	7672	7683	7694	7705	7716	7727	7738	7749	7760	2	4	6	7	9
51	.7771	7782	7793	7804	7815	7826	7837	7848	7859	7869	2	4	5	7	9
52	.7880	7891	7902	7912	7923	7934	7944	7955	7965	7976	2	4	5	7	9
53	.7986	7997	8007	8018	8028	8039	8049	8059	8070	8080	2	3	5	7	8
54	.8090	8100	8111	8121	8131	8141	8151	8161	8171	8181	2	3	5	7	8
55	.8192	8202	8211	8221	8231	8241	8251	8261	8271	8281	2	3	5	7	8
56	.8290	8300	8310	8320	8329	8339	8348	8358	8368	8377	2	3	5	6	8
57	.8387	8396	8406	8415	8425	8434	8443	8453	8462	8471	2	3	5	6	8
58	.8480	8490	8499	8508	8517	8526	8536	8545	8554	8563	2	3	4	6	8
59	.8572	8581	8590	8599	8607	8616	8625	8634	8643	8652	1	3	4	6	7
60	.8660	8669	8678	8686	8695	8704	8712	8721	8729	8738	1	3	4	6	7
61	.8746	8755	8763	8771	8780	8788	8796	8805	8813	8821	1	3	4	6	7
62	.8829	8838	8846	8854	8862	8870	8878	8886	8894	8902	1	3	4	5	7
63	.8910	8918	8926	8934	8942	8949	8957	8965	8973	8980	1	3	4	5	6
64	.8988	8996	9003	9011	9018	9026	9033	9041	9048	9056	1	3	4	5	6
65	.9063	9070	9078	9085	9092	9100	9107	9114	9121	9128	1	2	4	5	6
66	.9135	9143	9150	9157	9164	9171	9178	9184	9191	9198	1	2	3	5	6
67	.9205	9212	9219	9225	9232	9239	9245	9252	9259	9265	1	2	3	4	6
68	.9272	9278	9285	9291	9298	9304	9311	9317	9323	9330	1	2	3	4	5
69	.9336	9342	9348	9354	9361	9367	9373	9379	9385	9391	1	2	3	4	5
70	.9397	9403	9409	9415	9421	9426	9432	9438	9444	9449	1	2	3	4	5
71	.9455	9461	9466	9472	9478	9483	9489	9494	9500	9505	1	2	3	4	5
72	.9511	9516	9521	9527	9532	9537	9542	9548	9553	9558	1	2	3	4	4
73	.9563	9568	9573	9578	9583	9588	9593	9598	9603	9608	1	2	2	3	4
74	.9613	9617	9622	9627	9632	9636	9641	9646	9650	9655	1	2	2	3	4
75	.9659	9664	9668	9673	9677	9681	9686	9690	9694	9699	1	1	2	3	4
76	.9703	9707	9711	9715	9720	9724	9728	9732	9736	9740	1	1	2	3	3
77	.9744	9748	9751	9755	9759	9763	9767	9770	9774	9778	1	1	2	3	3
78	.9781	9785	9789	9792	9796	9799	9803	9806	9810	9813	1	1	2	2	3
79	.9816	9820	9823	9826	9829	9833	9836	9839	9842	9845	1	1	2	2	3
80	.9848	9851	9854	9857	9860	9863	9866	9869	9871	9874	0	1	1	2	2
81	.9877	9880	9882	9885	9888	9890	9893	9895	9898	9900	0	1	1	2	2
82	.9903	9905	9907	9910	9912	9914	9917	9919	9921	9923	0	1	1	2	2
83	.9925	9928	9930	9932	9934	9936	9938	9940	9942	9943	0	1	1	1	2
84	.9945	9947	9949	9951	9952	9954	9956	9957	9959	9960	0	1	1	1	2
85	.9962	9963	9965	9966	9968	9969	9971	9972	9973	9974	0	0	1	1	1
86	.9976	9977	9978	9979	9980	9981	9982	9983	9984	9985	0	0	1	1	1
87	.9986	9987	9988	9989	9990	9990	9991	9992	9993	9993	0	0	0	1	1
88	.9994	9995	9995	9996	9996	9997	9997	9997	9998	9998	0	0	0	0	0
89	.9998	9999	9999	9999	9999	1·000	1·000	1·000	1·000	1·000	0	0	0	0	0
90	1·000														

NATURAL SINES

Degrees	0' 0°.0	6' 0°.1	12' 0°.2	18' 0°.3	24' 0°.4	30' 0°.5	36' 0°.6	42' 0°.7	48' 0°.8	54' 0°.9	1	2	3	4	5
0	·0000	0017	0035	0052	0070	0087	0105	0122	0140	0157	3	6	9	12	15
1	·0175	0192	0209	0227	0244	0262	0279	0297	0314	0332	3	6	9	12	15
2	·0349	0366	0384	0401	0419	0436	0454	0471	0488	0506	3	6	9	12	15
3	·0523	0541	0558	0576	0593	0610	0628	0645	0663	0680	3	6	9	12	15
4	·0698	0715	0732	0750	0767	0785	0802	0819	0837	0854	3	6	9	12	15
5	·0872	0889	0906	0924	0941	0958	0976	0993	1011	1028	3	6	9	12	14
6	·1045	1063	1080	1097	1115	1132	1149	1167	1184	1201	3	6	9	12	14
7	·1219	1236	1253	1271	1288	1305	1323	1340	1357	1374	3	6	9	12	14
8	·1392	1409	1426	1444	1461	1478	1495	1513	1530	1547	3	6	9	12	14
9	·1564	1582	1599	1616	1633	1650	1668	1685	1702	1719	3	6	9	12	14
10	·1736	1754	1771	1788	1805	1822	1840	1857	1874	1891	3	6	9	12	14
11	·1908	1925	1942	1959	1977	1994	2011	2028	2045	2062	3	6	9	11	14
12	·2079	2096	2113	2130	2147	2164	2181	2198	2215	2232	3	6	9	11	14
13	·2250	2267	2284	2300	2317	2334	2351	2368	2385	2402	3	6	8	11	14
14	·2419	2436	2453	2470	2487	2504	2521	2538	2554	2571	3	6	8	11	14
15	·2588	2605	2622	2639	2656	2672	2689	2706	2723	2740	3	6	8	11	14
16	·2756	2773	2790	2807	2823	2840	2857	2874	2890	2907	3	6	8	11	14
17	·2924	2940	2957	2974	2990	3007	3024	3040	3057	3074	3	6	8	11	14
18	·3090	3107	3123	3140	3156	3173	3190	3206	3223	3239	3	5	8	11	14
19	·3256	3272	3289	3305	3322	3338	3355	3371	3387	3404	3	5	8	11	14
20	·3420	3437	3453	3469	3486	3502	3518	3535	3551	3567	3	5	8	11	14
21	·3584	3600	3616	3633	3649	3665	3681	3697	3714	3730	3	5	8	11	14
22	·3746	3762	3778	3795	3811	3827	3843	3859	3875	3891	3	5	8	11	14
23	·3907	3923	3939	3955	3971	3987	4003	4019	4035	4051	3	5	8	11	14
24	·4067	4083	4099	4115	4131	4147	4163	4179	4195	4210	3	5	8	11	13
25	·4226	4242	4258	4274	4289	4305	4321	4337	4352	4368	3	5	8	11	13
26	·4384	4399	4415	4431	4446	4462	4478	4493	4509	4524	3	5	8	10	13
27	·4540	4555	4571	4586	4602	4617	4633	4648	4664	4679	3	5	8	10	13
28	·4695	4710	4726	4741	4756	4772	4787	4802	4818	4833	3	5	8	10	13
29	·4848	4863	4879	4894	4909	4924	4939	4955	4970	4985	3	5	8	10	13
30	·5000	5015	5030	5045	5060	5075	5090	5105	5120	5135	3	5	8	10	13
31	·5150	5165	5180	5195	5210	5225	5240	5255	5270	5284	3	5	7	10	12
32	·5299	5314	5329	5344	5358	5373	5388	5402	5417	5432	2	5	7	10	12
33	·5446	5461	5476	5490	5505	5519	5534	5548	5563	5577	2	5	7	10	12
34	·5592	5606	5621	5635	5650	5664	5678	5693	5707	5721	2	5	7	10	12
35	·5736	5750	5764	5779	5793	5807	5821	5835	5850	5864	2	5	7	9	12
36	·5878	5892	5906	5920	5934	5948	5962	5976	5990	6004	2	5	7	9	12
37	·6018	6032	6046	6060	6074	6088	6101	6115	6129	6143	2	5	7	9	12
38	·6157	6170	6184	6198	6211	6225	6239	6252	6266	6280	2	5	7	9	11
39	·6293	6307	6320	6334	6347	6361	6374	6388	6401	6414	2	4	7	9	11
40	·6428	6441	6455	6468	6481	6494	6508	6521	6534	6547	2	4	7	9	11
41	·6561	6574	6587	6600	6613	6626	6639	6652	6665	6678	2	4	6	9	11
42	·6691	6704	6717	6730	6743	6756	6769	6782	6794	6807	2	4	6	9	11
43	·6820	6833	6845	6858	6871	6884	6896	6909	6921	6934	2	4	6	8	11
44	·6947	6959	6972	6984	6997	7009	7022	7034	7046	7059	2	4	6	8	10

NATURAL COSINES

[Numbers in difference columns to be subtracted, not added.]

Degrees	0' 0°·0	6' 0°·1	12' 0°·2	18' 0°·3	24' 0°·4	30' 0°·5	36' 0°·6	42' 0°·7	48' 0°·8	54' 0°·9	Mean Differences				
											1	2	3	4	5
45	·7071	7059	7046	7034	7022	7009	6997	6984	6972	6959	2	4	6	8	10
46	·6947	6934	6921	6909	6896	6884	6871	6858	6845	6833	2	4	6	8	11
47	·6820	6807	6794	6782	6769	6756	6743	6730	6717	6704	2	4	6	9	11
48	·6691	6678	6665	6652	6639	6626	6613	6600	6587	6574	2	4	7	9	11
49	·6561	6547	6534	6521	6508	6494	6481	6468	6455	6441	2	4	7	9	11
50	·6428	6414	6401	6388	6374	6361	6347	6334	6320	6307	2	4	7	9	11
51	·6293	6280	6266	6252	6239	6225	6211	6198	6184	6170	2	5	7	9	11
52	·6157	6143	6129	6115	6101	6088	6074	6060	6046	6032	2	5	7	9	12
53	·6018	6004	5990	5976	5962	5948	5934	5920	5906	5892	2	5	7	9	12
54	·5878	5864	5850	5835	5821	5807	5793	5779	5764	5750	2	5	7	9	12
55	·5736	5721	5707	5693	5678	5664	5650	5635	5621	5606	2	5	7	10	12
56	·5592	5577	5563	5548	5534	5519	5505	5490	5476	5461	2	5	7	10	12
57	·5446	5432	5417	5402	5388	5373	5358	5344	5329	5314	2	5	7	10	12
58	·5299	5284	5270	5255	5240	5225	5210	5195	5180	5165	3	5	8	10	12
59	·5150	5135	5120	5105	5090	5075	5060	5045	5030	5015	3	5	8	10	13
60	·5000	4985	4970	4955	4939	4924	4909	4894	4879	4863	3	5	8	10	13
61	·4848	4833	4818	4802	4787	4772	4756	4741	4726	4710	3	5	8	10	13
62	·4695	4679	4664	4648	4633	4617	4602	4586	4571	4555	3	5	8	10	13
63	·4540	4524	4509	4493	4478	4462	4446	4431	4415	4399	3	5	8	11	13
64	·4384	4368	4352	4337	4321	4305	4289	4274	4258	4242	3	5	8	11	14
65	·4226	4210	4195	4179	4163	4147	4131	4115	4099	4083	3	5	8	11	13
66	·4067	4051	4035	4019	4003	3987	3971	3955	3939	3923	3	5	8	11	14
67	·3907	3891	3875	3859	3843	3827	3811	3795	3778	3762	3	5	8	11	14
68	·3746	3730	3714	3697	3681	3665	3649	3633	3616	3600	3	5	8	11	14
69	·3584	3567	3551	3535	3518	3502	3486	3469	3453	3437	3	5	8	11	14
70	·3420	3404	3387	3371	3355	3338	3322	3305	3289	3272	3	5	8	11	14
71	·3256	3239	3223	3206	3190	3173	3156	3140	3123	3107	3	6	8	11	14
72	·3090	3074	3057	3040	3024	3007	2990	2974	2957	2940	3	6	8	11	14
73	·2924	2907	2890	2874	2857	2840	2823	2807	2790	2773	3	6	8	11	14
74	·2756	2740	2723	2706	2689	2672	2656	2639	2622	2605	3	6	8	11	14
75	·2588	2571	2554	2538	2521	2504	2487	2470	2453	2436	3	6	8	11	14
76	·2419	2402	2385	2368	2351	2334	2317	2300	2284	2267	3	6	8	11	14
77	·2250	2233	2215	2198	2181	2164	2147	2130	2113	2096	3	6	8	11	14
78	·2079	2062	2045	2028	2011	1994	1977	1959	1942	1925	3	6	8	11	14
79	·1908	1891	1874	1857	1840	1822	1805	1788	1771	1754	3	6	9	11	14
80	·1736	1719	1702	1685	1668	1650	1633	1616	1599	1582	3	6	9	12	14
81	·1564	1547	1530	1513	1495	1478	1461	1444	1426	1409	3	6	9	12	14
82	·1392	1374	1357	1340	1323	1305	1288	1271	1253	1236	3	6	9	12	14
83	·1219	1201	1184	1167	1149	1132	1115	1097	1080	1063	3	6	9	12	15
84	·1045	1028	1011	0993	0976	0958	0941	0924	0906	0889	3	6	9	12	15
85	·0872	0854	0837	0819	0802	0785	0767	0750	0732	0715	3	6	9	12	15
86	·0698	0680	0663	0645	0628	0610	0593	0576	0558	0541	3	6	9	12	15
87	·0523	0506	0488	0471	0454	0436	0419	0401	0384	0366	3	6	9	12	15
88	·0349	0332	0314	0297	0279	0262	0244	0227	0209	0192	3	6	9	12	15
89	·0175	0157	0140	0122	0105	0087	0070	0052	0035	0017	3	6	9	12	15
90	·0000														

NATURAL COSINES

[Numbers in difference columns to be subtracted, not added.]

Degrees	0' 0°·0	6' 0°·1	12' 0°·2	18' 0°·3	24' 0°·4	30' 0°·5	36' 0°·6	42' 0°·7	48' 0°·8	54' 0°·9	Mean Differences				
											1	2	3	4	5
0	1·000	1·000	1·000	1·000	1·000	1·000	·9999	9999	9999	9999	0	0	0	0	0
1	·9998	9998	9998	9997	9997	9997	9996	9996	9995	9995	0	0	0	0	0
2	·9994	9993	9993	9992	9991	9990	9990	9989	9988	9987	0	0	0	0	1
3	·9986	9985	9984	9983	9982	9981	9980	9979	9978	9977	0	0	0	1	1
4	·9976	9974	9973	9972	9971	9969	9968	9966	9965	9963	0	0	1	1	1
5	·9962	9960	9959	9957	9956	9954	9952	9951	9949	9947	0	1	1	1	2
6	·9945	9943	9942	9940	9938	9936	9934	9932	9930	9928	0	1	1	1	2
7	·9925	9923	9921	9919	9917	9914	9912	9910	9907	9905	0	1	1	2	2
8	·9903	9900	9898	9895	9893	9890	9888	9885	9882	9880	0	1	1	2	2
9	·9877	9874	9871	9869	9866	9863	9860	9857	9854	9851	0	1	1	2	2
10	·9848	9845	9842	9839	9836	9833	9829	9826	9823	9820	1	1	2	2	3
11	·9816	9813	9810	9805	9803	9799	9796	9792	9789	9785	1	1	2	2	3
12	·9781	9778	9774	9770	9767	9763	9759	9755	9751	9748	1	1	2	3	3
13	·9744	9740	9736	9732	9728	9724	9720	9715	9711	9707	1	1	2	3	3
14	·9703	9699	9694	9690	9686	9681	9677	9673	9668	9664	1	1	2	3	4
15	·9659	9655	9650	9646	9641	9636	9632	9627	9622	9617	1	2	2	3	4
16	·9613	9608	9603	9598	9593	9588	9583	9578	9573	9568	1	2	2	3	4
17	·9563	9558	9553	9548	9542	9537	9532	9527	9521	9516	1	2	3	3	4
18	·9511	9505	9500	9494	9489	9483	9478	9472	9466	9461	1	2	3	4	5
19	·9455	9449	9444	9438	9432	9426	9421	9415	9409	9403	1	2	3	4	5
20	·9397	9391	9385	9379	9373	9367	9361	9354	9348	9342	1	2	3	4	5
21	·9336	9330	9323	9317	9311	9304	9298	9291	9285	9278	1	2	3	4	5
22	·9272	9265	9259	9252	9245	9239	9232	9225	9219	9212	1	2	3	4	6
23	·9205	9198	9191	9184	9178	9171	9164	9157	9150	9143	1	2	3	5	6
24	·9135	9128	9121	9114	9107	9100	9092	9085	9078	9070	1	2	4	5	6
25	·9063	9056	9048	9041	9033	9026	9018	9011	9003	8996	1	3	4	5	6
26	·8988	8980	8973	8965	8957	8949	8942	8934	8926	8918	1	3	4	5	6
27	·8910	8902	8894	8886	8878	8870	8862	8854	8846	8838	1	3	4	5	7
28	·8829	8821	8813	8805	8796	8788	8780	8771	8763	8755	1	3	4	6	7
29	·8746	8738	8729	8721	8712	8704	8695	8686	8678	8669	1	3	4	6	7
30	·8660	8652	8643	8634	8625	8616	8607	8599	8590	8581	1	3	4	6	7
31	·8572	8563	8554	8545	8536	8526	8517	8508	8499	8490	2	3	5	6	8
32	·8480	8471	8462	8453	8443	8434	8425	8415	8406	8396	2	3	5	6	8
33	·8387	8377	8368	8358	8348	8339	8329	8320	8310	8300	2	3	5	6	8
34	·8290	8281	8271	8261	8251	8241	8231	8221	8211	8202	2	3	5	7	8
35	·8192	8181	8171	8161	8151	8141	8131	8121	8111	8100	2	3	5	7	8
36	·8090	8080	8070	8059	8049	8039	8028	8018	8007	7997	2	3	5	7	9
37	·7986	7976	7965	7955	7944	7934	7923	7912	7902	7891	2	4	5	7	9
38	·7880	7869	7859	7848	7837	7826	7815	7804	7793	7782	2	4	5	7	9
39	·7771	7760	7749	7738	7727	7716	7705	7694	7683	7672	2	4	5	7	9
40	·7660	7649	7638	7627	7615	7604	7593	7581	7570	7559	2	4	6	8	9
41	·7547	7536	7524	7513	7501	7490	7478	7466	7455	7443	2	4	6	8	10
42	·7431	7420	7408	7396	7385	7373	7361	7349	7337	7325	2	4	6	8	10
43	·7314	7302	7290	7278	7266	7254	7242	7230	7218	7206	2	4	6	8	10
44	·7193	7181	7169	7157	7145	7133	7120	7108	7096	7083	2	4	6	8	10

NATURAL TANGENTS

Degrees	0' 0°.0	6' 0°.1	12' 0°.2	18' 0°.3	24' 0°.4	30' 0°.5	36' 0°.6	42' 0°.7	48' 0°.8	54' 0°.9	1	2	3	4	5
45	1·0000	0035	0070	0105	0141	0176	0212	0247	0283	0319	6	12	18	24	30
46	1·0355	0392	0428	0464	0501	0538	0575	0612	0649	0686	6	12	18	25	31
47	1·0724	0761	0799	0837	0875	0913	0951	0990	1028	1067	6	13	19	25	32
48	1·1106	1145	1184	1224	1263	1303	1343	1383	1423	1463	7	13	20	27	33
49	1·1504	1544	1585	1626	1667	1708	1750	1792	1833	1875	7	14	21	28	34
50	1·1918	1960	2002	2045	2088	2131	2174	2218	2261	2305	7	14	22	29	36
51	1·2349	2393	2437	2482	2527	2572	2617	2662	2708	2753	8	15	23	30	38
52	1·2799	2846	2892	2938	2985	3032	3079	3127	3175	3222	8	16	24	31	39
53	1·3270	3319	3367	3416	3465	3514	3564	3613	3663	3713	8	16	25	33	41
54	1·3764	3814	3865	3916	3968	4019	4071	4124	4176	4229	9	17	26	34	43
55	1·4281	4335	4388	4442	4496	4550	4605	4659	4715	4770	9	18	27	36	45
56	1·4826	4882	4938	4994	5051	5108	5166	5224	5282	5340	10	19	29	38	48
57	1·5399	5458	5517	5577	5637	5697	5757	5818	5880	5941	10	20	30	40	50
58	1·6003	6066	6128	6191	6255	6319	6383	6447	6512	6577	11	21	32	43	53
59	1·6643	6709	6775	6842	6909	6977	7045	7113	7182	7251	11	23	34	44	56
60	1·7321	7391	7461	7532	7603	7675	7747	7820	7893	7966	12	24	36	48	60
61	1·8040	8115	8190	8265	8341	8418	8495	8572	8650	8728	13	26	38	51	64
62	1·8807	8887	8967	9047	9128	9210	9292	9375	9458	9542	14	27	41	55	68
63	1·9626	9711	9797	9883	9970	2·0057	2·0145	2·0233	2·0323	2·0413	15	29	44	58	73
64	2·0503	0594	0686	0778	0872	0965	1060	1155	1251	1348	16	31	47	63	78
65	2·1445	1543	1642	1742	1842	1943	2045	2148	2251	2355	17	34	51	68	85
66	2·2460	2566	2673	2781	2889	2998	3109	3220	3332	3445	18	37	55	73	92
67	2·3559	3673	3789	3906	4023	4142	4262	4383	4504	4627	20	40	60	79	99
68	2·4751	4876	5002	5129	5257	5386	5517	5649	5782	5916	22	43	65	87	108
69	2·6051	6187	6325	6464	6605	6746	6889	7034	7179	7326	24	47	71	95	119
70	2·7475	7625	7776	7929	8083	8239	8397	8556	8716	8878	26	52	78	104	131
71	2·9042	9208	9375	9544	9714	9887	3·0061	3·0237	3·0415	3·0595	29	58	87	116	145
72	3·0777	0961	1146	1334	1524	1716	1910	2106	2305	2506	32	64	96	129	161
73	3·2709	2914	3122	3332	3544	3759	3977	4197	4420	4646	36	72	108	144	180
74	3·4874	5105	5339	5576	5816	6059	6305	6554	6806	7062	41	81	122	163	204
75	3·7321	7583	7848	8118	8391	8667	8947	9232	9520	9812	46	93	139	186	232
76	4·0108	0408	0713	1022	1335	1653	1976	2303	2635	2972	53	107	160	213	267
77	4·3315	3662	4015	4374	4737	5107	5483	5864	6252	6646					
78	4·7046	7453	7867	8288	8716	9152	9594	5·0045	5·0504	5·0970					
79	5·1446	1929	2422	2924	3435	3955	4486	5026	5578	6140					
80	5·6713	7297	7894	8502	9124	9758	6·0405	6·1066	6·1742	6·2432					
81	6·3138	3859	4596	5350	6122	6912	7720	8548	9395	7·0264					
82	7·1154	2066	3002	3962	4947	5958	6996	8062	9158	8·0285					
83	8·1443	2636	3863	5126	6427	7769	9152	9·0579	9·2052	9·3572	Mean differences cease				
84	9·5144	9·677	9·845	10·02	10·20	10·39	10·58	10·78	10·99	11·20	to be sufficiently				
85	11·43	11·66	11·91	12·16	12·43	12·71	13·00	13·30	13·62	13·95	accurate.				
86	14·30	14·67	15·06	15·46	15·89	16·35	16·83	17·34	17·89	18·46					
87	19·08	19·74	20·45	21·20	22·02	22·90	23·86	24·90	26·03	27·27					
88	28·64	30·14	31·82	33·69	35·80	38·19	40·92	44·07	47·74	52·08					
89	57·29	63·66	71·62	81·85	95·49	114·6	143·2	191·0	286·5	573·0					
90	∞														

NATURAL TANGENTS

Degrees	0' 0°.0	6' 0°.1	12' 0°.2	18' 0°.3	24' 0°.4	30' 0°.5	36' 0°.6	42' 0°.7	48' 0°.8	54' 0°.9	1	2	3	4	5
0	·0000	0017	0035	0052	0070	0087	0105	0122	0140	0157	3	6	9	12	15
1	·0175	0192	0209	0227	0244	0262	0279	0297	0314	0332	3	6	9	12	15
2	·0349	0367	0384	0402	0419	0437	0454	0472	0489	0507	3	6	9	12	15
3	·0524	0542	0559	0577	0594	0612	0629	0647	0664	0682	3	6	9	12	15
4	·0699	0717	0734	0752	0769	0787	0805	0822	0840	0857	3	6	9	12	15
5	·0875	0892	0910	0928	0945	0963	0981	0998	1016	1033	3	6	9	12	15
6	·1051	1069	1086	1104	1122	1139	1157	1175	1192	1210	3	6	9	12	15
7	·1228	1246	1263	1281	1299	1317	1334	1352	1370	1388	3	6	9	12	15
8	·1405	1423	1441	1459	1477	1495	1512	1530	1548	1566	3	6	9	12	15
9	·1584	1602	1620	1638	1655	1673	1691	1709	1727	1745	3	6	9	12	15
10	·1763	1781	1799	1817	1835	1853	1871	1890	1908	1926	3	6	9	12	15
11	·1944	1962	1980	1998	2016	2035	2053	2071	2089	2107	3	6	9	12	15
12	·2126	2144	2162	2180	2199	2217	2235	2254	2272	2290	3	6	9	12	15
13	·2309	2327	2345	2364	2382	2401	2419	2438	2456	2475	3	6	9	12	16
14	·2493	2512	2530	2549	2568	2586	2605	2623	2642	2661	3	6	9	12	16
15	·2679	2698	2717	2736	2754	2773	2792	2811	2830	2849	3	6	9	13	16
16	·2867	2886	2905	2924	2943	2962	2981	3000	3019	3038	3	6	9	13	16
17	·3057	3076	3096	3115	3134	3153	3172	3191	3211	3230	3	6	10	13	16
18	·3249	3269	3288	3307	3327	3346	3365	3385	3404	3424	3	6	10	13	16
19	·3443	3463	3482	3502	3522	3541	3561	3581	3600	3620	3	7	10	13	16
20	·3640	3659	3679	3699	3719	3739	3759	3779	3799	3819	3	7	10	13	17
21	·3839	3859	3879	3899	3919	3939	3959	3979	4000	4020	3	7	10	13	17
22	·4040	4061	4081	4101	4122	4142	4163	4183	4204	4224	3	7	10	14	17
23	·4245	4265	4286	4307	4327	4348	4369	4390	4411	4431	3	7	10	14	17
24	·4452	4473	4494	4515	4536	4557	4578	4599	4621	4642	4	7	11	14	18
25	·4663	4684	4706	4727	4748	4770	4791	4813	4834	4856	4	7	11	14	18
26	·4877	4899	4921	4942	4964	4986	5008	5029	5051	5073	4	7	11	15	18
27	·5095	5117	5139	5161	5184	5206	5228	5250	5272	5295	4	7	11	15	18
28	·5317	5340	5362	5384	5407	5430	5452	5475	5498	5520	4	8	11	15	19
29	·5543	5566	5589	5612	5635	5658	5681	5704	5727	5750	4	8	12	15	19
30	·5774	5797	5820	5844	5867	5890	5914	5938	5961	5985	4	8	12	16	20
31	·6009	6032	6056	6080	6104	6128	6152	6176	6200	6224	4	8	12	16	20
32	·6249	6273	6297	6322	6346	6371	6395	6420	6445	6469	4	8	12	16	20
33	·6494	6519	6544	6569	6594	6619	6644	6669	6694	6720	4	8	13	17	21
34	·6745	6771	6796	6822	6847	6873	6899	6924	6950	6976	4	9	13	17	21
35	·7002	7028	7054	7080	7107	7133	7159	7186	7212	7239	4	9	13	18	22
36	·7265	7292	7319	7346	7373	7400	7427	7454	7481	7508	5	9	14	18	23
37	·7536	7563	7590	7618	7646	7673	7701	7729	7757	7785	5	9	14	18	23
38	·7813	7841	7869	7898	7926	7954	7983	8012	8040	8069	5	9	14	19	24
39	·8098	8127	8156	8185	8214	8243	8273	8302	8332	8361	5	10	15	19	24
40	·8391	8421	8451	8481	8511	8541	8571	8601	8632	8662	5	10	15	20	25
41	·8693	8724	8754	8785	8816	8847	8878	8910	8941	8972	5	10	16	21	26
42	·9004	9036	9067	9099	9131	9163	9195	9228	9260	9293	5	11	16	21	27
43	·9325	9358	9391	9424	9457	9490	9523	9556	9590	9623	6	11	17	22	28
44	·9657	9691	9725	9759	9793	9827	9861	9896	9930	9965	6	11	17	23	29

Index

References in *italics* are to material in
the Answers and Discussion Section